Edited by
Purushottam D. Gujrati
and Arkadii I. Leonov

Modeling and Simulation in Polymers

Related Titles

Pascault, J.-P., Williams, R. J. J. (eds.)

Epoxy Polymers

New Materials and Innovations

2010

ISBN: 978-3-527-32480-4

Mittal, V. (ed.)

Optimization of Polymer Nanocomposite Properties

2010

ISBN: 978-3-527-32521-4

Mathers, R. T., Meier, M. A. R. (eds.)

Green Polymerization Methods

Renewable Starting Materials, Catalysis and Waste Reduction

2010

ISBN: 978-3-527-32625-9

Xanthos, M. (ed.)

Functional Fillers for Plastics

2010

ISBN: 978-3-527-32361-6

Matyjaszewski, K., Müller, A. H. E. (eds.)

Controlled and Living Polymerizations

From Mechanisms to Applications

2009

ISBN: 978-3-527-32492-7

Elias, H.-G.

Macromolecules

Series: Macromolecules (Volume 1-4)

2009

ISBN: 978-3-527-31171-2

Martín, N., Giacalone, F. (eds.)

Fullerene Polymers

Synthesis, Properties and Applications

2009

ISBN: 978-3-527-32282-4

Severn, J. R., Chadwick, J. C. (eds.)

Tailor-Made Polymers

Via Immobilization of Alpha-Olefin Polymerization Catalysts

2008

ISBN: 978-3-527-31782-0

Matyjaszewski, K., Gnanou, Y., Leibler, L. (eds.)

Macromolecular Engineering

Precise Synthesis, Materials Properties, Applications

2007

ISBN: 978-3-527-31446-1

Edited by Purushottam D. Gujrati and Arkadii I. Leonov

Modeling and Simulation in Polymers

WILEY-VCH

WILEY-VCH Verlag GmbH & Co. KGaA

The Editors

Dr. Purushottam D. Gujrati
The University of Akron
Department of Polymer Science
302 Buchtel Common
Akron, OH 44325-3909
USA

Dr. Arkady I. Leonov
The University of Akron
Department of Polymer Engineering
Polymer Engineering Academic Center
Akron, OH 44325-0301
USA

Library of Congress Card No.: applied for

British Library Cataloguing-in-Publication Data
A catalogue record for this book is available from the British Library.

Bibliographic information published by the Deutsche Nationalbibliothek
The Deutsche Nationalbibliothek lists this publication in the Deutsche Nationalbibliografie; detailed bibliographic data are available on the Internet at http://dnb.d-nb.de.

Cover Design Formgeber, Eppelheim
Typesetting Thomson Digital, Noida, India
Printing and Bookbinding Strauss GmbH, Mörlenbach

Printed in the Federal Republic of Germany
Printed on acid-free paper

ISBN: 978-3-527-32415-6

Contents

Modeling and Simulation in Polymers. Edited by P.D. Gujrati and A.I. Leonov

ISBN: 978-3-527-32415-6

Preface

Polymers are now one of the major materials used in many industrial applications. The prediction of their behavior depends on our understanding of these complex systems. As happens in all experimental sciences, our understanding of complex physical phenomena requires modeling the system by focusing on only those aspects that are supposedly relevant to the observed behavior. Once a suitable model has been identified, we have to validate it by solving it and comparing its predictions with experiments. Solving the model usually requires some approximation or approximations. The resulting solutions are known as theories under those approximations. Indeed, the interpretation and analyses of experimental data depend in a crucial way on the way a system is modeled and the availability of corresponding reliable theories. This is also true for polymers, whose behavior can be very complex. After more than 70 years of research and development, polymer science and engineering has reached a high level of sophistication. Application of modern computational tools and methods, and modeling and theories of observed behavior, made it possible to solve, thereby understand, many basic and applied problems, and along with that helped to resolve some fundamental problems in polymer physics, physicochemistry, and polymer mechanics. This book illustrates the progress achieved in using computational methods to understand the behavior and reliability of various models in polymer science and engineering. The editors wanted a well-balanced presentation from scientists and engineers. Accordingly, their attempt was to seek contributions from universities, industries, and national laboratories so that the book could represent a wide array of topics of interest in the field. The main desire was to have a book that will not only function as a resource for people active in the field but will also help educate them and graduate students.

One of the tough jobs the editors had was of limiting the number of contributions to keep the size of the book manageable. Their job was made easy though by the contributors, who are well recognized in their respective fields. The editors are pleased to present their reviews for the benefit of the reader. The selection of the topics covered in this book in no way reflects their bias; rather, it reflects the strengths of the contributors. The topics cover a range of problems in polymers,

Modeling and Simulation in Polymers. Edited by P.D. Gujrati and A.I. Leonov

ISBN: 978-3-527-32415-6

including liquid crystals and biopolymers. Since in many cases the science and engineering are not well distinguishable, the editors decided to use a "mixed" approach in presenting the contributions in the book in alphabetic order of the authors.

The chapter by Beris and Housiadas ("Computational Viscoelastic Fluid Mechanics and Numerical Studies of Turbulent Flows of Dilute Polymer Solutions") aims at resolving the famous long-standing problems of turbulent drag reduction. This contribution describes recent efforts and achievements in direct computations of near-wall turbulent flows of dilute polymer solutions and comparisons with experimental data.

In spite of many efforts, attempts to model complicated properties of liquid crystalline polymers (LCPs) are far from being complete. The constitutive equations of continuum type for thermotropic LCPs were proposed only last year. Multiparametric character of these equations is the challenging problem for LCP simulations. The chapter by Chen and Leonov ("Liquid Crystalline Polymers: Theories, Experiments, and Nematodynamic Simulations of Shearing Flows") reviews the major findings in this field, describes new continuum theory valid for thermotropic LCPs, and illustrates simulations of their shearing flows.

Glasses form an important part not only of industry but also of our daily life. A satisfactory understanding of glasses is not within our reach at present, mainly because it represents a nonequilibrium state of matter, which seems to not follow Nernst's postulate or the third law of thermodynamics. Gujrati in his chapter ("Energy Gap Model of Glass Formers") addresses the issue by elevating to a fundamental level the observed fact that the energy of a glass is much higher than that of the corresponding crystal (an energy gap) even at absolute zero. The chapter therefore deals with glass formation in a supercooled liquid for which a more thermodynamically stable state exists in the form of a crystal. The resulting metastability is studied thermodynamically by treating it as constrained equilibrium. The theoretical model is also supported by performing simple exact calculations.

Polymer nanocomposites are prospective new materials. Nevertheless, their properties are still not well understood. The chapter by Heinz, Patnaik, Pandey, and Farmer ("Modeling of Polymer Matrix Nanocomposites") demonstrates the application of modern computational methods for investigating dispersion of various nanofillers in polymer matrices under the action of microscopic forces. In addition, an attempt is made to calculate the thermal conductivity in a model system of nanotubes in polymer matrix.

Predicting flow properties of polymers such as interfacial slip is of paramount importance in industries and poses a major challenge at present. It truly requires a multiscale attack. Ilg, Mavrantzas, and Öttinger provide in their contribution ("Multiscale Modeling and Coarse Graining of Polymer Dynamics: Simulations Guided by Statistical Beyond-Equilibrium Thermodynamics") a comprehensive treatment of polymer flow dynamics by borrowing ideas from nonequilibrium thermodynamics. They develop a multiscale modeling approach, which successfully bridges microscopic and macroscopic scales. By using GENERIC formalism, they attempt to avoid thermodynamic inconsistency that is present in most coarse-grained models. They achieve this by carefully separating timescales.

Studying charged polymers in aqueous solutions provides another example of a major challenge in polymer technology, and is considered by Kundagrami, Kumar, and Muthukumar ("Simulations and Theories of Single Polyelectrolyte Chains"). Only single chains are considered. Chain connectivity and topological considerations are the sources of complication in understanding the interactions between the solute and the solvent. They consider two different kinds of theoretical methods, variational and self-consistent, and employ Langevine dynamics for their simulations.

Studies of polymerization kinetics have a long history. Nevertheless, many problems in this field remain unresolved. Using computational methods, several of these problems are clarified in the chapter by Litvinenko ("Computational Studies of Polymer Kinetics"). Special attention is paid to the effect of chain transfer reactions on polymer molecular weight and applications to different types of polymerization methods.

Modeling of polymer processing has more than a 60-year history. Computational methods gave many possibilities for optimization of processing operations. The chapter by Mitsoulis ("Computational Polymer Processing") provides a comprehensive review of simulations and computational efforts for a majority of polymer processing operations and also forecasts the future development in this important part of polymer industry.

Multicomponent melts that are commonly found in such varied situations as material fabrification, reinforcement, blending, and so on are discussed in the chapter by Müller ("Computational Approaches for Structure Formation in Multicomponent Polymer Melts"). Only equilibrium properties are discussed along with computational approaches for coarse-grained models in the mean field approximation. Both hard-core and soft-core models are used to cover a multitude of scales of length, time, and energy. Attention is also paid to methods that go beyond the mean field approximation.

The chapter by Poldneff and Heinstein ("Computational Mechanics of Rubber and Tires") reviews the latest achievements in using finite element analysis for solving highly nonlinear problems of rubber and tire mechanics, with several illustrative examples of industrial importance.

Application of ideas from polymers has begun to revolutionize bio-related disciplines. Therefore, this review book will not be complete without a chapter detailing such an application. The last chapter ("Modeling the Hydrodynamics of Elastic Filaments and its Application to a Biomimetic Flagellum") by Stark attempts to model Nature's successful strategies for propulsion such as of sperm cells and fluid transport such as of mucus. The artificial cilium is based on a superparamagnetic filament, actuated by an external magnetic field; the latter allows one to explore the filament's capacity to transport fluid.

The editors hope that the collection of reviews will be beneficial to graduate students, scientists, and engineers, whether practicing or just eager to familiarize themselves with new models and computational tools. The editors also take full responsibility for any shortcoming of the book.

Finally, the editors offer their sincere thanks to Manfred Kohl for inviting us to take on this project and to Stefanie Volk and Claudia Nussbeck, all at Wiley-VCH, for their patience and help to ensure the completion of this project.

August 2009
Akron, OH, USA

Puru Gujrati and Arkady Leonov

List of Contributors

Antony N. Beris
University of Delaware
Department of Chemical Engineering
Newark, DE 19716
USA

Hongyan Chen
The University of Akron
Department of Polymer Engineering
Akron, OH 44325-0301
USA

Barry L. Farmer
Wright-Patterson Air Force Base
Air Force Research Laboratory
Materials and Manufacturing
Directorate
Dayton, OH 45433
USA

Puru D. Gujrati
The University of Akron
The Departments of Physics and
Polymer Science
Akron, OH 44325
USA

Martin W. Heinstein
Sandia National Laboratory
Computational Solid Mechanics and
Structural Dynamics
P.O. Box 5800 MS 0380
Albuquerque, NM 87185-0380
USA

Hendrik Heinz
The University of Akron
Department of Polymer Engineering
Akron, OH 44325
USA

Kostas D. Housiadas
University of the Aegean
Department of Mathematics
Karlovassi
Samos
Greece

Patrick Ilg
Polymer Physics, ETH Zürich
Department of Materials
CH-8093 Zürich
Switzerland

Modeling and Simulation in Polymers. Edited by P.D. Gujrati and A.I. Leonov

ISBN: 978-3-527-32415-6

Rajeev Kumar
University of Massachusetts
Polymer Science and Engineering
Department
Amherst, MA
USA

and

University of California
Materials Research Laboratory
Santa Barbara, CA
USA

Arindam Kundagrami
University of Massachusetts
Polymer Science and Engineering
Department
Amherst, MA
USA

Arkady I. Leonov
The University of Akron
Department of Polymer Engineering
Akron, OH 44325-0301
USA

Galina Litvinenko
Karpov Institute of Physical Chemistry
Vorontsovo Pole 10
Moscow 105064
Russia

Vlasis Mavrantzas
University of Patras and FORTH-ICE/
HT
Department of Chemical Engineering
Patras GR 26504
Greece

Evan Mitsoulis
National Technical University of Athens
School of Mining Engineering &
Metallurgy
Zografou
157 80 Athens
Greece

Marcus Müller
Georg-August-Universität
Institut für Theoretische Physik
37077 Göttingen
Germany

Murugappan Muthukumar
University of Massachusetts
Polymer Science and Engineering
Department
Amherst, MA
USA

Hans Christian Öttinger
Polymer Physics, ETH Zürich
Department of Materials
CH-8093 Zürich
Switzerland

Ras B. Pandey
University of Southern Mississippi
Department of Physics and Astronomy
Hattiesburg, MS 39406
USA

Soumya S. Patnaik
Thermal and Electrochemical Branch
Energy Power Thermal Devision
Propulsion Directorate
Air Force Research Laboratory
Dayton, OH 45433
USA

Michael J. Poldneff
External Science & Technology
Programs
The Goodyear Tire & Rubber Company
Innovation Center, P.O. Box 3531
D/410A
Akron, OH 44309-3531
USA

Holger Stark
Technische Universität Berlin
Institut für Theoretische Physik
Hardenbergstr. 36
10623 Berlin
Germany

1
Computational Viscoelastic Fluid Mechanics and Numerical Studies of Turbulent Flows of Dilute Polymer Solutions

Antony N. Beris and Kostas D. Housiadas

1.1
Introduction and Historical Perspective

According to the late A.B. Metzner [1], the phenomenon of polymer-induced drag reduction was independently discovered by two researchers, K.J. Mysels in May 1945, as reported by the discoverer at an AIChE symposium on drag reduction in 1970, and B.A. Toms in the summer of 1945, as reported by the discoverer at the IUTAM symposium on the structure of turbulence and drag reduction held in Washington, DC in 1976. The original fluids studied in these two first experimental investigations were micellar aluminum disoaps or rubber in gasoline (K.J. Mysels) and polymethyl methacrylate in monochlorobenzene (B.A. Toms). However, due to the war, the first records in an accessible publication were found later [2, 3] with journal contributions even much later [4, 5]. Since that time, the field has literally exploded with 500 papers until the seminal review by Virk [6] and 4900 papers by 1995 [1].

The first attribution of drag reduction to fluid viscoelasticity was by Dodge and Metzner [7], whereas the first description of drag reduction as "Toms effect" was by Fabula at the Fourth International Congress on Rheology, held in 1996 [8]. The first measurements of viscoelasticity of drag-reducing fluids were performed by Metzner and Park [9] and Hershey and Zakin [10]. The first articulations of a "maximum drag reduction asymptote" for dilute polymer solutions were reported by Castro and Squire [11], Giles and Pettit [12], and Virk *et al.* [13]. As far as the first proposed mechanisms of drag reduction due to fluid mechanical effects are concerned, Lumley attributed drag reduction to molecular stretching in the radial flow patterns in turbulent flows [14]. Simultaneously, Seyer and Metzner [15] clarified it even further, as due to high extensional deformation rates in radial flow patterns in turbulent flows and high resistance to stretching of viscoelastic fluids. More recently, Lumley and Blossey [16] elaborated further by arguing that polymer additives, by boosting the extensional viscosity of the fluid, affect especially the structure of the turbulent bursts; see also Ref. [17]. This same mechanism is also suspected to be operative when other additives are employed, such as micellar surfactant solutions [2, 4, 18–20]

Modeling and Simulation in Polymers. Edited by P.D. Gujrati and A.I. Leonov

ISBN: 978-3-527-32415-6

and fiber suspensions [21]. Ever since Virk [6] made the first thorough review of drag reduction, the interest in the phenomenon continues uninterrupted – see, for example, the experimental works [22–26], the more recent reviews by Sellin and Moses [27, 28] and Bushnell and Hefner [29], the book by Gyr and Bewersdorff [30], and the latest reviews by Nieuwstadt and Den Toonder [31], Graham [32], and White and Mungal [33].

The phenomenon of polymer-induced drag reduction [5, 6, 34, 35] describes the effect of relatively small quantities, as small as on the order of ppm by weight, of a high molecular weight polymer, which when added to a low molecular weight solvent, such as water or crude oil, reduce the turbulent drag; see Figure 1.1 for a representation of some early but standard experimental results as adapted from Virk [6]. According to Metzner [1], Lummus *et al.* [36] recorded what appears to be the first commercial exploitation of turbulent drag reduction, using polymeric additives to increase oil well drilling rates in 50 wells in the United States and Canada. Increases in drilling rate were tabulated. The first publicly recorded usage of drag reduction in pipelines appears to be that of Burger *et al.* [37], who reported results for 1 and 2 in.

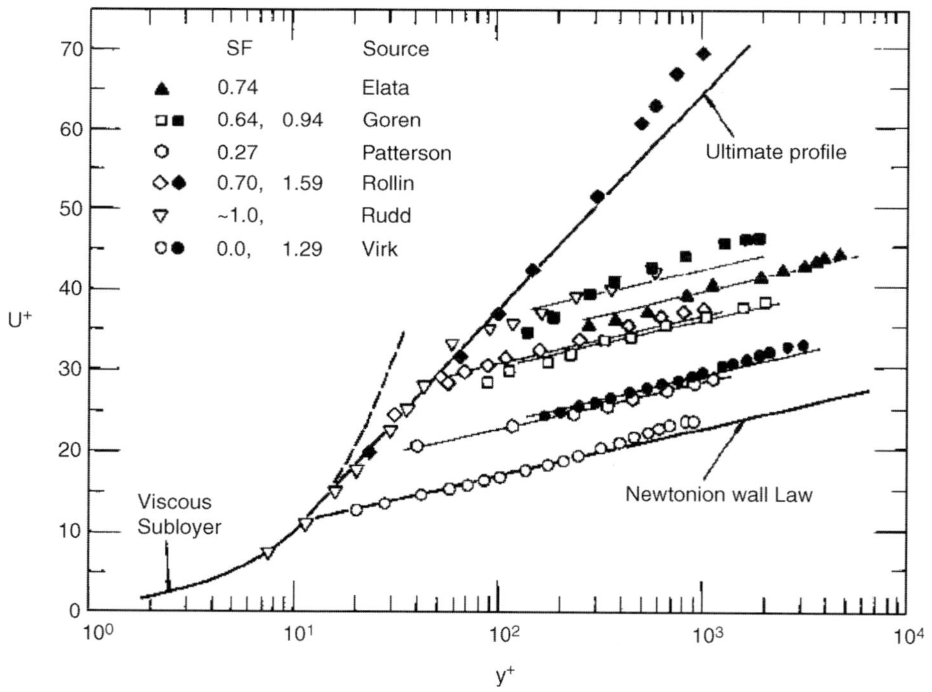

Figure 1.1 Velocity profiles in flows with drag reduction. The different symbols represent different polymer solutions; the open circles provide the pure solvent (water) base data. As polymer additives are introduced, and for the same pressure drop (equivalently, the same wall stress, i.e., the same velocity slope at the wall), the flow rate increases, from the base line to the maximum (Virk) asymptote, called "Ultimate Profile" in the figure [6]. (Reprinted with permission from John Wiley & Sons)

tubes as well as 14 and 48 in. pipelines. The best available estimates of usage in petroleum transportation in 2004, obtained from industry sources, state that drag reducers are employed to treat approximately 20 million barrels of hydrocarbons per day, which may be compared with a worldwide crude oil production rate of about 80 million bpd. A major application is in facilitating the oil transfer through the Alaskan pipeline [38]. Commercial-scale tests of drag-reducing additives in municipal heating and cooling systems are described by Zakin *et al.* [18], with emphasis on surfactant additives – there is also a reference to an application to the heating system of pipelines in offshore drilling [20]. It is interesting to note that the drag reductions achieved by using surfactant (colloidal) additives are very large and may exceed those of the "maximum drag-reducing asymptote" for dilute polymer solutions, as is also the case when fiber–polymer mixtures are employed [39]. We are unaware of any regular commercial use of these materials. Incidentally, marine applications of turbulent drag reduction have been studied for many decades and measurements performed of dilute solutions of fish mucus that is polymeric in nature in turbulent pipe flow exhibited as much as 66% reduction in friction drag [40]. Even snails were found to produce a mucus that reduces drag [40].

During the past 15 years, the most significant progress undoubtedly came in the theoretical front from the development of direct numerical simulations (DNS) due to advances in computational viscoelastic fluid mechanics, models, methods, and computer hardware (parallel computing). The goal was to theoretically understand in more depth the nature and the underlying mechanisms of polymer-induced drag reduction. The simulations, with the exception of some limited work on homogeneous viscoelastic turbulence [41], have primarily focused on the analysis of inhomogeneous turbulent viscoelastic flows. Even there, with the exception of some limited work on boundary layer flows [42, 43] – see also the very recent review by White and Mungal [33] – and pipe flow [44], the bulk of the work has focused on simulations of channel flow. The very first simulations (those reported by Orlandi [45] and Den Toonder *et al.* [44]) used ad hoc constitutive relationships for the stress that attempted to empirically capture the increase in the resistance to extensional deformation so characteristic to polymeric systems. As a result, they did get some encouraging results, exhibiting the right trends with increasing viscoelasticity in the flow, in agreement with experimental observations [23, 30, 44]. For example, they noticed a decrease in the strength of longitudinal structures accompanied by an increase in their spacing with increasing polymer concentration [45] and drag reduction with the right changes in the root mean square (rms) values of the velocity fluctuations and velocity probability distribution functions (pdf), exhibiting increasing intermittency [44].

However, the first DNS based on a microscopically originated constitutive equation for the polymer dynamics (the finite extensibility nonlinear elastic with the Peterlin approximation (FENE-P) model [46]) was conducted by Sureshkumar *et al.* [47]. In this work, for a fixed friction Reynolds number and other rheological parameters, drag reduction was observed as the Weissenberg number increased beyond a critical onset value. Moreover, accompanying drag reduction, characteristic changes were observed in the velocity and vorticity mean and rms values, the Reynolds stress, and the energy spectra all in qualitative agreement with experimental observations obtained with drag-reducing dilute polymer solutions [22, 23, 48]. Shortly afterward, DNS based on a network-based constitutive model, the Giesekus model [49], also produced similar results [50]. These large-scale (three-dimensional and time-dependent) simulations under turbulent conditions are similar to the ones carried out for Newtonian fluids [51, 52] and also account for the presence of the polymer additives through a full modeling of the coupled effect of additives on the flow.

Since these initial works, we have seen a flourishing of DNS works by our own [53–58] and many other [59–66] research groups. We include in the discussion here only those works that involve the fully coupled flow and polymer concentration problem; there are many more works that monitored the polymer molecular extension against a fixed flow field obtained from Newtonian DNS (see, for example, Ref. [67]) – this is a much simpler computational problem, many of the numerical stability problems emerge only when the fully coupled problem is considered. Of interest is the spectral DNS work by De Angelis *et al.* [59] showing as main effect of viscoelasticity a quite concentrated action on bursting phenomena and a stabilization of the streaks resulting in fewer, bigger, coherent structures, reinforcing previous results [47]. Later work [60] involving the application of a Karhunen–Loéve (K-L) decomposition for the extraction of information on larger coherent structures in the flow (following Refs [68–70] and others) has reported that the main change in viscoelasticity is primarily an energy redistribution among the K-L modes, while the spatial profile of the most energetic modes stays the same. This is in contrast to the findings in Ref. [58] where a similar K-L decomposition has also revealed substantial changes in the spatial structure of the most energetic modes and an overall enhancement of the energy content of the first few modes of the decomposition upon addition of viscoelasticity. However, this may be just due to the higher elasticity of flows in direct numerical simulations in Ref. [58] allowing to more clearly see these effects.

Another work worth mentioning as a measure of the activity in the field is that of Ptasinski *et al.* [62]. The authors in this work asserted that they have simulated cases close to maximum drag reduction by using model parameters (such as a high Weissenberg number, 50, a high extensibility parameter value, 33, and, primarily, a small solvent to total viscosity ratio, of order 0.5) implying a high polymer contribution. The results indeed seem to be in agreement with expectations (such as a law-of-the-wall with a significantly high slope) with drag reductions on the order of 60–65% for the higher elasticity cases. However, there still remain questions primarily due to the small computational domain size used (minimal channel of $(3 \times 2 \times 2)$ made dimensionless by the half channel-height and small mesh resolution of $(48 \times 32 \times 100)$ in the x, y, and z directions, respectively, along with mixed spectral (along the period directions, x and y) and finite difference (along the wall-normal direction, z) approximations). Given the fact that, as mentioned above, one dominant effect of viscoelasticity is to lead to substantially larger organized coherent structures than those seen in Newtonian flows, the use of this minimal computational domain size places questions on the validity of the final results. Even more questions are placed based on the small mesh resolution used, especially along the shear direction. For the Reynolds number used (friction Reynolds number of 180), a mesh size of at least 96 ought to have been used. Moreover, the technique used was not immune to flow instabilities (as also the standard spectral method), and an extra conformation diffusion was also needed to be used (let alone the diffusion introduced by the use of the finite difference approximation along the sheatwise (z) direction). So, although progress is definitely made, still there are questions regarding the validity of this and other maximum drag reduction results that make that state still, in our opinion, unsolved. To achieve a solution in this regime is a very demanding task both from a computational perspective and from a polymer modeling perspective.

Other works used, instead of spectral, lower order accuracy finite difference approximations, but they have to be noted here since they employed more suitable numerical formulations for the constitutive equations that explicitly avoided the introduction of artificial diffusivity in the numerical solution [61, 63–66]. Those works also employed the FENE-P model to simulate dilute polymer solutions [61, 63, 64] or the Giesekus model for surfactant turbulent flow [19, 65, 66].

Moreover, we have recently seen, in addition to straight DNS computations, some very substantial theoretical analyses, as very elegant analyses based on the examination of exact coherent states in plane shear flow, by Stone *et al.* [71]. These represent unstable solutions of the laminar flow problem in terms of traveling waves – see also the excellent review by Graham [32]. The analysis of the effects of viscoelasticity on these states has shown [72] that they mirrored the changes observed in experiments and simulations of fully turbulent flows of polymer solutions: this similarity establishes again the importance of large-scale structures in turbulent flow. Their modification by viscoelasticity elegantly reveals most of the drag-reducing effects. This work followed a similar analysis performed only relatively recently for Newtonian turbulent flows [73]. An exciting development was also the recent experimental discovery of the dominant traveling wave mode in a Newtonian turbulent pipe flow field [74] indicating the physical significance in further theoretical investigations of large coherent motion structures in turbulent flows. These results are also consistent with other methods of analysis of large-scale coherent motion such as the K-L decomposition ([68–70] and references therein). As already mentioned, K-L analysis of viscoelastic turbulent channel flow has also been conducted by de Angelis *et al.* [60], Housiadas *et al.* [58], and, more recently, by Handler *et al.* [75] and Samanta *et al.* [76], indicating a significant enhancement of the energy contained in the large scales as viscoelasticity sets in. The relevance of these results in improving our understanding makes these types of analyses highly desirable in future work, along with more traditional analysis of the turbulence statistics. In parallel, there is considerable recent work, taking sometimes advantage of the insight developed thanks to DNS results, aimed at developing $\varkappa-\varepsilon$-type turbulent models applicable for drag-reducing polymer flows – see, for example, Refs [17, 77] and references therein. However, due to space limitations, we will not discuss any further these and related investigations.

This chapter aims to present, in a concise way, the major elements and results from applications of computational viscoelastic fluid mechanics in numerical studies (DNS) of turbulent channel flows of homogeneous, dilute, polymer solutions under drag-reducing conditions. In the next section, we present a summary and outline of governing equations with emphasis on polymer modeling. In Section 1.3, we describe key elements of the spectral method algorithm as the principal numerical method used in DNS investigations. Section 1.4 overviews the main results of the DNS with an emphasis on the principal effects of the variation of key dimensionless numbers and rheological characteristics on the most important turbulent statistics. A discussion on the influence of numerical parameters is also included. Finally, Section 1.5 contains the conclusions and some thoughts for future work.

1.2
Governing Equations and Polymer Modeling

The first among the governing equations used in DNS are the mass and momentum balances, which for an incompressible, isothermal, and constant polymer concentration flow in a channel are given in dimensionless form as

$$\underline{\nabla} \cdot \underline{u} = 0, \tag{1.1}$$

$$\frac{D\underline{u}}{Dt} = -\underline{\nabla} p + \frac{\beta_0}{Re_{\tau 0}} \underline{\nabla}^2 \underline{u} + \frac{1-\beta_0}{Re_{\tau 0}} \underline{\nabla} \cdot \underline{\underline{\tau}} + \underline{e}_x, \tag{1.2}$$

where $D/Dt \equiv (\partial/\partial t) + \underline{u} \cdot \underline{\nabla}$ denotes the material derivative, $\beta_0 \equiv (\eta_s^*/\eta_0^*) = (\eta_s^*/(\eta_s^* + \eta_{p0}^*))$ is the ratio of the solvent viscosity η_s^* to the total zero shear rate viscosity of the solution η_0^*, and $Re_{\tau 0}$ is the zero shear rate friction Reynolds number, $Re_{\tau 0} \equiv (u_\tau^* h^*/\nu_0^*)$, where u_τ^* is the friction velocity defined as $u_\tau^* \equiv \sqrt{\tau_w^*/\varrho^*}$, where τ_w^* is the total wall shear stress and ϱ^* is the constant density of the solution, h^* is the half channel width, and $\nu_0^* \equiv (\eta_0^*/\varrho^*)$ is the zero shear rate kinematic viscosity of the solution. In Eqs. (1.1) and (1.2), $\underline{u} = u_x \underline{e}_x + u_y \underline{e}_y + u_z \underline{e}_z$ is the (dimensionless) velocity vector, where $\underline{e}_x, \underline{e}_y, \underline{e}_z$ are the unit vectors along the axes, p is the (dimensionless) periodic part of the pressure, and $\underline{\underline{\tau}}$ is the (dimensionless) extra stress tensor, introduced due to the presence of polymer in the flow. The momentum balance, Eq. (1.2), has been formulated for pressure-driven flow along the streamwise direction x, and this is how the constant term $\underline{e}_x$ arises. Equations (1.1) and (1.2) have been nondimensionalized by using the characteristic scales reported in the first column in Table 1.1, usually referred to as "computational scales," as opposed to the "zero shear rate wall scales" and the "actual wall scales" reported in columns 2 and 3, respectively. In Table 1.1, u_b^* is the average (bulk) velocity, η_{p0}^* is the zero shear rate polymer viscosity, k_B^* is the Boltzmann constant, T^* is the absolute temperature, and K^* is the characteristic elastic spring strength of the nonlinear elastic dumbbell. Note that the superscript asterisk denotes a dimensional quantity. More details can be found in Ref. [78].

Note that in viscoelastic turbulent flows, because of the shear thinning effect [35, 79], we have to distinguish between two different types of wall units. One is based on the zero shear properties and the other, applicable for channel and boundary layer

Table 1.1 Characteristic computational and wall scales for fixed wall stress except for the last column that is for fixed flux conditions.

Quantity	Computational scale	Reference wall scale	Actual wall scale	Computational scale (fixed flux)
Length	h^*	ν_0^*/u_τ^*	ν^*/u_τ^*	h^*
Velocity	u_τ^*	u_τ^*	u_τ^*	u_b^*
Time	h^*/u_τ^*	ν_0^*/u_τ^{*2}	ν^*/u_τ^{*2}	h^*/u_b^*
Pressure	τ_w^*	τ_w^*	τ_w^*	$\varrho^* u_b^{*2}$
Polymer stress	$\eta_{p0}^* u_\tau^*/h^*$	$\eta_{p0}^* u_\tau^{*2}/\nu_0^*$	$\eta_{p0}^* u_\tau^{*2}/\nu^*$	$\eta_{p0}^* u_b^*/h^*$
Conformation	$k_B^* T^*/K^*$	$k_B^* T^*/K^*$	$k_B^* T^*/K^*$	$k_B^* T^*/K^*$

flows, on the stationary wall conditions [78]. The advantage of the first is that it gives rise to quantities that are a priori known so that it makes easier the setting up of numerical computations. In fact, in practice, the numerical simulations are set only in terms of such nondimensionalization. However, because of shear thinning, the actual viscous shear viscosity next to the wall is different from its zero shear rate value. As a result, the proportionality of the velocity and the distance that it is expected to hold in the viscous sublayer next to the wall, when both are expressed in terms of wall units, $u^+ = y^+$ [34], no longer holds when the zero shear rate viscosity is used to scale the length distance. Therefore, to allow a physical interpretation of the results, it is necessary to use the actual wall viscosity value in order to construct the dimensionless wall length and time units. However, a disadvantage of this approach is that since with viscoelastic systems the total shear viscosity is not a material property but essentially an "effective" quantity that can be determined only a posteriori after the full solution is known, such nondimensionalization can also be performed only a posteriori. For dilute solutions, since the bulk of the solution shear viscosity is due to the solvent and stays constant, we only have small changes due to this rescaling from zero shear rate (nominal) to the actual wall values. Moreover, as DNS information indicates [78], the correction can be approximated on the basis of the laminar steady shear flow model predictions, albeit the shear flow that is established in the shear sublayer next to the wall is neither laminar nor steady in time. For brevity, we leave such a correction out of the reported calculations here, all of which are therefore reported in terms of zero shear rate quantities except for the drag reduction where those corrections can be both important and essential for an accurate estimation (see Section 1.4.1).

Equations (1.1) and (1.2) are not closed because of the presence of the extra stress tensor $\underline{\underline{\tau}}$. Therefore, one more equation is required, which is provided by a viscoelastic constitutive model [35, 49, 79]. According to nonequilibrium thermodynamics, the most thermodynamically consistent way to describe the constitutive model is in terms of internal (structural) variables for which separate evolution equations are to be described [49]. The simplest case is when a single, second-order and positive definite, tensor, the conformation tensor, $\underline{\underline{c}}$, is used. In that case, the most general constitutive equation is given, in dimensionless form, as [80]

$$\frac{D\underline{\underline{c}}}{Dt} = \underline{\underline{c}} \cdot \underline{\nabla u} + \underline{\nabla u}^{\mathrm{T}} \cdot \underline{\underline{c}} + \underline{\underline{g}} + \frac{D_0^+}{Re_{\tau 0}} \nabla^2 \underline{\underline{c}}, \tag{1.3}$$

where the superscript T denotes the transpose and $\underline{\underline{g}} = \underline{\underline{g}}(\underline{\underline{c}})$ is a second-order tensor that models the relaxation effects and can be interpreted as a restoring force to equilibrium.

On the right-hand side of the constitutive equation, Eq. (1.3), a diffusion term has been added, as proposed by Sureshkumar and Beris [81], so that in turbulent simulations the high wavenumber contributions of the conformation tensor do not diverge during the numerical integration of this equation in time. This parallels the introduction of a numerical diffusion term in any scalar advection equation (e.g., a concentration equation with negligible molecular diffusion) that is solved along with the flow equations under turbulent conditions [82]. In Eq. (1.3), D_0^+ is the dimensionless numerical diffusivity [54–56]. The issue of the numerical diffusivity is further discussed in Sections 1.3.2 and 1.4.3.

For the cases considered here, the relaxation term $\underline{\underline{g}}$ simplifies into the following expression:

$$\underline{\underline{g}} = -\underline{\underline{\tau}} - \alpha We \underline{\underline{\tau}} \cdot \underline{\underline{\tau}}, \tag{1.4}$$

where the Weissenberg number We is defined as

$$We \equiv \frac{\lambda^* u_\tau^*}{h^*} = \left(\frac{\lambda^* (u_\tau^*)^2}{\nu_0^*}\right)\left(\frac{\nu_0^*}{h^* u_\tau^*}\right) \equiv \frac{We_{\tau 0}}{Re_{\tau 0}},$$

in terms of the zero shear rate friction Weissenberg number, $We_{\tau 0} \equiv \lambda^*((u_\tau^*)^2/\nu_0^*)$, and the zero shear rate friction Reynolds number, $Re_{\tau 0}$, where λ^* is the relaxation time of the polymer chains. The polymer extra stress tensor is directly connected to $\underline{\underline{c}}$, through

$$\underline{\underline{\tau}} = \underline{\underline{c}} \cdot \frac{\delta H}{\delta \underline{\underline{c}}}, \tag{1.5}$$

where H is the Hamiltonian (total energy of the system) [49, 80, 83]. For the cases considered here, Eq. (1.5) simplifies in the following dimensionless expression:

$$\underline{\underline{\tau}} = \frac{f(\underline{\underline{c}})\underline{\underline{c}} - \underline{\underline{I}}}{We}. \tag{1.6}$$

The conformation tensor has a definite physical origin and interpretation, typically associated with the second moment of a suitably defined chain end-to-end distribution function [84]

$$\underline{\underline{c}} \equiv \langle \underline{RR} \rangle, \tag{1.7}$$

where $\underline{R}$ is the chain end-to-end distance vector. It is important to note here that it is exactly this association that also induces a very special property in the conformation tensor, its positive definiteness. In fact, this property is absolutely essential for its proper interpretation and the thermodynamic consistency (i.e., the requirement of a nonnegative entropy production) of the accompanying evolution equations [49, 83, 85]. Indeed, the eigenvalues of $\underline{\underline{c}}$ and their corresponding eigenvectors have the physical meaning of the square of the average macromolecular size along the primary three directions and the orientation of these directions in space, respectively. More discussion on the implication of the positive definiteness requirements on the numerical methods follows in Section 1.3.2.

When $We \to 0$, then $f \to 1$ and $\underline{\underline{c}} \to \underline{\underline{I}}$, which give that $\underline{\underline{\tau}} = \underline{\nabla v} + (\underline{\nabla v})^{T}$, and Eqs. (1.1) and (1.2) are reduced to the Navier–Stokes equations. Neglecting the numerical diffusion term and depending on the rheological parameters, α, β_0, and the specific form for function $f = f(\underline{\underline{c}})$, entering Eqs. (1.5), (1.2) and (1.6), respectively, the following well-known constitutive models are recovered:

$$f = 1, \quad a = 0, \quad \beta_0 = 0: \quad \text{Maxwell}, \tag{1.8a}$$

$$f = 1, \quad a = 0, \quad 0 < \beta_0 < 1: \quad \text{Oldroyd-B}, \tag{1.8b}$$

$$f = 1, \quad a > 0, \quad 0 < \beta_0 < 1: \quad \text{Giesekus}, \tag{1.8c}$$

$$f = \frac{L^2 - 3}{L^2 - tr(\underline{\underline{c}})}, \quad a = 0, \quad 0 < \beta_0 < 1: \quad \text{FENE} - \text{P}, \tag{1.8d}$$

where L is the maximum polymer chain extensibility parameter. Note that $f \to 1$ as $L \to \infty$; thus, in this case the FENE-P reduces to the Oldroyd-B model. So far, in DNS of viscoelastic turbulent flows, basically two constitutive models were employed. First, to describe the finite chain length effects in dilute solutions of long flexible macromolecules, the FENE-P model [35, 49, 84] was used. This model arises as an averaging approximation of a microscopic kinetic theory description of the end-to-end distribution of the macromolecular chain deformation with the conformation tensor representing the second moment of that distribution [84]; see Eq. (1.1). Notice that for this model, due to the inherent finite chain extensibility, there is an upper bound for the conformation tensor in addition to its positive definiteness constraint [35]. Second, the network-based Giesekus model [35, 49, 84] has also been used to describe drag-reducing turbulent channel flows of surfactant solutions [19, 65, 66] and viscoelastic effects in concentrated polymer solutions.

Finally, the appropriate boundary conditions are the standard nonslip conditions at the channel walls and periodicity conditions along the homogeneous directions:

$$\underline{u}(x, y = \pm 1, z) = 0, \tag{1.9a}$$

$$\underline{u}, p, \underline{\underline{c}} \text{ periodic in } x,\ z \text{ directions}. \tag{1.9b}$$

No boundary conditions are applied at the walls for the conformation (or the stress) tensor since even in the presence of numerical diffusion, the numerical diffusivity D_0^+ is nonzero only in the bulk; for more details, see Refs [50, 54, 78].

1.3
Numerical Methods for DNS

Outside the spectral method-based DNS work, special mention needs to be made for two finite difference works. First, Min *et al.* [64] used a special upwind finite difference formulation through which they managed to simulate without additional diffusivity for the stress a highly elastic turbulent viscoelastic case (corresponding to Oldroyd-B model) for which maximum drag reduction is obtained. Although the method used has admittedly some distinct advantages over the spectral approach in terms of enhanced stability, one should still be cautious of the fact that the lower order finite difference approximations used in Min's work (as well as in above-mentioned other finite difference works, to a greater extent) also introduce diffusion in the results based on the diffusive character of the finite difference approximations (in Min's work, all approximations are second-order finite differences except for a fourth-order scheme used to approximate the stress divergence term in the momentum equation). Thus, the results of the maximum drag reduction need further independent corroboration.

Second, we need to mention the work by Yu and Kawaguchi [65] that, within a finite difference approximation, employed a special MINMOD finite difference formulation that they demonstrated to be more accurate and more stable than the corresponding formulation that employs an artificial diffusion term. However, it should be noted that since they patterned the value of the numerical diffusivity after the work of Dimitropoulos *et al.* [50], they used significant larger values for it than the ones used in subsequent spectral works [54], also reported here. Yu and Kawaguchi used the Giesekus constitutive model for various flow and rheological parameter values. Whenever those values happen to be close to the ones used in spectral simulations [55], the results are similar. Yu and Kawaguchi [65] also ran simulations at much lower viscosity ratios, corresponding to much higher maximum extensional viscosities, and under these conditions they reported very high drag reductions (more than 70%).

We believe that as far as investigating the large-scale features of turbulence is concerned, lower order approximations are powerful enough, if formulated suitably, to give valuable results. However, we also believe that if one wants to investigate in detail turbulence in many scales of length and time, it is hard to beat the accuracy and efficiency of spectral methods, albeit one has to abide by their limitations in simple geometry applications. Since most of our emphasis is on developing a better understanding of the mechanisms of drag reduction through a detailed investigation of interactions between viscoelasticity and turbulence at multiple scales of length and time, we believe that our primary emphasis on the spectral methods is fully justified.

1.3.1 Spectral Methods: Influence Matrix Formulation

In a fully spectral representation, each primary dependent variable, $S = S(x, y, z, t)$, is approximated through a triple series expansion in terms of its spectral coefficient, $\hat{s}_{rkl} = \hat{s}_{rkl}(t)$. This involves a double Fourier series along the two periodic directions (streamwise and spanwise, using N_x and N_z Fourier modes, respectively) and $N_y + 1$ Chebyshev orthogonal polynomial series along the nonperiodic (shear) direction:

$$S(x, y, z, t) = \sum_{r=-(Nx/2)}^{(Nx/2)-1} \sum_{k=-(Nz/2)}^{(Nz/2)-1} \sum_{l=0}^{Ny} \hat{s}_{rkl}(t) T_l(y) \exp\left\{2\pi i\left(\frac{x}{L_x} r + \frac{z}{L_z} k\right)\right\}, \quad (1.10)$$

where the three subscripts "r", "k", and "l" of the generic spectral coefficient $\hat{s}_{rkl}$ are used to denote the corresponding mode in the streamwise, spanwise, and shearwise directions, respectively. The Chebyshev orthogonal polynomials are defined in Refs [86, 87].

The time integration of Eqs. (1.1)–(1.3), where the $\underline{\underline{g}}$ tensor is given by Eq. (1.4) and the stress tensor $\underline{\underline{\tau}}$ by Eq. (1.6), is performed by first formally integrating these equations with respect to time from $t = t_n$ to $t = t_{n+1}$. Then, the following discretized equations are derived:

$$\underline{\nabla} \cdot \underline{u}_{n+1} = 0, \quad (1.11)$$

$$\underline{u}_{n+1} - \underline{u}_n + \int_{t_n}^{t_{n+1}} \underline{u} \cdot \underline{\nabla}\, \underline{u}\, dt = -\underline{\nabla} p_{\text{eff}} + \frac{\beta_0}{Re_{\tau 0}} \nabla^2 \int_{t_n}^{t_{n+1}} \underline{u}\, dt + \frac{1-\beta_0}{Re_{\tau 0}} \int_{t_n}^{t_{n+1}} \underline{\nabla} \cdot \underline{\underline{\tau}}\, dt + \Delta t \underline{e}_x, \quad (1.12)$$

$$\underline{\underline{c}}_{n+1} - \underline{\underline{c}}_n = \int_{t_n}^{t_{n+1}} \underline{\underline{F}}\, dt + \frac{D_0^+}{Re_{\tau 0}} \nabla^2 \int_{t_n}^{t_{n+1}} \underline{\underline{c}}\, dt, \quad (1.13)$$

where $\Delta t = t_{n+1} - t_n$, and, for clarity, we have defined the effective pressure, $p_{\text{eff}} \equiv \int_{t_n}^{t_{n+1}} p\, dt$, and a second-order tensor: $\underline{\underline{F}} \equiv -\underline{u} \cdot \underline{\nabla}\, \underline{\underline{c}} + \underline{\underline{c}} \cdot \underline{\nabla}\, \underline{u} + \underline{\nabla}\, \underline{u}^{\mathrm{T}} \cdot \underline{\underline{c}} - \underline{\underline{\tau}} - \alpha We \underline{\underline{\tau}} \cdot \underline{\underline{\tau}}$. Note that neither any boundary condition is imposed for the pressure nor any additional approximation is used as in the fractional step method [87, 88]. Instead, to solve for the pressure variable, an iterative approach, the influence matrix or Green's function method, described in Refs [87, 89], is followed.

We describe next two slightly different solution procedures used to find the solution at the time step "$n + 1$".

1.3.1.1 The Semi-Implicit/Explicit Scheme

The first scheme is a classical mixed semi-implicit/explicit scheme [51, 52]. According to this scheme, all linear terms in Eqs. (1.11)–(1.13) are treated implicitly and nonlinear terms explicitly based on the second-order Adams–Basforth formula:

$$\underline{u}_{n+1}-\underline{u}_n+\frac{\Delta t}{2}(3\underline{u}_n\cdot\underline{\nabla}\,\underline{u}_n-\underline{u}_{n-1}\cdot\underline{\nabla}\,\underline{u}_{n-1})$$
$$=-\underline{\nabla}\,p_{\text{eff}}+\frac{\Delta t\beta_0}{2Re_{\tau 0}}\underline{\nabla}^2(\underline{u}_{n+1}+\underline{u}_n)$$
$$+\frac{\Delta t(1-\beta_0)}{2\text{Re}_{\tau 0}}\underline{\nabla}\cdot(\underline{\underline{\tau}}_{n+1}+\underline{\underline{\tau}}_n)+\Delta t\underline{e}_x, \tag{1.14}$$

$$\underline{\underline{c}}_{n+1}-\underline{\underline{c}}_n=\frac{\Delta t}{2}(3\underline{\underline{F}}_n-\underline{\underline{F}}_{n-1})+\frac{\Delta t\,D_0^+}{2Re_{\tau 0}}\underline{\nabla}^2(\underline{\underline{c}}_{n+1}+\underline{\underline{c}}_n). \tag{1.15}$$

Equation (1.11) remains unchanged. Then, the following steps are used to calculate the required solution at the new time step.

Step 1: The quantity $\underline{\underline{c}}_{n+1/2}=\underline{\underline{c}}_n+\frac{\Delta t}{2}(3\underline{\underline{F}}_n-\underline{\underline{F}}_{n-1})$ is calculated first, and Eq. (1.15) is rearranged as follows:

$$\underline{\nabla}^2(\underline{\underline{c}}_{n+1}+\underline{\underline{c}}_n)-\frac{2Re_{\tau 0}}{\Delta t\,D_0^+}(\underline{\underline{c}}_{n+1}+\underline{\underline{c}}_n)=-\frac{2Re_{\tau 0}}{\Delta t\,D_0^+}(\underline{\underline{c}}_{n+1/2}+\underline{\underline{c}}_n). \tag{1.16}$$

Equation (1.16) is a Helmholtz equation with the unknown sum $\underline{\underline{c}}_{n+1}+\underline{\underline{c}}_n$. This is first transformed in the spectral domain where, due to the separability of the Helmholtz equation, the equations for each pair of Fourier modes (r, k) are fully decoupled with the only coupling appearing among the Chebyshev modes. Thus, by solving the Helmholtz equation with a fast solver [86], the updated solution for the conformation tensor $\underline{\underline{c}}_{n+1}$ is obtained, from which, with the aid of Eq. (1.6), the extra stress tensor $\underline{\underline{\tau}}_{n+1}$ is calculated.

Step 2: By taking the divergence of Eq. (1.13) and demanding the continuity equation, Eq. (1.11), to be satisfied, a Poisson equation for p_{eff} is obtained:

$$\underline{\nabla}^2 p_{\text{eff}}=\underline{\nabla}\cdot\left(-\frac{\Delta t}{2}(3\underline{u}_n\cdot\underline{\nabla}\,\underline{u}_n-\underline{u}_{n-1}\cdot\underline{\nabla}\,\underline{u}_{n-1})+\frac{1-\beta_0}{Re_{\tau 0}}\underline{\nabla}\cdot(\underline{\underline{\tau}}_{n+1}+\underline{\underline{\tau}}_n)\right). \tag{1.17}$$

Equation (1.17) is solved with the same solver as the Helmholtz equation for the conformation tensor by using the influence matrix method [87, 89].

Step 3: The quantity

$$\underline{u}_{n+1/2}=\underline{u}_n-\frac{\Delta t}{2}(3\underline{u}_n\cdot\underline{\nabla}\,\underline{u}_n-\underline{u}_{n-1}\cdot\underline{\nabla}\,\underline{u}_{n-1})$$
$$-\underline{\nabla}\,p_{\text{eff}}+\frac{1-\beta_0}{Re_{\tau 0}}\underline{\nabla}\cdot(\underline{\underline{\tau}}_{n+1}+\underline{\underline{\tau}}_n)+\Delta t\,\underline{e}_x$$

is evaluated first. Then, Eq. (1.14) is rearranged to give another Helmholtz equation for the sum $\underline{u}_{n+1}+\underline{u}_n$:

$$\underline{\nabla}^2(\underline{u}_{n+1}+\underline{u}_n)-\frac{2Re_{\tau 0}}{\Delta t\,\beta_0}(\underline{u}_{n+1}+\underline{u}_n)=-\frac{2Re_{\tau 0}}{\Delta t\,\beta_0}(\underline{u}_{n+1/2}+\underline{u}_n), \tag{1.18}$$

which is solved by using the fast solver and by applying no-slip and no-penetration conditions at the walls. These wall divergence values are all that is needed in order to obtain the correct boundary conditions for the effective pressure by enforcing the requirement that the actual velocity field is divergent free on the boundaries, $y = \pm 1$ (influence matrix technique [47, 89]). This is achieved by considering the final pressure and velocity as the linear superposition of the homogeneous solution and a weighted sum over appropriate Green's functions [47, 89]. The solution of Eq. (1.17) completes the update of all the dependent variables. Then, the same procedure is applied in order to further advance in the next time step.

1.3.1.2 The Fully Implicit Scheme

The fully implicit scheme is based on the second-order accurate Adams–Moulton formulae, that is, at each time step, $n + 1$, all terms (both linear and nonlinear) in Eqs. (1.11)–(1.13) are integrated implicitly as follows:

$$\int_{t=t_n}^{t=t_{n+1}} g(t)\,\mathrm{d}t \approx \frac{\Delta t}{2}(g_n + g_{n+1}), \qquad n > 0. \tag{1.19}$$

In this case, Eqs. (1.12) and (1.13) are given as

$$\underline{u}_{n+1} - \underline{u}_n + \frac{\Delta t}{2}(\underline{u}_n \cdot \underline{\nabla u}_n + \underline{u}_{n+1} \cdot \underline{\nabla u}_{n+1}) = -\underline{\nabla} p_{\mathrm{eff}} + \frac{\Delta t \beta_0}{2 Re_{\tau 0}} \underline{\nabla}^2(\underline{u}_{n+1} + \underline{u}_n) + \frac{\Delta t(1-\beta_0)}{2 Re_{\tau 0}} \underline{\nabla} \cdot (\underline{\underline{\tau}}_{n+1} + \underline{\underline{\tau}}_n) + \Delta t\, \underline{e}_x, \tag{1.20}$$

$$\underline{\underline{c}}_{n+1} - \underline{\underline{c}}_n = \frac{\Delta t}{2}(\underline{\underline{F}}_n + \underline{\underline{F}}_{n+1}) + \frac{\Delta t\, D_0^+}{2 Re_{\tau 0}} \underline{\nabla}^2(\underline{\underline{c}}_{n+1} + \underline{\underline{c}}_n). \tag{1.21}$$

Since the governing equations are nonlinear, the implementation of the implicit algorithm requires an iterative method. A direct iterative scheme based on Newton's method is computationally too demanding. Instead, as an alternative, a predictor-corrector scheme can be used with the corrector to be iteratively applied until a convergence criterion is met. To this end, in Eqs. (1.19) and (1.20) $\underline{u}_{n+1}$, $\underline{\underline{\tau}}_{n+1}$, p_{eff}, and $\underline{\underline{c}}_{n+1}$ are replaced by $\underline{u}^{j+1}_{n+1}$, $\underline{\underline{\tau}}^{j+1}_{n+1}$, p^{j+1}_{eff}, and $\underline{\underline{c}}^{j+1}_{n+1}$, respectively, whenever they appear in a linear way in Eqs. (1.19) and (1.20) and by $\underline{u}^{j}_{n+1}$, $\underline{\underline{\tau}}^{j}_{n+1}$, p^{j}_{eff}, and $\underline{\underline{c}}^{j}_{n+1}$, respectively, whenever they appear in a nonlinear way, where $j = 0, 1, 2, \ldots$. Therefore, Eqs. (1.19) and (1.20) become

$$\underline{u}^{j+1}_{n+1} - \underline{u}_n + \frac{\Delta t}{2}(\underline{u}_n \cdot \underline{\nabla}\, \underline{u}_n + \underline{u}^{j}_{n+1} \cdot \underline{\nabla}\, \underline{u}^{j}_{n+1}) = -\underline{\nabla}\, p_{\mathrm{eff}} + \frac{\Delta t\, \beta_0}{2 Re_{\tau 0}} \underline{\nabla}^2(\underline{u}^{j+1}_{n+1} + \underline{u}_n) + \frac{\Delta t(1-\beta_0)}{2 Re_{\tau 0}} \underline{\nabla} \cdot (\underline{\underline{\tau}}^{j+1}_{n+1} + \underline{\underline{\tau}}_n) + \Delta t\, \underline{e}_x, \tag{1.22}$$

$$\underline{\underline{c}}_{n+1}^{j+1}-\underline{\underline{c}}_{n}=\frac{\Delta t}{2}(\underline{\underline{F}}_{n}+\underline{\underline{F}}_{n+1}^{j})+\frac{\Delta t\, D_0^+}{2Re_{\tau 0}}\underline{\nabla}^2(\underline{\underline{c}}_{n+1}^{j+1}+\underline{\underline{c}}_{n}). \tag{1.23}$$

The initial guess for the solution that is required to start the iterative procedure at each time step is obtained either by using a second-order extrapolation scheme, that is, $X_{n+1}^0 \approx 2X_n - X_{n-1}$ where $X = \underline{u}, \underline{\underline{\tau}}, \underline{\underline{c}}$, or by simply applying the semi-implicit/explicit scheme described in Section 1.3.1.1. Note that at each step of the iterative procedure, all nonlinear terms are known. Therefore, a procedure similar to the one described for the semi-implicit/explicit scheme can be applied. It is also worth noting that the last step in the full implicit scheme is the satisfaction of the divergence-free condition, which in spectral space can be enforced to machine accuracy. The implicit use of the continuity in the development of the Poisson equation for the pressure is not enough to guarantee with machine accuracy the satisfaction of the divergence-free condition due to approximation error involved in the solution of the Poisson equation caused by the coupling of the Chebyshev modes. However, the structure and linearity of the continuity equation is such to allow a posteriori correction of the velocity field (in spectral space) so that it is identically satisfied.

The advantage of the full implicit scheme and the direct enforcement of the continuity equation can be more clearly seen when we formulate a suitable Poisson equation for the pressure by taking the divergence of the momentum equation, Eq. (1.2):

$$\underline{\nabla}^2 p = \left(-\frac{\partial(\underline{\nabla}\cdot\underline{u})}{\partial t}+\frac{\beta_0}{Re_{\tau 0}}\underline{\nabla}^2(\underline{\nabla}\cdot\underline{u})\right)+\left(-\underline{\nabla}\cdot(\underline{u}\cdot\underline{\nabla}\,\underline{u})+\frac{1-\beta_0}{Re_{\tau 0}}\underline{\nabla}\cdot(\underline{\nabla}\cdot\underline{\underline{\tau}})\right). \tag{1.24}$$

With the semi-implicit/explicit scheme, the terms in the first parentheses of the right-hand side of Eq. (1.18) are not identically zero. However, with the full implicit scheme in which the continuity equation is satisfied with machine accuracy, these terms are zero and therefore the periodic part of the pressure is more accurately evaluated.

Finally, it should be noted that the spectral coefficient for nonlinear terms can be optionally evaluated using the three half rule for dealiasing along all directions [86]. According to this procedure, and starting from the original $(N_x) \times (N_y + 1) \times N_z$ spectral coefficients, all nonlinear terms are evaluated in an extended physical space with $(3N_x/2) \times (3N_y/2 + 1) \times (3N_z/2)$ points. This is efficiently accomplished by first extending the spectral coefficients, for all dependent variables, by a half in each direction. The additional spectral coefficients are set equal to zero and the extended spectral information is transformed into the extended physical space and all nonlinear terms are evaluated. Then, the results are transformed back into the extended spectral space and are truncated to the original spectral space. The three half rule employed here makes the dealiasing complete (i.e., all the aliasing error is removed) for all quadratic terms and partial for higher order nonlinear terms. In fact, the only partial dealiasing occurs in the calculation of the Peterlin function when the FENE-P model is used (see Eq. (1.6)) since all other nonlinearities are quadratic.

1.3.1.3 Typical Simulation Conditions

Most of the simulation results that are reported in the literature and all those considered here are for zero shear rate friction Reynolds numbers, $Re_{\tau 0}$, 125, 180, 395, and 590. For viscoelastic simulations, the zero shear rate friction Weissenberg number $We_{\tau 0}$ varies from 6 to 125, the viscosity ratio β_0 varies from 0.6 to 0.99, and the Giesekus molecular extensibility parameter α is 1/900. A typical viscoelastic case, for which many reliable results are available, is $Re_{\tau 0} = 180$, $We_{\tau 0} = 50$, $\beta_0 = 0.9$, and $L = 30$ (when the FENE-P model is used) or $a = 1/900$ (when the Giesekus model is used). In this case, an adequate computational domain to capture all the main turbulent events is $L_x \times L_y \times L_z = 9 \times 2 \times 4.5$, which means that in wall units is $0 \leq x^+ \leq L_x Re_{\tau 0}$, $-Re_{\tau 0} \leq y^+ \leq Re_{\tau 0}$, and $0 \leq z^+ \leq L_z Re_{\tau 0}$. Each dependent variable is approximated by N_x and N_z Fourier modes along the streamwise and spanwise directions, respectively, and $N_y + 1$ Chebyshev polynomials along the shear direction, and for this particular friction Reynolds number and computational domain size, are $N_x \times N_y \times N_z = 96 \times 96 \times 96$. The time step of numerical integration Δt in computational units is 5×10^{-4} for viscoelastic case and 10^{-3} for Newtonian case in the fully implicit method; the semi-implicit and explicit schemes require much smaller Δt, that is, $\Delta t = 2 \times 10^{-4}$ and 5×10^{-4}, respectively.

1.3.2 The Positive Definiteness of the Conformation Tensor

As mentioned above, the conformation tensor $\underline{\underline{c}}$ is a second-order internal structural parameter that has a definite physical origin and interpretation, typically associated with the second moment of a suitably defined chain end-to-end distribution function [84]. It is exactly this association that also induces a very special property in the conformation tensor, its positive definiteness. In fact, this property is absolutely essential for its proper interpretation. The eigenvalues of $\underline{\underline{c}}$ and their corresponding eigenvectors have the physical meaning of the square of the average macromolecular size along the primary three directions and the orientation of those directions in space, respectively. For this physical interpretation to be possible, it is clear that all those eigenvalues need to be positive, that is, that $\underline{\underline{c}}$ is positive definite. This is therefore a property that all models need to preserve and indeed most of them do, provided their corresponding governing equations are exactly solved [49, 90].

However, in numerical simulations of viscoelastic flows the positive definiteness of the conformation tensor can be occasionally lost due to the accumulation of numerical error [54, 55]. Moreover, under these conditions the evolutionarity of the models is not guaranteed [90, 91]. As a consequence, under certain circumstances this can lead to catastrophic Hadamard instabilities [90]. It is therefore advantageous if a general way existed to guarantee the preservation of the positive definiteness in numerical simulations. This is not dissimilar to concentration profile calculations where the nonnegative character of various concentration variables is an essential feature to allow a physical interpretation of the results. Indeed, various techniques have been proposed to circumvent the numerical loss of nonnegativeness, most often through the use of an exponential mapping [92]. However, the development of the exponential mapping here is a more delicate situation due to the tensorial character of the conformation tensor.

Upwind techniques can and have been developed that preserve the positive definiteness and simultaneously do the smoothing of high frequencies [64, 65]. However, their application is restricted to low-order finite difference schemes. For higher order schemes, such as spectral methods, the only alternative to achieve the necessary smoothing is the explicit addition of artificial numerical diffusion. This idea was first developed by Crochet and coworkers [93, 94] with regard to laminar viscoelastic flows where the equivalence of artificial diffusion to upwind approximations was demonstrated. It was further utilized in a spectral method developed by Sureshkumar and Beris [95] for viscoelastic turbulent simulations. A drawback of the latter method is that its application may destroy the positive definiteness of the conformation tensor. In addition, the conformation tensor can grow beyond its physical bounds. Although these aphysical results appear typically in a small fraction of the computational domain, they still hinder the physical interpretation and cast doubt on the accuracy of the results. However, other alternatives exist, as discussed in the following.

Indeed, new variable representations have recently been developed [41, 96–98] that explicitly avoid the loss of positive definiteness of the conformation tensor. Vaithianathan and Collins [41] use either a continuous eigen decomposition of the conformation tensor (which, when used to numerically solve for eigenvalues, may require an occasional corrective action so that all eigenvalues of C are nonnegative) or a Cholesky decomposition $\underline{\underline{c}} = \underline{\underline{L}} \cdot \underline{\underline{L}}^{\mathrm{T}}$ that automatically guarantees by construction that when one numerically follows $\underline{\underline{L}}$, and for any value of $\underline{\underline{L}}$, $\underline{\underline{c}}$ will retain its positive definiteness. Explicit ways of constructing equations for new variables and implementing their numerical solution were also given. This scheme allowed the solution to be obtained in homogeneous viscoelastic turbulence up to high *We* numbers, the spectra though showed a sharp increase in the magnitude of the high-frequency modes. On the other hand, Fattal and Kupferman [96] proposed an exponential decomposition, $\underline{\underline{c}} = \exp(\underline{\underline{A}})$, so that, again, when one numerically follows the evolution of $\underline{\underline{A}}$ (which is the logarithm of $\underline{\underline{c}}$) and for any value of $\underline{\underline{A}}$, the corresponding $\underline{\underline{c}}$ remains positive definite. Also, here explicit directions were offered as to how to construct the evolution equation for $\underline{\underline{A}}$.

So far, this approach has been followed in several works. We here note only the first two. The first is a finite difference simulation of the lid-driven flow of an Oldroyd-B fluid for which the results appeared convergent up to $We = 2$ with the appearance of oscillations at higher *We* values [97]. The second work concerns the finite element simulation of the flow of a viscoelastic fluid past a cylinder using either an Oldroyd-B or a Giesekus model [49]; convergence results were obtained up to a higher *We* number than before. The convergence is ultimately limited only because of artifacts of the model (in the case of Oldroyd-B) or lack of sufficient resolution to resolve the small scales of the Giesekus model. Either one of these techniques is promising in making spectral (and other macroscopic) simulations equally compatible with hybrid micro–macro simulations as far as the capability of either method to produce physical results is concerned. In addition, adopting the simpler exponential representation for the conformation tensor of Fattal and Kupferman [96], Housiadas *et al.* [99] have developed a simpler mapping that offers numerical advantages. This most recent idea is based on the Cayley–Hamilton theorem that is used to analytically relate the conformation tensor $\underline{\underline{c}}$ to its logarithm tensor $\underline{\underline{A}}$. In fact, the process of updating $\underline{\underline{c}}$ employs three simple successive mappings, each one of which is introduced to preserve a specific property of $\underline{\underline{c}}$: its boundness, its positive definiteness, and (for computational convenience) the avoidance of underflow or overflow error in calculations. Although the method has already produced promising full DNS results [99], more work is required to improve its computational efficiency before it can see more widespread utilization. The most important outcome so far is the verification of previous DNS spectral results without the drawback of any physical violations on the conformation tensor.

1.4 Effects of Flow, Rheological, and Numerical Parameters on DNS of Turbulent Channel Flow of Dilute Polymer Solutions

1.4.1 Drag Reduction Evaluation

One of the most significant quantities to be evaluated in viscoelastic turbulent flow simulations is undoubtedly the achieved drag reduction [47, 54, 78]. The drag reduction is properly defined on the basis of the ratio of the drag observed after the introduction of the polymer additives versus the drag obtained with pure solvent, while keeping the same bulk Reynolds number [34], taking into account that all transitional effects are eliminated so that to avoid miscounting for drag reduction changes, if any, occurring related to the onset of turbulence [14] are determined as [78]

$$\mathrm{DR} = 1 - \left(\frac{Re_\tau^{(\mathrm{visc})}}{Re_\tau^{(\mathrm{Newt})}} \right)^2_{Re_b}, \qquad (1.25)$$

where the bulk Reynolds number Re_b is defined in terms of the average streamwise velocities $\bar{\upsilon}^*_{av}$ and the wall shear rate (effective) kinematic viscosity ν^* as $Re_b \equiv (2h^*\bar{\upsilon}^*_{av}/\nu^*)$. Now, when the viscoelastic system resulting from the addition of polymers to the Newtonian solvent is really very dilute, the bulk shear viscosity hardly changes, and the bulk Reynolds numbers in both cases (the viscoelastic and the Newtonian one) are simply proportional to the bulk (average) flow velocity; keeping that constant while one measures the pressure drop is therefore sufficient for a drag reduction measurement. In reality though, we do have changes in the effective solution viscosity, and these changes need to be taken into account when calculating the bulk Reynolds number. However, still this effect is typically low.

Most importantly for computational viscoelastic fluid mechanics, most of the channel DNS calculations are not performed for a constant flux (which would have naturally resulted in a constant bulk Reynolds number) but for a constant pressure drop per unit length that results in a constant zero shear rate friction Reynolds number. These runs lead to substantial variations in the (instantaneous and average) bulk Reynolds number from which the drag reduction needs to be estimated. Knowing roughly the relationship between the friction and the average bulk Reynolds number for a Newtonian fluid (from the experimentally determined and DNS confirmed empirical relationships for the skin friction factor – see, for example, Ref. [34]), one can extract such a relationship that also takes into account the already mentioned (in Section 1.2) shear thinning effect in association with viscoelastic results [78].

The final expression, relying on the average streamwise velocities $\bar{\upsilon}_{\mathrm{av}}^{(\mathrm{visc})}$ and $\bar{\upsilon}_{\mathrm{av}}^{(\mathrm{Newt})}$ evaluated from the viscoelastic and the Newtonian simulations, respectively, at the same $Re_{\tau 0}$, is given [78] as

$$\mathrm{DR} = 1-\mu_{\mathrm{w}}^{2(1-n)/n}\left(\frac{\bar{\upsilon}_{\mathrm{av}}^{(\mathrm{visc})}}{\bar{\upsilon}_{\mathrm{av}}^{(\mathrm{Newt})}}\right)_{Re_{\tau 0}}^{-2/n} = 1-\mu_{\mathrm{w}}^{-2}\left(\frac{Re_{\mathrm{b}}^{(\mathrm{visc})}}{Re_{\mathrm{b}}^{(\mathrm{Newt})}}\right)_{Re_{\tau 0}}^{-2/n}, \quad n \approx 1.14775, \tag{1.26}$$

where

$$\mu_{\mathrm{w}} = \frac{\beta_0}{1+\dfrac{1-\beta_0}{2Re_{\tau 0}}\Delta(\bar{\tau}_{xy})} \tag{1.27}$$

and $\Delta(\bar{\tau}_{xy})$ is the average wall shear stress at the two walls. Note that for the Oldroyd-B constitutive model that does not show shear thinning behavior, the mean viscosity ratio at the wall μ_{w} is 1. By an overbar, we denote x, z, and time-averaged quantities:

$$\bar{g} \equiv \frac{1}{L_x L_z T_f}\int\limits_{x=0}^{x=L_x}\int\limits_{z=0}^{z=L_z}\int\limits_{t=t_0}^{t=t_0+T_{\mathrm{f}}} g(x, y, z, t)\,\mathrm{d}x\,\mathrm{d}z\,\mathrm{d}t, \tag{1.28}$$

where t_0 is the required integration time to reach stationary state (beyond which the statistics are taken), T_{f} is the integration time in stationary state, and Δ denotes the difference of the measured values at the two walls, $\Delta(\bullet) = (\bullet)_{\text{top wall}} - (\bullet)_{\text{bottom wall}}$, used in order to average the wall results for antisymmetric quantities such as τ_{xy}. For dilute systems, μ_{w} can be very well approximated by considering the steady-state simple shear flow model predictions corresponding to the average wall shear rate [55, 56, 78]. Equation (1.26) describes the drag reduction for runs performed at constant zero shear rate friction Reynolds number, while Eq. (1.25) describes the drag reduction when the bulk Reynolds number is kept constant. Note that this means that if the flux is the quantity that is instead maintained constant between runs, there is a corresponding correction that also needs to be implemented to accommodate potential viscosity changes.

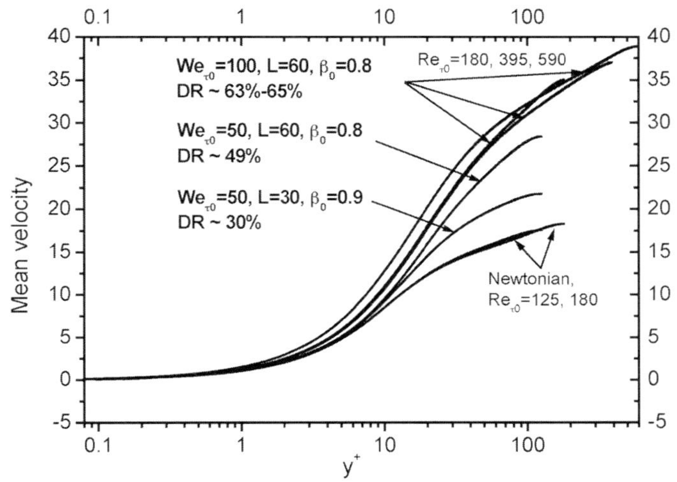

Figure 1.2 DNS results for the mean velocity profiles, obtained for various parameter values of the FENE-P constitutive model, the Newtonian representing the pure viscous solvent base case. (Adapted from Ref. [56].)

1.4.2
Effects of Flow and Rheological Parameters

From our computational viscoelastic DNS work and others (see, for example, Ref. [66]), a rich information has been accumulated concerning the effects that various rheological and flow parameters have on key turbulence statistics and the drag reduction. We will like to present here, drawing from our own work for convenience, some indicative results that we believe have especially led us to improve our understanding of the polymer-induced drag reduction. First, in Figure 1.2, we have collected some representative data on the mean streamline velocity profiles corresponding to a variety of models and model parameter values. It is reassuring to see there that the numerical results look very similar to the experimental data sample presented in Figure 1.1. More specifically, we see in Figure 1.2 that under conditions little departing from the Newtonian reference case (i.e., corresponding to either small Weissenberg number or small extensional parameter L, and/or small polymer concentration that is proportional to $1-\beta_0$), we have a small departure from the Newtonian turbulent results, consisting of a simple parallel translation to higher values of the log linear law segment, in agreement with experimental results [23]. In contrast, under conditions corresponding to higher viscoelasticity, we see that the velocity log-law profile moves further away, and eventually even its slope increases, almost reaching the level of Virk's maximum drag reduction asymptote, exactly in parallel to the trends exhibited by the most drag reduction experimental data shown in Figure 1.1. Therefore, we have confirmed here the main experimental finding that with viscoelasticity the shape of the viscous sublayer remains practically unchanged; however, the buffer layer widens and the log-law layer is pushed higher.

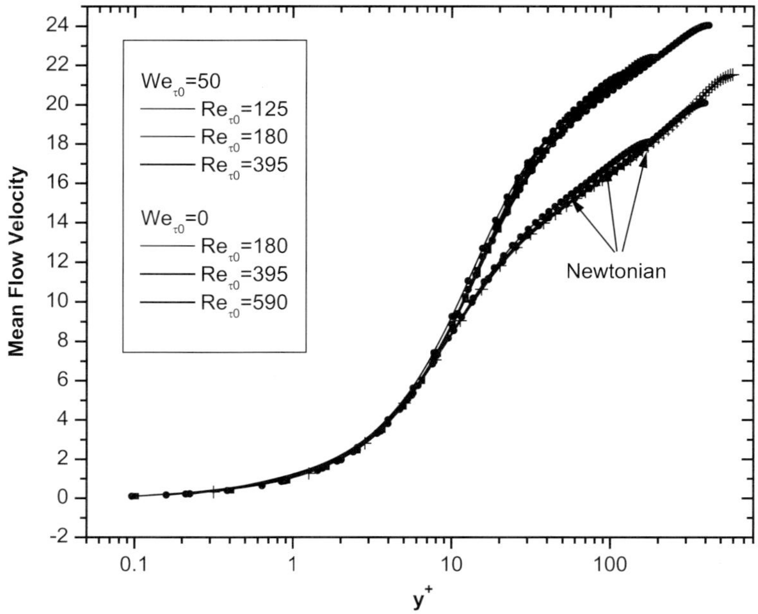

Figure 1.3 Effect of the friction Reynolds number on the mean velocity profile. (Adapted from [54].)

Already in Figure 1.2, by comparing results collected at various Reynolds numbers, we can see that the effect of the Reynolds number, as far as the mean velocity profile (and therefore the drag reduction) is concerned, is minimal. We can also see that even more clearly in Figure 1.3, where results have been collected for fixed rheological parameters (corresponding to a moderate drag reduction case, calculated with the FENE-P model) and at different friction Reynolds numbers. We see that the primary effect of increasing the friction Reynolds number is to be able to allow the log-linear layer to develop more, as well as to push further from the wall as small downturn on the profile that can be attributed to a centerline effect. Note that exactly similar behavior is obtained with Newtonian DNS, also reported in the same figure for comparison purposes. Needless to say those Newtonian DNS results, collected with the same code as in the viscoelastic case, are in perfect agreement with the literature [100].

The effect of the Reynolds and Weissenberg numbers on drag reduction can be most clearly seen in Figure 1.4, where we present the Weissenberg number dependence of the drag reduction, calculated according to the expressions provided in Section 1.4.1. The various curves in Figure 1.4 correspond to different combinations of other rheological parameters and the friction Reynolds number, as indicated in that figure. First, it is important to note that for the range of Reynolds numbers considered, there is minimal, if any, effect that they have on drag reduction. In fact, this parallels what we saw in Figures 1.2 and 1.3 before. As it turns out, one will have

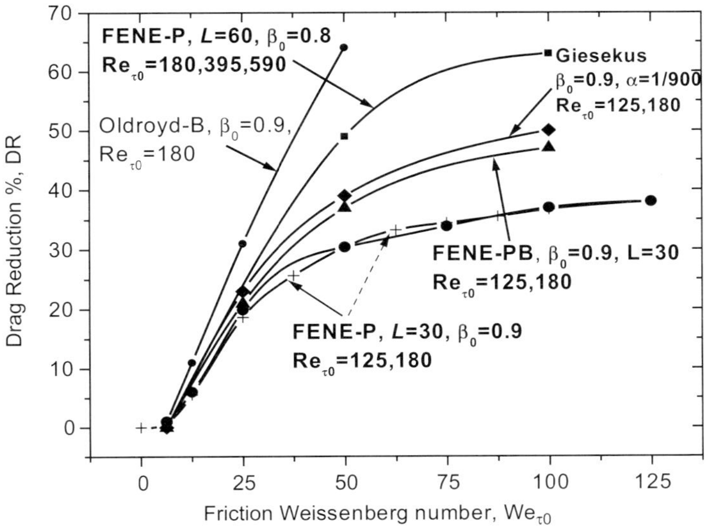

Figure 1.4 Drag reduction predictions as a function of *We* for various values of model parameters. (Adapted from [56] and unpublished data.)

to go to higher order turbulent statistics to see a clear Reynolds number effect – see below for an example and Ref. [58] for more details – on the exception, possibly of results obtained at the smallest Reynolds number, $Re_{\tau 0} = 125$, which being very close to the turbulent transition regime (that can change, depending on viscoelasticity) can introduce some bias in calculations. Similar Reynolds effects are also observed with Newtonian fluids based on experimental [101] and DNS [100] data. Thus, we focus here on discussing results obtained at higher $Re_{\tau 0}$, 180 and more.

Second, note that the various curves shown in Figure 1.4, each one corresponding to a different model with all rheological parameters other than the $We_{\tau 0}$ number fixed, are sigmoidal in shape, with all curves exhibiting a clear onset, around $We_{\tau 0} = 6{-}8$ and saturation for $We_{\tau 0} > 100$. The main difference between the various curves is then just a question of scale, with the scale determined by the maximum drag reduction obtained at saturation in the limit of very high $We_{\tau 0}$. This saturation value appears to be predominantly dictated by the maximum extensional viscosity that is predicted for a steady extensional flow by the corresponding model and model parameter values: the higher the maximum extensional viscosity, the higher the saturation value for DR. Therefore, an extensional viscosity that significantly increases above the Newtonian limit is identified as the most significant prerequisite for additive-induced drag reduction to take place; the higher the extensional viscosity, the higher the effect. This is most appropriately measured by the Trouton ratio, defined as the ratio of the extensional to shear viscosity [79]. This is consistent with the very high values attributed for the Trouton ratio of even dilute drag-reducing polymer solutions [102] also mirrored, although not to the same high levels, by the Trouton ratio of drag-reducing surfactant solution [103]. For the FENE-P, the maximum extensional viscosity ratio to the shear viscosity is proportional to $(1-\beta_0)L^2$, where β_0 is the ratio of the solvent viscosity to the total solution viscosity and L is the maximum extensibility parameter. Then, the monotonic increase in DR is indeed observed as we change accordingly the relevant parameter values – compare the FENE-P curves in Figure 1.4 corresponding to $L = 30$, $\beta_0 = 0.9$ (low extensional viscosity) and $L = 60$ and $\beta_0 = 0.8$ (high extensional viscosity). However, when the models vary, there are further secondary effects that enter the picture – one of the main results from DNS is the relatively high sensitivity of the results to details of the rheology and the rheological modeling.

A secondary dependency of DR, beyond that on the extensional viscosity and the Trouton ratio, on the particular constitutive model used is to be expected. This is because various other rheological characteristics (e.g., the second normal stress difference in simple shear flow) that may also be important in the mechanisms underlying drag reduction [104] critically depend on the model. Indeed, in Figure 1.4 we note a significant increase in drag reduction as we move from FENE-P ($L = 30$) to Giesekus ($a = 1/900$), even when the maximum extensional viscosity predicted under these conditions from both models is the same. This enhancement of drag reduction with the Giesekus model has also been observed by other investigators [65], and it is consistent with the hypothesis on the effect of second normal stress difference (present only in the Giesekus model) in drag-reducing flows [104].

It is of interest that even smaller differences between the models, for example, only affecting the detail way that the extensional viscosity depends on the extensional rate (at steady state) and/or the time (during transients), without affecting other rheological parameters, such as the second normal stress difference, may also contribute to significant changes in the predicted drag reduction. Such is the case when a variant of the FENE-P model, the FENE-PB, is introduced in the calculations [57]. Note that both the FENE-P and the FENE-PB predict zero second normal stress difference in simple shear flow and for the same L value ($L = 30$ here) the same maximum extensional viscosity [57]. However, the fact that the latter corresponds to a higher steady-state extensional viscosity at intermediate extensional rates reaching the asymptotic maximum value at smaller extensional rates [57], and also corresponds to faster transients, is sufficient to lead to significantly higher drag reductions, almost as high as those observed with the Giesekus model ($a = 1/900$), as seen in Figure 1.4.

Even more striking than the above observations is the fact that additional simulations show that there is also a secondary effect to the flow keeping the same model and the same maximum extensional viscosity! Indeed, when the combination of the L and β_0 parameters is varied in the FENE-P model, in such a way as to keep the maximum extensional viscosity constant, $(1-\beta_0)L^2 = 90$, we still see some changes in the solution and in particular in the drag reduction, as shown in Figure 1.5. As we can see there, the drag reduction seems to significantly increase for small values of the extensibility parameter L, corresponding to smaller values of the solvent ratio, β_0. A possible explanation of this effect can be again the detail way the extensional viscosity responds to intermediate extensional rates at steady state and/or to intermediate times during transients, albeit the differences in this case are much smaller than those seen before between the FENE-PB and the FENE-P cases. This

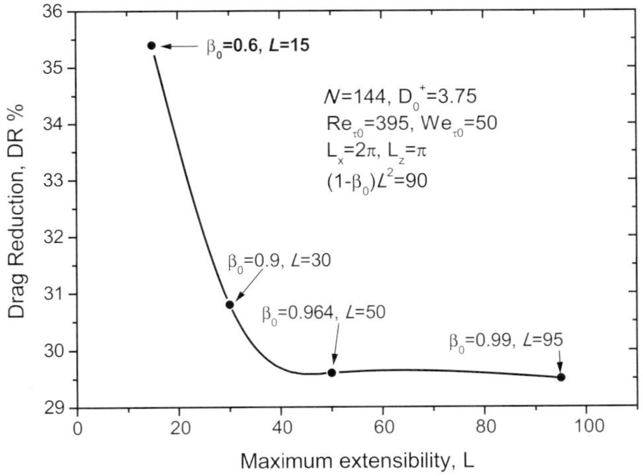

Figure 1.5 Effect of the maximum extensional viscosity to the drag reduction, FENE-P model.

further reinforces our belief that although the extensional viscosity is the primary factor that controls drag reduction, one needs to consider supplemental information in addition to its maximum predicted (saturation) value at infinite extensional rates. This may be explained since despite the fact that typical extensional rates encountered in turbulent DNS are large, they are not infinite, and they definitely vary with time. Thus, it is also important to examine the dependence of the extensional viscosity on the extensional rate and/or time, and the current data seem to support such a view.

Beyond this correlation of model parameters with average macroscopic effects, such as represented by the overall drag reduction, one can find in the detailed numerical results a wealth of additional information. For example, looking at the root mean square of the velocity fluctuation profiles, such as those shown in Figure 1.6, one can clearly see that with increasing viscoelasticity the turbulent velocity fluctuations become more anisotropic. While the rms values of the streamline component increase, those in the shearwise and spanwise directions decrease. The increased anisotropicity has also been confirmed recently through the calculation from DNS of the Lumley anisotropic tensors [105] as has been measured experimentally [106, 107]; it is also interesting to note that this anisotropy tensor information may be used to better understand the drag reduction mechanism [108]. As also seen for Figure 1.6, these changes are accompanied with a clear shift of the streamline rms maximum, typically occurring in the buffer layer, further away from the wall, offering another indication (beyond that seen with the mean streamline profiles in Figure 1.2) of the widening of the wall eddies structures with viscoelasticity. Incidentally, this is also further corroborated by a recent analysis of the large, coherent structures in the flow (eddies) [58, 75].

Form the above DNS findings, it became obvious that the coherence of the large-scale structures is significantly enhanced in the presence of polymer additives. Simultaneously, the eddies become considerably weaker and therefore less capable

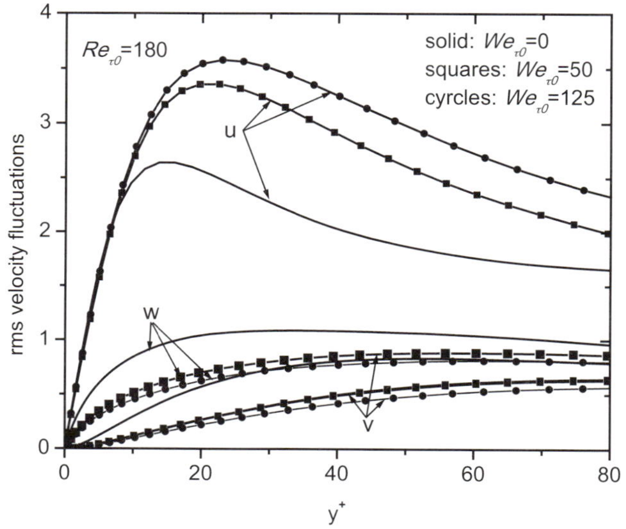

Figure 1.6 The rms velocity fluctuations at friction Reynolds 180. (Adapted from Ref. [54].)

of transferring momentum from the wall. This effect emerges, therefore, naturally as the main mechanism through which turbulence increases drag. This effect has been unequivocally connected through DNS to the enhanced resistance to extension offered by polymer molecules in their stretched state. For example, in Figure 1.7

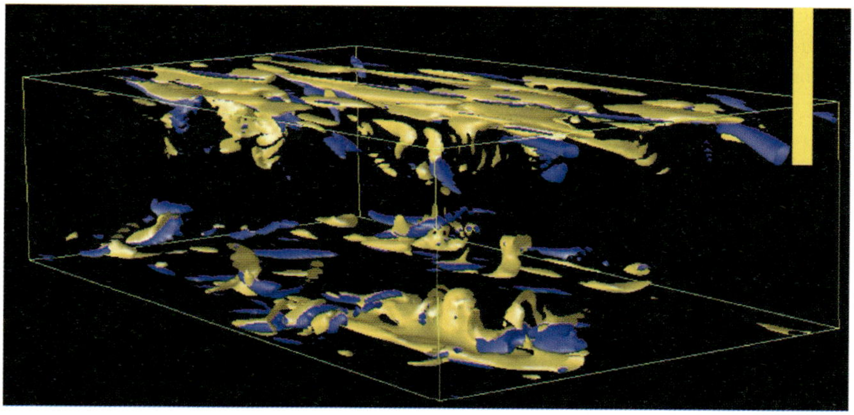

Figure 1.7 Vorticity (with blue) and trace of conformation tensor (with yellow) isosurfaces, close to their maximum values, obtained in a FENE-P polymer simulation of turbulent channel flow. The flow is from back to front with the bottom and top surfaces representing the channel walls. The close-to-maximum vorticity isosurfaces paint the eddies, whereas the close-to-maximum trace of the conformation tensor isosurfaces the location of the most stretched polymer chains in the flow – their close correlation is an indication of the role of the polymers in eddies formation. Alternative rendering of the data also shown in Ref. [58].

we see that the most stretched molecules (as documented from the isosurfaces corresponding to close to the maximum value obtained for the trace of the conformation tensor that can be used to characterize the macromolecular deformation) are to be found near the surface of those eddies (as represented by vorticity isosurfaces at values close to the maximum streamline vorticity). Also, in that figure one can notice the large size of the eddies, much larger than the corresponding Newtonian turbulent channel flow, that closely parallels the increase in the streak spacing also observed in experiments with dilute polymer solutions [23]. In fact, from the first DNS publication [47] it has been noted that the empirically derived linear relationship between the increased streak spacing λ^+ and the drag reduction DR, at small values of drag reduction for dilute polymer solutions [23], $\lambda^+ = 1.9\mathrm{DR} + 99.7$, has also been confirmed from the DNS results [47]. This substantial increase in the streaky structures has also been confirmed in subsequent DNS works with both FENE-P and Giesekus models [50].

The main outcome of the DNS studies has, therefore, been the ample supply of evidence (a small sample of which is provided here) in favor of an extensional viscosity thickening-driven drag reduction mechanism exactly as foreseen by Metzner and coworkers [15] and also independently postulated by Lumley [14]: as the eddies form at the wall, they see with viscoelasticity an increased resistance due to the predominantly extensional character of the velocity deformation and the enhanced resistance to extensional deformation offered by polymers. As a result, the eddies are larger in length, have smaller vorticity values, and are more sluggish; therefore, they become less effective in transferring momentum from the wall, resulting in drag reduction. This is also illustrated in the significantly depressed, by viscoelasticity, values for the streamwise vorticity as seen in Figure 1.8. In the same figure, we see the changes in the vorticity close to the wall because of the changes in the friction Reynolds number. Most of the changes happen when the Reynolds number increases from 125 to 180, then some changes still occur as the Reynolds number increases to 395, from which point there are no perceptible changes as the Reynolds number increases further to 590. The vorticity is one of the quantities most sensitive to Reynolds number change. Figure 1.8 provides evidence that for Reynolds number larger than 395 the results next to wall have converged.

Finally, a very important quantity for determining the wall friction is the Reynolds stress. As numerous DNS studies [47, 54–57, 59, 60, 62–65, 109] have shown, a major viscoelastic effect is a significant lowering in the Reynolds stress in agreement with experimental evidence [22, 23, 25, 26, 62, 110]. In addition, results from a more detailed quadrant analysis have also been reported [58, 111] and show most changes due to viscoelasticity to be concentrated with the fourth and second quadrant events that are found to be significantly decreased in magnitude due to viscoelasticity in the viscous sublayer and the buffer layers, as has been observed experimentally [22, 23]. Furthermore, for a constant friction Reynolds number, we have shown [53, 58] that viscoelasticity is responsible for significantly lowering the production in the Reynolds stresses, in agreement with experiments [23, 48]. A similar decrease is observed in the magnitudes of the terms representing the pressure–velocity gradient correlations, transport, and the overall (viscous and viscoelastic) diffusion and dissipation.

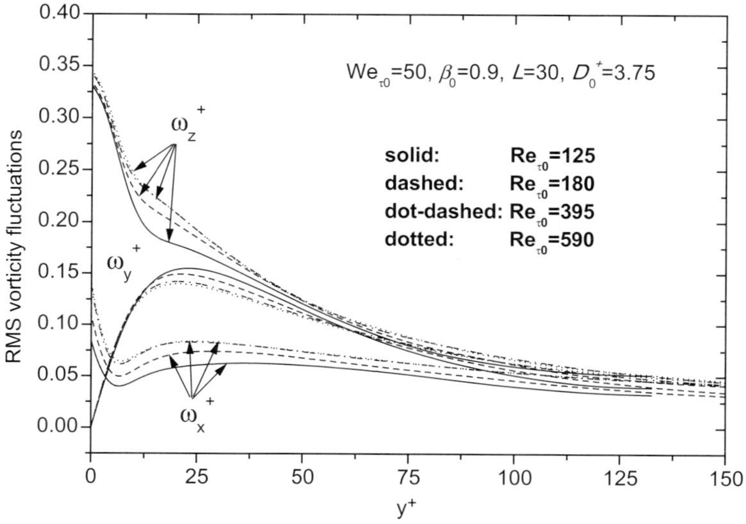

Figure 1.8 Effect of the friction Reynolds number on the vorticity components. (Adapted from Ref. [54].)

Simultaneously, the peaks in the profiles widen and shift toward the centerline. The changes are on the order of 30–70% and are localized in the near-wall boundary layer ($y^+ < 75-100$), leaving the region far away from the wall virtually unaffected. Regarding the terms appearing on the enstrophy budgets, the changes observed due to viscoelasticity are qualitatively similar but much more dramatic in magnitude with reductions of 80–90% being typical. In fact, it is this drastic annihilation of the enstrophy in the wall boundary that may explain the saturation of the drag reduction at a level of about 30–40% obtained with the FENE-P dilute polymer solution constitutive model [58].

1.4.3 Effects of Numerical Parameters

It is important to mention several computational and numerical issues related to the sensitivity of the results to

- time integration scheme;
- different values for the numerical diffusivity;
- different mesh sizes;
- different computational domain sizes;
- long time integration (both in transient and in stationary state).

First, we should comment on the semi-implicit/explicit and fully implicit schemes. The fully implicit scheme gives more physical results and is more accurate and more

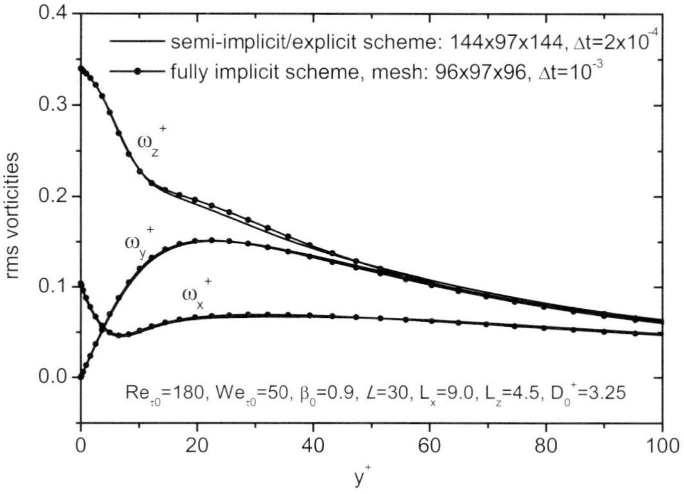

Figure 1.9 Comparison between the semi-implicit/explicit and the fully implicit scheme for the rms vorticity fluctuations. (Adapted from Ref. [55].)

stable allowing considerably larger time steps of integration to be used. On the other hand, it requires both more computational memory and more arithmetic operations per time step. A comparison between these two schemes is offered in Figure 1.9, in which the average vorticity components are given as a function of the distance from the wall. It is clearly seen there that the fully implicit scheme with $\Delta t = 10^{-3}$ and mesh size $96 \times 97 \times 96$ yields the same statistics with the semi-implicit/explicit scheme with $\Delta t = 2 \times 10^{-4}$ and mesh size $144 \times 145 \times 144$.

Second, we present some thoughts about the role of artificial diffusivity in relation to spectral simulations. So far, all indications are that some diffusivity in viscoelastic turbulence simulations is unavoidable (added either explicitly, as is done in spectral methods, or implicitly by using diffusive low-order approximations, such as finite differences, especially when upwind formulations or linearizations in particle Lagrangian formulations [112] are used) as we have seen in all successful simulations of complex viscoelastic flows. Indeed, it is believed that the need to have a conformation diffusion term is the same as the need to add a diffusive term in solving for a passive scalar advection equation in a turbulent flow: it is due to the feature that any chaotic flow field has in creating finer and finer features in the distribution of any passive scalar that it is advected upon its action ([82] and references therein). We stipulate that this is exactly what is happening with the conformation tensor. This becomes even more obvious when one examines the equation for each one of the conformation tensor eigenvalues (Eq. (16) in Ref. [41]) that is exactly like a passive scalar advection equation plus two extra scalar terms, one representing velocity gradient stretching and the other polymer relaxation.

However, the magnitude of the artificial diffusivity has to be kept small, and, when probing convergence of the results with mesh refinement, for consistency,

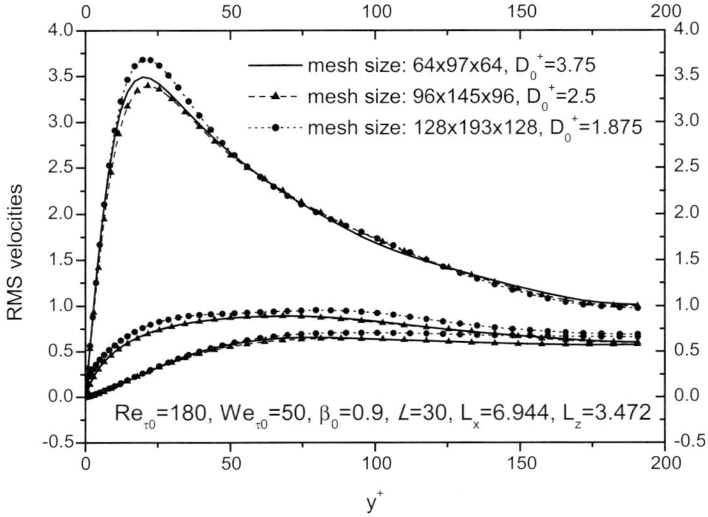

Figure 1.10 Effect of the mesh size and the numerical diffusivity. (Adapted from Ref. [54].)

as the mesh resolution increases, the numerical diffusivity should also decrease. Indeed, this is happening in the convergence study shown in Figure 1.10. The fact that after two successive refinements of the mesh resolution (accompanied by corresponding decreases in the numerical diffusivity) the results for the rms velocities are almost the same is a strong evidence that at least in that case we have reached convergence in the results to a solution independent of the value of the artificial diffusivity. Although every effort has been made in the reported calculations to use diffusivity values that were judged as similarly not altering the physical content of the results, it is to be noted that as the parameter values and/or models change, the role of numerical diffusivity needs to be carefully reexamined. In particular, this is one of the most important factors as extreme range of the parameter values is reached, such as when we approach maximum drag reduction.

In view of the observed enhancement of the larger scales of turbulence by the viscoelasticity of the flow, special attention has also to be paid to viscoelastic simulations to allow a large enough computational domain so that even the larger scales of turbulence can develop. So far, only limited investigations of this effect have been carried out mainly due to the significant computational cost associated with it. Such an example is offered here for the rather small friction Reynolds number 125. In this case, as shown in Figure 1.11, the results obtained after a significant enlargement of the domain by 50% in each direction are almost the same. In this figure, it is interesting to point out another characteristic result arising from the interaction of viscoelasticity with the turbulent flow. In particular, note that the average profile for the trace of the conformation tensor shows a pronounced peak not at the wall (where the maximum shear rates are obtained) but

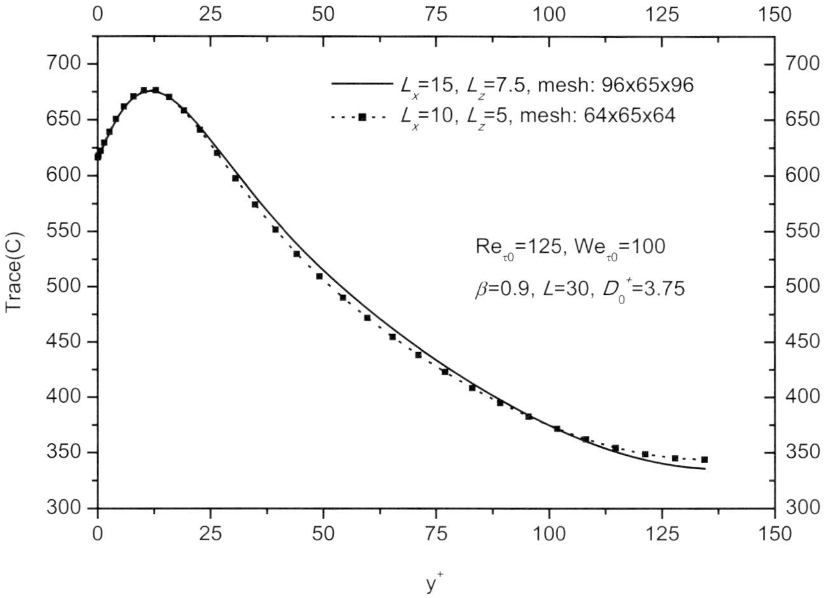

Figure 1.11 Effect of the computational domain.

rather at the beginning of the buffer layer where most of the eddies action takes place. This is a clear indication that the most significant molecular deformation occurs due to the transient extensional characteristics of the flow that accompany the formation of eddies. Note that the issue of identification of the velocity structures that give rise to that pronounced peak in macromolecular deformation (and therefore is critical in the elucidation of details of the drag reduction mechanism) is still a subject of active investigation [105].

1.5 Conclusions and Thoughts on Future Work

We offered in this chapter an overview of some of the most important elements of recent viscoelastic computational work on turbulent channel flow under drag-reducing conditions. After a brief history on the polymer-induced drag reduction phenomenon and presentation of recent computational works, we focused on spectral DNS of turbulent channel flow for dilute polymer solutions. We first presented the governing equations, with emphasis on the two most popular models (FENE-P and Giesekus). Then, we presented the key elements of the spectral method-based numerical algorithms. Following that we discussed in some detail the most important changes to the turbulence induced by viscoelasticity. We can summarize those here as a significant enhancement of the size of the buffer layer, the streamline velocity fluctuations, and the size of the main eddies that form there, while there is a significant weakening of the streamwise vorticity, and therefore the intensity of the eddies, the Reynolds stress, and the shearwise and spanwise velocity fluctuations. All these were found to contribute to drag reduction through the displacement of the log-law component of the mean streamline velocity profile in agreement with experimental observations. The sensitivity of the above findings to flow and rheological parameters provided significant evidence for the primary role of the extensional viscosity in drag reduction, exactly as originally proposed by Metzner and Lumley.

At present, the numerical simulation activity in this area is focusing on deepening our understanding of the changes induced by viscoelasticity in the structure of turbulence flows. For example, we can mention here the continuing work on the exact coherent states [72], on large-scale structure dynamics through K-L analysis in time [105], and, most recently, on the investigation of the effects of viscoelasticity on small (dissipative) scales of turbulence and intermittency through an analysis of the probability density functions of the velocity and velocity derivatives [113]. These investigations, in addition to further elucidating the underlying mechanism for drag reduction, can also contribute possibly to the development of low-dimensional models of turbulence from first principles. In parallel, as the work to develop various $\varkappa - \varepsilon$-based empirical models of drag reduction [17] is continuing, the DNS data are expected to be further utilized to test those ideas.

Still considerable challenges remain as the strength of viscoelasticity increases and the region of maximum drag reduction is approached due to the very steep computational requirements to evaluate mesh-converged numerical solutions under those conditions. Issues related to polymer inhomogeneities are also important – see, for example, the experimental data of Refs [26, 114] suggesting polymer aggregation and filament formation under conditions of maximum drag reduction. In addition, there is a considerable room for improvement in the development of LES and averaged equations for general viscoelastic turbulent flows, moving beyond channel and pipe flows. One thing is sure that the polymer-induced drag reduction problem is a very challenging one that will keep numerical analysts occupied for a long time.

Acknowledgments

ANB would like to acknowledge the help of the late A.B. Metzner in sharing most of the historical information that appears in the beginning of this work and for the many stimulating discussions and guidance throughout this work. The authors would like to acknowledge the financial support provided by DARPA, Grant No. MDA972-01-1-0007. We are also grateful to the National Center for Supercomputing Applications (NCSA) for providing the computational resources needed for this work through a sequence of NSF-supported supercomputing time allocation grants, the last of which is TG-MCA96N005. Dr. Robert A. Handler and Dr. Gaurab Samanta are also thanked for their contributions. KDH would like to acknowledge the Department of Mathematics, at the University of the Aegean, Greece, for the sabbatical leave of absence in the fall semester of the academic year 2008–2009 during which this work was done.

References

1 Metzner, A.B. (2004) Turbulent drag reduction. Presentation in a Symposium in Honor of Jan Mewis, Leuven University, Leuven, Belgium, September 22, 2004.

2 Mysels, K.J. (1949) Flow of thickened fluids. US Patent 2,492,173.

3 Toms, B.A. (1949) Some observations on the flow of linear polymer solutions through straight tubes at large Reynolds numbers, in *Proceedings of the International Rheological Congress, Holland, II*, North-Holland, pp. 135–141.

4 Agostan, G.A., Harte, W.H., Hottel, H.C., Klemm, W.A., Mysels, K.J., Pomeroy, H.H., and Thompson, J.M. (1954) Flow of gasoline thickened by Napalm. *Ind. Eng. Chem.*, **46**, 1017–1019.

5 Toms, B.A. (1977) On the early experiments on drag reduction by polymers. *Phys. Fluids*, **20**, S3–S5.

6 Virk, P.S. (1975) Drag reduction fundamentals. *AIChE J.*, **21**, 625–656.

7 Dodge, D.W. and Metzner, A.B. (1959) Turbulent flow of non-Newtonian systems. *AIChE J.*, **5**, 189–203.

8 Fubala, A.G. (1965) The Toms phenomenon in the turbulent flow of very dilute polymer solutions. Proceedings of the 4th International Congress on Rheology, Part 3, pp. 455–479.

9 Metzner, A.B. and Park, M.G., III (1964) Turbulent flow characteristics of viscoelastic fluids. *J. Fluid Mech.*, **20**, 291–303.

10 Hershey, H.C. and Zakin, J.L. (1967) Molecular approach to predicting the onset of drag reduction in the turbulent flow of dilute polymer solutions. *Chem. Eng. Sci.*, **22**, 1847–1857.

11 Castro, W. and Squire, W. (1967) The effect of polymer additives on transition in pipe flow. *Appl. Sci. Res.*, **18**, 81–96.

12 Giles, W.B. and Pettit, W.T. (1967) Stability of dilute viscoelastic flows. *Nature*, **216** (11), 470–472.

13 Virk, P.S., Merrill, E.W., Mickley, H.S., Smith, K.A., and Molo-Christensen, E.L. (1967) The Toms phenomenon: turbulent pipe flow of dilute polymer solutions. *J. Fluid Mech.*, **30**, 305–328.

14 Lumley, J.L. (1969) Drag reduction by additives. *Ann. Rev. Fluid Mech.*, **1**, 367–384.

15 Seyer, A. and Metzner, A.B. (1969) Turbulence phenomena in drag-reducing systems. *AIChE J.*, **15**, 426.

16 Lumley, J.L. and Blossey, P. (1998) Control of turbulence. *Ann. Rev. Fluid Mech.*, **30**, 311–327.

17 Mehrabadi, M.A. and Sadeghy, K. (2008) Simulating drag reduction phenomenon in turbulent pipe flows. *Mech. Res. Commun.*, **35**, 609–613.

18 Zakin, J.L., Lu, B., and Bewersdorff, H.W. (1998) Surfactant drag reduction. *Rev. Chem. Eng.*, **14** (4–5), 253–320.

19 Yu, B. and Kawaguchi, Y. (2003) Effect of Weissenberg number on the flow structure: DNS study of drag reducing flow with surfactant additives. *Int. J. Heat Fluid Trans.*, **24**, 491–499.

20 Zhang, Y., Schmidt, J., Talmon, Y., and Zakin, J.L. (2005) Co-solvent effects on drag reduction, rheological properties and micelle microstructures of cationic surfactants. *J. Colloid Interface Sci.*, **286**, 696–709.

21 Mewis, J. (1974) The rheological properties of suspensions of fibers in Newtonian fluids subjected to extensional deformations. *J. Fluid Mech.*, **62**, 593–600.

22 Luchik, T.S. and Tiederman, W.G. (1988) Turbulent structure in low-concentration drag-reducing channel flows. *J. Fluid Mech.*, **190**, 241–263.

23 Tiederman, W.G. (1990) The effect of dilute polymer solutions on viscous drag and turbulent structure, in *Structure of Turbulence and Drag Reduction* (ed. A. Gyr) Springer, pp. 187–200.

24 Virk, P.S. and Wagger, D.L. (1989) Aspects of mechanisms in Type B drag reduction, in *Structure of Turbulence and Drag Reduction* (ed. A. Gyr) Springer, pp. 201–213.
25 Warholic, M.D., Massah, H., and Hanratty, T.J. (1999) Influence of drag-reducing polymers on turbulence: effects of Reynolds number, concentration and mixing. *Exp. Fluids*, **27**, 461–472.
26 Vlachogiannis, M., Liberatore, M.W., McHugh, A.J., and Hanratty, T.J. (2003) Effectiveness of a drag reducing polymer: relation to molecular weight distribution and structuring. *Phys. Fluids*, **15**, 3786–3794.
27 Sellin, R.H.J. and Moses, R.T. (1984) Proceedings of the International Conference on Drag reduction, Bristol University, Bristol.
28 Sellin, R.H.J. and Moses, R.T. (1989) Drag reduction in fluid flows, in *Proceedings of the IAHR International Conference, Davos*, Ellis Horwood, Chichester.
29 Bushnell, D.M. and Hefner, J.N. (1990) Viscous drag reduction in boundary layers. *Prog. Astronaut. Aeronaut.*, Vol. 123, AIAA, New York.
30 Gyr, A. and Bewersdorff, H.-W. (1995) *Drag Reduction of Turbulent Flows by Additives*, Kluwer Academic Publishers.
31 Nieuwstadt, F. and Den Toonder, J. (2001) Drag reduction by additives: a review, in *Turbulence Structures and Motion* (eds A. Soldati and R. Monti) Springer, New York, pp. 269–316.
32 Graham, M.D. (2004) Drag reduction in turbulent flow of polymer solutions, in *Rheology Reviews 2004* (eds D.M. Binding and K. Walters) The British Society of Rheology, pp. 143–170.
33 White, C.M. and Mungal, M.G. (2008) Mechanics and predictions of turbulent drag reduction with polymer additives. *Annu. Rev. Fluid Mech.*, **40**, 235–56.
34 McComb, W.D. (1990) *The Physics of Fluid Turbulence*, Oxford University Press.
35 Tanner, R.I. (2000) *Engineering Rheology*, 2nd edn, Oxford University Press.
36 Lummus, J.L., Fox, J.E., Jr., and Anderson, D.B. (1961) New low solids polymer mud cuts drilling costs for Pan American. *J. Oil Gas.*, **87**, 1211
37 Burger, E.D., Chorn, L.G., and Perkins, T.K. (1980) Studies of drag reduction conducted over a broad range of pipeline conditions when flowing Prudhoe Bay crude oil. *J. Rheol.*, **24**, 603–626.
38 Alyeska-pipeline (2005) Internet online information at http://www.alyeska-pipe.com/Pipelinefacts/PipelineOperations.html.
39 Lee, W.K., Vaseleski, R.C., and Metzner, A.B. (1974) Turbulent drag reduction in polymeric solutions containing suspended fibers. *AIChE J.*, **20**, 128–133.
40 Hoyt, J.W. (1975) Hydrodynamic drag reduction due to fish slimes. Jn T.Y. Wu, C.J. Brokaw, and C. Brennen (eds.) *Swimming and Flying in nature*, Vol. 2, Plenum Press, New York, pp. 653–672.
41 Vaithianathan, T. and Collins, L.R. (2003) Numerical approach to simulating turbulent flow of a viscoelastic polymer solution. *J. Comput. Phys.*, **187**, 1–23.
42 Dubief, Y., White, C.M., Terrapon, V.E., Shaqfeh, E.S.G., Moin, P., and Lele, S.K. (2004) On the coherent drag-reducing turbulence-enhancing behavior of polymers in wall flows. *J. Fluid Mech.*, **514**, 271–280.
43 Dimitropoulos, C.D., Dubief, Y., Shaqfeh, E.S.G., and Moin, P. (2006) Direct numerical simulations of polymer-induced drag reduction in turbulent boundary layer flow of inhomogeneous polymer solutions. *J. Fluid Mech.*, **566**, 153–166.
44 Den Toonder, J.M., Hulsen, M.A., Kuiken, G.D.C., and Nieuwstadt, F.T.M. (1997) Drag reduction by polymer additives in a turbulent pipe flow: numerical and laboratory experiments. *J. Fluid Mech.*, **337**, 193–231.
45 Orlandi, P. (1995) A tentative approach to the direct simulation of drag reduction by polymers. *J. Non-Newtonian Fluid Mech.*, **60**, 277–301.
46 Wedgewood, L.E. and Bird, R.B. (1988) From molecular models to the solution of flow problems. *Ind. Eng. Chem. Res.*, **27**, 1313–1320.

47 Sureshkumar, R., Beris, A.N., and Handler, R.A. (1997) Direct numerical simulation of the turbulent channel flow of a polymer solution. *Phys. Fluids*, **9**, 743–755.

48 Walker, D.T. and Tiederman, W.G. (1990) Turbulent structure in a channel flow with polymer injection at the wall. *J. Fluid Mech.*, **218**, 377–403.

49 Beris, A.N. and Edwards, B.J. (1994) *Thermodynamics of Flowing Systems*, Oxford University Press, New York.

50 Dimitropoulos, C.D., Sureshkumar, R., and Beris, A.N. (1998) Direct numerical simulation of viscoelastic turbulent channel flow exhibiting drag reduction: effect of the variation of rheological parameters. *J. Non-Newtonian Fluid Mech.*, **79**, 433–468.

51 Orszag, S.A. and Kells, L.C. (1980) Transition to turbulence in plane Poiseuille and plane Couette flow. *J. Fluid Mech.*, **96**, 159–205.

52 Moin, P. and Kim, J. (1980) On the numerical solution of time-dependent viscous incompressible fluid flows involving solid boundaries. *J. Comput. Phys.*, **35**, 381–392.

53 Dimitropoulos, C.D., Sureshkumar, R., Beris, A.N., and Handler, R.A. (2001) Budgets of Reynolds stress, turbulent kinetic energy and vorticity production in the turbulent channel flow of a dilute polymer solution. *Phys. Fluids*, **13**, 1016–1027.

54 Housiadas, K.D. and Beris, A.N. (2003) Polymer-induced drag reduction: effects of the variations in elasticity and inertia in turbulent viscoelastic channel flow. *Phys. Fluids*, **15**, 2369–2384.

55 Housiadas, K.D. and Beris, A.N. (2004) An efficient fully implicit spectral scheme for DNS of turbulent viscoelastic channel flow. *J. Non-Newtonian Fluid Mech.*, **122**, 243–262.

56 Housiadas, K.D. and Beris, A.N. (2005) Direct Numerical Simulations of viscoelastic turbulent channel flows at high drag reduction. *Korea-Aust. Rheol. J.*, **17** (3), 131–140.

57 Housiadas, K.D. and Beris, A.N. (2006) Extensional behavior influence on viscoelastic turbulent channel flow. *J. Non-Newtonian Fluid Mech.*, **140**, 41–56.

58 Housiadas, K.D., Beris, A.N., and Handler, R.A. (2005) Viscoelastic effects on higher order statistics and on coherent structures in turbulent channel flow. *Phys. Fluids*, **17**, 035106, 20 pp.

59 De Angelis, E., Casciola, C.M., and Piva, R. (2002) DNS of wall turbulence: dilute polymers and self-sustaining mechanisms. *Comput. Fluids*, **31**, 495–507.

60 De Angelis, E., Casciola, C.M., L'vov, V.S., Piva, R., and Procaccia, I. (2003) Drag reduction by polymers in turbulent channel flows: energy redistribution between invariant empirical modes. *Phys. Review E*, **67**, 056312 (11).

61 Sibilla, S. and Baron, A. (2002) Polymer stress statistics in the near-wall turbulent flow of a drag-reducing solution. *Phys. Fluids*, **14** (3), 1123–1136.

62 Ptasinski, P.K., Boersma, B.J., Nieuwstadt, F.T.M., Hulsen, M.A., Van den Brule, B.H.A.A., and Hunt, J.C.R. (2003) Turbulent channel flow near maximum drag reduction: simulations, experiments and mechanisms. *J. Fluid Mech.*, **490**, 251–291.

63 Min, T., Yoo, J.Y., Choi, H., and Joseph, D.D. (2003) Drag reduction by polymer additives in a turbulent channel flow. *J. Fluid Mech.*, **486**, 213–238.

64 Min, T., Choi, H., and Yoo, J.Y. (2003) Maximum drag reduction in a turbulent channel flow by polymer additives. *J. Fluid Mech.*, **492**, 91–100.

65 Yu, B. and Kawaguchi, Y. (2004) Direct numerical simulations of viscoelastic drag-reducing flow: a faithful finite difference scheme. *J. Non-Newtonian Fluid Mech.*, **116**, 431–466.

66 Yu, B. and Kawaguchi, Y. (2006) Parametric study of surfactant-induced drag-reduction by DNS. *Int. J. Heat Fluid Trans.*, **27**, 887–894.

67 Zhou, Q. and Akhavan, R. (2003) A comparison of FENE and FENE-P dumbbell and chain models in turbulent flow. *J. Non-Newtonian Fluid Mech.*, **109**, 115–155.

68 Moin, P. and Moser, R.D. (1989) Characteristic eddy decomposition of turbulence in a channel. *J. Fluid Mech.*, **200**, 471–509.

69 Sirovich, L., Ball, K.S., and Keefe, L.R. (1990) Plane waves and structures in turbulent channel flow. *Phys. Fluids A*, **2**, 2217–2226.

70 Sirovich, L., Ball, K.S., and Handler, R.A. (1991) Propagating structures in wall bounded turbulent flow. *Theor. Comput. Fluid Dyn.*, **2**, 307–317.

71 Stone, P.A., Roy, A., Larson, R.G., Waleffe, F., and Graham, M. (2004) Polymer drag reduction in exact coherent structures of plane shear flow. *Phys. Fluids*, **16**, 3470–3482.

72 Li, W. and Graham, M.D. (2007) Polymer induced drag reduction in exact coherent structures of plane Poiseuille flow. *Phys. Fluids*, **19**, 083101. (15 pages).

73 Waleffe, F. (1998) Three-dimensional coherent states in plane shear flows. *Phys. Rev. Lett.*, **81**, 4140–4143.

74 Hof, B., van Doorne, C.W.H., Westerweel, J., Nieuwstadt, F.T.M., Faisst, H., Eckhardt, B., Wedin, H.H., Kerswell, R.R., and Waleffe, F. (2004) Experimental observation of nonlinear traveling waves in turbulent pipe flow. *Science*, **305**, 1594–1598.

75 Handler, R.A., Housiadas, K.D., and Beris, A.N. (2006) Karhunen–Loeve representations of drag reduced turbulent channel flows using the method of snapshots. *Int. J. Numer. Meth. Fluids*, **52**, 1339–1360.

76 Samanta, G., Oxberry, G., Beris, A.N., Handler, R.A., and Housiadas, K.D. (2008) Time-evolution K-L analysis of coherent structures based on DNS of turbulent Newtonian and viscoelastic flows. *J. Turbulence*, **9** (41), 1–25.

77 Pinho, F.T. (2003) A GNF framework for turbulence flow models of drag reducing fluids and proposal for a $\kappa-\varepsilon$ type closure. *J. Non-Newtonian Fluid Mech.*, **114**, 149–184.

78 Housiadas, K.D. and Beris, A.N. (2004) Characteristic scales and drag reduction evaluation in turbulent channel flow of nonconstant viscosity viscoelastic fluids. *Phys. Fluids*, **16**, 1581–1586.

79 Bird, R.B., Armstrong, R.C., and Hassager, O. (1987) *Dynamics of Polymeric Fluids*, vol. **1**, 2nd edn, John Wiley & Sons, Inc., New York.

80 Beris, A.N. (2003) Simple non-equilibrium thermodynamics applications to polymer rheology, in *Rheology Reviews* (eds D.M. Binding and K. Walters), The British Society of Rheology.

81 Sureshkumar, R. and Beris, A.N. (1995) Linear stability analysis of viscoelastic Poiseuille flow using an Arnoldi orthogonalization algorithm. *J. Non-Newtonian Fluid Mech.*, **56**, 151–182.

82 Fox, R.O. (2003) *Computational Models for Turbulent Reacting Flows*, Cambridge University Press, New York.

83 Leonov, A.I. and Prokunin, A.N. (1994) *Nonlinear Phenomena in Flows of Viscoelastic Polymer Fluids*, Springer.

84 Bird, R.B., Curtiss, C.F., Armstrong, R.C., and Hassager, O. (1987) *Dynamics of Polymeric Fluids*, vol. 2, 2nd edn, John Wiley & Sons, Inc., New York.

85 Simhambhatla, M. and Leonov, A.I. (1995) On the rheological modeling of viscoelastic polymer liquids with stable constitutive equations. *Rheol. Acta*, **34**, 259–273.

86 Canuto, C., Houssaini, M.Y., Quarteroni, A., and Zang, T.A. (2006) *Spectral Methods. Fundamentals in Single Domains*, Springer.

87 Deville, M.O., Fischer, P.F., and Mund, E.H. (2002) *High-Order Methods for Incompressible Fluid Flow*, Cambridge University Press.

88 Chorin, A.J. (1968) Numerical solution of the Navier–Stokes equations. *Math. Comput.*, **22**, 745–762.

89 Phillips, T.N. and Soliman, I.M. (1991) The influence matrix technique for the numerical spectral simulation of viscous incompressible flows. *Num. Meth. PDE*, **7**, 9–24.

90 Joseph, D.D. (1990) *Fluid Dynamics of Viscoelastic Liquids*, Springer, New York.

91 Dupret, F. and Marchal, J.M. (1986) Loss of evolution in the flow of viscoelastic fluids. *J. Non-Newtonian Fluid Mech.*, **20**, 143–171.

92 Apostolakis, M.V., Mavrantzas, V.G., and Beris, A.N. (2002) Stress gradient-

induced migration effects in the Taylor–Couette flow of a dilute polymer solution. *J. Non-Newtonian Fluid Mech.*, **102**, 409–445.

93 Crochet, M.J. and Legat, V. (1992) The consistent streamline upwind Petrov–Galerkin method for viscoelastic flow revisited. *J. Non-Newtonian Fluid Mech.*, **42**, 283–299.

94 Marchal, J.M. and Crochet, J.M. (1987) A new mixed finite-element for calculating viscoelastic flow. *J. Non-Newtonian Fluid Mech.*, **26**, 77–114.

95 Sureshkumar, R. and Beris, A.N. (1995) Effect of artificial stress diffusivity on the stability of the numerical calculations and the flow dynamics of time-dependent viscoelastic flows. *J. Non-Newtonian Fluid Mech.*, **60**, 53–80.

96 Fattal, R. and Kupferman, R. (2004) Constitutive laws for the matrix-logarithm of the conformation tensor. *J. Non-Newtonian Fluid Mech.*, **123** (2–3), 281–285.

97 Fattal, R. and Kupferman, R. (2005) Time-dependent simulation of viscoelastic flows at high Weissenberg number using the log-conformation representation. *J. Non-Newtonian Fluid Mech.*, **126** (1), 23–37.

98 Hulsen, M.A., Fattal, R., and Kupferman, R. (2004) Flow of viscoelastic fluids past a cylinder at high Weissenberg number: stabilized simulations using matrix logarithms. *J. Non-Newtonian Fluid Mech.*, **127**, 27–39.

99 Housiadas, K.D., Wang, L., and Beris, A.N. (2010) A new method preserving the positive definiteness of a second order tensor variable in flow simulations with application to viscoelastic turbulence. *Comput. Fluids*, **39**, 225–241.

100 Kim, J., Moin, P., and Moser, R. (1987) Turbulence statistics in fully developed channel flow at low Reynolds number. *J. Fluid Mech.*, **177**, 133–166.

101 Antonia, R.A., Teitel, M., Kim, J., and Browne, L.W.B. (1992) Low-Reynolds-number effects in a fully developed turbulent channel flow. *J. Fluid Mech.*, **236**, 579–605.

102 James, D.F. and Yogachandran, N. (2006) Filament-breaking length: a measure of elasticity in extension. *Rheol. Acta*, **46**, 161–170.

103 Nguyen, H.-P., Ishihara, K., Susuki, H., Kawaguchi, Y., and Usui, H. (2005) Biaxial extensional characteristics of drag-reducing surfactant solution. *J. Soc. Rheol. Jpn.*, **33**, 145–150.

104 Renardy, M. (1995) On the mechanism of drag reduction. *J. Non-Newtonian Fluid Mech.*, **59**, 93–101.

105 Samanta, G., Beris, A.N., Handler, R.A., and Housiadas, K.D. (2009) Velocity and conformation statistics based on reduced Karhunen–Loeve projection data from DNS of viscoelastic turbulent channel flow. *J. Non-Newtonian Fluid Mech.*, **160**, 55–63.

106 Resende, P.R., Escudier, M.P., Presti, F., Pinho, F.T., and Cruz, D.O.A. (2006) Numerical predictions and measurements of Reynolds normal stresses in turbulent pipe flow of polymers. *Int. J. Heat Fluid Flow*, **27**, 204–219.

107 Jovanovic, J., Pashtrapanska, M., Frohnapfel, B., Durst, F., Koskinen, J., and Koskinen, K. (2006) On the mechanism responsible for turbulent drag reduction by dilute addition of high polymers: theory, experiments, simulations, and predictions. *J. Fluids Eng.*, **128**, 118–130.

108 Frohnapfel, B., Lammers, P., Jovanovíc, J., and Durst, F. (2007) Interpretation of the mechanism associated with turbulent drag reduction in terms of anisotropy invariants. *J. Fluid Mech.*, **577**, 457–466.

109 Beris, A.N. and Dimitropoulos, C.D. (1999) Pseudospectral simulation of turbulent viscoelastic channel flow. *Comput. Meth. Appl. Mech. Eng.*, **180** (3–4), 365–392.

110 Kawaguchi, Y., Segawa, T., Feng, Z., and Li, P. (2002) Experimental study on drag-reducing channel flow with surfactant additives: spatial structure of turbulence investigated by PIV system. *Int. J. Heat Fluid Flow*, **23**, 700–709.

111 Kim, K., Li, C.-F., Sureshkumar, R., Balachandar, S., and Adrian, R.J. (2007)

Effects of polymer stresses on eddy structures in drag-reduced turbulent channel flow. *J. Fluid Mech.*, **584**, 281–299.

112 Wapperom, P., Keunings, R., and Legat, V. (2000) The backward-tracking Lagrangian particle method for transient viscoelastic flows. *J. Non-Newtonian Fluid Mech.*, **91**, 273–295.

113 Samanta, G., Housiadas, K.D., Handler, R.A., and Beris, A.N. (2009) Effects of viscoelasticity on the probability density functions in turbulent channel flow. *Phys. Fluids*, **21** article 115106.

114 Shetty, A.M. and Solomon, M.J. (2009) Aggregation in dilute solutions of high molar mass poly(ethylene) oxide and its effect on polymer turbulent drag reduction. *Polymer*, **50**, 261–270.

2
Modeling of Polymer Matrix Nanocomposites

Hendrik Heinz, Soumya S. Patnaik, Ras B. Pandey, and Barry L. Farmer

2.1
Introduction

Nanocomposites are materials made of nanometer-sized particles dispersed in a matrix, typically but not exclusively, composed of a polymeric material. Although the term "nanocomposites" has been fashionable for the last few decades, these materials have been around for a long time. Besides the synthetic nanocomposites of recent interest, similarly constructed materials are also found in nature – spider silk, bone, and nacre from abalone represent a few cases of organic, inorganic, and organic/inorganic nanocomposites in the biological world. The multifunctional nature of these biological nanocomposites stems from the presence of different structures and compositions. In spider silk, the exceedingly high flexibility and toughness is attributed to the presence of alternating alanine-rich crystalline blocks that impart hardness and glycine-rich amorphous blocks that provide elasticity. Together, the nanostructured morphology of these hard and soft components accounts for the incredible mechanical properties of dragline spider silk (five times tougher than steel by weight and up to 30% stretch without breaking). The multifunctional nature of bone leading to its high strength is accomplished by a hierarchical structure consisting of carbonated apatite platelets and collagen fibrils. Similarly, the nacreous layers in red abalone consist of aragonite plates in between layers of organic matrix resulting in a composite structure with fracture toughness about 3000 times greater than that of inorganic aragonite alone.

In addition to these natural examples, synthetic nanocomposites have been in use since early civilization. The concept of enhancing the properties of materials by introducing nanofillers can be traced back to as early as 100 BC. The bright color and corrosion resistance of the Maya blue paint has been attributed to its nature as an organoclay nanocomposite.

The second half of the last century has seen many applications of nanoscale organoclays to control the flow of polymer solutions and gels. More recently, nanocomposites have been the subject of a very rapidly growing body of research. Along with advances in tools to characterize and control structure at ever-smaller size

Modeling and Simulation in Polymers. Edited by P.D. Gujrati and A.I. Leonov

ISBN: 978-3-527-32415-6

scales, the advantages of new physical properties that can be achieved by controlling nanoscale morphology and dispersion, and thereby effectively changing the nature of the parent matrix, have generated great interest in these materials. Nanocomposites are being developed for a wide range of applications. The conventional definition of nanoparticles has included particles that are less than 100 nm in at least one dimension. Due to their extremely large surface to volume ratio, in most cases, the nanoparticles have dramatically different properties compared to their bulk size equivalents, and the presence of nanoparticles distributed within a host matrix at the nanometer scale has provided extraordinary improvement in properties. These properties, however, depend not only on the properties of the individual constituents but also on their morphology and interfacial characteristics.

Polymer matrix nanocomposites represent a subclass of nanocomposites consisting of nanoparticulates in a polymer matrix. The discovery of carbon nanotubes in the 1990s, the developments in processing of nanoparticles and nanocomposites, and the realization that sometimes only small loading of the nanoparticles can provide exceptional improvement in properties have all led to an unprecedented increase in polymer nanocomposite research [1] in recent years. Although it is now understood that along with the properties of the matrix and the filler, the filler size and distribution, and the nature of its interfacial characteristics with the matrix dictate the properties of the nanocomposites, a thorough understanding of the underlying physics is still in its infancy. For polymer matrix nanocomposites, the presence of nanofillers not only introduces interfacial area but also modifies the properties of the host matrix material over several radii of gyration from the particle surface. The observed enhancement in properties has contributions both from the confined polymer layers close to the filler particle and from the effective entanglements that arise when the average distance between the particles is comparable to the radius of gyration of the polymer chains. To maximize the enhancement in properties, the relative effects of these various factors need to be better understood. With rapid advances in computer power over the last few decades and the tantalizing similarity of the important size scale that dictates nanocomposite properties and the size of computational models that hardware and software can now address, computer modeling and simulations are playing an increasing role in providing this fundamental understanding, along with guiding synthesis and characterization of nanocomposites [2].

Comprehensive modeling of polymer nanocomposites calls for description of the material at many size scales – from chemical bonding to chain aggregates spanning sizes from angstrom to millimeters in length. The multiphysics aspects of their multifunctional properties require the study of molecular processes from bond vibrations to collective motion of polymer chains ranging from femtoseconds to seconds in timescale. Thus, the hierarchical structure and multifunctional properties of nanocomposites span multiple length and timescales, requiring a multiscale modeling approach. Although both sequential and concurrent multiscale approaches are being developed, by far the majority of development to date has been in the area of parameter-passing sequential approaches where different scale computational methods are linked such that quantities calculated from simulations at one scale are used as parameters for the next larger scale. Such hierarchical methods have been found to be particularly effective when material behavior can be separated into different length and timescales. An excellent review of the applications of various multiscale methods for studying polymer nanocomposites has recently been provided by Zeng *et al.* [3]. Rather than providing another comprehensive review of the application of modeling and simulation techniques to polymer nanocomposites in this chapter, we shall address some additional important aspects that have been the subjects of our own ongoing efforts in modeling nanocomposites. We have utilized Monte Carlo (MC) modeling to augment molecular and coarse-grain dynamics methods, and we have developed force fields that suitably represent the interactions of organic materials (host matrix) and inorganic materials (the nanoparticulate components). Using these approaches, we have focused on two important aspects of modeling polymer nanocomposites: factors responsible for good dispersion of nanoparticles in the polymer matrix and the effects of polymer chain and filler interactions on material properties. These are strongly interrelated – chain filler interaction is one of the important factors controlling dispersion behavior and good dispersion is needed to enhance the macroscopic properties. However, these two aspects present unique modeling challenges. Traditional homogenization theories that were developed for microcomposites have been found lacking in describing the dispersion behavior of nanocomposites, primarily because the representative volume element can no longer be considered to be homogeneous. The comparable dimensions of the nanoparticles and the polymer chains require the use of nonlocal constitutive equations. Furthermore, the interplay between the change in conformational entropy of the polymer chains close to the particles and the polymer–nanoparticle interaction energies determine the polymer configuration and its properties close to the nanoparticles. Therefore, any model describing the dispersion behavior should take all these details into consideration. A full atomistic representation of such systems is not always practical. A reasonable size polymer clay system with 5% loading can require up to a billion atoms [3] and even a simple nanocomposite made with 5% loading of carbon nanotubes dispersed in a thermoset polymers can require half a million atoms. Multiscale modeling studies [4–6] have presented arguments for the validity of coarse-grained models and these models represent a growing area of nanocomposite modeling. In the first section of this chapter, we summarize some of our own efforts in modeling polymer clay nanocomposites using coarse-grained Monte Carlo simulations. In the second section, we present some of our work in atomistic molecular dynamics (MD) modeling of interfaces between polymer and nanotube, as well as clay and metals. The detailed nature of the property of interest dictates whether a strong or weak interface is desirable and atomic simulations are helping us shape our understanding of how to achieve the desired interactions and properties. Although computationally costly, atomistic simulations play a special role in the study of nanocomposites, and MD simulations have been widely used to provide detailed insight of various nanoscale interfaces. The reliability of results from atomistic simulations strongly depends on the quality of the forcefields and molecular models used. We include a description of our efforts in developing accurate forcefields for inorganic–organic interfaces and molecular models for cross-linked polymers.

We conclude the chapter with selected examples of applications of nanocomposites where computational models might be especially informative, summarizing the results of our studies thereon. Nanostructured materials may offer advantageous thermal transport, but such properties have not received nearly as much attention as mechanical properties. Thus, we have included a section on thermal properties across interfaces.

2.2
Polymer Clay Nanocomposites and Coarse-Grained Models

Nanoclay composites [7] may consist of a range of organic and inorganic components [8, 9], in addition to clay platelets. The spatial distribution of these components is crucial in determining physical properties of these nanocomposites. In particular, the distribution of clay platelets (the largest component) in a polymer matrix is very important in designing nanoclay composites with desirable thermal and mechanical characteristics [7, 8, 10–17], such as reducing gas permeability, reducing flame propagation, and enhancing strength with respect to tensile response. A typical clay platelet (e.g., montmorillonite) is of the order of hundreds of nanometers in lateral size but only one nanometer in thickness. These platelets stack together with an interlayer spacing of the order of a nanometer to form a layered structure. Controlling the distribution of individual clay platelets in an appropriate polymer matrix is, therefore, crucial to achieving enhanced thermomechanical characteristics.

Enormous efforts have been made in recent years [7, 8, 10–28] to understand the dispersion of platelets in a range of host matrices. Critical to dispersion are controlling or achieving an intercalated or exfoliated structure. In an intercalated structure, clay plates are separated by an interleaving layer of (typically) organic material, but the platelets retain a stacked morphology. In an exfoliated material, individual clay platelets are distributed in the host matrix, with little or no correlation between the positions of the platelets. Processing may still result in some correlation of the orientations of the platelets even though they may be isolated from each other. Achieving exfoliation and a uniform distribution of the individual clay platelets remains a technical challenge due to their unique characteristics. Understanding the exfoliation and intercalation of the layers of clay platelets is a key to finding ways to control their distribution. In laboratory experiment, it is difficult to monitor the basic mechanisms that lead to a unique dispersion of the components of the nanocomposite. Computer simulations [25–28] have become valuable learning tools in such systems [18–23, 25, 26] where it is easier to invoke hypotheses and constraints and explore a range of parameters usually inaccessible in laboratory experiments.

Many of the previous studies [18, 19, 23–26] deal with the intercalation of polymer in a slit or "gallery" formed by two constrained surfaces/platelets with little or no dynamics. While some of the theoretical investigations [23, 24] are too crude to incorporate relevant fluctuations, others lack ample dynamics of its components. For example, Lee *et al.* have used MD simulation to study the intercalation of polymer chains in a slit with fixed surfaces [25] and surfaces with constrained movement along the longitudinal direction for variable slit spacing [26]. Sinsawat *et al.* [27] have examined the intercalation and exfoliation processes of clay sheets using an off-lattice MD simulation. It was difficult to investigate desired structural evolution (in particular, long-time exfoliation) due to the relatively long relaxation times in these materials – a typical bottleneck in computer simulations of such complex systems with off-lattice methods. Thus, the utility of a powerful computational tool such as off-lattice simulations is severely limited despite its strength in probing the microscopic structural details in depth.

We have proposed a Monte Carlo bond-fluctuating model [29, 30] for clay platelets with ample degrees of freedom on a discrete lattice to capture some of the pertinent details including multiscale mode dynamics [28, 31–33]. Unlike the kink analysis [34, 35] of platelets, we have studied the exfoliation of a stack (layer) of sheets by the bond fluctuation model. With a coarse-grained model, effects of solvent on the exfoliation and dispersion of platelets can be examined in a host matrix with effective solvent [28] by Monte Carlo simulations. While this study provides insight into how the exfoliation is enhanced by increasing the temperature and changing the quality of solvent, the solvent fluctuations are neglected as in a typical mean field approximation. Very recently, we have examined [36] the exfoliation of a stack of platelets in a matrix with explicit solvent particles. With a relatively strong attractive interaction between sheets and the solvent particles, we find that the stack of sheets remains intact via intercalation of solvent particles into the interstitial spacing (gallery) between the sheets. For weaker attractive-to-repulsive interactions between sheets and solvent particles, the sheets easily exfoliate. We have also extended [37] this study to incorporate polymer molecules (i.e., replacing solvent particles with chains) in the presence of a stack of platelets and examined the effect of the molecular weight of polymer chains and their interaction with sheets on exfoliation.

Attempts are thus being made to push forward the computer simulation of nanoclay composites to incorporate the dynamics of each component, that is, platelets, polymer, and solvent and their cooperative and competing effects via a coarse-grained description. With the efficiency of the coarse-grained approach, it is possible to cover much larger time and possibly spatial scales. We must, however, caution that the coarse-grained system considered thus far is still far from reality but represents a step toward developing a model closer to realistic systems. In the following, we point out how the constituents of composite components are described in a coarse-grained approximation. A brief remark is made on relevant methods to cover appropriate characteristic details at different timescales. Modeling of a platelet is then described and its multiscale dynamics is pointed out via an interconnection to particles and chains. Models for clay platelets in solvent particles and in a polymer matrix are considered and results of the distribution of constituents and their dynamics are discussed systematically. We conclude with a remark on the future outlook.

2.2.1
Coarse-Grained Components

In computer simulation and in the interpretation of the results of laboratory measurements, constituents of a nanocomposite are conceptualized by such representations as particles, chains, sheets and aggregates. These constituents are assembled in a composite via such basic mechanisms as attractive (noncovalent) interactions, covalent bonding, physical entanglement or simply hardcore confinement, and combinations thereof. The morphology of nanocomposites depends not only on the concentration and composition of their basic components but also on their assembly processes (e.g., compaction, consolidation via shaking and tapping, annealing, and quenching). The fabrication (assembly) involves thermodynamics and kinetic aspects such as changes in and out of equilibrium, and pinning (arresting) of evolving structures at local and global scales.

Understanding the fundamental issues from the atomic or molecular scale to macroscopic morphology in such a complex system is challenging. Most analytical theories [11] are severely limited for such complex systems that exhibit linear and nonlinear response properties on different spatial and temporal scales. Computer simulations remain the primary choice to probe multiscale phenomena from microscopic characteristics of constituents to macroscopic observables in such complex systems. Most real systems [9] are still too complex to be fully addressed by computing and computer simulations alone. Coarse-grained descriptions are almost unavoidable in developing models for such nanocomposite systems.

What are the basic building blocks and how are they incorporated in a model to describe the assembly of a nanocomposite and characterize its physical properties? The concept of a particle, the smallest unit, can be used to describe constituents in nearly any physical, chemical, and biological system. Basic building blocks such as a chain, sheet, microaggregate, or microgel, for example, can be designed by tethering the particles, incorporating appropriate interactions among them, and imposing other constraints such as prescribed neighborhood (i.e., in a covalent bonding of specific constituents), some of which will be described in this chapter in a hierarchical fashion. We consider here the nanoclay composite and focus on the exfoliation and dispersion of clay platelets (sheets) in a solvent and a polymer matrix represented by particles and chains, respectively.

2.2.2
Methods and Timescales

In computer simulation [30] by standard Monte Carlo and MD methods, one generally considers off-lattice, discrete lattice, and their combination if feasible, depending on the system and properties of interest. Advantages and disadvantage of these for describing a particular system are more transparent with specific examples; it is, however, worth pointing out some general features here.

2.2.2.1 Off-Lattice (Continuum) Approach

A particle can execute its stochastic motion with consecutive hops in any direction in a continuum host space, that is, in virtually unlimited directions as opposed to a limited number of possible moves to neighboring grid points on a fixed grid (see below). Because of the large degrees of freedom accessible to a particle, the off-lattice approach is extremely valuable in probing the short timescale, for example, vibration spectra of a simple liquid. Let us consider a polymer chain [30] designed by tethering consecutive particles (particles or beads) with spring bonds, that is, bead-spring model. The conformations and dynamics of a single polymer chain in free space or dilute solution can be described very well by the bead-spring model in a continuum host lattice, that is, the short time Rouse dynamics and long time diffusion of the chain can be verified with good accuracy along with the scaling of its radius of gyration with molecular weight. The problem becomes much more complex in the melt when many chains are included in the simulation box. Ideally, the dynamics of a polymer chain is expected to exhibit different characteristics over the range of longer timescales, that is, Rouse, reptation, postreptation, and diffusion. In large-scale simulations of melts with the bead-spring model of the polymer chains, one can recover short time Rouse dynamics and reptation in part. Since they require both large chain sizes and high concentration, it is not feasible to recover longer time dynamics, that is, postreptation and diffusion beyond reptation. Thus, the feasibility of this *bottom-up*, off-lattice approach for the bead-spring chain systems in a complex melt breaks down beyond a certain timescale. In order to cover longer scales, one has to resort to another level of coarse-grained approach.

2.2.2.2 Discrete Lattice Approach

Consider the random motion of a particle on a cubic lattice. The particle occupies a lattice node and can hop from one lattice node to another nearest-neighbor node. Since there are only six nearest-neighbor sites available for the particle to hop to, its degrees of freedom are severely limited in comparison to a particle in a continuum space described above. However, if one calculates the root mean square (rms) displacement of the particles on the discrete cubic lattice as a function of time steps, one can show that its asymptotic behavior is diffusive, similar to that of the particle in a continuum space. The degrees of freedom of a particle can be improved by describing the particle as a cube (say the unit cube) on a cubic lattice where the particle (cube) can hop to one of its 26 adjacent cubic spaces (sites). The polymer chain [30] can be described by tethering the consecutive particles (cubes) by bonds of fluctuating length (between 2 and $\sqrt{10}$ with an exception of $\sqrt{8}$, expressed in multiples of the length of the edge of the unit cube). If the size of the polymer chain is long enough, one can recover both the short time Rouse dynamics and the long time diffusion with this bond fluctuation model of the polymer chain. Note the difference between the bond-fluctuating and bead-spring model of the polymer chain. The degrees of freedom of the polymer chain (for both particle and bond) are much larger for a chain described by the bead-spring model in a continuum space than that of the bond-fluctuating model on a discrete lattice. Therefore, one needs a relatively long chain length to recover the short time Rouse dynamics with the bond-fluctuating chain. There is no problem in recovering the asymptotic diffusive motion with relatively smaller bond-fluctuating chains. In this *top-down* approach, one needs longer chain lengths to approach shorter timescales; the longer chain length provides larger degrees of freedom for the chain. The advantage of this approach is the simplicity and the computational efficiency of the discrete lattice. The equilibrium and asymptotic properties of a complex melt can be studied more effectively and efficiently with the bond-fluctuating chains.

2.2.2.3 **Hybrid Approach**

In principle, one can exploit the effectiveness of both top-down and bottom-up approaches by implementing both methods in appropriate systems. For example, one may use discrete lattice simulation to reach around a particular point in thermodynamic phase space (large-scale stirring) followed by off-lattice simulation (small-scale stirring) to capture the microscopic details. This is an efficient method to reach closer to equilibrium while capturing appropriate short-scale details. This has been used [38] to show how a system of polymer chains assembles into short-range folds and long-range aggregates. The disadvantage of such method is the lack of universal applications – one has to develop such a method for a specific system.

For simplicity and consistency, we restrict ourselves to a discrete-lattice system with the bond-fluctuation method to describe basic constituents such as polymer chains and clay platelets in this section. As mentioned above, we focus on the distribution of clay platelets, for example, their exfoliation and dispersion, in a complex dynamic polymer and solvent matrix. Dynamics of particles [39] (hardcore, interacting) and polymer chains [30] (dilute solution, melt) have been extensively studied in the last few decades. Self-organizing structural evolutions leading to phase separation, segregation, and mixing resulting from the cooperative and competing characteristics of these constituents have also been studied in detail over the years. Despite a large volume of work, such studies for clay platelets in the presence of polymer chains and solvent particles with computer simulations are limited [9, 11] particularly due to difficulties encountered in extending the simulations to model platelets. Our goal is to understand the structural evolution in a complex mixture of interacting solvent (represented by particles), polymer chains, and clay platelets (represented by flexible sheets). It would be prudent to describe the model for the platelet and point out its unique conformation and dynamics first.

2.2.3 **Coarse-Grained Sheet**

The morphology and shape of a sheet play crucial roles in its specific functions. A platelet or tethered membrane [33, 40–44] exhibits unique conformational characteristics as it wrinkles and crumples under various physical constraints, such as quality of solvent and temperature, while entropy dissipates from its open boundaries. As pointed out above, one can tether particles – the smallest units in a coarse-grained description – to form both sheets and other hierarchical units such as chains, aggregates, and microgel particulates. To appreciate the collective dynamics of the tethered particles in a sheet [33, 40, 45], it would be appropriate to briefly summarize the characteristics of an untethered particle first.

We know that the stochastic motion of a particle exhibits a range of dynamics (anomalous diffusion, diffusion, and drift) from short to long asymptotic time regime. Let us consider the motion of a random walk of a particle on a cubic lattice in which a particle hops to one of its nearest-neighbor sites (six on a cubic lattice) randomly at each time step. The probability that the random walker reaches a distance R from the origin (the starting point) in t hops (or time steps) is $P(t,R) = A\exp(-R^2/C_1 t)$, where A and C_1 are constants. The root mean square displacement (R) can be evaluated by integrating it with this distribution and provides $R = C_2 t^{1/2}$ with a diffusion constant C_2. The motion can be explicitly monitored in a computer simulation and the power-law relation between the rms displacement (R) and the time step (t) can be verified,

$$R = Dt^{\nu}, \tag{2.1}$$

where D is a constant (diffusion) and ν is the exponent. The exponent $\nu = 1/2$ characterizes the diffusive motion of the random walker.

The characteristic dynamics [45] of the particle changes if a direction, say $+x$, is selected with a biased probability B $(0 \leq B \leq 1)$. In the presence of bias, the motion is diffusive $(R \approx Dt^{1/2})$ in the short time and drift $(R \approx Bt)$ in the long time. The crossover from diffusion to drift occurs around the time step $t_c \sim (D/B)^2$ where the crossover time t_c depends on the bias. This simple example illustrates how the characteristic multiple dynamics of a mobile particle depends on the bias and the time regimes. The motion of the particle becomes more complex [39] if we place the mobile particle in a heterogeneous porous matrix, that is, a percolating medium [46]. The restricted movement of particles leads to well-defined long time dynamics characterized by anomalous diffusion [39, 46] at the percolation threshold where $\nu = 0.20$ for the random walk motion of a particle at the percolation threshold in three dimensions. The crossover from short to long time dynamics depends on the heterogeneity (porosity) of the underlying matrix. Applying a bias (B) to the stochastic motion of the random walker in a percolating medium adds more complexity with the exponent ν depending on B and the fraction of the conducting sites [39, 46]. Thus, the characteristic dynamics of a simple particle executing its stochastic motion is very rich and depends on the underlying host matrix, bias, and time range.

When particles are tethered together in a string, that is, a polymer chain, their motion becomes more complex. A polymer chain [30] is described by a set of particles (nodes) tethered together in a string via appropriate bonds in a coarse-grained model. In a dilute solution, the stochastic motion of an interior particle is restricted due to constraints imposed by adjacent particles (connected bonds) and excluded volume effects. The rms displacement of the particle varies with the time step (t) as described by Eq. (2.1) but with a different power-law exponent ν. The short time motion is known to be slow with $\nu = 1/4$ (Rouse dynamics) [30] while the long time motion is still diffusive $(\nu = 1/2)$. In the long time (asymptotic) regime, the polymer chain behaves like a particle (random walker) while the internal structure of the coarse-grained chain becomes crucial in the short time dynamics. The motion of the particle in the chain becomes much more complex when the chain is placed in a complex environment, for example, in the melt where it exhibits multiple dynamics, namely, rouse, reptation, postreptation, diffusion in order of increasing timescales [30]. The crossovers from one dynamics to another depends on the molecular weight, environment, and so on and play a key role in orchestrating the viscoelastic properties of the polymer systems [30]. The fundamental characteristics of a particle in an assemblage is thus important in understanding the basic characteristics of such building blocks and their cooperative and/or competing behavior in the morphological structures in composites.

In a coarse-grained description, a sheet is a set of particles tethered together in a plane, in analogy with and in extension of the concept used for modeling a polymer chain. Accordingly, one can use off-lattice [25–27] or on-lattice bond fluctuation models for a sheet [33, 40]. Despite the apparent strength in exploring the microscopic structural details, the large degrees of freedom (for particles and bonds) with off-lattice simulations create a bottleneck in computing [27] in probing many complex issues in nanocomposites due to long relaxation time. In order to accelerate the relaxation process, a bond-fluctuating model on a discrete lattice becomes a viable choice [33].

A platelet (sheet) is described [33] by a set of L_s^2 particles tethered together by bonds with an initial configuration on a square grid (see Figure 2.1) on a cubic lattice of size L^3. A particle is represented by a unit cube (i.e., eight sites) of the lattice. In the bond fluctuation model for a polymer chain [30], the minimum bond length between particles is twice the lattice constant due to excluded volume. The bond length between particles can fluctuate between 2 and $\sqrt{10}$ with an exception of $\sqrt{8}$ (expressed as multiples of the length of the edge of the unit cube) as the particle

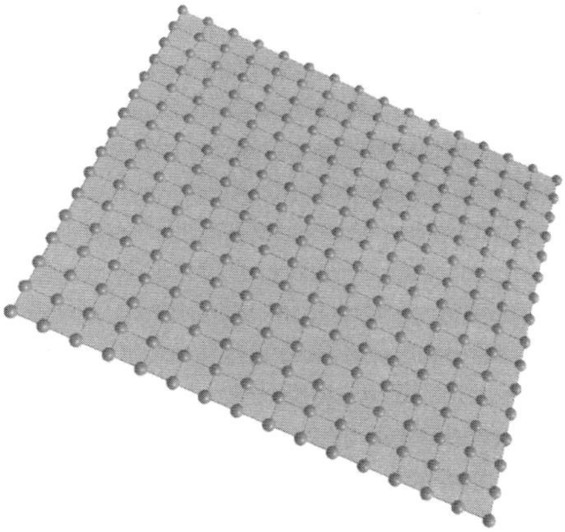

Figure 2.1 Model of a sheet in its initial configuration where nearest-neighbor particles are bonded with fluctuating bond length. Size of the sheet 16^2 (particles are represented by spheres) on a 128^3 lattice. (From Ref. [40].)

executes its stochastic move to one of its 26 adjacent cubes (referred to as sites) similar to the bond fluctuation model for a polymer chain [29, 30].

A stochastic move is defined as follows: select a particle at a site i and select one of its 26 neighboring sites, say j. If site j is empty and the constraint on the changes in the bond lengths (due to the proposed move) to all connected particles is satisfied, then the particle is moved from site i to site j. On the other hand, if either of the criteria is not met, the particle remains at its original site i. Attempts to move each particle once defines the unit Monte Carlo step (MCS) time. A simulation then consists of a large number of such (MCS) steps, typically millions.

The excluded volume, self-avoiding sheet (SAS) deforms and wrinkles propagate as its particles execute their stochastic motion. The conformation and dynamics of the sheet can be studied by keeping track of the rms displacement of the center of mass of the sheet and that of its center (interior) particle, radius of gyration, as a function of molecular weight and rigidity of the covalent bonds. Effects of the quality of the solvent and entanglement barriers (e.g., polymer chains matrix) can be incorporated (see below). A few observations from our computer simulations are briefly discussed in the next section.

2.2.3.1 **Conformation and Dynamics of a Sheet**

In addition to excluded volume, the conformation and dynamics of a sheet can be controlled by introducing (intrasheet) interactions among its particles as well as interactions and constraints with other constituents in the host matrix. The physical properties of the sheet (clay platelet or tethered membrane) depend on the type of interaction with the underlying matrix (e.g., solvent), concentration, temperature, and a host of external and internal parameters.

In a simple model [33, 40], one can use empty lattice sites to represent an effective solvent medium in which a particle can exchange its position with the solvent sites. The interaction energy of a particle can be described by

$$E = \Sigma_{ij} J(i, j), \tag{2.2}$$

where i runs over each particle and j over neighboring sites within a range r. Different types of particle–particle ($J(n,n) = \varepsilon_{nn}$) and particle–solvent ($J(n,s) = \varepsilon_{ns}$) interactions (e.g., square-well, Lennard-Jones, etc.) can be considered within a range (r) that itself can be varied. Let us consider a square-well potential with $r = \sqrt{8}$, $\varepsilon_{nn} = -1$ or $+1$, and $\varepsilon_{ns} = 0.5$ for simplicity. Each particle executes its stochastic move to one of its neighboring sites based on the Metropolis algorithm – that is, determined by the change in energy ΔE due to a possible change in configuration (caused by the move from site i to site j) via hopping probability $\exp(-\Delta E/T)$, within excluded volume and bond fluctuation constraints mentioned above; T is the reduced temperature measured in an arbitrary unit (involving Boltzmann constant and interaction energy). Periodic boundary conditions are used along each direction.

Wrinkles are caused by the local movements of particles and propagate along their connected pathways of the membrane. How the global dynamics emerge from the interplay between the competition and cooperation of these local dynamical modes is one of the fundamental questions to be addressed.

One of the unique features of a tethered membrane is the dissipation of propagating wrinkles from the open edges (best seen in visual animations). The number of configurations due to local modes (due to changes in positions of particles) is a measure of the configurational entropy. Since the entropy dissipates from the edges, it is rather difficult to retain many structural changes for a long time. As a result, the global conformation of the sheets retains its planar conformation that can be verified from the scaling of its radius of gyration (R_g) with the size, $R_g \approx L_s$. The sheet can be crumpled, of course, if one imposes severe constraints. Regardless, a sheet expands (repulsive) or contracts (attractive) at low and high temperatures depending on the particle–particle interaction (attractive, repulsive). From systematic simulations [28], one can predict how the radius of gyration varies with the temperature.

The collective dynamics of the sheet can be examined by tracking the motion (rms displacement) of its center of mass. It is often diffusive in simple (unconstrained) systems in the asymptotic time regime. The dynamics of the center of mass, however, does not elucidate the effect of internal structure succinctly due to mode averaging. Such details can be captured by monitoring the motion of an interior particle, say, the center one.

The mean square displacement of the center particle of the sheet with the time step shows well-defined modes from short to long time regimes (see Figures 2.2 and 2.3). From Figure 2.2, we can recognize three types of power-law dynamics in sequence: (i) short time dynamics, $R_n \propto t^{\nu_1}$, with $\nu_1 \sim 1/8$, followed by (ii) a faster dynamics $R_n \propto t^{\nu_2}$ with $\nu_2 \sim 1/3$ before slowing down substantially, $R_n \propto t^{\nu_3}$ with $\nu_3 \sim 1/10$. Although a closer examination of the data in Figure 2.2 ($t \sim 10^7$) shows the onset of

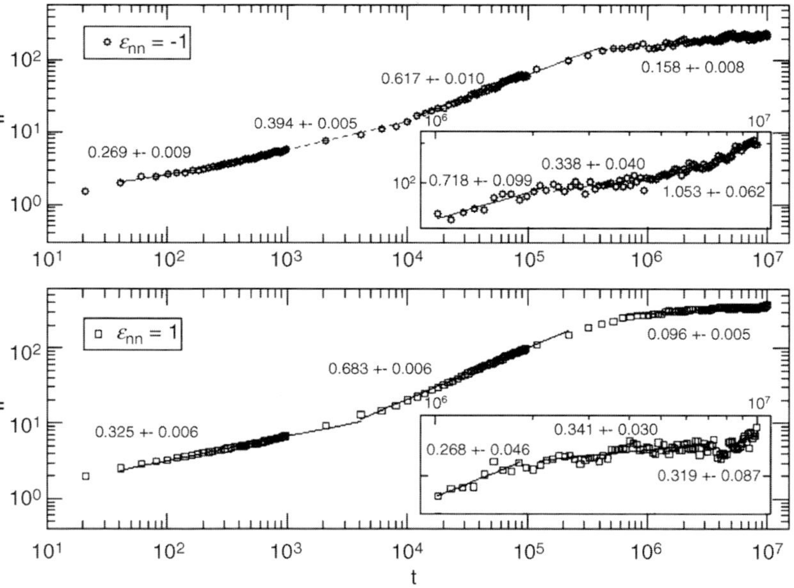

Figure 2.2 Variation of mean square displacement of the center particle (R_n^2) with the time step t of the membrane of size 64^2 on a 200^3 lattice at time steps with interactions $\varepsilon_{nn} = -1$, 1 and $\varepsilon_{ns} = -0.5$ and $T = 2$; 10–1000 independent samples are used. (From Ref. [33].)

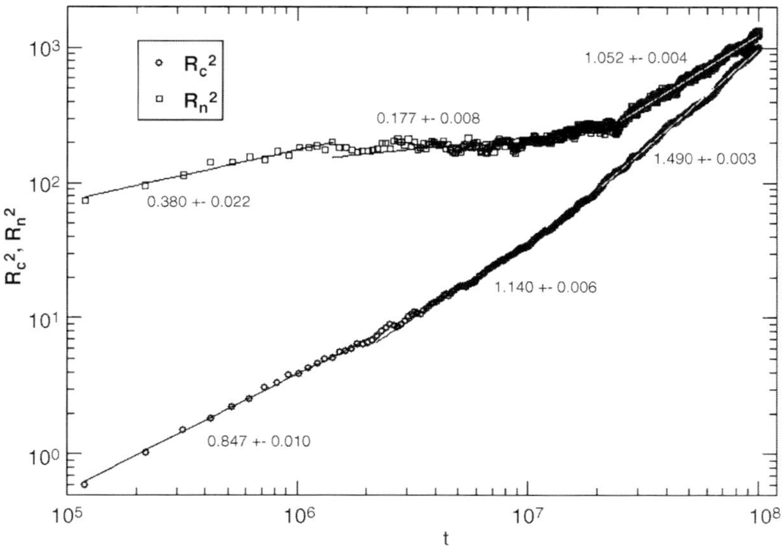

Figure 2.3 Variation of mean square displacement of the center of mass (R_c^2) of the membrane of size 64^2 and that of its center particle (R_n^2) with the time step t on a 200^3 lattice at time steps with interactions $\varepsilon_{nn} = -1$ and $\varepsilon_{ns} = -0.5$ and $T = 2$. About 50–100 independent samples are used. (From Ref. [33].)

diffusive behavior in the long time despite large fluctuations (see the figure inset), the larger scale simulation results shown in Figure 2.3 clarify it with much better accuracy, $R_n \propto t^{\nu_4}$ with $\nu_4 \sim 1/2$. Note that the short time dynamics ($\nu_1 \sim 1/8$, analogue of Rouse dynamics of a polymer chain but with half the size of its exponent) and long time asymptotic diffusion ($\nu_4 \sim 1/2$) are similar to that of a polymer chain. However, the intermediate dynamic modes (described by the exponent ν_2, for example) are unique features of the sheet. Thus, the segmental motion of a tethered membrane exhibits multiscale dynamics much richer than that of a chain. Furthermore, the type and range of the multiscale dynamics depends on the type of sheet (ε_{nn}) and the matrix in which it is embedded. One would expect the intermediate dynamics to exhibit its consequences in the viscoelastic properties of a nanoclay composite, analogous to that of the polymer melt (reptation) [30].

In order to cover the full range of multiscale dynamics, one needs a relatively large dimension of the sheet in our bond fluctuation (*top-down*) approach as in studying the multiscale dynamics of polymer chains. As mentioned above, a composite consists of a number of components represented by particles, chains, and sheets. Multiscale characteristics of each component, for example, particles, chains, and sheets, are further modified when a large number of these constituents are placed in a simulation box, that is, by their concentrations (volume fractions). Interaction and physical constraints at higher concentrations introduce multiple relaxation times for composites to reach equilibrium or steady state. In order to carry out a systematic investigation and draw meaningful conclusions, we restrict ourselves to constituents of smaller sizes and appropriately lower concentrations to capture their cooperative response properties.

2.2.4
Coarse-Grained Studies of Nanocomposites

We consider a cubic lattice of size L^3 as a host matrix. The components of the composites such as particles, chains, and sheets (described above) are then placed into the box (Figure 2.4).

Before dropping in all components at once and studying their stochastic and steady-state properties, it is good to systematically examine one component at a time, that is, clay platelets in a solvent, polymer matrix, and so on. Since clay platelets are the largest constituents of the composite and they stack together in layers, understanding their distribution (exfoliation and dispersion) in an appropriate matrix is crucial. Therefore, we place a layer of stacked platelets, each of size L_s^2 first in the center of the simulation box [28] when preparing the sample. Subsequently, we add solvent particles and/or polymer chains each of length (molecular weight) L_c as desired. As mentioned above, a coarse-grained polymer chain is a set of L_c particles tethered together by fluctuating bonds [29, 30] while a platelet (sheet) is described by a set of L_s^2 particles tethered together by flexible bonds on a square grid. Initially, four sheets, each in a planar square grid configuration, are stacked together with a small

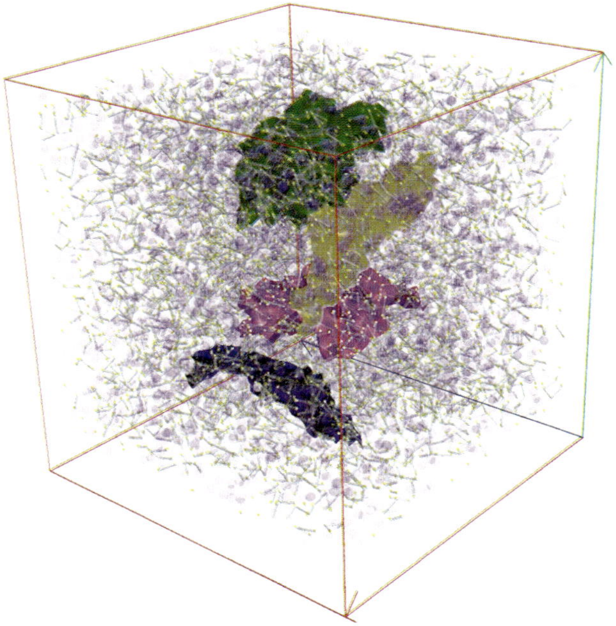

Figure 2.4 Representation of a coarse-grained model for mixture of clay platelets, polymer chains, and solvent particles in a cubic box.

separation of distance of 4 (lattice constants) in the center of the lattice. Attempts are made to insert solvent particles (each of the same size as an individual particle of a sheet) and/or chain particles randomly to occupy a fraction ϕ of the lattice sites. These matrix components, that is, solvent (individual) particles and chains of particles, are moved randomly (keeping the sheets in the same initial position) for about a million time steps to prepare the initial state of the sample (see below).

In addition to excluded volume, solvent particles and polymer chains interact with sheets with a short-range square-well pair potential. The interaction energy of a particle (representing particle, chain, or sheet) is described by

$$E = \Sigma_{ij} V(i,j), \tag{2.3}$$

where i runs over each particle and j over neighboring sites within a range $r = \sqrt{6}$. Clay–solvent ($V_{(c,s)} = \varepsilon$) and clay polymer interactions ($V_{(p,s)} = \varepsilon$) are taken to be the same for all particles within the range (r). We have used a range of values of ε to examine the effect of interaction. For example, $V_{(c,s)} = \varepsilon = -1$ and 1 represent sheets in the polymer matrix with attractive and repulsive interactions, respectively. Additional interactions, for example, sheet–sheet and polymer–polymer, can be included as desired. However, it is better to start with simpler interactions, for example, purely excluded volume, and add more interactions systematically. The distance r is measured in units of the lattice constant and the interaction energy is in arbitrary units as before [28, 31–33, 36, 37]. The Metropolis algorithm is used to move each particle (solvent particle, particle of sheets, or of polymer chains) stochastically. Periodic boundary conditions are used along each direction. Attempts to move each particle once (on average) define the unit Monte Carlo step time constant [30].

After the sample has been prepared (solvent particles and polymer chains moved randomly while keeping the sheets in the initial stacked configuration mentioned above), the Monte Carlo simulation is performed for a sufficiently long time (typically of the order of million time steps here) to evaluate morphological changes in the distribution of sheets and polymer chains. We study the density profiles of sheets, chains, and solvent particles as well as variations of the root mean square displacements and radius of gyration of chains and sheets with time steps. As in most simulations, a number of independent samples are used to obtain statistical averages of these quantities. Further, different lattice sizes are used to assure that the qualitative results are independent of the finite size effects. In the following, we focus on the effect on exfoliation and intercalation of the interaction between the matrix (solvent, polymer) and clay platelets and molecular weight of polymer.

2.2.4.1 **Probing Exfoliation and Dispersion**

One of the main questions of interest in nanocomposites is how can one disperse platelets of common clay (e.g., montmorillonite)? Instead of considering all the components of a composite (Figure 2.4), let us first consider only a stack of self-avoiding (excluded volume) sheets with a small interlayer spacing (as shown in Figure 2.5) placed on a cubic lattice in an otherwise empty box. If these sheets are allowed to execute their stochastic motion, they will obviously disperse (see

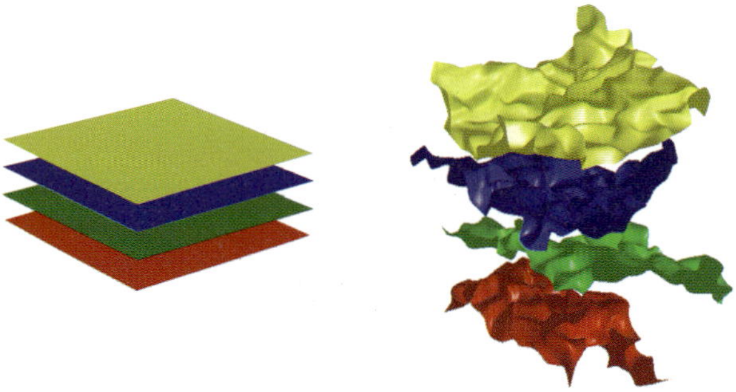

Figure 2.5 A stack of four SAS sheets (left) each of size 16^2 and their dispersion (right) in an empty space after 10^5 steps on a 128^3 sample. (From Ref. [40].)

Figure 2.5), a basic principle of flow from high to low concentration. How fast they disperse depends on the size of sheets and the thickness of the interlayer spacing. However, even in simple clay composite systems, platelets are exposed to an interacting environment, and therefore incorporation of solvent and other constraints is highly desirable.

2.2.5 Platelets in Composite Matrix

Solvent constitutes one of the simple host matrices for a layer of clay platelets to exfoliate and disperse. How can solvent be incorporated? Consider a layer of four sheets as shown in Figure 2.5 with a small interstitial separation (four lattice constant) – the minimum number of sheets (four) to incorporate the effect of external (exposed) and internal surfaces in the stack. In an effective medium approximation, the empty lattice sites may constitute the effective solvent medium if each particle interacts with the solvent (empty) sites. Each particle can exchange its position with the neighboring solvent site depending on the change in its energy via Boltzmann probability distribution in addition to excluded volume and bond fluctuation constraints (described above). Note that the effective medium representation [28] of the solvent is a crude approximation as there is no fluctuation in its concentration and its movement is highly correlated with that of the sheet particles. However, it is possible to incorporate the effects of temperature and quality of solvent on the exfoliation [28]. For example, the quality of the solvent (type and strength of the solvent–sheet interactions) can constrain the platelets to retain a stacked structure or drive them to adopt a fully exfoliated morphology. Raising the temperature enhances the mobility and dispersion [28].

2.2.5.1 Solvent Particles

Describing the effective solvent medium by empty lattice sites is efficient, but it is difficult to identify mobility, fluctuation, and correlations. Representing the solvent by mobile particles (say, each with the same size as a particle of the sheet) on a fraction

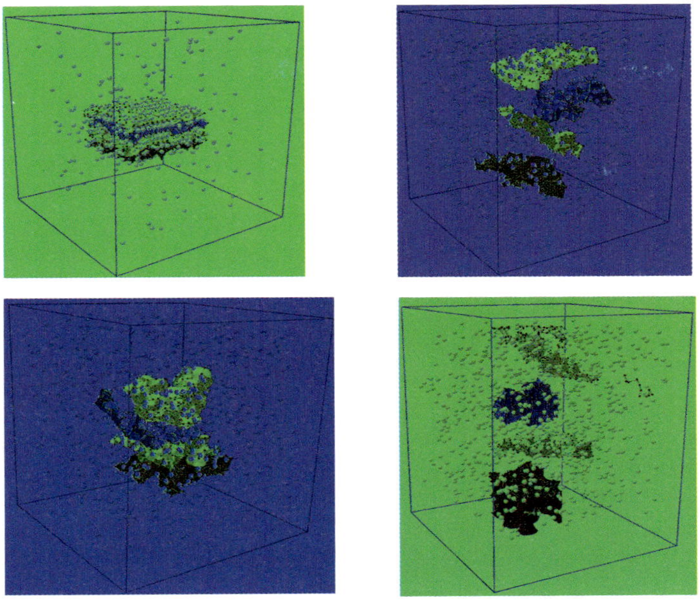

Figure 2.6 Snaps of stacked platelets (each of size 16^2) after 10^6 time steps on 64^3 lattice with clay–solvent interaction $\varepsilon_{ns} = -2$ ($T = 1$, left top, $T = 5$ right top), $\varepsilon_{ns} = -1$ ($T = 1$, left bottom), $\varepsilon_{ns} = 1$ ($T = 1$, right bottom). Solvent particle concentration is 0.2. (From Ref. [36].)

of lattice sites could be a viable approach. A layer of platelets is now immersed in such a solvent, the concentration of which can be varied (Figure 2.6). Interactions between the solvent and the particles of the sheets describe the quality of the solvent. Both solvent and particles of the clay platelets can perform their stochastic motion. The interaction between solvent particles and platelets and the temperature orchestrate their spatial distribution. In addition to dispersion of platelets, one can investigate the intercalation of the solvent particles into the interstitial spacing between the sheets. Extensive computer simulations are performed to investigate the density profile and mobility of platelets and solvent particles. Figure 2.6 shows the snapshots toward the end of the simulation to illustrate the effect of the quality of solvent and temperature on the dispersion of the layered platelets. We find that the attractive interaction between the solvent and the sheets leads to intercalation of solvent particles between interstitial platelet layers. The solvent-mediated interaction between sheets (via interstitial particles) tends to retain the layered morphology. Lowering the interaction strength reduces their cohesion, and sheets begin to disperse; a repulsive interaction causes the platelets to fully exfoliate. Dispersion of layered sheets is also enhanced by raising the temperature.

Density profiles of clay and solvent provide some insight into the distribution of clay and solvent particles that help clarify the effects of temperature and that of the quality of solvent (particle–solvent interaction). Let us define the y-axis (normal to the initial platelet planes) as the longitudinal direction; the z- and x-axes constitute transverse directions. The planar density (d_i) is

$$d_i = \frac{1}{L^2} \sum_{j,k} \varrho_{i,j,k}, \tag{2.4}$$

where $\varrho_{i,j,k} = 1$ if the lattice site i, j, k (at the Cartesian coordinate x, y, z) is occupied by a particle and $\varrho_{i,j,k} = 0$ if it is empty. The transverse density profile (d_x) for clay and solvent particles at $T = 1, 2, 5, 10$ is presented in Figure 2.7 for the particle–solvent interactions $\varepsilon = 1, -1$, and -2. At the low temperature $T = 1$ and $\varepsilon = -2$, the clay density profile shows that the platelets remain nearly static in the center of the lattice.

The corresponding density profile of the solvent particles follows a remarkably similar pattern with a lower magnitude which implies the solvent particles follow the clay platelets. As expected, the temperature ($T = 1$) is too low and the attractive interaction energy ($\varepsilon_{ns} = -2$) between the platelets and solvent is too strong to separate the platelets. Reducing the attractive interaction energy to $\varepsilon = -1$ leads to a considerable change in the density profiles for both clay and solvent. While there is a considerable change in the structural distribution of clay platelets, the corresponding shift in solvent density is not as large. This is because the overall solvent density is low. However, the shape of the solvent density profile suggests that the distribution of intercalated solvent preserves its memory of the close proximity of clay platelets. By changing the interaction further to $\varepsilon = 1$, we see that the distribution of solvent

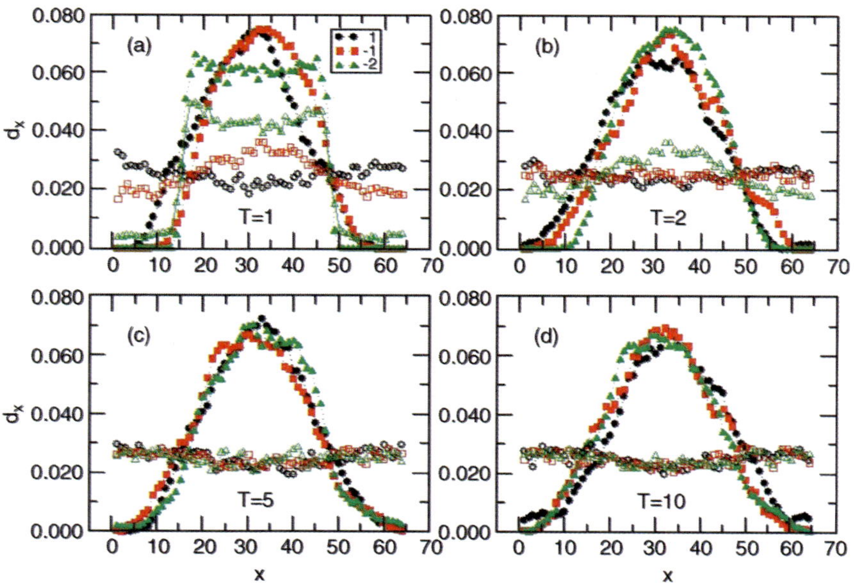

Figure 2.7 Transverse planar density (d_x) of clay (filled) and solvent (open) particles versus x at $T = 1$–10 with clay–solvent interaction $\varepsilon_{ns} = 1, -1, -2$; plates are in the zx-plane initially (Figure 2.6). A stack of four platelets each a size of 16^2 is used on a lattice of size 64^3 each with 10 independent runs. (From Ref. [36].)

particles is nearly homogeneous throughout while the clay platelets remain in the center of the lattice in distorted configurations. This is consistent with the snapshots seen in Figure 2.6. When the temperature is raised to $T = 2$, the distribution of solvent particles is nearly homogeneous except with $\varepsilon = -2$. At high temperature ($T = 5$, 10), the distribution of solvent remains homogeneous and the clay platelets somewhat exfoliated. Note that it is difficult to reach a fully exfoliated state even with our efficient computer simulation model despite the long simulation runs. However, the trend is clear.

2.2.5.2 **Polymer Matrix**

The preceding study is extended [37] by replacing solvent particles with polymer chains. A set of typical snapshots at the end of simulations is presented in Figures 2.8 and 2.9 with attractive and repulsive interactions between the polymers and the sheets with the polymer matrix density $\phi = 0.2$. Let us examine these figures to probe the effect of interaction between the sheets and the polymer chains and their molecular weight. Because of the entropic constraints on the conformations, shorter chains can move and intercalate more easily than the chains with higher molecular weights.

Snapshots with the lowest molecular weight ($L_c = 4$) polymer matrix show that the sheets maintain their layered configuration with $\varepsilon = -1$ (attractive) interaction

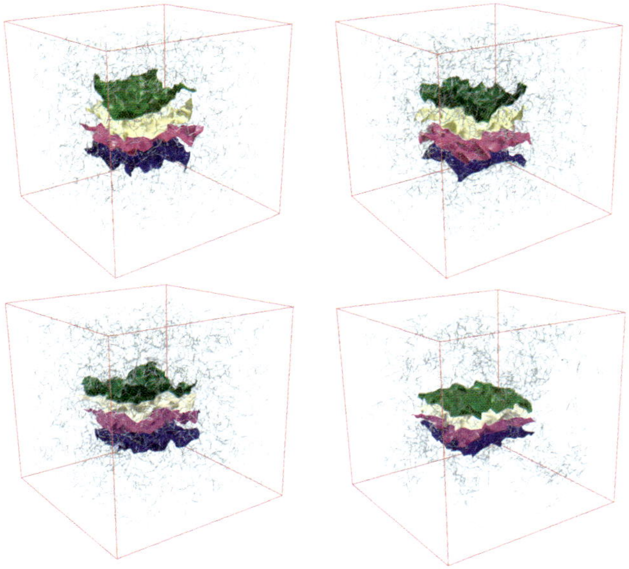

Figure 2.8 Snapshots of the final configurations with the attractive ($\varepsilon = -1$) interaction between polymer matrix and sheets) after 10^6 time steps. Sample size 64^3 is used with the polymer volume fraction $\phi = 0.2$ and polymer chain lengths $L_c = 4$ (first row, first column), 16 (first row, second column), 32 (second row, first column), and 64 (second row, second column). (From Ref. [37].)

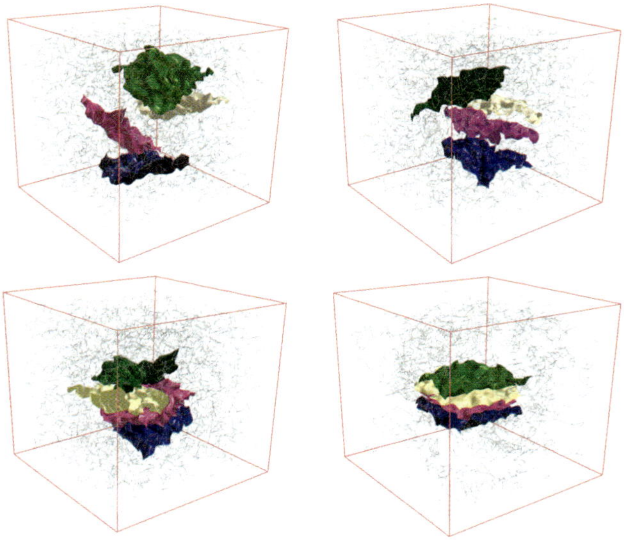

Figure 2.9 Snapshots of the final configurations with the repulsive $\varepsilon = 1$ interaction between polymer matrix and sheets) after 10^6 time steps. Sample size 64^3 is used with the polymer volume fraction $\phi = 0.2$ and polymer chain lengths $L_c = 4$ (first row, first column), 16 (first row, second column), 32 (second row, first column) and 64 (second row, second column). (From Ref. [37].)

(Figure 2.8) and disperse well with $\varepsilon = +1$ (repulsive) interaction (Figure 2.9). One may draw an immediate conclusion that the type of interaction between the polymer matrix and the sheets affects their exfoliation. The concept of a gallery between the sheets and intercalation of polymer chains therein is obviously more suitable for the layered configuration with $\varepsilon = -1$ where sheets are held together via interactions mediated by the interstitial polymer chains. The repulsive polymer matrix ($\varepsilon = +1$) leads to a well-exfoliated configuration (Figure 2.9 with $L_c = 4$). There is no well-defined layering or gallery at the end of the equilibration. The intercalation of chains into the initial stacked structure with a gallery separates the sheets in time and the concept of gallery becomes irrelevant.

Apart from conformational constraints, interactions between sheets and polymer or solvent particles determine the behavior: interstitial polymer chains (solvent) maintain the layered structure when $\varepsilon = -1$ and exfoliate the stacked sheets (with relatively isotropic dispersion) when $\varepsilon = +1$. The effect of dynamic polymer matrix on the exfoliation becomes somewhat smeared at higher molecular weights as the constraints on sheet mobility are enhanced. For example, with $L_c = 64$, the dispersion of sheets is constrained; their stacked (layered) morphology persists due to an entangled polymer matrix even with the repulsive interaction ($\varepsilon = +1$). Thus, the entropic trapping (cage) of sheets becomes more clear with $L_c = 64$ especially at higher polymer volume fraction $\phi = 0.2$ (see below). Note that the size of higher molecular weight polymer chains (e.g., $L_c = 64$) is much larger than the linear dimension of the sheets that may not be representative of a laboratory sample. However, it accentuates the physical mechanism, for example, entropic trapping without increasing the concentration of the polymer matrix for a theoretical analysis.

In general, the exfoliation is suppressed by the attractive polymer matrix and its higher molecular weights; the repulsive interaction causes them to exfoliate in a relatively low molecular weight of polymer at least at these volume fractions ($\phi = 0.1, 0.2$). These observations seem consistent with theoretical studies by Balazs *et al.* [23, 47]. Note that the polymer entanglements become dominant with the high molecular weight polymer (e.g., $L_c = 64$) even at the polymer volume fraction $\phi = 0.2$ where the layering persists regardless of the type of interaction between the polymer matrix and the platelets. Thus, the layering is favorable in an attractive matrix and further re-enforced by higher molecular weight of the polymer chains (Figure 2.8). In contrast, the exfoliation is enhanced in repulsive matrix but can be suppressed by higher molecular weight of the polymer matrix (Figure 2.9). Identifying the trend in exfoliation of sheets becomes complex at the intermediate molecular weight of the polymer matrix due to the interplay between the interaction-driven thermodynamics and the underlying structural entropy. Attempts are also made to quantitatively confirm these observations [37], some of which are briefly discussed here.

Figure 2.10 shows the longitudinal density profile of sheets in an attractive polymer matrix with molecular weight $L_c = 4$–64 at the volume fraction $\phi = 0.2$. The

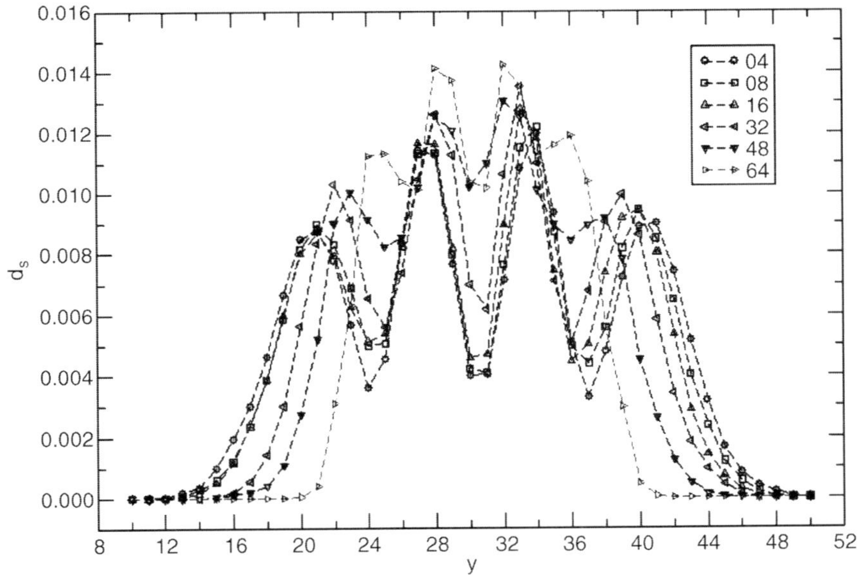

Figure 2.10 Variation of the density of sheets with the longitudinal distance. Statistics: sample size 64^3, polymer volume fraction $\phi = 0.2$, $\varepsilon = -1$, 10 independent samples, 10^6 time steps. (From Ref. [37].)

oscillatory profile with four maxima in the density of sheets (location of sheets) shows that the layered structure is sustained at long times. The interlayer spacing, however, decreases when the molecular weight of the polymer is increased. Sheets are held together via attractive interaction with the interstitial (intercalated) polymer chains especially at low molecular weights. Intercalation of polymer chains is more prevalent at lower molecular weight ($L_c \leq L_S$) due to lower energy. As a result, the intersheet distance is higher due to intercalation of shorter chains. Intercalation becomes increasingly difficult with the higher molecular weight polymer ($L_c > L_S$). The probability of adsorption of polymer chains to the external surfaces of the layered sheets increases with L_c while their entanglement becomes more dominant. For example, at $L_c = 64$ ($\gg L_S$), polymer chains seem to form a cage surrounding the sheets via (i) adsorption at the sheets and (ii) entanglement of chains. Both the entanglement of the polymer matrix and its attractive interaction with the sheets push the sheets closer.

The corresponding longitudinal density profile of polymer chains is presented in Figure 2.11. One can immediately note the complementary oscillation in polymer density in the same regions of the lattice, for example, zx planes around $y = 20$–44. The complementary variation of the polymer density to that of the platelets is rather easy to see by comparing Figures 2.10 and 2.11. For example, the dominant polymer density maxima peaks for $L_c = 4$ are around $y = 24$, 30, 36 where the platelet density

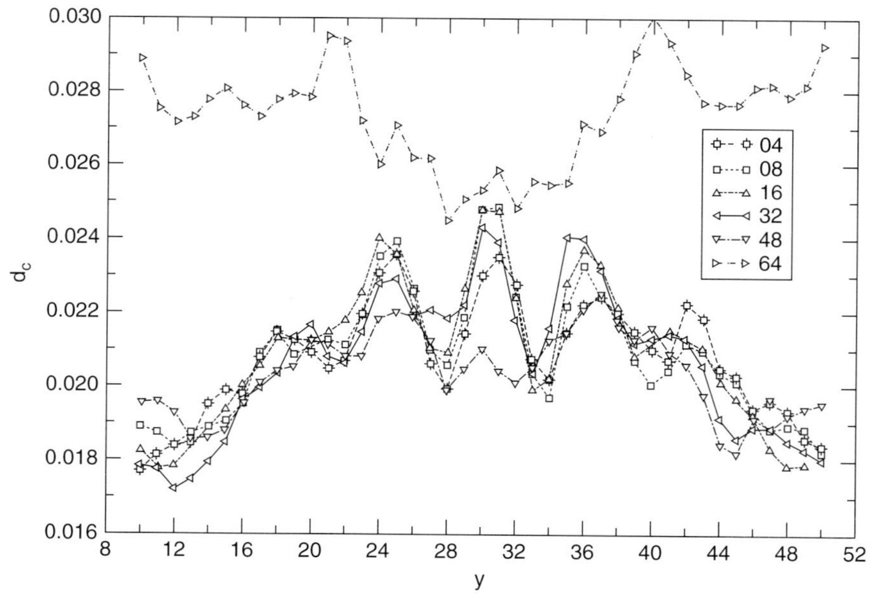

Figure 2.11 Variation of the polymer density with the longitudinal distance. Statistics: sample size 64^3, polymer volume fraction $\phi = 0.2$, $\varepsilon = -1$, 10 independent samples, 10^6 time steps. (From Ref. [37].)

shows its minima. The locations of maxima peaks of the platelet density are around $y = 20$, 28, 34, 42 where polymer density has its minima. The polymer density is higher between the sheet layers. Increasing the molecular weight of polymer smears their distribution with larger fluctuations in contrast to well-defined and systematic shifts in density peaks for sheets (Figures 2.10 and 2.11). A close examination of Figure 2.11, however, reveals a systematic trend: the magnitude of the maxima peaks increases on increasing the molecular weight, for example, from $L_c = 4$ to $L_c = 16$, followed by a decreasing (reverse) trend with $L_c = 32$ to $L_c = 64$, a nonmonotonic dependence of interlayer polymer density on the molecular weight. Now, we will try to explain this trend.

Since the interaction between the polymer chains and the sheets is attractive, the energy is lower with a larger number of chain particles close to the sheet surface including the interstitial spacing between the sheets. Increasing the molecular weight from $L_c = 4$ allows more particles to interact with the sheets that reduces the energy until it reaches the linear dimension of the sheet, that is, $L_c = L_s = 16$. A chain (in the interstitial gallery) with the conformation comparable with the two neighboring sheets has lower energy than chains with both lower molecular weight ($L_c < L_s$) and higher molecular weight ($L_c > L_s$). Chains with lower molecular weights have lower number of particles surrounding the sheets since the packing of chains with lower molecular weight in the gallery is not as efficient as those with the higher molecular weight and vice versa. On increasing the molecular weight (i.e., $L_c = 32$, 48, 64), the probability of intercalation of the whole chain in the gallery (interstitial spacing between sheets) becomes lower due to larger fluctuations with larger radius of gyration. The number of sheet particles surrounding the polymer chain within the range of interaction is, therefore, lower for larger chains ($L_c > L_s$). Chains in the gallery become less close-packed, resulting in a lower density. It is extremely difficult to intercalate large chains ($L_c = 64$) due to a comparatively large radius of gyration and its fluctuation and enhanced entanglement of the polymer matrix at $\phi = 0.2$ in comparison to a lower polymer density (follows). As a result, the polymer density profile with $L_c = 64$ exhibits lower density in the sheet regions, dispersion of platelets is arrested by the entangled or spanning network of the polymer cage – an entropic-induced layering.

We have also examined [37] a number of such physical quantities as variation of the rms displacement of chains and sheets, and their radius of gyration, to supplement our general observations. For example, the effect of molecular weight on the dynamics of the polymer matrix is more apparent [37] when we analyze the variation of the rms displacement of the center of mass of the polymer chains with time steps for both an attractive and a repulsive interaction matrix. A systematic slowing down is seen on increasing the molecular weight regardless of interaction. At a high molecular weight, the motion of the polymer chains becomes so slow that the free volume for sheets to move is arrested in a nearly quasistatic state over simulation-accessible timescales. The dynamics of the polymer matrix slows down considerably on increasing the molecular weight of the polymer chains. We know that percolation of chains [48] depends strongly on their molecular weight and that the percolation threshold decreases on increasing the molecular weight of the polymer chains. Because of the spanning network of polymer chains, with and without their entanglement, the relaxation of free volume becomes very slow.

2.2.6
Conclusions and Outlook

We have shown examples of investigations of exfoliation and dispersion of clay platelets and intercalation of solvent particles and polymer chains by coarse-grained computer Monte Carlo simulations. As pointed out, many of the early studies on intercalation of polymer chains deal with confined surfaces with little or no dynamics, particularly with the off-lattice molecular dynamics simulations [25, 26]. Although attempts [27] have been made more recently to incorporate the dynamics of clay platelets to examine the intercalation and exfoliation, covering sufficient timescales has not been feasible. Despite enormous literature on the subject, we are not aware of any computer simulation to date that shows intercalation and exfoliation so clearly with freely mobile stacked sheets, solvent particles, or polymer chains concurrently. In the presence of solvent particles or low molecular weight polymer ($L_c \leq L_s$), the stacked platelet configuration persists with attractive matrix ($\varepsilon = -1$) via an indirect intersheet interaction mediated by the intercalated solvent particles or polymer chains. The stacked platelets exfoliate in the presence of repulsive solvent or polymer matrix ($\varepsilon = +1$) with low molecular weights. In a polymer matrix with high molecular weight, the stacked (layered) platelet configuration is trapped in a cage via entanglement and/or percolation of surrounding polymer chains, an entropic trapping. Although the entropic cage formed by mobile chains is dynamic, the relaxation time for the cage renewal is too large to provide free volume for sheets to exfoliate. In the polymer matrix with intermediate molecular weight, the interplay between the interaction-controlled thermodynamics and structural constraints leads to a complex dispersion behavior with varying degrees of layering and exfoliation depending on the molecular weight of the matrix and type of interaction – a subject of continued interest.

Regardless of insights gained, the systems considered so far, clay platelets in solvent particles and clay platelets in a polymer matrix were still simplified and subject to assumptions of interaction parameters. A more realistic scenario in agreement with available experimental surface data would involve attractive clay-clay interactions, attractive polymer-polymer interactions, and attractive polymer-clay interactions tuned such that a small positive interface tension results upon dispersion. Specific surface modification of the clay minerals could then be represented by further simulations with modified interaction parameters for clay-clay attraction and clay-polymer attraction. Such semi-quantitative correlations with experiment might explain exfoliation processes more insightfully and remain as a future challenge. A natural extension is to consider clay platelets in the simultaneous presence of both solvent particles and polymer chains as shown in Figure 2.4. Incorporating particles having the characteristics of amino acids and peptides in the presence of clay platelets [49] would be a step toward understanding biofunctionalized nanomaterials and biomineralization.

2.3
All-Atom Models for Interfaces and Application to Clay Minerals

Force fields to describe the interactions between organic molecules and inorganic components such as clay minerals have heretofore been associated with substantial uncertainties. In this section, we will describe necessary strategies to derive compatible force fields for reliable simulations of surfactants, biopolymers, and synthetic polymers in contact with oxidic minerals and explain characteristic properties of clay-organic interfaces relevant to exfoliation in nanocomposites.

The complexity of interfacial regions in polymer composites with organically modified clay minerals or metal particles necessitates the understanding of interactions at the molecular level up to length scales of micrometers. As outlined in the introductory sections, nanocomposites are bulk materials of interfaces and the design of interfacial regions allows control over structural, mechanical, electrical, optical, and barrier properties. A major challenge that remains is reliably and reproducibly achieving a homogeneous dispersion of layered silicates (Figure 2.12), termed exfoliation, in conventional polymer matrices instead of an intercalated, partially exfoliated, or agglomerated layers of clay mineral in the matrix.

Miscibility is related to interface tensions, and such interfacial processes could only recently be simulated quantitatively with chemical details. Classical MD simulation, in combination with coarse-grained Monte Carlo approaches to generate

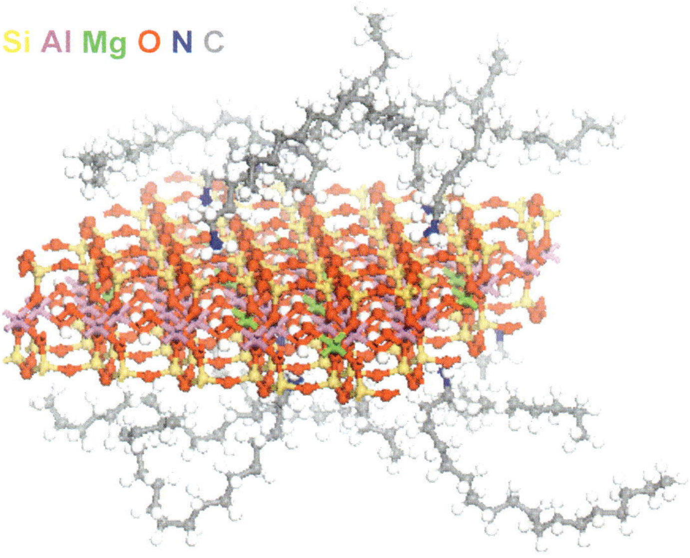

Figure 2.12 Part of a montmorillonite lamella (cation exchange capacity = 91 meq/100 g) modified with n-$C_{18}H_{37}$–NH_3^+ surfactants. Corresponding chemical compositions of the natural and modified mineral are $Na_{0.333}[Si_4O_8][Al_{1.667}Mg_{0.333}O_2(OH)_2]$ and $(n\text{-}C_{18}H_{37}\text{–}NH_3)_{0.333}[Si_4O_8][Al_{1.667}Mg_{0.333}O_2(OH)_2]$, respectively. Interfaces of such modified fillers with polymers influence a range of nanocomposite properties.

thermodynamic equilibrium structures, allows the analysis of structural, dynamic, thermal, and mechanical properties of up to 10^6 atoms at timescales up to 1 μs. For the desired quantitative estimates, however, the quality of the energy expression is of primary importance, namely, the quality of the force field in classical semiempirical simulations. The strength of such force fields is the full chemical detail and the ability to simulate nonbonded interactions and dynamic phenomena at a low computational cost, that is, 10^6–10^9 times faster than with *ab initio* methods. A limitation is the difficulty in simulating the formation of new covalent bonds, for example, the reaction of end-functionalized surfactants attached to the clay mineral surface with functional groups on the polymer chain, which requires a combination with quantum-mechanical methods or direct adjustments of the molecular connectivity in the course of the simulation. Therefore, the combination of classical models both with electronic structure methods for covalent effects and with coarse-grained models to access larger macroscopic length and timescales is also important.

We focus in this section on improvements in energy models for the inorganic components to allow reliable simulations of inorganic–organic interfaces at the nanometer scale, the self-assembly and cohesive energies of various alkylammonium surfactants on montmorillonite surfaces, and on a relation between packing density, tilt angle, and the occurrence of thermal phase transitions of surface-grafted alleyl chain. A focus throughout this discussion is the close agreement between experimental data and simulation results that helps establish a realistic understanding of the interfacial structure and dynamics.

2.3.1
Force Fields for Inorganic Components

Force field parameters for inorganic constituents have been a major hurdle in the quantitative analysis of interfaces in nanocomposites while energy models for surfactants, polymers, and biopolymers have been developed in sufficient reliability to simulate densities, cohesive energies, and interface energies. At the classical atomistic level, several force fields are in use, such as AMBER [50], CHARMM [51], CVFF [52], COMPASS [53], OPLS-AA [54], PCFF [55], and UFF [56]. For example, the energy expressions of CVFF [52], and PCFF [55] (same as COMPASS [53]), are as follows:

CVFF

$$\begin{aligned} E_{\text{pot}} &= \sum_{ij\,\text{bonded}} K_{r,ij}(r_{ij}-r_{0,ij})^2 + \sum_{ijk\,\text{bonded}} K_{\theta,ijk}(\theta_{ijk}-\theta_{0,ijk})^2 \\ &\quad + \sum_{ijkl\,\text{bonded}} V_{\phi,ijkl}[1+\cos(n\phi_{ijkl}-\phi_{0,ijkl})] + E_{\text{oop}} \\ &\quad + \frac{1}{4\pi\varepsilon_0\varepsilon_r} \sum_{\substack{ij\,\text{nonbonded}\\(1,2\,\text{and}\,1,3\,\text{excl})}} \frac{q_i q_j}{r_{ij}} + \sum_{\substack{ij\,\text{nonbonded}\\(1,2\,\text{and}\,1,3\,\text{excl})}} \varepsilon_{ij}\left(\left(\frac{r_{0,ij}}{r_{ij}}\right)^{12} - 2\left(\frac{r_{0,ij}}{r_{ij}}\right)^{6}\right). \end{aligned} \tag{2.5}$$

PCFF (and COMPASS)

$$\begin{aligned} E_{\text{pot}} &= \sum_{ij\text{ bonded}} \sum_{n=2}^{4} K_{rn,ij}(r_{ij}-r_{0,ij})^n + \sum_{ijk\text{ bonded}} \sum_{n=2}^{4} K_{\theta n,ijk}(\theta_{ijk}-\theta_{0,ijk})^n \\ &+ \sum_{ijkl\text{ bonded}} \sum_{n=1}^{3} V_{n,ijkl}[1-\cos(n\phi_{ijkl}-\phi_{0n,ijkl})] + E_{\text{oop}} + E_{\text{cross}} \\ &+ \frac{1}{4\pi\varepsilon_0\varepsilon_r} \sum_{\substack{ij\text{ nonbonded}\\(1,2\text{ and }1,3\text{ excl})}} \frac{q_i q_j}{r_{ij}} + \sum_{\substack{ij\text{ nonbonded}\\(1,2\text{ and }1,3\text{ excl})}} \varepsilon_{ij}\left(2\left(\frac{r_{0,ij}}{r_{ij}}\right)^9 - 3\left(\frac{r_{0,ij}}{r_{ij}}\right)^6\right). \end{aligned} \tag{2.6}$$

The term "force field" refers to the set of adjustable parameters in such models, such as equilibrium bond lengths $r_{0,ij}$ and angles $\theta_{0,ijk}$, vibration constants $K_{rn,ij}$, $K_{\theta n,ijk}$, or atomic charges q_i, which are used in the simulation. The majority of such force fields were developed for small organic and large biological macromolecules (AMBER, CHARMM, CVFF, and OPLS-AA), some force fields can be applied to a wider range of material structures (CVFF, COMPASS, and PCFF), and some are useful only for specific systems (UFF). A common theme of biologically oriented force fields is the focus on protein structure, which has led to data structures that are difficult to apply to bonded frameworks of inorganic structures and inorganic–biological or organic–inorganic interfaces. Furthermore, the transferability of parameters between force fields is limited by virtue of their use of different scaling factors for nonbonded interactions between 1,4-bonded atoms and differences in combination rules to obtain 12-6 or 9-6 Lennard-Jones interaction parameters between pairs of different atom types.

These differences complicate somewhat the development of a unified, accurate force field for inorganic, organic, and biological chemical species. A present functional solution to this problem involves the extension of CVFF and PCFF for accurate inorganic parameters, using the existing (bio)molecular parameters and focusing on their continued improvement. Inorganic parameters can be transferred to biomolecular force fields such as Lennard-Jones parameters for fcc metals. We will thus outline a method to derive reliable force fields for inorganic components using the examples of layered silicates and metals.

In accordance with the typical energy expressions shown in Eqs. (2.5) and (2.6), atomic charges, van der Waals parameters, bonded properties, and the compatibility of the parameters for inorganic components with existing force fields for organic molecules are essential for reliable simulations of the minerals themselves and an unlimited scope of hybrid interfaces (Table 2.1) [57]. Inorganic minerals often also include defects, for example, $AlOOH \rightarrow MgOOH^- \cdots Na^+$ in montmorillonite, which require suitable model descriptions to ensure the success of a simulation. In the following sections, we describe the derivation of an accurate force field parameterization for the examples of layered silicates and fcc metals that also applies to other chemical species.

Table 2.1 Influence of force field parameters on the simulation of inorganic components and inorganic–organic interfaces, for example, in composite materials.

Parameters in a classical simulation	Effect
Atomic charges	Interface structure, energy, and long-range (polar) interactions
Van der Waals parameters (LJ well depth)	Surface and interface energies
Vibrational constants	Elastic properties
Compatibility of the energy expression with FFs for (bio)organic compounds	Scope of application
Distribution and parametrization of charge defects	Interface structure and dynamics

2.3.1.1 Atomic Charges

A major parameter in the simulation of both inorganic and organic components is the magnitude and distribution of polarity in the models. Atomic charges reflect the distribution of the valence electron density among the atoms and arguably are most important for meaningful results from molecular simulations which has been discussed by Heinz and Suter (Figure 2.13) [58]. Atomic charges perform best in simulations when they coincide with atomic charges derived from experimentally measured electron deformation densities through mapping on spherical atomic basins. For example, this approach works well for simple molecules such as H_2O in the condensed phase with an experimental O charge $-0.74 \pm 0.1e$ compared to $-0.82 \pm 0.05e$ in simulations, or for SiO_2 in various tetrahedrally coordinated silicates with an experimental Si charge $+1.18 \pm 0.15e$ compared to $+1.1 \pm 0.1e$ as a suitable value in molecular simulation. By considering atomic charges first in the

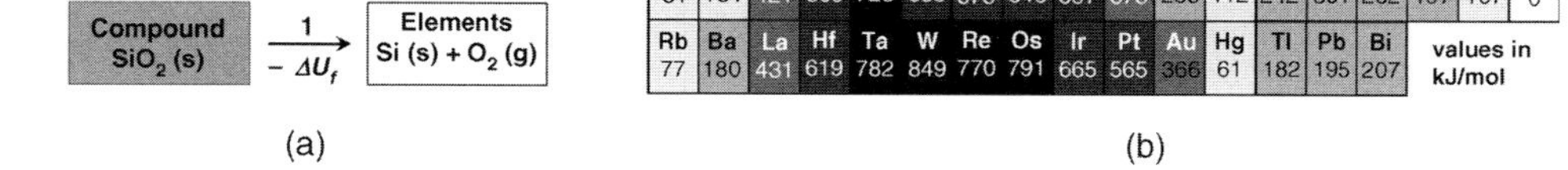

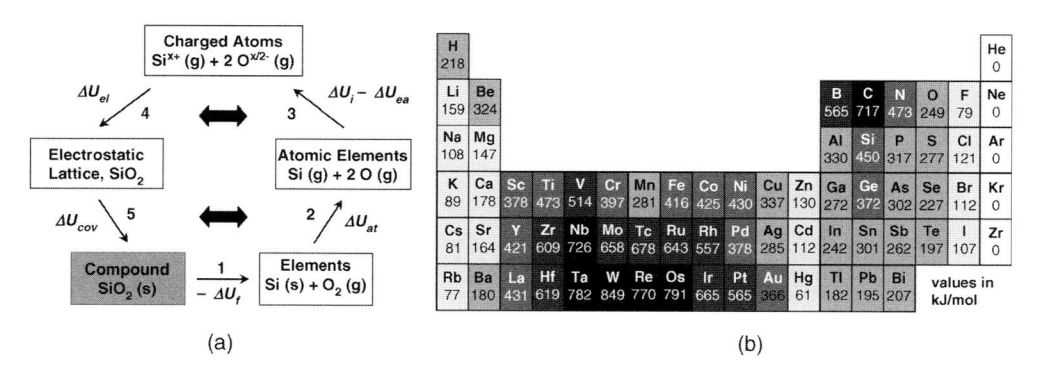

Figure 2.13 (a) The extended Born model assumes ionization to partial atomic charges and distinguishes covalent (step 2, step 5) and ionic contributions (step 3, step 4) to chemical bonding. (b) For example, covalent bonding contributions are represented by atomization energies of the elements (up to 0.8 MJ/atom) and often dominate over ionic contributions represented by ionization potentials/electron affinities (up to 0.7 MJ/atom). Darker shading means higher potential for covalent bonding and a tendency toward lower atomic charges. (After Ref. [58].)

process of parameterization, it is possible to obtain the best agreement between computed densities, surface tensions (electrostatic and van der Waals contributions), and vaporization energies (when applicable) with experimental data after identification of the rest of the parameters. The sensitivity of molecular models to atomic charges is exemplified for computed surface tensions and cleavage energies of layered silicates in Table 2.2. Using other atomic charges, agreement with experiment cannot be obtained, although it is also very important to use appropriate Lennard-Jones parameters that contribute the van der Waals portion to the surface tension. On the basis of electron deformation densities, experimental values for atomic charges have become available for an increasing number of minerals and organic crystals with a typical uncertainty of $\pm 0.1e$ or less [58]. In a similar way, estimates for atomic charges can be derived from dipole moments with the same accuracy. For example, the dipole moment of H_2CO in the gas phase equals a C charge $+0.40e$ compared to $+0.40 \pm 0.05e$ as a suitable value in molecular simulation. Measurements of dipole moments, however, often relate to the gaseous state and may differ somewhat from the dipole moment in the liquid or solid state. The use of electron deformation densities or dipole moments, in most cases, leads to suitable estimates of atomic charges that are up to an order of magnitude more accurate compared to charges derived from *ab initio* approaches that strongly depend on basis sets, exchange, and correlation energies, and often assume nonintuitive partition schemes of the electron density [58]. When experimental data on the electron distribution are not available for a given system, access to physically justified atomic charges is provided by an extended Born model that takes into account the atomization energies, ionization energies, and electron affinities of the constituting atoms (Figure 2.13). This model weights covalent and ionic contributions to chemical bonding in the correct proportion and gives excellent estimates of atomic charges when employed relative to one or two reference compounds nearby in the periodic table, particularly when the reference compounds have similar atomization and ionization potential. For example, the most suitable Si charge of $+1.1 \pm 0.1e$ in tetrahedral oxygen coordination has been initially derived only on the basis of this model and is firmly supported by various comparisons across the periodic table [58].

2.3.1.2 **Lennard-Jones Parameters**

In conjunction with atomic charges, the parameters $r_{0,ii}$ and ε_{ii} in the 12-6 or 9-6 Lennard-Jones potential are critical for reproducing surface energies, interface energies, and solvation energies of inorganic components. In particular, the well depth ε_{ii} (Eqs. (2.1) and (2.2)) has a profound influence on surface and cleavage energies of layered silicates due to the contribution of van der Waals energy (Table 2.2). The discussion of computed surface energies for inorganic solids in relation to experimental data is fairly recent and provides an important test of a model with regard to interfacial interactions. The first such comparison for Spinel, $Mg_2Al_2O_4$ [59] showed an overestimate of 50% in the computation, related to shortcomings both in atomic charges and in Lennard-Jones parameters. Even greater deviations of 50–500% are associated with other models (Table 2.2). In these cases, the subtle interplay of interfacial interactions in nanocomposites during the disper-

Table 2.2 Comparison of experimental and computed cleavage energies for pyrophyllite, montmorillonite, and mica in mJ/m^2 according to Ref. [57].

References in ref. [57]	Pyrophyllite			Montmorillonite ΔE_{cleav}	Mica ΔE_{cleav}	Principal charges (e)		Principal well depths (kcal/mol)		
	γ^{tot}	γ^{el}	γ^{vdW}			Si^{tet}	Al^{oct}	Si	Al	O
Exptl.	39.7	5.8	33.9	50–200	375	1.2	1.45			
This work	40	8	32	140	380	1.1	1.45	0.03	0.03	0.015
[13, 15]		[155]				4	3	0	0	49
[16]		[>30]		[>300]	[>500]	2.4				0.47
[17]	−515	2	−517	−3000	−433	0.52	1.33	0	0	0
[18]	−1094	13	−1107	−433	−162	1.4	1.68	0	0	6.86
[19]		[>15]				1.2	3	0	0	
[20]		[>15]				1.2	2.8			
[22] ← CLAYFF	81	30	51	167	484	2.1	1.58	10^{-6}	10^{-6}	0.155
[23]	265	155	110	251	683	4	3	0.04	9.04	0.228
[25]	260	8	252	340	631	1.1	1.45	0.40	0.50	0.06

The strong influence of principal atomic charges and Lennard-Jones well depths on computed surface tensions and cleavage energies can be seen. Si^{tet} refers to Si in tetrahedral oxygen coordination and Al^{oct} refers to Al in octahedral oxygen coordination.

sion of inorganic minerals, or the interaction with solvents, cannot be captured. Therefore, a new rationale to assign parameters for van der Waals interactions was introduced by Heinz *et al.* [57]. Lennard-Jones parameters $r_{0,ii}$ and ε_{ii} can be assigned in two steps, leading to computed interfacial tensions within 5% of experimental values. First, the van der Waals equilibrium distances $r_{0,ii}$ are assigned on the basis of experimentally known crystallographic or van der Waals values across the periodic table [60]. With an uncertainty of approximately ±10%, these distances are directly suited for force field parameterizations and adjust the computed density. Second, well depths ε_{ii} are assigned. The parameterization of well depths is more demanding compared to atomic charges since there is less experimental justification of these quantities. Accurate well depth values are known for rare gases, and individual values across the periodic table approximately increase toward the end of each row. However, additional adjustments are needed according to polarizability and charge of the atom, and the extent of covalent bonding plays a role due to nonbonded exclusions between 1,2, 1,3, and sometimes 1,4 bonded atoms. These trends can be derived from existing force fields [57] and, ultimately, the well depths ε_{ii} are fitted so that the computed surface tension agrees with experiment. In the final model, computed solvation energies with water or organic solvents should also coincide with experimental values, as well as the computed vaporization energy if the compound boils within ±200 K of the chosen reference temperature (force fields parameterized at 298 K will typically need changes when applied at temperatures outside a ±200 K range). More information on the physical interpretation of the Lennard-Jones parameters $r_{0,ii}$ and ε_{ii} can also be found in the context of fcc metals [61].

2.3.1.3 **Bonded Parameters**

Equilibrium bond lengths, bond angles, and vibration constants are known from X-ray data and spectroscopy, and may be implemented in molecular models without major modifications [57]. Initial values of vibration constants can be obtained from compilation of interatomic force constants and then optimization by the following procedure [57]: (1) Short MD trajectories of 5 ps with snapshots every 3 fs are computed, followed by analysis of the Fourier transform of the velocity autocorrelation function. The transform equals the computed superposition of IR and Raman spectra. (2) As long as differences with experiment persist, vibration constants are iteratively adjusted and step (1) repeated. In addition to suitable vibration constants, this approach also provides insight both into the coupling between internal degrees of freedom and into limitations of the accuracy of vibration modes in the force field model. Dihedral angles and rotation barriers can be set to zero for mineral structures. In contrast, for chain molecules, dihedral angles and rotation barriers are important for the dynamics, and current force fields often overestimate torsion barriers. For example, the eclipsed barrier in *n*-butane was measured at 3.95 kcal/mol [62] while several force fields assume 5–6 kcal/mol on the basis of *ab inito* results. Thus, chain rotation in simulations is sometimes slowed down by a factor up to 30 (2 kcal/mol $\approx$ 3.4 RT) compared to experiment so that adjustments in force fields for polymers and biopolymers are also important.

In brief, we have described a physical basis for a new generation of force fields that encompass inorganic, organic, and biomolecular constituents on a common modeling platform. To date, accurate models for layered silicates (mica, montmorillonite, and pyrophyllite) [57] and fcc metals (Ag, Al, Au, Cu, Ni, Pb, Pd, and Pt) [61] are available for quantitative simulation of nanocomposites using material-oriented (extended CVFF, extended PCFF) and biologically oriented force fields.

2.3.2 Self-Assembly of Alkylammonium Ions on Montmorillonite: Structural and Surface Properties at the Molecular Level

The capability of atomistic simulation in comparison to experiment is nicely documented through the understanding of molecular-level properties related to the dispersion of organically modified clay minerals in polymer matrices. The following considerations explain relevant factors for exfoliation and may lend themselves to all-atom simulations of assemblies with polymers, although the simulation of such large assemblies is yet the domain of coarse-grained models (Section 2.2). A substantial body of experimental data, including X-ray, TEM, NEXAFS, IR, NMR, DSC, dielectric measurements, and surface energies is available for alkylammonium-modified clay minerals. However, the data provide only indirect evidence of the actual structure, surface properties, and thermal behavior [18, 63–78]. MD simulations with the models described in the previous section use the same unit definitions as in experiment and agree closely with available data, in most cases quantitatively [10, 57, 58, 79–87], which provides the link to molecular-level details.

Organically modified clay minerals are synthesized through the incubation of montmorillonite in water/ethanol solution with surfactants below the critical micelle concentration. Upon completion of the ion exchange reaction and removal of solvent, alkylammonium ions arrange on the surface in characteristic patterns (Figure 2.14). Hereby, the presence of $AlO(OH) \rightarrow MgO(OH)^{-} \cdots Na^{+}$ charge defects in the montmorillonite octahedral sheet leads to the positioning of positively charged ammonium head groups in locations on the surface where previously Na^{+} ions have resided [10, 40, 57, 79, 88]. The alkyl tails of the surfactants are forced by van der Waals interactions into characteristic interlayer structures upon removal of the solvent. At low cation exchange capacity (CEC), layer-by-layer alkyl structures are formed with increasing chain length (Figure 2.14a) whereas at higher CEC, the packing density increases, the layering effect diminishes, and the character of the surfactants shifts toward self-assembled monolayers (Figure 2.14b). Therefore, the surfactants tilt to a greater extent relative to the surface plane as the CEC increases [10].

In the assembly process, the distinction between primary and quaternary ammonium head groups plays a major role (Figure 2.15). Both primary and quaternary ammonium ions are attached to the surface primarily through an electrostatic bond caused by the charge defects $AlO(OH) \rightarrow MgO(OH)^{-} \cdots NR_4^{+}$. In addition, primary ammonium head groups form hydrogen bonds with oxygen in the montmorillonite surface that consists of characteristic [Si,O] dodecacycles (Figure 2.15a). Each

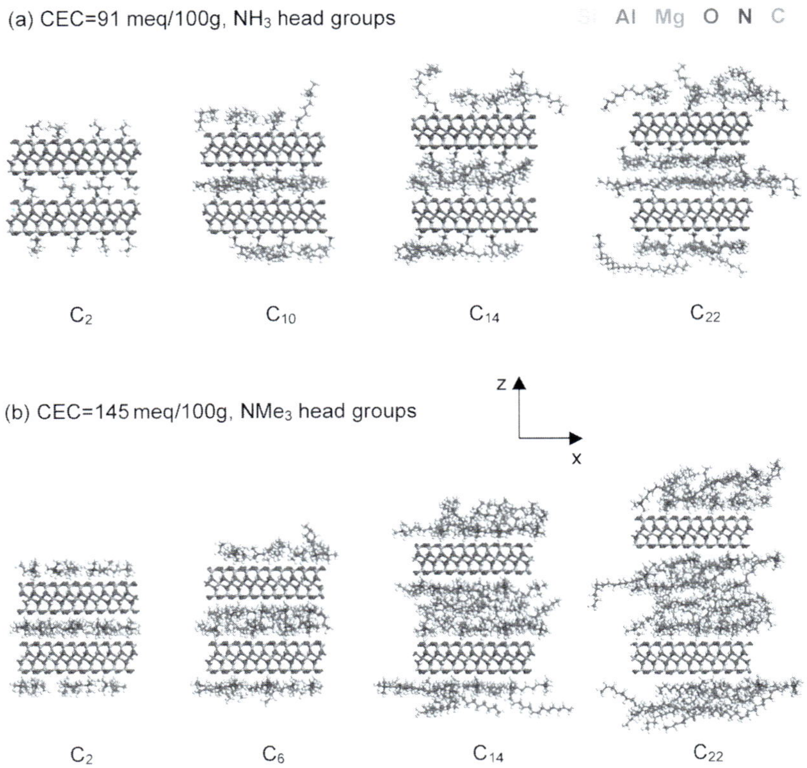

Figure 2.14 Representative MD snapshots of alkylammonium montmorillonites, viewed along the y-direction. (a) CEC = 91 meq/100 g and NH_3^+–C_n chains. The difference between partially formed layers (C_2, C_{14}) and completely formed layers (C_{10}, C_{22}) can be seen. (b) CEC = 145 meq/100 g and NMe_3^+–C_n chains. The successive formation of layers with decreasing order can be seen.

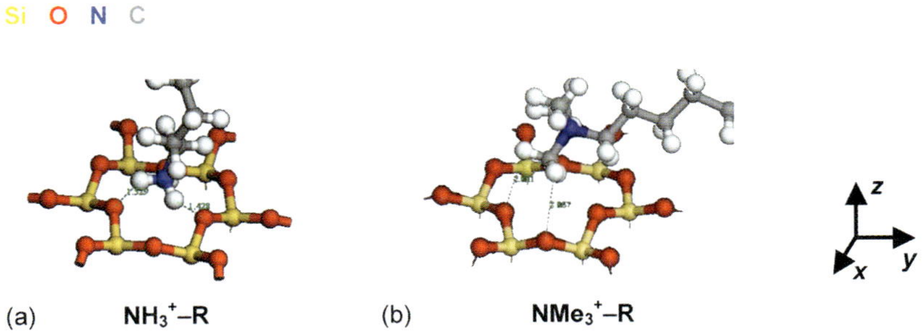

Figure 2.15 The organic–inorganic interface. (a) R–NH_3^+ head groups form hydrogen bonds with oxygen on the silicate surface. The average O···H distance is approximately 150 pm and the mean N to O distance is 245 pm. (b) R–$N(CH_3)_3^+$ head groups do not form hydrogen bonds with the surface. The average O···H distance is almost twice as large, approximately 290 pm, and the mean N to O distance is 390 pm. As a result, the mobility of R–NMe_3^+ surfactants on the surface is higher than that of R–NH_3^+ surfactants.

primary ammonium cation can form up to three hydrogen bonds and assumes a position a short distance from the montmorillonite surface plane, resulting in an approximately 3.0–3.5 kcal/mol stronger bond to the surface and less lateral flexibility compared to a quaternary ammonium ion. The latter are positioned substantially further above the superficial cavities and cannot form hydrogen bonds on the surface (Figure 2.15b) [10, 57].

The successive formation of alkyl layers with increasing surfactant length at low CEC can be seen in the computed gallery spacing (Figure 2.16a). Stepwise increases in the computed basal plane spacing of the organically modified layered silicates are found as the chain length increases, and the curve for $R-N(CH_3)_3{}^+$ surfactants is slightly shifted to the left due to a higher surfactant volume compared to $R-NH_3{}^+$ surfactants (three additional CH_2 groups). The gallery spacing increases steeply when one alkyl layer is densely packed and additional interlayer material forms an additional, frustrated alkyl layer on top. In these regions, we also observe a drop in interlayer density ($NMe_3{}^+-C_{10}$ and $NH_3{}^+-C_{12}$ in Figure 2.16b) that gradually recovers until the next alkyl layer is densely packed (see also Figure 2.16). The periodic changes in interlayer density are also accompanied by changes in the percentage of gauche conformations of the surfactants (Figure 2.16c). Thereby, the two different head groups show distinctive behavior. For the $R-N(CH_3)_3{}^+$ system that is less strongly bound to the surface, the lowest energy state is all-*anti* (all-*trans*). The required lateral space for such extended chains is available only in weakly packed alkyl layers that accordingly show the lowest percentage of gauche conformations, about 15%. When the alkyl layers are densely packed, the surfactant chains adjust by folding and the percentage of gauche conformations increases to 35%. For the $R-NH_3{}^+$ system that is more strongly bound to the surface due to hydrogen bonds, the preferred tripod arrangement on the surface (Figure 2.15a) leads to kinks in the first N-terminal torsion angles due to confinement of the alkyl chains between the stacked silicate layers. Therefore, the percentage of gauche conformations is very high at 40% for short chains and responds less to changes in interlayer density as the chain length increases (Figure 2.16c). For chains longer than C_{20}, the percentage of gauche torsions converges to 20–25% for both head groups that resembles a liquid-like state. In comparison, <5% gauche torsions are found in crystalline C_{18} surfactants and 39% gauche torsions in unconstrained, liquid octadecylamine (at 100 °C) [10, 57].

A comparatively smoother increase in basal plane spacing and weaker steps in the basal plane spacing as a function of chain length are found at higher CEC (Figure 2.17a). The curve for $R-N(CH_3)_3{}^+$ head groups is also slightly offset to the left due to higher surfactant volume relative to the $R-NH_3{}^+$ head groups. A softer, undulating increase in gallery spacing leads to weaker fluctuations in interlayer density at high CEC (Figure 2.17b). The fluctuations are larger for the formation of the first and second alkyl layers and converge with the formation of pseudomultilayers to a steady value. Then, the structure becomes reminiscent of polymer brushes (Figure 2.14b). The steady interlayer density in the semiliquid state of the alkyl chains is computed at 725 kg/m^3, in reasonable agreement with the experimental density of crystalline long-chain alkanes approximately 800 kg/m^3,

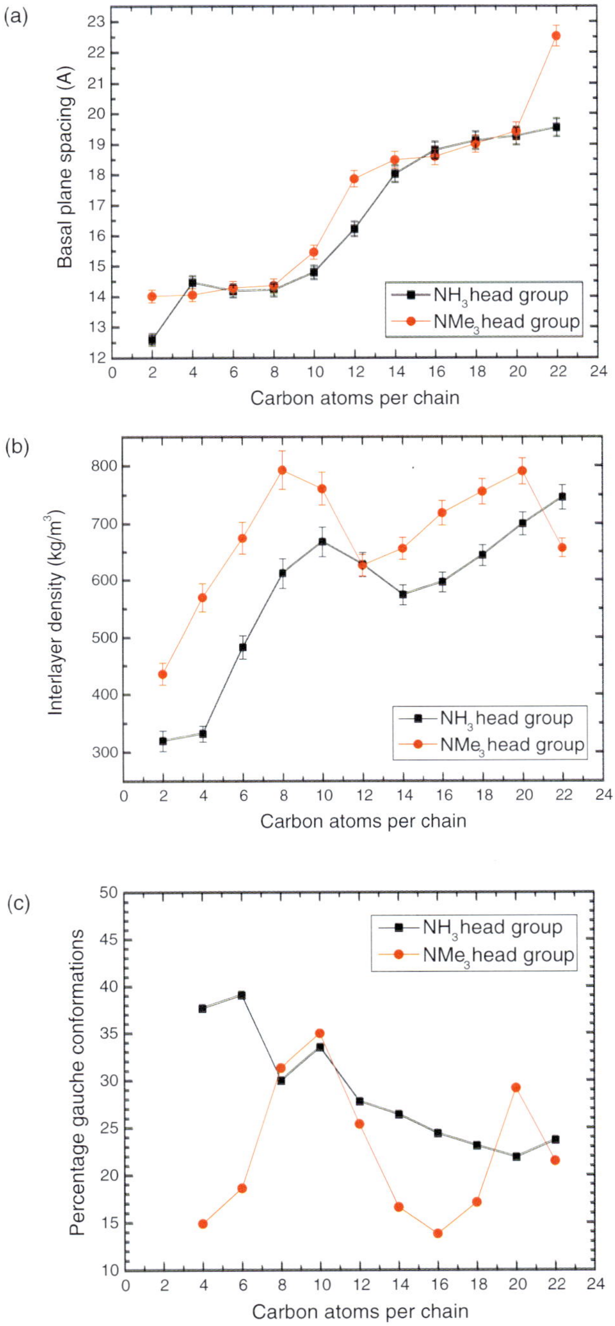

Figure 2.16 (a) Computed basal plane spacing, (b) interlayer density, and (c) average conformation of the alkyl chains between the silicate layers for alkylammonium-modified montmorillonite with CEC = 91 meq/100 g. The series $NH_3^+-C_nH_{2n+1}$ and N $(CH_3)_3^+-C_nH_{2n+1}$ with $n = 2, 4, \ldots, 22$ are shown.

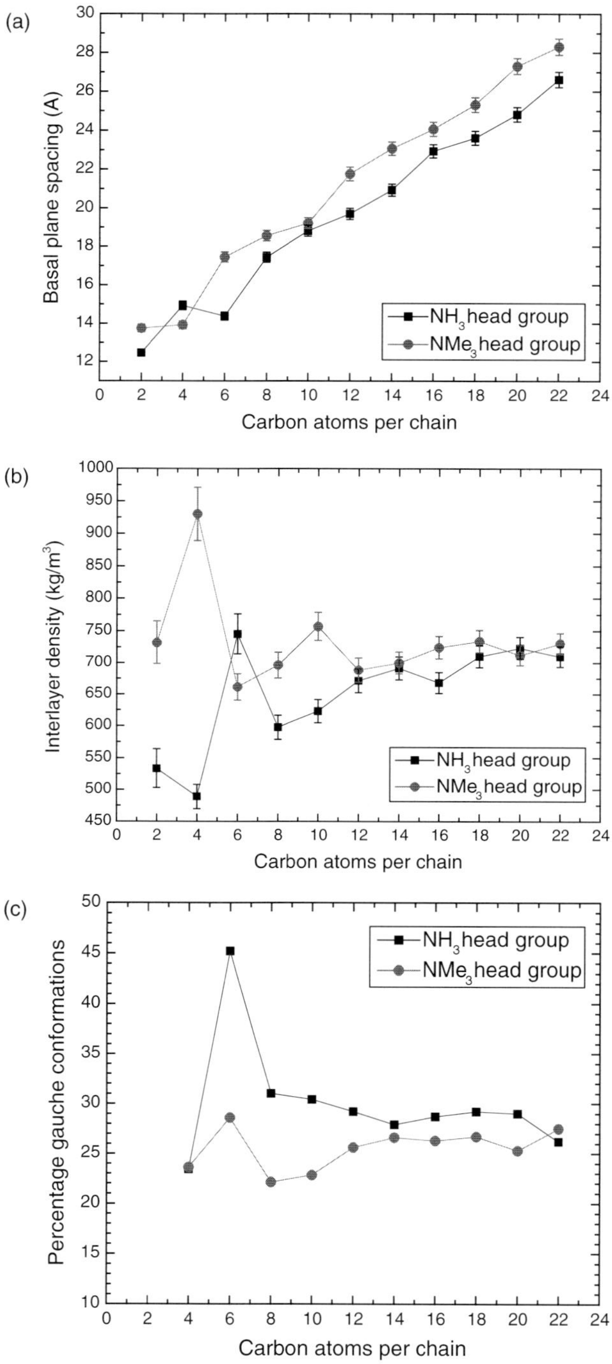

Figure 2.17 (a) Computed basal plane spacing, (b) interlayer density, and (c) average conformation of the alkyl chains between the silicate layers for alkylammonium-modified montmorillonite with CEC = 145 meq/100 g. The series $NH_3^+{-}C_nH_{2n+1}$ and $N(CH_3)_3^+{-}C_nH_{2n+1}$ with $n = 2, 4, \ldots, 22$ are shown.

since a somewhat higher density is expected for the crystalline phase. The percentage of gauche conformations in the alkyl backbones is nearly invariant in a 25–30% range for different chain length and different head group structure (Figure 2.17c), although peaks up to 45% are seen at short chain length for both head groups. Particularly for $R{-}NH_3^+$, the preference for a tripod arrangement of the head group on the surface (Figure 2.15a) leads to extra gauche conformations due to confinement between the montmorillonite lamellae. The relative weight of this effect on the percentage of gauche conformations diminishes as the chain length increases. The evidence for the confined head group tripod arrangement as a cause of the high percentage of gauche conformation at short chain length comes from the analysis of the percentage of gauche conformations of the alkyl chains on the outer montmorillonite surfaces (to vacuum) that are not confined (Figure 2.14) and exhibit only between 15 and 25% gauche conformations at short chain length [10].

These computational findings are consistent with a variety of experimental data. Foremost, the gallery spacing measured by X-ray diffraction has been known for decades [18, 63–65, 67, 69, 78] and is reproduced quantitatively by simulation (Table 2.3) [10, 79, 86, 88]. The difference between experiment and simulation is approximately 5%, and both approaches have associated errors. Uncertainties in experimental gallery spacing are associated with a distribution of interlayer environments due to heterogeneities within the natural montmorillonite, uncertainty in CEC (±5 mequiv/100 g), and nonequilibrium structures in the interlayer space depending on the process history of the sample. Residual amounts of water may also be present in the sample. Uncertainties in computed gallery spacing may include an overestimate by 2–3% owing to the limited accuracy of the energy model and an open simulation box in the z-direction during molecular dynamics simulation. In cases where the CEC was very accurately known, differences of only 2–3% are seen [10].

Table 2.3 Comparison of experimental and computed basal plane spacings (nm).

	Expt CEC = 90 Me_3NR^+	Sim CEC = 91 Me_3NR^+	Expt CEC = 90 H_3NR^+	Sim CEC = 91 H_3NR^+	Expt CEC = 150 H_3NR^+	Sim CEC = 145 H_3NR^+
C_4	1.42 (2)	1.40 (2)	1.36 (3)	1.45 (2)		
C_6			1.36 (3)	1.42 (2)	1.50 (3)	1.44 (2)
C_8	1.42 (2)	1.43 (2)	1.36 (3)	1.42 (2)		
C_{10}			1.45 (3)	1.48 (3)	1.80 (3)	1.89 (3)
C_{12}	1.66 (9)	1.78 (3)	1.70 (3)	1.62 (3)	1.87 (3)	1.97 (3)
C_{14}	1.77 (10)	1.85 (3)	1.75 (3)	1.80 (3)	2.03 (3)	2.09 (3)
C_{16}	1.81 (10)	1.86 (3)	1.75 (3)	1.88 (3)	2.28 (3)	2.29 (3)
C_{18}	1.85 (3)	1.90 (3)	1.85 (2)	1.91 (3)	2.30 (3)	2.36 (3)
Std dev to Expt		0.07		0.08		0.08

Standard deviations of the last digit are given in brackets. (After Ref. [10].)

Infrared and NMR spectroscopy have been further employed to probe the backbone conformation of alkylammonium surfactants attached to montmorillonite and mica surfaces [22, 25, 26, 30, 74]. The symmetric and asymmetric CH_2 stretching vibrations exhibit shifts from ~2848 to ~2854 cm^{-1} and from ~2917 to ~2928 cm^{-1}, respectively, which indicate conformational transitions from predominantly *anti* (s-*trans*) to gauche. The conformational details presented in Figures 2.16c and 2.17c are confirmed by IR data, including the sharp increase in the percentage of gauche conformations for short $R{-}NH_3{}^+$ chains at CEC = 143 meq/100 g. Similarly, ^{13}C NMR chemical shifts are suitable to monitor transitions from an ordered array of chains to disorder, which is indicated by a shift from a single peak at 33 pm to a combination of peaks at both 33 and 30 ppm. NMR data are also in agreement with observations in the simulation, particularly in the description of thermal transitions as discussed in Section 3.3. For a detailed comparison of experimental IR and NMR data with the simulation, the reader may refer to Ref. [10].

Another interesting aspect arises from the different modes of binding of the surfactants to the clay mineral surface (Figure 2.15). Diffusion of surfactants on the montmorillonite surface is possible by hopping across the cavities by Brownian motion. When ion exchange with surfactants is incomplete, hopping of surfactants may also be caused by local concentration gradients of defect sites. The self-diffusion of the surfactants by Brownian motion, that is, the exchange rate between adjacent surfactants, has been analyzed in MD simulation. A correlation within a group of at least two surfactants is characteristic of the motion of the surfactants. This results from their desire to remain close to charge defects.

1) The surfactants confined in the interlayer space between two silicate layers show virtually no lateral mobility on a timescale of 10 ns in the simulation, independent of CEC, head group, and chain length. 2D diffusion constants at room temperature are $\leq 10^{-9}$ cm^2/s, related to close packing of the chains. Self-diffusion then requires conformational changes in the backbone and concerted movements of alkylammonium ions (Figure 2.14), which is associated with high-energy barriers. Qualitatively, we expect the lowest diffusion constants (and highest energy barriers) for high CEC, H-bonded NH_3 head groups, and long alkyl chains.
2) Significantly higher diffusion rates can be found only on single, unconfined surfaces such as on the outside of the duplicate structures (Figure 2.14). The order of magnitude of self-diffusion is still sensitive to the CEC, head group chemistry, and chain length. Surfactants with NH_3 head groups (both at low and at high CEC, any chain length) yield either no net movement on the surface or occasional jumps to a neighbor cavity at a timescale of 1 ns. Alkyl chains with NMe_3 head groups exhibit more frequent hopping on the surface. At CEC = 145 meq/100 g for any chain length and at CEC = 91 meq/100 g for chains longer than C_6, 2D diffusion constants are still $< 10^{-7}$ cm^2/s. Rapid motion on the surface is seen only for the smallest quaternary ammonium ions at low CEC. Computed 2D diffusion constants are 5.0×10^{-6} cm^2/s for Me_3NEt^+ and 1.5×10^{-6} cm^2/s for Me_3NBu^+ at CEC = 91 meq/100 g, which amounts to

approximately 20% and 6% of the 3D self-diffusion constant of liquid water at room temperature ($2.3 \times 10^{-5}\ cm^2/s$) [10, 84].

In addition to conventional alkylammonium surfactants as compatibilizers between the clay mineral and a polymer matrix, other functionalized surfactants may be employed. Examples are ammonium-functionalized, azobenzene-containing surfactants with light-responsive properties. Azobenzene can then act as a nanoscale actuator during a photon-induced *trans–cis* isomerization (Figure 2.18). MD simulation has explained changes in the basal plane spacing upon photoisomerization in agreement with X-ray diffraction and UV/Vis spectral data. The *trans–cis* isomerization takes place quantitatively in the constrained geometries, and the isomerization reaction was examined for flexible, semiflexible, and rigid azobenzene derivatives. Conformational flexibility of intercalated azo dyes facilitates molecular rotation in the interlayer space and causes small to no changes in gallery spacing upon photoisomerization, unless differences in solvation energy between the two isomers are exploited by adding a cointercalate. For example, 6-(4-phenylazophenyl)hexylammonium ions on montmorillonite do not lead to changes in the gallery spacing upon photoconversion due to the flexibility of the alkyl-containing backbone. This leads to a parallel orientation of the azobenzene substituents to the surface with a preference toward perpendicular orientation of the phenyl rings relative to the surface. In contrast, conformationally rigid azo dyes support reversible actuation in the interlayer space upon isomerization, particularly in near-upright orientation of the azobenzene unit on the surface. For example, (4-phenylazophenyl)ammonium ions lead to a change of up to 11% in gallery spacing upon *trans–cis* isomerization (Figure 2.18). Rigid *trans*-isomers are

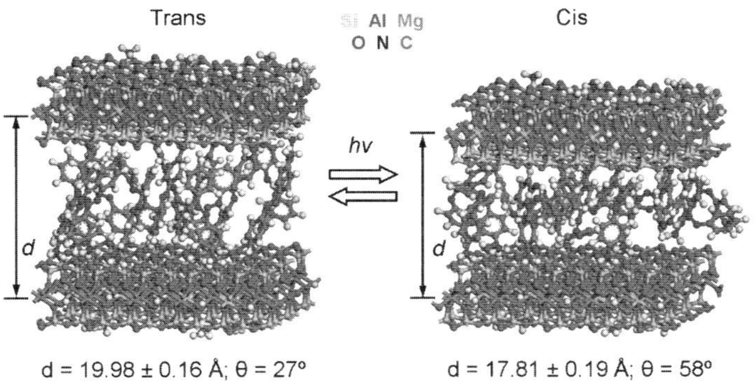

Figure 2.18 Snapshots of montmorillonite (CEC = 143 mequiv/100 g) modified with (4-phenylazophenyl)ammonium ions in near-upright orientation on the surface. The basal plane spacing changes by 2.17 Å (11%) on photoisomerization. The average tilt angle of the connectors between the ammonium N atoms and the phenyl 4′-C atoms of the surfactants relative to the surface normal increases from 27 to 58° upon *trans* → *cis* conversion. The *cis*-configured surfactant assumes a higher interlayer density. (From Ref. [86].)

found to be sterically more demanding than the corresponding *cis*-isomers that flexibly arrange into dense interlayer structures. The use of rigid dications, such as (4,4′-phenylazophenyl)diammonium ions, to achieve changes in gallery spacing upon photoisomerization is more effective when they act as cross-links between two montmorillonite layers and less effective when they are doubly bonded to the same montmorillonite layer. Significant controllable actuation also depends on, or is supported by, the presence of cointercalates that can reversibly enter into and exit from the interlayer gallery space. Simulation results indicate that experimentally achieved reversible changes in gallery spacing of 0.9 Å (4%) may be improved up to 2.8 Å (14%) through (1) the presence of a cointercalate to compensate, and possibly overcompensate, associated changes in interlayer density; (2) conformational rigidity of the azobenzene-containing moieties, and (3) upright orientation of the dye molecules on the surface. A moderate-to-high cation exchange capacity, the absence of flexible alkyl spacers in the surfactants, and the use of rigid macrocyclic "pedestals" support this objective. To date, few rigid azobenzene derivatives and cointercalates have been explored in the interlayer space, and further advances in this direction appear feasible [86].

From the computational perspective, a classical molecular dynamics approach to simulate the *trans–cis* isomerization of azobenzene and its derivatives was developed. The procedure is based on a temporary modification of the torsion potential to model the impact of photon energy and is compatible with any force field containing three-term torsion potentials (Eq. (2.6)) or one-term torsion potentials (Eq. (2.5)). By incorporating UV spectroscopic data and *ab initio* understanding of the isomerization reaction into the model, it accounts for the input of excitation energy, the timescale of the reaction, the relative energies of the *trans*- and the *cis*-isomers, as well as for the barrier to thermal conversion. Testing on simple azobenzene systems indicates that it is possible to retrieve details about the reaction dynamics and to probe the sensitivity to external pressure, temperature, approximate excitation time (duration and number of laser pulses), and the molecular environment in large systems [86].

Simulations have also provided quantitative understanding of known surface energies, cleavage energies [68, 70, 73, 89–91], and reconstruction processes upon separation of clay mineral layers, including the principal difference between natural alkali clay and organically modified clay minerals that is critical for understanding separation processes of montmorillonite layers in nanocomposites (Figure 2.19). The cleavage energy of layers of the pure minerals mica and montmorillonite in vacuum (equivalent to an inert medium) comprises mostly electrostatic contributions that are four–five times stronger than dispersion contributions (Figure 2.19a). In the range 0.2–0.8 nm separation, alkali ions partition between the silicate sheets and rearrange on the surface with a tendency to stay close to the charge defects (Figure 2.19b). Upon completion of the (equilibrium) separation process, silicate layers remain as electrically neutral units and feel no significant Coulomb interactions between them. Rearrangements of K^+ ions on the surface of the silicate layers (jump from one cavity to another) are associated with a high barrier ($\sim$20 kcal/mol K^+ in mica).

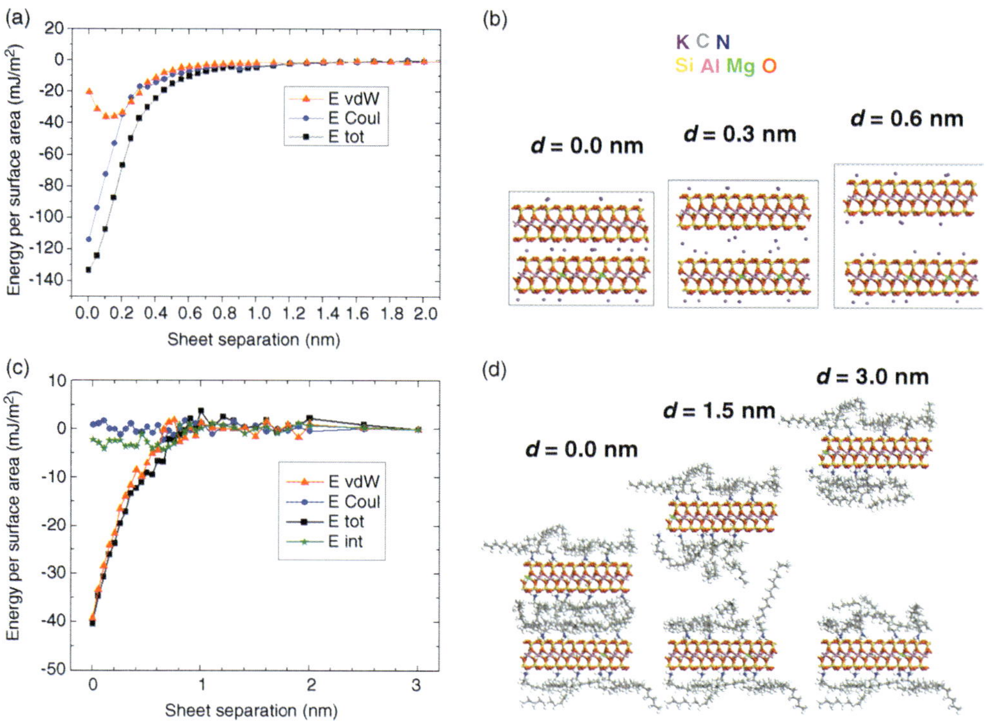

Figure 2.19 Cleavage of montmorillonite layers by stepwise separation in MD simulation. Energy components (van der Waals, Coulomb, total, internal) and representative structures are shown for (a and b) potassium montmorillonite and for (c and d) octadecylammonium-modified montmorillonite (CEC = 91 mequiv/100 g). (After Ref. [84].)

As a consequence, a layer of organic surfactant of a thickness >0.8 nm can effectively shield polar interactions and reduce the high cleavage energy of alkali clays up to 90%. Attached organic layers are flexible upon separation of the clay mineral layers and modify their superficial arrangement at medium separation (1–2 nm) to maintain van der Waals interactions (Figure 2.19c and d). Total interaction energies become smaller than 1 mJ/m^2 at 1.5 nm separation for alkali clay and at 3 nm separation for octadecylammonium (C_{18})-clay. Therefore, long-range effects between separated layers of montmorillonite or organically modified montmorillonite are minor. Such residual interactions are related to lateral head group mobility on the clay mineral layer and of a Coulomb multipole nature; no contributions from van der Waals interactions are seen. Cleavage energies as well as their Coulomb and van der Waals contributions are obtained in very good agreement with experiment (Table 2.4) and complement experimental data of surface forces and surface tensions for mica, montmorillonite, C_{18}-mica, and C_{18}-montmorillonite [84].

The possible computational accuracy for cleavage energies has been ± 1 mJ/m^2 for alkali clays and ± 3 mJ/m^2 for octadecylammonium-modified clays. In principle, this

Table 2.4 Cleavage energy (mJ/m^2) and its components according to experiment and simulation.

	Mica (CEC 251)		C_{18}-Mica		Montm. (CEC 91)		C_{18}-Montm.	
	Sim	Expt	Sim	Expt	Sim	Expt	Sim	Expt[a)]
Coulomb	298		-1 ± 2		113		-1 ± 2	1 ± 1
VdW	85		45 ± 2		20		39 ± 2	40 ± 1
Internal	0		1 ± 2		0		2 ± 2	
Total	383	375	45 ± 3		133	50–200	40 ± 3	41

a) For C_{18}-montmorillonite, surface tensions are given (see text).

performance is sufficient to study interactions with chemically different solvents and polymers that often have surface tensions in the range 20–60 mJ/m^2. Future challenges include more accurate approaches to sample the phase space and to analyze thermodynamic quantities, particularly entropic contributions. Future models will also allow the direct simulation of interfaces of organically modified clay minerals and polymer matrices in full atomistic detail, in which the CEC, the surfactant chain length and architecture, the chemistry of the polymer, the polymer chain length, and the volume fraction of organoclay can be explicitly varied. Computed interface tensions [92] indicate the thermodynamic potential for exfoliation: the closer to zero, or ideally negative, the higher the tendency for dispersion in the polymer matrix.

2.3.3
Relationship Between Packing Density and Thermal Transitions of Alkyl Chains on Layered Silicate and Metal Surfaces

The detailed information on specific systems has also led to the conceptual understanding of structural and surface properties of a wide range of alkyl modified clay mineral and metal surfaces, including the occurrence of thermal-phase transitions of the tethered surfactants in some instances (Figure 2.20). Extensive experimental characterization and simulation using molecular dynamics and Monte Carlo methods suggest a model on the basis of the packing density as a single geometry parameter that explains the average segmental tilt angle of the surfactants, the likely surface structure, and the occurrence of thermal order–disorder transitions in good approximation (Figure 2.20) [87]. As a central quantity, the packing density λ_0 is defined as the ratio between the average cross-sectional area of an all-*anti* configured surfactant chain $A_{C,0}$ and the available surface area per alkyl chain A_S:

$$\lambda_0 = \frac{A_{C,0}}{A_S}. \tag{2.7}$$

In this model, we consider a minimum chain length of 10 carbon atoms since shorter chains do not form sufficiently distinctive patterns on the surface and do not yield significant thermal transitions. Furthermore, we assume a typical temperature

$0 \leq \lambda_0 \leq 0.2$	$0.2 \leq \lambda_0 \leq 0.75$	$0.75 \leq \lambda_0 \leq 1$
Amorphous	Semicrystalline	Quasicrystalline
(up to 40% gauche)		
No phase transitions	Reversible phase	No phase transitions
(a)	(b)	(c)

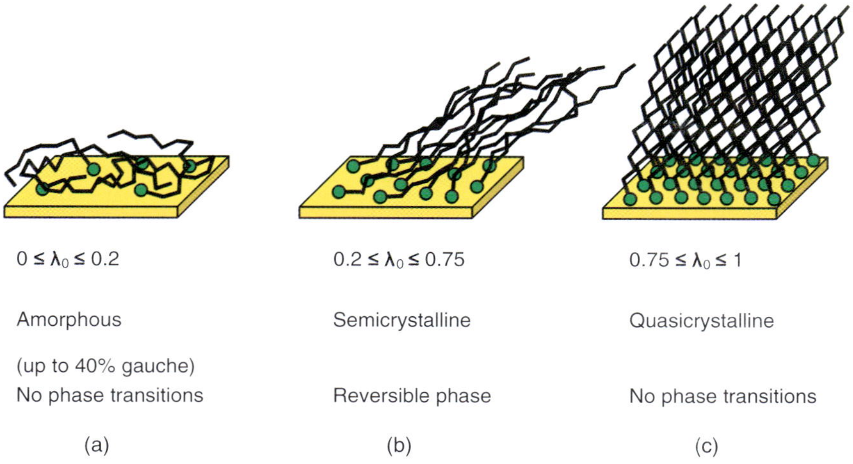

Figure 2.20 The range of homogeneous alkyl layers (chain length $\geq C_{10}$) on even surfaces includes (a) disordered chains oriented parallel to the surface, (b) intermediately ordered chains with an intermediate collective tilt angle, and (c) nearly vertically oriented, all-*anti* configured chains. Significant reversible thermal transitions are found only in case (b).

range of 250–400 K for the thermal transitions and the absence of cross-links between the alkyl chains as cross-linked chains have a higher resistance to order–disorder transitions. Under these conditions, the key observation is that the packing density determines the surface structure and thermal behavior while the chemical details of the surface, the surfactant head group, and the surfactant chains only fine-tune specific interactions, chemical functionality, and actual temperature of thermal transitions (Figure 2.20).

The packing density can be easily modified for layered silicates such as montmorillonites and micas by changing the CEC or the type of surfactant while such control is still a challenge on metal and oxide surfaces. The packing density λ_0 further correlates with the collective segmental tilt angle θ_0 of the alkyl chains relative to the surface normal [79, 87]

$$\lambda_0 = \cos \theta_0. \tag{2.8}$$

More details of this concept, which also applies to surfactants grafted on metal surfaces and curved surfaces with a distance-dependent radius of curvature, are described in Ref. [87].

2.4
Interfacial Thermal Properties of Cross-Linked Polymer–CNT Nanocomposites

Over the past decade, a significant amount of research has been carried out on carbon nanotube reinforced polymers. The high modulus, high strength, and high thermal and electrical conductivities of CNTs have made it an ideal candidate to be explored for use in reinforced polymer nanocomposites for structural and multifunctional applications. The extraordinary properties of CNTs have, however, not always translated into similar properties of the polymer nanocomposites and extensive research has pointed out that both good dispersion and good interfacial properties are extremely critical to realize ultimate composite properties. The effective bulk properties have been found to depend on several factors that include CNT–polymer bonding, alignment of CNTs, and uniformity of dispersion. The dispersion in turn depends on factors such as diameter, chirality, and length of CNTs [93]. Due to the difficulty in individually controlling these factors experimentally, modeling and simulations have been proven to be very useful in providing insight into their relative significance. Multiscale modeling studies have focused on different aspects – from mesoscale simulations of the dispersion process [6, 94] to finite element-based simulation of the dependence of mechanical properties on interfacial strength [95] and continuum-based micromechanical models with effective interfaces [96].

Experimental and theoretical studies on interfacial characterization have found improvement in mechanical properties with better bonding between the CNT and the polymer that subsequently improves the interfacial strength and stress transfer between them. An atomic-level understanding of the physics and chemistry of the interfaces is essential to tailor these interfaces for various multifunctional applications, and atomistic simulations can provide insight into the molecular structure. Several theoretical studies have looked at the structure and mechanical properties of these interfaces [97–99] and a few have looked specifically at bonding energies. Panhuis *et al.* [100] have presented one of the earliest studies on selective interaction of polymer backbone and CNTs, and Elliot and coworkers [101] have reported MD studies that show large deviations from results expected from simple rule-of-mixture models for cases with strong interfacial interactions. One of the effective techniques to improve the interfacial bonding has been found to be chemical functionalization of CNTs. Although chemical functionalization has been found to negatively affect the mechanical properties of individual CNTs [102–104], it can enhance that of the interface [96]. Functionalization can also modify the stacking and solvation properties that in turn can improve the dispersion. The functional groups can also act as tethers that increase the interaction with the polymer matrix.

Along with good mechanical properties, good thermal conductance in composites is also of primary interest for many thermal management applications [105, 106]. However, unlike electrical conductivity where a filler volume fraction above a percolation threshold is found to provide a dramatic increase in the electrical conductivity of the polymer matrix, the thermal conductivity is orders of magnitude less than that predicted by an engineering rule of mixtures [105]. Although the thermal conductivity of multiwall CNTS is in the range of 3000 W/(m K) and even higher than that for single-wall CNTs, the experimental values for polymer composites are around 0.3–0.4 W/(m K), a very modest improvement from that of the pristine matrix [105]. This has been attributed to a very high interfacial Kapitza resistance [107], and Keblinski *et al.* have reported an exceptionally small interface thermal conductance of $G \approx 12$ MW/(m^2 K) [108] for a nanotube submerged in octane liquid. They have calculated the thermal resistance between two nanotubes to be equivalent to the resistance of a 20 nm thick layer of polymer with a standard conductivity of 0.25 W/(m K), implying that the thermal conductivity is primarily controlled by the interface thermal conductance. Similar to improving the mechanical properties, chemical functionalization is also being considered as a means to improve the interfacial bonding and thereby the thermal conductance of polymer–CNT composites. There are two main competing factors – the reduction in the intrinsic thermal conductivity due to introduction of scattering centers and the strengthening of the interfacial bonding, both of which depend on parameters such as the grafting density and chain length of functional groups, and CNT tube lengths [109, 110].

Thermoset polymers are network-forming polymers widely used in composites and as adhesives and coatings. One of the important classes of thermosetting polymers are epoxy-based thermosets that are the primary polymers used in advanced aerospace applications due to their high modulus and fracture strength, low creep, and stable high temperature performance. Nanostructured reinforcement of these polymers was initially investigated for their potential to provide better mechanical properties but are now also being explored for multifunctional applications in the area of thermal management [105, 106]. Atomistic simulations of thermoset–CNT interfaces with and without functionalization of CNTs provide insight into the nature of thermal transport at these interfaces and help with our understanding of controlling them.

2.4.1 Model Building

Building atomistic models for highly cross-linked thermoset polymers is a challenge, and there has been a steady progress by several research groups in creating representative molecular models. Some of the earlier work focused on understanding the reaction kinetics of the cross-linking process and the effect of polymerization, molecular weight distribution, polydispersity, and sol–gel transition, using lattice [111] as well as nonlattice [112] Monte Carlo simulations. Subsequent atomistic studies did consider topological information. Doherty *et al.* [113] created networks using lattice-based simulations and polymerization molecular dynamics scheme, whereas Yarovsky and Evans [114] have presented a methodology to cross-link epoxy resins in a single-step procedure and Xu and coworkers [115] have performed cross-linking simulations for epoxy resin using an iterative MD/MM procedure. This [115] approach was used to build small model systems carrying out one cross-link per iteration. For lightly cross-linked polymers (PDMS networks), Heine *et al.* [116] formulated a method using a united atom model and a dynamic cross-linking approach based on cutoff distance criteria. Here, the newly formed topology is relaxed based on a modified potential that is linear at large distances and quadratic at short distances. A more robust approach to model network polymers was recently presented [117] that provides a stepwise procedure to build a highly cross-linked system of epoxy-based networks. It combines Heine's dynamic

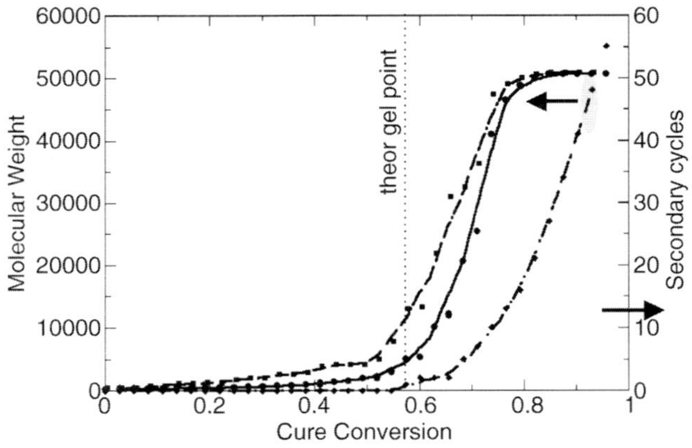

Figure 2.21 Plot of weight average molecular weight (circles), largest molecular weight (squares), and secondary cycles (diamonds) as a function of cure conversion. The dashed lines are guide to the eye. The dotted line suggests theoretical gel point [117].

cross-linking concept [116] at an atomistic level with Xu's iterative MD/MM concept [118] and a new multistep relaxation procedure for relaxing the molecular topology during cross-linking. This dynamic cross-linking approach has been applied to an epoxy-based thermoset (EPON-862/DETDA) and several material properties such as density, glass transition temperature, thermal expansion coefficient, and volume shrinkage during curing were calculated and found to be in good agreement with experimental results [117]. The simulations also highlight the distribution of molecular weight build-up and inception of gel point (Figures 2.21 and 2.22) during the network formation, paving the way for characterizing both thermodynamic and structural properties as a function of temperature and degree of curing.

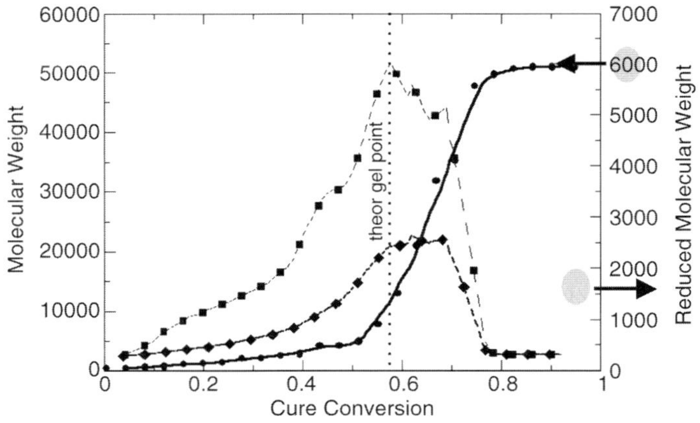

Figure 2.22 Plot of largest molecular weight (circles), second-largest molecular weight (squares), and weight averaged reduced molecular weight (diamonds) as a function of cure conversion. The dashed lines are guide to the eye. The dotted line suggests theoretical gel point [117].

2.4.2
Thermal Conductivity

MD simulations provide not only a convenient method of estimating thermal conductivity but also additional atomistic information that is useful in relating to the nanoscale structure and conformations [119]. Although classical MD cannot simulate electron–electron or electron–photon interactions, the dominance of phonon contributions to thermal conductivity of CNTs has made this approach suitable for studying thermal transport behavior in CNTs for a wide range of temperatures [120, 121]. In general, there are two methods to compute thermal conductivity in a solid [122, 123]. Nonequilibrium molecular dynamics (NEMD) [124] is widely used to calculate temperature profiles in molecular systems. It is also known as the direct method because it is analogous to experimental conditions and provides a direct physical representation of heat flow. The general approach is to apply a constant heat flux and calculate the thermal conductivity from the resulting temperature gradient using Fourier's law that states that under steady-state conditions, the amount of heat flow per unit area in unit time is directly proportional to the temperature gradient at the cross section. The thermal conductivity λ can be defined as

$$\lambda = \frac{\dfrac{Q}{A\,\Delta t}}{\dfrac{\mathrm{d}T}{\mathrm{d}z}}, \tag{2.9}$$

where Q is heat flow through the cross section, A is the cross-sectional area, Δt is the time for which heat is flowing, and $\mathrm{d}T/\mathrm{d}z$ is the steady-state temperature gradient. In a representative case, the system of interest is first built with desired periodic boundary conditions depending on the specific problem of interest and is then equilibrated at the desired temperature and pressure. After equilibration, a heat flux is imposed on the system by adding a fixed amount of energy to the atoms at the hot end of the model system and removing the same amount of energy from atoms at the cold end at every time step. In doing so, a temperature gradient is established across the system and heat flows from the hot to the cold region. The temperature of each region can be calculated as

$$T_i = \frac{1}{3N_i k_\mathrm{B}} \sum_{k=1}^{N_i} m_k v_k^2, \tag{2.10}$$

where N_i is number of atoms in ith region and the temperature gradient is calculated by the slope of the resulting temperature profile.

Similarly, the heat flux per unit area, $Q/A\ \Delta t$ is calculated as

$$\frac{Q}{A\,\Delta t} = \frac{1}{A\,\Delta t}\left\langle \frac{1}{2}\sum_{k=1}^{N_\mathrm{B}} m_k\left(v_k^2 - vp_k^2\right)\right\rangle, \tag{2.11}$$

where vp_k and v_k are the velocities of the atoms before and after rescaling to a desired temperature, respectively, and N_B is the number of atoms in the boundary layers.

Once the temperature gradient and the heat flux are known, the thermal conductivity is subsequently calculated.

Another approach to calculate thermal conductivity is equilibrium molecular dynamics (EMD) [125] that uses the Green–Kubo relation derived from linear response theory to extract thermal conductivity from heat current correlation functions. The thermal conductivity λ is calculated by integrating the time autocorrelation function of the heat flux vector and is given by

$$\lambda = \frac{1}{k_B T^2 V} \int_0^{\infty} \langle \mathbf{J}(t) \cdot \mathbf{J}(0) \rangle \, dt. \tag{2.12}$$

Here, $\mathbf{J}(t)$ is the heat flux vector at time t and is defined as

$$\mathbf{J}(t) = \frac{d}{dt} \sum_{i=1}^{N} \mathbf{r}_i E_i \tag{2.13}$$

and

$$E_i = \frac{1}{2} m_i \mathbf{v}_i^2 + \frac{1}{2} \sum_{j \neq i}^{N} u(r_{ij}). \tag{2.14}$$

m_i and $\mathbf{v}_i$ represent the mass and velocity of atom i, $u(r_{ij})$ is the total potential energy of atom i, and r_{ij} is the distance between atom i and j. A 12-6 Lennard-Jones potential for nonbonded van der Waals interactions along with an Ewald summation for electrostatic interactions yield the heat current vector $\mathbf{J}(t)$ [126] to be

$$\mathbf{J}(t) = \frac{1}{2} \sum_{i=1}^{N} \left[m_i \mathbf{v}_i^2 + \sum_{j \neq i}^{N} u(r_{ij}) \right] \mathbf{v}_i + \frac{1}{2} \sum_{i=1}^{N} \sum_{j \neq i}^{N} \left(\mathbf{r}_{ij} \mathbf{F}_{ij}^R \right) \cdot \mathbf{v}_i + \frac{1}{2} \sum_{i=1}^{N} \sum_{j=1}^{N} \mathbf{v}_i \cdot \overleftrightarrow{\mathbf{S}}_{ij} \tag{2.15}$$

$\mathbf{F}_{ij}$ represents the short-range van der Waals force and real part of the Ewald–Coulomb force (calculated within certain cutoff distance). In addition, it also includes forces due to bonded interaction terms such as bond stretching, angle bending, and so on. On the other hand, tensor $\mathbf{S}$ represents the forces due to electrostatic interactions beyond the cutoff distance.

The thermal conductance across nanointerfaces is mostly studied using constant flux simulations [107–109, 127–129]. The high thermal interfacial resistance between tube and matrix, and between two tubes, indicates that even for high loading, the thermal conductivity of these nanocomposites will be limited by the interfacial resistance [107]. Chemical functionalization has been found to reduce this interfacial resistance [106, 109]. The radial temperature profile from molecular dynamics simulation of two adjacent carbon nanotubes submerged in a thermoset polymer (Figure 2.23a) shows a significant temperature difference between the tubes that improves on adding covalent bonds between the tubes (Figure 2.23b) [106]. The temperature profile of two (10,0) CNTs placed with a separation distance of 11.18 Å between the tube axes was monitored. To analyze heat conduction between the nanotubes, one of the tubes was heated and its effect on the adjacent nanotube was

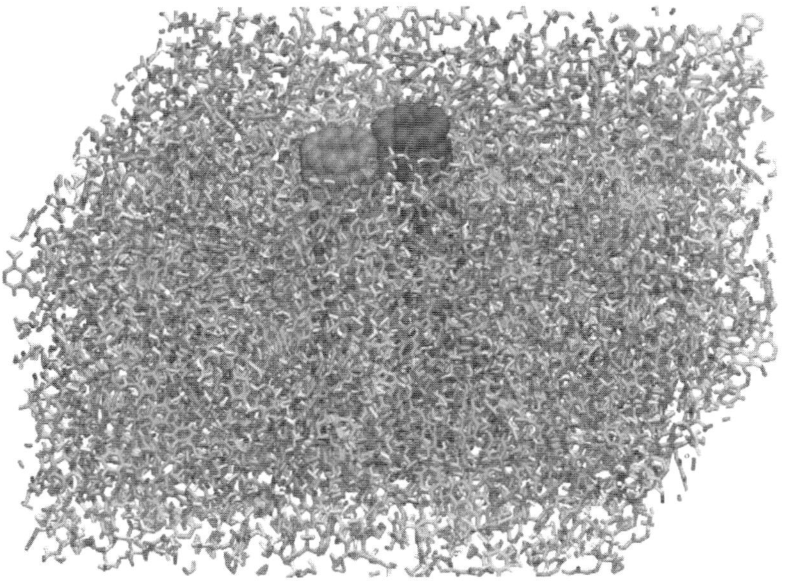

(a)

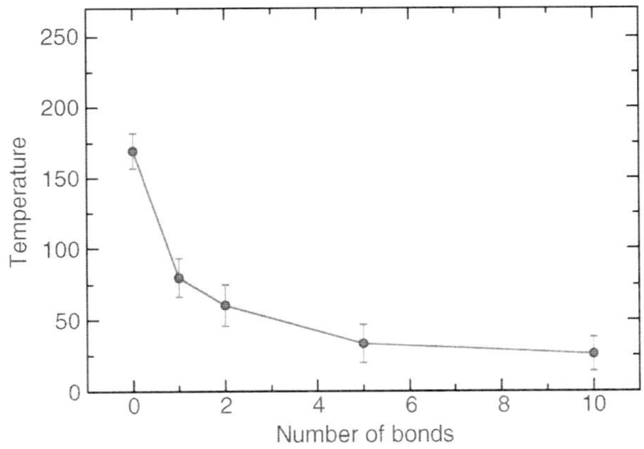

(b)

Figure 2.23 (a) Schematic of two SWCNTs aligned sidewise in epoxy (EPON-862 cross-linked with DETDA) based matrix. Heat is pumped into one SWCNT (shown in red) and taken out from the other SWCNT (shown in blue). (b) Temperature difference (in °C) between two simulated nanotubes. The number of bonds indicates the number of covalent bonds between the tubes (diameter = 7.83 Å and length = 42.6 Å).

monitored as a function of increasing number of single covalent bonds between the tubes. The temperature difference between the tubes was found to significantly decrease by the introduction of only a few bonds indicating the importance of covalent bonding between the tubes toward improving the transverse thermal conductance across nanotubes.

2.5
Conclusion

Interfacial Thermal Properties of Cross-Linked Polymer–CNT Nanocomposites Computer modeling and simulations are playing an ever increasing role in helping us control and design synthetic polymer matrix composites. Along with providing a fundamental understanding of the underlying physics responsible for their unique characteristics, simulations are also guiding synthesis and characterization of these materials. This chapter describes some of our own on-going efforts in modeling nanocomposites, using coarse-grained and all-atom models. Methods, timescales, and comparison to experimental data are discussed. The application of coarse-grained models to platelet dispersion in polymer matrices using on-lattice Monte Carlo simulation qualitatively reveals the influence of interaction strength on agglomeration versus exfoliation of the inorganic filler material. Novel force fields in full atomic resolution facilitate accurate exploration of interfacial regions in the 1–10 nm nanometer range and reproduce surface and interface properties in excellent agreement with experiment. Computation of thermal conductance across carbon nanotube and epoxy matrix interface provides better understanding of the nature of interfacial thermal resistance and the effect of covalent bonding in improving this resistance.

References

1 Schadler, L.S. (2003) Polymer-based and polymer-filled nanocomposites, in *Nanocomposite Science and Technology* (eds P.M. Ajayan, L.S. Schadler, and P.V. Braun), Wiley-VCH Verlag GmbH, Weinheim, pp. 77–153.

2 Picu, C.R. and Keblinski, P. (2003) Modeling of nanocomposites, in *Nanocomposite Science and Technology* (eds P.M. Ajayan, L.S. Schadler, and P.V. Braun), Wiley-VCH Verlag GmbH, Weinheim.

3 Zeng, Q.H., Yu, A.B., and Lu, G.Q. (2008) Multiscale modeling and simulation of polymer nanocomposites. *Prog. Polym. Sci.*, **33**, 191–269.

4 Scocchi, G. *et al.* (2007) Polymer-clay nanocomposites: a multiscale molecular modeling approach. *J. Phys. Chem. B*, **111**, 2143–2151.

5 Scocchi, G. *et al.* (2007) To the nanoscale, and beyond! Multiscale molecular modeling of polymer-clay nano-composites. *Fluid Phase Equilib.*, **261**, 366–374.

6 Fermeglia, M. and Pricl, S. (2007) Multiscale modeling for polymer systems of industrial interest. *Prog. Org. Coat.*, **58**, 187–199.

7 Pinnavaia, T.J. and Beall, G.W. (eds) (2001) *Polymer Clay Nanocomposites*, John Wiley & Sons, Inc., New York.

8 Krishnamoorti, R. and Vaia, R.A. (eds) (2002) *Polymer Nanocomposites: Synthesis, Characterization and Modeling, ACS Symposium Series*, vol. 804, American Chemical Society, Washington, DC.
9 Vaia, R.A. and Maguire, J.F. (2007) Polymer nanocomposites with prescribed morphology: Going beyond nanoparticle-filled polymers. *Chem. Mater.*, **19**, 2736–2751.
10 Heinz, H. *et al.* (2007) Self-assembly of alkylammonium chains on montmorillonite: effect of chain length, headgroup structure, and cation exchange capacity. *Chem. Mater.*, **19**, 59–68.
11 Balazs, A.C., Bicerano, J., and Ginzburg, V.V. (2008) Chapter 15, in *Polyolefin Composites* (eds D. Nwabunma and T. Kyu), John Wiley & Sons, Inc., New York.
12 Vermogen, K. *et al.* (2005) Evaluation of the structure and dispersion in polymer-layered silicate nanocomposites. *Macromolecules*, **38**, 9661–9669.
13 Gelfer, M. *et al.* (2004) Thermally induced phase transitions and morphological changes in organoclays. *Langmuir*, **20**, 3746–3758.
14 Lin-Gibson, S. *et al.* (2004) Shear-induced structure in polymer-clay nanocomposite solutions. *J. Colloids Interface Sci.*, **274**, 515–525.
15 Liu, J. *et al.* (2006) A review on morphology of polymer nanocomposites reinforced by inorganic layer structures. *Mater. Manuf. Process*, **21**, 143–151.
16 Gopalkumar, T.G. *et al.* (2002) Influence of clay exfoliation on the physical properties of ontmorillonite/polyethylene composites. *Polymer*, **43**, 5483–5491.
17 Jiankun, L. *et al.* (2000) Study on intercalation and exfoliation behavior of organoclays in epoxy resin. *J. Polym. Sci. Part B: Polym. Phys.*, **39**, 115–120.
18 Vaia, R.A. and Giannelis, E.P. (1997) Polymer melt intercalation in organically-modified layered silicates: Model predictions and experiment. *Macromolecules*, **30**, 8000–8009.
19 Vaia, R.A. and Giannelis, E.P. (1997) Lattice model of polymer melt intercalation in organically modified layered silicates. *Macromolecules*, **30**, 7990–7999.
20 Vaia, R.A. *et al.* (1997) Relaxations of confined chains in polymer nanocomposites: Glass transition properties of poly(ethylene oxide) intercalated in montmorillonite. *Polym. Sci. Part B: Polym. Phys.*, **35**, 59–67.
21 Vaia, R.A. *et al.* (1995) Kinetics of polymer melt intercalation. *Macromolecules*, **28**, 8080–8085.
22 Manias, E. *et al.* (2000) Intercalation kinetics of long polymers in 2 nm confinements. *Macromolecules*, **33**, 7955–7966.
23 Balazs, A.C., Singh, C., and Zhulina, E. (1998) Modeling the interactions between polymers and clay surfaces through self-consistent field theory. *Macromolecules*, **31**, 8370–8381.
24 Ginzburg, V.V., Singh, C., and Balazs, A.C. (2000) Theoretical phase diagrams of polymer/clay composites: The role of grafted organic modifiers. *Macromolecules*, **33**, 1089–1099.
25 Lee, J.Y., Baljon, A.R.C., and Loring, R.F. (1998) Simulation of polymer melt intercalation in layered nanocomposites. *J. Chem. Phys.*, **109**, 10321–10330.
26 Lee, J.Y., Baljon, A.R.C., and Loring, R.F. (1999) Spontaneous swelling of layered nanostructures by a polymer melt. *J. Chem. Phys.*, **111**, 9754–9760.
27 Sinsawat, A. *et al.* (2003) Influence of polymer matrix composition and architecture on polymer nanocomposite formation: Coarse-grained molecular dynamics simulation. *J. Polym. Sci. Part B: Polym. Phys.*, **41**, 3272–3276.
28 Pandey, R.B., Anderson, K., and Farmer, B.L. (2006) Exfoliation of stacked sheets: Effects of temperature, platelet size, and quality of solvent by a Monte Carlo simulation. *J. Polym. Sci. Part B: Polym. Phys.*, **44**, 3580.
29 Carmesin, I. and Kremer, K. (1988) The bond fluctuation method-a new effective algorithm for the dynamics of polymers in all spatial dimensions. *Macromolecules*, **21**, 2819–2823.
30 Binder, K. (ed.) (1995) *Monte Carlo and Molecular Dynamics Simulations in*

Polymer Science, Oxford University Press, New York.

31 Pandey, R.B. *et al.* (2005) Conformation and dynamics of a self-avoiding sheet: Bond-fluctuation computer simulation. *J. Polym. Sci. Part B: Polym. Phys.*, **43**, 1041–1050.

32 Pandey, R.B., Anderson, K.L., and Farmer, B.L. (2005) Effect of temperature and solvent on the structure and transport of a tethered membrane: Monte Carlo simulation. *J. Polym. Sci. Part B: Polym. Phys.*, **75**, 061913-1–061913-5.

33 Pandey, R.B., Anderson, K.L., and Farmer, B.L. (2007) Multiscale mode dynamics of a tethered membrane. *Phys. Rev. E*, **75**, 061913-1–061913-5.

34 Ginzburg, V.V., Gendelman, O.V., and Manevitch, L.I. (2001) Simple "kink" model of melt intercalation in polymer-clay nanocomposites. *Phys. Rev. Lett.*, **86**, 5073–5075.

35 Gendelman, O.V., Manevitch, L.I., and Manevitch, O.L. (2003) Solitonic mechanism of structural transition in polymer-clay nanocomposites. *J. Chem. Phys.*, **119**, 1066–1069.

36 Pandey, R.B. and Farmer, B.L. (2008) Effect of temperature and solvent on dispersion of layered platelets studied by Monte Carlo simulation. *Macromol. Theory Simul.*, **17**, 208–216.

37 Pandey, R.B. and Farmer, B.L. (2008) Exfoliation of a stack of platelets and intercalation of polymer chains: Effects of molecular weight, entanglement, and interaction with the polymer matrix. *J. Polym. Sci. Part B: Polym. Phys.*, **46**, 2696–2710.

38 Foo, G.M. and Pandey, R.B. (1999) Effects of field on temperature-induced segregation and folding of polymer chains. *J. Chem. Phys.*, **110**, 5993–5997.

39 Ben-Avraham, D. and Havlin, S. (2000) *Diffusion and Reaction in Fractals and Disordered Systems*, Cambridge University Press.

40 Pandey, R.B., Anderson, K.L., and Farmer, B.L. (2008) Sheets: Entropy dissipation, multiscale dynamics, dispersion, and intercalation. *Comput. Sci. Eng.*, **10**, 80–88.

41 Kantor, Y., Kardar, M., and Nelson, D.R. (1986) Statistical mechanics of tethered surfaces. *Phys. Rev. Lett.*, **57**, 791–794.

42 Abraham, F.F., Rudge, W.E., and Plischke, M. (1989) Molecular dynamics of tethered membranes. *Phys. Rev. Lett.*, **62**, 1757–1759.

43 Nelson, D.R., Piran, T., and Weinberg, S. (eds) (1989) *Statistical Mechanics of Membranes and Interfaces*, World Scientific, Singapore.

44 Ho, J.-S. and Baumgartner, A. (1989) Self-avoiding tethered membranes. *Phys. Rev. Lett.*, **63**, 1324–1324.

45 Pandey, R.B., Anderson, K.L., and Farmer, B.L. (2007) Morphing structures and dynamics with particles, chains and sheets. SAMPE Fall Technical Conference, Cincinnati.

46 Stauffer, D. and Aharony, A. (1994) *Introduction to Percolation Theory*, 2nd edn, Taylor & Francis.

47 Balazs, A.C. *et al.* (1999) Modeling the phase behavior of polymer/clay composites. *Acc. Chem. Res.*, **32**, 651–657.

48 Becklehimer, J.L. and Pandey, R.B. (1994) Percolation of chains and jamming coverage in 2 dimensions by computer-simulation. *J. Stat. Phys.*, **75**, 765–771.

49 Nakayama, H., Wada, N., and Tsuhako, M. (2004) Intercalation of amino acids and peptides into Mg-Al layered double hydroxide by reconstruction method. *Int. J. Pharm.*, **269**, 469–478.

50 Pearlman, D.A. *et al.* (1995) AMBER, a package of computer programs for applying molecular mechanics, normal mode analysis, molecular dynamics, and free energy calculations to simulate the structural and energetic properties of molecules. *Comp. Phys. Commun.*, **91**, 1–41.

51 Mackerell, J.A.D. *et al.* (1998) CHARMM: the energy function and its parameterization with and overview of the program, in *The Encyclopedia of Computational Chemistry* (ed. R. Scheleyer), John Wiley & Sons, Inc., New York, pp. 271–277.

52 Dauber-Osguthorpe, P. *et al.* (1988) Structure and energetics of ligand

binding to proteins: *E. coli* dihydrofolate reductase-trimethoprim, a drug-receptor system. *Proteins: Struct. Funct. Genet.*, **4**, 31–47.
53 Sun, H. (1998) Compass: an *ab initio* force-field optimized for condensed-phase applications – overview with details on alkane and benzene compounds. *J. Phys. Chem.*, **102**, 7338–7364.
54 Jorgensen, W.L., Maxwell, D.S., and Tirado-Rives, J. (1996) Development and testing of the OPLS all-atom force field on conformational energetics and properties of organic liquids. *J. Am. Chem. Soc.*, **118**, 11225–11236.
55 Sun, H. *et al.* (1994) An *ab initio* CFF93 all-atom forcefield for polycarbonates. *J. Am. Chem. Soc.*, **116**, 2978–2987.
56 Rappé, A.K. *et al.* (1992) UFF, a full periodic-table force-field for molecular mechanics and molecular dynamics simulations. *J. Am. Chem. Soc.*, **114**, 10024–10035.
57 Heinz, H. *et al.* (2005) Force field for phyllosilicates and dynamics of octadecylammonium chains grafted to montmorillonite. *Chem. Mater.*, **17**, 5658–5669.
58 Heinz, H. and Suter, U.W. (2004) Atomic charges for classical simulations of polar systems. *J. Phys. Chem. B*, **108**, 18341–18352.
59 Fang, C.M., Parker, S.C., and DeWith, G. (2000) Atomistic simulation of the surface energy of spinel $MgAl_2O_4$. *J. Am. Ceram. Soc.*, **83**, 2082–2084.
60 Batsanov, S.S. (2001) Van-der-Waals radii of elements. *Inorg. Mater.*, **37**, 871–885.
61 Heinz, H. *et al.* (2008) Accurate simulation of surfaces and interfaces of FCC metals using 12-6 and 9-6 Lennard-Jones potentials. *J. Phys. Chem. C*, **112**, 17281–17290.
62 Herrebout, W.A. *et al.* (1995) Enthalpy difference between conformers of n-butane and the potential function governing conformational interchange. *J. Phys. Chem.*, **99**, 578–585.
63 Vaia, R.A., Teukolsky, R.K., and Giannelis, E.P. (1994) Interlayer structure and molecular environment of alkylammonium layered silicates. *Chem. Mater.*, **6**, 1017–1022.
64 Lagaly, G., Weiss, A., and Kolloid, Z.Z. (1971) Arrangement and orientation of cationic surfactants on silicate surfaces. IV. Arrangement of n-alkylammonium ions on weakly charged layer silicates. *Polymer*, **243**, 48–55.
65 Lagaly, G., Weiss, A., and Kolloid, Z.Z. (1970) Anordnung und Orientierung Kationischer Tenside auf Silicatoberflaechen. II. Paraffinähnliche Strukturen bei den n-Alkylamonium-Schichtsilicaten mit hoher Schichtladung (Glimmer). *Polymer*, **237**, 364–368.
66 Brovelli, D., Caseri, W.R., and Hahner, G. (1999) Self-assembled monolayers of alkylammonium ions on mica: Direct determination of the orientation of the alkyl chains. *J. Colloids Interface Sci.*, **216**, 418–423.
67 Osman, M.A., Seyfang, G., and Suter, U.W. (2000) Two-dimensional melting of alkane monolayers ionically bonded to mica. *J. Phys. Chem. B*, **104**, 4433–4439.
68 Giese, R.F. and van Oss, C.J. (2002) *Colloid and Surface Properties of Clays and Related Minerals*, Dekker, New York.
69 Osman, M.A. *et al.* (2002) Structure and molecular dynamics of alkane monolayers self-assembled on mica platelets. *J. Phys. Chem. B*, **106**, 653–662.
70 Yariv, S. and Cross, H. (eds) (2002) *Organo-Clay Complexes and Interactions*, Dekker, New York.
71 Zhu, J.X. *et al.* (2005) Characterization of organic phases in the interlayer of montmorillonite using FTIR and ^{13}C NMR. *J. Colloids Interface Sci.*, **286**, 239–244.
72 Osman, M.A., Ploetze, M., and Skrabal, P. (2004) Structure and properties of alkylammonium monolayers self-assembled on montmorillonite platelets. *J. Phys. Chem. B*, **108**, 2580–2588.
73 Lewin, M., Mey-Marom, A., and Frank, R. (2005) Surface free energies of polymeric materials, additives and minerals. *Polym. Adv. Technol.*, **16**, 429–441.
74 He, H.P. *et al.* (2005) Thermal characterization of surfactant-modified montmorillonites. *Clay Miner.*, **53**, 287–293.

75 Lagaly, G. and Dekany, I. (2005) Adsorption on hydrophobized surfaces: Clusters and self-organization. *Adv. Colloid Interface*, **114**, 189–204.
76 Jacobs, J.D. *et al.* (2006) Dynamics of alkyl ammonium intercalants within organically modified montmorillonite: dielectric relaxation and ionic conductivity. *J. Phys. Chem. B*, **110**, 20143–20157.
77 Drummy, L.F. *et al.* (2005) High-resolution electron microscopy of montmorillonite and montmorillonite/ epoxy nanocomposites. *J. Phys. Chem. B*, **109**, 17868–17878.
78 Osman, M.A., Rupp, J.E.P., and Suter, U.W. (2005) Gas permeation properties of polyethylene-layered silicate nanocomposites. *J. Mater. Chem.*, **15**, 1298–1304.
79 Heinz, H., Castelijns, H.J., and Suter, U.W. (2003) Structure and phase transitions of alkyl chains on mica. *J. Am. Chem. Soc.*, **125**, 9500–9510.
80 Zeng, Q.H. *et al.* (2003) Molecular dynamics simulation of organic-inorganic nanocomposites: layering behavior and interlayer structure of organoclays. *Chem. Mater.*, **15**, 4732–4738.
81 Hackett, E., Manias, E., and Giannelis, E.P. (1998) Molecular dynamics simulations of organically modified layered silicates. *J. Chem. Phys.*, **108**, 7410–7415.
82 He, H.P., Galy, J., and Gerard, J.F. (2005) Molecular simulation of the interlayer structure and the mobility of alkyl chains in DTMA+/montmorillonite hybrids. *J. Phys. Chem. B.*, **109**, 13301–13306.
83 Heinz, H. and Suter, U.W. (2004) Surface structure of organoclays. *Chem. Int. Ed.*, **43**, 2239–2243.
84 Heinz, H., Vaia, R.A., and Farmer, B.L. (2006) Interaction energy and surface reconstruction between sheets of layered silicates. *J. Chem. Phys.*, **124**, 224713:1–224713:9.
85 Gardebien, F. *et al.* (2004) Molecular dynamics simulations of intercalated poly(epsilon-caprolactone)-montmorillonite clay nanocomposites. *J. Phys. Chem. B*, **108**, 10678–10686.
86 Heinz, H. *et al.* (2008) Photoisomerization of azobenzene grafted to montmorillonite: simulation and experimental challenges. *Chem. Mater.*, **20**, 6444–6456.
87 Heinz, H., Vaia, R.A., and Farmer, B.L. (2008) Relation between packing density and thermal transitions of alkyl chains on layered silicate and metal surfaces. *Langmuir*, **24**, 3727–3733.
88 Heinz, H. and Suter, U.W. (2004) Surface structure of organoclays. *Angew. Chem. Int. Ed.*, **43**, 2239–2243.
89 Rimai, D.S., DeMejo, L.P., and Mittal, K.L. (eds) (1995) *Fundamentals of Adhesion and Interfaces*, VSP, Utrecht.
90 Chassin, P., Jouany, C., and Quiquampoix, H. (1986) *Clay Mineral*, **21**, 899–907.
91 Adamson, A.W. and Gast, A.P. (1997) Chapter VII, in *Physical Chemistry of Surfaces*, 6th edn, John Wiley & Sons, Inc., New York.
92 Heinz, H. (2007) Calculation of local and average pressure tensors in molecular simulation. *Mol. Simul.*, **33**, 747–758.
93 Fiedler, B. *et al.* (2006) Fundamental aspects of nano-reinforced composites. *Compos. Sci. Technol.*, **66**, 3115–3125.
94 Maiti, A., Wescott, J., and Kung, P. (2005) Nanotube–polymer composites: insights from Flory–Huggins theory and mesoscale simulations. *Mol. Simul.*, **31** (2–3), 143–149.
95 Namilae, S. and Chandra, N. (2005) Multiscale model to study the effect of interfaces in carbon nanotube: based composites. *Trans. ASME*, **127**, 222–232.
96 Odegard, G.M., Clancy, T.C., and Gates, T.S. (2005) Modeling of the mechanical properties of nanoparticle/polymer composites. *Polymer*, **46** (2), 553–562.
97 Tatke, S.S., Renugopalakrishnan, V., and Prabhakaran, M. (2004) Interfacing biological macromolecules with carbon nanotubes and silicon surfaces: a computer modeling and dynamic simulation study. *Nanotechnology*, **15**, S684–S690.
98 Maiti, A. and Ricca, A. (2004) Metal–nanotube interactions: binding energies and wetting properties. *Chem. Phys. Lett.*, **395**, 7–11.

99 Wei, C. (2006) Radius and chirality dependent conformation of polymer molecule at nanotube interface. *Nano Lett.*, **6** (8), 1627–1631.

100 Panhuis, M.I.H. *et al.* (2003) Selective interaction in a polymer-single-wall carbon nanotube composite. *J. Phys. Chem. B*, **107** (2), 478–482.

101 Han, Y. and Elliott, J. (2007) Molecular dynamics simulations of the elastic properties of polymer/carbon nanotube composites. *Comp. Mater. Sci.*, **39**, 315–323.

102 Garg, A. and Sinnott, S.B. (1998) Effect of chemical functionalization on the mechanical properties of carbon nanotubes. *Chem. Phys. Lett.*, **295**, 273–278.

103 Kuzmany, H. *et al.* (2004) Functionalization of carbon nanotubes. *Synth. Met.*, **141**, 113–122.

104 Sinnott, S.B. (2002) Chemical functionalization of carbon nanotubes. *J. Nanosci. Nanotechnol.*, **2** (2), 113–123.

105 Biercuk, M.J. (2002) Carbon nanotube composites for thermal management. *Appl. Phys. Lett.*, **80**, 2767–2769.

106 Varshney, V. *et al.* (2008) Thermal transport in composite materials and their interfaces. Eighth World Conference on Nanocomposite, San Diego, CA.

107 Shenogin, S. *et al.* (2004) Role of thermal boundary resistance on the heat flow in carbon-nanotube composites. *J. Appl. Phys.*, **95** (12), 8136–8144.

108 Huxtable, S.T. *et al.* (2003) Interfacial heat flow in carbon nanotube suspensions. *Nat. Mater.*, **2**, 731–734.

109 Shenogin, S. *et al.* (2004) Effect of chemical functionalization on thermal transport of carbon nanotube composites. *Appl. Phys. Lett.*, **85** (12), 2229–2231.

110 Clancy, T.C. (2006) Modeling of interfacial modification effects on thermal conductivity of carbon nanotube composites. *Polymer*, **47**, 5990–5996.

111 Rohr, D. and Klein, M.T. (1990) modeling diffusion and reaction in cross-linking epoxy-amine cure kinetics: a dynamic percolation approach. *Ind. Eng. Chem. Res.*, **29** (7), 1210–1218.

112 Leung, Y.K. and Eichinger, B.E. (1984) Computer simulation of end-linked elastomers. 1. Trifunctional networks cured in the bulk. *J. Chem. Phys.*, **80**, 3877–3879.

113 Doherty, D.C. *et al.* (1998) Polymerization molecular dynamics simulations. I. Cross-linked atomistic models for poly (methacrylate) networks. *Comput. Theor. Polym. Sci.*, **8** (1–2), 169–178.

114 Yarovsky, I. and Evans, E. (2002) Computer simulation of structure and properties of crosslinked polymers: application to epoxy resins. *Polymer*, **43** (3), 963–969.

115 Wu, C. and Xu, W. (2007) Atomistic simulation study of absorbed water influence on structure and properties of crosslinked epoxy resin. *Polymer*, **48**, 5440–5448.

116 Heine, D.R. *et al.* (2004) Atomistic simulations of end-linked poly (dimethylsiloxane) networks: structure and relaxation. *Macromolecules*, **37** (10), 3857–3864.

117 Varshney, V. *et al.* (2008) A molecular dynamics study of epoxy-based networks: cross-linking procedure and prediction of molecular and materials properties. *Macromolecules*, **41**, 6837–6842.

118 Wu, C. and Xu, W. (2006) Atomistic molecular modelling of crosslinked epoxy resin. *Polymer*, **47**, 6004–6009.

119 Varshney, V. *et al.* (2009) Heat transport in epoxy networks: A molecular dynamics study. *Polymer*, **50**, 3378–3385.

120 Che, J., Cagin, T., and Goddard, W.A. (2000) Thermal conductivity of carbon nanotubes. *Nanotechnology*, **11**, 65–69.

121 Lukes, J.R. and Zhong, H. (2007) Thermal conductivity of individual single wall carbon nanotubes. *J. Heat Transfer*, **129**, 705–716.

122 Allen, M.P. and Tidesley, D.J. (1987) *Computer Simulation of Liquids*, Clarendon Press, Oxford.

123 Lukes, J.R. *et al.* (2000) Molecular dynamics study of solid thin-film thermal conductivity. *Trans. ASME*, **122**, 536–543.

124 Hoover, W.G. and Ashurst, W.T. (1975) Nonequilibrium molecular dynamics, in *Theoretical Chemistry: Advances and Perspectives* (eds H. Eyring and D.G.

Henderson), Academic Press, New York, pp. 1–51.

125 Frenkel, D. and Smit, B. (2002) *Understanding molecular simulations: From Algorithms to Applications,* 2nd edn, Academic Press, San Diego.

126 Varshney, V. *et al.* (2007) Molecular dynamics study of thermal conductivity of curing agent-W (Detda). SAMPE Technical Conference, Cincinnati, OH.

127 Abramson, A.R., Tien, C.-L., and Majumdar, A. (2002) Interface and strain effects on the thermal conductivity of heterostructures: a molecular dynamics study. *J. Heat Transfer,* **124**, 963–970.

128 Zhong, H. and Lukes, J.R. (2006) Interfacial thermal resistance between carbon nanotubes: molecular dynamics simulations and analytical thermal modeling. *Phys. Rev. B,* **74**, 125403-1–125403-10.

129 Diao, J., Srivastava, D., and Menon, M. (2008) Molecular dynamics simulations of carbon nanotube/silicon interfacial thermal conductance. *J. Chem. Phys.,* **128**, 164708-1–164708-5.

3
Computational Studies of Polymer Kinetics

Galina Litvinenko

3.1
Introduction

An important problem in the chemistry of high molecular weight compounds is obtaining polymers with predetermined properties that, in turn, are defined by their molecular weight characteristics and conditions of synthesis. In general, whether it is a free radical (FR), cationic, or anionic polymerization, the kinetic scheme of the process includes the following steps: chain initiation, propagation, termination, and various chain transfer reactions. In principle, in order to calculate the kinetics and molecular weight distribution (MWD), one should solve an enormous number of kinetic equations for the time evolution of the concentrations $R(l)$ of active macromolecules (or free radicals) of each value of chain length l. It is clear that brute-force programs may be applied only for short chains, $l \leq 10$ (oligomerization), whereas for high molecular weight polymers special approaches and approximations that simplify calculations should be used.

In most situations when, for example, such reactions as chain transfer or termination take place, obtaining analytical expressions for the MWD is rather complicated or even impossible. However, it is often sufficient to know only average degrees of polymerization (DP)

$$\text{number average DP } \bar{P}_\text{n} = \sum lR(l) / \sum R(l),$$

$$\text{weight average } \bar{P}_\text{w} = \sum l^2 R(l) / \sum lR(l),$$

$$z\text{-average } \bar{P}_z = \sum l^3 R(l) / \sum l^2 R(l) \text{ etc.}$$

and the polydispersity index (PDI), $\bar{P}_\text{w}/\bar{P}_\text{n}$.

The most common way to calculate average DPs is to use the well-known mathematical method of statistical moments. By definition, the MWD moment of the ith order is given by $\mu_i = \sum l^i R(l)$. The use of this method decreases the number of equations to solve from almost infinite set of equations for $R(l)$ to several differential equations for μ_i. Therefore, in order to calculate $\bar{P}_\text{n}$, $\bar{P}_\text{w}$, and $\bar{P}_z$, one has to solve only four equations since $\bar{P}_\text{n} = \mu_1/\mu_0$, $\bar{P}_\text{w} = \mu_2/\mu_1$, and $\bar{P}_z = \mu_3/\mu_2$. At the same time, the zeroth and the first MWD moments have a clear physical sense.

Modeling and Simulation in Polymers. Edited by P.D. Gujrati and A.I. Leonov

ISBN: 978-3-527-32415-6

Thus, $\mu_0 = \sum R(l)$ is the total concentration of chains and $\mu_1 = M_0 - M$ is the amount of the polymer formed.

Sometimes, the use of generating functions $G(y) = \sum_{l=1}^{\infty} R(l) y^l$ is very helpful in calculating MWD. The generating function transformation converts the infinite set of differential equations for $R(l)$ to a single ordinary equation and in many cases (but not always!) can provide the total MWD. In this case, $R(l)$ is proportional to the lth derivative of $G(y)$ over y,

$$R(l) = \frac{1}{l!} \cdot \frac{\mathrm{d}^l G}{\mathrm{d}y^l}_{|y=0}.$$

Statistical moments, in turn, are related to $G(y)$ as

$$\mu_0 = G_{|y=1}, \quad \mu_1 = \frac{\mathrm{d}G}{\mathrm{d}y}_{|y=1}, \quad \mu_2 = \mu_1 + \frac{\mathrm{d}^2 G}{\mathrm{d}y^2}_{|y=1.}$$

For high molecular mass polymers, the continuum approximation was proved very useful. It is assumed that chain length may continuously change rather than stepwise. The approximation consists in replacing the differences such as $R(l) - R(l-1)$ with the derivatives $\mathrm{d}R(l)/\mathrm{d}l$ and the sums over l by the corresponding integrals. This approach allows one to obtain results in a much simpler and closed form. This is the common approximation in modeling free radical polymerization. For nonterminating polymerization, a comparison of the exact solutions and the results obtained under the continuum approximation was undertaken by several authors for chain transfer to monomer and slow initiation [1, 2]. It is recognized that some fine details can be lost. For example, instead of the Poisson (or Gauss) MWD of living polymers with the polydispersity index $\mathrm{PDI} = 1 + \bar{P}_\mathrm{n}^{-1}$, the continuum approximation gives the Dirac δ-function with a polydispersity strictly equal to unity. Nevertheless, the difference between the discrete and the continuum approaches is getting smaller for higher $\bar{P}_\mathrm{n}$. For $\bar{P}_\mathrm{n} > 50$, it is already negligible.

In fact, nowadays, the appearance of powerful computers and special software programs, such as MATLAB, CHEMKIN, and so on, and especially those intended exactly for kinetic calculations in polymerization, such as PREDICI (polyreaction distributions by countable system integration), has considerably facilitated calculations. For example, PREDICI [3] applies the Galerkin approximation [4] for the representation of chain length distributions and the adaptive Rothe method [5] for time discretization. When using PREDICI, the engineer should only correctly choose reactions in a special window, and all other jobs will be done by the program package itself. Numerous comparisons performed by the author of this review of the results obtained using this software with the results obtained with other methods confirm the validity of PREDICI for most, if not all possible situations, of the homogeneous as well as emulsion and suspension polymerization. Nevertheless, this does not mean that all other methods of MWD and kinetic calculations are outdated. Of course, numerical calculations are very helpful, but they hardly give the possibility to qualitatively analyze and to predict polymer characteristics. Much more helpful in this regard are the good old methods especially when analytical solutions can be obtained. The area of polymerization processes is extremely extensive, and it is impossible to discuss all calculation problems in a single review. Thus, most attention will be paid to the nonterminating polymerization.

3.2
Batch Polymerization

3.2.1
Ideal Living Polymerization

The simplest from the viewpoint of kinetics and MWD – although, probably, complex in chemical mechanism – is the classical scheme of formation of living polymers [6]. It includes only one type of reactions, namely, chain propagation:

$$R(l) + M \xrightarrow{k_p} R(l+1). \tag{3.1}$$

Here, M is the monomer and $R(l)$ is the growing polymer chain containing l monomer units. The initiation in such systems is presumed to be instantaneous. This means that the total concentration of growing chains $R = \sum R(l)$ is constant from the very beginning of polymerization and is equal to the initiator concentration I_0.

If the dependence of $R(l)$ on length l is known, it is possible to calculate the average DPs and the polydispersity.

According to Eq. (3.1), the dependence of monomer concentration and $R(l)$ on time are to be calculated from the set of equations

$$\begin{aligned} \frac{dM}{dt} &= -k_p RM \\ \frac{dR(l)}{dt} &= k_p M[R(l-1) - R(l)] \quad (i = 1, 2, \ldots) \end{aligned} \tag{3.2}$$

with the initial conditions:

$$M = M_0; \quad R(l) = I_0 \delta_{l,1},$$

where $\delta_{i,j}$ is the Kronecker δ-symbol, $\delta_{ij} = \begin{cases} 1, & i = j \\ 0, & i \neq j. \end{cases}$

The set of Eq. (3.2) is easy to integrate and it results in the following expression for the dependence of monomer conversion $x = (M_0 - M)/M_0$ on time

$$x = 1 - e^{-k_p I_0 t} \tag{3.3}$$

and for the dependence of fraction of chain length l, $f_n(l) \equiv R(l)/I_0$, on conversion

$$f_n(l) = \frac{e^{-\beta x} (\beta x)^{l-1}}{(l-1)!}. \tag{3.4}$$

Here, $\beta = M_0/I_0$.

Equation (3.3) may be expressed in a different form as the dependence of polymerization rate on time

$$-\frac{1}{M_0}\cdot\frac{\mathrm{d}M}{\mathrm{d}t}=k_\mathrm{p}I_0.$$

In this form, it represents a straight line with a slope $k_\mathrm{p}I_0$. Any deviation from the straight line indicates to proceed further reactions that affect the concentration of active centers.

Function (3.4) represents the Poisson distribution. This expression for the MWD of living polymers was obtained by Flory for the first time many years ago [7].

From Eq. (3.4), it follows that number average DP linearly increases with conversion, $\bar{P}_\mathrm{n}=\beta x$, and at full conversion $\bar{P}_n=\beta$. This is why the parameter β is often designated as "theoretical degree of polymerization."

At a sufficiently high β, a characteristic of most polymerization processes, the discrete Poisson distribution may be very good approximated by the continuous Gauss distribution

$$f_\mathrm{n}(l)\approx\frac{\exp\left(-\frac{(l-\bar{P}_\mathrm{n})^2}{2\bar{P}_\mathrm{n}}\right)}{\sqrt{2\pi\bar{P}_\mathrm{n}}}.$$

For this case, the polydispersity index characterizing the broadness of distribution is equal to

$$\frac{\bar{P}_\mathrm{w}}{\bar{P}_\mathrm{n}}=\frac{1+1}{\bar{P}_\mathrm{n}}.$$

As can be seen, $\bar{P}_\mathrm{w}/\bar{P}_\mathrm{n}$ decreases from 2 at $t=0$ to $1+1/\beta$ at full monomer conversion.

The Poisson distribution is the narrowest MWD that may be obtained. Any additional reaction will result in broadening MWD. First, the initiation may be noninstantaneous

$$M+I\xrightarrow{k_\mathrm{i}}R(1).$$

The effect of finite initiation rate on the kinetics and MWD was studied in Ref. [8], where it was shown that slow initiation may increase the polydispersity index to maximum 4/3. This result is very clear from the following considerations, without excessive mathematics as in Ref. [8].

In the case of very slow initiation, new chains will be generated at an almost constant rate, thus the distribution will be proportional to the step function, $f(l)=\theta(L_\mathrm{max}-l)/L_\mathrm{max}$, where the step function $\theta(x)$ is 0 for $x\leq 0$ and 1 for $x>0$, and $L_\mathrm{max}=k_\mathrm{p}\int M\,\mathrm{d}t$ is the maximum chain length at time t. According to the definition of DPs, the number average DP for such $f(l)$ is equal to $L_\mathrm{max}/2$ and the weight average to $2L_\mathrm{max}/3$; hence, the polydispersity index is 4/3. It can be shown that the criterion of instantaneous initiation is the condition $\beta k_\mathrm{i}/k_\mathrm{p}\gg 1$ rather than

$k_i/k_p \gg 1$. Thus, taking into account that usually $\beta \gg 1$ initiation can be considered as instantaneous, even for small ratios of k_i/k_p.

Other important reactions affecting kinetics and MWD are various chain transfer reactions. They are discussed in the next section.

3.2.2 Effect of Chain Transfer Reactions

Chain transfer reactions are considered here irrespective of the exact chemical mechanisms of transfer that are discussed in detail, for example, in the comprehensive review of Glasse [9]. It should be stressed that all calculations were performed bearing in mind anionic polymerization, so the main concern in all theoretical papers reviewed is the absence of kinetic termination rather than the anionic mechanism of chain propagation. Thus, the conclusions drawn are valid for all processes that satisfy this condition.

3.2.3 Chain Transfer to Solvent

Chain transfer to solvent is the best investigated reaction among all chain transfer reactions. The first evidence of chain transfer to solvent based on the formation of low molecular weight polymers and on the direct detection of solvent fragments in macromolecules was obtained more than half a century ago for various combinations of monomers and solvents. These include the polymerization of butadiene in toluene [9] or styrene in liquid ammonia [10]. Later on, chain transfer to aromatic solvents was reported for many other systems. Therefore, the most important is not a qualitative result (whether chain transfer to solvent takes place or not) but rather quantitative one (to what extent it goes). That is why this reaction deserves to be considered in detail.

The effects of chain transfer to solvent on molecular masses and MWDs of polymers formed in nonterminating polymerizations have been theoretically studied in a number of papers [11–18]. A detailed review is given in Ref. [19]. Most studies have examined common types of polymerization, that is, homopolymerization by monofunctional initiators (of the type RMt with a single active center). The calculations have been based on the kinetic scheme 3.1 that was proposed initially by Higginson and Wooding [11].

In this scheme, S is the solvent, S^* is an intermediate active species arising due to chain transfer, and $R(l)$ and $P(l)$ are the growing and dead macromolecules, respectively, containing l monomer units. Later on, a similar scheme was independently proposed in Refs [12, 15] for spontaneous transfer

$$(b') \quad R(l) \xrightarrow{k_{sp}} P(l) + S^*.$$

It should be said that in spite of that chemistry of chain transfer to solvent and spontaneous transfer is different, the two mechanisms are identical both in mathematics and in results at $k_{sp} = k_{ts}S$.

Instantaneous primary initiation was assumed in all papers. In combination with the absence of chain termination, this means that the total concentration of active species $R+S^*$ is constant from the very beginning of polymerization and is equal to the concentration of initiator I_0. As was discussed previously, this is satisfied for $\beta k_i/k_p \gg 1$.

The effect of chain transfer on the average degrees of polymerization was studied much earlier for free radical polymerization. For this case, the number average degree of polymerization of the polymer formed during a short time interval from t to $t+\mathrm{d}t$ (instantaneous $\bar{P}_n$) is calculated as the amount of monomer converted into polymer during $\mathrm{d}t$ divided by the number of newly formed macromolecules. The simple expression was obtained that relates $\bar{P}_n$ to the constants of chain transfer [20, 21]

$$\frac{1}{\bar{P}_n} = \frac{1}{\bar{P}_n^0} + C_m + C_s\frac{S}{M} + C_i\frac{1}{M}. \tag{3.5}$$

Here, $C_m = k_{tm}/k_p$, $C_s = k_{ts}/k_p$, and $C_i = k_{ti}/k_p$ are the relative constants of chain transfer to monomer, solvent, and initiator, respectively, and $\bar{P}_n^0$ is the number average DP in the absence of chain transfer reactions. Using an expression of this type at low conversions ($M \approx M_0$) when the difference between instantaneous $\bar{P}_n$ and the experimentally measured time-averaged DP is small, and from Eq. (3.5), it is easy to evaluate C_m, C_s, and C_i. However, this procedure usually cannot be applied to living polymerizations. First, this expression does not take into account the contribution from growing macromolecules. For nonterminating polymerizations with long-living active centers, it is correct only if chain transfer is extremely extensive. Second, in some cases, it is difficult to isolate the polymer at low conversions due to very high polymerization rates. Moreover, because the ratio of the chain transfer rate, $k_{ts}SR$, to the propagation rate, k_pMR, increases in the course of polymerization, the effect of chain transfer can be small at low conversions but becomes significant toward the end of the process. Therefore, it is necessary to derive expressions that can be used over the entire range of conversions.

In accordance with Scheme 3.1, the evolution of the system in time can be described by the following set of differential equations:

$$\begin{aligned}
\frac{\mathrm{d}M}{\mathrm{d}t} &= -M(k_pR + k_{ri}S^*),\\
\frac{\mathrm{d}R(l)}{\mathrm{d}t} &= k_pM[R(l-1)-R(l)]-k_{ts}SR(l) + \delta_{l,1}k_{ri}S^*M,\\
\frac{\mathrm{d}P(l)}{\mathrm{d}t} &= k_{ts}SR(l),\\
\frac{\mathrm{d}S^*}{\mathrm{d}t} &= k_{ts}SR-k_{ri}S^*M.
\end{aligned} \tag{3.6}$$

The first term on the right side of Eq. (3.6) describes the increase in chain length due to propagation, the second is the disappearance of $R(l)$ due to chain transfer, and the third denotes the formation of short active chains as a result of reinitiation.

(a)	Propagation	$R(l) + M \xrightarrow{k_p} R(l+1)$
(b)	Chain transfer to solvent	$R(l) + S \xrightarrow{k_s} P(l) + S^*$
(c)	Reinitiation	$S^* + M \xrightarrow{k_{ri}} R(1)$

Scheme 3.1

The set (3.6) was used to calculate average DPs and full MWDs. Except for one publication [11], the calculations were performed assuming that reinitiation is fast compared to chain transfer and, consequently, the total concentration of growing chains R is equal to I_0. Then, the dependence of monomer conversion $x = (M_0 - M)/M_0$ on time is described by the expression slightly different from Eq. (3.3)

$$x = (1 + \alpha)(1 - e^{-k_p I_0 t}).$$

Here, the parameter $\alpha = k_{ts} S / k_p M_0$ characterizes the intensity of chain transfer. Normally, $\alpha \ll 1$, otherwise no polymer would be formed.

Although the expressions for $\bar{P}_n$ and $\bar{P}_w$ in Refs [15, 18, 22] differ from each other in the form, this difference is not important and disappears if one takes into account that $\alpha \ll 1$ and the "theoretical" DP $= \beta = M_0/I_0 \gg 1$. Here, β is the degree of polymerization of the living polymer formed in the absence of side reactions. Then, the dependence of the number and weight average DPs on conversion can be expressed as

$$\bar{P}_n = \frac{\beta x}{1 - \gamma_s \ln\left(\dfrac{1 - x + \alpha}{1 + \alpha}\right)}, \tag{3.7}$$

$$\bar{P}_w = \frac{2\beta}{(\gamma_s - 1)x}\left[x - \frac{x^2}{2} - \frac{1 - (1 - x)^{\gamma_s + 1}}{\gamma_s + 1}\right], \tag{3.8}$$

where $\gamma_s = \alpha\beta = k_t S / k_p I_0$.

It should be mentioned that expression (3.7) for the number average DP can be easily derived without solving the full set (3.6). By definition, $\bar{P}_n$ is equal to the total number of moles of monomer converted into polymer, $M_0 x$, divided by the total number of macromolecules. The latter increases in time due to chain transfer at a constant rate $k_{ts} SR$ and hence, at time t it is $I_0 + k_{ts} S I_0 t$. Replacing time with conversion yields formula (3.7).

A typical dependence of the average DPs on conversion is shown in Figure 3.1. The deviation from the straight line (living polymer) is more significant for higher values of α. The occurrence of chain transfer results in the decrease of $\bar{P}_n$ at high conversions even at small intensities of chain transfer when $\bar{P}_w$ increases with conversion. However, at higher α, $\bar{P}_w$ also begins to fall after some maximum value. This is a crucial difference between chain transfer to solvent and all other chain transfer reactions. This difference can be used to elucidate the mechanism of transfer.

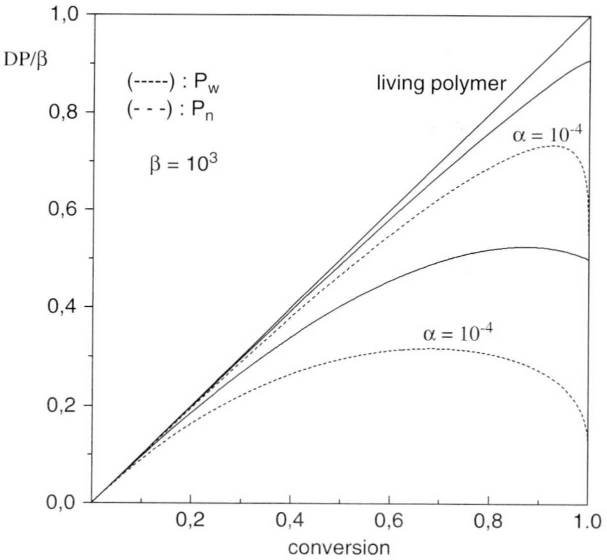

Figure 3.1 Dependence of average DPs of polymer formed under conditions of chain transfer to solvent on conversion.

At full conversion, the average DPs are [15, 22]

$$\bar{P}_n = \frac{\beta}{1-\gamma_s \ln \alpha}, \quad \bar{P}_w = \frac{\beta}{\gamma_s + 1}, \quad \bar{P}_z = \frac{2\beta}{\gamma_s + 2}. \tag{3.9}$$

While the average DPs of the final polymer decrease with increasing intensity of chain transfer, the behavior of the polydispersity index PDI $= \bar{P}_w/\bar{P}_n$ is more complex. As can be shown, the ratio $\bar{P}_w/\bar{P}_n$ as a function of α has a maximum. The position of the maximum is defined by the condition $-\ln \alpha = \gamma_s + 2$. Thus, for typical values $\beta = 10^3$, the maximal value of the polydispersity index $\bar{P}_w/\bar{P}_n \sim 4.62$ is reached at $\alpha \approx 3.62 \times 10^{-3}$. For very intensive chain transfer ($\gamma_s \gg 1$), $\bar{P}_w/\bar{P}_n \sim -\ln \alpha$. The ratio $\bar{P}_z/\bar{P}_w$ varies over a shorter interval from about 1 for $\gamma_s \ll 1$ to 2 for $\gamma_s \gg 1$.

Several words should be said about the importance of taking into account the reinitiation step for the calculation of average DPs. Sometimes, similar to free radical polymerization, the consumption of monomer at this step is neglected leading to the following expression for $\bar{P}_n$:

$$\bar{P}_n = \frac{\beta x}{1-\gamma_s \ln(1-x)}.$$

This expression almost coincides with the exact solution (3.7) for the whole conversion range except for values of x close to 1 because it gives a senseless limiting value $\bar{P}_n \to 0$ at $x \to 1$. At high monomer conversion, the contribution of reinitiation to monomer consumption becomes comparable with that of propagation and should be taken into account, whereas the simplified scheme of chain transfer does not exclude the transfer of active center from one solvent molecule to another when no new polymer chains are formed.

Some observations should also be made on the reinitiation kinetics. For free radical polymerization, only reactions with fast reinitiation are regarded as chain transfer. The processes with slow reinitiation are called inhibition. In this case, the kinetics and molar masses are described by equations quite different from those for chain transfer to solvent [20]. Nevertheless, our calculations demonstrated that in nonterminating polymerization, in contrast to free radical polymerization, Eqs. (3.7) and (3.8) derived originally for fast initiation remain also valid for slow reinitiation ($k_{ri}/k_{ts} \ll 1$). In fact, the main condition for the validity of Eqs. (3.7) and (3.8) is not the high rate of reinitiation compared to transfer but the quasisteady-state approximation with regard to S^*, which is valid when $\beta k_{ri}/k_p \gg 1$. On the other hand, it was shown in Ref. [11] that slow reinitiation leads to the second-order kinetics with respect to monomer. This effect was not, however, observed in anionic polymerization of nonpolar monomers.

It should also be noted that while Eqs. (3.7) and (3.8) were obtained for chain transfer to solvent, they may be successfully applied to chain transfer to impurities or to a specially added transfer agent if the transfer constant $C_s = k_{ts}/k_p$ is small ($C_s \leq 0.1$). However, when the rate constants of chain transfer and propagation are comparable, the consumption of chain transfer agent should be taken into account, $S/S_0 = (1-x)^{C_s}$; hence, the effect of this transfer type on the average DPs is smaller than that of chain transfer to solvent (the consumption of the latter is negligible). In this case, one can obtain the dependence of $\bar{P}_n$ on conversion as

$$\bar{P}_n = \frac{\beta x}{1 + \frac{S_0}{I_0}\left[1-(1-x)^{C_s}\right]}.$$

If $C_s \ll 1$, then, $(1-x)^{C_s} \approx 1 - C_s \ln(1-x)$, and this expression transforms into Eq. (3.7).

It is possible to obtain from set (3.6) the full MWD functions. This can be done in different ways. One is the direct solution of the equations for $R(l)$ and $P(l)$ [18]. However, this method is tedious and the resulting expressions are represented in the form of sums over l, which is extremely inconvenient for analysis. As was already mentioned, the use of the direct approach is reasonable only for oligomerization ($l \leq 10$). The weight fraction of macromolecules of length l

$$f_w(l) = l\frac{R(l) + P(l)}{\int_0^\infty (R(l) + P(l))\,dl}$$

was calculated independently in Refs [12, 13] under condition of constant monomer concentration and is given by

$$f_w(l) = lp^{l-1}(1-p)^2.$$

Here, $p = k_p M/(k_p M + k_{ts} S)$ is the probability of propagation. However, the process usually proceeds at a variable monomer concentration. Nanda and Jain

[15] were the first who tried to derive MWD functions for arbitrary conversions, but they did not succeed in obtaining analytical expressions. Such an expression was derived in Ref. [22] for the general case of initiators of arbitrary functionality that will be considered later. In particular, for a monofunctional initiator, the weight distribution is described by a relatively simple expression

$$f_{\mathrm{w}}(z) = \frac{\gamma_{\mathrm{s}} z}{x}\left[1 - \ln\frac{1-x+\alpha}{1+\alpha}\right]\delta(z) + \frac{\gamma_{\mathrm{s}}}{x}\left[(1-z)^{\gamma_{\mathrm{s}}-1} - z\left(\frac{1-x}{1-x+z}\right)\right]\theta(x-z)$$
$$+\left(\frac{1-x+\alpha}{1+\alpha}\right)^{\gamma_{\mathrm{s}}}\delta(x-z).$$

Here, $z = l/\beta$ is the normalized length and $\delta(z)$ is the Dirac delta function. At full conversion, the above expression is simplified to

$$f_{\mathrm{w}}(z) = \gamma_{\mathrm{s}}(1-z)^{\gamma_{\mathrm{s}}-1}\theta(1-z).$$

3.2.4
Multifunctional Initiators

Along with the simplest case of one-ended growing chains discussed above, more complicated processes also occur. For example, electron transfer initiators such as sodium naphthalene, widely used in the study of living polymers, form growing chains with two living ends [6]. Initiators with two or more active centers are also used, in particular, for the synthesis of ABA-type block copolymers or star-shape polymers. In the case of living polymers, the use of multifunctional initiators has no effect on the polymer MWD (except, of course, for an m-fold increase in DP for the initiator of functionality m). But in the presence of chain-breaking reactions, the functionality of growing chains greatly affects the MWD. It was observed long ago that the effect of impurities on the MWD is different for one- or two-ended living polymers [6]. MWD calculations of polymers formed with multifunctional initiators under conditions of chain transfer to solvent were considered in detail in Ref. [22]. The calculations were based on Scheme 3.1 generalized for the case when a growing chain may contain an arbitrary number of active centers.

In such a case, it is convenient to subdivide polymer chains into two types: primary, or direct chains $D(i,l)$, containing the fragmentation of the initiator molecule, and the secondary, or transfer chains $T(i,l)$, formed during chain transfer to solvent. Such a subdivision can be done, of course, even if the initiator is monofunctional [15], the difference between chains of different types greatly increases with increasing initiator functionality. Therefore, reactions for chain transfer and reinitiation can now be written as follows:

In Scheme 3.2, the symbol i denotes the number of active centers on the macromolecule. For an initiator of functionality m, the value of i can lie between 0 and m for direct chains, whereas for transfer chains it is either 0 or 1. If m is greater than 2, the direct macromolecules are branched, whereas the secondary are always linear.

$$D(i,l) + S \xrightarrow{k_{ts}} D(i-1,l) + S^*$$

$$T(1,l) + S \xrightarrow{k_{ts}} T(0,l) + S^*$$

$$S^* + M \xrightarrow{k_{ri}} T(1,1)$$

Scheme 3.2

The same simple considerations made in the preceding section afford the possibility of calculating the number average DP without solving the full set of differential equations. If we denote now by I_0 the total concentration of active centers (i.e., the molar concentration of initiator I_0/m), the dependence of $\bar{P}_n$ on conversion is described by expression similar to Eq. (3.7).

$$\bar{P}_n = \frac{m\beta x}{1 - m\gamma_s \ln\left(\dfrac{1-x+\alpha}{1+\alpha}\right)}. \tag{3.10}$$

Here, $\beta = M_0/I_0$ and $\gamma_s = \alpha\beta$ are the same as for monofunctional initiators. Weight- and z-average DPs of the total polymer and of the primary and secondary fractions were also calculated [22]. The description of conversion for average DPs of the total polymer is similar to that for monofunctional initiator. The behavior of the direct polymer is the least sensitive to chain transfer.

Expressions for the average DPs of the final polymer are provided in Table 3.1. The characteristics of the secondary polymer obviously do not depend on the functionality of initiator.

An analysis of these expressions shows that the dependence of the number average DP of the total polymer on the initiator functionality is already small for $\gamma_s > 0.1$, and at $\gamma_s \gg 1$ it ceases to depend on the initiator functionality, and

$$\bar{P}_n \approx \frac{1}{\alpha \ln \alpha}$$

whereas $\bar{P}_w$ increases almost linearly with m even at $\gamma_s \sim 1$ (Figure 3.2). As a result, under otherwise similar conditions, the polydispersity index in case of multifunctional

Table 3.1 Average degrees of polymerization for a multifunctional initiator.

Polymer	Degree of polymerization		
	Number average	**Weight average**	**z-average**
Total	$\dfrac{\beta m}{1 - m\gamma_s \ln \alpha}$	$\beta \dfrac{\gamma_s + m}{(\gamma_s + 1)^2}$	$\beta \dfrac{2\gamma_s^2 + (m^2 + 3m)\gamma_s + 2m^2}{(\gamma_s + 1)(\gamma_s + 2)(\gamma_s + m)}$
Primary	$\dfrac{\beta m}{\gamma_s + 1}$	$\beta \dfrac{(m+1)\gamma_s + 2m}{(\gamma_s + 1)(\gamma_s + 2)}$	$\beta \dfrac{\gamma_s^2(m+1)(m+2) + \gamma_s(m+2)(5m-1) + 6m^2}{(\gamma_s + 1)(\gamma_s + 3)[(m+1)\gamma_s + 2m]}$
Transfer	$-\dfrac{\beta}{(\gamma_s + 1)\ln \alpha}$	$\dfrac{\beta}{\gamma_s + 2}$	$\dfrac{2\beta}{\gamma_s + 3}$

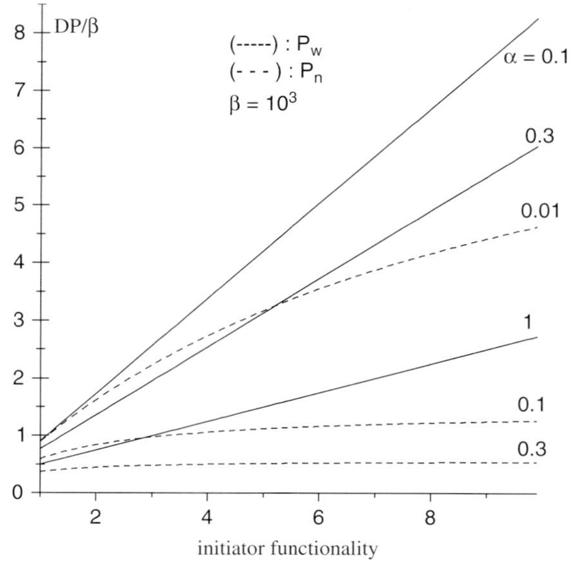

Figure 3.2 Dependence of average DPs of the total polymer on the initiator functionality.

initiators is higher than for monofunctional ones. For example, at $\beta = 10^3$ and $\alpha = 10^{-4}$ ($\gamma_s = 0.1$), the polydispersity index is equal to 1.74 for monofunctional, 2.47 for bifunctional, and 4.0 for tetrafunctional initiators.

The form of the MWD curves is determined by superposing the contributions from the direct and transfer fractions. As far as the total rate of chain transfer is concerned, $k_{ts}MI_0$ is proportional to the total concentration of active centers and hence does not depend on initiator functionality, the relationship between these fractions does not depend on the functionality either, and the weight fraction of direct polymer, $\omega = W_{direct}/W_{total}$, is given by

$$\omega = \frac{1-(1-x)^{\gamma_s+1}}{x(\gamma_s+1)}. \tag{3.11}$$

At full conversion ($x = 1$), this ratio converts into $\omega = (\gamma_s + 1)^{-1}$. For example, for $\gamma_s = 1$, only a half polymer is formed on the initiator molecules.

The change in the initiator functionality at a constant intensity of chain transfer will cause a shift in the DP of the primary polymer to higher values, whereas the distribution of the secondary polymer does not depend on m. As a result, the MWD of the total polymer has a distinct bimodality. Typical MWD curves for various initiator functionalities are shown in Figure 3.3.

Therefore, polymerization with multifunctional initiators is the only case of a one-state polymerization (i.e., polymerization where all active centers are of the same nature and reactivity) leading to a bimodal MWD. It is worth noting that the MWD cannot have more than two maxima independent of the functionality of initiator. The same is valid for all other kinds of chain transfer.

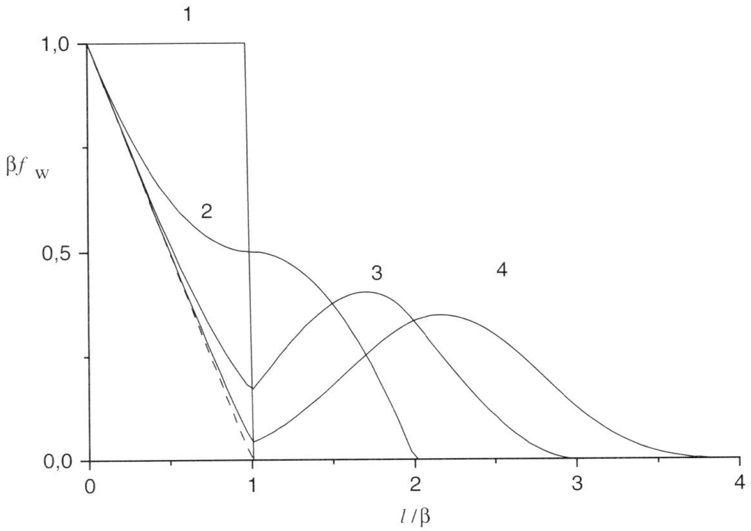

Figure 3.3 MWD of final polymer obtained with initiators of different functionalities. $\gamma_s \sim 1$.

3.2.5
Chain Transfer to Polymer

Chain transfer to polymer is another important reaction that can greatly affect the structure of macromolecules and, hence, the properties of the polymers. Moreover, contrary to chain transfer to solvent that can be eliminated or at least suppressed by the use of another solvent, transfer to polymer is an inherent property of the polymerizing system. The effect of this reaction on molecular weight characteristics of polymers has been studied only for free radical polymerization [23–25]. In particular, it was shown that transfer to polymer may cause gelation if termination proceeds via combination. Principal results for living polymerization were obtained in Ref. [26].

The occurrence of chain transfer to polymer gives rise to the formation of macromolecules that simultaneously contain more than one active center even if the initiator is monofunctional. If one denotes $R(i,l)$ the concentration of macromolecules containing i active centers and l monomer units, the reaction of chain transfer to polymer can be represented as follows:

$$R(i,l) + R(j,n) \xrightarrow{k_{tp}} \begin{cases} R(i-1,l) + R(j+1,n) \\ R(i+1,l) + R(j-1,n). \end{cases} \quad (3.12)$$

This reaction does not affect the rate of monomer consumption; hence, the dependence of monomer conversion on time is given by Eq. (3.3). There is no effect on the number average DP, either, because the number of chains does not change due to this reaction type. Nevertheless, chain transfer to polymer leads to the formation of branched polymer, thus broadening the MWD. The number of side branches N_{br} increases with time at a rate

$$\frac{dN_{br}}{dt} = k_{tp}\sum_{i,j}\sum_{l,n} inR(i,l)R(j,n) = k_{tp}I_0M_0x.$$

The average degree of branching per monomer unit, $\varrho_m = N_{br}/M_0x$, is given by the well-known Flory expression [27] that is valid for all addition polymerization processes independent of their mechanisms:

$$\varrho_m = -C_p\left[1 + \frac{\ln(1-x)}{x}\right]. \tag{3.13}$$

Here, C_p is the relative constant of chain transfer to polymer, $C_p = k_{tp}/k_p$. When chain transfer to solvent proceeds along with chain transfer to polymer, this expression for nonterminating polymerization changes to [26]

$$\varrho_m = -C_p\left[1 + \frac{1}{x}\ln\frac{1-x+\alpha}{1+\alpha}\right].$$

More informative is the average degree of branching per polymer chain $\varrho_p = N_{br}/N_{ch}$. These two parameters are connected by the obvious relationship: $\varrho_p = \varrho_m\bar{P}_n$. Therefore, for the polymerization under conditions of combined chain transfer to polymer and solvent,

$$\varrho_p = -\gamma_p\frac{x + \ln\dfrac{1-x+\alpha}{1+\alpha}}{1-\gamma_s\ln\dfrac{1-x+\alpha}{1+\alpha}}. \tag{3.14}$$

Here, $\gamma_p = \beta C_p$ is the intensity of chain transfer to polymer.

Equations (3.13) and (3.14) show that in the absence of chain transfer to solvent, both ϱ_m and ϱ_p go to infinity when $x \to 1$. Two approaches [26] were suggested to avoid this singularity. The reinitiation step similar to Scheme 3.1 enables one to exclude from the total number of branching N_{br} those sites of branching that do not add a single monomer unit. This procedure leads to the following finite limiting values of ϱ_m and ϱ_p

$$\varrho_m(x=1) = -C_p(1 + \ln C_p),$$
$$\varrho_p(x=1) = -\gamma_p(1 + \ln C_p).$$

On the other hand, as shown in Ref. [26] the formation of relatively long side branches containing, say, 10–20 units, ends at $x = 90$–95%. Branches formed at higher conversions have only a small effect on polymer properties. So, it was suggested to use the values of ϱ_m and ϱ_p calculated at this conversion as final.

It is interesting to compare the effects of chain transfer to polymer in nonterminating and free radical polymerizations. Calculations [26] were performed for an ideal free radical polymerization, that is, in the absence of the gel effect. At equal C_p and $\bar{P}_n$, the number of branches per macromolecule for nonterminating and free radical

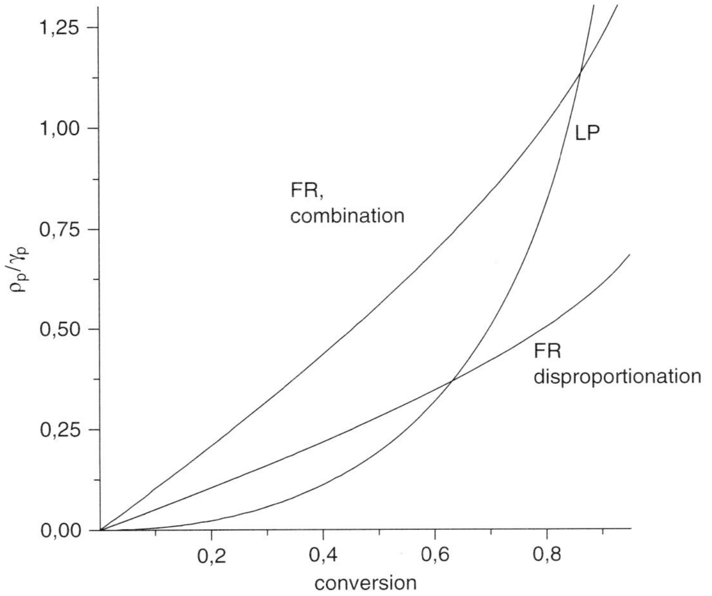

Figure 3.4 Dependence of the average number of branches per macromolecule on conversion for living (LP) and free radical polymerization.

polymerizations becomes comparable to moderate conversions, and at high conversions, anionic ϱ_p is even higher than that for the free radical polymerization (Figure 3.4).

Another important feature of the polymer formed under conditions of chain transfer to polymer is the distribution of polymer between the main chains and the side branches. The weight fraction of polymer in the side branches ω_s is defined by the expression [26]

$$\omega_s \equiv \frac{W}{M_0 x} = 1 - \frac{1}{x}\int_0^x (1-y)^{\gamma_p}\, e^{\gamma_p y}\, dy. \tag{3.15}$$

The increase in ω_s with conversion calculated for different values of γ_p is shown in Figure 3.5.

As seen, the final values of living and free radical ω_s calculated at equal C_p and $\bar{P}_n$ are comparable, but the conversion behavior is different. For anionic polymerization, a marked increase in ω_s is observed only at high conversion, whereas for free radical polymerization ω_s is relatively high even at small conversions. This difference can be explained by different mechanisms of side branch formation. In free radical polymerization, each newly formed side radical quickly grows to a long chain whereas in nonterminating polymerization, the growth of both main and side chains continues during the whole process.

One more distinction between living and free radical polymerizations is the relationship between the average length of the backbone L_b and the side branches

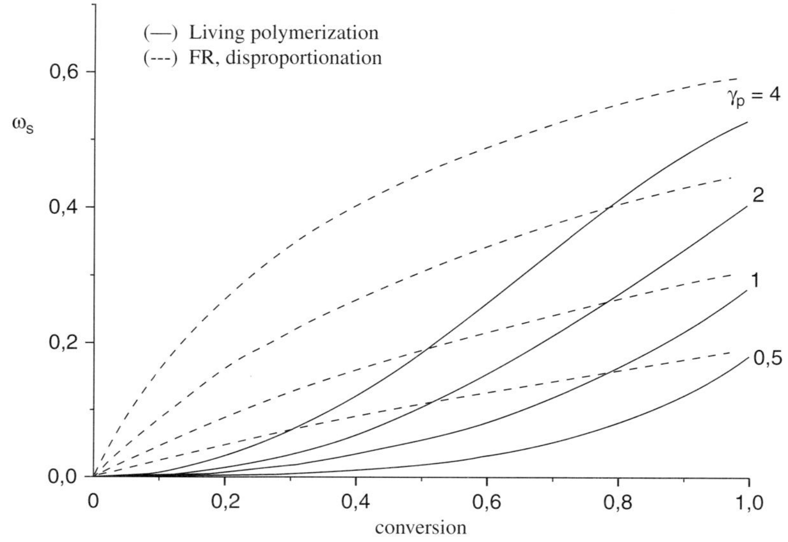

Figure 3.5 Dependence of weight fraction of polymer in side branches on conversion.

L_s. These parameters can be easily calculated if ω_s is known:

$$L_b = \frac{M_0 x(1-\omega_s)}{N_{ch}} = (1-\omega_s)\bar{P}_n,$$
$$L_s = \frac{M_0 x \omega_s}{N_{br}} = \frac{\omega_s}{\varrho_p}\bar{P}_n. \qquad (3.16)$$

The analysis of expressions (3.16) shows that side branches formed in anionic polymerization are much shorter than the backbone. Thus, for typical values ($\beta = 10^3$ and $C_p = 10^{-3}$), the ratio L_b/L_s is about 5 at $x = 0.9$ and increases up to about 20 by full conversion. On the other hand, in free radical polymerization with short-living active centers, the average lifetimes of active centers in the side and main chains are equal, that is, $L_b \approx L_s$.

The weight and z-average DPs for a polymer formed under conditions of chain transfer to polymer and simultaneous chain transfer to polymer and solvent were numerically calculated using the method of statistical moments [26]. Effect of chain transfer to polymer on the MWD broadening in living and free radical polymerization is shown in Figure 3.6. The results show that for living polymerization, the contribution of transfer to polymer in the increase in polydispersity index is much smaller than that for transfer to solvent. This is explained by the fact that main effect that broadens MWDs is a sharp decrease in $\bar{P}_n$ under chain transfer to solvent, whereas transfer to polymer does not change $\bar{P}_n$.

As seen in Figure 3.6, in nonterminating polymerizations the average DPs depend on the intensity of chain transfer to polymer to a significantly less extent than in free radical polymerization. As discussed in the following section, the situation is quite different, however, for continuous mode of polymerization.

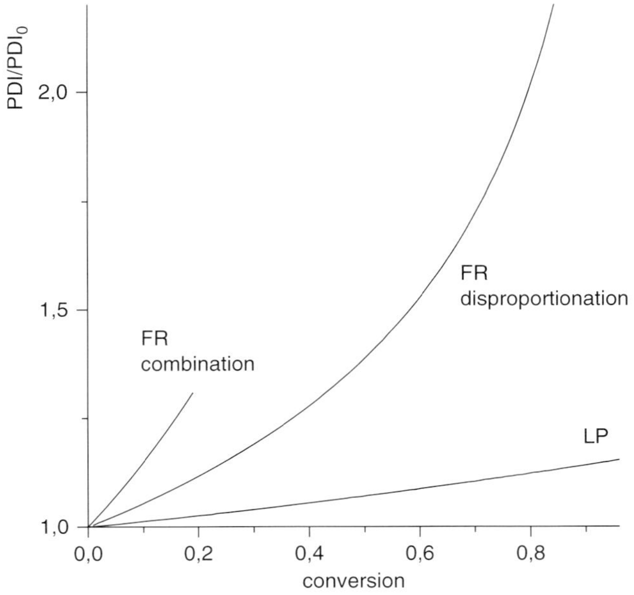

Figure 3.6 Effect of chain transfer to polymer on the polydispersity index. $\gamma_p = 1$; PDI$_0$ is the polydispersity index in the absence of chain transfer.

3.2.6
Chain Transfer to Monomer

The first theoretical papers on anionic chain transfer to monomer were published long ago [2]. Later on, the effect of chain transfer to monomer on the molecular characteristics of the polymers formed in nonterminating polymerization was considered in a number of publications [28–33]. Usually, chain transfer is described according to a scheme that is similar to the free radical polymerization [21]

$$R(l) + M \xrightarrow{k_{tm}} P(l) + R(1). \tag{3.17}$$

According to Eq. (3.17), the evolution of growing $R(l)$ and dead $P(l)$ macromolecules in time is described by differential equations:

$$\begin{aligned} \frac{dR(l)}{dt} &= k_p M[R(l-1) - R(l)] - k_{tm} M R(l) + \delta_{l,1} k_{tm} I_0 M, \\ \frac{dP(l)}{dt} &= k_{tm} M R(l). \end{aligned} \tag{3.18}$$

Since the rate constant (reactivity) of transfer k_{tm} is much less than the propagation rate k_p, the consumption of monomer due to the chain transfer may be neglected. Then, the dependence of monomer conversion on time is given by the same relation (3.3), $x = 1 - \exp(-k_p I_0 t)$, as in the absence of transfer. The solution of Eq. (3.18) gives the following dependence of the number and weight average DPs on conversion

$$\bar{P}_n = \frac{\beta x}{1 + \gamma_m x}, \qquad \bar{P}_w = \frac{2\beta}{\gamma_m^2 x}\left[\gamma_m x - 1 + e^{-\gamma_m x}\right]. \tag{3.19}$$

Here, $\gamma_m = \beta k_{tm}/k_p = \beta C_m$ characterizes the intensity of chain transfer to monomer. As follows from formulas (3.19), at small and moderate values of γ_m, both $\bar{P}_n$ and $\bar{P}_w$ increase with conversion. At high γ_m ($\gamma_m > 10$), the average DPs quickly approach even at low conversions of the stationary values $\bar{P}_n = C_m^{-1}$ and $\bar{P}_w = C_m^{-1}$, and after that they do not change. The polydispersity index cannot exceed 2 however large γ_m is. These results are different from the chain transfer to solvent when $\bar{P}_w/\bar{P}_n$ can exceed 4 and a decrease of $\bar{P}_n$ is observed at high conversion. In addition, as the ratio of propagation and transfer rates is constant, the average DPs under chain transfer to monomer do not depend on the monomer concentration, whereas in the case of chain transfer to solvent, this dependence manifests itself through the parameter α.

The effect of simultaneous chain transfer to monomer and solvent was considered in Refs [28, 29]. Since the number of polymeric chains due to the two reactions increases additively, it is easy to calculate $\bar{P}_n$

$$\bar{P}_n = \frac{\beta x}{1 + \gamma_m x - \gamma_s \ln\left(\dfrac{1 - x + \alpha}{1 + \alpha}\right)}.$$

The effect of chain transfer to monomer for multifunctional initiators was considered in Ref. [34]. Because of the complexity of the original differential equations, the author managed to derive analytical expressions without approximations only for $\bar{P}_n$. However, taking into account that k_m/k_p is much less than unity and using methods of statistical moments, it was rather easy to derive the following analytical expressions for the evolution of average DPs with conversion [35]

$$\bar{P}_n = \frac{m\beta x}{m + \gamma_m x}, \qquad \bar{P}_w = \beta\frac{(m-1)(1 - e^{-\gamma_m x})^2 - 2(1 - e^{-\gamma_m x}) + 2\gamma_m x}{\gamma_m^2 x}. \tag{3.20}$$

Here, m is the functionality of initiator.

The chain length distribution of the polymer formed under conditions of chain transfer to monomer was calculated in a number of papers [2, 34, 36]. However, the continuum approximation was not used and the results are represented in the form of very complicated sums over the chain length. This complexity was the reason for erroneous MWD curves of the polymer formed in the presence of chain transfer to monomer for the case of initiator of functionality $m = 3$, calculated in Ref. [34] by a graphical method. Using the continuum approximation, and neglecting k_{tm}/k_p, one can obtain [35] a fairly simple expression for the weight fraction of macromolecules of

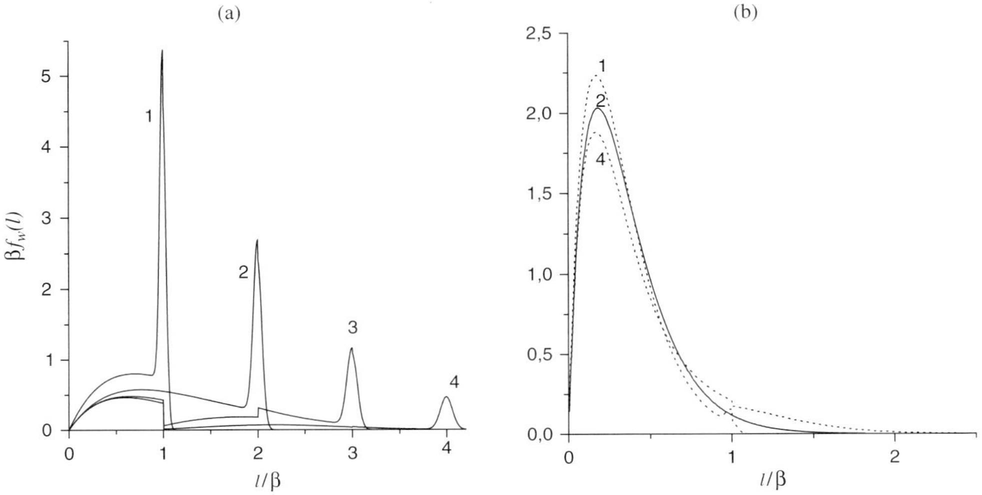

Figure 3.7 Final ($x = 1$) MWD of polymer obtained in polymerization under conditions of chain transfer to monomer for initiators of various functionalities (numbers at the curves). The intensity of transfer $\gamma_m = 0.5$ (a) and 5 (b).

normalized length $z = l/\beta$, at monomer conversion

$$f_w(z,x) = \frac{z\,e^{-\gamma_m z}}{mx}\Bigg\{\delta(z-mx) + m\gamma_m[1+\gamma_m(x-z)]\theta(x-z)$$

$$+\sum_{i=1}^{m}\sum_{k=0}^{i}(-1)^k C_m^i C_i^k \frac{\gamma_m^i[z-(m-i+k)x]^{(i-1)}}{(i-1)!}\theta(z-(m-i+k)x)\Bigg\},$$

where $z = l/\beta$ and $C_m^i = \dfrac{m!}{i!(m-i)!}$.

For monofunctional initiator, this expression becomes much simpler

$$f_w(z,x) = \frac{z\,e^{-\gamma_m z}}{x}\{\delta(x-z) + \gamma_m^2(x-z)\theta(x-z)\}. \tag{3.21}$$

Figure 3.7 shows typical MWD curves for polymer obtained under conditions of transfer to monomer for multifunctional initiator. As seen from this figure, at m higher than 2 and not very high γ_m, the MWD can be multimodal.

3.3 Continuous Polymerization

The main distinction of polymerization in continuous stirred tank reactors (CSTRs) from batch/plug-flow polymerization is the distribution of reactor residence times. In a single CSTR, the distribution of molecules in residence time (i.e., the probability to be in the reactor during time period t is given by [37, 38]

$$f(t) = \frac{1}{\tau}\exp\left(-\frac{t}{\tau}\right).$$

Here, τ is the average residence time defined by the reactor volume and flow rates of the components. If the volume of a separate reactor is V and the volumetric rate is q, then $\tau = V/q$. In the processes with short-living active species, for example, in free radical polymerization, this distinction does not affect MWDs. On the contrary, in the processes where the average lifetime of active species is large compared to the mean residence time τ, the transition from batch polymerization to CSTR gives rise to the distribution in chain growth times. This, in turn, results in crucial changes in the MWD, even in the absence of any chain transfer. The chain transfer reactions should lead to even more significant differences.

Obviously, the increase in the number of reactors in a series approaches the process to the batch one because residence time distribution becomes narrower; thus, in the kth reactor [37]

$$f_k(t) = \frac{(t/\tau)^{k-1}\exp(-t/\tau)}{k!\tau}.$$

However, in commercial processes the number of reactors is limited (usually three to five), thus the difference between MWD of batch and continuous processes may be considerable. These conclusions were qualitatively formulated long ago by Denbigh [39]. Nevertheless, they seem to be underestimated in spite of a wide use of CSTR in commercial polymer synthesis. Till the late 1980s, such processes remained poorly studied except for modeling of few particular polymerization systems [40, 41]. There were also general indications that the polymer formed in a CSTR has the most probable Flory chain length distribution with the polydispersity index equal to 2 instead of the very narrow Poisson distribution, characteristic of polymer produced in living batch polymerization [6, 38]. Later on, papers [42–47] reported results of modeling of the structure and MWD of living polymers formed in CSTRs under various conditions. Some publications took into account chain transfer to solvent [48], impurity [44], or modifier [49, 50]. However, only single CSTR was considered. A detailed theoretical study of nonterminating polymerization in CSTRs, including chain transfer to solvent and polymer, was undertaken by the author [51–53].

A series of r identical continuous stirred tank reactors is considered

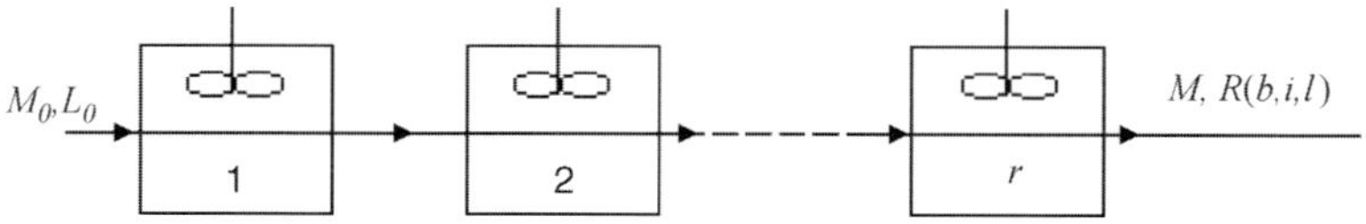

Monomer, solvent, initiator, and other components, if appropriate, are fed into the first reactor at a constant volumetric rate. The concentrations of monomer, solvent, and initiator in the input stream are M_0, S, and I_0, respectively.

The kinetic scheme of polymerization in CSTR is the same as for batch process, but differential equations for the concentrations of monomer M_k and macromolecules $R_k(b,i,l)$ (b, i, and l denote, respectively, the number of branches, the number of active centers, and the number of monomer units) in the kth reactor of the series comprise additional terms related to enter and exit from the reactor. In the most general case when chain transfer reactions to monomer, polymer, and solvent are possible, the corresponding equations take the following form [53]:

$$\frac{dM_k}{dt} = \frac{M_{k-1}-M_k}{\tau} - k_{p,k}M_k I_0, \tag{3.22}$$

$$\begin{aligned}\frac{dR_k(b,i,l)}{dt} &= \frac{R_{k-1}(b,i,l)-R_k(b,i,l)}{\tau} - ik_{p,k}M_k\frac{\partial R_k(b,i,l)}{\partial l}\\ &\quad + k_{tp,k}lI_0[R_k(b-1,i-1,l)-R_k(b,i,l)]\\ &\quad + (k_{tp,k}(M_0-M_k)+k_{ts,k}S+k_{tm}M)[(i+1)R_k(b,i+1,l)-iR_k(b,i,l)]\\ &\quad + \frac{I_0}{\tau}\delta_{k,1}\delta_{b,0}\delta_{i,1}\delta_{l,1} + (k_{ts,k}SI_0+k_{tm}MI_0)\delta_{b,0}\delta_{i,1}\delta_{l,1}.\end{aligned} \tag{3.23}$$

The first term on the right-hand side of Eq. (3.23) describes the input and output from the kth reactor, the second describes the increase in chain length due to propagation, the next two are responsible for the variation in the number of branching b and active centers i due to chain transfer, and two last terms describe the formation of short ($l=1$) linear ($b=1$) chains due to instantaneous initiation and chain transfer to solvent. In the general case, the temperature in different reactors can be varied; therefore, the reactivities also bear index k.

This set of equations gives the possibility to calculate the dynamic behavior of a CSTR and, what is more important, the characteristics of the steady state. In contrast to free radical polymerization where multiple steady states and even oscillating regimes can exist owing to the second-order chain termination reaction and the gel effect [54], Eqs. (3.22) and (3.23) have only one stationary solution. Hereinafter, steady-state parameters of the MWD are discussed.

3.3.1 MWD of Living Polymers Formed in CSTR

First, the simplest case of ideal living polymerization (the absence of termination and any chain transfer reaction) is considered. The kinetic scheme of such process contains only propagation step 1.1. The polymer formed in such a case is linear ($b=0$), and each macromolecule contains one active center (monofunctional initiator)

$$R_k(b,i,l) = R_k(l)\delta_{i,1}\delta_{b,0}.$$

Then, Eq. (3.23) simplifies to

$$\frac{dR_k(l)}{dt} = \frac{R_{k-1}(l)-R_k(l)}{\tau} - k_{p,k}M_k\frac{\partial R_k(l)}{\partial l} + \frac{I_0}{\tau}\delta_{k,1}\delta(l) = 0. \tag{3.24}$$

Equation (3.24) shows that changing temperature in the reactors and, hence, reactivities of propagation, results in the possibility, at least in principle, of obtaining a desired conversion in the kth reactor, $x_k = 1 - M_k/M_0$,

$$x_k = \frac{\varphi_k + x_{k-1}}{\varphi_k + 1}, \tag{3.25}$$

where $\varphi_k = k_{p,k} I_0 \tau$.

Using the stationary solution of Eq. (3.24) makes it easy to calculate the average DPs in the kth reactor:

$$\bar{P}_{n,k} = \beta x_k, \quad \bar{P}_{w,k} = \frac{2\beta}{x_k} \sum_{j=1}^{k} x_j (x_j - x_{j-1}). \tag{3.26}$$

Here, as before, $\beta = M_0/I_0$ is the "theoretical" DP of living polymer. The chain length distribution calculated from stationary Eq. (3.24) is described by the expression

$$R_k(l) = \frac{(-1)^{k+1} I_0}{\beta} \left(\prod_{j=1}^{k} \varepsilon_j \right) \sum_{j=1}^{k} \frac{\exp(-\varepsilon_j^l/\beta)}{\prod_{\substack{n=1 \\ n \neq j}}^{k} (\varepsilon_j - \varepsilon_n)}, \tag{3.27}$$

where $\varepsilon_j = (x_j - x_{j-1})^{-1}$.

As mentioned, in a single CSTR the MWD represents Flory distribution $R(l) = (I_0/\beta)\, e^{-l/\beta x}$ with the number average DP, $\bar{P}_n = \beta x_1$ and the polydispersity index equal to 2. The behavior of MWD at increasing number of reactors depends on the distribution of conversions over reactors. The conversion mode that gives the narrowest MWD at a given conversion at the outlet of the last reactor, $x_r = X$, may be determined from the condition $\left.\frac{\partial(\bar{P}_{w,r}/\bar{P}_{n,r})}{\partial x_k}\right|_{x_r = X} = 0$. This condition and expressions (3.26) suggest that in this mode the increment of conversion in each reactor should be equal, $x_k - x_{k-1} = X/r$, that is, $x_k = X(k/r)$ (the so-called mode of equal distribution of conversions). In this case, Eq. (3.27) considerably simplifies,

$$R_k(l) = \frac{I_0}{\beta x_1} \cdot \frac{\exp(-l/\beta x_1)}{(k-1)!} \cdot \left(\frac{l}{\beta x_1} \right)^{k-1},$$

and PDI is equal to $\bar{P}_{w,r}/\bar{P}_{n,r} = 1 + r^{-1}$. Thus, it depends only on the number of reactors and is independent of final conversion.

Under nonstationary conditions, in particular at periodic change of input concentration, it is possible to obtain even more narrow MWD [42–44] than $1 + r^{-1}$, obtained for stationary process. However, such regime, similar to the regime of equal distribution of conversions, is hardly practicable on a commercial scale. More realistic is the isothermal regime (temperature in all reactors is equal). For such a case, the conversion in the kth reactor is described by a known function [39]

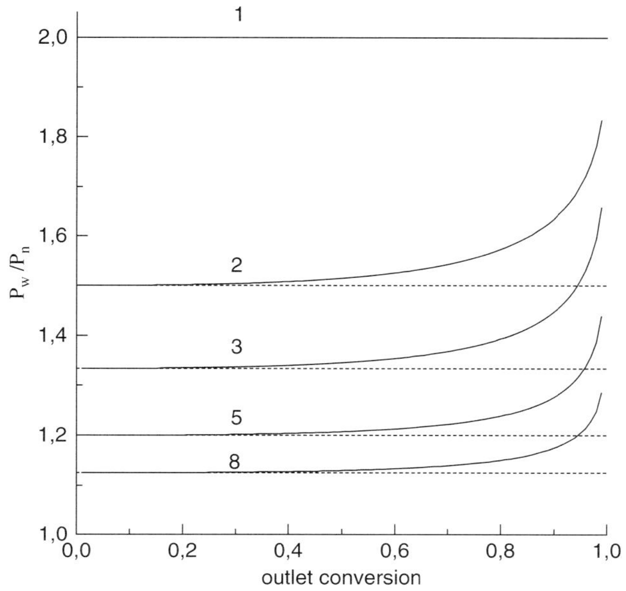

Figure 3.8 Dependence of polydispersity index on final conversion for the regime of equal distribution of conversion (dashed line) and isothermal regime.

$$x_k = 1-(\varphi + 1)^{-k}, \tag{3.28}$$

and Eq. (3.26) for the number and weight average DPs transforms into

$$\bar{P}_{n,k} = \beta x_k, \quad \bar{P}_{w,k} = 2\beta\left\{1-\frac{1+(\varphi+1)^{-k}}{\varphi+2}\right\}. \tag{3.29}$$

Equation (3.29) shows that in isothermal regime, contrary to the regime of equal distribution of conversions, the polydispersity does depend on the outlet conversion. However, as seen from Figure 3.8, the difference becomes noticeable only at high conversions at the outlet of the series.

Thus, contrary to batch polymerization, obtaining narrow MWDs in continuous polymerization is rather a complicated problem even in the absence of secondary reactions. The narrowing MWD is favored by the increasing number of reactors (this is trivial because the process approaches the batch polymerization) and, in the isothermal regime, maximum possible decrease in conversion in the first reactor. For example, in order to obtain polymer with $\bar{P}_{w,r}/\bar{P}_{n,r} < 1.3$-$1.4$ at the outlet conversion at least 90%, it is necessary to maintain the conversion in the first reactor x_1 not higher than 50% and use at least four to five reactors (see Figure 3.9).

An important practical consequence of MWD broadening compared to batch polymerization is the appearance of a high molecular weight fraction, that is, the polymers with DPs several times higher than the average values. This is also the result of the residence time distribution. In real commercial reactors, the fraction of high

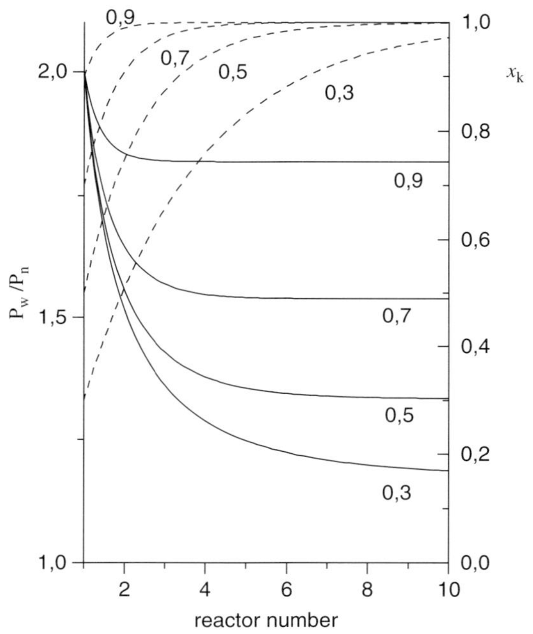

Figure 3.9 Dependence of PDI (solid line) and conversion (dashed line) on the reactor number for isothermal regime at the given conversion in the first reactor (indicated in the plot).

molecular weight polymer will be even higher due to imperfect stirring, existence of stagnant zones, and so on. A considerable contribution to the MWD broadening is made by chain transfer reaction.

3.3.2
Chain Transfer to Solvent

Using stationary Eqs. (3.22) and (3.23), the method of statistical moments enables to obtain the following expression for the number average degree of polymerization in the isothermal regime:

$$\begin{aligned} \bar{P}_{n,k} &= \frac{\beta x_k}{1 + k\gamma_s\varphi}, \\ \bar{P}_{w,k} &= \frac{2\beta}{1-\gamma_s}\left\{\frac{1-(1-x_k)(1+\gamma_s\varphi)^{-1}}{x_k(\gamma_s\varphi+\gamma_s+1)} - \frac{2-x_k}{\varphi+2}\right\}. \end{aligned} \tag{3.30}$$

Here, similar to the batch process, $\gamma_s = k_{t,k}S/k_{p,k}I_0$ represents the intensity of chain transfer to solvent in the kth reactor, and the conversions x_k are given by formula (3.28).

The comparison of the average DPs of polymer formed in batch [22] and continuous [52, 53] polymerization at equal conversions shows that in a CSTR $\bar{P}_n$

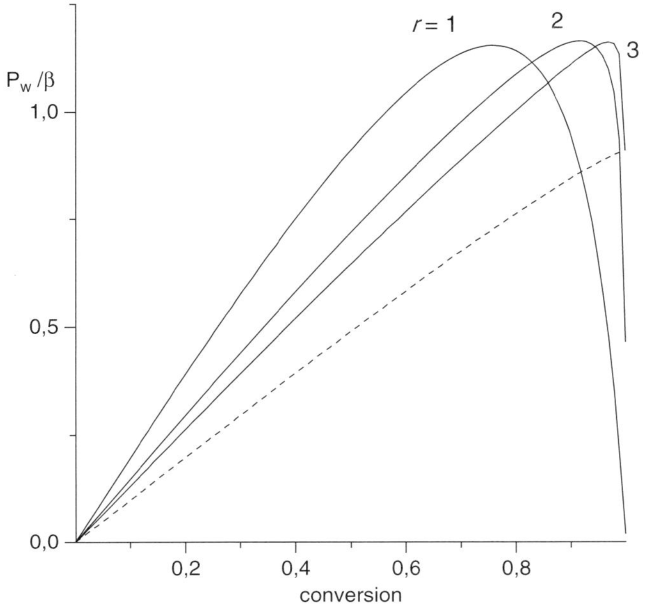

Figure 3.10 Average degree of polymerization versus conversion at the outlet for $\gamma_s = 0.1$ and different numbers of reactors.

is always smaller, especially at very high conversions. At the same time, at moderate values of γ_s and final conversion X, the weight average DP can be even greater than that formed in batch polymerization (Figure 3.10). For three to five reactors in a series and $\gamma_s < 0.5$, the average DPs are the highest if $X = 80$–90%.

The polydispersity index in the first reactor is always equal to 2, independent of conversion and intensity of chain transfer. This is in contrast to batch polymerization, where $\bar{P}_w/\bar{P}_n$ can exceed 4. The MWD of the polymer formed in the first reactor is the most probable Flory distribution, as for living polymerization in CSTR, but with a considerably smaller $\bar{P}_n$. Typical MWD curves of the polymer formed in CSTRs are shown in Figure 3.11.

Due to chain transfer reactions, the concentration of reactive direct chains, that is, the macromolecules formed via primary initiation, decreases. This worsens the conditions for subsequent modification reactions, for example, functionalization. The weight fraction q_w of active polymer in the kth reactor is

$$q_{w,k} = \frac{(1 + \gamma_s\varphi)^{-k} - (1 - x_k)}{(1 - \gamma_s)x_k}. \tag{3.31}$$

Equation (3.31) and Figure 3.12 show that at $X = 0.9$ a considerable part of macromolecules in the output flow is already inactive at small $\gamma_s \sim 0.1$. For example, for one CSTR $q_w = 0.53$ and for three reactors $q_w = 0.77$. In this case, it is possible to obtain 90% weight fraction of active polymer at 90% conversion only at $\gamma_s < 0.0123$ when one reactor is used and $\gamma_s < 0.04$ for a series consisting of three reactors.

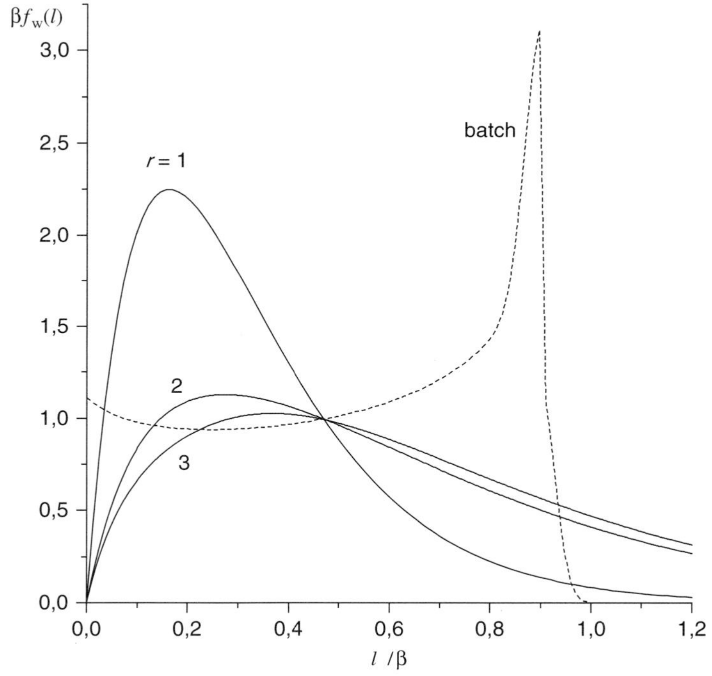

Figure 3.11 Comparison of MWDs formed in continuous and batch processes at 90% conversion. $\gamma_s = 0.5$.

Therefore, in nonterminating polymerization the use of CSTR instead of batch process greatly enhances the role of chain transfer reactions. This is explained by much smaller operating concentrations of the monomer even in the first reactor. In some cases, this might be useful, for example, the use of such weak chain-transferring solvent as ethylbenzene (which has almost no effect on batch polymerization of styrene) in continuous process, which enables one to obtain high molecular weight polystyrene with fourfold decreased amount of butyllithium [48]. Similarly, the control of polybutadiene molecular weight by the addition of 1,2-butadiene is more pronounced in CSTR than in batch polymerization [40]. And, of course, the use of CSTR is favorable for manufacturing low molecular weight polymers (e.g., nonfunctional liquid rubbers) in chain-transferring solvents. On the other hand, for the synthesis of high molecular weight polymers even relatively weak chain transfer can cause certain problems.

3.3.3 Chain Transfer to Monomer

For isothermal conditions, the following expressions were obtained for the average degree of polymerization in the kth reactor

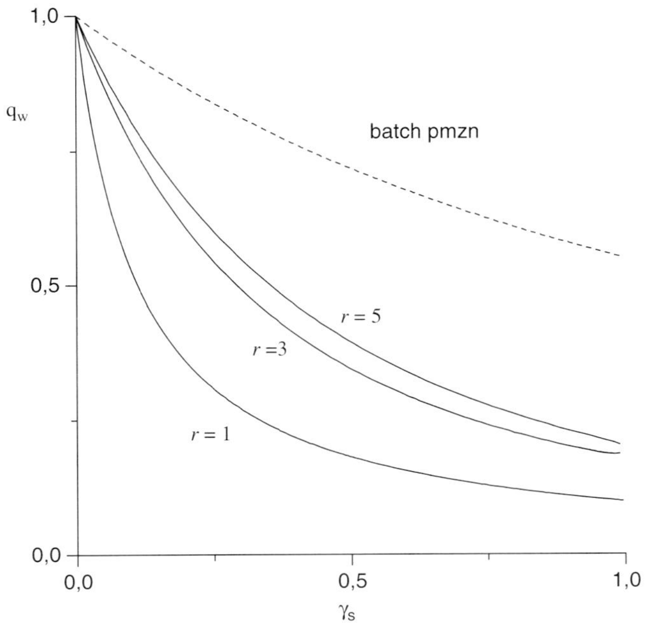

Figure 3.12 Dependence of weight fraction of active chains on the intensity of chain transfer to solvent γ_s. Final conversion 90%.

$$\bar{P}_{n,k} = \frac{\beta x_k}{1+\gamma_m x_k}, \quad \bar{P}_{w,k} = \frac{2\beta}{\gamma_m}\left\{1 - \frac{\varphi}{x_k}\sum_{j=1}^{k}\frac{(1-x_j)}{\prod_{n=j}^{k}(1+\gamma_m\varphi(1-x_n))}\right\}. \tag{3.32}$$

Here, as in the batch process, $\gamma_m = \beta\, k_{tm}/k_p$ represents the intensity of chain transfer to monomer.

Similar to transfer to solvent, the polydispersity index in the first reactor is equal to 2; the increasing number of reactors makes the MWD narrower.

By comparing Eqs. (3.30) and (3.32), one can conclude that at equal final conversions and intensity of transfer $\gamma_m = \gamma_s$, the effect of chain transfer to monomer on $\bar{P}_n$ and $\bar{P}_w$ is not so pronounced as in the case of transfer to solvent (see Figure 3.13). For example, for a single reactor operating at 90% conversion and at $\gamma_m = \gamma_s = 0.5$, the number average degree is 0.16β for chain transfer to solvent and 0.62β for chain transfer to monomer. One more distinction of these two reactions is that $\bar{P}_n$, in contrast to chain transfer to solvent, is independent of the number of reactors in the series being defined only by final conversion X.

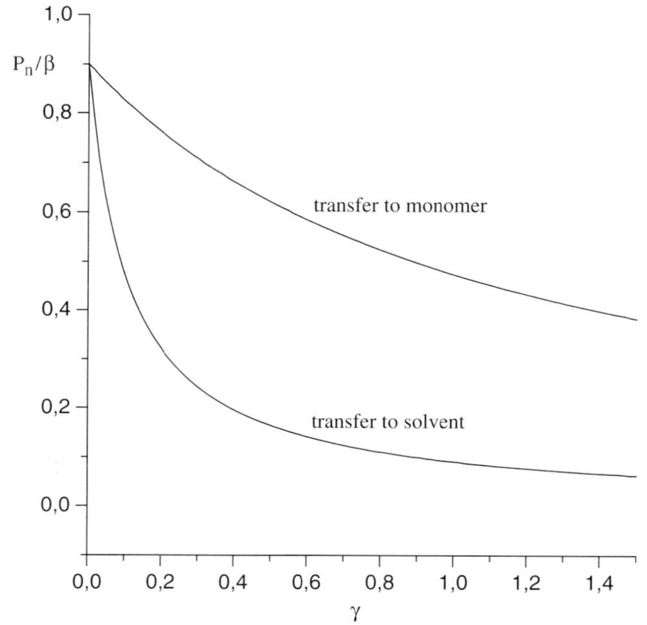

Figure 3.13 Comparison of the effect of chain transfer to monomer and solvent in a single CSTR. Outlet conversion $X = 0.9$.

3.3.4
Chain Transfer to Polymer

The most dramatic difference between batch and CSTR polymerization can be observed under conditions of chain transfer to polymer. Similar to batch polymerization, this reaction does not change $\bar{P}_n$, but leads to the formation of branched polymer. The average number of branching per macromolecules, ϱ_p, calculated from Eq. (3.23) is

$$\varrho_{p,k} = \frac{\gamma_p(k\varphi - x_k)}{1 + k\gamma_s\varphi + \gamma_m x_k},$$

where $\gamma_p = \beta k_{tp}/k_p$ characterizes the intensity of chain transfer to polymer.

For a single CSTR, the dependence of ϱ_p on conversion and γ_p is compared in Figure 3.14 with batch polymerization. As seen, at equal conversions and intensity of chain transfer to polymer, ϱ_p is much higher in CSTRs. This difference is the most significant for a single reactor. Nevertheless, in continuous polymerization, even for three reactors in the series, ϱ_p is 1.5–2 times higher than in batch polymerization.

The relationship between average lengths of backbone and side branches L_b and L_s also differs from that of batch polymerization. While in batch polymerization, side branches are considerably shorter than main chains as shown in Section 3.1, in the first CSTR of the series $L_b = L_s$, because here all active centers, whether in the main or

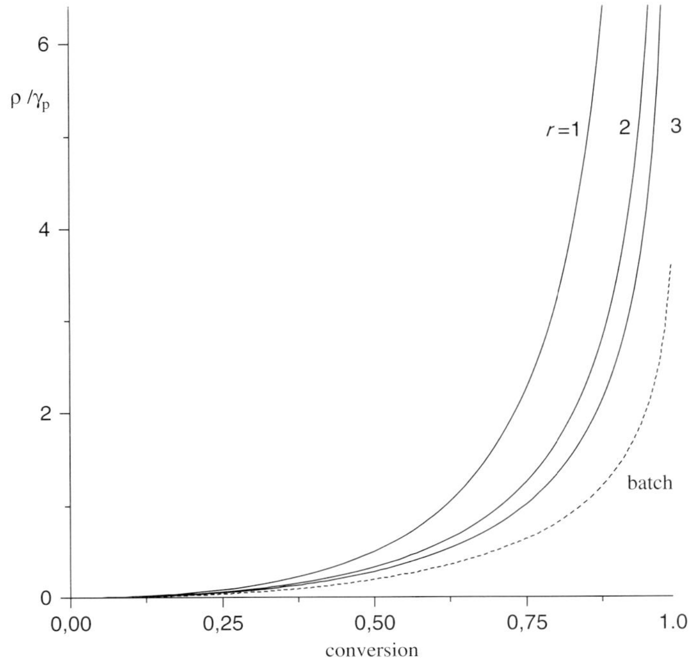

Figure 3.14 Dependence of the reduced average number of branchings arising due to chain transfer to polymer on conversion.

side chains, have equal probability of growth. In subsequent reactors, the ratio L_b/L_s decreases but still remains higher than in batch polymerization.

The weight fraction of polymer in side branches, $\omega_s = W_s/M_0x$, for the first reactor is

$$\omega_s = \frac{\gamma_p \varphi x}{1 + \varphi(\gamma_p x_1 + \gamma_s)}.$$

As follows from this expression, ω_s is much higher than in batch polymerization [26], and at high conversions main portion of polymer is located in side branches even at small γ_p (see Figure 3.15).

The difference between batch and continuous polymerization is the most pronounced for MWD. The weight average degree of polymerization in the first reactor, calculated using the method of statistical moments, is

$$\bar{P}_{w,1} = \frac{2\beta x_1[1 + 2\varphi(\gamma_p x_1 + \gamma_s)]}{[1 + \varphi(\gamma_p x_1 + \gamma_s)][1 - 2\varphi(\gamma_p x_1 - \gamma_s)]}. \tag{3.33}$$

The dependence of the polydispersity index $\bar{P}_w/\bar{P}_n$ on conversion in the first reactor is shown in Figure 3.16.

While in batch polymerization the increase in $\bar{P}_w/\bar{P}_n$ by the end of the process is only 10–20% even at a relatively high $\gamma_p = 1$, the polydispersity in continuous

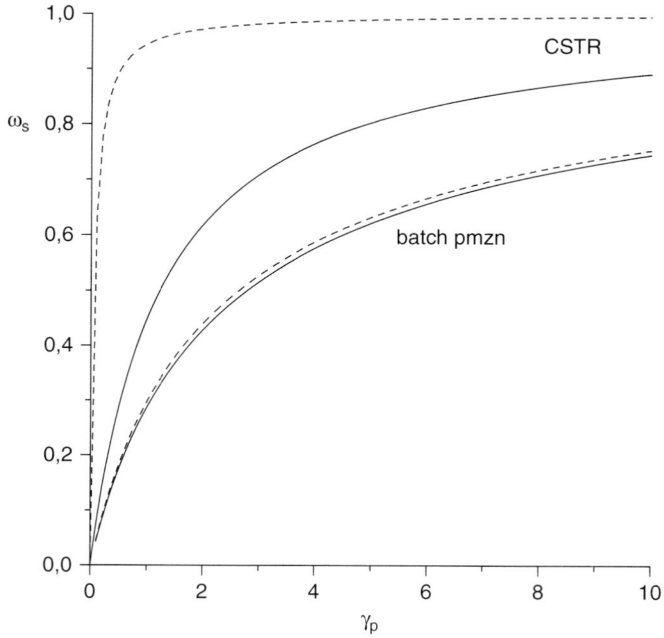

Figure 3.15 Weight fraction of polymer in side branches versus γ_p. Solid lines $x = 0.9$, dashed lines $x = 0.95$.

processes can achieve very high values even at small intensities of chain transfer ($\gamma_p < 0.01$). Moreover, the analysis of Eq. (3.33) shows that at certain values of parameters and conversions $\bar{P}_w$ goes to infinity, which means gelation. Equation (3.33) shows that the critical conversion of gel formation in the first reactor is to be calculated from the condition

$$x_{1,cr} = \frac{2\gamma_s - 1 + [(2\gamma_s - 1)^2 + 8\gamma_p]^{1/2}}{4\gamma_p} \tag{3.34}$$

or, in the absence of chain transfer to solvent,

$$x_{1,cr} = \frac{(8\gamma_p + 1)^{1/2} - 1}{4\gamma_p}.$$

A detailed analysis shows that most dangerous from the viewpoint of gelation is the first reactor of the series. If $\bar{P}_w$ of the polymer formed in the first reactor is finite, it cannot go to infinity in subsequent reactors. The occurrence of chain transfer to solvent (or spontaneous transfer) reduces the danger of gelation and, as follows from Eq. (3.34), at $\gamma_s > \gamma_p$ no gelation can occur at any conversion. In this regard, it should be noted that most patents propose to add some quantity of chain transfer agent to suppress gel formation.

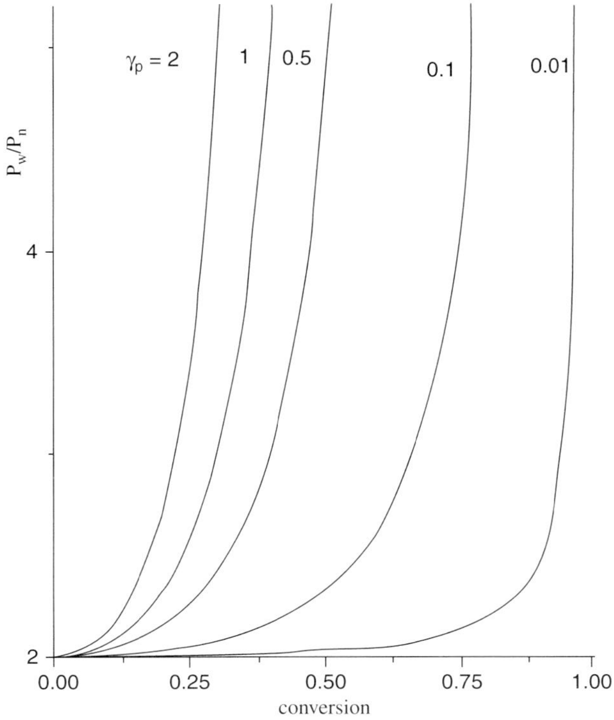

Figure 3.16 Dependence of polydispersity index on conversion in a single CSTR.

3.4 Conclusions

This chapter distinctly shows that the results obtained in laboratories in batch polymerization cannot be uncritically applied to continuous processes. Some complications can arise even for such simple systems as living polymers, but the difference in the structure of macromolecules formed in batch and continuous polymerization is most pronounced for chain transfer to polymer. This is explained by the different conditions of polymer formation. In batch polymerization, the probability of branching increases with increasing conversion; therefore, the main part of side branches is formed at high conversion and thus cannot grow to high lengths owing to lack of free monomer. On the contrary, in a CSTR the monomer concentration and conversion are constant, and each active center, independent of its position in the chain, has equal probability of growth. In addition, due to the distribution in residence time, the degree of polymerization of a small part of macromolecules, with their age much higher than τ, is much larger than $\bar{P}_n$. As a result, chain transfer will occur mainly on these macromolecules. They will contain a large number of active centers that favors gelation.

The appearance of an increasing number of processes with long-living active species, such as generation of Ziegler–Natta or lanthanide-based catalysts or new cationic and certain free radical polymerizations with controlled chain propagation, makes the results obtained especially important for polymerization engineering.

Nomenclature

B	number of branches in a macromolecule
$C_i = k_{ti}/k_p$	constant of chain transfer to initiator
$C_m = k_{tm}/k_p$	constant of chain transfer to monomer
$C_s = k_{ts}/k_p$	constant of chain transfer to solvent
$f_w(l)$	weight fraction of polymer of length l
i	number of active centers in a macromolecule
I_0	initiator concentration
k_i	rate constant of initiation
k_p	rate constant of propagation
k_{ri}	rate constant of reinitiation
k_{sp}	rate constant of spontaneous chain transfer
k_{tm}	rate constant of chain transfer to monomer
k_{tp}	rate constant of chain transfer to polymer
k_{ts}	rate constant of chain transfer to solvent
l	number of monomer units in a macromolecule
L_b	average length of backbone
L_s	average length of side branches
m	initiator functionality
M	monomer concentration
M_0	initial monomer concentration
$P(l)$	concentration of dead macromolecules with l monomer units
$\bar{P}_n$	number average degree of polymerization
$\bar{P}_w$	weight average degree of polymerization
$\bar{P}_z$	z-average degree of polymerization
r	number of reactors in a series of CSTR
$R_k(b,i,l)$	macromolecule with b branching, i active centers and l monomer units
$R(l)$	concentration of growing macromolecules with l monomer units
S	solvent concentration
$x = \dfrac{M_0 - M}{M_0}$	monomer conversion
x_k	conversion in the kth reactor
X	conversion at the outlet from the series of CSTR
$\alpha = \dfrac{k_{ts} S}{k_p M_0}$	parameter characterizing rate of chain transfer to solvent
$\beta = M_0/I_0$	degree of polymerization of ideal living polymer at full conversion

$\gamma_m = \beta C_m$	intensity of chain transfer to monomer
$\gamma_s = \alpha\beta$	intensity of chain transfer to solvent
$\gamma_p = \beta C_p$	intensity of chain transfer to polymer
$\varphi = k_p I_0 \tau$	effective parameter defining monomer conversion in a CSTR
ϱ_m	average number of branching per monomer unit
ϱ_p	average number of branching per polymer chain
τ	average residence time in one reactor

References

1 Nanda, V.S. and Jain, R.K. (1963) *J. Chem. Phys.*, **39**, 1363.
2 Bresler, S.E., Korotkov, A.A., Mosevitskii, M.I., and Poddubnyi, I.Y. (1958) *Zh. Tekh. Fiz.*, **28**, 114.
3 For the description of the program, see http://www.cit-wulkow.de.
4 Wulkow, M. and Deuflhard, P. (1995) *Mathematik in der Praxis*, Springer, Heidelberg.
5 Wulkow, M. (1992) *IMPACT Comput. Sci. Eng.*, **4**, 153.
6 Szwarc, M. (1968) *Carbanions, Living Polymers and Electron Transfer Processes*, Interscience, New York.
7 Flory, P.J. (1940) *J. Am. Chem. Soc.*, **62**, 1561.
8 Litt, M. (1962) *J. Polym. Sci.*, **58**, 429.
9 Glasse, M.D. (1983) *Prog. Polym. Sci.*, **9**, 133.
10 Robertson, R.E. and Marion, L. (1948) *Can. J. Res.*, **26B**, 657.
11 Higginson, W.C.E. and Wooding, N.S. (1952) *J. Chem. Soc.*, 760.
12 Burton, R.E. and Pepper, D.C. (1961) *Proc. Rov. Soc.*, **A-263**, 58.
13 Kume, S. (1966) *Makromol. Chem.*, **98**, 120.
14 Nanda, V.S. and Jain, R.K. (1968) *J. Chem. Phys.*, **48**, 1858.
15 Nanda, V.S. and Jain, S.C. (1970) *Eur. Polym. J.*, **6**, 1605.
16 Livshits, I.A. and Podol'nyi, Yu.B. (1970) *Vysokomol. Soedin., Ser. A*, **12**, 2655.
17 Largo-Cabrerizo, J. and Guzman, J. (1979) *Macromolecules*, **12**, 526.
18 Yuan, C. and Yan, D. (1987) *Makromol. Chem.*, **188**, 341.
19 Litvinenko, G.I. and Arest-Yakubovich, A.A. (1996) *Prog. Polym. Sci.*, **21**, 335.
20 Barson, C.A. (1989) *Comprehensive Polymer Science*, vol. 3 (eds G. Allen and J.C. Bevington), Pergamon Press, Oxford, p. 171.
21 Bamford, C.H. (1988) *Encyclopedia of Polymer Science and Engineering*, vol. 13, 2nd edn (ed J.J. Kroshwitz), John Wiley & Sons, Inc., New York, p. 708.
22 Litvinenko, G.I., Arest-Yakubovich, A.A., and Zolotarev, V.L. (1987) *Vysokomol. Soedin., Ser. A*, **29**, 732.
23 Kume, S., Saka, H., Takahashi, A., Nishikava, G., Hatana, M., and Kambara, S. (1966) *Makromol. Chem.*, **98**, 109.
24 Proni, A., Corno, C., Roggero, A., Santi, G., and Gandini, A. (1979) *Polymer*, **20**, 116.
25 De Chirico, A., Proni, A., Roggero, A., and Bruzzone, M. (1979) *Angew. Makromol. Chem.*, **79**, 185.
26 Litvinenko, G.I. andArest-Yakubovich, A.A.(1988)*Vysokomol. Soedin., Ser. A*,**30**,1218.
27 Flory, P.J.(1953)*Principles of Polymer Chemistry*,Cornell University Press,New York.
28 Jain, S.C. andNanda, V.S.(1977)*Eur. Polym. J.*,**13**,137.
29 Yuan, C. andYan, D.(1986)*Makromol. Chem.*,**187**,2641.
30 Yuan, C. andYan, D.(1988)*Eur. Polym. J.*,**24**,729.
31 Yuan, C. andYan, D.(1988)*Polymer*, **29**,924.
32 Litvinenko, G.I. andArest-Yakubovich, A.A.(1992)*Makromol. Chem. Theory Simul.*,**1**,321.
33 Arest-Yakubovich, A.A. andLitvinenko, G.I.(1993)*Makromol. Chem. Macromol. Symp.*,**67**,277.
34 Yan, D.(1986)*J. Macromol. Sci. Chem.*, **A 23**,129.

35 Litvinenko, G.I. (1996) Dissertation (Dr. Sci.). Karpov Institute of Physical Chemistry, Moscow.

36 Nanda, V.S. (1964) *Trans. Faraday Soc.*, **60**, 949.

37 Sebastian, D.H. and Biesenberger, J.A. (1983) *Principles of Polymerization Engineering*, John Wiley & Sons, Inc., New York.

38 Reichert, K.-H. and Moritz, H.-U. (1989) *Comprehensive Polymer Science*, vol. 3 (eds G. Allen and J.C. Bevington), Pergamon Press, Oxford, p. 327.

39 Denbigh, K.G. (1947) *Trans. Faraday Soc.*, **43**, 648.

40 Lander, B. and Bandermann, F. (1978) *Angew. Makromol. Chem.*, **71**, 101.

41 Ceausescu, E. (1981) *Not Cercetari in Domentiul Compusilar Macromoleculari*, Editura Acad. Republicii Socialista Romania, Bucharest.

42 Frontini, G.L., Elicabe, G.E., Couso, D.A., and Meira, G.R. (1986) *J. Appl. Polym. Sci.*, **31**, 1019.

43 Frontini, G.L., Elicabe, G.E., and Meira, G.R. (1987) *J. Appl. Polym. Sci.*, **33**, 2165.

44 Alassia, L.M., Frontini, G.L., Vega, J.R., and Meira, G.R. (1988) *J. Polym. Sci. Polym. Lett.*, **26**, 201.

45 Chang, C.C., Miller, J.W., and Schorr, G.R. (1990) *J. Appl. Polym. Sci.*, **39**, 2395.

46 Chang, C.C., Halasa, A.F., and Miller, J.W. (1993) *J. Appl. Polym. Sci.*, **47**, 1589.

47 Chang, C.C., Halasa, A.F., Miller, J.W., and Hsu, W.L. (1994) *Polym. Int.*, **33**, 151.

48 Priddy, D.B. (1994) *Adv. Polym. Sci.*, **111**, 67.

49 Puskas, J.E. (1993) *Makromol. Chem.*, **194**, 187.

50 Ermakov, I., De Clercq, R., and Goethals, E. (1994) *Macromol. Theory Simul.*, **3**, 427.

51 Litvinenko, G.I. and Arest-Yakubovich, A.A. (1989) *Teor. Osnovy Khim. Tekhnol. (in Russian)*, **23**, 469.

52 Litvinenko, G.I., Arest-Yakubovich, A.A., and Zolotarev, V.L. (1991) *Vysokomol. Soedin., Ser. A*, **33**, 1410.

53 Litvinenko, G.I. (1996) *Chem. Eng. Sci.*, **51**, 2471.

54 Sebastian, D.H. and Biesenberger, J.A. (1983) *Principles of Polymerization Engineering*, John Wiley & Sons, Inc., New York.

4
Computational Polymer Processing

Evan Mitsoulis

4.1
Introduction

4.1.1
Polymer Processing

Polymer processing is defined as the "engineering activity concerned with operations carried out on polymeric materials or systems to increase their utility" [1]. Primarily, it deals with the conversion of raw polymeric materials into finished products, involving not only shaping but also compounding and chemical reactions leading to macromolecular modifications and morphology stabilization, and thus to "value-added" structures [2]. The subject matter of polymer processing in general has been superbly laid out in the 1979 monograph by Tadmor and Gogos and has been updated significantly in the 2006 second edition [2]. Notable additions to the subject are the two very recent books by C.D. Han on rheology and processing of polymers [3]. They provide a wealth of information on rheological modeling of various polymer processes and an extensive list of relevant references. An excellent overview article has also appeared [4], where important up-to-date information on the polymer-processing industry is given, together with the most important developments in various polymer processes, and a most informative list of references and textbooks on the subject for further reading.

Synthetic polymers can be classified into two categories. *Thermoplastics* (by far the largest volume) can be melted by heating, solidified by cooling, and remelted repeatedly. Major types are polyethylene (PE), polypropylene (PP), polystyrene (PS), polyvinyl chloride (PVC), polycarbonate (PC), polymethyl methacrylate (PMMA), polyethylene terephthalate (PET), and polyamide (PA, nylon). *Thermosets* are hardened by the application of heat and pressure, owing to cross-linking, that is, the creation of permanent three-dimensional networks. They cannot be softened by heating for reprocessing. Bakelite, epoxies, and most polyurethanes are thermosets.

Modeling and Simulation in Polymers. Edited by P.D. Gujrati and A.I. Leonov

ISBN: 978-3-527-32415-6

This chapter is devoted to the processing of thermoplastics and its computations. Thermoplastics are usually processed in the molten state. Molten polymers have very high viscosity values and exhibit shear-thinning (*pseudoplastic*) behavior. As the rate of shearing increases, the viscosity decreases, owing to alignments and disentanglements of the long molecular chains. The viscosity also decreases with increasing temperature. In addition to the viscous behavior, molten polymers exhibit *elasticity*. Elasticity is responsible for a number of unusual rheological phenomena [5–7], including stress relaxation and normal stresses. Slow stress relaxation causes frozen-in stresses in injection-molded and extruded products. Normal stress differences are responsible for some flow instabilities during processing and also extrudate swelling, that is, the significant increase in cross-sectional area when a molten material is extruded out of a die.

The most important polymer processing operations are *extrusion* and *injection molding* (Vlachopoulos and Strutt, 2003). Extrusion is material-intensive and injection molding is labor-intensive. Both these processes involve the following sequence of steps: (a) heating and melting the polymer, (b) pumping the polymer to the shaping unit, (c) forming the melt into the required shape and dimensions, and (d) cooling and solidification. Other processing methods include calendering, roll coating, wire coating, fiber spinning, film casting, film blowing, blow molding, thermoforming, compression molding, and rotational molding (Figure 4.1).

4.1.2
Historical Notes on Computations

All these processes have been analyzed computationally in varying degrees of complexity. The mathematical analysis of the processes has followed the development of high-speed digital computers. Computational polymer processing was first tackled in the 1950s using analytical solutions for a few tractable problems, such as flows in channels with the power-law model of pseudoplasticity [8]. Then, the advent of digital computers saw numerical solution of simple problems in simple geometries, using mainly the finite difference method (FDM) due to its simplicity. With an increase in computer power, the 1970s saw the utilization of the more involved finite element method (FEM), which proved more capable in handling complicated geometries and boundary conditions (BCs) [9]. Two-dimensional (2D) flow problems in polymer processing were then handled for the first time [10]. The 1980s saw an overwhelming majority of computational works dealing with *viscoelasticity* (see book by Crochet *et al.* [11], and a review article by Crochet and Walters [12]). The efforts were directed toward overcoming the high Weissenberg number (Wi) problem (HWNP) [13], which did not allow solution of the viscoelastic models above a critical Wi number on the order of 1 (see below), where most phenomena were not very different from their inelastic counterparts. As expected, and due to a gigantic effort by many researchers around the world, this problem was eventually resolved successfully by using numerical schemes best suited for hyperbolic equations [14] in the late 1980s and early 1990s. Crochet's group and others then managed to reach highly viscoelastic numerical solutions in the range $1 < Wi < 100$ [15].

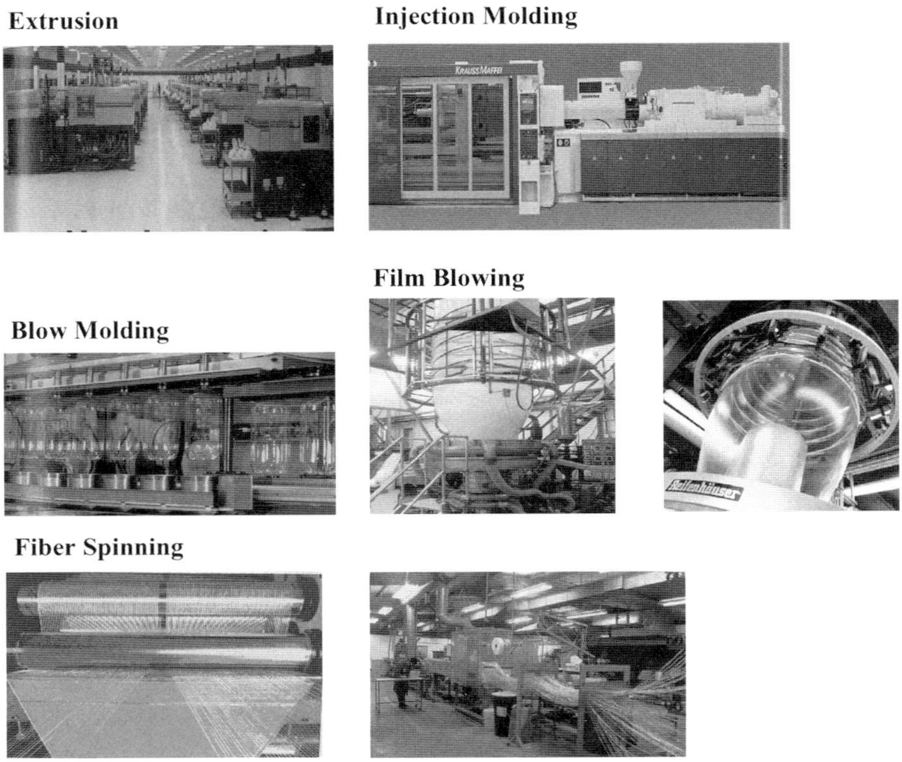

Figure 4.1 Photographs of various polymer processes. (From various issues of *Plastics Engineering*.)

From these significant efforts on viscoelasticity, the first 2D and then 3D FEM codes developed into commercial packages, with varying degrees of success, in the 1990s. Among these, some software particularly suited for polymer processing were POLYFLOW [16] and POLYCAD [17]. More recently, polymer processing software were made available from www.compuplast.com [18] and www.polyXtrue.com [19]. Other software were more generic in nature for viscous fluids, such as FIDAP, FLUENT, NEKTON, PHOENICS, and so on [20]. All these software use sophisticated mesh generation schemes and solvers for solving the governing differential equations along with appropriate boundary conditions. The FEM is the numerical method of choice for most packages, while inroads have also been made with the finite volume method (FVM) and the boundary element method (BEM).

This chapter reviews some major contributions regarding polymer melt flows that appear in polymer processing and discusses several issues (usually still unresolved) and their influence in polymer processing. Due to the vastness of the subject matter, the outlay of what follows is rather personal and subjective, but it is hoped that it will add some focus to the present state of affairs in polymer processing computations.

4.2 Mathematical Modeling

4.2.1 Governing Conservation Equations

In order to study polymer flows in processing equipment, it is essential to consider first the governing flow equations. The flow of incompressible fluids (such as polymer solutions and melts, at least in situations where they are considered as incompressible for pressures below 100 MPa) is governed by the conservation equations of mass, momentum, and energy [2, 6, 21], that is,

$$\nabla \cdot \boldsymbol{v} = 0, \tag{4.1}$$

$$\varrho \boldsymbol{v} \cdot \nabla \boldsymbol{v} = -\nabla p + \nabla \cdot \boldsymbol{\tau}, \tag{4.2}$$

$$\varrho c_{\mathrm{p}} \boldsymbol{v} \cdot \nabla T = k \nabla^2 T + \boldsymbol{\tau} : \nabla \boldsymbol{v}, \tag{4.3}$$

where $\boldsymbol{v}$ is the velocity vector, p is the scalar pressure, τ is the extra stress tensor, ϱ is the density, c_{p} is the heat capacity, k is the thermal conductivity, and T is the temperature.

The above system of conservation equations is usually called the *Navier–Stokes equations* in fluid mechanics.

4.2.2 Constitutive Equations

The above system of conservation equations is not closed for non-Newtonian fluids due to the presence of the stress tensor τ. The required relationship between the stress tensor τ and the kinematics (velocities and velocity gradients) must be given by appropriate *rheological constitutive equations*, and this is an eminent subject in theoretical rheology [5, 6, 22]. A cartoon showing the importance of stresses for polymers has been put forward in Figure 4.2, where a zebra is losing its stripes under stress. The implicit message is that polymers under stress exhibit unusual, unexpected, and counterintuitive behavior or that the wrong constitutive equation may give stresses totally inappropriate for a polymer undergoing deformation and flow.

For purely viscous fluids, the rheological constitutive equation that relates the stresses τ to the velocity gradients is the generalized Newtonian model [5, 6, 21] and is written as

$$\boldsymbol{\tau} = \eta(\dot{\gamma})\dot{\gamma}, \tag{4.4}$$

where $\dot{\gamma} = \nabla \boldsymbol{v} + \nabla \boldsymbol{v}^{\mathrm{T}}$ is the rate-of-strain tensor and $\eta(\dot{\gamma})$ is the apparent viscosity given in its simplest form by the power-law model [5]

$$\eta(\dot{\gamma}) = K\dot{\gamma}^{n-1}, \tag{4.5}$$

Figure 4.2 The stress state of a polymer melt is essential for any computation. A cartoon showing a zebra losing its stripes is a good analogue for the computation of a polymer melt with the wrong model.

where K is the consistency index and n is the power-law index (usually $0 < n < 1$, representing a degree of shear-thinning). Another popular model for viscosity computations, among others, is the Carreau model [5] given by

$$\eta(\dot{\gamma}) = \eta_\infty + \eta_0[1 + (\lambda_C\dot{\gamma})^2]^{\frac{n-1}{2}} \tag{4.6}$$

and the Cross model [23] given by

$$\eta(\dot{\gamma}) = \eta_\infty + \frac{\eta_0}{1 + (\lambda_C\dot{\gamma})^{1-n}}. \tag{4.7}$$

In the above equations, η_0 is the zero shear rate viscosity, η_∞ is the infinite shear rate viscosity, λ_C is a time constant, and n is again the power-law index. The magnitude $\dot{\gamma}$ of the rate-of-strain tensor is given by

$$\dot{\gamma} = \sqrt{\frac{1}{2}\mathrm{II}_{\dot{\gamma}}} = \sqrt{\frac{1}{2}(\dot{\gamma} : \dot{\gamma})}, \tag{4.8}$$

where $\mathrm{II}_{\dot{\gamma}}$ is the second invariant of the rate-of-strain tensor. The Carreau model describes well the shear-thinning behavior of polymer solutions and melts for all shear rates and exhibits two plateaus for low and for high shear rates, while for intermediate to high shear rates it represents well the power law. For example, for a low-density polyethylene (LDPE) melt, experimental data at different temperatures are fitted well with the Carreau model, as evidenced in Figure 4.3.

The effect of temperature on the viscosity is of primordial importance in polymer processing, where tight control of temperatures is required for a successful

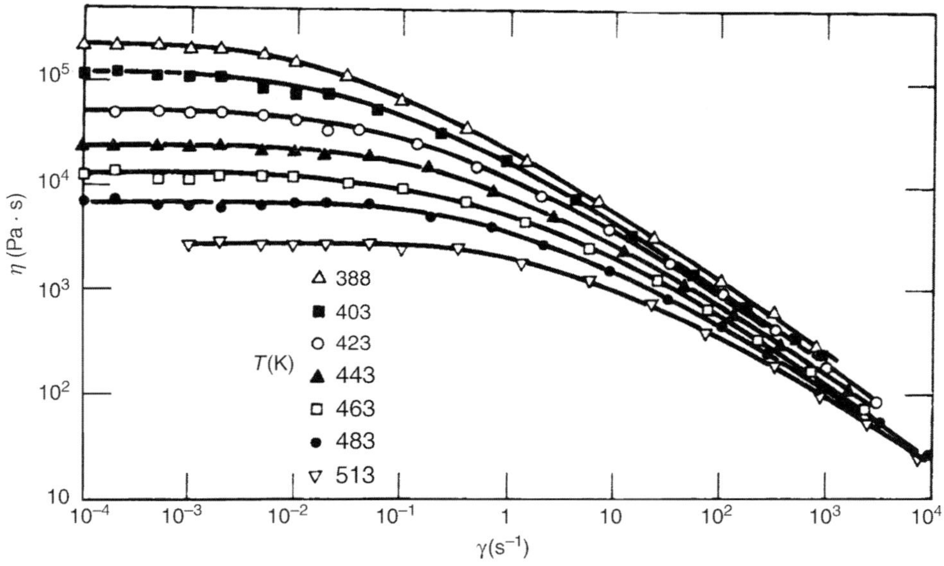

Figure 4.3 Non-Newtonian viscosity of an LDPE melt at several temperatures [5].

operation. The viscosity as a function of temperature is given by an exponential relationship, according to

$$\eta_T = \eta_0 \exp\left[-\beta_T (T - T_0)\right], \tag{4.9}$$

where β_T is a temperature-shift factor in the expression that relates viscosity to temperature and η_0 is the viscosity at a reference temperature, T_0. The values of β_T for polymers are usually in the range of 0.01–0.04/°C, but occasionally they may reach 0.1/°C or more for some polymers.

Another expression for the temperature-dependence of the viscosity is the Arrhenius law [21]:

$$\eta_T = \eta_0 \exp\left[\frac{E_a}{R_g}\left(\frac{1}{T} - \frac{1}{T_0}\right)\right], \tag{4.10}$$

where R_g is the ideal gas constant (= 8.13 J/(K mol)), E_a is the activation energy (J/mol), T is the absolute temperature (K), and T_0 is the absolute reference temperature (K).

Then combining Eqs. (4.8) and (4.9) yields

$$\beta_T = \frac{E_\alpha}{R_g\, T\, T_0}. \tag{4.11}$$

Non-Newtonian fluids (polymer solutions and melts) are rheologically complex materials, which exhibit both viscous and elastic effects, and are therefore called *viscoelastic* [6]. Regarding viscoelasticity, a plethora of constitutive equations exist with varying degrees of success and popularity. Standard textbooks on the subject [5, 6, 22, 24–26] list categories of these equations and their predictions in several types of flow and deformation. There are constitutive equations of differential type, of integral type, molecular models, and so on. From the *differential models*, eminent in the early 1980s were the upper convected Maxwell (UCM) and the Oldroyd-B models. Then, in the 1990s the Phan-Thien/Tanner (PTT) and the Giesekus models were among the most popular, while in the 2000s the "pom-pom" model has been the model of choice for computations. From the *integral models*, the K-BKZ model (from the initials of Kaye, Bernstein, Kearsley and Zapas) has been by far the most popular. This subject matter is more fully explored in Chapter 1.

As an example of a popular viscoelastic constitutive equation used in the past 25 years, which possesses enough degree of complexity so as to capture as accurately as possible the complex nature of polymeric liquids, we present here the K-BKZ integral constitutive equation with multiple relaxation times proposed by Papanastasiou *et al.* [27] and further modified by Luo and Tanner [28]. This is often referred to in the literature as K-BKZ/PSM model (from the initials of Papanastasiou, Scriven, Macosko) and is written as

$$\tau = \frac{1}{1-\theta}\int_{-\infty}^{t}\sum_{k=1}^{N}\frac{G_k}{\lambda_k}\exp\left(-\frac{t-t'}{\lambda_k}\right)H(\mathrm{I}_{C^{-1}},\mathrm{II}_{C^{-1}})\left[C_t^{-1}(t')+\theta C_t(t')\right]\,\mathrm{d}t', \tag{4.12}$$

where τ is the stress tensor for the polymer, λ_k and G_k are the relaxation times and relaxation moduli, respectively, N is the number of relaxation modes, θ is a material constant, C_t is the Cauchy–Green tensor, C_t^{-1} is the Finger strain tensor, and $\mathrm{I}_{C^{-1}}$, $\mathrm{II}_{C^{-1}}$ are its first and second invariants. The function H is a strain-memory (or damping) function, and the following formula was proposed by Papanastasiou *et al.* [27]:

$$H(\mathrm{I}_{C^{-1}},\mathrm{II}_{C^{-1}}) = \frac{\alpha}{(\alpha-3)+\beta\mathrm{I}_{C^{-1}}+(1-\beta)\mathrm{II}_{C^{-1}}}, \tag{4.13}$$

where α and β are nonlinear model constants to be determined from shear and elongational flow data, respectively. The θ-parameter (a negative number) relates the second normal stress difference $N_2=\tau_{22}-\tau_{33}$ to the first N_1 according to

$$\frac{N_2}{N_1} = \frac{\theta}{1-\theta}. \tag{4.14}$$

The linear viscoelastic storage and loss moduli, G' and G'', can be expressed as a function of frequency ω as follows:

$$G'(\omega) = \sum_{k=1}^{N} G_k \frac{(\omega\lambda_k)^2}{1+(\omega\lambda_k)^2}, \tag{4.15a}$$

$$G''(\omega) = \sum_{k=1}^{N} G_k \frac{(\omega\lambda_k)}{1+(\omega\lambda_k)^2}. \tag{4.15b}$$

These functions are independent of the strain-memory function, and only λ_k and G_k can be determined from dynamic data of the viscoelastic moduli. As an example, we give in Figure 4.4a and Table 4.1 the fitting of experimental G' and G'' data for another popular benchmark LDPE melt (the IUPAC-LDPE, melt A) [29], where it is shown that a spectrum of eight relaxation times, ranging from 10^{-4} to 10^{3} s is able to capture well the data at all frequencies.

The strain-memory function is derived from the first and second invariants of the Finger strain tensor. For simple shear flow, the strain-memory function is given as

$$H(\mathrm{I}_{C^{-1}}, \mathrm{II}_{C^{-1}}) = \frac{\alpha}{\alpha + \gamma^2}, \tag{4.16}$$

where γ is the shear strain. The strain-memory function in simple shear flow depends on α but not on β. This is expected since α is viewed as a *shear* parameter, while β is viewed as an *elongational* parameter.

The above constitutive equation has been successfully used for fitting the data for many polymer solutions and melts [27, 30]. Apart from fitting well the G' and G'', the model can give a good fit for other rheological data, as shown in Figure 4.4b, with the values of the model for parameters α, β, and θ given in Table 4.1. The full fitting involves the determination of the relaxation spectrum (parameters N, λ_k, and G_k) from experimental data on the storage and loss moduli, G' and G''. Then, the nonlinear parameters, α and β, are determined from shear and elongational data, in this case from shear viscosity, η_S, first normal stress difference, N_1, and uniaxial elongational viscosity, η_E. The value of θ is usually set to be a small negative number (around -0.1) according to experimental evidence [28]. Other extensional viscosities in planar extension, η_P, and in biaxial extension, η_B, are predicted by the model. These predictions can also be extended to transient effects for all rheological functions at different times [27].

This integral model has been used in numerical flow simulations for a number of flow problems more or less successfully (see Refs [28, 31–34]). A recent review [35] on the subject gives a list of problems solved with this model through numerical simulation, including many flows from polymer-processing operations. Other flows solved with a number of different constitutive equations can be found in a recent book on computational rheology [36].

4.2.3
Dimensionless Groups

Before proceeding with the boundary conditions, it is interesting to examine the relevant dimensionless numbers in polymer processing. The dimensionless groups are calculated at a reference temperature, here taken as the temperature of the process, T_0. As a characteristic length, it is usually assumed the smallest dimension, for example, in a capillary tube its radius, R. As a characteristic speed, it is usually assumed the average velocity defined by

$$V = Q/\pi R^2. \tag{4.17}$$

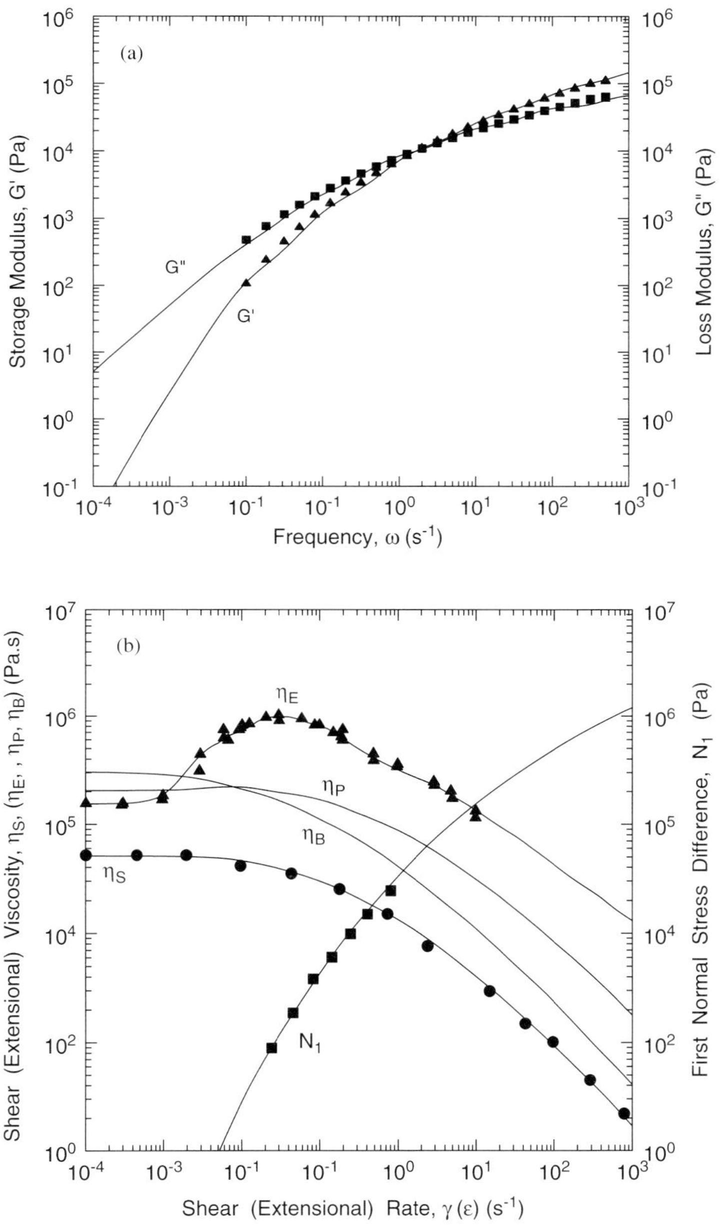

Figure 4.4 Rheological data and their best fit for the IUPAC-LDPE melt-A using the K-BKZ/PSM integral constitutive equation with eight relaxation modes and the data of Table 4.1 [32]. Symbols correspond to experimental data [29], solid lines correspond to their best fit.

Table 4.1 Material parameter values used in Eq. (4.12) for fitting the data of the IUPAC-LDPE (sample A) melt at 150 °C ($\alpha = 14.38$, $\theta = -1/9$) [28].

k	λ_k (s)	G_k (Pa)	β_k (−)
1	10^{-4}	1.29×10^5	0.018
2	10^{-3}	9.48×10^4	0.018
3	10^{-2}	5.86×10^4	0.08
4	10^{-1}	2.67×10^4	0.12
5	10^{0}	9.80×10^3	0.12
6	10^{1}	1.89×10^3	0.16
7	10^{2}	1.80×10^2	0.03
8	10^{3}	1.00×10^0	0.002

A characteristic apparent shear rate is then defined according to

$$\dot{\gamma}_{\mathrm{a}} = V/R \tag{4.18}$$

and a characteristic viscosity is given as a function of apparent shear rate and reference temperature, that is,

$$\bar{\eta} = \eta(\dot{\gamma}_{\mathrm{a}}, T_0). \tag{4.19}$$

For a power-law model, the characteristic viscosity can be found as

$$\bar{\eta}(T_0) = \frac{\tau}{\dot{\gamma}_a} = K\dot{\gamma}_{\mathrm{a}}^{n-1}, \tag{4.20}$$

where the material parameters K and n are calculated at T_0.

The relative importance of inertia forces in the equation of momentum is assessed by the Reynolds number, defined for Newtonian fluids by

$$Re = \frac{\varrho V D}{\mu}, \tag{4.21}$$

where D is the characteristic diameter ($= 2R$).

For power-law fluids, Boger and Walters [7] gives a generalized Re^*

$$Re^* = \frac{\varrho V^{2-n} D^n}{8^{n-1} K} \left(\frac{4n}{3n+1}\right)^n. \tag{4.22}$$

It is noted that for most polymer melt flows the Re number is usually small, in the range of 0.0001–0.01. Therefore, these flows are inertialess or creeping.

For *viscoelastic* fluids with a relaxation time λ, several dimensionless groups can be defined, but these can be seen as being equivalent [21]. For example, the Deborah number (De) is defined as

$$De = \frac{\lambda}{t_{\mathrm{p}}} = \lambda\dot{\gamma}, \tag{4.23}$$

where λ is a material relaxation time, t_p is a process relaxation time usually taken to be equal to $1/\dot{\gamma}$, and $\dot{\gamma}$ is a shear rate usually evaluated at the channel wall. The Weissenberg number (*Wi*, also written as *We* or *Ws*) is defined as

$$Wi = \lambda \frac{V}{R}. \tag{4.24}$$

The recoverable shear or stress ratio (S_R) is defined as

$$S_R = \frac{N_{1,w}}{2\tau_w}, \tag{4.25}$$

where $N_{1,w} = \tau_{11} - \tau_{22}$ is the first normal stress difference and τ_w is the shear stress, both evaluated at the channel wall. The equivalence is evident when we take

$$\dot{\gamma} = V/R, \quad N_{1,w} = \Psi_1 \dot{\gamma}^2, \quad \tau_w = \eta\dot{\gamma}, \quad \lambda = \Psi_1/2\eta, \tag{4.26}$$

where Ψ_1 is the first normal stress difference coefficient and η is the shear viscosity. The case of $De = Wi = S_R = 0$ corresponds to inelastic fluids ($\lambda = \Psi_1 = 0$), while it is understood that $De = Wi = S_R = 1$ corresponds to the elastic effects being as important as the viscous effects, and for $De = Wi = S_R > 1$ the elastic effects dominate the flow over the viscous effects.

The relative importance of surface tension effects (usually for polymer solutions) is assessed by the capillary number defined by

$$\mathrm{Ca} = \frac{\mu V}{\gamma}, \tag{4.27}$$

where γ is the surface tension. For very viscous fluids, such as polymer melts, the surface tension effects are negligible ($Ca \to \infty$), and the boundary terms, including a force balance with the capillary forces, can be set to zero.

The relative importance of each term in the energy equation is assessed through a variety of dimensionless groups [37, 38].

The Peclet number is defined by

$$Pe = \frac{\varrho C_p V D}{k}. \tag{4.28}$$

The Peclet number is a measure of convective heat transfer with regard to conductive heat transfer. High *Pe* values indicate a flow dominated by convection. From a numerical point of view, these flows are notorious because of instabilities that manifest themselves in the form of spurious oscillations in the temperature field. Special upwinding techniques must then be used to remedy the oscillations [37].

Another group related to *Pe* is the Graetz number defined by

$$\mathrm{Gz} = \frac{\varrho C_p V D^2}{kL} = Pe\frac{D}{L}, \tag{4.29}$$

where L is the axial length of the die. The Graetz number can be understood as the ratio of the time required for heat conduction from the center of the capillary to the wall and the average residence time in the capillary. As with *Pe*, a large value of *Gz* means that heat convection in the flow direction is more important than conduction toward the walls.

The Nahme number is defined by

$$Na = \frac{\beta_T \bar{\eta} V^2}{k}. \tag{4.30}$$

The Nahme number is a measure of viscous dissipation effects compared to conduction; hence, it is an indicator of coupling of the energy and momentum equations. For values of $Na > 0.1$–0.5 (depending on geometry and thermal boundary conditions), the viscous dissipation leads to considerable coupling of the conservation equations, and a nonisothermal analysis is necessary.

The relative importance of heat transfer mode at the boundaries is expressed in terms of the dimensionless *Biot* number defined by

$$Biot = \left(\frac{\partial T}{\partial r}\right)_w \frac{r_i}{(T_s - T_w)}, \tag{4.31}$$

where r_i is the local radius (gap), T_s is some temperature of the surroundings, and T_w is the local boundary wall temperature. A high value of *Biot* (>100) approaches isothermal conditions ($Biot = \infty$), while a low value of *Biot* (<l) describes poor heat transfer to the surroundings (nearly adiabatic case, for which $Biot = 0$). Usual *Biot* values in highly viscous flows inside dies or other processing equipment range between 10 and 100 [38].

For engineering calculations, the specific heat flux $\boldsymbol{q}$ is often described by the Nusselt number

$$Nu = \frac{h_T D}{k}, \tag{4.32}$$

where h_T is a heat transfer coefficient. However, as explained by Winter [38], the Nusselt number is not adequate for describing the wall heat flux in flows with significant viscous dissipation.

4.2.4
Boundary Conditions

The solution of the conservation Eqs. (4.1–4.3) and constitutive Eq. (4.4) (or Eq. (4.12)) is possible only after a set of boundary conditions has been imposed on the flow domain. Boundary conditions for flow analysis are highly dependent on the problem at hand, and as such they defy a complete description for all polymer processing applications. However, a rough guide encompassing most of the types of boundary conditions used in the past follows.

For *steady-state* problems, the set of equations is elliptic for viscous flows and elliptic-hyperbolic for viscoelastic flows. Elliptic problems have boundary conditions everywhere in the perimeter of the domain, while in hyperbolic problems boundary conditions are more difficult to determine and may need some degree of trial and error. This is especially true at the outflow boundaries, which are more often than not "artificial" or "computational" boundaries, arbitrarily set to reduce the computational domain.

For *unsteady-state* problems, the set of equations is parabolic due to time, and it also requires initial conditions at time $t=0$. Then, the solution proceeds in time until a steady-state is reached (if such exists).

The boundary conditions can be either "fixed" or "natural." These types refer, respectively, to the primary variables (velocities, pressures, and temperatures) or variables involving their derivatives (stresses, surface tractions, heat fluxes, etc.).

The usual *flow boundary conditions* in polymer processing are as follows:

(a) Along the domain *entry*, a fully developed velocity profile is imposed, according to the assumed constitutive model and corresponding to the apparent shear rate $\dot{\gamma}_a$, which is in turn related to the volumetric flow rate Q. If viscoelasticity is involved through some viscoelastic constitutive model, then the fully developed stress profiles have to be imposed as well.
(b) Along the *center line* (if one exists, as is the case in axisymmetric flows), and because of symmetry, the radial velocity component is set to zero, as well as the shear stresses.
(c) Along *solid walls*, usually the no-slip velocity boundary condition is imposed, which states that the velocity of the fluid is that of the boundary, that is, zero if the boundary is stationary or nonzero if the boundary is moving. In cases where the fluid slips at the wall, as is the case with some polymers [39], then a slip law has to be assumed based on measurements, which relates the tangential velocity to the tangential components of the stress tensor, while the normal velocity is set to zero.
(d) Along the domain *exit*, it is not clear what are the correct or physical boundary conditions for all situations. The best candidate appears to be the "open" or "free" boundary condition advocated by Papanastasiou *et al.* [40], which basically assumes an extrapolation of the governing equations to the artificial exit boundary. However, the vast majority of the computations assume a long enough domain, where they impose zero surface tractions and zero transverse velocity (assuming implicitly a fully developed profile at exit). For viscoelastic models, fully developed stresses based on the model at hand are used, unless there is a known force or velocity imposed at the exit boundary.
(e) In problems with *free surfaces*, and especially for polymer melts due to their very viscous character (zero surface tension assumed), zero surface tractions are imposed along with a *kinematic* boundary condition of no flow normal to the surface, that is, $\boldsymbol{n}\cdot\boldsymbol{v}=0$, where n is the unit outward normal vector to the surface.

For the *thermal boundary conditions*, the situation is even more difficult to describe due to the many types of thermal conditions used in polymer processing. However, in the vast majority of computations, the following set of boundary conditions is a good representation of what has been applied, based on the previous flow sets:

(a) Along the domain *entry*, a set (quite often constant) temperature profile is assumed, sometimes based on measurements or settings.

(b) Along the *center line* (if one exists, as is the case in axisymmetric flows), and because of symmetry, the heat flux is set to zero.

(c) Along *solid walls*, either isothermal walls (a set temperature) are assumed or adiabatic walls (heat flux set to zero) or a heat balance between the fluid and the solid boundary [37, 38]. In the latter case, a local *Biot* number can then be calculated, which is neither 0 (adiabatic walls) nor ∞ (isothermal walls) but may range between 10 and 100. Other types of wall thermal boundary conditions involve a known heat flux at the wall based on measurements of an effective heat transfer coefficient, but this is also tantamount to having a nonzero local *Biot* number.

(d) Along the domain *exit*, the same comments apply as above, regarding the flow BCs, with the best candidate being the "open" or "free" BC. However, the majority of computational problems have assumed a long enough domain so as to impose a zero heat flux at exit.

(e) In problems with *free surfaces*, a zero heat flux is usually imposed, that is, $\boldsymbol{n}\cdot\boldsymbol{q}=0$, or a heat balance according to

$$q_n = h_{\mathrm{T}}(T_{\mathrm{f}} - T_{\mathrm{a}}), \tag{4.33}$$

where h_{T} is a heat transfer coefficient to the ambient cooling medium (e.g., air) of temperature T_{a}, and T_{f} is the unknown free surface temperature. A more elaborate set of conditions has been applied to the process of film blowing, where turbulent air blowing outside and inside the film bubble requires a sophisticated coupling and interaction between the two fluids (blowing air and stretched polymer melt) [41–43].

For unsteady-state processes, and in particular in injection molding, initial conditions are also needed, while the boundary conditions applied inside the mold must invariably take into account the interaction between the mold and the flowing and solidifying polymer melt. The variety of such conditions makes a full description beyond the scope of this chapter. The interested reader can find useful information in the manuals of custom-made commercial computer software, such as MOLDFLOW [44], and in a recent thesis [20].

4.3
Method of Solution

As mentioned in Section 4.1, computational polymer processing was first tackled in the 1950s using analytical solutions, and since the 1960s numerical solutions have been made possible with a variety of numerical methods. The FEM is the numerical method of choice for most works and it is included in most commercial packages, which are able to handle polymer-processing flows. We will give here in some more detail the requirements for solving viscous and viscoelastic polymer flow problems with FEM.

Implementation of a finite element formulation for the mass and momentum equations uses the *primitive variables approach*, that is, velocities and pressure (in 2D, this is called *u-v-p formulation*). For viscoelastic formulations, the stresses are also needed as discrete variables, and this leads to a *mixed variable approach* (called *u-v-p-τ formulation* or *EVSS* for elastic–viscous split stress). A better formulation also requires the rates of strain as extra variables, thus leading to an *enhanced mixed variable approach* (called *u-v-p-τ-g formulation* or *DEVSS-G* for discontinuous elastic–viscous split stress–strain rate). Variations of these also exist. For the energy formulation, temperature T is the single primitive field variable. Special streamline upwind/Petrov–Galerkin (SU/PG) techniques are used for highly convective flows (high *Pe* numbers) [37]. More details about the SU/PG scheme can be found in the landmark paper by Marchal and Crochet [14].

At this point, it is perhaps instructive to give a feel of the complexity of solving flow problems in polymer processing with viscous or viscoelastic constitutive models, and in the latter case the differences between using differential and integral models to describe polymer melts. We take as an example a 2D flow problem and the FEM employing a typical 9-node Lagrangian quadrilateral element, as shown in Figure 4.5. A *viscous problem*, based on the *u-v-p* formulation, would require 22 dof per element ($9u$, $9v$, and $4p$ variables). If the flow problem is also coupled with the thermal problem, then we have 31 dof/element (9T variables). A serendipity element is cheaper as it has only 8 nodes (it lacks the centroid node), thus giving 20(28) dof/element (flow/ + thermal problem).

Now for a *viscoelastic problem*, with one (1) relaxation mode, the stresses and the strain rates have to be added to the nodal unknowns. In 2D flows, we distinguish between planar ($\tau_{xx}, \tau_{yy}, \tau_{xy}; g_{xx}, g_{yy}, g_{xy}$) and axisymmetric flows ($\tau_{rr}, \tau_{zz}, \tau_{rz}, \tau_{\theta\theta}; g_{rr}, g_{zz}, g_{rz}, g_{\theta\theta}$). These extra nodal unknowns are defined at the corner nodes (due to linear interpolation), thus adding $4 \times 3 = 12$ stresses and $4 \times 3 = 12$ strain rates, a total of 24 extra dof (planar); and $4 \times 4 = 16$ stresses and $4 \times 4 = 16$ strain rates, a total of 32 dof (axisymmetric). Thus, the total dof/element are $22 + 24 \times 46(55)$ dof/element

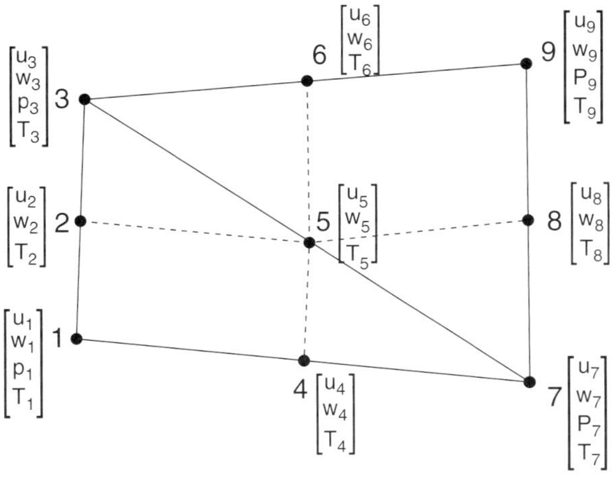

Figure 4.5 A 9-node Lagrangian isoparametric finite element used in the *u-v-p-T* formulation of 2D polymer melt flows [14, 37].

Table 4.2 Degrees of freedom analysis needed for differential versus integral constitutive models for the 2D finite element of Figure 4.5.

Integral models	Differential models				
u-v-p(-T) (dof/element)	*u-v-p(-T)* (dof/element)	τ_k (dof/element)		g_k (dof/element)	
		Planar	Axisym.	Planar	Axisym.
$8u$	$9u$	$4\tau_{xx}$	$4\tau_{rr}$	$4g_{xx}$	$4g_{rr}$
$8v$	$9v$	$4\tau_{yy}$	$4\tau_{zz}$	$4g_{yy}$	$4g_{zz}$
$4p$	$4p$	$4\tau_{xy}$	$4\tau_{rz}$	$4g_{xy}$	$4g_{rz}$
$8T$	$9T$	—	$4\tau_{\theta\theta}$	—	$4g_{\theta\theta}$
20(28)	22(31)	12	16	12	16
	1-mode total (planar): 22(31) + 12 + 12 = 46(55)				
	1-mode total (axisym.): 22(31) + 16 + 16 = 54(63)				
	8-mode total (planar): 22(31) + 8 × 12 + 8 × 12 = 214(223)				
	8-mode total (axisym.): 22(31) + 8 × 16 + 8 × 16 = 278(287)				

(flow/ + thermal planar problem) and 22 + 32 = 54(63) dof/element (flow/ + thermal axisymmetric problem). Note that these numbers are for a *single* relaxation model.

For the eight relaxation modes describing the IUPAC-LDPE of Figure 4.4, the corresponding numbers become 214(223) dof/element (flow/ + thermal planar problem) and 278(287) dof/element (flow/ + thermal axisymmetric problem). These numbers are collectively given in Table 4.2, and are necessary for solving any problem with a differential viscoelastic model, for example, the popular "pom-pom" model.

It is then obvious that even for a sparse FEM mesh of a 1000 elements the total numbers of dof climbs to $O(10^5)$. So, it is not surprising that even in today's computers many viscoelastic problems have not been solved with the full spectrum and differential viscoelastic models (such as the "pom-pom") even for simple 2D flows.

The situation is different with integral constitutive models, such as the K-BKZ/PSM model of Eq. (4.12). In this case, the formulation remains *u-v-p(-T)* with the same number of dof/element as in viscous flows. The stresses are then calculated a posteriori along streamlines according to the integral of Eq. (4.12) using a 15-point Gauss–Laguerre quadrature suitable for exponentially fading functions [45]. The summation used in the integration is very fast with digital computers, and it does not make much difference in computational time by using either one or eight relaxation modes. Most of the computational time is used for the solution of the *u-v-p* problem, while the a posteriori computation of the stresses takes approximately an equal amount of CPU time. Thus, it was possible to solve many 2D flow problems by the year 2000 in the personal computers (PCs) of the day, as evidenced in a series of papers by the author and his coworkers [35].

As an example, we give here the case of flow through a contraction of the IUPAC-LDPE melt-A, using Eq. (4.12) with eight relaxation modes and the data of Table 4.1, and employing the computational method of streamline integration developed by Luo and Mitsoulis [34]. In the year 2000, working on a PC Pentium-II at 300 MHz with 128 MB RAM, for a mesh with 640 elements it was taking 20 CPU s for the frontal solution of the *u-v-p* system of equations and another 20 CPU s for the calculation of stresses for each iteration. To get a solution at an elevated flow rate (corresponding to an apparent shear rate $\dot{\gamma}_a = 100\,s^{-1}$), about 1000 iterations were needed for a total of 11 CPU h. In today's computers (Intel Core2 Duo at 2×2.66 GHz with 2 GB RAM), the same problem can be solved in the same overall time but with 10 times the number of finite elements (6400) and the results are the same, just smoother. Thus, it has been possible to solve viscoelastic problems of multimode fluids, such as polymer melts, in more complex geometries, such as those encountered in polymer processing, as will be discussed in the following section.

4.4 Polymer Processing Flows

Polymer processing flows may be classified into *steady-state flows* (e.g., extrusion and its subsequent processing flows, such as roll- or wire-coating, calendering, fiber spinning, film blowing, film casting, etc.) and *unsteady-state flows* (e.g., injection molding and other molding processes, such as blow molding, thermoforming, etc.). In the first classification, we deal with continuous processing, where time is not of essence, and is not part of the solution. In the second classification, we deal with discontinuous (or batch) processes, where a process cycle is completed, and where time is of primary essence, as each plastic object is produced in a set time interval.

4.4.1 Extrusion

4.4.1.1 Flow Inside the Extruder

The flow inside the extruder has been a challenge computationally from the early 1960s when it was possible to solve numerically simple problems, not amenable to analytical solutions. The complicated nature of the extruder channel with its helical screw configuration and the different zones of solids-conveying, melting, and melt-conveying (see Figure 4.6) have rendered its modeling and simulation a formidable intellectual task. Accordingly, some brilliant minds have come to the fore and answered the challenge, more or less successfully. The early model of Tadmor [2] has seen many successful applications, and it has been built upon to render it more complete and user-friendly by many others [2]. It is still used in the industry for the analysis and design of screws and extruders due to its simplicity and ease of calculations. For example, the software package NEXTRUCAD [17] uses the finite volume method for simulating solids transport, melting, and metering in single-screw extruders. It supports conventional single-flighted screws, barrier screws of Maillefer/Uniroyal types and parallel-flight designs, and screws with mixing elements. The simulation results include solid-bed profile, pressure build-up, and bulk

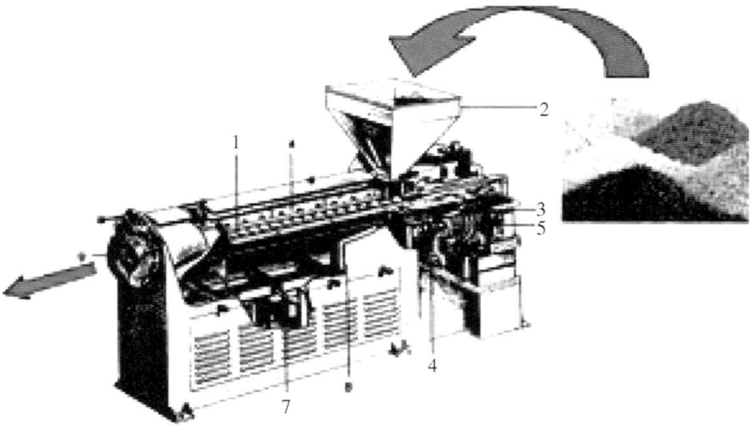

(a)

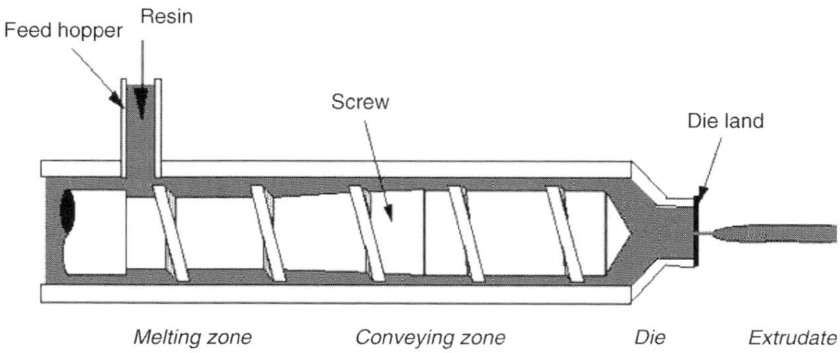

(b)

Figure 4.6 Single-screw extrusion: (a) photograph of an extruder [16] and (b) schematic of the extruder and its zones.

temperature along the screw axis, along with extruder output, power requirement, torque, and average residence time. Typical results are of the form shown in Figure 4.7. These results can be had in a matter of a few minutes in modern PCs.

More complicated models have been published in the open literature [47], while the first fully 3D computational effort of the whole extruder was made in 1984 [48] based on the assumption of a very viscous fluid even for the solids-conveying zone. Since then, more computational effort has been expended on the subject. For example, the work of Moysey and Thompson [49] uses the discrete element method to study interactions of polymer pellets as they flow in the solids-conveying zone. These authors have shown interesting patterns in the extruder, which can be used for analysis and design, albeit on a much more demanding basis due to the full 3D nature of the geometry.

The related subject of modeling and computations in a twin-screw extruder (Figure 4.8), its kneading elements, and its mixing characteristics has been undertaken by a notable number of researchers. Starting with the monograph of

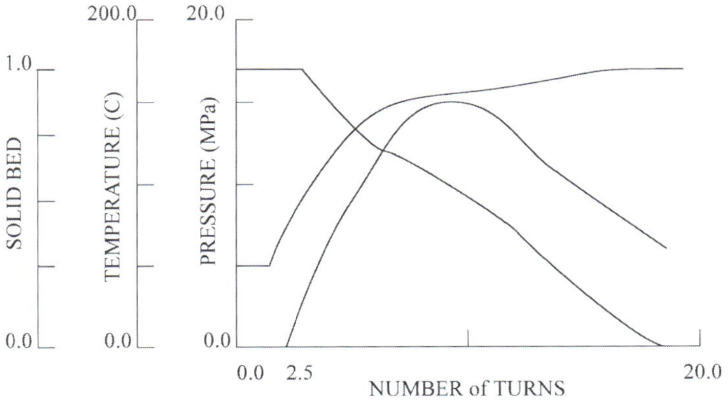

Figure 4.7 Typical results from a commercial software (EXTRUCAD) for extruder simulation showing the development of solids bed, pressure, and temperature as a function of the number of turns of the screw [17].

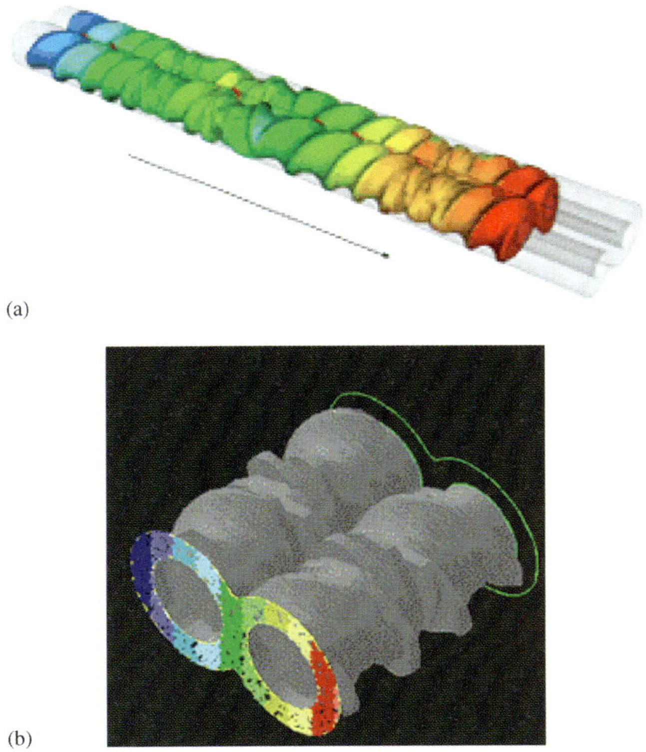

Figure 4.8 Twin-screw extrusion: (a) simulation of the pressure drop in the system and (b) simulation of particle tracking in the intermeshing chamber of the extruder [16].

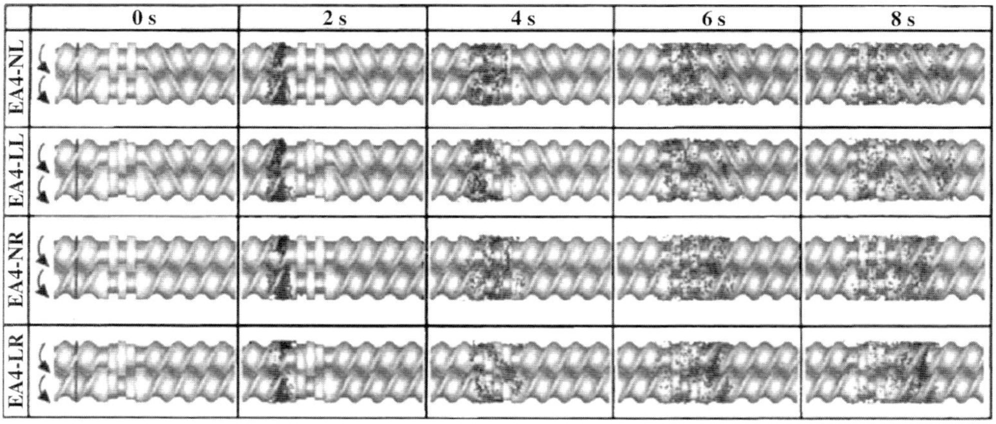

Figure 4.9 Twin-screw extrusion. Simulation of marker tracking at various time instants for different configurations and a given flow rate and screw speed [52].

Janssen [50], White and his group [51] and Funatsu and his group [52] have worked extensively on the subject and provided many clues about the relative features of twin-screw extrusion and its advantages and drawbacks. For example, Figure 4.9 shows the progression of particles in a fully 3D simulation at different time frames [52]. The important issue of deformation and mixing has been undertaken by Manas-Zloczower and her group [53, 54] who in a series of papers have analyzed what happens inside a twin-screw extruder and how its design and operating conditions affect the polymer melt under consideration.

In all these cases of simulation inside extruders, the major focus has been the development of efficient geometry modules to describe in a quick and user-friendly way the complicated geometrical characteristics of screws and extruder channels. The question of adequately describing the rheology of the polymer melt is usually resolved with good viscosity data as a function of shear rate and temperature (Eq. (4.19)). The Carreau model (Eq. (4.6)) for the former and the exponential model (Eq. (4.9)) or Arrhenius model (Eq. (4.10)) for the latter are sufficient in these computations, where viscoelasticity does not seem to be of importance or has not been attempted in any meaningful way. The predominance of shear flow inside the extruder seems to be the justification for that.

4.4.1.2 **Flow in an Extruder Die (Contraction Flow)**

Few flows have received so much attention both experimentally and computationally than the flow in a contraction. It is related to the flow in the final stage of the extruder, that of the forming die (see Figure 4.6b), which, in the simplest of cases is just a tubular orifice through which the polymer exits to the atmosphere from a larger reservoir. It is also the main feature in rheometry for measurements of viscosity. A simplified version of the problem is depicted in Figure 4.10. The polymeric fluid passes from a larger reservoir of diameter D_{res} to a smaller tube of diameter D_0. In the analogous sudden *planar* contraction, there is a step change in slit width. The

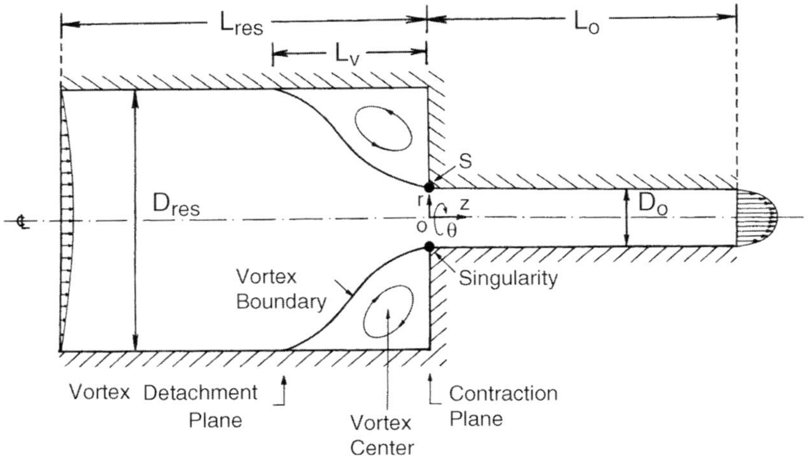

Figure 4.10 Schematic representation of an abrupt contraction [31].

presence of the abrupt contraction gives rise to many interesting flow patterns, including secondary flows in the reservoir, while the singularities at the salient corners render the problem extremely difficult to solve with any of the known numerical techniques. Contrary to the situation inside the extruder, here viscoelasticity may indeed play a significant role and distinguish the behavior among different polymer melts.

For any fluid, Newtonian or non-Newtonian, there is in these geometries a region of extensional flow near the center line and the contraction plane, while shear dominates near the walls. The well-known and different entry patterns of LDPE and HDPE melts (Figure 4.11) in a 20-to-1 circular contraction with their distinct behavior have beguiled the scientific community for more than 40 years, since they appeared in the work of Bagley and Birks [55]. While for Newtonian fluids fluid inertia can change the flow and the shape of the salient corner vortex [56], much more interesting transitions occur for viscoelastic fluids, such as polymer melts, as the flow rate is increased and flow becomes more elastic. The early studies by Giesekus have been followed by studies by Boger and coworkers [7, 56–60].

Using the rheological data for typical LDPE and HDPE melts and their best fit using the K-BKZ/PSM model [34] has made it possible to successfully simulate these distinct flow patterns, as shown in Figure 4.12 [34]. Simulations have shown that a strain-hardening elongational viscosity exhibited by the branched LDPE is primarily responsible for big vortices in the contraction. On the other hand, the linear HDPE melt exhibiting strain-thinning in elongation, generates no vortices for the same wall shear stresses. Further results have been obtained for different polymers and different conditions, including stress birefringence [61, 62]. It is interesting to note that while the flow of *polymer melts* through contractions has more or less been resolved successfully from a computational point of view, there are still difficulties with some *polymer solutions* [31]. For the latter, experiments show a pulsating flow with a lip vortex that oscillates near the entry to the die, while simulations show a

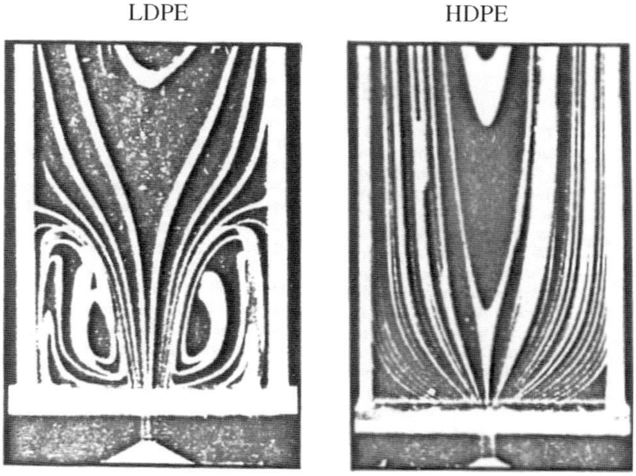

Figure 4.11 Flow patterns of LDPE and HDPE melts flowing through a 20 : 1 axisymmetric contraction before unstable conditions set in. Note the big vortex for LDPE and the absence of any vortex activity for HDPE [55].

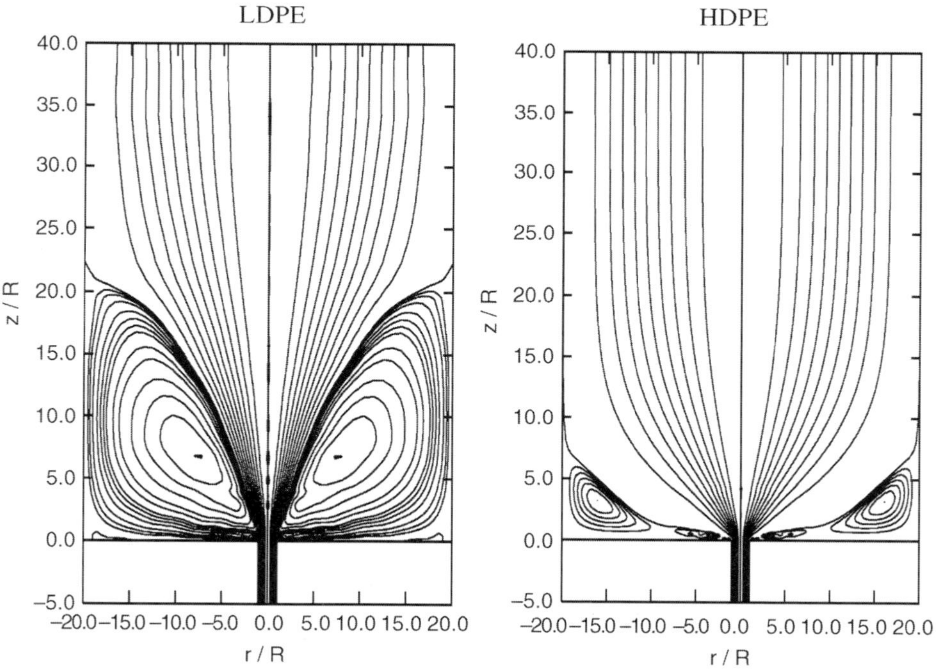

Figure 4.12 Viscoelastic simulation of flow patterns for the melts of Figure 4.11 using the K-BKZ integral model (Eq. (4.12)) [34].

stable flow with strong vortex growth [31]. Obviously, any steady-state, two-dimensional, axisymmetric simulations are incapable of reproducing three-dimensional, time-dependent effects, such as the pulsating flow patterns exhibited by some polymer solutions in contractions.

4.4.1.3 Flow Outside the Extruder – Extrudate Swell

As much interest as has been shown for contraction flows, an equal amount has been devoted to the flow just outside the extruder, where the accompanying phenomenon of "extrudate swell" (formerly known as "die swell") [63] is a major preoccupation in the polymer processing industry. An extra computational difficulty arises from the free surfaces present at the exit, which are unknown a priori, and must be found iteratively as part of the solution, by satisfying extra free-surface BCs, which have long been established to be of the kinematic type for polymer melts [64]. Thus, the Newtonian extrudate swell problem was solved first, and it has been since a benchmark problem against which all computer software for polymer processing are checked. While the small Newtonian swelling of 13% for tubular dies and 18% for slit dies is easy to get, the viscoelastic case is much harder to solve and to achieve good agreement with experiments. It requires sophisticated numerical schemes, a thorough rheological characterization of the melt at hand, and a careful increase in viscoelastic loading (or equivalently a careful increase in extrusion flow rate) to achieve the large swelling ratios experimentally observed. Although this has been feasible for some 20 years now, it is still not a trivial matter for 3D problems and inverse die design.

Some typical examples of successful extrudate swell simulations are given in Figure 4.13 in extrusion from orifice dies for the IUPAC-LDPE melt-A with the data of Table 4.1 [32]. As the flow rate increases, the vortex inside the reservoir increases and so does the extrudate swell outside the die, which can reach as much as 100%, in agreement with the experiments by Meissner [29]. Another example is given in Figure 4.14 for the flow of an HDPE melt through tapered annular dies, where the effect of die design is obvious and in agreement with experiments [65]. Namely, for the same flow rate a diverging annular die produces the least swelling, followed by a straight annular die, and then by a converging annular die, which produces the most swelling due to enhanced memory phenomena of the melt during its flow [66].

It is obvious that these successful simulations are feasible because of (a) thorough rheological characterization, (b) fitting with a sophisticated multimode integral model, and (c) simplicity of geometry. As mentioned above, for 3D problems, this is still not an easy matter. For example, Baird and Collias [23] show extrudate cross sections for two polymer melts (polyethylene and rigid PVC) and the die cross sections that produced them (Figure 4.15). Extrudate swell is responsible for a mismatch between die used and polymer produced. This leads then to the *inverse design*, as shown in Figure 4.16. Namely, what should be the die shape for a required product? Ellwood *et al.* [67] have shown how this is possible for Newtonian fluids, and modern software packages claim that they can handle this problem, as shown in Figure 4.17 from www.polyflow.be [16]. For complicated geometries and viscoelastic polymer melts, this is still though a formidable task!

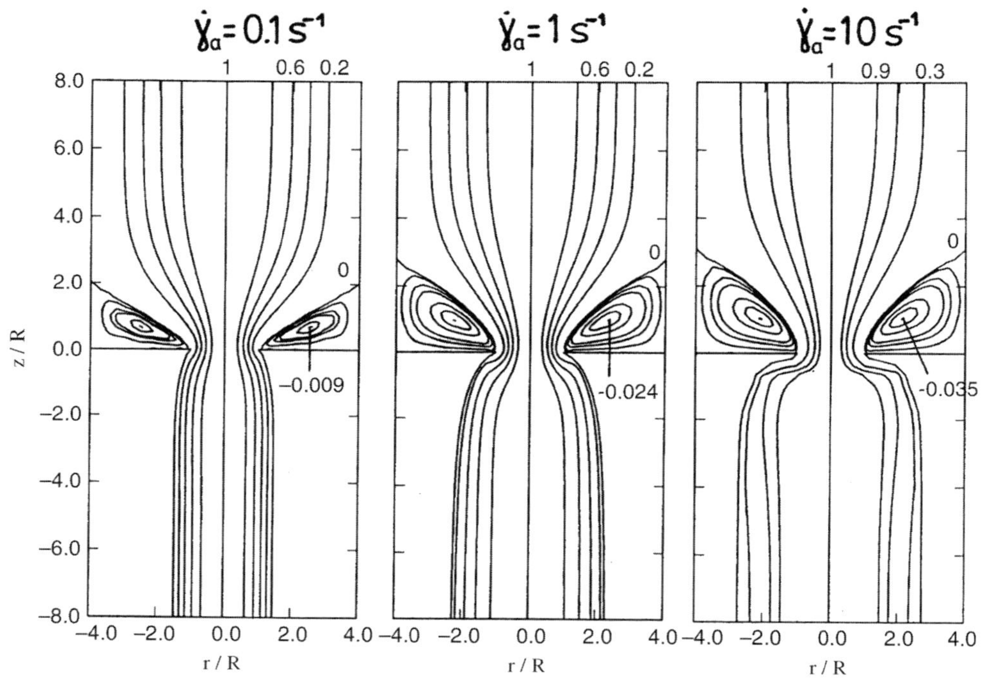

Figure 4.13 Simulation of vortex growth and enhanced extrudate swell as the apparent shear rate increases for extrusion of the IUPAC-LDPE melt-A through a circular orifice. Viscoelastic simulations with the K-BKZ integral model (Eq. (4.12)) with data given in Table 4.1 [32].

4.4.1.4 Coextrusion Flows

An interesting variation of extrusion is coextrusion of two or more fluids through dies (Figure 4.18). Coextrusion is practiced increasingly in the polymer processing industry to impart special qualities on films, especially in packaging (food and beverage) and elsewhere [68]. In the coextrusion process, two or more fluids come together in a die and flow having a common interface. Computationally, this is a difficult problem because the interfaces are not known a priori and must be found iteratively like the free surfaces of extrudate swell. A typical example of coextrusion is given here. The flow inside the dies of two polymer solutions is shown from experiments in Figure 4.19 [69], while the simulations are shown in Figure 4.20 [70]. There is good agreement of the flow patterns, including the presence of recirculation in the lower fluid. To obtain such a good agreement, it was necessary to know (a) the viscous rheological properties of the two polymers (Carreau model sufficed), (b) the geometry of the coextrusion die, and (c) the flow rate for each polymer solution. This is the case because inside the coextrusion die, the flow is dominated by shear, therefore a purely viscous model is sufficient.

On the other hand, viscoelastic coextrusion simulations (Figure 4.21) show the relative importance of the stratification of two polymer melts of different viscoelastic strength in the extrudate swelling. Namely, a PS/HDPE (inner/outer) and a HDPE/PS

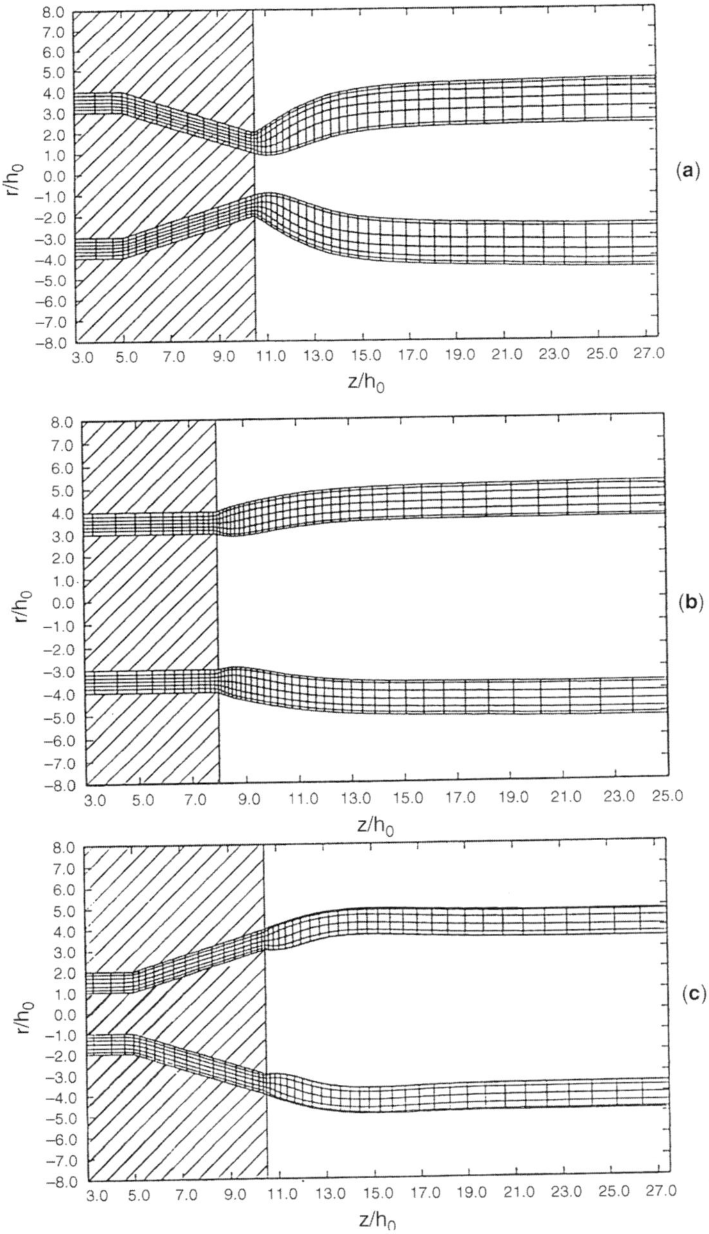

Figure 4.14 Simulation of extrudate swell from tapered annular dies for an HDPE melt. Memory phenomena are present manifested in enhanced swelling from converging dies. Viscoelastic simulations with the K-BKZ integral model (Eq. (4.12)) [66].

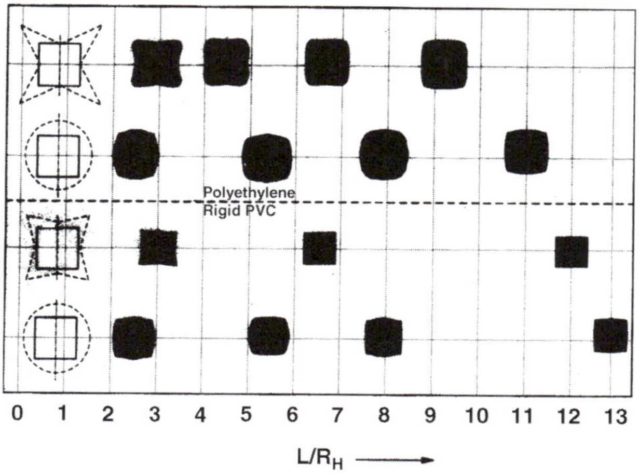

Figure 4.15 Die design for polymer melts. A square die will not produce a square extrudate due to viscoelastic phenomena. The right design must be a star-shaped die [23].

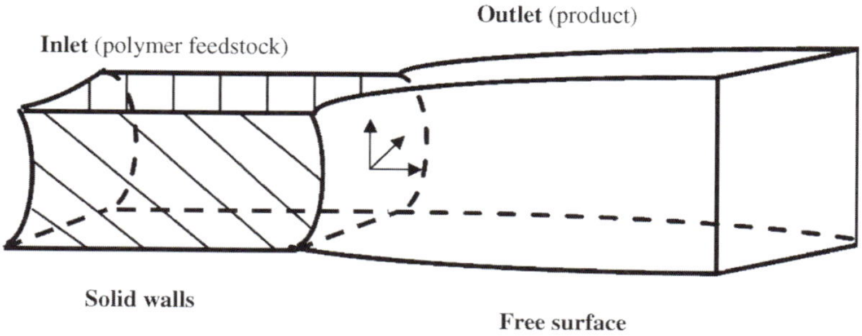

Figure 4.16 Schematic representation for the inverse problem of designing a die that produces a squared section polymer product [67].

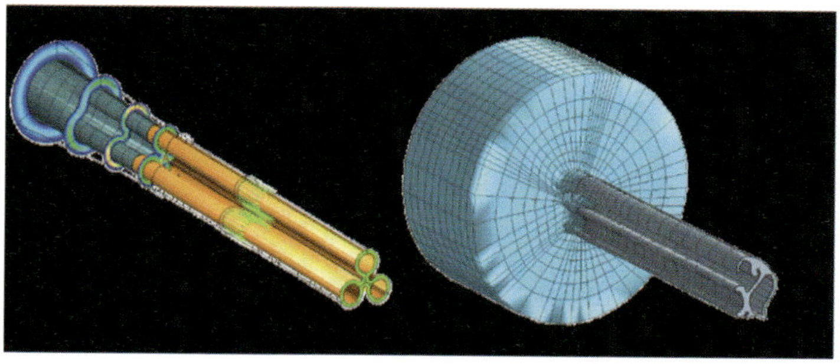

Figure 4.17 3D simulations for die design using the commercial software POLYFLOW: (a) extrusion of a bundle of three fibers and (b) extrusion of a "parrot" toy extrudate [16].

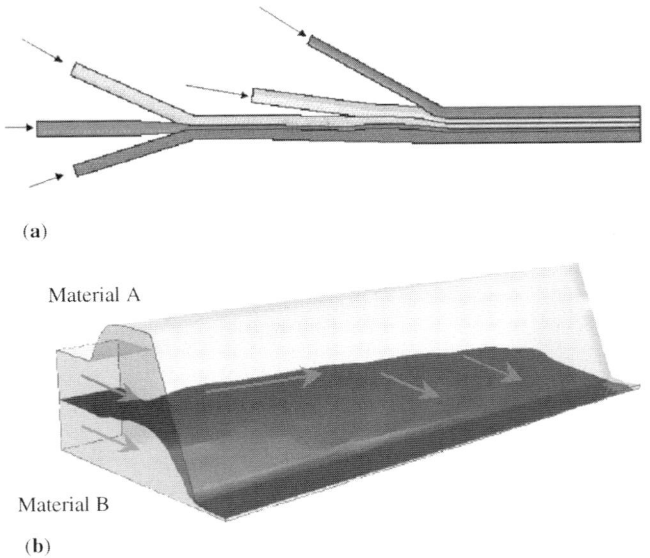

Figure 4.18 Multilayer coextrusion simulations using the commercial software POLYFLOW: (a) 5-layer 2D coextrusion and (b) 2-layer 3D coextrusion in a coat-hanger die [16]. In both cases, there is a simultaneous determination of the unknown a priori interfaces.

configuration will give different swelling behavior for the same flow rate, according to their elastic nature, something which cannot be predicted a priori [46].

Computations for many layers (multilayer coextrusion) and different configurations are best handled by the lubrication approximation theory (LAT) for quick engineering calculations [68]. However, FEM 2D simulations are also feasible and provide good insight for industrial applications as shown by the author [71]. It is understood that 3D viscoelastic effects for polymer melts flowing in a coat-hanger die, such as the one shown in Figure 4.18b, are again not a trivial matter, even for two fluids. A great piece of work for viscoelastic 2D and 3D effects in multilayer polymer coextrusion is the thesis by Dooley [72] with many references therein, and the author's own papers (see Ref. [73]).

4.4.1.5 Extrusion Die Design

While in most cases the extruder can be regarded as a "black box" with no extra effort required for its analysis, this is not the case for die design. The latter remains in many industries, both small and large, a major challenge, for producing useful products of different shapes and dimensions. This is accomplished with profile dies of an amazing variety of shapes, of which the coat-hanger die represents one of the most basic die designs. In all these cases, the No. 1 problem is balancing the flow at the die exit so that no uneven thickness of film or sheet or extrudate distortion will ensue, and the product will have everywhere precise dimensions. To that effect, appropriate software has been developed (e.g., PROFILECAD from POLYDYNAMICS, Inc. [17]),

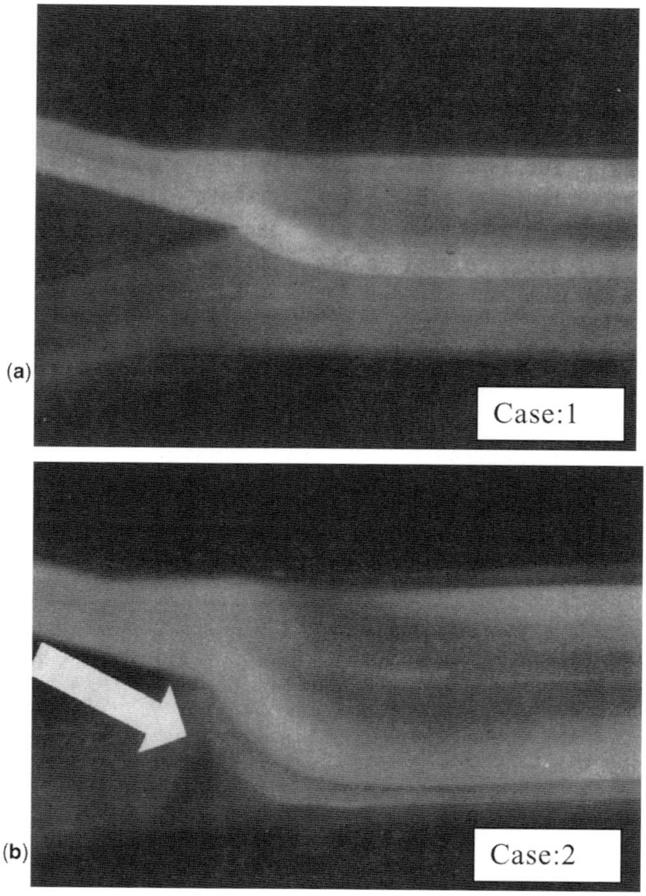

Figure 4.19 Flow patterns in coextrusion dies: (a) stable configuration and (b) unstable configuration with recirculation [69].

which significantly cuts down on trials used in machine shops for prototype designs. The principles behind such profile design are pressure balancing and cross-flow minimization [74, 75]. The problem is also amenable to the inverse design methodology, mentioned above (see Figures 4.16 and 4.17), as explained by Marchal [76] and Debbaut and Marchal [77]. This area of modeling is still nascent and needs further attention to improve profile die design.

4.4.2 Postextrusion Operations

The term applies to many different continuous shaping operations after the polymer has exited the extruder, in the vast majority of cases as a hot homogeneous melt. The postextrusion operations can be classified into two types: (i) *shear dominated*, such as

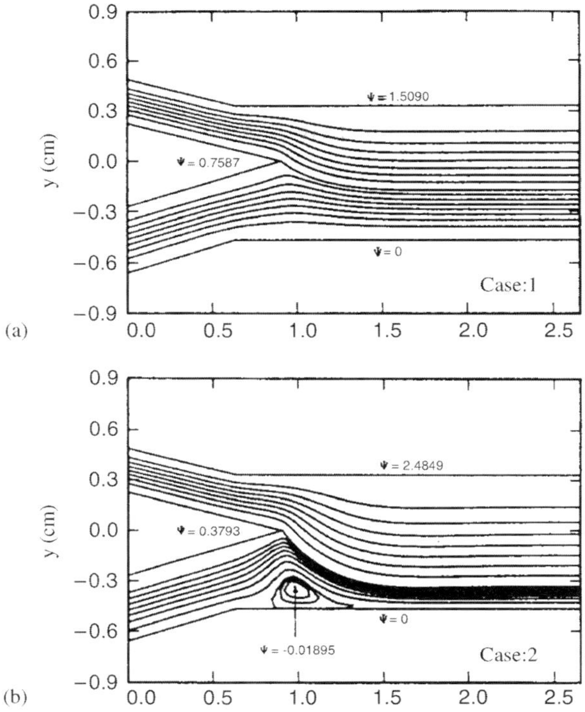

Figure 4.20 Simulated flow patterns in coextrusion dies corresponding to cases (a) and (b) of Figure 4.19 [70].

calendering, roll coating, wire coating, and so on, where the polymer melt is guided by walls, and (ii) *extension dominated*, such as fiber spinning, film casting, film blowing, and so on, where the polymer melt is stretched or pulled, so that extension is the major mode of deformation and flow. In shear-dominated flows, viscoelastic effects have been shown to be minimal, so computationally purely viscous, non-isothermal models are capable of correctly simulating the processes and predict flow patterns and engineering quantities of interest (pressures, torques, forces, etc.). This is not the case for extension-dominated flows, where viscoelastic effects are dominant, and usually a purely viscous model gives wrong and misleading predictions. In these cases, a thorough viscoelastic simulation is necessary, as has been found in the literature and will be shown below.

4.4.2.1 **Calendering**

In the process of calendering, a molten polymer enters usually as a sheet on one of the two rotating rolls and leaves on the other with a reduced thickness. The process is schematically shown in Figure 4.22. It is seen that due to the reduced area, a melt bank is created before the nip region. In this melt bank, a very interesting flow pattern develops with multiple recirculation regions, as shown in Figure 4.23 from

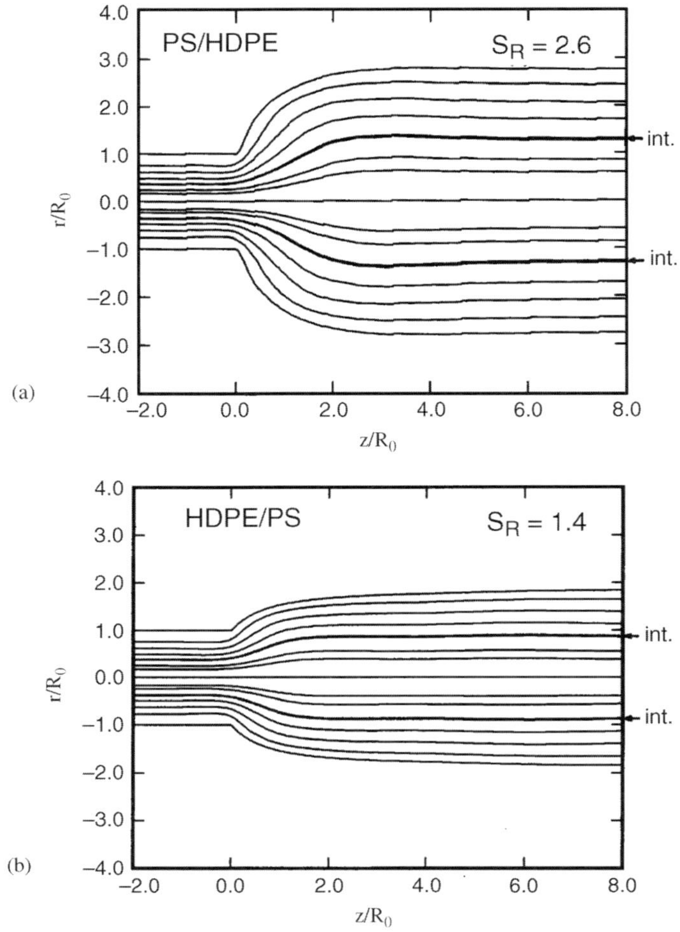

Figure 4.21 Simulation of extrudate swell from coextrusion circular dies for a combination of HDPE and PS melts (PS/HDPE = inner/outer). A different configuration produces different swelling for the same flow rate due to different viscoelastic properties of the melts, hence different stress ratio, S_R. Viscoelastic simulations with the K-BKZ integral model (Eq. (4.12)) [46].

experiments on rigid PVC by Agassant and Espy [78, 79]. A huge vortex appears in the melt bank, while smaller ones develop near the entering sheet and near the nip region.

A purely viscous nonisothermal model (power law with slip at the wall) for PVC is able to capture this vortex behavior, as evidenced in Figure 4.24a [80, 81]. The simulations also provide the temperature field (Figure 4.24b) [80, 81] and the pressure distribution (Figure 4.25), which is in good agreement with experiments when slip is included. When the calendering process is stable (i.e., when the melt bank is stable), the experimental streamlines are very similar to the computed ones (the presence of two stable recirculating regions). When the calendering process is

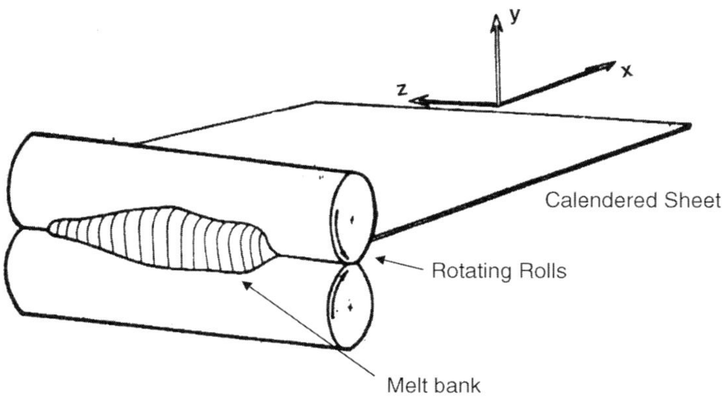

Figure 4.22 Schematic representation of calendering a plastic sheet [80, 81].

unstable (high output rate or too big a melt bank), the bank and the recirculating regions vary in the third direction, and a spiral flow ensues. The spiral flow in the third direction shown in Figure 4.26 was observed by Unkrüer [82] and gives rise to several instabilities as pointed out also by Agassant and Espy [78]. Computationally, the stability of calendering has not been tackled yet and remains an unsolved problem.

A 3D analysis of calendering was performed by Luther and Mewes [83] and reveals the spiral motion in the third or z-direction as shown in Figure 4.27. Although this work represents the only 3D effort to date in the open literature, it is not a trivial matter to reproduce it or use it iteratively for design purposes. Therefore, a very interesting 3D computation of calendering still remains a challenge.

4.4.2.2 **Roll Coating**

In the process of roll coating, a sheet is produced much as in calendering but usually for polymeric solutions. In such flows, surface tension becomes important. Both forward and reverse roll-coating operations are used in practice. The important work by Coyle *et al.* [84–86] has done much to increase our understanding of the fluid dynamics in both forward and reverse roll coating. The reverse process is schemat-

Figure 4.23 Experimental flow pattern in the melt bank of calendered rigid PVC [78, 79].

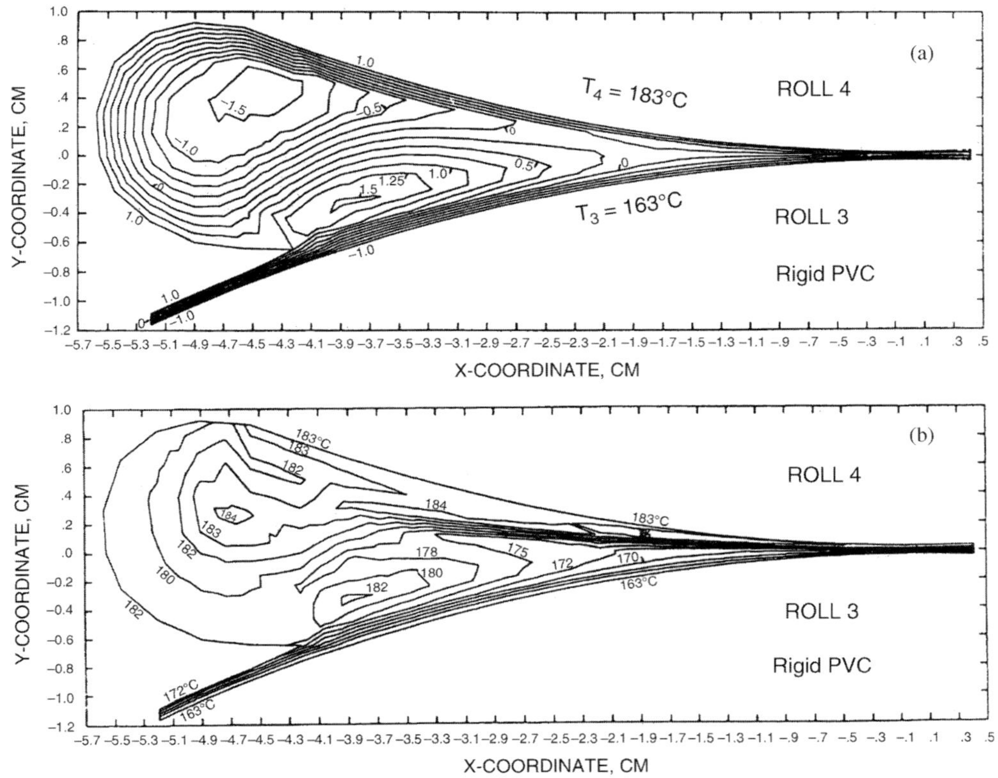

Figure 4.24 Nonisothermal simulations of flow in calendering rigid PVC with slip at the wall: (a) flow pattern seen through streamlines and (b) isotherms [80, 81].

ically depicted in Figure 4.28 (from Ref. [86]). Experiments coupled with theory show interesting flow patterns with vortices and various types of instabilities, such as cascade, ribbing, seashore, and so on. For example, in reverse roll coating of Newtonian and shear-thinning inelastic fluids, vortices appear as evidenced in Figure 4.29 [86]. Typical roll-coating instabilities experimentally observed are shown schematically in Figure 4.30 [21]. The stability analysis was based on Newtonian fluids with the capillary number as the nonlinear effect causing instabilities. Experimental work based on shear-thinning inelastic fluids has shown that shear-thinning also plays an important role, as evidenced in Figure 4.31 [86]. For viscoelastic polymer solutions, the experimental work of Coyle *et al.* [86] showed that the ribbing phenomenon becomes irregular and time-dependent. There is no sharp transition to cascade instability, accompanied by the steep upturn in coating thickness as speed ratio is increased. Rather, the coated film becomes mottled in appearance and the average coating thickness stays relatively constant [86]. These instabilities are due to secondary helical flows in the third dimension and must be resolved before a better coating is obtained. Computationally, this still remains a formidable task and has not been successfully addressed yet.

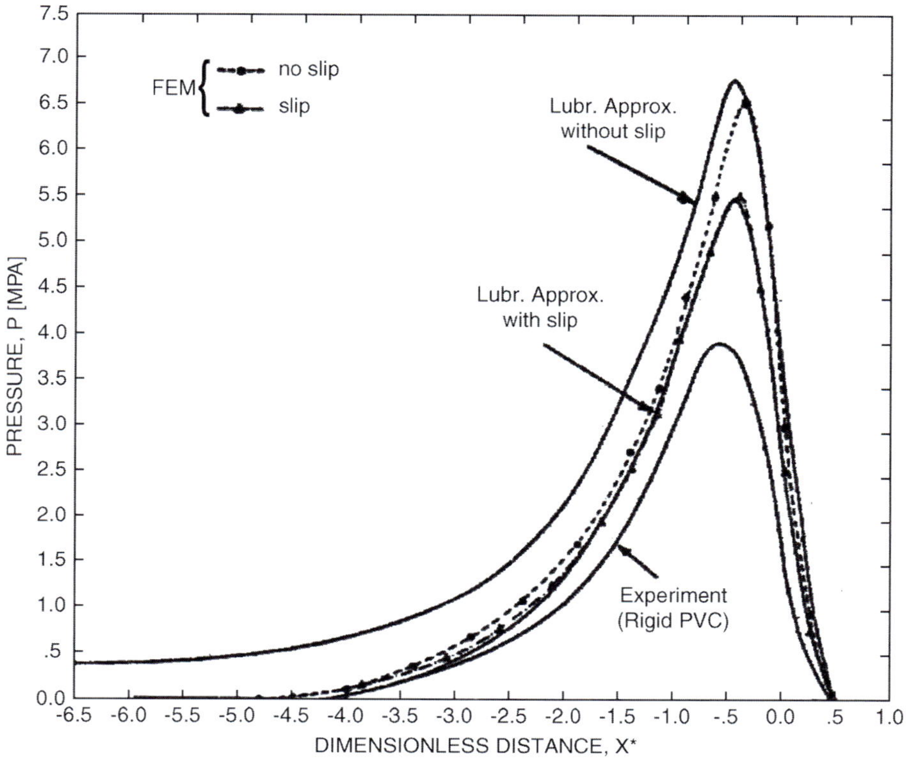

Figure 4.25 Nonisothermal simulations of flow in calendering rigid PVC with slip at the wall. Pressure distribution as found by the LAT, the FEM, and experimentally [80, 81].

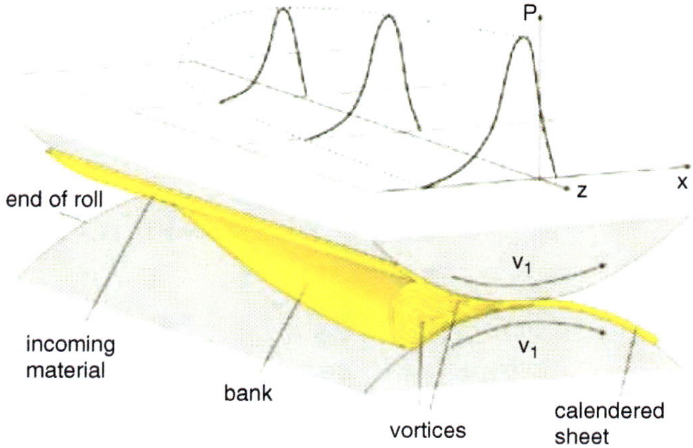

Figure 4.26 Schematic representation of 3D flow in calendering, showing the vortex formation, the melt bank development in the *z*-direction, and the pressure distributions along various positions in the *z*-direction [83].

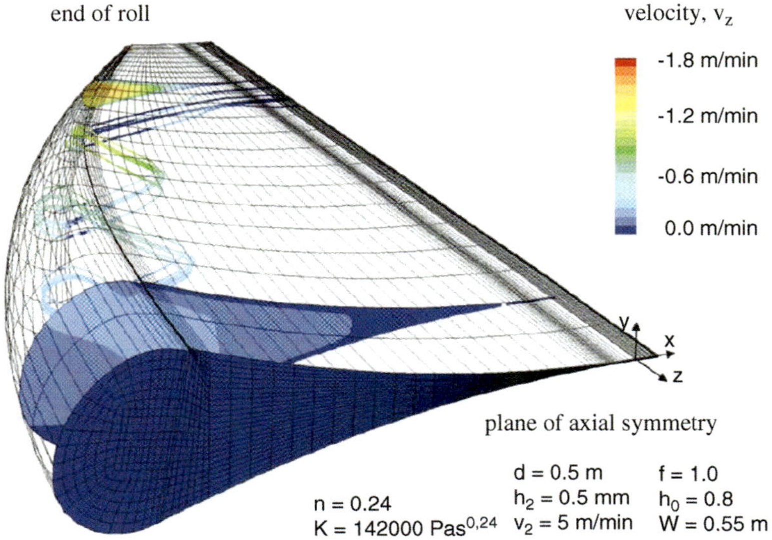

Figure 4.27 3D simulation results in calendering power-law fluids, showing the finite element grids used together with the spiral motion in the melt bank in the *z*-direction [83].

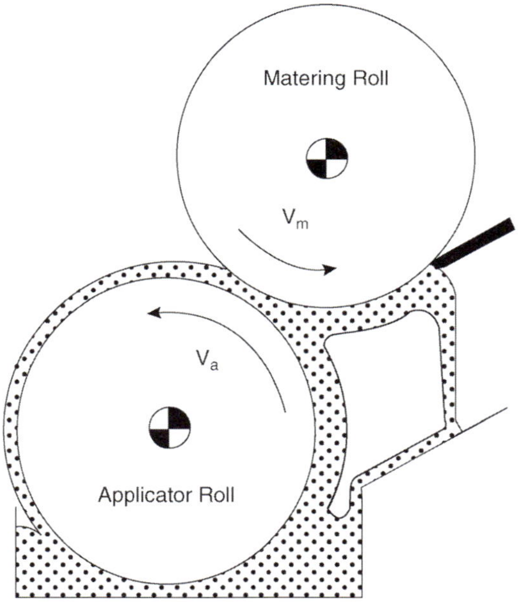

Figure 4.28 Schematic representation of reverse roll coating to study the flow in the metering gap [85, 86].

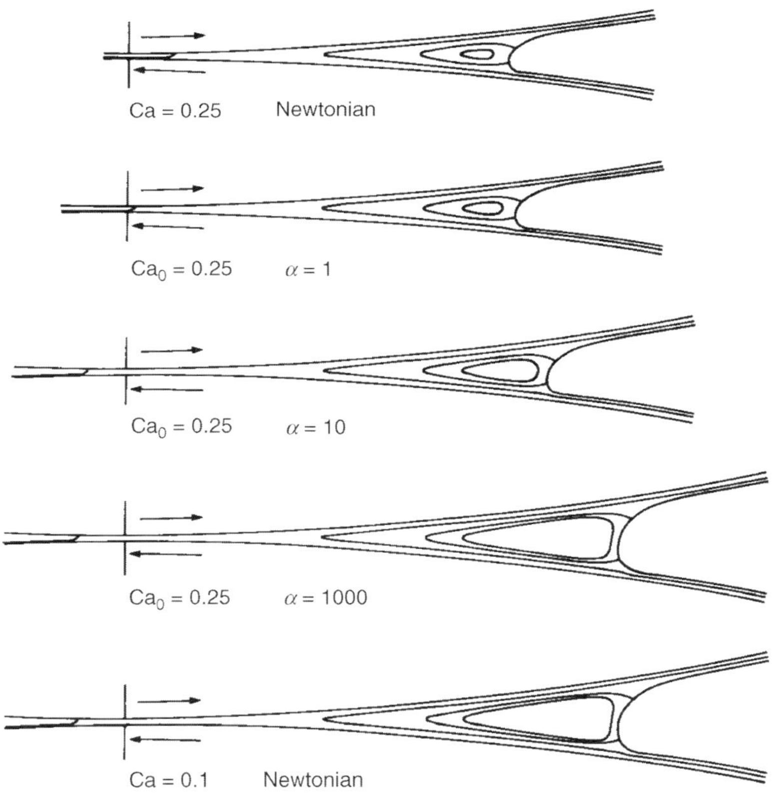

Figure 4.29 Predicted streamlines in flow for Newtonian and shear-thinning inelastic fluids in reverse roll coating [85].

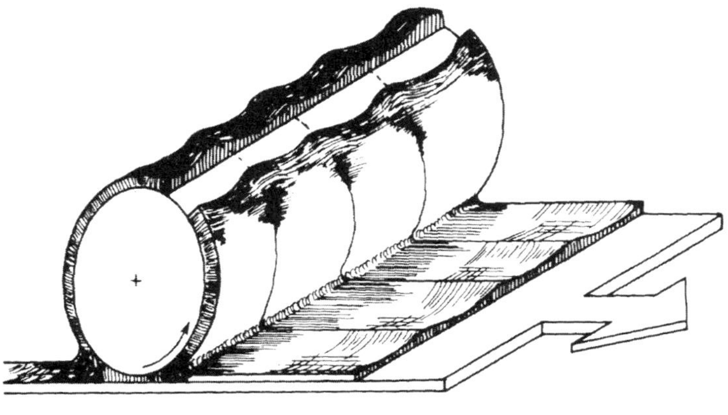

Figure 4.30 Schematic instabilities in roll coating [21].

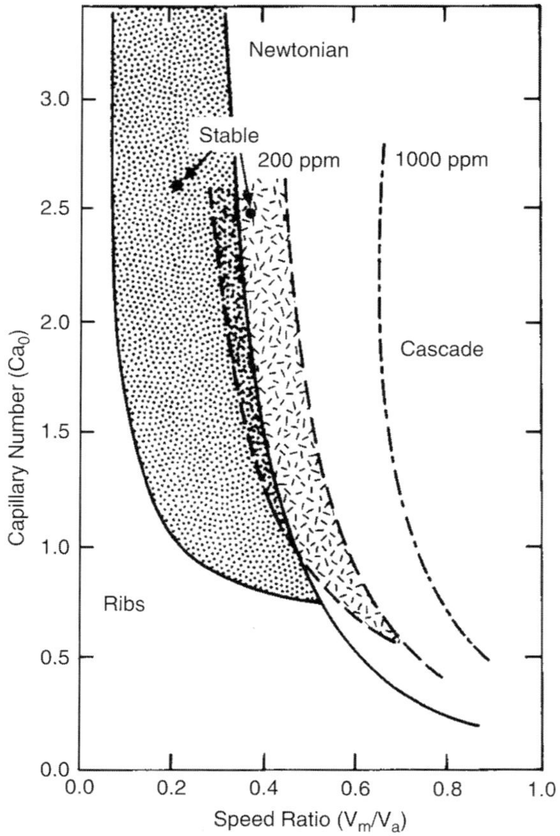

Figure 4.31 Effect of shear-thinning on the stability in reverse roll coating [86].

4.4.2.3 **Wire Coating**

The process of wire coating is depicted in Figure 4.32. A molten polymer, usually under the influence of a pressure-driven flow, is extruded through a wire-coating die having inside a "torpedo" (guider) to guide the wire. The position of the torpedo and its guider tip is crucial for a good design that avoids recirculation and limits the level of stresses generated in the melt as it enters from the annular channel into the die region under the dragging action of the moving wire [87].

A design that generates a big vortex for a Newtonian fluid may not generate one for a polymer melt. This is evidenced in Figure 4.33, where the full solution of the steady-state flow shows just that [88]. On the other hand, for a viscous polymer melt, a bad design that leaves enough "gum space" between torpedo and die wall may generate a vortex that is detrimental to the exiting melt [87]. Good designs based on the appropriate "gum space" have been put forward, and these eliminate both the unwanted vortices and the stress "jumps" along the die walls [87].

Successful simulations for a coating-grade LDPE melt have been performed by Mitsoulis *et al.* [37]. These simulations were based on an industrial high-speed wire-

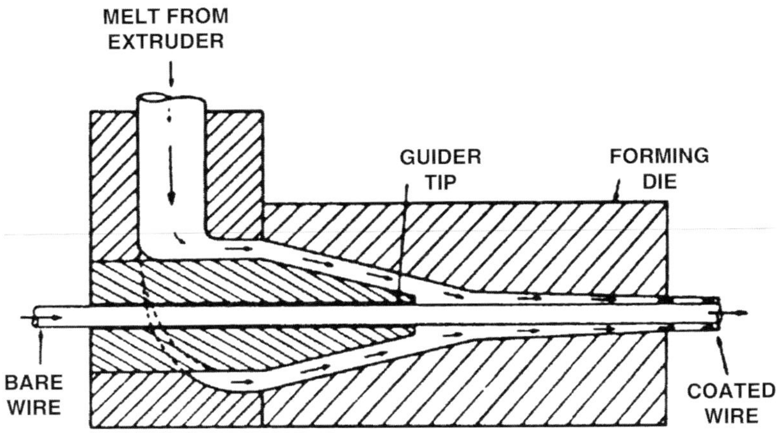

Figure 4.32 Schematic representation of flow in a wire-coating die [88].

coating die design, while for the polymer viscosity data were used for a wide range of shear rates and temperatures. Purely viscous, nonisothermal simulations produced the smooth streamline patterns inside and outside the die of Figure 4.34 and correctly predicted the pressure drop as a function of wire speed, as evidenced in Figure 4.35. Thus, successful simulations for wire coating need (a) good viscosity data as a function of shear rate and temperature, (b) geometry data for the die design, (c) correct boundary conditions for the nonisothermal analysis, and (d) operating conditions, such as the final coating thickness and wire speed. Nonisothermal effects were found to be of major importance for good predictions, rather than slip effects as it has been surmised by some people in the field.

Vortices may also be produced in wire-coating coextrusion, depending on the viscosities of the two fluids, as evidenced in Figure 4.36 [89]. Accordingly, they can be eliminated by a judicious choice of the polymer viscosities and/or temperatures [89]. The analysis of Tadmor and Bird [90] has shown that a nonzero second normal stress difference N_2 exhibited by polymer melts has a stabilizing effect on the wire-coating process, by reducing the eccentricity that may appear in high-speed operations. A full stability analysis of the process is still lacking.

4.4.2.4 **Fiber Spinning**

The fiber-spinning process is used throughout the plastics industry to manufacture synthetic fibers. The process is schematically shown in Figure 4.37. The fibers are produced by the extrusion of the polymer through a die, usually of circular cross section, and taken up downstream at a higher velocity by the chill roll. The ratio of take-up velocity to extrusion velocity is known as the *draw ratio* (D_R).

The fiber-spinning process is a prime example of *uniaxial extension*. The process consists of two regions: (i) the first is the *extrudate swell region*, where normal forces accumulated during extrusion suddenly relax to cause swelling; (ii) the second is the *draw-down region*, where the fiber diameter decreases according to the velocity

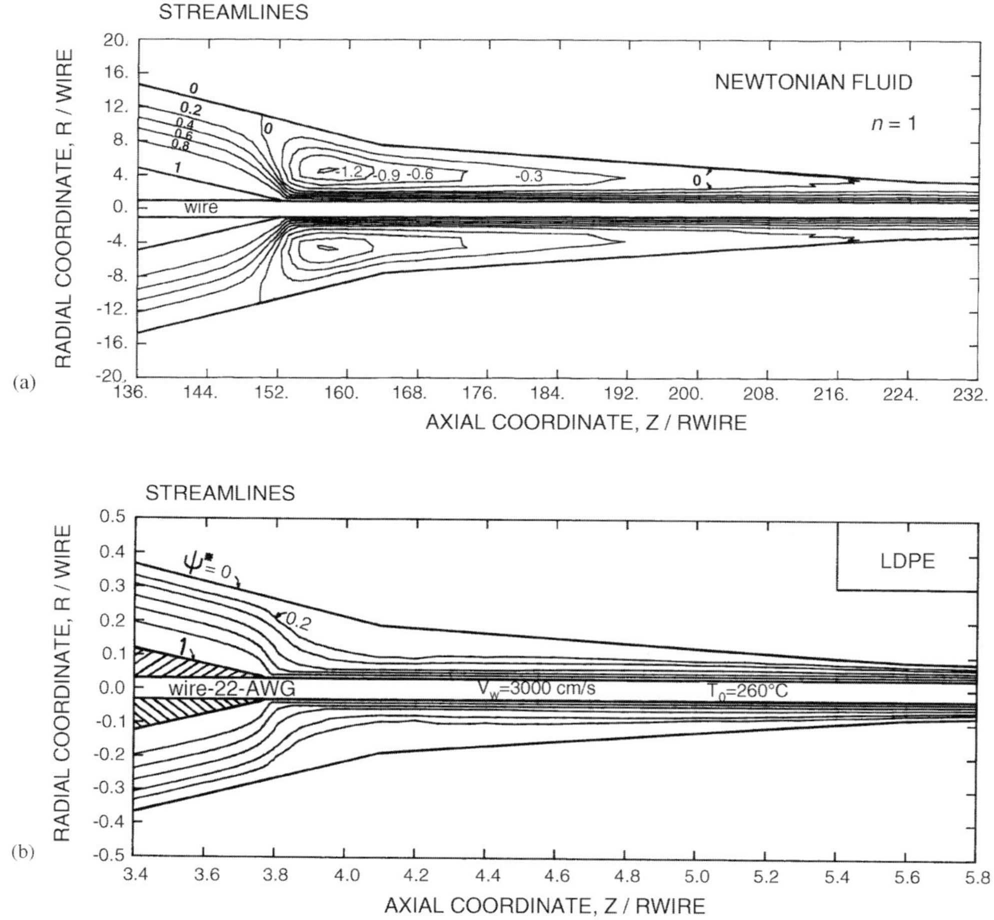

Figure 4.33 Simulated flow patterns of (a) Newtonian and (b) shear-thinning polymer melts (LDPE) in a wire-coating die. The Newtonian fluid exhibits a big vortex, while in the same geometry the LDPE melt flows without recirculation [88].

increase. Usually, cross-flow air is used to aid in the solidification of the fiber, which occurs between the die and the chill roll. Spinning speeds range from 30 to 7000 m/min, with fiber diameters ranging from 4 μm to 2 mm. At high spinning speeds (4000 m/min and greater), stress-induced crystallization can become a factor for solidification and hence it can affect fiber properties [23, 91].

Numerical simulation of the fiber-spinning process began with the early work of Matovich and Pearson [92], who analyzed the spinning of a Newtonian liquid and arrived at an analytical solution. Attempts were then made to analyze the process with *differential constitutive models*. Early work by Denn *et al.* [93] considered the upper convected Maxwell model, including nonisothermal effects [94]. Later, Gagon and Denn [95] used the PTT model and included nonisothermal effects to simulate

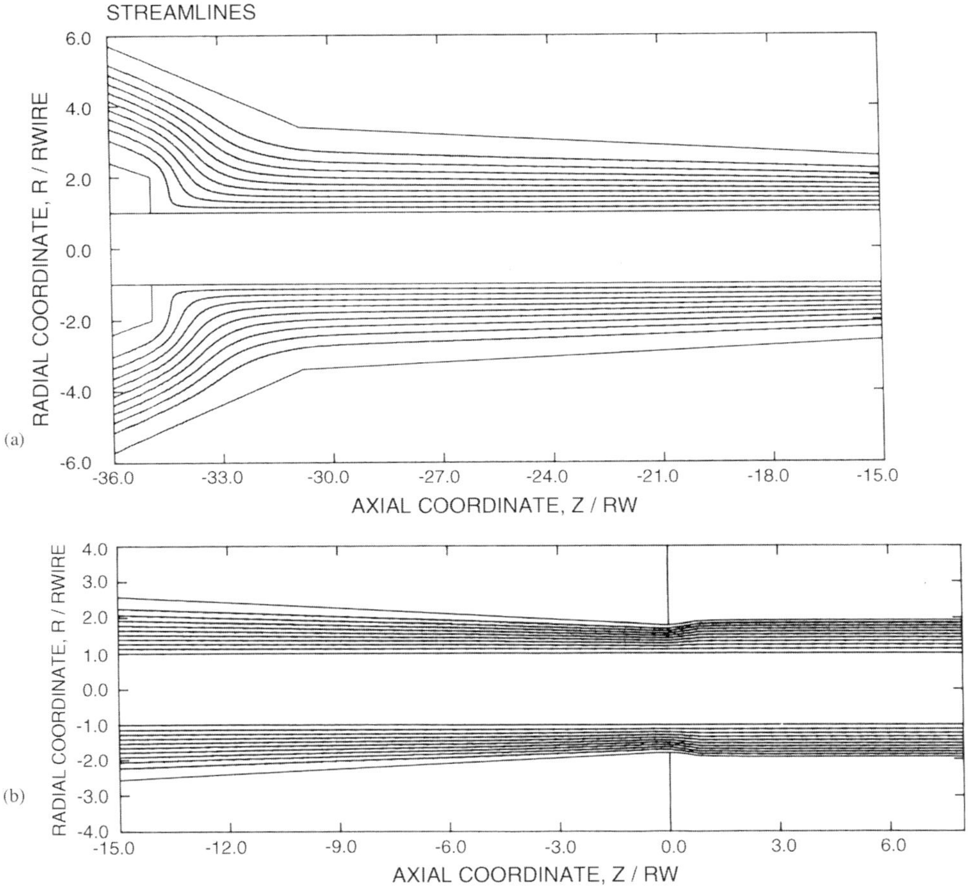

Figure 4.34 Simulated flow patterns of an LDPE melt in an industrial wire-coating die. The polymer melt flows smoothly in a streamlined flow both (a) inside and (b) outside the die, verifying the good die design [37].

experimental data given by George [96] for a PET melt. They also included the extra forces acting on the fiber, namely, inertia, gravity, and air drag. Keunings *et al.* [97] used the UCM, Oldroyd-B, and PTT models to simulate short fibers using a two-dimensional finite element method. The major findings on the process up to 1983 were reviewed by Denn [98].

Attempts were also made to analyze the fiber-spinning process with *integral constitutive models*. Papanastasiou *et al.* [99] used the K-BKZ/PSM model to analyze and simulate experimental data from Zeichner [100] for a polystyrene melt. More recently, analyses have been performed by Fulchiron *et al.* [101], using the Wagner integral model, and by Rauschenberger and Laun [102], using the K-BKZ model to simulate spinning data obtained with the IUPAC-LDPE melt-A. The K-BKZ/PSM model has also been used by Mitsoulis and Beaulne [103] for full rheological

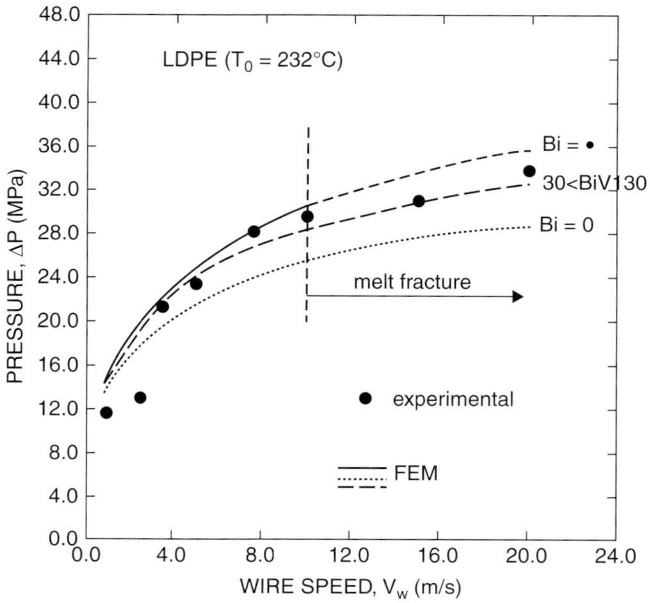

Figure 4.35 Pressure drop versus wire speed for flow of an LDPE melt in an industrial wire-coating die. The results from the simulations (lines) are in good agreement with experimental results (symbols) when a full nonisothermal analysis is used with appropriate thermal boundary conditions at the wall and the wire [37].

characterization and simulation of the process, including both the nonisothermal effects and the effects of inertia, gravity, and air drag.

As an example of fiber-spinning simulations, we present results from this latter work. The results have been obtained by a quick and approximate method of analysis with the use of an efficient code, F-SPIN [104]. The one-dimensional formulation is used instead of a 2D or 3D formulation since they require much more complicated numerical techniques, a great deal of computing time, and greater computing power. Starting with the rheological characterization and fitting of data with the K-BKZ/PSM model (Eq. (4.12)) for a PET melt, the velocity and temperature profiles are predicted in Figure 4.38, in good agreement with the experiments by George [96].

The time-dependent simulation of the process has been performed by Beris and Liu [105] and Liu and Beris [106]. Another development of importance in fiber spinning is the inclusion of crystallinity. This has been addressed by Doufas *et al.* [107], and the model is rather involved. Figure 4.39 shows a schematic of the fiber due to crystallization with the distinct neck-in phenomenon. The simulations by Doufas *et al.* [108] show good agreement for the velocity and the temperature profiles with experiments under a wide range of conditions for nylon fibers (Figure 4.40). The problem still remains though about getting correct values for all the parameters needed in such a sophisticated model, and careful measurements have to be made to obtain them.

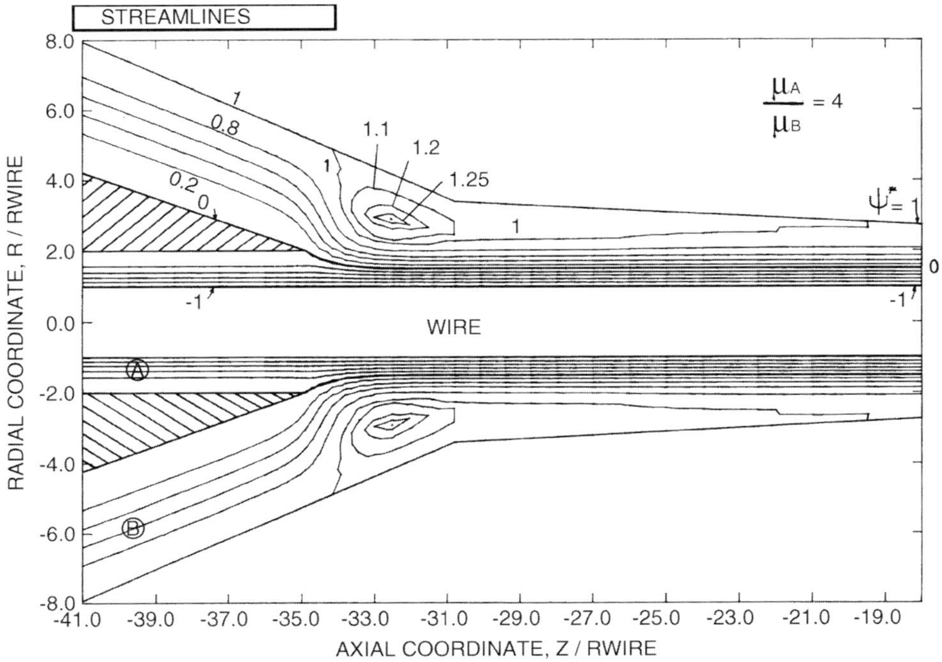

Figure 4.36 Flow pattern of two Newtonian fluids in wire-coating coextrusion. Depending on the viscosity ratio, a big vortex can be produced in this die design [88].

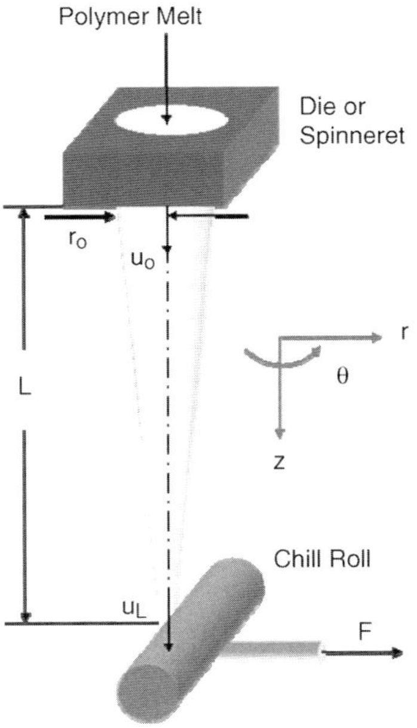

Figure 4.37 Schematic representation of the fiber-spinning process.

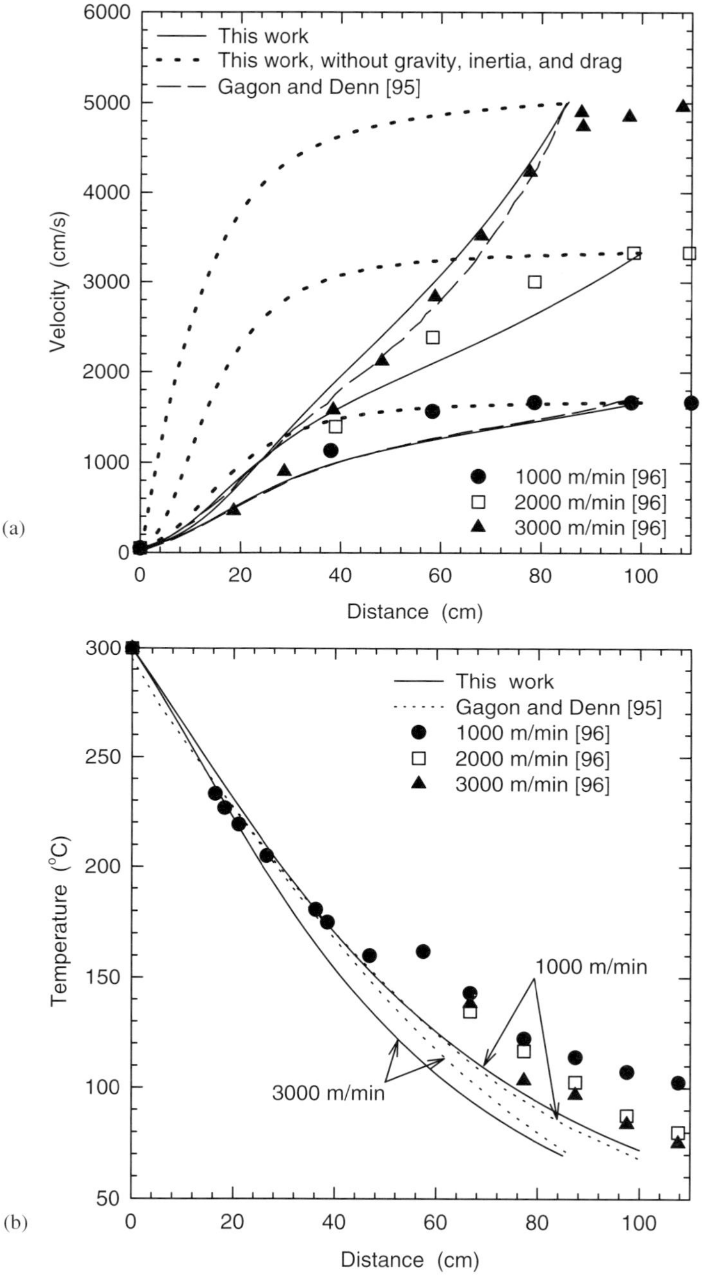

Figure 4.38 (a) Velocity and (b) temperature profiles for the spinning of a PET melt at 300 °C. Symbols represent experimental measurements by George [96], dashed lines represent simulations by Gagon and Denn [95], while full lines are viscoelastic predictions using the K-BKZ integral model (Eq. (4.12)) [103].

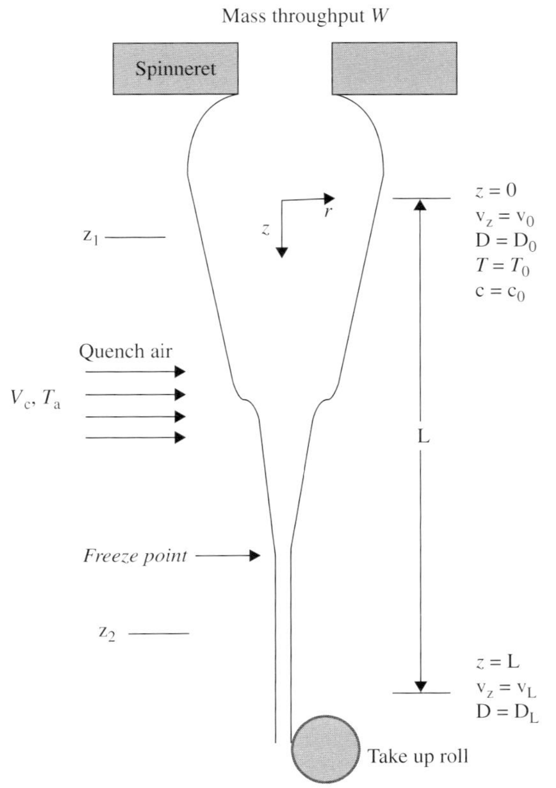

Figure 4.39 Schematic representation of the single filament melt-spinning model [107, 108].

4.4.2.5 **Film Casting**

The film-casting process is used industrially to manufacture thin flat sheets or films. The process itself is portrayed in Figure 4.41, along with the respective Cartesian coordinate system used in simulations. For this process, a polymeric material is extruded through a rectangular die or slit and taken up at the drum or chill roll. Afterward, the film is subjected to additional processes, such as biaxial extension or thermoforming, to increase the tensile strength. Typical thicknesses can vary from 10 to 2500 μm, whereas lateral dimensions range from 40 to 320 cm. For untreated films, the film-casting process has operating speeds ranging from 120 to 400 m/min, whereas if the film undergoes biaxial orientation, speeds range from 280 to 350 m/min. Cast films are primarily used in the packaging industry for either foodstuffs or other consumer products, and other uses include magnetic strips for audio and video tapes. Film casting is very similar to the fiber-spinning process, except where a fiber is drawn uniaxially (stretching of a cylindrical rod), the film or sheet is drawn planarly (stretching of a flat surface). After the film reaches the drum or chill roll, it is usually trimmed at the sides to remove the thick edges (*edge beads*).

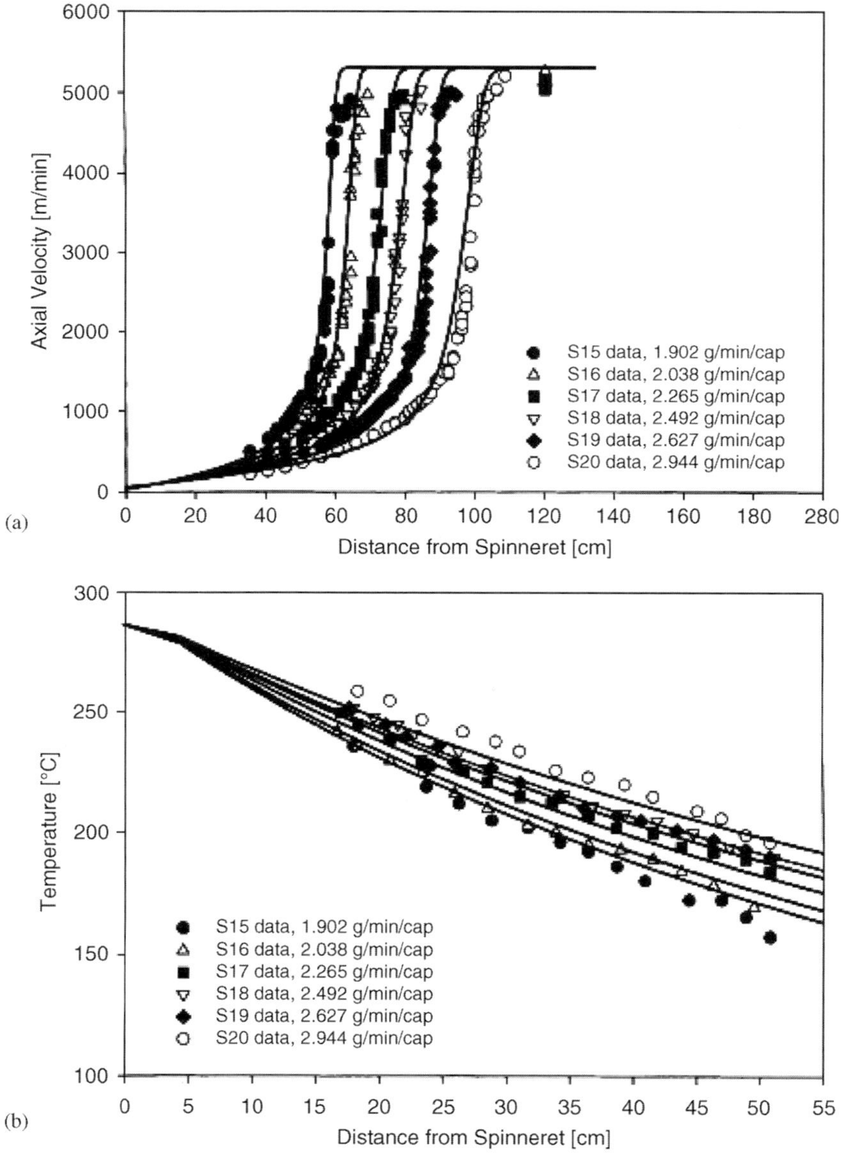

Figure 4.40 (a) Velocity and (b) temperature profiles for the spinning of industrial nylon melts at 290 °C. Symbols represent experimental measurements, while full lines are simulation predictions using a Giesekus model with crystallization [108].

In contrast to the fiber-spinning process, the film-casting process has received less attention in the literature. Dobroth and Erwin [109] examined the causes of thick edges or the *edge bead effect* (also called the *bone effect*) and attributed it to the change from planar extension in the center of the film to uniaxial extension at the edges.

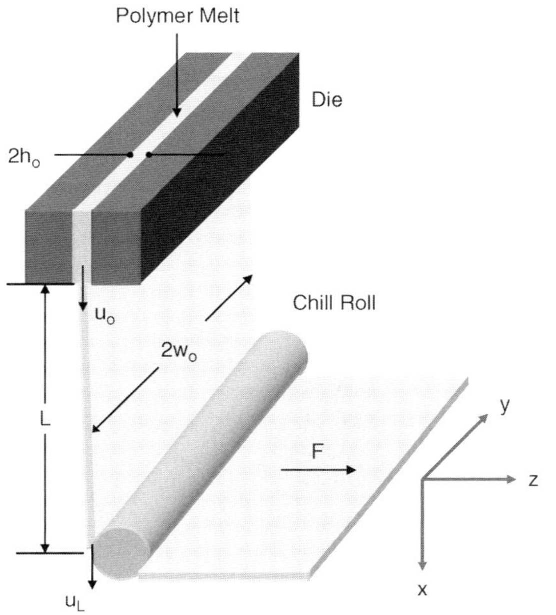

Figure 4.41 Schematic representation of the film-casting process.

Many studies have been conducted by the Co group [110–113] analyzing the stability of the film-casting process with numerous constitutive models (UCM, Carreau, Giesekus). Another major source of studies on the film-casting process has emerged from Agassant's group [114–117], which has analyzed stability and performed simulations with the Newtonian and UCM models. An attempt was made by Alaie and Papanastasiou [118] to simulate the data of Kase [119] employing a one-dimensional approximation with no *neck-in effect* (thinning in the third direction) and using the modified K-BKZ integral constitutive model. Debbaut *et al.* [120] performed a two-dimensional analysis of the film-casting process with the use of the Newtonian and UCM models. Sakaki *et al.* [121] performed a full three-dimensional analysis of the film-casting process using only the Newtonian model.

Along the lines of their previous work on fiber spinning, Beaulne and Mitsoulis [122] studied steady-state film casting with the Newtonian, UCM, and integral K-BKZ/PSM constitutive equations under isothermal and nonisothermal conditions based on simplified 1D equations. Typical results from the simulations are shown in Figure 4.42 and are compared against the previous simulations of two- and three-dimensional type [120], for both the Newtonian and UCM models. It is seen that the 1D results compare well with the 2.5D results for all models regarding the dimensionless film thickness (Figure 4.42b), while they compare well for the UCM but not for the Newtonian model regarding the dimensionless film width (Figure 4.42a). This is explained in Figure 4.43, where the Newtonian model has big curvatures in the width direction, contrary to the Maxwell model that does not. Next, results are shown from viscoelastic simulations based on experiments conducted by Kase and

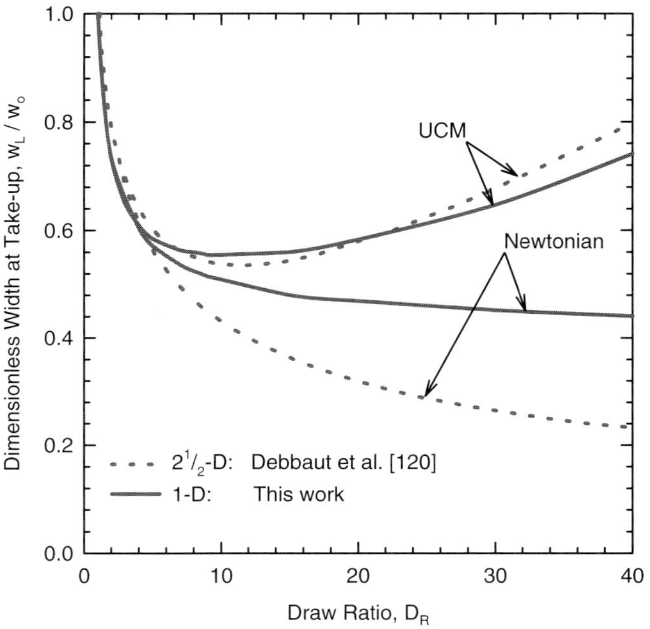

(a)

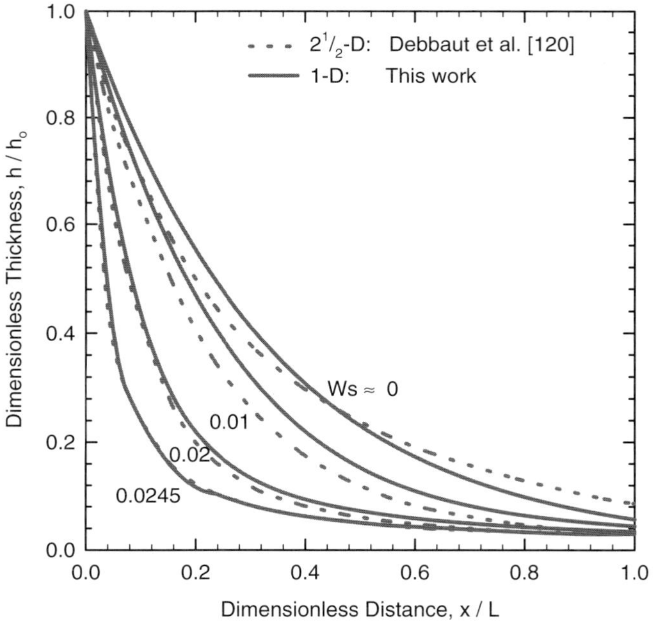

(b)

Figure 4.42 (a) Width (neck-in effect) and (b) thickness profiles for the casting of Newtonian and upper convected Maxwell (UCM) fluids in film casting. Dashed lines represent simulations by Debbaut *et al.* [120] using a 2.5D-approach, while full lines are predictions using a 1D approach with the K-BKZ integral model (Eq. (4.12)), appropriately reduced to the Newtonian and UCM models [122]. Surprisingly, the 1D UCM model gives similar results to the more thorough 2.5D-approach.

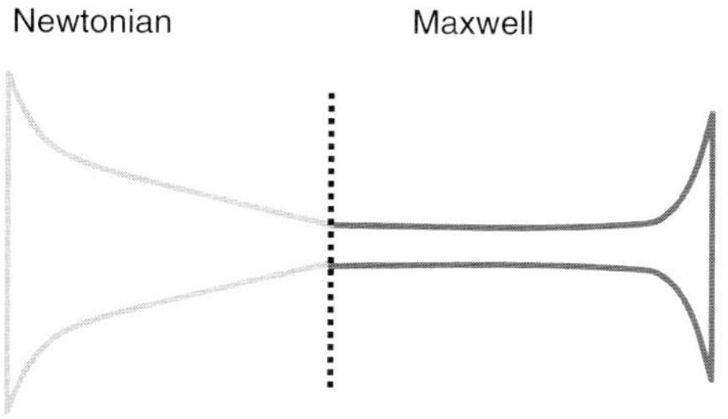

Figure 4.43 Edge bead effect for the Newtonian and the Maxwell (UCM) models. At a given cross section, the Newtonian edge bead occupies a much larger area than the UCM edge bead, at the same flow rate [120, 122].

Matsuo [123] for a polypropylene melt. Fitting the rheological data with the K-BKZ/PSM integral multimode model and knowing the operating conditions for the process, the film thickness (Figure 4.44a) and its temperature (Figure 4.44b) are well predicted by the 1D model. These results have been obtained by a quick and approximate method of analysis with the use of an efficient code, F-CAST [124]. Such a code can be used for parametric studies where there is a lack of experimental data for the process. Beaulne and Mitsoulis [122] show such a parametric study for PET and LDPE melts because they are commonly used in film casting.

4.4.2.6 **Film Blowing**

The film-blowing process is used industrially to manufacture plastic films that are biaxially oriented. Many attempts have been made to predict and model this complex but important process, which continues to mystify rheologists and polymer processing engineers worldwide. A constitutive equation, able to predict well the polymer melt in all forms of deformation, is required to model the process, together with the standard conservation equations of continuity, momentum, and energy. Pearson and Petrie [125, 126] were the first to predict the forces within the blown film by the use of the thin-shell approximation, force balances, and the Newtonian constitutive equation. The use of the thin-shell approximation and force balances is standard in any attempt to model the film-blowing process, and it has been used in the vast majority of subsequent studies.

The process itself is portrayed in Figure 4.45, where a polymer melt is extruded through an annular die, and biaxial extension is effected by slight internal pressurization and axial drawing. Cooling air is supplied by air ring jets surrounding the mid- to upper portion of the bubble. The height above the die at which solidification occurs, also known as *freeze line*, can be controlled by the cooling air. Both the deformation of the bubble and the changes in velocity and temperature are negligible above the freeze line in most processes. The bubble dimensions are measured in terms of the *blow-up ratio* (BUR), the *draw ratio* (DR), and the *thickness reduction* (TR). The BUR, which is the ratio of the bubble radius at the freeze line to the inner die radius, is typically in the range of 1–4. The DR is the ratio of the velocity at the freeze line to that of the average velocity at the die, and is typically in the range of 10–40. The TR is the

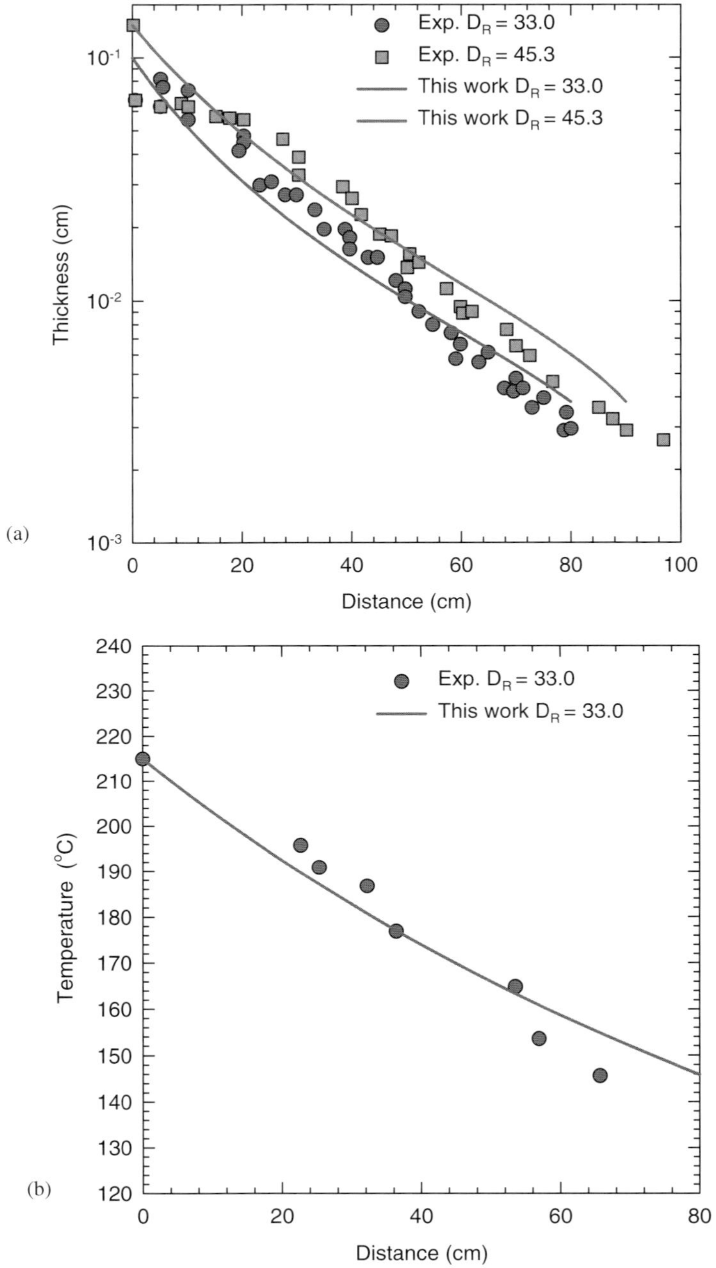

Figure 4.44 (a) Thickness and (b) temperature profiles for the casting of a polypropylene melt at 215 °C and different draw ratios, D_R. Symbols represent experimental measurements by Kase [119], while full lines are viscoelastic predictions using the K-BKZ integral model (Eq. (4.12)) [122].

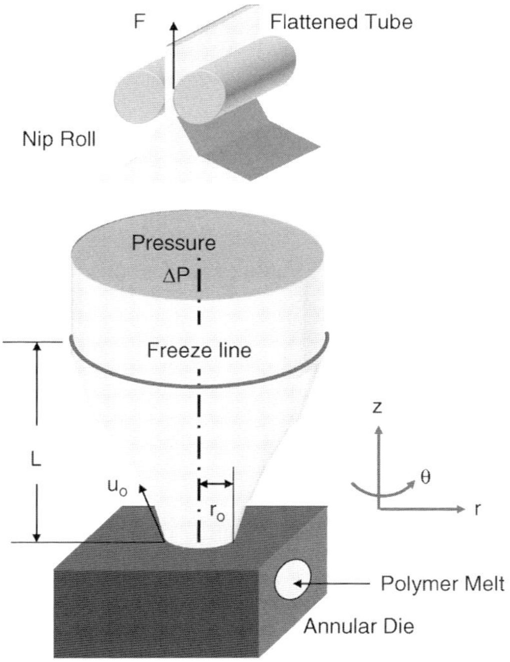

Figure 4.45 Schematic representation of the film-blowing process.

ratio of the die annular spacing to the thickness at the freeze line and is typically in the range of 20–200. The bubble is then flattened by a set of guide rolls and taken up by a set of nip rolls that form an airtight seal at the upper end of the bubble, thus forming a double-layered collapsed tube or sheet. Finally, the film is wound onto reels and sold as "lay-flat" tubing or trimmed at the edges and wound into two reels of flat film. While the film is being drawn and blown, it undergoes a nonuniform biaxial deformation. The biaxial extension of the film is the primary attraction of the film-blowing process, which increases the strength of the film in two directions and allows precise control over the mechanical, shrink, and optical properties of the finished product.

There have been numerous studies on the film-blowing process. Since the initial thin-shell approximation proposed by Pearson and Petrie [125, 126] with the Newtonian model assumed for deformation, various rheological models have been incorporated in simulations, such as the power-law model [127, 128], a crystallization model [129], the Maxwell model [130–133], the Leonov model [133], a viscoplastic–elastic model [134], the K-BKZ/PSM model [135–137], and a nonisothermal viscosity model [138]. A complete set of experimental data was reported by Gupta [139] for the Styron 666 polystyrene and by Tas [140] for three different grades of LDPE.

An approach different from that of Pearson and Petrie [125, 126] was used by Housiadas and Tsamopoulos [141, 142] according to a mapping technique. Thus, they were able to solve the time-dependent flow problem of the bubble formation and examined various aspects of the effects of forces at play in film blowing. They included viscoelasticity, nonisothermal effects, gravity, surface tension, air drag, in a series of parametric studies. The results give the bubble shape and the temperature and forces along the bubble, but no comparison was made with experimental results. Doufas and McHugh [143] addressed the subject of crystallinity in the blown film along similar lines to their previous work in fiber spinning. The stability and dynamic analysis of the process has also been recently addressed by Shin *et al.* [144].

A lack of agreement with experimental results despite the use of sophisticated viscoelastic models [136, 137] has put into question the analysis of the process based solely on rheological models and the force balance on a thin film. The work of Wolf *et al.* [41] showed that air cooling outside and inside the film are of primary importance to correctly account for the forces exerted on the blown film. This line of work was taken up in a series of papers by Sidiropoulos and Vlachopoulos [42, 43, 145, 146], who unequivocally showed that the blown air determines in a large measure the process. In all these cases, though, a known bubble shape is assumed and the simulations are based on turbulent air flow around and inside the bubble (see Figure 4.46). A full interaction of the viscoelastic film with the blowing air is still lacking and presents another difficult task for the numerical analysis of the process.

4.4.3
Unsteady-State Processes

Unsteady-state processes are very important in polymer processing, as the bulk of distinct plastic objects are manufactured in batch mode, within a time cycle. Injection molding is the most widely practiced process, followed by blow molding and its close relation, thermoforming. This has been a vast subject of research and development for the past 50–60 years, and it is impossible here to list all major developments in the field. Some highlights will be given, with key references for the interested reader to search deeper in these fascinating subjects.

4.4.3.1 Blow Molding

Extrusion blow molding or injection (stretch) blow molding is the method of choice to form any kind of hollow product, such as bottles, gas tanks, car bumpers, and so on. The whole process can be broken into four stages: (a) extrusion, (b) pinch-off, (c) blowing, and (d) cooling. The variation of the injection (stretch) blow molding is shown in Figure 4.47 [147], while the variation of extrusion blow molding is shown in Figure 4.48, along with FEM computations of a blown plastic bottle [16]. The final goal of the extrusion blow molding process consists of a blown bottle with a uniform thickness or at least a formed container where the minimum thickness of the blown product is above the minimum thickness everywhere else.

The demand for hollow products with geometric complexity and increased functionality is growing. Hence, together with the process operating parameters and the polymer rheology, the geometrical shape of a new hollow product brings its

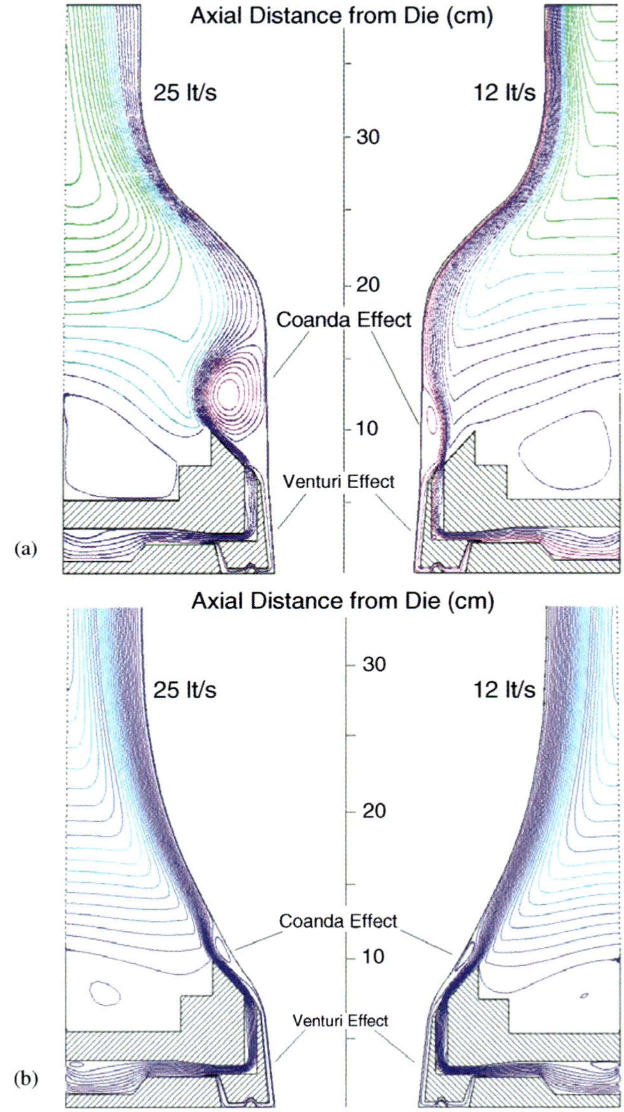

Figure 4.46 Cooling air streamlines around (a) a long-neck and (b) a shot-neck blown-film bubble for different operating setups of the adjustable air ring. Turbulent air simulations around a known and fixed bubble shape giving rise to the Coanda and Venturi effects [42, 43].

own additional difficulties. Consequently, the design and development of a new product, a new shape, remains a challenging task for the engineer since several technological and economic requirements must be met. The monograph by Rosato and Rosato [148] presents an updated survey of the complete blow molding operation. Additional modeling background is given by DeLorenzi and Nied [149].

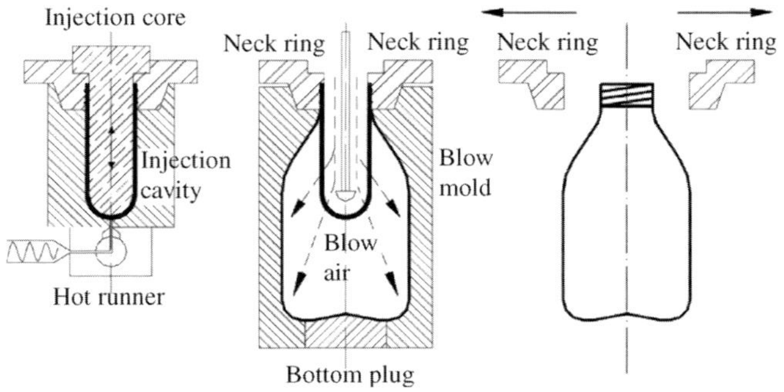

Figure 4.47 Schematic representation of the injection stretch blow-molding process [147].

In both types of blow molding, the process is a time-dependent moving boundary flow problem with free surfaces, where the polymer melt comes to an eventual contact with the walls, and after taking the shape of the mold it solidifies. This is indeed a very difficult problem, and different types of approximation have been used. DiRaddo and Garcia-Rejon [150] have used a neural network computing approach that predicts the thickness of the part from the initial configuration and the operating properties, thus getting rid of the classical fluid dynamics approach. Another study addresses the problem of parison formation [151], which is really a time-dependent variant of the annular extrudate swell problem. In this way, viscoelastic models can be used for predicting the parison thickness. The whole inflation process with a viscoelastic fluid was modeled by Schmidt *et al.* [152], who used sophisticated mesh-generation techniques to track the evolution of the moving boundaries in time. McEvoy *et al.* [153] used the commercial code ABAQUS and elastoviscoplastic models from solid mechanics to model the process. Viscoplasticity was also used by Wang *et al.* [154] to obtain fully 3D bottle shapes. Viscoelasticity with the integral multimode K-BKZ constitutive model (Eq. (4.12)) was employed by Debbaut and Homerin [155]. A novel approach based on FEM with a dynamic explicit procedure was used by Marckmann *et al.* [156] and more recently by Erchiqui [157], who also used the K-BKZ integral model (Eq. (4.12)). Nonisothermal effects have been addressed by Yang *et al.* [158]. Most works deal with PET due to its predominance in manufacturing plastic bottles [147, 158]. The degree of complexity of the models used has increased, and it appears now that a good understanding of predictive capabilities for the thickness, temperature, stress, and strain variations can be had.

4.4.3.2 **Thermoforming**

Thermoforming is the process of shaping a heated thermoplastic sheet by applying a positive air pressure, a vacuum, mechanical drawing, or combinations of these operations (see Figure 4.49). Thermoforming is the method of choice to form any kind of large flat product, such as a car door panel, a truck wind deflector, a yogurt cup, food trays, and so on. Various types of thermoforming techniques have been developed, such as plug-assist forming, drape forming, and matched mold

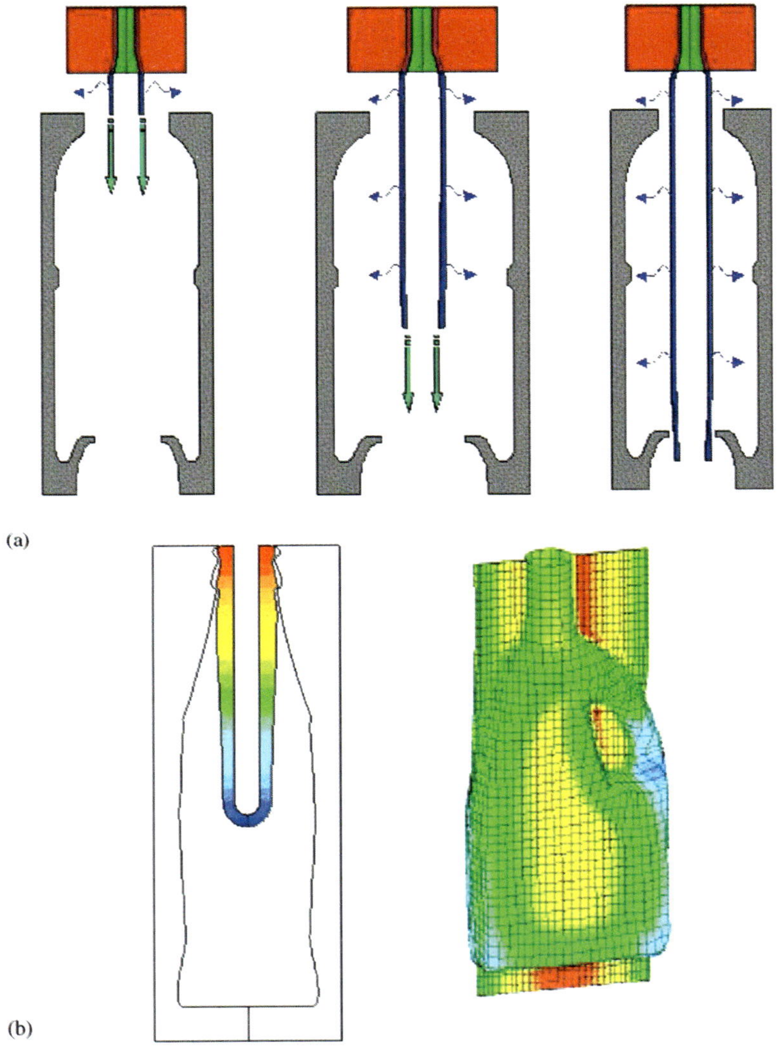

Figure 4.48 Schematic representation of the blow-molding process [16]: (a) simulation results at different time instants of parison extrusion (time-dependent annular extrudate swell) and (b) simulation results of the blow-molding process color-coded for thickness (red: thickest; blue: thinnest).

forming. The objective is to achieve good thickness distribution and to limit peripheral waste.

The whole process can be broken down into three stages: (a) extrusion, (b) inflation, and (c) cooling.

Mathematical modeling can provide valuable insights into mold design and process improvement. The objective of computer simulation of thermoforming is

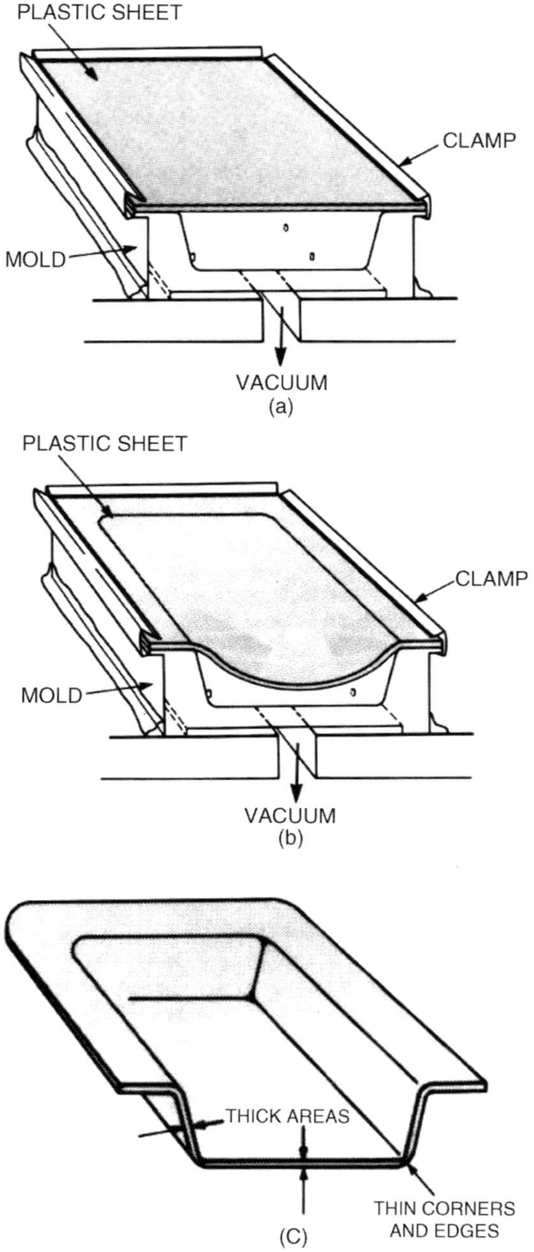

Figure 4.49 Schematic representation of straight vacuum thermoforming in three steps: (a) a thermoplastic sheet is clamped to the mold, is heated well above its glass transition temperature, and vacuum is applied; (b) the sheet is quickly bent (inflated) under vacuum; and (c) the finished hollow product is cooled and taken out of the mold [161].

an accurate determination of thickness distribution throughout the entire area of a finished product. This is of great importance in thermoforming in complex three-dimensional geometries. In such situations, it is possible to have corners, edges, bumps, dents, and other highly stretched areas with thicknesses below an acceptable limit. In production, such articles will be discarded. Thus, computer simulation may lead to a rational mold design without the need to perform expensive trial-and-error procedures. The prediction of thickness distribution via computer simulations will enable the mold designer and the process engineer to select an optimum from many possible alternatives. The monograph by Throne [159] presents an updated survey of the complete thermoforming operation.

Three types of modeling techniques have been used in the past: (a) *simple mass balances*. These models involve only geometrical considerations and are independent of material behavior [160]; (b) finite element analysis *with the membrane approximation*. This is applicable for thin-walled parts for which bending resistance is negligible [149, 161]; (c) finite element analysis *without the membrane approximation*. This is applicable for thick-walled parts and multilayer thermoforming [162, 163].

Biaxial extensional behavior is important. Strain-hardening, that is, increasing resistance of the resin to extension as deformations increase, cannot be neglected anymore. Otherwise, inaccurate results would be calculated. POLY*FLOW* [16] has implemented differential viscoelastic models in 2D and K-BKZ viscoelastic models in 3D in order to take this behavior into account.

Most of the time a membrane approximation for the sheet is considered (fast computations). The sheet is modeled as a surface deforming in a 3D environment. However, in some specific situations where the temperature, velocity, or thermal gradient across the thickness cannot be neglected anymore, a full 3D simulation including volume elements across the sheet can be done as well.

Early work by Nied *et al.* [161] and by Kouba *et al.* [164] considered the membrane as an elastic material obeying the Ogden model for large deformation of rubber sheets. Typical results are shown in Figure 4.50 for a 3D sheet being inflated (from Ref. [165]). Another 3D work [166] has used the Mooney–Rivlin strain-energy function for describing the material. Erchiqui [167] has used the K-BKZ integral model, while in other works hyperelastic and viscoelastic models have been assumed [168, 169]. Nonisothermal effects have been included in a recent publication on PET sheets [170]. In this work, it is instructive to look at the table of data (35 values) needed to model the process to get a feel of the degree of its complexity.

The development of sophisticated meshing algorithms and the overall agreement with experiments show that thermoforming is now feasible for 3D computations of complicated shapes. However, more work is needed regarding solidification and crystallization of the melts.

4.4.3.3 **Injection Molding**

Injection molding is a major polymer processing operation for producing identical articles from a hollow mold. It is an intermittent cyclic process with the following steps: (i) *filling stage*: a polymer melt is injected into the cold-walled cavity where it spreads under the action of high pressures and fills the mold;

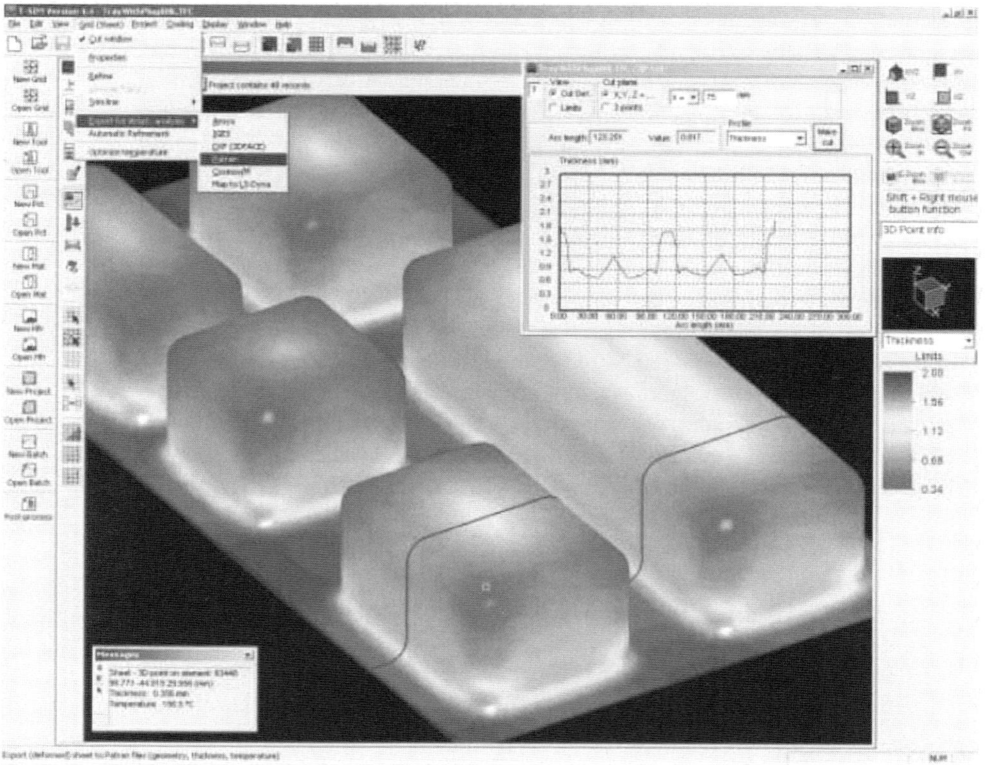

Figure 4.50 Simulation of thermoforming a hollow product showing the many drop-down menus of the software [165].

(ii) *packing stage*: after the mold is filled, high pressure is maintained, and additional melt flows into the cavity to compensate for density changes (shrinkage) during cooling; and (iii) *cooling stage*: the melt is cooled, and the shaped article is ejected. More information on the overall injection molding process is provided in Refs [2, 21, 171, 172].

During the filling–packing–cooling cycle, the polymer properties, mold design, and molding conditions interact to produce the thermomechanical history experienced by the polymer melt, which in turn determines the physical properties of the molded article. Understanding the links between process conditions and product properties is crucial for successful operation, and a predictive model enables process optimization for given product specifications.

The filling stage is the most complex and important step in the sequence and has attracted much attention in the literature. The implications and the problems related to the mathematical modeling of mold filling have been presented in a review by Mavridis *et al.* [173–175]. Earlier reviews have also been presented (see, Ref. [176]), while the latest review can be found in a most recent thesis [20].

The process in its entirety is a formidable task to tackle. The flows are time-dependent, 3D, compressible, and nonisothermal, with moving free boundaries, and solidification and crystallization phenomena taking place inside the molds. Despite (or perhaps because of) its complexity, injection molding has received the most attention in modeling and software development. Tremendous growth has been achieved by MOLDFLOW [42, 177, 178], which in the late 1970s and early 1980s used engineering ingenuity to produce useful approximate solutions in molds (it is instructive to quote the original developer, Colin Austin, telling the author in 1985: "I started it on my kitchen table!". Since then MOLDFLOW has grown incredibly and put down vast resources and efforts to address all these issues, more or less successfully. The software has moved from two-dimensional analysis to 2.5D and 3D, given the continued improvements in computer speed and memory. A typical example from a MOLDFLOW simulation on a PC window is given in Figure 4.51, where a great variety of drop-down menus is available. The whole process is broken into different stages and modules, and very fast solutions can be had with today's PCs [20]. However, since this is a proprietary software, all details are not available, and academics are still working toward developing more detailed models and computational tools for specific problems of the process. In particular,

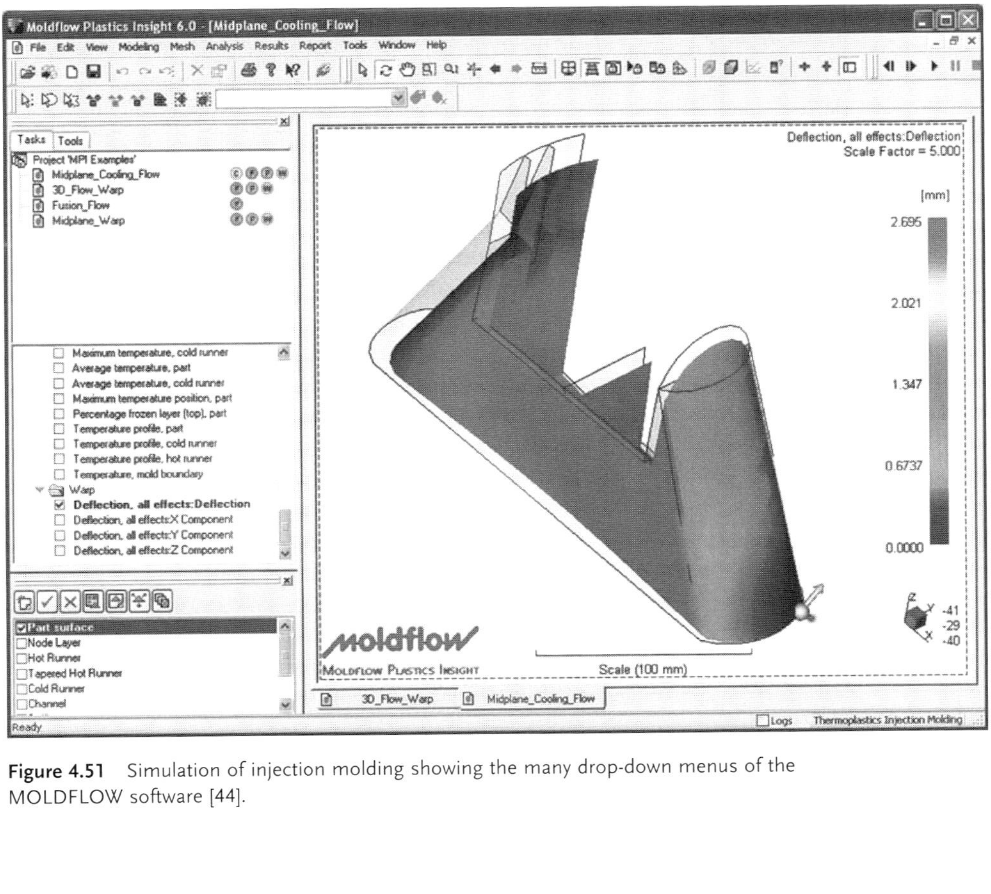

Figure 4.51 Simulation of injection molding showing the many drop-down menus of the MOLDFLOW software [44].

many challenges still remain with regard to warpage and shrinkage of the molded parts.

It is worth noting that most of the early developments in modeling and simulation came from Kamal's group at McGill University in Canada [176, 179, 180] and from Wang's group at Cornell University in the United States [181]. Important contributions were made by Vlachopoulos's group at McMaster University in Canada to the well-known "fountain flow" phenomenon of injection molding (Figure 4.52) [173–175, 182].

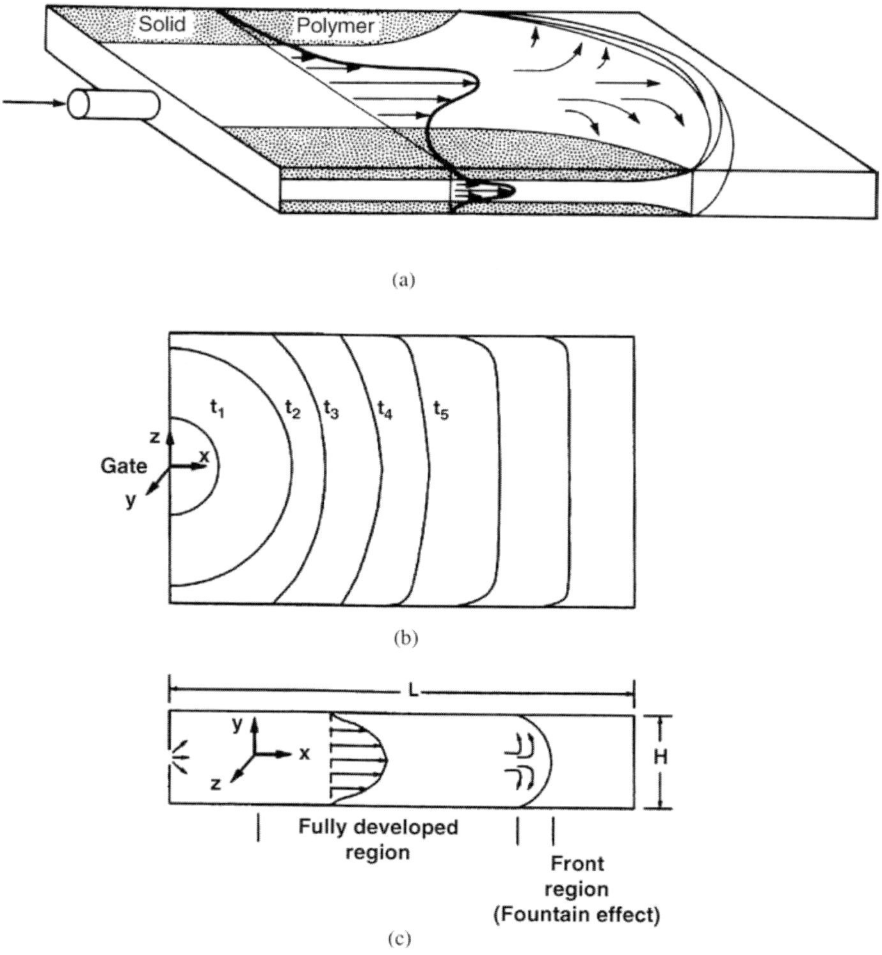

Figure 4.52 Schematic representation of flow patterns during the injection filling of an end-gated rectangular mold whose width is much greater than its thickness. (a) advancement of the flow front with fountain flow and solidification at the mold walls, (b) width direction flow fronts at various times, and (c) velocity profiles in the fully developed region and schematic representation of the fountain effect in the front region [2].

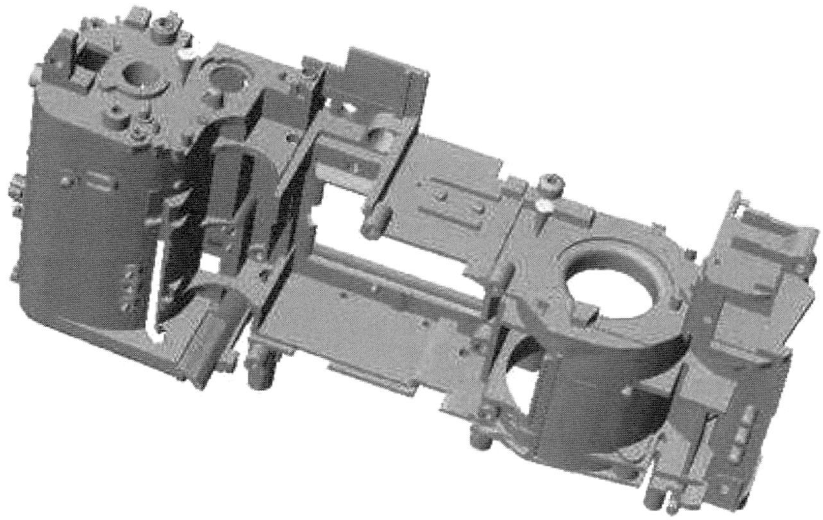

Figure 4.53 Modern injection-molding software, such as MOLDFLOW, can handle such complicated shapes as the Kodak Advantix camera [20].

Other people have also made contributions, too numerous to include in this chapter, so the interested reader may refer to Kennedy's thesis [20].

The process continues to be a challenging subject for academics, and every year at the Annual Technical Conference (ANTEC) of the Society of Plastics Engineers (SPE) in the United States, the injection molding sessions are numerous and well-attended. As mentioned in Kennedy's thesis [20], the subject of geometry (Figure 4.53), material properties, mechanical properties, additives, and so on need further work in order to predict the ultimate morphology and end-use properties of the molded product.

4.5 Conclusions

The focus of this chapter was based on the work done so far both in academia and in commercial software companies dedicated to polymer processing. Examples from some interesting and important polymer flows have been reviewed, such as those that arise from polymer processing operations with viscous (inelastic) and viscoelastic polymer melts. The topic is vast and this chapter deals exclusively with the continuum approach of flow problems. Although experimental evidence has accumulated over the years from various researchers and for various processes, the theory and predictions have lagged behind, mainly due to the complexity of the subject.

On the basis of what has been laid out in this chapter, an attempt is made in Table 4.3 to show the impact of computations on various polymer processes outlined

Table 4.3 Impact of simulations and computations on different polymer processes.

Process	Mark
Extrusion–extruder	A
Extrusion die design	A
Twin-screw extrusion	B
Coextrusion	B
Calendering	A
Roll coating	B +
Wire coating	A
Fiber spinning	B +
Film casting	B +
Film blowing	B
Blow molding	B
Thermoforming	A
Injection molding	A

Letter marking scheme corresponding to % (A + = 90–100, A = 85–89, A– = 80–84, B + = 75–79, B = 70–74, B– = 65–69, C = 60–69, D = 50–59, F = fail).

above. A professor's marking scheme has been adopted, and the marks are obviously subjective. Therefore, extrusion gets an A both for the extruder and the die design because many software exist that are routinely used by people in the field to design screws and dies rather successfully. The twin-screw extrusion gets a B because of its complexity and the relatively less attention it has received in software development. Coextrusion also gets a B because of its complexity, the unknown interfaces, the many layers, and data needed to feed into the computer (many of these have to be guessed).

The postextrusion processes follow. So, calendering gets an A because of the lubrication approximation (LAT), which has served the designers well. However, a full 3D analysis is not an easy matter. Roll coating is more difficult and gets a B + due to the splitting and unknown free surfaces present. Wire coating gets an A because the analysis and design do not seem to hide any major secrets, at least for the polymer melts used industrially. For the shear-free processes, fiber spinning gets a B + because despite many efforts, the effects of viscoelasticity and flow-induced crystallization need perhaps too many data not readily available for any meaningful computation. The same is true for film casting, although a 1D approach can give some insight into the process at hand. Film blowing appears to be worse off getting a B due to the severe interaction of turbulent blowing air with the stretched viscoelastic film.

The time-dependent or unsteady molding processes are more difficult and take longer computationally due to time effects. Thus, blow molding gets a B, again due to viscoelasticity and cooling effects and the many parameters needed for the full simulation. Thermoforming gets an A because of the thin-membrane approximation, which allows apparently fast solutions even for complicated 3D shapes. Finally, injection molding gets an A mainly due to the vast efforts of all those engineers and computer analysts who over the years have made MOLDFLOW a force to reckon with.

Another computational difficulty is related to the stability of polymer-processing flows. For example, many steady-state flows show vortices in stable operation, which may spiral out of control in an unstable operation. On top of viscoelasticity, which is still very much an active topic of research, one must consider the linear and nonlinear stability analysis, which can give operating windows, ever so crucial for good operations. This is still a much sought after subject, especially for such complicated flows as calendering and coating flows, shear-free flows, and flows through extruder channels and injection molds. It is guaranteed to keep the scientific community (and one might add, the academic community) busy for many years to come.

4.6 Current Trends and Future Challenges

Although the growth of the plastics industry is likely to continue, especially in developing countries, this industry is considered to have reached a stage of maturity [4]. Research and development efforts by major resin and machine producers have been severely curtailed in recent years, including their computational analysis and design departments. The plastic processors and original equipment manufacturers are not big enough to sustain major R&D programs that could lead to "quantum jumps" in technology.

At a recent workshop of university and industry experts [183], it was concluded that future efforts should go beyond machinery design and process analysis and optimization. The focus should be on predicting and improving the product properties of polymer-based products. The term *macromolecular engineering* was introduced as being more descriptive of future developments in the transformation of monomers into long-chain molecules and their subsequent shaping or molding into numerous useful products.

The prediction of end-use properties of polymeric products is faced with some huge challenges. The current polymer-processing simulation approach, which is based on the continuum mechanics of non-Newtonian fluids, such as was the focus of the present chapter [23, 184–187], must be combined with models describing macromolecular conformations, relaxation, and polycrystalline morphologies [188]. The various types of constitutive models, whether continuum [26], reptation [189], or pom-pom [190], have had limited success in predicting the unusual rheological phenomena exhibited by polymeric liquids, even under isothermal conditions. A few notable exceptions were achieved with the K-BKZ integral model, which has been successfully applied in polymer-processing flows, such as flows through extrusion dies, in fiber spinning, film casting, thermoforming, and so on. Determination of heat-transfer coefficients [42, 43] and modeling of flow-induced crystallization [107, 108, 191] are necessary for the eventual prediction of end-use properties of films and other extruded products. Viscoelasticity is important to predict stresses that influence crystallization, size of crystals, their properties, and so on. Numerous problems remain unresolved in other polymer processes, such as the prediction of shrinkage, warpage, and stress cracking in injection molding.

The goal of precise end-use property prediction from molecular characteristics of the polymer and the processing conditions is likely to remain the biggest challenge for a considerable time.

Acknowledgment

Financial assistance from the National Technical University of Athens (NTUA) is gratefully acknowledged.

References

1 Bernhardt, E.C. and McKelvey, J.M. (1958) Polymer processing – new engineering specialty. *Mod. Plast.*, **35**, 154–155.

2 (a) Tadmor, Z. and Gogos, C. G. (1979) *Principles of Polymer Processing, SPE Monograph Series*, 1st edn, John Wiley & Sons, Inc., New York; (b) Tadmor, Z. and Gogos, C.G. (2006) *Principles of Polymer Processing, SPE Monograph Series*, 2nd edn, John Wiley & Sons, Inc., New York.

3 (a) Han, C.D. (2007) *Rheology and Processing of Polymers. Vol. 1. Polymer Rheology*, Oxford University Press, Oxford; (b) Han, C.D. (2007) *Rheology and Processing of Polymers. Vol. 2. Polymer Processing*, Oxford University Press, Oxford.

4 Vlachopoulos, J. and Strutt, D. (2003) Overview: Polymer Processing. *Mat. Sci. Tech.*, **19**, 1153–1160.

5 Bird, R.B., Armstrong, R.C., and Hassager, O. (1987) Dynamics of polymeric liquids, in *Fluid Mechanics*, 2nd edn, vol. **1**, John Wiley & Sons, New York, pp. 1–52.

6 Barnes, H.A., Hutton, J.F., and Walters, K. (1989) *An Introduction to Rheology, Rheology Series*, Elsevier, Amsterdam, pp. 11–35.

7 Boger, D.V. and Walters, K. (1993) *Rheological Phenomena in Focus*, Elsevier, New York, pp. 35–72.

8 Bernhardt, E.C. (1959) *Processing of Thermoplastic Materials*, Reinhold, New York.

9 Huebner, K.M. and Thornton, E.A. (1982) *The Finite Element Method for Engineers*, John Wiley & Sons, Inc., New York.

10 Tanner, R.I., Nickell, R.E., and Bilger, R.W. (1975) Finite element methods for the solution of some incompressible non-Newtonian fluid mechanics problems with free surfaces. *Comput. Meth Appl. Mech. Eng.*, **6**, 155–174.

11 Crochet, M.J., Davies, A.R., and Walters, K. (1984) *Numerical Simulation of Non-Newtonian Flow*, Elsevier, Amsterdam.

12 Crochet, M.J. and Walters, K. (1983) Numerical methods in non-Newtonian fluid mechanics. *Ann. Rev. Fluid Mech.*, **15**, 241–260.

13 Keunings, R. (1986) On the high weissenberg number problem. *J. Non-Newtonian Fluid Mech.*, **20**, 209–226.

14 Marchal, J.M. and Crochet, M.J. (1987) A new mixed finite element for calculating viscoelastic flow. *J. Non-Newtonian Fluid Mech.*, **26**, 77–114.

15 Crochet, M.J. (1989) Numerical simulation of viscoelastic flow: a review. *Rubber Chem. Technol.*, **62**, 426–455.

16 www.polyflow.be (2008).

17 www.polydynamics.com (2008).

18 www.compuplast.com (2008).

19 www.polyXtrue.com (2008).

20 Kennedy, P.K. (2008) Practical and Scientific Aspects of Injection Molding Simulation. PhD Thesis, Technical University of Eindhoven (TUE), The Netherlands.

21 Middleman, S. (1977) *Fundamentals of Polymer Processing*, McGraw-Hill, New York, pp. 8–68.

22 Larson, R.G. (1988) *Constitutive Equations for Polymer Melts and Solutions*, Butterworths, Boston, pp. 59–89.

23 Baird, D.G. and Collias, D.I. (1995) *Polymer Processing: Principles and Design*, Butterworth-Heinemann, Boston, MA.

24 Larson, R.G. (1999) *The Structure and Rheology of Complex Fluids*, Oxford University Press, Oxford, pp. 1–56.
25 Dealy, J.M. and Wissbrun, K.F. (1990) *Melt Rheology and its Role in Polymer Processing. Theory and Applications*, Van Nostrand Reinhold, New York, pp. 1–41.
26 Tanner, R.I. (2000) *Engineering Rheology*, 2nd edn, Oxford University Press, Oxford, pp. 33–75.
27 Papanastasiou, A.C., Scriven, L.E., and Macosko, C.W. (1983) An integral constitutive equation for mixed flows: rheological characterization. *J. Rheol.*, **27**, 387–410.
28 Luo, X.L. and Tanner, R.I. (1988) Finite element simulation of long and short circular die extrusion experiments using integral models. *Int. J. Num. Meth Eng.*, **25**, 9–22.
29 Meissner, J. (1975) Basic parameters, melt rheology, processing and end-use properties of three similar low density polyethylene samples. *Pure Appl. Chem.*, **42**, 551–612.
30 Kajiwara, T., Barakos, G., and Mitsoulis, E. (1995) Rheological characterization of polymer solutions and melts with an integral constitutive equation. *Int. J. Polym. Anal. Char.*, **1**, 201–215.
31 Park, H.J. and Mitsoulis, E. (1992) Numerical simulation of circular entry flows of fluid M1 using an integral constitutive equation. *J. Non-Newtonian Fluid Mech.*, **42**, 301–314.
32 Barakos, G. and Mitsoulis, E. (1995) Numerical simulation of extrusion through orifice dies and prediction of bagley correction for an IUPAC-LDPE Melt. *J. Rheol.*, **39**, 193–209.
33 Sun, J., Phan-Thien, N., and Tanner, R.I. (1996) Extrudate swell through an orifice die. *Rheol. Acta*, **35**, 1–12.
34 Luo, X. L. and Mitsoulis, E. (1990) A Numerical Study of the Effect of Elongational Viscosity on Vortex Growth in Contration Flows of Polyethylene Melts *J. Rheol.*, **34**, 309–342.
35 Keunings, R. (2003) *Finite Element Methods For Integral Viscoelastic Fluids. Rheology Reviews 2003* (eds D.M. Binding and K. Walters), British Society of Rheology, London, pp. 167–195.
36 Owens, R.G. and Phillips, T.N. (2002) *Computational Rheology*, Imperial College Press, London, pp. 19–47.
37 Mitsoulis, E., Wagner, R., and Heng, F. L. (1988) Numerical Simulation of Wire-Coating Low-Density Polyethylene: Theory and Experiments. *Polym. Eng. Sci.*, **28**, 291–310.
38 Winter, H.H. (1977) Viscous dissipation in shear flows of molten polymers. *Adv. Heat Trans.*, **13**, 205–267.
39 Hatzikiriakos, S.G. and Dealy, J.M. (1992) Wall slip of molten high density polyethylenes. II. Capillary rheometer studies. *J. Rheol.*, **36**, 703–741.
40 Papanastasiou, T.C., Malamataris, N., and Ellwood, K. (1992) A new outflow boundary condition. *Int. J. Numer. Meth Fluids*, **14**, 587–608.
41 Wolf, D., Feron, B., and Wortberg, J. (1997) Numerical analysis of cooling air systems in film blowing. *Int. Polym. Proc.*, **12**, 38–44.
42 Sidiropoulos, V. and Vlachopoulos, J. (2000) An investigation of venturi and coanda effects in blown film cooling. *Int. Polym. Proc.*, **15**, 40–45.
43 Sidiropoulos, V. and Vlachopoulos, J. (2000) The effects of dual-orifice air-ring design on blown film cooling. *Polym. Eng. Sci.*, **40**, 1611–1618.
44 www.moldflow.com (2008).
45 Luo, X.L. and Tanner, R.I. (1986) A streamline element scheme for solving viscoelastic flow problems, Part II. Integral constitutive models. *J. Non-Newtonian Fluid Mech.*, **22**, 61–89.
46 Luo, X.L. and Mitsoulis, E. (1990) A finite element study of viscoelastic effects in double-layer coextrusion of polymer melts. *Adv. Polym. Technol.*, **10**, 47–54.
47 Agur, E.E. and Vlachopoulos, J. (1982) Numerical simulation of a single-screw plasticating extruder. *Polym. Eng. Sci.*, **2**, 1084–1094.
48 Viriyayuthakorn, M. and Kassahun, B. (1984) A three dimensional model for plasticating extrusion screw design. Proceedings of the ANTEC'84, Society of Plastics Engineers, vol. 30, pp. 81–84.
49 Moysey, P.A. and Thompson, A.R. (2004) Investigation of solids transport in a

single-screw extruder using a 3-D discrete particle simulation. *Polym. Eng. Sci.*, **44**, 2203–2215.

50 Janssen, L.P.B.M. (1978) *Twin Screw Extrusion*, Elsevier, Amsterdam.

51 Wang, Y., White, J.L., and Szydlowski, W. (1989) Flow in a modular intermeshing co-rotating twin screw extruder. *Int. Polym. Proc.*, **4**, 262–269.

52 Funatsu, K., Kihara, S.I., Miyazaki, M., Katsuki, S., and Kajiwara, T. (2002) 3-D numerical analysis on the mixing performance for assemblies with filled zone of right-handed and left-handed double-flighted screws and kneading blocks in twin-screw extruders. *Polym. Eng. Sci.*, **42**, 707–723.

53 Cheng, J.J. and Manas-Zloczower, I. (1989) Hydrodynamic analysis of a banbury mixer. *Polym. Eng. Sci.*, **29**, 701–708.

54 Yang, H.H. and Manas-Zloczower, I. (1992) Flow field analysis of the kneading disk region in a co-rotating twin screw extruder. *Polym. Eng. Sci.*, **32**, 1411–1417.

55 Bagley, E.B. and Birks, A.M. (1960) Flow of polyethylene into a capillary. *J. Appl. Phys.*, **31**, 556–561.

56 Boger, D.V., Hur, D.U., and Binnington, R.J. (1986) Further observations of elastic effects in tubular entry flows. *J. Non-Newtonian Fluid Mech.*, **20**, 31–49.

57 Cable, P.J. and Boger, D.V. (1978) A comprehensive experimental investigation of tubular entry flow of viscoelastic fluids: Part 1. Vortex characteristics in stable flow. *AIChE J.*, **24**, 869–879.

58 Cable, P.J. and Boger, D.V. (1978) A comprehensive experimental investigation of tubular entry flow of viscoelastic fluids: Part 2. The velocity field in stable flow. *AIChE J.*, **24**, 882–899.

59 Cable, P.J. and Boger, D.V. (1979) A comprehensive experimental investigation of tubular entry flow of viscoelastic fluids: Part 3. Unstable flow. *AIChE J.*, **25**, 152–159.

60 Nguyen, H. and Boger, D.V. (1979) The kinematics and stability of die entry flows. *J. Non-Newtonian Fluid Mech.*, **5**, 353–368.

61 Park, H.J., Kiriakidis, D.G., Mitsoulis, E., and Lee, K.-J. (1992) Birefringence studies in die flows of an HDPE melt. *J. Rheol.*, **36**, 1563–1583.

62 Kiriakidis, D.G., Park, H.J., Mitsoulis, E., Vergnes, B., and Agassant, J.-F. (1993) A study of stress distribution in contraction flows of an LLDPE melt. *J. Non-Newtonian Fluid Mech.*, **47**, 339–356.

63 Vlachopoulos, J. (1981) Extrudate swell in polymers. *Rev. Def. Beh. Mat.*, **3**, 219–248.

64 Tanner, R.I. (1973) Die-swell reconsidered: some numerical solutions using a finite element program. *Appl. Polym. Symp.*, **20**, 201–208.

65 Orbey, N. and Dealy, J.M. (1984) Isothermal swell of extrudate from annular dies; effects of die geometry, flow rate, and resin characteristics. *Polym. Eng. Sci.*, **24**, 511–518.

66 Luo, X.L. and Mitsoulis, E. (1989) Memory phenomena in extrudate swell simulations from annular dies. *J. Rheol.*, **33**, 1307–1327.

67 Ellwood, K.R.J., Papanastasiou, T.C., and Wilkes, J.O. (1992) Three-dimensional streamlined finite elements: design of extrusion dies. *Int. J. Numer. Meth Fluids*, **14**, 13–24.

68 Mitsoulis, E. (1988) Multilayer sheet coextrusion: analysis and design. *Adv. Polym. Technol.*, **8**, 225–242.

69 Strauch, T. (1986) Ein Betrag zur Rheologischen Auslegung von Coextrusionswerkzeugen. Doctoral Thesis, IKV, TH Aachen, Germany.

70 Hannachi, A. and Mitsoulis, E. (1993) Sheet coextrusion of polymer solutions and melts: comparison between simulation and experiments. *Adv. Polym. Technol.*, **12**, 217–231.

71 Mitsoulis, E. (2005) Multilayer film coextrusion of polymer melts: analysis of industrial lines with the finite element method. *J. Polym. Eng.*, **25**, 393–410.

72 Dooley, J. (2002) Viscoelastic Flow Effects in Multilayer Polymer Coextrusion. PhD Thesis, Technical University of Eindhoven (TUE), The Netherlands.

73 Anderson, P.D., Dooley, J., and Meijer, H.E.H. (2006) Viscoelastic effects in multilayer polymer extrusion. *Appl. Rheol.*, **16**, 198–205.

74 Koziey, B.L., Vlachopoulos, J., Vlcek, J., and Svabik, J. (1996) Profile die design by pressure balancing and cross flow minimization, ANTEC'96, Society of Plastics Engineers, Indianapolis, IN, pp. 247–252.

75 Carneiro, O.S. and Nobrega, J.M. (2004) Recent developments in automatic die design for profile extrusion. *Plast. Rubber Comp.*, **33**, 400–408.

76 Marchal, T. (2005) Challenges of modelling the extrusion process. *Plast. Rubber Comp.*, **34**, 265–270.

77 Debbaut, B. and Marchal, T. (2008) Numerical simulation of extrusion process and die design for industrial profile, using multimode pom-pom model. *Plast. Rubber Comp.*, **37**, 142–150.

78 Agassant, J.F. and Espy, M. (1985) Theoretical and experimental study of the molten polymer flow in the calender bank. *Polym. Eng. Sci.*, **25**, 118–121.

79 Agassant, J.F. (1980) Le Calandrage des Matières Thermoplastiques. Doctoral Thesis, Paris 6, France.

80 Mitsoulis, E., Vlachopoulos, J., and Mirza, F.A. (1985) Calendering analysis without the lubrication approximation. *Polym. Eng. Sci.*, **25**, 6–18.

81 Vlachopoulos, J. and Mitsoulis, E. (1988) *Fluid Flow and Heat Transfer in Calendering: A Review. Transport Phenomena in Polymeric Systems-2*, vol. VI (eds M.R. Kamal, R.A. Mashelkar, and A.S. Mujumdar), *Advances in Transport Processes*, Wiley Eastern, New Delhi, pp. 79–104.

82 Unkrüer, W. (1970) Beitrag zur Ermittlung des Drucksverlaufes und der Fliessvorgänge im Walzspalt bei der Kalanderverarbeitung von PVC Hart zu Folien. PhD Thesis, IKV, TU Aachen.

83 Luther, S. and Mewes, D. (2004) Three-dimensional polymer flow in the calender bank. *Polym. Eng. Sci.*, **44**, 1642–1647.

84 Coyle, D.J., Macosko, C.W., and Scriven, L.E. (1987) Film-splitting flows of shear-thinning liquids in forward roll coating. *AIChE J.*, **33**, 741–746.

85 Coyle, D.J., Macosko, C.W., and Scriven, L.E. (1990) The fluid dynamics of reverse roll coating. *AIChE J.*, **36**, 161–174.

86 Coyle, D.J., Macosko, C.W., and Scriven, L.E. (1990) Reverse roll coating of non-Newtonian liquids. *J. Rheol.*, **34**, 615–636.

87 Mitsoulis, E. (1986) Fluid flow and heat transfer in wire coating: a review. *Adv. Polym. Technol.*, **6**, 467–487.

88 Mitsoulis, E. (1986) Finite element analysis of wire coating. *Polym. Eng. Sci.*, **26**, 171–186.

89 Heng, F.L. and Mitsoulis, E. (1989) Numerical simulation of wire-coating coextrusion. *Int. Polym. Proc.*, **4**, 44–56.

90 Tadmor, Z. and Bird, R.B. (1974) Rheological analysis of stabilizing forces in wire-coating dies. *Polym. Eng. Sci.*, **14**, 124–136.

91 Ziabicki, A. and Kawai, H. (1985) *High-Speed Fiber Spinning*, John Wiley & Sons, Inc., New York.

92 Matovich, M.A. and Pearson, J.R.A. (1969) Spinning a molten threadline. *Ind. Eng. Chem. Fund*, **8**, 512–521.

93 Denn, M.M., Petrie, C.J.S., and Avenas, P. (1975) Mechanics of steady spinning of a viscoelastic fluid. *AIChE J.*, **21**, 791–799.

94 Fisher, R.J. and Denn, M.M. (1977) A theory of isothermal melt spinning and draw resonance. *AIChE J.*, **23**, 23–29.

95 Gagon, D.K. and Denn, M.M. (1981) Computer simulation of steady polymer melt spinning. *Polym. Eng. Sci.*, **21**, 844–853.

96 George, H.H. (1982) Model of steady-state melt spinning at intermediate take-up speeds. *Polym. Eng. Sci.*, **22**, 292–299.

97 Keunings, R., Crochet, M.J., and Denn, M.M. (1983) Profile development in continuous drawing of viscoelastic liquids. *Ind. Eng. Chem. Fund*, **22**, 347–355.

98 Denn, M.M. (1983) Fibre spinning, *Computational Analysis of Polymer Processing* (eds J.R.A. Pearson and S.M. Richardson), Applied Science Publishers, London, pp. 179–216.

99 Papanastasiou, A.C., Macosko, C.W., Scriven, L.E., and Chen, Z. (1987) Fiber spinning of viscoelastic liquid. *AIChE J.*, **33**, 834–842.

100 Zeichner, G.R. (1973) Spinnability of Viscoelastic Fluids. M.Sc.E. Thesis, Department of Chemical Engineering, University of Delaware, Newark, DE.

101 Fulchiron, R., Verney, V., Michel, A., and Roustant, J.C. (1995) Correlations between relaxation time spectrum and melt spinning behavior of polypropylene.

II: Melt spinning simulation from relaxation time spectrum. *Polym. Eng. Sci.*, **35**, 518–527.

102 Rauschenberger, V. and Laun, H.M. (1997) A recursive model for rheotens tests. *J. Rheol.*, **41**, 719–737.

103 Mitsoulis, E. and Beaulne, M. (2000) Numerical simulation of rheological effects in fiber spinning. *Adv. Polym. Technol.*, **19**, 155–172.

104 Mitsoulis, E. and Beaulne, M. (1997) *F-SPIN – A Software Package for Fiber Spinning of Plastics*, Department of Chemical Engineering, University of Ottawa, Canada.

105 Beris, A.N. and Liu, B. (1988) Time-dependent fiber spinning equations. 1. Analysis of the mathematical behavior. *J. Non-Newtonian Fluid Mech.*, **26**, 341–361.

106 Liu, B. and Beris, A.N. (1988) Time-dependent fiber spinning equations. 2. Analysis of the stability of numerical approximations. *J. Non-Newtonian Fluid Mech.*, **26**, 363–394.

107 Doufas, A.K., McHugh, A.J., Miller, C., and Immaneni, A. (2000) Simulation of melt spinning including flow-induced crystallization. Part I. Model development and predictions. *J. Non-Newtonian Fluid Mech.*, **92**, 27–66.

108 Doufas, A.K., McHugh, A.J., Miller, C., and Immaneni, A. (2000) Simulation of melt spinning including flow-induced crystallization. Part II. Quantitative comparisons with industrial spinline data. *J. Non-Newtonian Fluid Mech.*, **92**, 81–103.

109 Dobroth, T. and Erwin, L. (1986) Causes of edge beads in cast film. *Polym. Eng. Sci.*, **26**, 462–467.

110 Anturkar, N.R. and Co, A. (1988) Draw resonance in film casting of viscoelastic fluids: a linear stability analysis. *J. Non-Newtonian Fluid Mech.*, **28**, 287–307.

111 Vardarajan, R.I. and Co, A. (1996) Film casting of a modified giesekus fluid: stability analysis. *Chem. Eng. Sci.*, **9**, 1417–1430.

112 Piz-Lopez, M.E. and Co, A. (1996) Multilayer film casting of modified giesekus fluids. Part 1. Steady-state analysis. *J. Non-Newtonian Fluid Mech.*, **66**, 71–93.

113 Piz-Lopez, M.E. and Co, A. (1996) Multilayer film casting of modified giesekus fluids. Part 2. Linear stability analysis. *J. Non-Newtonian Fluid Mech.*, **66**, 95–114.

114 Barq, P., Haudin, J.M., Agassant, J.F., Roth, H., and Bourgin, P. (1990) Instability phenomena in film casting process. *Int. Polym. Proc.*, **4**, 264–271.

115 Barq, P., Haudin, J.M., and Agassant, J.F. (1992) Isothermal and anisothermal models for cast film extrusion. *Int. Polym. Proc.*, **7**, 334–349.

116 d'Halewyn, S., Agassant, J.F., and Demay, Y. (1990) Numerical simulation of the cast film process. *Polym. Eng. Sci.*, **30**, 335–340.

117 Silagy, D., Demay, Y., and Agassant, J.F. (1996) Study of the stability of the film casting process. *Polym. Eng. Sci.*, **36**, 2614–2625.

118 Alaie, S.M. and Papanastasiou, T.C. (1991) Viscoelastic film casting. *Polym. Eng. Sci.*, **31**, 67–75.

119 Kase, S. (1974) Studies on melt spinning. IV. On the stability of melt spinning. *J. Appl. Polym. Sci.*, **18**, 3279–3304.

120 Debbaut, B., Marchal, J.M., and Crochet, M.J. (1995) Viscoelastic effects in film casting. *Z Angew Math Phys.*, **46**, S679–S698.

121 Sakaki, K., Katsumoto, R., Kajiwara, T., and Funatsu, K. (1996) Three-dimensional flow simulation of a film-casting process. *Polym. Eng. Sci.*, **36**, 1821–1831.

122 Beaulne, M. and Mitsoulis, E. (1999) Numerical simulation of the film-casting process. *Int. Polym. Proc.*, **14**, 261–275.

123 Kase, S. and Matsuo, T. (1967) Studies on melt spinning. II. Steady-state and transient solutions of fundamental equations compared with experimental results. *J. Appl. Polym. Sci.*, **11**, 251–287.

124 Mitsoulis, E. and Beaulne, M. (1997) *F-CAST – A Software Package for Film Casting of Plastics*, Department of Chemical Engineering, University of Ottawa, Canada.

125 Pearson, J.R.A. and Petrie, C.J.S. (1970) The flow of a tubular film. Part 1. Formal mathematical representation. *J. Fluid Mech.*, **40**, 1–19.

126 Pearson, J.R.A. and Petrie, C.J.S. (1970) The flow of a tubular film. Part 2.

Interpretation of the model and discussion of solutions. *J. Fluid Mech.*, **42**, 609–625.

127 Han, C.D. and Park, J.Y. (1975) Studies on blown film extrusion II: analysis of the deformation and heat transfer process. *J. Appl. Polym. Sci.*, **19**, 3257–3277.

128 Gupta, R.K., Metzner, A.B., and Wissbrun, K.F. (1982) Modeling of polymeric film-blowing processes. *Polym. Eng. Sci.*, **22**, 172–181.

129 Kanai, T. and White, J.L. (1984) Kinematics, dynamics and stability of the tubular film extrusion of various polyethylenes. *Polym. Eng. Sci.*, **24**, 1185–1201.

130 Petrie, C.J.S. (1973) Memory effect in the non-uniform flow: a study of the behaviour of tubular film of viscoelastic fluid. *Rheol. Acta*, **12**, 92–98.

131 Wagner, M.H. (1976) Blown film extrusion as rheological-thermodynamic process. *Rheol. Acta.*, **15**, 40–51.

132 Cain, J.J. and Denn, M.M. (1988) Multiplicities and instabilities in film blowing. *Polym. Eng. Sci.*, **28**, 1527–1541.

133 Luo, X.L. and Tanner, R.I. (1985) A computer study of film blowing. *Polym. Eng. Sci.*, **25**, 620–629.

134 Cao, B. and Campbell, G.A. (1990) Viscoplastic-elastic modeling of tubular blown film processing. *AIChE J.*, **36**, 420–430.

135 Alaie, S.M. and Papanastasiou, T.C. (1993) Modeling of non-isothermal film blowing with integral constitutive equations. *Int. Polym. Proc.*, **8**, 51–65.

136 Beaulne, M. and Mitsoulis, E. (1998) Numerical simulation of the film-blowing process. *Int. J. Form. Proc.*, **1**, 451–484.

137 Beaulne, M. and Mitsoulis, E. (2007) Effect of viscoelasticity in the film-blowing process. *J. Appl. Polym. Sci.*, **105**, 2098–2112.

138 Sidiropoulos, V., Tain, J.J., and Vlachopoulos, J. (1996) Computer simulation of film blowing. *J. Plast. Film Sheeting*, **12**, 107–129.

139 Gupta, R.K. (1980) A New Non-Isothermal Rheological Constitutive Equation and its Application to Industrial Film Blowing Processes. PhD Thesis, Department of Chemical Engineering, University of Delaware, Newark, DE.

140 Tas, P.P. (1994) Film Blowing from Polymer to Product. PhD Thesis, Eindhoven University of Technology, Department of Mechanical Engineering, Eindhoven, The Netherlands.

141 Housiadas, K. and Tsamopoulos, J. (2000) Unsteady extrusion of a viscoelastic annular film – I. General model and its numerical solution. *J. Non-Newtonian Fluid Mech.*, **88**, 229–259.

142 Housiadas, K. and Tsamopoulos, J. (2000) Cooling of a viscoelastic film during unsteady extrusion from an annular die. *Rheol. Acta*, **39**, 44–61.

143 Doufas, A.K. and McHugh, A.J. (2001) Simulation of film blowing including flow-induced crystallization. *J. Rheol.*, **45**, 1085–1104.

144 Shin, D.M., Lee, J.S., Jung, H.W., and Hyun, J.C. (2007) Multiplicity, bifurcation, stability and hysteresis in dynamic solutions of film blowing process. *J. Rheol.*, **51**, 605–621.

145 Sidiropoulos, V. and Vlachopoulos, J. (2002) Numerical simulation of blown film cooling. *J. Reinf. Plast. Comp.*, **21**, 629–637.

146 Sidiropoulos, V. and Vlachopoulos, J. (2005) Temperature gradients in blown film bubbles. *Adv. Polym. Technol.*, **24**, 83–90.

147 Pham, X.T., Thibault, F., and Lim, L.T. (2004) Modeling and simulation of stretch blow molding of polyethylene terephthalate. *Polym. Eng. Sci.*, **44**, 1460–1472.

148 Rosato, D.V. and Rosato, D.V. (1989) *Blow Molding Handbook*, Hanser Publishers, Munich.

149 DeLorenzi, H.G. and Nied, H.F. (1991) Finite element simulation of thermoforming and blow molding, in *Progress in Polymer Processing* (ed. A.I. Isayev), Carl Hanser Verlag, Munich, p. 117.

150 DiRaddo, R.W. and Garcia-Rejon, A. (1993) On-line prediction of final part dimensions in blow molding: a neural network computing approach. *Polym. Eng. Sci.*, **33**, 653–664.

151 Tanoue, S., Kajiwara, T., Funatsu, K., Terada, K., and Yamabe, M. (1996) Numerical simulation of blow molding – prediction of parison diameter and

thickness distributions in the parison formation process. *Polym. Eng. Sci.*, **36**, 2008–2017.

152 Schmidt, F.M., Agassant, J.F., Bellet, M., and Desoutter, L. (1996) Viscoelastic simulation of PET stretch/blow molding process. *J. Non-Newtonian Fluid Mech.*, **64**, 19–42.

153 McEvoy, J.P., Armstrong, C.G., and Crawford, R.J. (1998) Simulation of the stretch blow molding process of PET bottles. *Adv. Polym. Technol.*, **17**, 339–352.

154 Wang, S., Makinouchi, A., and Nakagawa, T. (1998) Three-dimensional viscoplastic FEM simulation of a stretch blow molding process. *Adv. Polym. Technol.*, **17**, 189–202.

155 Debbaut, B. and Homerin, O. (1999) A comparison between experiments and predictions for the blow molding of an industrial part. *Polym. Eng. Sci.*, **39**, 1812–1822.

156 Marckmann, G., Verron, E., and Peseux, B. (2001) Finite element analysis of blow molding and thermoforming using a dynamic explicit procedure. *Polym. Eng. Sci.*, **41**, 426–439.

157 Erchiqui, F. (2006) A new hybrid approach using the explicit dynamic finite element method and thermodynamic law for the analysis of the thermoforming and blow molding processes for polymer materials. *Polym. Eng. Sci.*, **46**, 1554–1564.

158 Yang, Z.J., Harkin-Jones, E., Menary, G.H., and Armstrong, C.G. (2004) A non-isothermal finite element model for injection stretch-blow molding of PET bottles with parametric studies. *Polym. Eng. Sci.*, **44**, 1379–1390.

159 Throne, J.L. (1987) *Thermoforming*, Hanser Publishers, Munich.

160 Crawford, R.J. and Lui, S.K.L. (1982) Prediction of wall thickness distribution in thermoformed moldings. *Eur. Polym. J.*, **18**, 699–705.

161 Nied, H.F., Taylor, C.A., and DeLorenzi, H.G. (1990) Three-dimensional finite element simulation of thermoforming. *Polym. Eng. Sci.* **30**, 1314–1322.

162 Song, W.N., Mirza, F.A., and Vlachopoulos, J. (1991) Finite-element analysis of inflation of an axisymmetric sheet of finite thickness. *J. Rheol.*, **35**, 93–111.

163 Song, W.N., Mirza, F.A., and Vlachopoulos, J. (1992) Finite-element simulation of plug-assist forming. *Int. Polym. Proc.*, **7**, 248–256.

164 Kouba, K., Bartos, O., and Vlachopoulos, J. (1992) Computer simulation of thermoforming in complex shapes. *Polym. Eng. Sci.*, **32**, 699–704.

165 www.t-sim.com (2008).

166 Nam, G.J., Ahn, K.H., and Lee, J.W. (2000) Three-dimensional simulation of thermoforming process and its comparison with experiments. *Polym. Eng. Sci.*, **40**, 2232–2240.

167 Erchiqui, F. (2005) Thermodynamic approach of inflation process of K-BKZ polymer sheet with respect to thermoforming. *Polym. Eng. Sci.*, **45**, 1319–1335.

168 Erchiqui, F., Gakwaya, A., and Rachik, M. (2005) Dynamic finite element analysis of nonlinear isotropic hyperelastic and viscoelastic materials for thermoforming applications. *Polym. Eng. Sci.*, **45**, 124–134.

169 Karamanou, M., Warby, M.K., and Whiteman, J.R. (2006) Computational modelling of thermoforming processes in the case of finite viscoelastic materials. *Comput. Meth. Appl. Mech. Eng.*, **195**, 5220–5238.

170 Erchiqui, F., Souli, M., and Ben Yedder, R. (2007) Nonisothermal finite-element analysis of thermoforming of polyethylene terephthalate sheet: incomplete effect of the forming stage. *Polym. Eng. Sci.*, **47**, 2129–2144.

171 Pearson, J.R.A. (1975) *Mechanics of Polymer Processing*, Elsevier, London.

172 Bernhardt, E.C. (1983) *CAE: Computer Aided Engineering for Injection Molding*, Hanser, Munich.

173 Mavridis, H., Hrymak, A.N., and Vlachopoulos, J. (1986) Finite element simulation of fountain flow in injection molding. *Polym. Eng. Sci.*, **26**, 449–454.

174 Mavridis, H., Hrymak, A.N., and Vlachopoulos, J. (1986) Deformation and orientation of fluid elements behind an advancing flow front. *J. Rheol.*, **30**, 555–563.

175 Mavridis, H., Hrymak, A.N., and Vlachopoulos, J. (1986) Mathematical modeling of injection mold filling: a review. *Adv. Polym. Technol.*, **6**, 457–466.

176 Kamal, M.R. and Lafleur, P.G. (1982) Computer modeling of injection molding. *Polym. Eng. Sci.*, **22**, 1066–1074.

177 Austin, C. (1983) Filling of mold cavities, in *Computer Aided Engineering for Injection Molding* (ed. E.C. Bernhardt), Hanser, New York.

178 Austin, C.A. (1985) *Moldflow Design Principles*, Moldflow Pty Ltd, Melbourne, Australia.

179 Kamal, M.R., Chu, E., Lafleur, P.G., and Ryan, M.E. (1986) Computer simulation of injection mold filling for viscoelastic melts with fountain flow. *Polym. Eng. Sci.*, **26**, 190–196.

180 Kamal, M.R., Goyal, S.K., and Chu, E. (1988) Simulation of injection mold filling of viscoelastic polymer with fountain flow. *AIChE J.*, **34**, 94–106.

181 Wang, V.W., Hieber, C.A., and Wang, K.K. (1987) C-FLOW: A CAE package with high level interactive graphics, in *Applications of Computer Aided Engineering in Injection Molding* (ed. L.T. Manzione), Hanser, New York.

182 Mavridis, H., Hrymak, A.N., and Vlachopoulos, J. (1988) The effect of fountain flow on molecular orientation in injection molding. *J. Rheol.*, **32**, 639–663.

183 Final Workshop Report: Touchstones of Modern Polymer Processing. Polymer Processing Institute, NJIT, Newark, NJ, USA, June 2002.

184 Tucker, C.L., III (1989) *Fundamentals of Computer Modeling for Polymer Processing*, Hanser, Munich, Germany.

185 Isayev, A.I. (1991) *Modeling of Polymer Processing*, Hanser, Munich, Germany.

186 O'Brien, K.T. (1992) *Applications of Computer Modeling for Extrusion and Other Continuous Polymer Processes*, Hanser, Munich, Germany.

187 Covas, J.A., Agassant, J.F., Diogo, A.C., Vlachopoulos, J., and Walters, K. (1995) *Rheological Fundamentals of Polymer Processing*, Kluwer, Dordrecht, The Netherlands.

188 Theodorou, D.N. (2007) Hierarchical modeling of polymeric materials. *Chem. Eng. Sci.*, **62**, 5697–5714.

189 Macosko, C.W. (1994) *Rheology: Principles, Measurements and Applications*, VCH Publishers, New York.

190 McLeish, T.C.B. and Larson, R.G. (1998) Molecular constitutive equations for a class of branched polymers: the pom-pom polymer. *J. Rheol.*, **42**, 82–112.

191 Kanai, T. and Campbell, G.A. (1999) *Film Processing*, Hanser, Munich, Germany.

5
Computational Approaches for Structure Formation in Multicomponent Polymer Melts

Marcus Müller

5.1
Minimal, Coarse-Grained Models, and Universality

Blending different polymer species is a versatile strategy to fabricate new materials [1, 2]. A technologically important example is rubber-toughened polystyrene, where the brittle polystyrene is reinforced by a rubber material, polybutadiene. Unlike metallic alloys, however, multicomponent polymer blends often do not mix on a microscopic scale. Owing to the connectivity of monomeric units along the macromolecule, the translational entropy is greatly reduced and a minuscule repulsion χ between unlike monomeric units gives rise to local demixing. The different components segregate into domains, which are separated by interfaces, such that the material can be conceived as an assembly of interfaces. The morphology of the material – the size and spatial arrangement of domains – dictates the mechanical properties.

Phase separation in multicomponent polymer blends is technologically an important process. Thus, much effort has been directed toward controlling the morphology of a blend. In thermodynamic equilibrium, two incompatible polymers will separate into macroscopically large domains. Due to the slow dynamics of the long macromolecules and the small thermodynamic driving force for coarsening of the morphology, this ultimate equilibrium state is often not reached on experimentally relevant timescales. Both the thermodynamics and the kinetics of phase separation can be utilized to tailor the experimentally observed morphology. For instance, by adding copolymers to the blend, one can reduce the interfacial tension [3] and hinder the coalescence of domains [4] and, thereby, achieve a beneficial, finer dispersion of the two components.

Likewise, microphase separation [5] in block copolymer materials has attracted abiding interest. In a dibelock copolymer, two chemically different species are irreversibly bonded together to form a single macromolecule. The linkage of the two parts prevents macroscopic phase separation and the different species arrange into periodical spatial structures with a length scale that is dictated by the molecular extension, R_e. Depending on the volume fraction, f, of the different blocks, a variety of microphase-separated structures can be formed including lamellar structures,

Modeling and Simulation in Polymers. Edited by P.D. Gujrati and A.I. Leonov

ISBN: 978-3-527-32415-6

cylindrical domains arranged on a hexagonal lattice, or micellar structures that crystallize on a BCC lattice [5–7]. These fascinating materials have been employed, for example, to template structures at the nanoscale [8–10].

In both cases, a theoretical description has to deal with a multitude of length, time-, and energy scales, which range from the atomistic characteristics of the constituent chemical repeat units (length scale: Angstrom; timescale: 10^{-12}s; and energy scale: eV) to the long time and large length scales associated with the kinetics of macrophase separation, the annihilation of defects, or the motion of grain boundaries in microphase-separated structures (length scale: μm; timescale: hours or days; energy scale: $k_B T$). Up to now, there is no single computational technique for simultaneously describing all these different scales [11], and we will restrict ourselves to the discussion of coarse-grained models that are able to address large time- and length scales and yet retain the notion of molecular conformations. These models build a bridge between the atomistic description of chemical details and continuum models (e.g., phase-field models), where the relevant degrees of freedoms are collective variables (e.g., local density fields) rather than the microscopic coordinates of the individual molecules.

Coarse-grained models for dense polymeric systems are built upon the concept that the characteristics on short length scales and timescales can be parameterized by a small number of coarse-grained parameters or invariants. Systems that differ in their atomistic architecture but that are characterized by the same coarse-grained parameters will exhibit the same behavior on large length and time scales. This concept is illustrated in Figure 5.1. The universality of the behavior of multicomponent polymer melts is indeed confirmed by experiments. Its foundation lays in the Gaussian, self-similar structure of the polymers in a dense melt [12, 13]. Conceptually, a coarse-grained model can be obtained by successively lumping a small number of neighboring repeat units along the molecular backbone into effective interaction centers [11]. For instance, interactions between these effective interaction centers can be explicitly constructed by inverse Boltzmann sampling [14, 15]. Typically, they are softer than those between the original repeat units, they depend on the thermodynamic state of the system (i.e., temperature and density), and they are comprised of multibody interactions. If the polymers are very long, this coarse-graining procedure can be repeated over and over again. The limiting model that is obtained after many coarse-graining steps is characterized only by a small number of relevant interactions, that is, starting with systems that differ in their microscopic structure one arrives at a common, limiting, coarse-grained model that can be characterized by the strength of a small number of relevant interactions. The coarse-grained parameters are invariants that parameterize these relevant interactions [16]. The concept of successive elimination of the degrees of freedom is formally described by the renormalization group theory of polymers [13, 17–19], and it shares many aspects with critical phenomena. It rationalizes the experimental observation of universal behavior and justifies the use of minimal, coarse-grained models. Both different experimental systems and different computational models will exhibit the same behavior on large length scales if they converge toward the same system under successive elimination of degrees of freedom.

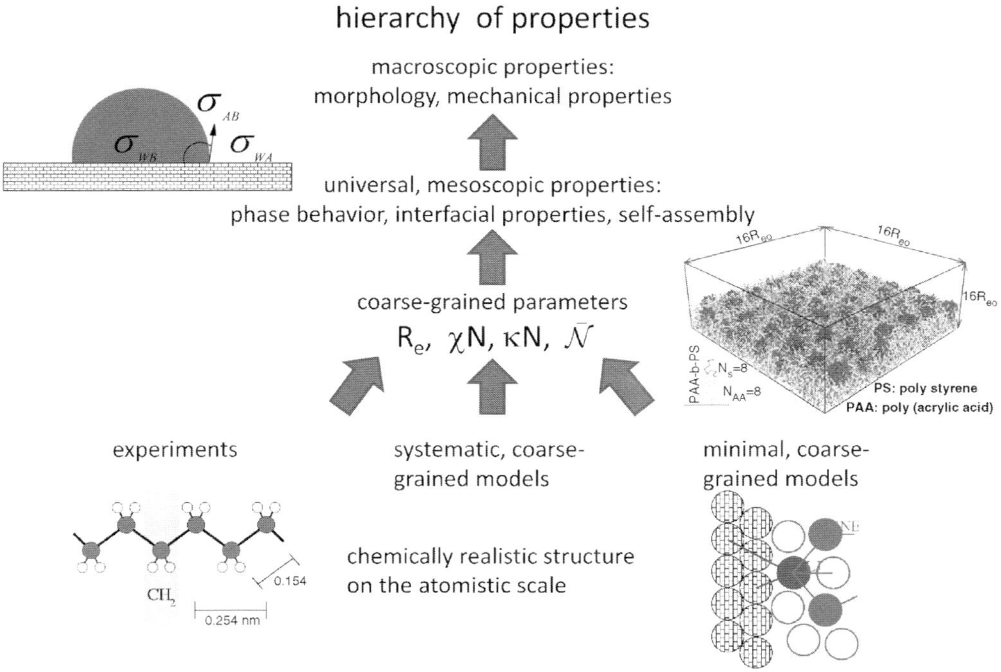

Figure 5.1 Illustration of properties on different length scales for a binary polymer blend. Coarse-grained models rely on the assumption that properties on large length scales depend on microscopic, chemical details only via a small number of invariants or coarse-grained parameters. For a symmetric, incompressible, polymer blend, the set of coarse-grained parameters comprises R_e, χN, $\bar{\mathcal{N}}$, characterizing length and energy scales and the strength of fluctuations, respectively. The set of coarse-grained parameters depends on the problem, for example, for asymmetric blends (cf. Figure 5.2) also the ratio of the chain extensions, $\eta = R_e^{(B)}/R_e^{(A)}$, or the ratio of the segmental volumes constitute invariants.

The limiting model for a multicomponent polymer melt is characterized by the following relevant interactions: (i) the connectivity of the long, flexible macromolecules, (ii) repulsive interactions between all segments in a dense liquid that restricts long wavelength density fluctuations, (iii) the density of molecules, and (iv) the thermal interactions between the different molecular species in a multicomponent melt, which drive macroscopic phase separation or self-assembly. The strength of these interactions can be described by the following coarse-grained parameters:

1) The shape of a long flexible macromolecule in a melt is Gaussian and solely described by its mean-squared end-to-end distance, R_e^2. This parameter serves to identify the unit of length. Obviously, it does not explicitly depend on the number of effective interaction centers that are used to describe the molecular contour.
2) The suppression of the density fluctuations can either be described by an incompressibility constraint or by a finite but small compressibility, $1/\varkappa$, of the dense polymeric liquid. It is important to realize that this constraint can be

enforced only after coarse-graining, that is, on a large length scale. On the length scale of the atomistic constituents, density fluctuations and packing effects always remain large.

3) The molecular density is parameterized by the invariant degree of polymerization $\bar{\mathcal{N}} \equiv \left(\frac{\varrho_0}{N} R_e^3\right)^2$ where ϱ_0 and N denote the density of coarse-grained segments and the number of segments per molecule, respectively. The Gaussian conformations in a melt give rise to $R_e \sim \sqrt{N}$ and thus $\bar{\mathcal{N}} \sim N$. This quantity remains invariant as one lumps successively more repeating units into one effective segment and, therefore, it is independent of the definition of a segment. $\sqrt{\bar{\mathcal{N}}}$ describes the number of neighboring partners a molecule interacts with, and it sets the free energy scale of composition fluctuations.
4) The incompatibility between molecules is described by the product of the Flory–Huggins parameter, χ, and the number of segments, N, per molecule. χ denotes the strength of repulsion between different segments. The product, χN, parameterizes the repulsion between molecules in a polymer blend or the distinct blocks of the diblock copolymer.

Some of the coarse-grained parameters, R_e and $\bar{\mathcal{N}}$, can be easily measured by experiments or in simulations. The other two parameters, χN and the suppression of density fluctuations, $\varkappa_0 N$, are thermodynamic characteristics, which are not directly related to the structure (i.e., they cannot be simply expressed as a function of the molecular coordinates). If density fluctuations of the polymeric liquid are small on the length scale of interest (e.g., width of an interface between domains), then the value of the compressibility has only a minor relevance and decreasing it even further will not significantly affect the behavior of the system. Thus, field-theoretic calculations often take the idealized limit of strict incompressibility. In particle-based simulations, however, one often softens the constraint in order to facilitate the motion of the interaction centers and, thereby, reduces the viscosity of the polymer liquid. The Flory–Huggins parameter, in turn, is a crucial coarse-grained parameter and different methods have been devised to extract it from experiments or simulations [16, 20–25]. We shall briefly discuss this important issue in Section 5.2.3, and further refer the reader to the literature, where computer simulations have been quantitatively compared with mean field predictions and where the role of fluctuations on the coarse-grained parameters is discussed [16, 22].

While the development of coarse-grained models for studying equilibrium properties has a rather well-understood conceptual foundation in its relation to renormalization group calculations and critical phenomena, the development of coarse-grained models for the kinetics of phase transformation is a wide-open field. It is tempting to postulate a similar universality for the dynamics as for the equilibrium properties but the comparison with phase transitions of simple liquids reveals that one equilibrium universality class can exhibit different dynamics depending on the quantities that are conserved [26]. Moreover, it is not obvious that there exists a single coarse-grained parameter that relates the time scale between different models. One may adopt the pragmatic point of view that the largest single-chain relaxation time, $\tau \equiv R_{\mathrm{eo}}^2/D$, where D denotes the self-diffusion coefficient, sets the relevant timescale.

This chapter is organized as follows: In Section 5.2, we describe how to derive a field-theoretic formulation of a particle-based model with spatially extended, pairwise interactions between segments and discuss the mean field approximation of the concomitant field-theoretic model [16, 27, 28]. In Section 5.3, we simplify this field-theoretic model and motivate a minimal, coarse-grained model for compressible, multicomponent, polymer blends. This model in conjunction with mean field theory has been extraordinarily successful in predicting the properties of dense, multicomponent, polymeric liquids [29]. Partial enumeration [30–34], Brownian dynamics [35, 36], and importance-sampling techniques [37, 38] to calculate the single-chain properties of non-Gaussian molecules in the mean field are briefly mentioned. Finally, computer simulation techniques of the discretized versions of the minimal, coarse-grained model are reviewed [39–42]. These techniques allow to account for fluctuation effects. Section 5.4 presents different methods to calculate free energies of soft matter system without invoking the mean field approximation [43–45]. The chapter closes with a brief summary and outlook.

5.2 From Particle-Based Models for Computer Simulations to Self-Consistent Field Theory: Hard-Core Models

5.2.1 Hubbard–Stratonovich Transformation: Field-Theoretic Reformulation of the Particle-Based Partition Function

Different particle-based models have been employed to computationally describe multicomponent polymer melts on a coarse-grained scale [46–48]. In these models, the coordinates of segments, $\{\mathbf{r}_\alpha(s)\}$, are the degrees of freedom. The index, α, runs over all polymers, $\alpha = 1, \ldots, n$, in the system and s parameterizes the location of the segment along the molecule, $s = 1, \ldots, N_\alpha$. Typically, segments interact via pairwise potentials. Bonded, intramolecular potentials, $\mathcal{H}_{\mathrm{b}}$, and nonbonded interactions, $\mathcal{H}_{\mathrm{nb}}$, can be distinguished. Segments along the backbone of the macromolecule are bonded together by anharmonic springs. Nonbonded interactions are applied between all segments, and they are typically comprised of a harsh repulsion at short distances and a softer attraction at intermediate distances. For computational convenience, nonbonded interactions are commonly cut off at a finite distance. The harsh repulsion between segments characterizes the segmental excluded volume. It gives rise to nontrivial liquid packing effects of the fluid of segments and, on large length scales, this harsh repulsion restricts density fluctuations and produces the typical behavior of a dense liquid. Often, the distance between bonded segments along the macromolecule and the typical distance of nonbonded nearest neighbors in the dense liquid, which can be identified by the first peak of the radial density–density pair correlation function, are comparable. In this case, the bonded interactions do not strongly perturb the liquid, and the structure of a liquid of nonbonded segments and polymers is quite similar. Often a maximal bond length is enforced, and the combination of harsh repulsion and finite bond length is chosen as to prevent the chain molecules from crossing through each other in the course of their motion [47, 49]. Then, the model captures the important condition of noncrossability of chain molecules, which has important consequences for the single-chain dynamics (i.e., entanglements) and results in a reptation-like motion of long macromolecules in a melt [12, 50].

Typical examples of this type of coarse-grained models are bead-spring models using a Lennard-Jones potential between effective segments and an FENE-potential or a bond length constraint as bonded interactions [47]. Repulsion between distinct segment species can be modeled through different Lennard-Jones parameters. A popular choice consists in using a purely repulsive Lennard-Jones potential, which is cut off and shifted at the minimum $\sigma_c = \sqrt[6]{2}\sigma$ [51, 52]. Another widely used, coarse-grained representation is the bond fluctuation model (BFM) [49, 53], where effective segments live on a cubic lattice and block all eight corners of a unit cube from further occupancy. Bonded segments along the chain are connected via one of the 108 bond vectors that can adopt the lengths $2, \sqrt{5}, \sqrt{6}, 3$, and $\sqrt{10}$ in units of the lattice spacing. The repulsion between distinct monomer species can be described by a simple square-well potential. Molecular dynamic simulations [47, 51] have been utilized to study off-lattice models and Monte Carlo techniques in conjunction with sophisticated reweighting techniques have been employed to extract information about the phase behavior and interface properties [54–56] of lattice models.

In the following discussion, we use the example of a symmetric binary polymer blend to develop the formalism of deriving a field-theoretic description, but the technique can be easily generalized to asymmetric systems, block copolymers, and their mixtures with homopolymers.

The starting point for a theoretical description is the partition function $\mathcal{Z}$ in the canonical ensemble. Specifically, we consider n_A polymers of species A and n_B macromolecules of species B in a volume, V, at temperature, T. Both species of the symmetric blend are comprised of the same number, $N = N_A = N_B$, of effective segments.

$$Z \propto \frac{1}{n_A! n_B!} \int \prod_{\alpha=1}^{n_A} \tilde{\mathcal{D}}_A[\mathbf{r}_\alpha(s)] \prod_{\beta=1}^{n_B} \tilde{\mathcal{D}}_B[\mathbf{r}_\beta(s)] \exp\left[-\frac{\mathcal{H}_{nb}[\{\mathbf{r}\}]}{k_B T}\right], \tag{5.1}$$

where k_B is Boltzmann constant. The factorials take account of the indistinguishability of the A- and B-polymers. The integration $\tilde{\mathcal{D}}_A[\mathbf{r}_\alpha]$ sums over all conformations of the αth A-polymer within the microscopic model using the appropriate statistical weight due to intramolecular interactions, that is,

$$\tilde{\mathcal{D}}_A[\mathbf{r}_\alpha] = \prod_{s=1}^{N_A} d^3\mathbf{r}_\alpha(s) \exp\left[-\frac{\mathcal{H}_b[\{\mathbf{r}_\alpha\}]}{k_B T}\right]. \tag{5.2}$$

The nonbonded interactions are comprised of two contributions, $\mathcal{H}_{nb} = \mathcal{H}_{melt} + \mathcal{H}_{ord}$. $\mathcal{H}_{melt}$ is the harsh, short-ranged excluded volume interaction, represented by the avoidance of double-occupancy of lattice sites in the bond fluctuation model. $\mathcal{H}_{\text{ord}}$ describes the longer-ranged, thermal interactions that distinguish the segment species. In the bond fluctuation model, these thermal interactions are described by the square-well potential between neighboring, effective segments. Typically, the interaction range is extended over the nearest 54 lattice sites, which constitute the first neighbor shell of the polymer liquid [20, 57]. In off-lattice models, the distinction between harsh repulsive interactions, dictating the liquid structure, and softer, longer-ranged interactions, which provide the cohesion of the liquid, is well established of liquid-state theory. In a model with purely repulsive Lennard-Jones potentials [58, 59], however, the quantitative separation of the two contributions (and, thus, the relation between the parameters of the Lennard-Jones potential and the Flory–Huggins parameter) is a challenge because the interaction between unlike species alters the local liquid structure.

$\mathcal{H}_{\text{melt}}$ and $\mathcal{H}_{\text{ord}}$ arise from pairwise interactions. For simplicity, we assume a simple, symmetric form

$$\mathcal{H}_{\text{melt}}[\{\mathbf{r}\}] = \sum_{\text{pairs}\{(\alpha,i)\ (\beta,j)\}} V_{\text{ev}}(\mathbf{r}_\alpha(i)-\mathbf{r}_\beta(j)), \tag{5.3}$$

where the sum is taken over all pairs of segment i on polymer α and segment j on polymer β, irrespective of their species, and

$$\mathcal{H}_{\text{ord}}[\{\mathbf{r}\}] = \sum_{\text{pairs}\{(\alpha,i)(\beta,j)\}} \pm V(\mathbf{r}_\alpha(i)-\mathbf{r}_\beta(j)), \tag{5.4}$$

where the plus-sign is used for interactions between unlike species, while the interaction energy is negative for contacts between species of the same type. Generalizations to nonsymmetric choices of interactions are straightforward. These types of coarse-grained representations are minimal, coarse-grained models, and the different interactions – bonding, harsh repulsion, longer-ranged interactions – and the density are related to the four coarse-grained parameters, $R_e, \varkappa N, \chi N$, and $\bar{\mathcal{N}}$.

This particle-based description can be reformulated in a field-theoretic language via a Hubbard–Stratonovich transformation. As the first step, we define the microscopic normalized density of A-segments, which depends on the positions of all segments, $\{\mathbf{r}_\alpha(s)\}$, of A-polymers [60]:

$$\hat{\phi}_A(\mathbf{r}) = \frac{1}{\varrho_0} \sum_{\alpha=1}^{n_A} \sum_{s=1}^{N} \delta(\mathbf{r}-\mathbf{r}_\alpha(s)). \tag{5.5}$$

A similar expression holds for the density of B-segments. The normalization is chosen such that the spatial average, $(1/V)\int d^3\mathbf{r}\,\hat{\phi}_A(\mathbf{r})$, equals the average composition, $0 \le \bar{\phi}_A = n_A N/\varrho_0 V \le 1$, of the system. The energy of excluded volume repulsion, $\mathcal{H}_{\text{melt}}$, then takes the form

$$\mathcal{H}_{\text{melt}}\left[\hat{\phi}_A, \hat{\phi}_B\right] = \frac{\varrho_0^2}{2} \int d^3\mathbf{r}\, d^3\mathbf{r}' \left[\hat{\phi}_A(\mathbf{r}) + \hat{\phi}_B(\mathbf{r})\right] V_{\text{ev}}(\mathbf{r}-\mathbf{r}') \left[\hat{\phi}_A(\mathbf{r}') + \hat{\phi}_B(\mathbf{r}')\right] - \frac{(n_A+n_B)N}{2} V_{\text{ev}}(0), \tag{5.6}$$

where V_{ev} is the excluded volume interaction defined in Eq. (5.3). The last term explicitly subtracts the self-interactions, which are included in the first term. Since they contribute only a constant to the energy, which is independent of the configuration, they are irrelevant for the following discussion and omitted; they merely give rise to a shift of the chemical potential in the grand-canonical ensemble. Similarly, one can rewrite the pairwise interactions, $\mathcal{H}_{ord}$, as a convolution of the interaction potential $V(\mathbf{r}-\mathbf{r}')$ with densities at positions $\mathbf{r}$ and $\mathbf{r}'$.

By choosing linear combinations of the densities, $\hat{\phi}_A$ and $\hat{\phi}_B$, we can eliminate the cross term proportional to $\hat{\phi}_A\hat{\phi}_B$ in the interactions. By virtue of the symmetry of the interactions, $V_{AA}(r) = V_{BB}(r)$, these linear combinations simply are the total segment density, $\hat{\varrho} \equiv \hat{\phi}_A + \hat{\phi}_B$, and the difference of densities or composition, $\hat{\phi} \equiv \hat{\phi}_A - \hat{\phi}_B$. Then, the interaction energy is quadratic in $\hat{\varrho}$ and $\hat{\phi}$ and the concomitant Boltzmann factor takes the form

$$\exp\left[-\frac{\mathcal{H}_{nb}}{k_B T}\right] = \exp\left[-\frac{\varrho_0}{2N}\int d^3\mathbf{r}\, d^3\mathbf{r}'\{\hat{\varrho}(\mathbf{r})V_+(\mathbf{r}-\mathbf{r}')\hat{\varrho}(\mathbf{r}') - \hat{\phi}(\mathbf{r})V_-(\mathbf{r}-\mathbf{r}')\hat{\phi}(\mathbf{r}')\}\right] \tag{5.7}$$

with

$$V_+(\mathbf{r}) = \frac{\varrho_0 N}{k_B T}V_{ev}(\mathbf{r}) \quad \text{and} \quad V_-(\mathbf{r}) = \frac{\varrho_0 N}{k_B T}V(\mathbf{r}). \tag{5.8}$$

Note that in models with harsh excluded volume interactions the repulsive part of the binary interaction, coupling to the total density, is of order $k_B T$ per segment, that is, $\int d^3\mathbf{r}\, V_+(\mathbf{r}) \sim \mathcal{O}(N)$. The longer-ranged interactions that distinguish the segment species and couple to the composition are only of order $1/N$ per segment. Thus, the integrated strength is of order unity, that is, $\int d^3\mathbf{r}\, V_-(\mathbf{r}) \sim \mathcal{O}(1)$. The two interactions also contribute with different signs to the energy. Physically, this is reflected in the qualitatively different behavior of $\hat{\varrho}$ and $\hat{\phi}$: a dense polymer melt is nearly incompressible, therefore $\hat{\varrho}$ will hardly vary in space and exhibit only minor thermal fluctuations. This is in marked contrast to the composition, $\hat{\phi}$. It distinguishes the A-rich and B-rich phase, and it exhibits strong thermal fluctuations in the vicinity of the second-order demixing transition.

As the second step, we use the Hubbard–Stratonovich formula

$$\exp\left[\frac{1}{2}x\alpha x\right] = \frac{1}{\sqrt{2\pi\alpha}}\int_{-\infty}^{+\infty} dy \exp\left[-\left(\frac{y^2}{2\alpha} + xy\right)\right] \tag{5.9}$$

at each point in space. The field W_- is introduced by identifying $\alpha = \varrho_0 V_-/N$, $x = \hat{\phi}$ and $y = \varrho_0 W_-/N$. The field W_+ is also inserted by the Hubbard–Stratonovich formula, but as $\alpha = -\varrho_0 V_+/N < 0$ we choose $x = \hat{\varrho}$ and $y = \varrho_0 i W_+/N$ with $i = \sqrt{-1}$ in order to make the integrals well behaved. Then, the field W_+, that is conjugated to the total density $\hat{\varrho}$, gives rise to an imaginary contribution. This Hubbard–Stratonovich transformation leads to the exact rewriting of the partition function:

$$Z \propto \frac{1}{n_A! n_B!} \int \prod_{\alpha=1}^{n_A} \tilde{\mathcal{D}}_A[\mathbf{r}_\alpha(s)] \prod_{\beta=1}^{n_B} \tilde{\mathcal{D}}_B[\mathbf{r}_\beta(s)] \int_{-\infty}^{\infty} \mathcal{D}W_+ \mathcal{D}W_-$$

$$\exp\left[-\int \mathrm{d}^3\mathbf{r}\,\mathrm{d}^3\mathbf{r}' \frac{\varrho_0}{2N} W_+(\mathbf{r}) V_+^{-1}(\mathbf{r}-\mathbf{r}') W_+(\mathbf{r}') - i \int \mathrm{d}^3\mathbf{r} \frac{\varrho_0}{N} W_+(\mathbf{r}) \hat{\varrho}(\mathbf{r}) \right]$$

$$\exp\left[-\int \mathrm{d}^3\mathbf{r}\,\mathrm{d}^3\mathbf{r}' \frac{\varrho_0}{2N} W_-(\mathbf{r}) V_-^{-1}(\mathbf{r}-\mathbf{r}') W_-(\mathbf{r}') - \int \mathrm{d}^3\mathbf{r} \frac{\varrho_0}{N} W_-(\mathbf{r}) \hat{\phi}(\mathbf{r}) \right]$$

$$\propto \frac{1}{n_A! n_B!} \int_{-\infty}^{\infty} \mathcal{D}W_+ \mathcal{D}W_- \, e^{\mathcal{S}[W_+, W_-]} \int \prod_{\alpha=1}^{n_A} \tilde{\mathcal{D}}_A[\mathbf{r}_\alpha(s)] \prod_{\beta=1}^{n_B} \tilde{\mathcal{D}}_B[\mathbf{r}_\beta(s)] \, e^{-\frac{\varrho_0}{N}\int \mathrm{d}^3\mathbf{r}\,(iW_+\hat{\varrho} + W_-\hat{\phi})} \tag{5.10}$$

with

$$S[W_+, W_-] = -\frac{\varrho_0}{2N} \int \mathrm{d}^3\mathbf{r}\,\mathrm{d}^3\mathbf{r}' \{ W_+(r) V_+^{-1}(\mathbf{r}-\mathbf{r}') W_+(\mathbf{r}') + W_-(\mathbf{r}) V_-^{-1}(\mathbf{r}-\mathbf{r}') W_-(\mathbf{r}') \}. \tag{5.11}$$

Here $V_+^{-1}(\mathbf{r})$ denotes the functional inverse of $V_+(\mathbf{r})$, which is defined by the equation

$$\int \mathrm{d}^3\mathbf{r}\, V_+^{-1}(\mathbf{r}''-\mathbf{r}) V_+(\mathbf{r}-\mathbf{r}') = \delta(\mathbf{r}''-\mathbf{r}') \tag{5.12}$$

and a similar definition holds for V_-^{-1}. At this stage, we tacitly assume that the functional inverses of the potential exist. This property imposes a restriction on the particle-based models, which can be converted into a field-theoretic description. With the use of the explicit form of the microscopic density, the argument of the last exponential in Eq. (5.10) takes the form

$$\varrho_0 \int \mathrm{d}^3\mathbf{r}\, \{iW_+\hat{\varrho} + W_-\hat{\phi}\} = \sum_{\alpha=1}^{n_A} \sum_{s=1}^{N_A} W_A(\mathbf{r}_\alpha(s)) + \sum_{\beta=1}^{n_B} \sum_{s=1}^{N_B} W_B(\mathbf{r}_\beta(s)) \tag{5.13}$$

with $W_A = iW_+ + W_-$ and $W_B = iW_+ - W_-$. Thus, the chains do not mutually interact, but we have reformulated the partition function in terms of independent chains subjected to the external, fluctuating fields, iW_+ and W_-. We define the partition function of a single A-polymer in the external field $W_A(\mathbf{r})$ as

$$\mathcal{Z}_A[W_A] = \int \tilde{\mathcal{D}}_A[\mathbf{r}(s)] \exp\left[-\frac{1}{N} \sum_{s=1}^{N_A} W_A(\mathbf{r}(s)) \right] \tag{5.14}$$

and a similar expression defines $\mathcal{Z}_B[W_B]$. Then, we obtain for the partition function of the multichain system:

$$
\begin{aligned}
Z &\propto \int DW_+ DW_- \, e^{\mathcal{S}[W_+,W_-]} \frac{(Q_A[iW_+ + W_-])^{n_A}}{n_A!} \frac{(Q_B[iW_+ + W_-])^{n_B}}{n_B!} \\
&\propto \int DW_+ DW_- \exp\left[-\frac{F[W_+, W_-]}{k_B T}\right],
\end{aligned}
\tag{5.15}
$$

where the free energy functional, $\mathcal{F}$, takes the form

$$
\begin{aligned}
\frac{\mathcal{F}[W_+, W_-]}{k_B T (V/R_{eo}^3)\sqrt{\mathcal{N}}} &= \bar{\phi}_A \ln\left(\frac{\bar{\phi}_A \varrho_0 V}{eN\mathcal{Q}_A[iW_+ + W_-]}\right) + \bar{\phi}_B \ln\left(\frac{\bar{\phi}_B \varrho_0 V}{eN\mathcal{Q}_B[iW_+ + W_-]}\right) \\
&+ \frac{1}{2V}\int d^3\mathbf{r}\, d^3\mathbf{r}' \{W_+(\mathbf{r}) V_+^{-1}(\mathbf{r}-\mathbf{r}') W_+(\mathbf{r}') + W_-(\mathbf{r}) V_-^{-1}(\mathbf{r}-\mathbf{r}') W_-(\mathbf{r}')\},
\end{aligned}
\tag{5.16}
$$

$\bar{\phi}_A \equiv n_A N/\varrho_0 V$ and $\bar{\phi}_B \equiv n_B N/\varrho_0 V$ denote the average compositions.

Up to this stage we have reformulated the original problem in terms of a problem of independent chains, in the fields iW_+ and W_-, without invoking any approximation. Note that the fields are collective degrees of freedom and thus we have arrived at a field-theoretic reformulation of the particle-based model. The initial difficulty due to the interactions between different molecules is now shifted to the equally formidable problem of the functional integration over the fluctuating fields, iW_+ and W_-.

5.2.2 Mean Field Approximation

Unfortunately, we cannot perform the functional integral over W_+ and W_-. To make further progress, we evaluate the functional integrals by a saddle-point approximation, that is, instead of integrating over all fields we estimate the integral by the most probable value of the integrand [6, 28, 60–62]. At this stage we neglect fluctuations of the external fields and, thereby, correlations between the chain molecules are ignored.

We split the saddle-point integration into two steps [27, 63]: First, we approximate the functional integral over W_+ by the most probable value of the integrand.

$$
\mathcal{Z} \propto \int \mathcal{D}W_+ \mathcal{D}W_- \exp\left[-\frac{\mathcal{F}[W_+, W_-]}{k_B T}\right] \approx \int \mathcal{D}W_- \exp\left[-\frac{\mathcal{G}_{EP}[W_-]}{k_B T}\right] \tag{5.17}
$$

with

$$
\mathcal{G}_{EP}[W_-] = \min_{W_+(\mathbf{r})} \mathcal{F}[W_+, W_-].
$$

The real field, W_+, gives rise to an imaginary contribution to W_A and W_B that, in turn, corresponds to a strongly oscillating behavior of the integrand. To evaluate those oscillating contributions, the standard procedure is to extend the auxiliary field W_+

into the complex plane and perform the integration parallel to the real axis. The imaginary shift perpendicular to the real axis can be chosen such that the functional derivative vanishes, $\frac{\delta\mathcal{F}}{\delta W_+(\mathbf{r})}=0$, at $W_+=w_+$, where w_+ is purely imaginary. Consequently, the integrand has a stationary phase at w_+ along the shifted path of integration, and this region yields the dominant contribution to the integral. From the condition of stationary phase, we obtain

$$\frac{1}{k_BT}\frac{\delta\mathcal{F}}{\delta W_+(\mathbf{r})}=-\frac{i\varrho_0}{N}\int d^3\mathbf{r}'\, V_+^{-1}(\mathbf{r}-\mathbf{r}')W_+(\mathbf{r}')-n_A\frac{\delta\ln\mathcal{Z}_A[W_A]}{\delta W_A(\mathbf{r})}-n_B\frac{\delta\ln\mathcal{Z}_B[W_B]}{\delta W_B(\mathbf{r})}\overset{!}{=}0 \tag{5.18}$$

and the field that fulfills the saddle-point equation is denoted by a lower case letter, w_+. The last terms are proportional to the densities, $\phi_A^*[W_A](\mathbf{r})$ and $\phi_B^*[W_B](\mathbf{r})$, that are created by a single A-polymer or B-polymer in the external field W_A or W_B, respectively. To demonstrate this explicitly, we use the definition of the single-chain partition function, Eq. (5.14) and calculate the functional derivative

$$\begin{aligned}\frac{\delta\ln\mathcal{Z}_A[W_A]}{\delta W_A(\mathbf{r})}&=\frac{1}{\mathcal{Z}_A}\frac{\delta}{\delta W_A(\mathbf{r})}\int\tilde{\mathcal{D}}_A[\mathbf{r}_\alpha(s)]\,e^{-\int d^3\mathbf{r}'\,W_A(\mathbf{r}')\frac{1}{N}\sum_{s=1}^{N_A}\delta(\mathbf{r}'-\mathbf{r}_\alpha(s))}\\&=-\frac{1}{\mathcal{Z}_A}\int\tilde{\mathcal{D}}_A[\mathbf{r}_\alpha(s)]\frac{1}{N}\sum_{s=1}^{N_A}\delta(\mathbf{r}-\mathbf{r}_\alpha(s))\,e^{-\int d^3\mathbf{r}'\,W_A(\mathbf{r}')\frac{1}{N}\sum_{s=1}^{N_A}\delta(\mathbf{r}'-\mathbf{r}_\alpha(s))}\\&\equiv-\left\langle\frac{1}{N}\sum_{s=1}^{N_A}\delta(\mathbf{r}-\mathbf{r}_\alpha(s))\right\rangle_{\text{single chain in external field, }W_A}\end{aligned} \tag{5.19}$$

$$\equiv-\frac{\varrho_0}{n_AN}\phi_A^*[W_A](\mathbf{r}). \tag{5.20}$$

The prefactor has been chosen such that $\int d^3\mathbf{r}\phi_A^*(\mathbf{r})=V\bar{\phi}_A$.
Inserting this expression into the saddle-point equation for w_+, we obtain

$$\int d^3\mathbf{r}'V_+^{-1}(\mathbf{r}-\mathbf{r}')iw_+(\mathbf{r}')=\phi_A^*(\mathbf{r})+\phi_B^*(\mathbf{r})$$

$$iw_+(\mathbf{r}'')=\int d^3\mathbf{r}V_+(\mathbf{r}''-\mathbf{r})[\phi_A^*(\mathbf{r})+\phi_B^*(\mathbf{r})], \tag{5.21}$$

where, in the last step, we have multiplied the expression by $V_+(\mathbf{r}''-\mathbf{r})$ and integrated over the volume using Eq. (5.12).

Substituting back the saddle-point value, w_+, into the partition function, Eq. (5.15), we find

$$\mathcal{Z}_{\rm EP} \equiv \int \mathcal{D}W_- \exp\left[-\frac{\mathcal{F}[w_+, W_-]}{k_B T}\right] \equiv \int \mathcal{D}W_- \exp\left[-\frac{\mathcal{G}_{\rm EP}[W_-]}{k_B T}\right]$$

with

$$\frac{\mathcal{G}_{\rm EP}[W_-]}{k_B T} = n_A \ln\left(\frac{n_A}{e\mathcal{Q}_A[iw_+ + W_-]}\right) + n_B \ln\left(\frac{n_B}{e\mathcal{Q}_B[iw_+ + W_-]}\right)$$
$$+ \frac{\varrho_0}{2N}\int d^3\mathbf{r}\, d^3\mathbf{r}'\, W_- V_-^{-1} W_- - \frac{\varrho_0}{2N}\int d^3\mathbf{r}\, d^3\mathbf{r}' (\phi_A^* + \phi_B^*) V_+ (\phi_A^* + \phi_B^*). \tag{5.22}$$

Since iw_+ is real, one has to evaluate the single-chain partition functions, Q_A and Q_B, and single-chain densities, ϕ_A^* and ϕ_B^*, of molecules subjected to real but fluctuating fields, $W_A = iw_+[W_-] + W_-$ and $W_B = iw_+[W_-] - W_-$. At this stage, we have eliminated fluctuations of the field, W_+, which is conjugated to the total density, but retained the fluctuations of the field that couples to the composition of the blend. We refer to this scheme as external potential theory [27, 63, 64]. The fluctuations can be sampled using real Langevin dynamic simulations [63] or Monte Carlo simulations [65]. In practice, it often turns out that retaining the fluctuations in W_- is sufficient to capture most of the long-wavelength composition fluctuations because the coupling between density and composition fluctuations is small in a symmetric, multicomponent polymer melt (cf. Eq. (5.7)).

We proceed in deriving the self-consistent field theory for the binary polymer blend, by also approximating the functional integral over the field d, W_-, by its saddle-point value. The condition for the extremum becomes

$$\frac{1}{k_B T}\frac{\mathcal{D}F}{\mathcal{D}W_-(\mathbf{r})} = +\frac{\varrho_0}{N}\int d^3\mathbf{r}' V_-^{-1}(\mathbf{r}-\mathbf{r}')W_-(\mathbf{r}') - n_A \frac{\mathcal{D}\ln \mathcal{Q}_A[W_A]}{\mathcal{D}W_A(\mathbf{r})} + n_B \frac{\mathcal{D}\ln \mathcal{Q}_B[W_B]}{\mathcal{D}W_B(\mathbf{r})} \stackrel{!}{=} 0, \tag{5.23}$$

$$w_-(\mathbf{r}'') = -\int d^3\mathbf{r}\, V_-(\mathbf{r}''-\mathbf{r})(\phi_A^*(\mathbf{r}) - \phi_B^*(\mathbf{r})). \tag{5.24}$$

The two Eqs. (5.21) and (5.24) relate the real saddle-point values, iw_+ and w_-, to $\phi_A^*[iw_+ + w_-](\mathbf{r})$ and $\phi_B^*[iw_+ - w_-](\mathbf{r})$, which are themselves functionals of the fields (cf. Eq. (5.20)). Solving the interacting multichain problem within mean field theory amounts to self-consistently fulfilling Eqs. (5.20),(5.21) and (5.24). The saddle-point values of the fields acting on A- and B-polymers take to form

$$w_A = iw_+ + w_- = \frac{R_{\rm eo}^3}{k_B T\sqrt{\bar{\mathcal{N}}}}\frac{\delta \mathcal{H}_{\rm nb}}{\delta \hat{\phi}_A(\mathbf{r})},$$
$$w_A = iw_+ + w_- = \frac{R_{\rm eo}^3}{k_B T\sqrt{\bar{\mathcal{N}}}}\frac{\delta \mathcal{H}_{\rm nb}}{\delta \hat{\phi}_B(\mathbf{r})}. \tag{5.25}$$

Substituting the saddle-point values, w_+ and w_-, back into the free energy functional, we arrive at the mean field estimate for the Helmholtz free energy:

$$\frac{F_{\mathrm{SCFT}}}{k_{\mathrm{B}}T} \equiv \frac{\mathcal{F}[w_+, w_-]}{k_{\mathrm{B}}T} \tag{5.26}$$

$$= n_{\mathrm{A}} \ln\left(\frac{n_{\mathrm{A}}}{e\mathcal{Z}_{\mathrm{A}}[iw_+ + w_-]}\right) + n_{\mathrm{B}} \ln\left(\frac{n_{\mathrm{B}}}{e\mathcal{Z}_{\mathrm{B}}[iw_+ + w_-]}\right)$$

$$-\frac{\varrho_0}{2N}\int \mathrm{d}^3\mathbf{r}\,\mathrm{d}^3\mathbf{r}'[\phi_{\mathrm{A}}^*(\mathbf{r}) + \phi_{\mathrm{B}}^*(\mathbf{r})]V_+(\mathbf{r}-\mathbf{r}')[\phi_{\mathrm{A}}^*(\mathbf{r}') + \phi_{\mathrm{B}}^*(\mathbf{r}')] \tag{5.27}$$

$$+\frac{\varrho_0}{2N}\int \mathrm{d}^3\mathbf{r}\,\mathrm{d}^3\mathbf{r}'[\phi_{\mathrm{A}}^*(\mathbf{r}) - \phi_{\mathrm{B}}^*(\mathbf{r})]V_-(\mathbf{r}-\mathbf{r}')[\phi_{\mathrm{A}}^*(\mathbf{r}') - \phi_{\mathrm{B}}^*(\mathbf{r}')].$$

To calculate the thermal average of the composition, $\langle\hat{\phi}\rangle$, we go back to the exact expression for the partition function in Eq. (5.10)

$$\langle\hat{\phi}(\mathbf{r})\rangle \equiv \frac{1}{\mathcal{Z}}\int \mathcal{D}W_+\mathcal{D}W_-\; \mathrm{e}^{\mathcal{S}[W_+,W_-]}\frac{1}{n_{\mathrm{A}}!n_{\mathrm{B}}!}\int\prod_{\alpha=1}^{n_{\mathrm{A}}}\tilde{\mathcal{D}}_{\mathrm{A}}[\mathbf{r}_\alpha(s)]$$

$$\times\prod_{\beta=1}^{n_{\mathrm{B}}}\tilde{\mathcal{D}}_{\mathrm{B}}[\mathbf{r}_\beta(s)]\,\hat{\phi}(\mathbf{r})\mathrm{e}^{-\frac{\varrho_0}{N}\int \mathrm{d}^3\mathbf{r}\{iW_+\hat{\varrho}+W_-\hat{\phi}\}}$$

$$= \frac{1}{\mathcal{Z}}\int \mathcal{D}W_+\mathcal{D}W_-\; \mathrm{e}^{\mathcal{S}[W_+,W_-]}\frac{1}{n_{\mathrm{A}}!n_{\mathrm{B}}!}\int\prod_{\alpha=1}^{n_{\mathrm{A}}}\tilde{\mathcal{D}}_{\mathrm{A}}[\mathbf{r}_\alpha(s)]$$

$$\times\prod_{\beta=1}^{n_{\mathrm{B}}}\tilde{\mathcal{D}}_{\mathrm{B}}[\mathbf{r}_\beta(s)]\left(-\frac{N}{\varrho_0}\frac{\mathcal{D}}{\mathcal{D}W_-(\mathbf{r})}\right)\mathrm{e}^{-\frac{\varrho_0}{N}\int \mathrm{d}^3\mathbf{r}\{iW_+\hat{\varrho}+W_-\hat{\phi}\}}$$

$$= \frac{1}{\mathcal{Z}}\int \mathcal{D}W_+\mathcal{D}W_-\; \mathrm{e}^{\mathcal{S}[W_+,W_-]}\left(-\frac{N}{\varrho_0}\frac{\mathcal{D}}{\mathcal{D}W_-(\mathbf{r})}\right)\frac{\mathcal{Z}_{\mathrm{A}}^{n_{\mathrm{A}}}\mathcal{Z}_{\mathrm{B}}^{n_{\mathrm{B}}}}{n_{\mathrm{A}}!n_{\mathrm{B}}!} \tag{5.28}$$

$$= \frac{\int \mathcal{D}W_+\mathcal{D}W_-\; \mathrm{e}^{\mathcal{S}[W_+,W_-]}\frac{\mathcal{Z}_{\mathrm{A}}^{n_{\mathrm{A}}}\mathcal{Z}_{\mathrm{B}}^{n_{\mathrm{B}}}}{n_{\mathrm{A}}!n_{\mathrm{B}}!}[\phi_{\mathrm{A}}^*(\mathbf{r})-\phi_{\mathrm{B}}^*(\mathbf{r})]}{\int \mathcal{D}W_+\mathcal{D}W_-\; \mathrm{e}^{\mathcal{S}[W_+,W_-]}\frac{\mathcal{Z}_{\mathrm{A}}^{n_{\mathrm{A}}}\mathcal{Z}_{\mathrm{B}}^{n_{\mathrm{B}}}}{n_{\mathrm{A}}!n_{\mathrm{B}}!}} \tag{5.29}$$

$$= \langle\phi_{\mathrm{A}}^*(\mathbf{r})-\phi_{\mathrm{B}}^*(\mathbf{r})\rangle_W,$$

where the last average $\langle\cdots\rangle_W$ is performed over all fields, W_+ and W_-, with the weight $\mathrm{e}^{\mathcal{S}[W_+,W_-]}\frac{\mathcal{Z}_{\mathrm{A}}^{n_{\mathrm{A}}}\mathcal{Z}_{\mathrm{B}}^{n_{\mathrm{B}}}}{n_{\mathrm{A}}!n_{\mathrm{B}}!}$. Within mean field approximation, this expression simplifies to

$$\langle\hat{\phi}(\mathbf{r})\rangle \equiv \langle\hat{\phi}_{\mathrm{A}}\rangle - \langle\hat{\phi}_{\mathrm{B}}\rangle \approx \phi_{\mathrm{A}}^*[w_{\mathrm{A}}] - \phi_{\mathrm{B}}^*[w_{\mathrm{B}}]. \tag{5.30}$$

Similarly, we arrive at

$$\langle \hat{\varrho}(\mathbf{r}) \rangle \equiv \langle \hat{\phi}_A \rangle + \langle \hat{\phi}_B \rangle \approx \phi_A^*[w_A] + \phi_B^*[w_B]. \tag{5.31}$$

These equations identify the density that a single chain in the external field creates as the thermodynamic average of the microscopic density.[1)]

5.2.3
Role of Compressibility and Local Correlations of the Fluid of Segments

The mean field approximation neglects fluctuations and correlations. Two types of fluctuation effects can be distinguished: (i) long-ranged fluctuations in dense polymer systems are small [16, 27]. Often their significance is controlled by the inverse of the invariant degree of polymerization, $Gi \sim 1/\bar{\mathcal{N}}$. This parameter is denoted as Ginzburg number. Equation (5.16) (and also Eq. (5.35)) demonstrates that the scale of the free energy functional is set by $\bar{\mathcal{N}} k_B T$. Fluctuations of the collective fields from their saddle-point values incur an increase of the free energy by an amount of $\bar{\mathcal{N}} k_B T$ and, thus, they are strongly suppressed for large $\bar{\mathcal{N}}$. (ii) In addition, particle-based models with harsh excluded volume interactions exhibit local, non-universal, liquid-like correlations that are important and cannot be controlled by a small parameter (such as Gi).

Since fluctuations and correlations are ignored in the mean field treatment, the results of the literal mean field theory are quantitatively inaccurate [23]. There are two possibilities to deal with this problem and extract accurate results in the limit $Gi \to 0$:

(a) The effect of local fluctuations is to renormalize the value of the coarse-grained parameters. Thus, rather than predicting the value of the coarse-grained parameters for a particle-based model, one should adjust those coarse-grained parameters to account for the effect of local packing in the dense liquid of segments or the avoidance of back-folding of the chain molecule due to the excluded volume of the segments, and local composition fluctuations [16, 22–25]. This approach is very much in the spirit of coarse-grained models and, in fact, when the coarse-grained parameters are treated as adjustable parameters, the mean field theory is able to simultaneously and accurately predict a variety of thermodynamic and structural quantities in multicomponent polymer melts [30, 55, 66–68].

(b) Alternatively, one can replace the microscopic interactions, $\mathcal{H}_{nb}$, as a function of the microscopic densities by an interaction free energy functional, $\mathcal{F}_{DFT}$, which depends on the collective densities. This interaction free energy functional is chosen, such that the minimization of the field-theoretic partition function with respect to the collective fields and densities will yield accurate results. Density

1) One can also start from Eq. (5.28) and integrate by parts to obtain a relation between the average of the microscopic density and the average of the field W_-. After the saddle-point approximation, $\langle W_-(\mathbf{r}) \rangle_W \approx w_-(\mathbf{r})$, and using Eq. (5.24), we obtain Eq. (5.31).

functional theory explains that there exists such a $\mathcal{F}_{\mathrm{DFT}}$, but its form is unknown. Liquid-state theory of simple fluids, however, offers much guidance in constructing suitable forms for polymeric systems. Particularly important is the distinction between harsh repulsive interactions and long-ranged attractions in the fluid of segments [52, 69–77].

Both strategies have been successfully applied to a variety of problems in multicomponent polymer melts and the predictions of the approximate mean field theory have been quantitatively compared with the exact results of computer simulations [16].

5.3 From Field-Theoretic Hamiltonians to Particle-Based Models: Soft-Core Models

5.3.1 Standard Model for Compressible Multicomponent Polymer Melts and Self-Consistent Field Techniques

In the previous sections, we have motivated the use of minimal coarse-grained models by their universal behavior on large length scales. Lumping successively more atoms into one coarse-grained segment, the size of an effective segment, $R_e/\sqrt{N}$, increases and the range of segmental interactions measured in units of $R_e/\sqrt{N}$ becomes smaller. Moreover, by virtue of universality, the detailed shape of the interactions does not matter and thus it is tempting to use the simplest possible form, which is sufficient to parameterize the relevant interactions that are required to reproduce the universal behavior of multicomponent polymer melts. Therefore, one often utilizes zero-ranged, pairwise interactions of the form

$$V_{\mathrm{ev}}(\mathbf{r},\mathbf{r}') = \varkappa_{\mathrm{o}}\frac{k_{\mathrm{B}}T}{\varrho_{\mathrm{o}}}\delta(\mathbf{r}-\mathbf{r}') \quad \text{and} \quad V(\mathbf{r},\mathbf{r}') = \frac{\chi_{\mathrm{o}}}{2}\frac{k_{\mathrm{B}}T}{\varrho_{\mathrm{o}}}\delta(\mathbf{r}-\mathbf{r}'). \tag{5.32}$$

Using these potentials, one can rewrite the nonbonded interactions, $\mathcal{H}_{\mathrm{nb}}$, in the form

$$\mathcal{H}_{\mathrm{melt}}[\phi_{\mathrm{A}},\phi_{\mathrm{B}}] = \varrho_{\mathrm{o}}k_{\mathrm{B}}T\frac{\varkappa_{\mathrm{o}}}{2}\int \mathrm{d}^3\mathbf{r}[\phi_{\mathrm{A}}+\phi_{\mathrm{B}}-1]^2, \tag{5.33}$$

$$\mathcal{H}_{\mathrm{ord}}[\phi_{\mathrm{A}},\phi_{\mathrm{B}}] = -\varrho_{\mathrm{o}}k_{\mathrm{B}}T\frac{\chi_{\mathrm{o}}}{4}\int \mathrm{d}^3\mathbf{r}[\phi_{\mathrm{A}}-\phi_{\mathrm{B}}]^2, \tag{5.34}$$

where we have again added/omitted terms in the integrand that are linear in density. These latter contributions are proportional to the number of polymers and can be adsorbed in a shift of the chemical potential. Using the simple interactions in conjunction with the Gaussian chain model, one arrives at the standard, field-theoretic model of a weakly compressible multicomponent polymer melt.

$$Z \propto \frac{1}{n_A! n_B!} \int \prod_{\alpha=1}^{n_A} \tilde{\mathcal{D}}_A[\mathbf{r}_\alpha(s)] \prod_{\beta=1}^{n_B} \tilde{\mathcal{D}}_B[\mathbf{r}_\beta(s)] \exp\left[-\frac{\sqrt{\bar{\mathcal{N}}}}{R_{eo}^3} \int d^3\mathbf{r} \left\{ \frac{\varkappa_o N}{Z} [\phi_A + \phi_B - 1]^2 \right.\right.$$

$$\left.\left. - \frac{\chi_o N}{4} [\phi_A - \phi_B]^2 \right\} \right]. \tag{5.35}$$

The parameters, $R_{eo}, \chi_o N, \bar{\mathcal{N}}$, and $\varkappa_o N$ are related to the coarse-grained parameters, $R, \chi N, \bar{\mathcal{N}}$, and $\varkappa N$, respectively, but they are not identical due to the effect of fluctuations and local correlations that have been discussed in Section 5.2.3. Only within the mean field approximation, the distinction between the bare parameters of the standard model, which are denoted by the subscript "o", and the coarse-grained parameters can be neglected.

In the standard model, the bonded interactions, which describe the Gaussian chain architecture, are given by the Wiener measure

$$\tilde{\mathcal{D}}_A[\mathbf{r}_\alpha(s)] = \mathcal{D}_A[\mathbf{r}_\alpha(s)] \exp\left[-\frac{3N}{2R_{eo}^2} \int_0^{N_\alpha} ds \left(\frac{d\mathbf{r}}{ds} \right)^2 \right], \tag{5.36}$$

that is, the polymer α is represented by a space curve, $\mathbf{r}(s)$, and the contour parameter, $0 \le s \le N_\alpha$, is a continuous variable.

Polymers in a melt exhibit self-similar structure from the length scale of the polymer extension, R_{eo}, down to a microscopic cutoff. The latter length scale is set by the range of interactions, the bond length, or the segment size. Using a continuous description of the chain architecture and zero-ranged interactions, the microscopic cutoff is eliminated, that is, R_e is the only relevant length scale and, formally, the self-similar structure is extended to arbitrarily small length scales.

For the simple interactions, Eqs. (5.33) and (5.34), the self-consistent condition that relates the fields to the density takes the form

$$\begin{aligned} w_A &= iw_+ + w_- = -\frac{\chi_o N}{2}[\phi_A - \phi_B] + \varkappa_o N[\phi_A + \phi_B - 1], \\ w_B &= iw_+ - w_- = +\frac{\chi_o N}{2}[\phi_A - \phi_B] + \varkappa_o N[\phi_A + \phi_B - 1]. \end{aligned} \tag{5.37}$$

Sophisticated numerical techniques have been devised to study this standard model within mean field approximation. They exploit that the mean field problem of a Gaussian chain in an external field can be described by a modified diffusion equation in an external field [60]. The latter leads to a partial differential equation that can be solved by efficient computational techniques. Advanced real-space, spectral, and pseudospectral algorithms have been devised to this end [28, 78–80].[2)]

Since the notion of segments is eliminated, the chain discretization, N, is obviously not a coarse-grained parameter. Only the combinations $\chi_o N$, $\varkappa_o N$, and $\bar{\mathcal{N}}$ occur in the

2) In practice, however, all these numerical techniques use a discretization of space, real-space and pseudospectral methods additionally discretize the chain contour. Thereby, a microscopic cutoff is introduced via the numerical methodology.

definition of the model. $\chi_o N$ describes the incompatibility between polymers and the corresponding length scale is the width, w, of the interface between A and B domains. Within the mean field approximation, one obtains

$$\frac{w_{\mathrm{SSL}}}{R_{\mathrm{eo}}} = \frac{1}{\sqrt{6\chi_o N}} \tag{5.38}$$

in the limit of strong segregation (SSL), $\chi_o N \to \infty$.

By the same token, incompressibility cannot be enforced on the length scale of a segment but only on a small fraction of R_{eo}. The correlation length, ξ_{ev}, of density fluctuations in the compressible standard model is given by

$$\frac{\xi_{\mathrm{ev}}}{R_{\mathrm{eo}}} \approx \frac{1}{\sqrt{12\varkappa_o N}} \tag{5.39}$$

within mean field approximation for a one-component system at $\chi_o N = 0$. The two length scales – the excluded volume screening length, ξ_{ev}, and the interfacial width, w – define the microscopic scales that correspond to the nonbonded interactions, $\mathcal{H}_{\mathrm{melt}}$ and $\mathcal{H}_{\mathrm{ord}}$, respectively.[3)] In order to avoid artifacts, these microscopic scales have to be larger than the microscopic length that arise from the discretization of space, ΔL, or chain contour, $b \equiv R_{\mathrm{eo}}/\sqrt{N-1}$.

The mean field approximation of the standard model of multicomponent polymer melts has been extraordinarily successful in describing the universal properties of melts of long polymers. Moreover, simple analytical expressions for various quantities can be obtained in the limit $\chi_o N \to \infty$ (strong segregation limit) or at the onset of ordering or demixing (weak segregation limit). In addition, much effort has been directed toward describing fluctuation effects within the field-theoretic description, and we direct the reader to Ref. [27].

5.3.2 Mean Field Theory for Non-Gaussian Chain Architectures

5.3.2.1 Partial Enumeration Schemes

One major advantage of the standard model is the absence of a microscopic cutoff length scale. In some circumstances, however, the interactions $\chi_o N$ or $\varkappa_o N$ become so strong that the corresponding length scales, w or ξ_{ev}, respectively, become comparable to the size of a chemical repeat unit and the local molecular structure on these short length scales is not well-described by the continuous Gaussian chain model. In this case, more detailed representations of the molecular architecture have to be used. Efficient numerical schemes have been developed for molecular representations on lattices by Scheutjens and Fleer [61] and the worm-like chain model [81–83]. Both approaches allow a modeling of stiffness of the chain contour on short-length scales and are able to describe the crossover from a rod-like behavior

3) The ratio of the coarse-grained parameters, R_{eo} and $\bar{\mathcal{N}}$, defines an additional microscopic scale, $p = R_{\mathrm{eo}}/\sqrt{\bar{\mathcal{N}}}$, which is denoted as packing length. This scale is proportional to the tube diameter of the reptation motion of the long, flexible macromolecules in a melt.

on short-length scales to the Gaussian behavior of flexible chains on large scales [83].

In the case of general chain architectures, however, the mean field problem of a single chain in an external field cannot be cast in the form of a modified diffusion equation, and the density that a single chain creates in the external field and the concomitant single-chain partition function have to be estimated by partial enumeration [30–34]. This methodology has been successfully applied to study the packing of short hydrocarbon chains in the hydrophobic interior of lipid bilayers [31, 32, 34] and polymer brushes [33] and to quantitatively compare the results of Monte Carlo simulations to the predictions of the mean field theory without adjustable parameters [30]. The latter application is illustrated in Figure 5.2.

In the partial enumeration scheme, we evaluate the single-chain problem by considering a large ensemble of single-chain configurations, $\{\mathbf{r}_\alpha(s)\}$. Typically, such an ensemble is comprised of $N_c = 10^7$ or more single-chain configurations (including different intramolecular conformations and translations of the molecule). Then, the density $\phi_A^*[w_A]$ of an A molecule subjected to the external field, w_A, and the ratio of the single-chain partition function and its value in the absence of an external field, $\tilde{\mathcal{Z}}_A[w_A]$, can be estimated according to

$$\phi_A^*[w_A] = \frac{1}{N_c \tilde{\mathcal{Z}}_A} \sum_{\alpha=1}^{N_c} \left(\exp\left[-\frac{1}{N} \sum_{s=1}^{N} w_A(\mathbf{r}_\alpha(s)) \right] \frac{n_A}{\varrho_o} \sum_{s=1}^{N} \delta(\mathbf{r} - \mathbf{r}_\alpha(s)) \right), \tag{5.40}$$

$$\tilde{\mathcal{Z}}_A[w_A] = \frac{\sum_{\alpha=1}^{N_c} \exp\left[-\frac{1}{N} \sum_{s=1}^{N} w_A(\mathbf{r}_\alpha(s)) \right]}{N_c}. \tag{5.41}$$

Given the single-chain density, $\phi_A^*[w_A]$, one obtains improved estimates for the external fields and this cycle of calculations is repeated until self-consistency of Eq. (5.25) (or Eq. (5.37)) is obtained. Typically, one uses the same set of single-chain conformations in each iteration such that the noise in $\phi_A^*[w_A]$ due to the finite number, N_c, of conformations does not interfere with the convergence of the self-consistency condition. Calculating the single-chain density, $\phi_A^*[w_A]$, is computationally expensive, but it can be easily implemented on parallel computers, where each processor evaluates the Boltzmann weight of a subset of chain conformations.

5.3.2.2 **Monte Carlo Sampling of the Single-Chain Partition Function and Self-Consistent Brownian Dynamics**

The number of single-chain conformations exponentially increases with the chain length. For instance, for a random walk with $N-1$ steps, the number of single-chain conformations increases like Vz_{eff}^{N-1}, where V denotes the volume of the system and this factor quantifies the translational degrees of freedom. z_{eff} is an estimate of the number of steps (or bonds) that connect subsequent steps (or segments) and the second factor counts the number of intramolecular conformations. Thus, for all but very short chains, the number of single-chain conformations, N_c, that are included in

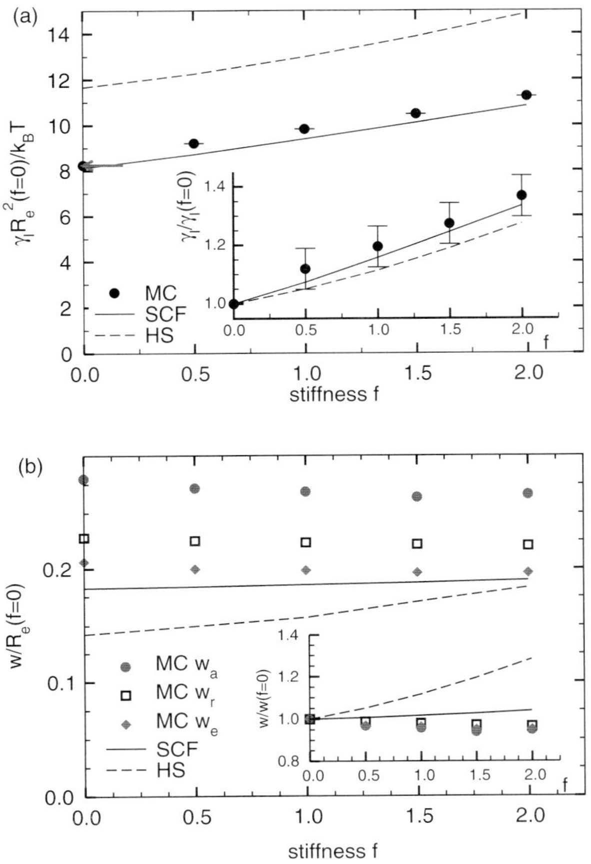

Figure 5.2 (a) Interface tension, γ_I, of an asymmetric polymer blend. One species is completely flexible, while the molecular architecture of the other component of the blend is semiflexible. The stiffness is modeled by a bond-angle potential of strength, *f*. Symbols correspond to Monte Carlo simulations of the bond fluctuation model for chain length $N = 32$, which have been obtained by analyzing the spectrum of interface fluctuations. The arrow on the left-hand side marks the interface tension of a symmetric blend obtained from semigrandcanonical Monte Carlo simulations. Solid lines correspond to the prediction of the self-consistent field theory using a partial enumeration scheme to incorporate the chain architecture on all length scales. The dashed line corresponds to the prediction of the Gaussian chain model in the strong segregation limit [158], $\frac{\gamma_{SSL} R_e^2}{k_B T} = \sqrt{\mathcal{N}}\sqrt{\frac{\chi N}{6}\frac{2(1+\eta+\eta^2)}{3(1+\eta)}}$. $\eta = R_e^{(B)}/R_e^{(A)}$ denotes the ratio of the chain extensions, which is an additional invariant for the asymmetric blend. The inset shows the relative increase of the interfacial tension due to stiffness disparity, *f*. (b) Estimates for the width of the interface as a function of stiffness on the lateral length scale $B = 3.8R_e$ (circles) and $B = 0.9R_e$ (squares). Diamonds correspond to an estimate of the intrinsic width from the excess energy of the interface, which is as an integral quantity less affected by interface fluctuations. The prediction of the self-consistent field theory using a partial enumeration scheme and the strong segregation limit for Gaussian chains, $\frac{w_{\text{intr,SSL}}}{R_e} = \frac{1}{\sqrt{6\chi N}}\sqrt{\frac{1+\eta^2}{2}}$, are shown by solid and dashed lines, respectively. The inset shows the relative variation as a function of stiffness, *f*. Reprinted with permission from Ref. [30].

the partial enumeration scheme is only a very small faction of the total number of molecular conformations.

If the external field is strong, the typical chain conformations in the external field will significantly differ from the original set of configurations. Under these conditions, the finite original set of single-chain conformations, which is representative of molecules in the absence of an external field, will only be a very poor representation of the physical system under investigation and severe sampling problems occur in Eq. (5.40). These problems are signaled by large fluctuations of the Boltzmann weights of the single-chain conformations in the external field, that is, very few chains dominate the weighted average.

One straightforward way to mitigate this problem is to calculate the density, $\phi_A^*[w_A]$, of a single-chain in the external field, w_A by computer simulation. Monte Carlo simulations or Brownian dynamics can be used to create a sample of single-chain conformations that are distributed according to the Boltzmann weight in Eq. (5.40). Using this importance-sampling scheme, the density is simply obtained by the nonweighted average over the generated single-chain conformations.

This procedure, however, is impractical because the generation of the representative set of single-chain conformations in the external field via importance sampling is computationally more expensive than partial enumeration and one would have to perform a computer simulation in each iteration cycle of the self-consistent adjustment of fields and densities.

Two solutions of this sampling problem have been proposed:

(a) Rather than requiring that the ensemble of single-chain conformations is a representative sample of conformations in the field, w_A, that fulfills the self-consistency condition, one can create a representative sample of single-chain conformations in a field $\tilde{w}_A$, which is sufficiently similar to the unknown field, w_A. Then, Eqs. (5.40) and (5.41) can be rewritten in the form [37, 38]:

$$\phi_A^*[w_A] \alpha \left\langle \exp\left[-\frac{1}{N}\sum_{s=1}^{N}\{w_A(\mathbf{r}_\alpha(s)) - \tilde{w}_A(\mathbf{r}_\alpha(s))\}\right] \frac{n_A}{\varrho_o}\sum_{s=1}^{N}\delta(\mathbf{r}-\mathbf{r}_\alpha(s))\right\rangle_{\tilde{w}_A}, \tag{5.42}$$

$$\tilde{\mathcal{Z}}_A[w_A] = \frac{\left\langle \exp\left[-\frac{1}{N}\sum_{s=1}^{N}\{w_A(\mathbf{r}_\alpha(s)) - \tilde{w}_A(\mathbf{r}_\alpha(s))\}\right]\right\rangle_{\tilde{w}_A}}{\left\langle \exp\left[\frac{1}{N}\sum_{s=1}^{N}\tilde{w}_A(\mathbf{r}_\alpha(s))\right]\right\rangle_{\tilde{w}_A}}, \tag{5.43}$$

where $\langle \cdots \rangle_{\tilde{w}_A}$ denotes the importance sampling with respect to the Boltzmann weight in the external field $\tilde{w}_A$. The technique is an analogue of histogram reweighting [84] for studying phase transitions. Typically, $\tilde{w}_A$ is obtained from one of the first iterations of the self-consistent adjustment of fields and densities or it is the self-consistent solution for a nearby set of control parameters (e.g., composition or incompatibility). If $\tilde{w}_A \approx w_A$, the single-chain conformations created by the importance-sampling procedure will also be representative for conformations in the field w_A and the same sample of conformations can be reused for different iterations of the self-consistent adjustment of fields and densities. The Boltzmann weights in Eq. (5.42) stem only from the difference, $w_A - \tilde{w}_A$, and they will not strongly fluctuate if this difference remains small. Thus, the sampling problem for the calculation of density is avoided. Apart from the single computer simulation used to generate the ensemble of conformations at the beginning, the computational cost is similar to the unbiased partial enumeration technique.

While this strategy mitigates the sampling problem for the density, it does not solve the problem for the estimation of the single-chain partition function, $\mathcal{Z}_A[w_A]$. In the denominator of Eq. (5.43), the Boltzmann factors will strongly fluctuate if there is little overlap between the representative sample for conformations generated in the field, $\tilde{w}_A \approx w_A$, and the typical conformations in the absence of a field. Thus, the sampling problem is shifted from the numerator to the denominator and the accurate calculation of the free energy remains to be difficult with this method.

Avalos *et al.* [37, 38] have devised this technique for studying the adsorption of macromolecules at surfaces. The underlying idea also shares common aspects with techniques employed in self-consistent PRISM calculations [85].

(b) The second strategy to avoid the cost of a complete computer simulation of the single-chain problem at each iteration of the self-consistent adjustment of fields and densities consists in relaxing the external fields toward their saddle-point values during the single-chain simulation [35, 36]. If the scheme converges to a stationary state, the mean field solution may be recovered. Previously, this numerical scheme has been employed in conjunction with Brownian dynamic simulations and is denoted self-consistent Brownian dynamics. Ganesan and coworkers [36] recommend to propagate the external fields on a timescale that is much slower than the motion of individual segments. The authors emphasize, however, that this scheme is purely phenomenological and lacks a firm basis at the molecular level. Similar to method (a), the overlap of the so-generated single-chain conformations with the original distribution in the absence of an external field cannot be easily estimated and thus one cannot directly compute the free energy.

5.3.3 Single-Chain-in-Mean-Field Simulations and Grid-Based Monte Carlo Simulation of the Field-Theoretic Hamiltonian

5.3.3.1 Single-Chain-in-Mean-Field Simulations

Within the field-theoretic formulation, the problem of mutually interacting chains is equivalent to that of a single-chain in fluctuating, complex fields. We emphasize again that this is an exact reformulation, which does not invoke any approximations. While the fluctuating external fields are complex, their saddle-point values, which make the dominant contribution to the integral over the fluctuating fields, are real. The mean field approximation consists in replacing the fluctuating, complex fields by their static, real saddle-point values. These saddle-point values of the external fields are related to the mean field value of the local densities via the self-consistency condition, Eqs. (5.21) and (5.24). They have a simple interpretation as the average interaction a segment experiences due to the surrounding, which is characterized by the average densities, $\phi_A^*(\mathbf{r})$ and $\phi_B^*(\mathbf{r})$.

Single-chain-in-mean-field (SCMF) simulation [40–42, 86] is an approximate, computational method that retains the computational advantage of self-consistent field theory but additionally includes fluctuation effects because, in contrast to self-consistent theory, SCMF simulations aim at preserving the instantaneous description of the fluctuating interactions of a segment with its environment. In this particle-based simulation technique, one studies an ensemble of molecules in fluctuating, real, external fields. The explicit particle coordinates are the degrees of freedom and not the collective variables, densities and fields.

An SCMF simulation cycle is comprised of two stages, which are illustrated in Figure 5.3. First, during a short Monte Carlo simulation, the chains are decoupled and are independently moved subjected to the external fields, $\hat{w}_A$ and $\hat{w}_B$. This decoupling allows an efficient implementation on parallel computers. In the second stage, after a small predetermined number of Monte Carlo moves of the ensemble of independent molecules in the external fields is accomplished, the changes in densities of the ensemble of chains due to the evolution of the molecules are calculated. The new local densities are obtained and the external fields updated according to Eqs. (5.21) and (5.24). It is important to note that in these equations the average densities, ϕ_A^* and ϕ_B^*, are replaced by the instantaneous values, $\hat{\phi}_A$ and $\hat{\phi}_B$, obtained from the ensemble of explicit molecular conformations. Then, a new cycle of SCMF simulations is commenced.

calculate the instantaneous, fluctuating and spatially varying density

recalculate the external fields from the fluctuating densities using

$$w_A = iw_+ + w_- = -\frac{\chi_o N}{2}[\phi_A - \phi_B] + \kappa_o N[\phi_A + \phi_B - 1]$$

$$w_B = iw_+ - w_- = +\frac{\chi_o N}{2}[\phi_A - \phi_B] + \kappa_o N[\phi_A + \phi_B - 1]$$

MC or BD simulation of single-chain molecules in fluctuating, external fields

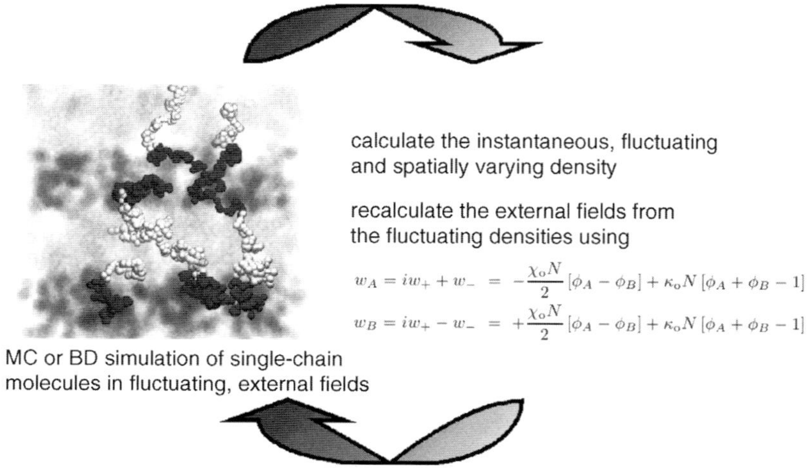

Figure 5.3 Illustration of the SCMF algorithm for a lamellar phase of a symmetric diblock copolymer. The snapshot on the left depicts five explicit chain configurations out of the large ensemble, which is comprised of thousands of diblock copolymer molecules. The fluctuating instantaneous densities of A- and B-segments are indicated by the shading of the background.

5.3.3.2 Minimal, Particle-Based, Coarse-Grained Model: Discretization of Space and Molecular Contour

In order to turn this prescription into a practical algorithm, one has to specify the molecular architecture and the way how to calculate the instantaneous densities from the explicit particle coordinates. Since the SCMF simulation algorithm is particle-based, it can be used with arbitrary single-chain models. In the following, we are interested in the universal behavior of multicomponent polymer melts. Therefore, we restrict ourselves to a bead-spring model, which is a computationally efficient representation of a Gaussian polymer chain. Generalizations to more complex molecular architectures are straightforward because the explicit molecular coordinates are the fundamental degrees of freedom.

$$\frac{\mathcal{H}_{\mathrm{b}}[\mathbf{r}_i(s)]}{k_{\mathrm{B}}T} = \sum_{s=1}^{N-1} \frac{3(N-1)}{2R_{\mathrm{eo}}^2}[\mathbf{r}_i(s)-\mathbf{r}_i(s+1)]^2. \tag{5.44}$$

While the bond lengths between atomistic monomeric units in a chemically realistic model are fixed, the harmonic potential between the coarse-grained segments stems from the Gaussian distribution of distances between sufficiently distant monomeric units along the backbone of a chemically realistic representation.

To relate the instantaneous densities of the ensemble of molecules to the external fields, one has to introduce a microscopic cutoff. Either one utilizes interactions with a finite range in Eqs. (5.21) and (5.24) or one regularizes the δ-function that appear in Eq. (5.5) by a smoothing function of finite support. We will show that both techniques are equivalent.

In the following discussion, we adopt zero-range, minimal interactions, Eqs. (5.33) and (5.34), and we regularize the instantaneous density by assigning it to a collocation lattice with spacing ΔL, that is, the instantaneous density, $\hat{\phi}_{\mathrm{A}}(\mathbf{c})$, of A-segments at a grid point, $\mathbf{c}$, is given by

$$\hat{\phi}_{\mathrm{A}}(\mathbf{c}) = \frac{1}{\Delta L^3}\int \mathrm{d}^3\mathbf{r}\,\Pi(\mathbf{r},\mathbf{c})\hat{\phi}_{\mathrm{A}}(\mathbf{r}) = \frac{1}{\varrho_{\mathrm{o}}\,\Delta L^3}\sum_{i=1}^{n_{\mathrm{A}}}\sum_{s=1}^{N}\Pi(\mathbf{r}_i(s),\mathbf{c}), \tag{5.45}$$

where $\Pi(\mathbf{r},\mathbf{c})$ denotes the assignment function onto the lattice. Often, a linear assignment of the segment position, $\mathbf{r}$, onto the point $\mathbf{c}$ of the collocation grid [39, 43, 87] is used, that is,

$$\Pi(\mathbf{r},\mathbf{c}) = w(r_x-c_x)w(r_y-c_x)w(r_z-c_z) \quad \text{with} \quad w(d) = \begin{cases} 1-\dfrac{|d|}{\Delta L}, & \text{for } |d| < \Delta L, \\ 0, & \text{otherwise,} \end{cases} \tag{5.46}$$

where d denotes the distance between the grid point and the segment position along a Cartesian direction. Similar schemes are used for particle-in-cell techniques in plasma physics [88] or particle-mesh methods in electrostatics [89, 90] in order to assign a particle-based density/charge distribution onto a lattice. The spatial integrals in Eqs. (5.33) and (5.34) are evaluated via the collocation grid

$$\frac{\mathcal{H}_{\mathrm{nb}}}{k_B T} = \sqrt{\bar{\mathcal{N}}} \frac{\Delta L^3}{R_{\mathrm{eo}}^3} \sum_{\mathbf{c}} \left\{ \frac{\varkappa_{\mathrm{o}} N}{2} [\hat{\phi}_A(\mathbf{c}) + \hat{\phi}_B(\mathbf{c}) - 1]^2 - \frac{\chi_{\mathrm{o}} N}{4} [\hat{\phi}_A(\mathbf{c}) - \hat{\phi}_B(\mathbf{c})]^2 \right\}, \tag{5.47}$$

where the sum is taken over all grid points. Alternatively, this interaction energy can also be obtained by using Eq. (5.5) for the instantaneous values of the microscopic densities in conjunction with the following, finite-ranged potentials:

$$V_{\mathrm{ev}}(\mathbf{r}, \mathbf{r}') = \varkappa_{\mathrm{o}} \frac{k_B T}{\varrho_{\mathrm{o}}} \nu(\mathbf{r}, \mathbf{r}') \quad \text{and} \quad V(\mathbf{r}, \mathbf{r}') = \frac{\chi_{\mathrm{o}}}{2} \frac{k_B T}{\varrho_{\mathrm{o}}} \nu(\mathbf{r}, \mathbf{r}') \tag{5.48}$$

with

$$\nu(\mathbf{r}, \mathbf{r}') \equiv \frac{1}{\Delta L^3} \sum_{\mathbf{c}} \Pi(\mathbf{r}, \mathbf{c}) \Pi(\mathbf{r}', \mathbf{c}). \tag{5.49}$$

The interactions are pairwise but not translationally invariant, that is, they do not only depend on the distance, $\mathbf{r}-\mathbf{r}'$, between the particles but also depend explicitly on their position relative to the collocation grid. The expression (5.49) also clarifies that ΔL plays the role of the range of interactions.

Unlike the pairwise interactions of coarse-grained models with harsh excluded volume interactions, the potentials in this coarse-grained model are soft, that is, effective segments can strongly overlap. This is quite a natural property because each segment represents the center of mass of a group of chemical repeat units. By the same token, the liquid of coarse-grained segments does not exhibit pronounced density–density correlations (i.e., packing effects).[4]

Note also that the pairwise interactions in Eq. (5.49) are density dependent. Thus, the interactions between coarse-grained entities are not potential energies but rather free energies that depend on the thermodynamic state. Again, this property is a natural consequence of the coarse-graining procedure that forms the basis of the model. The density-dependence is such that the energy per molecule remains finite in the limit of high density, $\sqrt{\bar{\mathcal{N}}} \sim \varrho_{\mathrm{o}} \to \infty$.

5.3.3.3 Monte Carlo Simulations and Advantages of Soft Coarse-Grained Models

The discretized Edwards–Hamiltonian in Eq. (5.44) and the nonbonded, pairwise interactions in Eq. (5.49) completely specify a coarse-grained, particle-based model for a multicomponent polymer system. Without resorting to any approximation, one can study the equilibrium properties of this coarse-grained, particle-based model by a variety of Monte Carlo simulation techniques, which can be chosen either to faithfully represent the single-chain dynamics [91] or to explore the configuration space of the model most efficiently. For instance, sophisticated and highly efficient rebridging techniques can be utilized to efficiently relax the macromolecular conformations [92]. There are two advantages of this class of soft, coarse-grained models:

4) In a particle-based model with harsh repulsions or the corresponding density functional theory, one would also use different assignment functions to describe the difference in the interaction range of V_{ev} and V.

(a) The increase in the molecular density, ϱ_0/N, is much more efficient to achieve experimentally relevant, high invariant degrees of polymerization, $\bar{\mathcal{N}}$, than the increase in the chain discretization, N.
Consider the example that one wants to describe a small patch of the lamellar phase with cubic geometry $(L = 5R_{eo})^3$, which is comprised of three lamellar sheets of a symmetric diblock copolymer with $\bar{\mathcal{N}} = 10\,000$ (cf. Figure 5.5 for a similar system). In a particle-based model with hard-core repulsion, the segment density is limited by $\varrho_o b^3 \approx 1$, where $b \equiv R_{eo}/\sqrt{N-1}$ denotes the statistical segment length. For higher densities, the liquid of segments freezes into a crystal or it arrests in a glassy structure. Since, $\bar{\mathcal{N}} = (\varrho_o b^3)^2 N$, the only way to achieve $\bar{\mathcal{N}} = 10\,000$ is to consider chains comprised of $N = 10^4$ beads. The total number of interaction centers in a box of size $(5R_{eo})^3$ is $n = N\sqrt{\bar{\mathcal{N}}}(L/R_{eo})^3 = 1.25 \cdot 10^8$. Since these molecules in a dense melt reptate, the single-chain relaxation time scales such as $\tau = \tau_o N^3 = \tau_o 10^{12}$, where τ_o is a microscopic time that characterizes the motion of an individual segment. The total number of segment motions to propagate the system by one single-chain relaxation time is $n\tau/\tau_o \approx 10^{20}$.
In a soft, coarse-grained model without harsh repulsion between segments, one can use a chain discretization of $N = 32$ in order to faithfully represent the Gaussian chain architecture. Using $\varrho_o b^3 = \sqrt{\bar{\mathcal{N}}/N} \approx 19$ one achieves the value $\bar{\mathcal{N}} = 10^4$. The system of size $(5R_{eo})^3$ is comprised of only $n = 4 \cdot 10^5$ segments. More importantly, since the single-chain dynamics obeys Rouse behavior, only $\tau/\tau_o = N^2$ segment movements are required to relax a chain conformation. Thus, the total effort to simulate the system amounts to $4 \cdot 10^8$ segment motions that are more than 10 orders of magnitude less than for models like the bond fluctuation model or Lennard-Jones bead-spring models.

(b) The computation of the interactions in Eq. (5.47) via the collocation grid is computationally very efficient in dense systems. In an off-lattice model with a potential of range, ΔL, the energy of a segment is computed by evaluating the interactions with all particles in a surrounding volume of size $(3\,\Delta L)^3$. For a typical parameter set, $N = 32$, $\bar{\mathcal{N}} = 10^4$, and $\Delta L/R_{eo} = 1/6$, the number of pairwise interactions to be computed amounts to $4 \cdot 10^2$. Assigning the particles to grid-based density, one can calculate the energy of a segment by a sum over the corresponding grid points. For a linear assignment (cf. Eq. (5.46)), there are eight sites of the collocation lattice to be considered. Thus, evaluating the interaction energy via the collocation grid saves another one or two orders of magnitude in computation time.

5.3.3.4 Comparison Between Monte Carlo and SCMF Simulations: Quasi-Instantaneous Field Approximation

It is important to realize that the external fields in SCMF simulations approximate the *instantaneous* interactions of a molecule with its surrounding. Therefore, they are frequently recalculated using the spatially inhomogeneous density distribution of the ensemble at that instant of time. Thus, in contrast to the mean field theory or self-consistent Brownian dynamics, SCMF simulations do not utilize the average fields but employ Eq. (5.37) to calculate the instantaneous, real, external fields, w_A and w_B, mimicking the instantaneous interactions of a segment with its fluctuating surrounding. The updating of the external fields utilizing the instantaneous density distribution introduces correlations between the molecules and allows a description of fluctuation effects [40, 41].

In the limit that the external fields follow the density distribution of the ensemble instantaneously the method becomes accurate. The quasi-instantaneous field approximation of the SCMF simulation consists in maintaining the external fields at their most recent values until a short, predetermined number of Monte Carlo moves has been accomplished, after which they are updated. The extent to which the external fields fail to mimic the interactions at each moment in time controls the quality of the quasi-instantaneous field approximation.

The comparison between SCMF simulations and Monte Carlo simulations of our soft, coarse-grained model allows us to quantify the accuracy of the quasi-instantaneous field approximation [41]. For simplicity, we consider a local Monte Carlo move, where one proposes to move an A-segment from position $\mathbf{r}$ to position $\mathbf{r}'$. The concomitant change in the grid-based A-density, $\delta\hat{\phi}_A(\mathbf{c})$, is of the order $1/\varrho_o \Delta L^3$. In the Monte Carlo simulations, one calculates the energy change ΔE^{MC} and accepts or rejects the proposed move via the Metropolis acceptance criterion, $p_{\mathrm{acc}} = \min[1, e^{-\Delta E}]$. The energy difference, ΔE, is comprised of bonded and nonbonded contributions, $\Delta E^{\mathrm{MC}} = \Delta\mathcal{H}_b + \Delta\mathcal{H}_{nb}^{\mathrm{MC}}$. The nonbonded contribution $\Delta\mathcal{H}_{nb}^{\mathrm{MC}}$ can be expanded as

$$\begin{aligned}\Delta\mathcal{H}_{nb}^{\mathrm{MC}} &= \mathcal{H}_{nb}[\hat{\phi}_A[\{\mathbf{r}'\}], \hat{\phi}_B[\{\mathbf{r}'\}]] - \mathcal{H}_{nb}[\hat{\phi}_A[\{\mathbf{r}\}], \hat{\phi}_B[\{\mathbf{r}\}]] \\ &= \sum_{\mathbf{c}} \left.\frac{\partial\mathcal{H}_{nb}}{\partial\hat{\phi}_A(\mathbf{c})}\right|_{\hat{\phi}_A(\mathbf{c})} \delta\hat{\phi}_A(\mathbf{c}) + \frac{1}{2}\sum_{\mathbf{cc}'} \left.\frac{\partial^2\mathcal{H}_{nb}}{\partial\hat{\phi}_A(\mathbf{c})\partial\hat{\phi}_A(\mathbf{c}')}\right|_{\hat{\phi}_A(\mathbf{c}),\hat{\phi}_B(\mathbf{c}')} \delta\hat{\phi}_A(\mathbf{c})\delta\hat{\phi}_A(\mathbf{c}') + \cdots\end{aligned} \tag{5.50}$$

In case of pairwise interactions (as discussed throughout this chapter), the expansion terminates after the term of second order. Higher order terms will become relevant only if one employs a density functional of higher orders for the interactions [93].

If we propose the same move in the SCMF simulations, the energy difference will take the form $\Delta E^{\mathrm{SCMF}} = \Delta\mathcal{H}_b + \Delta\mathcal{H}_{nb}^{\mathrm{SCMF}}$. The bonded contribution is identical to the value in the Monte Carlo simulations but the nonbonded contributions differ.

$$\Delta\mathcal{H}_{nb}^{\mathrm{SCMF}} = \frac{\varrho_o \Delta L^3}{N} \sum_{\mathbf{c}} \hat{w}_A(\mathbf{c})\delta\hat{\phi}_A(\mathbf{c}). \tag{5.51}$$

The choice

$$\hat{w}_A(\mathbf{c}) = \frac{N}{\varrho_o \Delta L^3} \frac{\partial\mathcal{H}_{nb}}{\partial\hat{\phi}_A(\mathbf{c})} \tag{5.52}$$

makes an energy change in the SCMF simulations equal to the energy change in the Monte Carlo simulations to first order in $\delta\hat{\phi}$. This is exactly the grid-based analogue of the saddle-point condition, Eq. (5.25), that we utilize to calculate the external field from the instantaneous density distribution of the ensemble of chains. Thus, the quasi-instantaneous field approximation of the SCMF simulation amounts to approximating the full change of the energy due to a proposed Monte Carlo move (cf. Eq. (5.50)) by its linear approximation (5.51) with respect to the concomitant change of density. To leading order, the difference, $\delta E = \Delta\mathcal{H}_{\text{nb}}^{\text{MC}} - \Delta\mathcal{H}_{\text{nb}}^{\text{SCMF}}$, between energy changes entering the Metropolis criterion in the SCMF simulations and the Monte Carlo simulation can be estimated by the second-order term in Eq. (5.50). Since it is quadratic in the change of density, $\delta\hat{\phi}_A(\mathbf{c})$, the quasi-instantaneous field approximation becomes the more accurate, the denser the system is and the smaller the density change between updates of the field is. For the specific model and move considered, the error of the energy change is given by

$$\delta E = \frac{\varrho_o \, \Delta L^3}{N}\left(\varkappa_o N - \frac{\chi_o N}{2}\right)\delta\hat{\phi}_A(\mathbf{c})^2 = \left(\varkappa N - \frac{\chi_o N}{2}\right)\varepsilon, \tag{5.53}$$

where

$$\varepsilon = \frac{V}{nN^2 \, \Delta L^3} = \frac{1}{N^2\sqrt{\bar{\mathcal{N}}}}\left(\frac{R_{\text{eo}}}{\Delta L}\right)^3 \tag{5.54}$$

is a small parameter that controls the accuracy of the quasi-instantaneous field approximation [41]. In the limit, $\varepsilon \ll 1/\max(\varkappa_o N, \chi_o N)$, and frequent updating of the fields (strictly after every accepted MC move), the external fields accurately describe the interactions with the surrounding segments and SCMF simulations become quantitatively accurate.

The parameter, ε, plays the same role as the Ginzburg parameter, Gi, does for the mean field theory. The essential difference is that $Gi = 1/\bar{\mathcal{N}}$ is a coarse-grained parameter, that is, it is a property of the physical system. The parameter, ε, that controls the accuracy of the quasi-instantaneous field approximation, however, also depends on the discretization of space, ΔL, and molecular contour, N. By a careful choice of these discretization parameters of the soft, coarse-grained model, one can reduce ε even if Gi will not be small and fluctuation effects will be important.

A suitable choice of the spatial discretization is $\Delta L \simeq R_{\text{eo}}/\sqrt{N-1}$ because the statistical segment length, $b = R_{\text{eo}}/\sqrt{N-1}$, sets the microscopic length scale of the discretized, Gaussian, single-chain structure. Then, ε takes the simple form:

$$\varepsilon = \frac{1}{\sqrt{N\bar{\mathcal{N}}}} = \sqrt{\frac{Gi}{N}}. \tag{5.55}$$

The quality of the quasi-instantaneous field approximation and the ability of SCMF simulations to capture nontrivial correlation effects is illustrated by studying deviations from the Gaussian chain statistics due to the correlation hole effect in polymer solutions [94] and melt [95, 96]. These intermolecular correlations result in a power-law decay of the intramolecular, bond–bond correlation function [95, 96] in a one-component melt as shown in Figure 5.4.

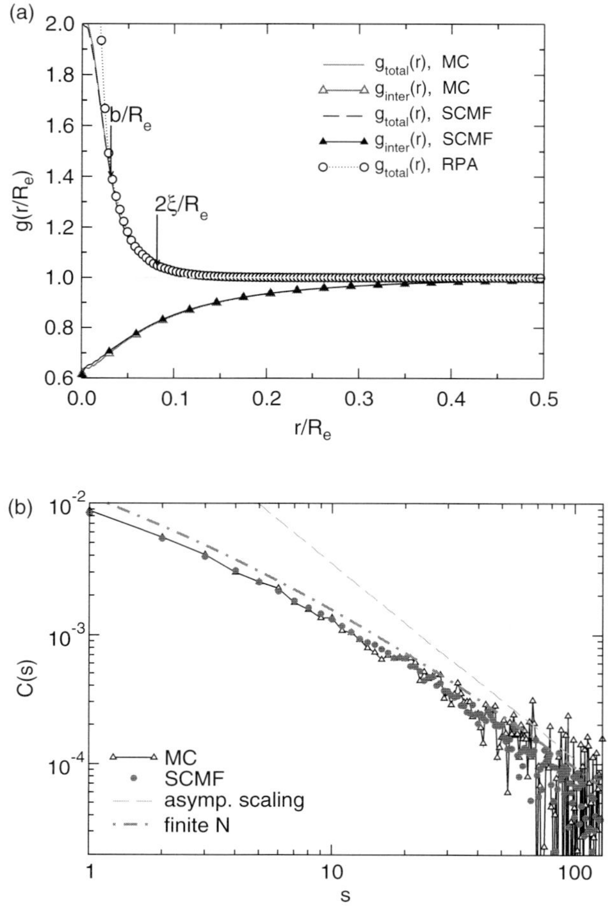

Figure 5.4 (a) Segment–segment radial distribution functions, $g(r)$, for a one-component melt with $\sqrt{\bar{\mathcal{N}}} = 32$, $\varkappa_o N = 50$, $\chi_o N = 0$, $N = 1024$, and $\Delta L = R_{eo}/32$. The random phase approximation, Monte Carlo, and SCMF simulation results for the total pair correlation function, $g_{\text{tot}}(r)$, are shown with open circles, solid lines, and dashed lines, respectively. The intermolecular part, $g_{\text{inter}}(r)$, is shown for Monte Carlo and SCMF simulations with open and solid triangles. The average bond length and the screening length, ξ_{ev}, of density fluctuations is indicated by arrows. The good agreement demonstrates the accuracy of the quasi-instantaneous field approximation. (b) Logarithmic plot of the decay of bond–bond correlations along the chain contour for a homopolymer melt (using the same parameters as in panel a). The results of Monte Carlo and SCMF simulations are represented by open triangles and circles, respectively. The dashed line marks the asymptotic power law [95] $C(s) \equiv \langle \mathbf{b}(i) \cdot \mathbf{b}(i+s) \rangle \rightarrow \sqrt{\frac{3}{8\pi^2 \bar{\mathcal{N}}}} \frac{\langle b^2 \rangle^2}{R_{eo}^2} \left(\frac{s}{N}\right)^{-3/2}$ in the limit $N \rightarrow \infty$. $\mathbf{b}(i) = \mathbf{r}(i+1) - \mathbf{r}(i)$ is the bond vector connecting two subsequent segments along the molecule [95]. The solid line, $C(s) = \frac{\sqrt{48}}{4\pi^2 N \sqrt{\bar{\mathcal{N}}}} \left(\frac{N}{g}\right)^{3/2} \left[\sqrt{\frac{\pi g}{2s}} - \pi\, e^{\frac{2s}{g}} \operatorname{erfc}\left(\sqrt{\frac{2s}{g}}\right)\right]$ with g being the number of segments contained inside an excluded volume blob of size ξ_{ev}, captures finite chain discretization effects. Reprinted with permission from Ref. [41]. © 2006, American Institute of Physics.

5.3.4
Off-Lattice, Soft, Coarse-Grained Models

While the evaluation of the interactions in a dense system is computationally beneficial, the underlying lattice structure requires the usage of special simulation techniques to accurately calculate the contribution of the nonbonded interactions to the pressure. These difficulties can be mitigated by using a soft, coarse-grained, off-lattice model. Since forces are well defined in off-lattice models, one can use Brownian dynamics or dissipative particle dynamics methods [97–103]. Also, simulations under constant pressure or surface tension are feasible.

One possibility to motivate such an off-lattice description is to use the pairwise interactions in Eq. (5.49) in conjunction with a linear assignment function Π and average the potential $v(\mathbf{r},\mathbf{r}')$ over a uniform distribution of grid positions keeping the particle positions, $\mathbf{r},\mathbf{r}'$, fixed.

$$\bar{v}(\mathbf{r},\mathbf{r}') \equiv \frac{1}{\Delta L^6}\int \mathrm{d}^3\,\Delta\mathbf{c}\sum_{\mathbf{c}}\Pi(\mathbf{r}-\Delta\mathbf{c},\mathbf{c})\Pi(\mathbf{r}'-\Delta\mathbf{c},\mathbf{c}). \tag{5.56}$$

This procedure restores translational invariance [41, 104] and results in a potential of the form

$$\Delta L^3\,\bar{v}(\mathbf{r},\mathbf{r}') = \bar{w}(|r_x-r'_x|)\bar{w}(|r_y-r'_y|)\bar{w}(|r_z-r'_z|) \quad \text{with}$$

$$\bar{w}(d) = \begin{cases} \dfrac{2}{3}-\dfrac{d^2}{\Delta L^2}+\dfrac{d^3}{2\,\Delta L^3}, & \text{for } |d|<\Delta L,\\[2ex] \dfrac{1}{3}-\dfrac{d^3}{6\,\Delta L^3}, & \text{for } 1\le |d|<2\,\Delta L,\\[2ex] 0, & \text{otherwise.}\end{cases} \tag{5.57}$$

These interactions with cubic anisotropy can be well approximated by an isotropic Gaussian form

$$\Delta L^3\,\bar{v}(\mathbf{r},\mathbf{r}') \approx \left(\frac{3}{2\pi}\right)^{3/2}\exp\left[-\frac{3(\mathbf{r}-\mathbf{r}')^2}{2\,\Delta L^2}\right]. \tag{5.58}$$

Such a model with a Gaussian form of the interaction, v, has been utilized by Zuckermann and coworkers [105] in order to investigate polymer brushes. The DPD-model of Groot and collaborators [100–102] also shares many features with this soft coarse-grained model. In this model, the conservative force is derived from a pairwise potential of the form

$$v_{\mathrm{DPD}}(\mathbf{r},\mathbf{r}') \propto \begin{cases} \dfrac{1}{2}\left(1-\dfrac{|\mathbf{r}-\mathbf{r}'|}{\Delta L}\right)^2, & \text{for } |\mathbf{r}-\mathbf{r}'|<\Delta L,\\[2ex] 0, & \text{otherwise.}\end{cases} \tag{5.59}$$

While the calculation of the pairwise interactions in the off-lattice model is computationally more demanding than in the grid-based version, this off-lattice model retains the capability of investigating dense polymer systems with an experimentally relevant invariant degree of polymerization, $\bar{\mathcal{N}}$. This benefit crucially relies on the ability to increase the density without introducing strong density–density correlations or crystallization in the fluid of segments.

The structure of the fluid depends on the form of the pairwise potential, v (or, in the lattice-based description, on the assignment function, Π). Our strategy is to choose the soft, pairwise potential, v, between effective segments such that the density–density pair correlation function of the fluid of nonbonded segments, $g(r)$, which measures the probability of finding two segments a distance r apart, does not exhibit much structure. Conceptually, inverse Boltzmann iteration [14] could be utilized to construct such an interaction numerically. The so-generated potential would, however, depend on the thermodynamic state, that is, it is optimized for a certain density (or $\bar{\mathcal{N}}$).

More generally, one may wish to ensure that the choice of the soft, intersegment potential, v, results in a liquid state. The phase diagram of fluids of soft particles has been explored and exhibits quite a rich behavior [106–108]. In addition to the liquid and crystal phases that occure in fluids with harsh repulsions exhibit at low temperatures, these fluids exhibit either re-entrant melting or formation of cluster crystals at high densities. The latter phases are thermodynamically stable crystal phases that feature a lattice constant that is independent of density and, in turn, the occupancy of a lattice site increases linearly with ϱ_0. In the present context, a coarse-grained segment describes a chain segment comprised of a small number of atomistic, monomeric repeat units, and a melt of these short chains does not exhibit inhomogeneous structure. Therefore, the soft potential, v, should be chosen to avoid the formation of cluster crystal.

It has been shown by computer simulation [109–111] and density functional theory [106, 108] that the soft, purely repulsive, radially symmetric potential, $V(r)$, will form cluster crystals at sufficiently high density if its Fourier transform, $\tilde{V}(k)$, becomes negative for a range of wave vectors. Within mean field approximation, the stability limit of the homogeneous liquid is given by the λ-line [108]

$$1+\varrho_0 \frac{\tilde{V}(k^*)}{k_B T} = 0, \tag{5.60}$$

where k^* denotes the wave vector at which the Fourier transform, $\tilde{V}$, of the potential attains its minimum. The condition (5.60) can be fulfilled only if $\tilde{V}(k^*) < 0$. Then, $2\pi/k^*$ sets the lattice spacing of the cluster crystal.

The Gaussian form of the interaction potential, V, in Eq. (5.58) and the DPD-potential in Eq. (5.59) are particularly suitable because their Fourier transforms are nonnegative. Thus, Eq. (5.60) cannot be fulfilled and the liquid structure is stable against the formation of cluster crystals.

5.4 An Application: Calculating Free Energies of Self-Assembling Systems

5.4.1 Crystallization in Hard Condensed Matter Versus Self-Assembly of Soft Matter

One important application of self-consistent field theory is the calculation of free energies of self-assembled morphologies or spatially inhomogeneous structures. In order to obtain a mean field approximation of the free energy, one substitutes the saddle-point values of the fields back into the free energy functional, $\mathcal{F}$, of Eq. (5.16). Comparing the free energy of different morphologies, phase diagrams as a function of composition or molecular architecture can be constructed. With the same technique, the free energy of interfaces between macroscopic domains in a homopolymer blend or the free energy of grain boundaries in self-assembling copolymer systems, where morphologies with different orientations meet, can be accurately determined. The calculation of free energies in self-assembling systems without invoking the mean field approximation is a difficult task and, only recently, computational strategies to accomplish this purpose have been devised.

In order to illustrate the challenge, it is interesting to draw a comparison between self-assembly in soft matter systems and crystallization in simple, hard condensed matter systems (e.g., a Lennard-Jones solid). The local difference in volume fraction, $\phi(\mathbf{r})$, of the two species of the amphiphilic system plays a similar role as the density of a hard crystal. Its dominant Fourier mode is the order parameter of the transition. In a well-ordered hard crystal, each particle fluctuates little around its corresponding crystal lattice position. Thus, the system resembles an Einstein crystal, in which noninteracting particles are tethered by harmonic springs to their ideal lattice positions. Frenkel and Ladd have used thermodynamic integration from this Einstein crystal to the well-ordered solid for calculating the absolute free energy of a hard crystal [112].

In a self-assembling soft matter system, the composition also fluctuates little around the ideally ordered value; however, the molecules are in a liquid state, that is, they diffuse and are not "tethered" to ideal positions. Therefore, there is no simple reference state of the particle-based model with a known free energy, and previous simulation techniques for calculating the absolute free energy of hard crystals do not straightforward carry over to self-assembling soft matter systems.

In fact, even calculating the free energy of a homogeneous melt without repulsion between the two different monomeric species (i.e., $\chi_o N = 0$), which would be the analogue of an ideal gas in a simple, hard condensed matter system, is a formidable task. In a liquid, the (bonded and nonbonded) interactions of a segment with its surrounding are on the order $k_B T$. Thus, the free energy per molecule is proportional to $k_B TN$. In order to accurately determine the location of phase boundaries, to calculate the free energy costs of defects or grain boundaries, or to assess the thermodynamic stability of morphologies, one needs to know the free energy per molecule with an accuracy of $\mathcal{O}(10^{-3} k_B T)$. Therefore, the absolute free energy would be required to be known with a precision of the order $10^{-3}/N \approx 10^{-5}$. The free energy difference between the disordered phase and a self-assembled structure, however, is only of the order $\chi_o N k_B T \sim k_B T$. Thus, it is advantageous to directly calculate the free energy difference between the disordered and the ordered state rather than to determine it as a difference of two large absolute free energies.

5.4.2
Field-Theoretic Reference State: The Einstein Crystal of Grid-Based Fields

The property that the collective fields in the self-assembled state fluctuate little around their saddle-point values in the self-assembled state has been exploited by Fredrickson and coworkers [45] using the field-theoretic formulation of the particle-based model in terms of the collective fields, W_+ and W_- (cf. Eq. (5.15)). When one employs a collocation lattice, these fields become a discrete set of continuous variables, $W_+(\mathbf{c})$ and $W_-(\mathbf{c})$. The integral over the fluctuating fields is performed via a complex Langevin sampling, which mitigates the sign problem due to the incompressibility field, W_+. Both fields fluctuate around their saddle-point values, which characterize the mean field approximation. Drawing an analogy between the fluctuation of a particle in a hard crystal around their ideal lattice position and the fluctuation of the grid-based fields around the mean field solution, Fredrickson and coworkers transferred the Frenkel–Ladd method [112] from particles to discretized fields [45]. The reference state of the Einstein crystal of grid-based fields is described by the free energy functional

$$\mathcal{F}_{\mathrm{ref}}[W_+, W_-] \equiv F_{\mathrm{SCFT}} + \frac{\Delta L^3}{2}\sum_{\mathbf{c}}\{\alpha_+|\Delta W_+(\mathbf{c})|^2 + \alpha_-|\Delta W_-(\mathbf{c})|^2\}, \tag{5.61}$$

where $\Delta W_\pm \equiv W_\pm - w_\pm$ denote the deviations of the local fields from their saddle-point values, and $\alpha_\pm > 0$ are real "spring" constants that dictate the magnitude of fluctuations. $F_{\mathrm{SCFT}} \equiv F[w_+, w_-]$ is the mean field approximation of the free energy. The free energy of this reference state can be easily calculated by performing the Gaussian integrals over the fields at the different grid points and one obtains

$$F_{\mathrm{ref}} \equiv -k_B T \ln \int \mathcal{D}W_+ \mathcal{D}W_- \exp\left[-\frac{\mathcal{F}_{\mathrm{ref}}[W_+, W_-]}{k_B T}\right], \tag{5.62}$$

$$= F_{\mathrm{SCFT}} - \frac{k_B T V}{2\Delta L^3}\left\{\ln\frac{2\pi}{\alpha_+ \Delta L^3} + \ln\frac{2\pi}{\alpha_- \Delta L^3}\right\}, \tag{5.63}$$

where $V/\Delta L^3$ denotes the number of points the grid is comprised of. Then, one uses a coupling parameter, λ, to relate this reference system to the original system. To this end, one performs simulations of a system characterized by the free energy functional

$$\mathcal{F}_\lambda[W_+, W_-] \equiv \lambda \mathcal{F}[W_+, W_-] + (1-\lambda)\mathcal{F}_{\mathrm{ref}}[W_+, W_-], \tag{5.64}$$

where $\mathcal{F}[W_+, W_-]$ is the grid-based analogue of Eq. (5.16). The free energy difference, $F - F_{\text{ref}}$, between the original and the reference system can be estimated via standard thermodynamic integration with respect to the parameter, λ.

$$F - F_{\text{ref}} = \int_0^1 \mathrm{d}\lambda \left\langle \frac{\mathrm{d}\mathcal{F}_\lambda}{\mathrm{d}\lambda} \right\rangle_\lambda \tag{5.65}$$

$$= \int_0^1 \mathrm{d}\lambda \left\langle \mathcal{F}[W_+, W_-] - \left(\mathcal{F}[w_+, w_-] + \frac{\Delta L^3}{2} \sum_{\mathbf{c}} \{\alpha_+ |W_+ - w_+|^2 + \alpha_- |W_- - w_-|^2\} \right) \right\rangle_\lambda . \tag{5.66}$$

The "spring" constants, $\alpha_\pm$, should be chosen as to minimize the absolute value of integrand of Eq. (5.66) as a function of λ [112]. Note that in simple crystals in hard condensed matter, all particles fluctuate around the ideal lattice position by the same amount. The fluctuations of the fields, $\Delta W_\pm(\mathbf{c})$, in a self-assembled structure, however, depend on the spatial position, $\mathbf{c}$.

In principle, the method is able to calculate the absolute free energies. In practice, however, only differences of free energies between systems using the same spatial discretization, ΔL^3, are meaningful because the value of the free energy sensitively depends on ΔL. This ultraviolet divergency [113] has to be regularized, for example, by subtracting the free energy of the disordered state simulated on the same lattice. The method has successfully been employed to determine fluctuation effects on the phase diagram of copolymers [45].

5.4.3
Particle-Based Approach: Reversible Path in External Ordering Field

5.4.3.1 How to Turn a Disordered Melt into a Microphase-Separated Morphology Without Passing Through a First-Order Transition?

Within a particle-based model, there is no well-defined reference state for the self-assembled structure. However, one can try to relate the self-assembled structure to a disordered melt (or a different self-assembled morphology) via a reversible path and calculate the change of the free energy by thermodynamic integration. Typically, transitions between disordered and ordered morphologies or between different self-assembled structures are of first order. Thus, in an analogy to crystallization of hard condensed matter, there is no path in the space of physical intensive variables – for example, temperature, incompatibility, or composition – that reversibly connects disordered and ordered structures.

Such a reversible path, however, can be constructed with the help of an external ordering field that is adapted to the spatial structure of the self-assembled morphology. Our method for self-assembling systems is inspired by the work of Sheu, Mou, and Lovett calculating the absolute free energy of a Lennard-Jones solid [114]. Related reversible integration paths between a solid and a liquid have been used by Grochola [115] and Eike *et al.* [116] for calculating the free energy difference between a solid and a liquid.

To illustrate the computational method, we consider the free energy difference between a disordered phase at $\chi_{\text{init}}N = \chi_o N = 0$ and a lamellar, microphase-separated morphology at $\chi_{\text{final}}N = \chi_o N = 20$ of a symmetric diblock copolymer. Specifically, we use the minimal, coarse-grained model for a compressible multicomponent polymer melt with grid-based interactions as defined by Eqs. (5.44) and (5.47). Generalizations to off-lattice representations or different molecular architectures are straightforward.

The reversible path, which relates the disordered phase to the lamellar one, is comprised of two branches. First, as illustrated in Figure 5.5, we structure the homogeneous, disordered melt at low incompatibility, $\chi_{\text{init}}N$, by applying an external ordering field, $h(\mathbf{r})$, that is conjugated to the composition, $\phi = \phi_A - \phi_B$. The nonbonded energy of the system in the external ordering field takes the form:

$$\mathcal{H}_{\text{nb}} = \mathcal{H}_{\text{melt}} + \mathcal{H}_{\text{ord}} + \mathcal{H}_{\text{ext}}, \tag{5.67}$$

where $\mathcal{H}_{\text{melt}}$, defined in Eq. (5.33), denotes the nonbonded interactions in the melt without repulsion between different species. $\mathcal{H}_{\text{ord}} \sim \chi_o$, given in Eq. (5.34), represents the thermal repulsion giving rise to self-assembly into the lamellar microphase, and $\mathcal{H}_{\text{ext}}$ is the contribution of the external ordering field:

$$\frac{\mathcal{H}_{\text{ext}}}{k_B T (V/R_{\text{eo}}^3)\sqrt{\bar{\mathcal{N}}}} = -\frac{\lambda N}{V} \int d^3\mathbf{r}\, f_{\text{ext}}(\mathbf{r})\phi(\mathbf{r}), \tag{5.68}$$

where λN characterizes the strength and f_{ext} the spatial variation of the external ordering field.

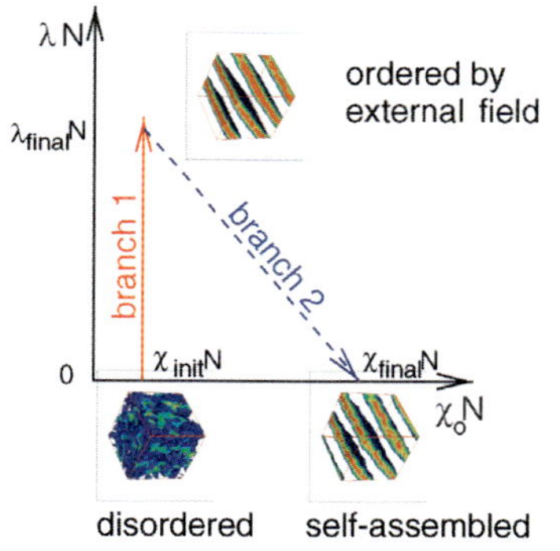

Figure 5.5 Sketch of the reversible path that connects the homogeneous, disordered state, the externally ordered and the self-assembled state. Configurational snapshots of a symmetric diblock melt illustrate the different states. In the snapshots three-dimensional contour plots of the composition are shown. The B-rich component is removed for clarity and the interface between the different components is colored blue. $\chi_{\text{init}}N = 0$, $\chi_{\text{final}}N = 20$, and the maximal strength of the ordering field is $\lambda_{\text{final}}N = 10$. The SCMF simulations correspond to $\bar{\mathcal{N}} = 14\,884$ and use a chain discretization, $N = 32$. The linear extension of the simulation cell is $L = 4.77R_{\text{eo}}$ and the spacing of the lamellar structure is $L_o = L/(2\sqrt{2}) = 1.686R_{\text{eo}}$. Reprinted with permission from Ref. [43]. © 2008, American Institute of Physics.

Gradually increasing the external ordering field, $h(\mathbf{r}) = \lambda N f_{\text{ext}}(\mathbf{r})$, from zero to h_{final}, along the first branch, we do not encounter any thermodynamic singularity because the ordering is not collective, that is, the molecules gradually arrange into the lamellar structure in response to the external ordering field but not by virtue of pairwise interactions.

Along the second branch, we gradually replace the effect of the external ordering field by pairwise interactions, that is, λN is decreased to zero and, in turn, $\chi_o N$ increases from $\chi_{\text{init}} N$ to $\chi_{\text{final}} N$. Reference [114] demonstrated that the optimal choice of the ordering field, $h(\mathbf{r})$, is such that the composition, $\phi(\mathbf{r})$, at $\chi_{\text{init}} N$ in the presence of the ordering field, h_{final} (i.e., at the end of the first branch) closely mimics the morphology at the final, self-assembled state, $\chi_{\text{final}} N$, in the absence of the ordering field. The second branch, $\chi_o N(\lambda N)$, of the transformation path has to be chosen such that the variation of the structure along the branch is minimized. The absence of any abrupt changes of the structure, $\phi(\mathbf{r})$, along the second branch indicates that it is also free of any thermodynamic singularity. Thus, the two branches reversibly connect the disordered and the ordered state in the expanded parameter space that includes a spatially varying, external ordering field. Thus, one turns a disordered system into a spatially ordered one without passing through a first-order transition [114].

The self-consistent field theory provides an accurate estimate of the ordering field, h_{final}, at the end of the first path and an appropriate choice of the second branch, $\chi_o N(\lambda N)$. The system at $\chi_o N = 0$ in the static, external ordering field, $h(\mathbf{r})$, closely resembles the single-chain problem of the mean field theory. h plays the same role as $-w_-$ in the mean field approximation. The difference between the system along the first branch and the single-chain problem of the mean field theory is that density fluctuations are not suppressed by a mean field, iw_+, but by the pairwise interactions, $\mathcal{H}_{\text{melt}}$. If density and composition fluctuations do not strongly couple, this difference is negligible and we can use the self-consistent field theory to obtain an accurate approximation of the external field that creates a given composition, $\phi_{\text{final}}(\mathbf{r})$. Within the mean field approximation, we obtain

$$h_{\text{final}}(\mathbf{r}) \approx \frac{\chi_{\text{final}} N}{2} \phi_{\text{final}}(\mathbf{r}), \tag{5.69}$$

that is, $\lambda_{\text{final}} N = \chi_{\text{final}} N/2$ and $f_{\text{ext}}(\mathbf{r}) = \phi_{\text{final}}(\mathbf{r})$.

If we apply the mean field theory to the system along the second branch of the transformation path, then we obtain for the saddle-point value of the field, w_-, coupling to the composition:

$$w_-(\mathbf{r}) = -\frac{\chi_o N}{2} \phi(\mathbf{r}) - h(\mathbf{r}). \tag{5.70}$$

Within the mean field approximation, the structure will not change along the second branch of the integration path if $w_-(\mathbf{r})$ remains constant. Thus, mean field theory predicts

$$h(\mathbf{r}) \approx \frac{\chi_{\text{final}} N - \chi_o N}{2} \phi_{\text{final}}(\mathbf{r}) \tag{5.71}$$

for the optimal, second branch of the path. More elaborated choices could be envisioned if the self-consistent field theory were not accurate and the configurations were to change significantly along this branch.

5.4.3.2 Thermodynamic Integration Versus Expanded Ensemble and Replica-Exchange Monte Carlo Simulation

The free energy change along the transformation path in the two-dimensional parameter space spanned by $\chi_o N$ and λN can be calculated via thermodynamic integration. Let $\chi_o N(\lambda N)$ with $\lambda_1 N \le \lambda N \le \lambda_{N_p} N$ denote such a path, then the free energy change along this path is given by

$$\frac{\Delta F}{k_B T} = -\int_{\lambda_1 N}^{\lambda_{N_p} N} \mathrm{d}\lambda N \left(\langle \mathcal{I}_{\mathrm{ext}} \rangle + \frac{\mathrm{d}\chi_o N}{\mathrm{d}\lambda N} \langle \mathcal{I}_{\mathrm{ord}} \rangle \right), \tag{5.72}$$

where we have introduced

$$\mathcal{I}_{\mathrm{ord}} \equiv -\frac{1}{k_B T} \frac{\partial \mathcal{H}_{\mathrm{ord}}}{\partial \chi_o N} = \frac{1}{4} \frac{\sqrt{\bar{\mathcal{N}}} \Delta L^3}{R_{eo}^3} \sum_{\mathbf{c}} [\hat{\phi}_A(\mathbf{c}) - \hat{\phi}_B(\mathbf{c})]^2, \tag{5.73}$$

$$\mathcal{I}_{\mathrm{ext}} \equiv -\frac{1}{k_B T} \frac{\partial \mathcal{H}_{\mathrm{ext}}}{\partial \lambda N} = \frac{\sqrt{\bar{\mathcal{N}}} \Delta L^3}{R_{eo}^3} \sum_{\mathbf{c}} h(\mathbf{c}) [\hat{\phi}_A(\mathbf{c}) - \hat{\phi}_B(\mathbf{c})]. \tag{5.74}$$

Advanced Monte Carlo algorithms can be utilized to accurately compute the free energy difference [117]. We discretize the integration path into a set of sampling points, $\lambda_i N$, with $i = 1, \ldots, N_p$. These sampling points are grouped into M_r intervals, I_j with $j = 1, \ldots M_r$, which share a common boundary point, that is, the first interval, I_1, is comprised of the sampling points $\lambda_1 N, \ldots, \lambda_{N_p/M_r} N$, the second interval, I_2, contains the points $\lambda_{N_p/M_r} N, \ldots, \lambda_{2N_p/M_r} N$, and so on. This partitioning of the integration path is illustrated in Figure 5.6.

Within each interval, we use an expanded ensemble technique to explore the sampling points and configurations between neighboring intervals are swapped via replica-exchange Monte Carlo moves [118]. The partition function, $\mathcal{Z}$, takes the form:

$$\mathcal{Z} \sim \prod_{j=1}^{M_r} \sum_{\lambda_i N \in I_j} \frac{e^{\eta(\lambda_i N)}}{n!} \int \prod_{\alpha=1}^{n} \tilde{\mathcal{D}}[\mathbf{r}_\alpha(s)] \exp\left(-\frac{\mathcal{H}_{\mathrm{melt}} + \mathcal{H}_{\mathrm{ord}} + \mathcal{H}_{\mathrm{ext}}}{k_B T} \right), \tag{5.75}$$

where $\eta(\lambda_i N)$ are the weighting factors of the expanded ensemble. These factors have to be appropriately chosen in order to achieve approximately uniform sampling of the different sampling points in each interval.

The Monte Carlo simulation comprises three distinct moves: (i) Canonical Monte Carlo moves update the molecular conformations in the M_r replica. In this specific application, we employ a Smart Monte Carlo algorithm [119] that utilizes strong bonded forces to propose a trial displacement [43, 87]. The amplitude of the trial displacement has been optimized in order to maximize the mean-square displacement of molecules [91], and the single-chain dynamics closely resembles the Rouse-dynamics of unentangled macromolecules [120]. (ii) Since each replica is an

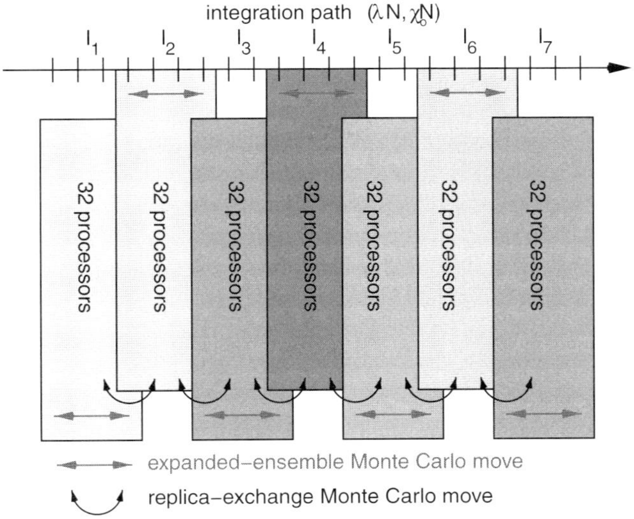

Figure 5.6 Illustration of the discretization of the integration path into $M_r = 7$ overlapping intervals. A configuration (replica) is associated with each interval. Each interval, in turn, is comprised of four states of an expanded ensemble. Expanded-ensemble Monte Carlo moves that change the values of λN and $\chi_o N$ within one interval are indicated by horizontal arrows. Replica-exchange Monte Carlo moves are represented by bent arrows. The hierarchical parallelization scheme – combining a moderate parallel program for a single replica with replica-exchange Monte Carlo simulation – allows an efficient usage of massively parallel computers. Adapted from Ref. [146]. Reproduced by permission of the PCCP Owner Society.

expanded ensemble [121], we attempt to change the control parameters, $\lambda_i N$ and $\chi_i N \equiv \chi_o N(\lambda_i N)$, to neighboring sampling points along the integration path. These Monte Carlo moves are accepted according to the Metropolis criterion

$$\mathrm{acc}(\lambda_i N \to \lambda_k N) = \min\left(1, \mathrm{e}^{\mathcal{T}_{\mathrm{ext}}[\lambda_k N - \lambda_i N] + \mathcal{T}_{\mathrm{ord}}[\chi_k N - \chi_i N] + \eta(\lambda_k N) - \eta(\lambda_i N)}\right), \tag{5.76}$$

where both $\lambda_i N$ and $\lambda_k N$ are within the same interval. Special care has to be taken at the interval boundaries to fulfill detailed balance. (iii) The third Monte Carlo move consists in swapping configurations at $\lambda N \in I_i$ and $\lambda' N \in I_l$, which are located in neighboring intervals, I_i and I_l. These replica-exchange moves [122–124] are accepted with probability

$$\mathrm{acc}(\lambda N \rightleftharpoons \lambda' N) = \min\left(1, \mathrm{e}^{-[\mathcal{T}'_{\mathrm{ext}} - \mathcal{T}_{\mathrm{ext}}][\lambda' N - \lambda N] - [\mathcal{T}'_{\mathrm{ord}} - \mathcal{T}_{\mathrm{ord}}][\chi'_o N - \chi_o N]}\right). \tag{5.77}$$

In order to sample all states of the expanded ensembles with uniform probability, the weighting factors, $\eta(\lambda_i N)$, should closely resemble the free energy

$$\eta(\lambda_i N) - \eta(\lambda_1 N) \approx \frac{F(\lambda_i)N - F(\lambda_1 N)}{k_B T} = -\int_{\lambda_1 N}^{\lambda_i N} \mathrm{d}\lambda N \left(\langle \mathcal{T}_{\mathrm{ext}} \rangle + \frac{\mathrm{d}\chi_o N}{\mathrm{d}\lambda N} \langle \mathcal{T}_{\mathrm{ord}} \rangle \right). \tag{5.78}$$

An initial estimate for the weighting factors can be obtained by performing simulations at a small number of fixed points along the path. Interpolating the integrands, $\langle \mathcal{I}_{\text{ext}} \rangle$ and $\langle \mathcal{I}_{\text{ord}} \rangle$, one obtains an initial estimate for the weighting factors according to Eq. (5.78). Then, the weighting factors are iteratively refined. There are several schemes available for this purpose (e.g., Wang-Landau sampling [125], Transition-Matrix techniques [126–129], multicanonical recursion [130], or successive umbrella sampling [131]). Note that a nonoptimal choice of weighting factors, η, does not affect the properties at fixed λN and $\chi_o N$ but the different sampling points are visited with different rates. Thus, we use Eq. (5.78) and continuously improve the estimate for $\langle \mathcal{I}_{\text{ext}} \rangle$ and $\langle \mathcal{I}_{\text{ord}} \rangle$. This strategy avoids discarding information that has been generated with a different choice of η [132]. Note that compared to other reweighting techniques (utilized, for example, to calculate phase diagrams, interface tensions, or potentials of mean force) [117], the free energy difference along the integration path is large, $\mathcal{O}(10^4 k_B T)$, and a systematic method for obtaining/improving the weighting factors is required.

If all states of the expanded ensembles are sampled, the probability, $P(\lambda_i N)$ of finding the system at the sampling point, $\lambda_i N$ and $\chi_o N(\lambda_i N)$ will be related to the free energy according to

$$\frac{F(\lambda_i N)}{k_B T} = \eta(\lambda_i N) - \ln P(\lambda_i N) + \text{const}_i, \tag{5.79}$$

where the additive normalization constants differ for the different intervals, I_i. Alternatively to Eq. (5.72), one can obtain the integrand of the thermodynamic integration from

$$\frac{1}{k_B T}\frac{dF}{d\lambda N} = \frac{d\eta}{d\lambda N} - \frac{d\ln P}{d\lambda N}. \tag{5.80}$$

The combination of expanded ensemble and replica-exchange techniques has several advantages: (i) since every integration path is discretized into many sampling points, exploring the entire integration path by replica-exchange alone is impractical. If we used a significantly smaller number of sampling points, however, the distribution functions of neighboring replica would not overlap. (ii) The expanded ensemble technique is useful because the probability distribution, $P(\lambda_i N)$, provides a very sensitive error estimate for the weighting factors and thus for the free energy. A nonoptimal choice of the reweighting factor results in an extremely nonuniform sampling of the integration path because the free energy change is large, $\Delta F \sim \mathcal{O}(10^4 k_B T)$, and, therefore, already small inaccuracies give rise to large Boltzmann factors. In this preliminary stage, the replica-exchange technique ensures that there is at least one sampling point visited in each interval. (iii) The combination of a moderately parallel program, for example, SCMF simulation, Brownian or molecular dynamics, for a single configuration, and replica-exchange allows to efficiently utilize massively parallel computers with more than 1024 compute cores. The simulations, which are presented in the following section have been performed at the Julich supercomputer Center (JSC) and the Norddentsche Verbund for Hoch– and Hochstleistungsrechen (HLRN).

The method can be readily applied for locating phase transitions between distinct self-assembled morphologies or to calculate the free energy of defects and interfaces between morphologies with different orientation (e.g., T-junctions of a lamellar phase). Generalizations to liquid crystalline phases with an orientational order parameter can be envisioned by utilizing an external ordering director field that couples to the orientational degrees of freedom. This generalization will also allow the calculation of the free energy differences between different phases of coarse-grained models for biological membranes.

The thermodynamic integration scheme can be applied to different models including coarse-grained, particle-based models of amphiphilic systems and membranes [133, 134] (e.g., soft DPD-models [135–137], Lennard-Jones models [138, 139], or solvent-free models [140–142] of membranes) as well as field-theoretic representations [28]. It can be implemented in Monte Carlo or molecular dynamic simulations, as well as SCMF simulations [40–42, 86], field-theoretic simulations [28], and external potential dynamics [27, 63, 64] or dynamic density functional theory [143, 144].

5.4.3.3 **Selected Applications**

In order to calculate the free energy difference between the disordered state, $\chi_o N = 0$, and the lamellar ordered structure at $\chi_{\text{final}} N = 20$, we discretize both branches of the integration path such that the distributions of the integrands, $\mathcal{I}_{\text{ord}}$ and $\mathcal{I}_{\text{ext}}$, overlap for neighboring points along the path of integration.

Qualitatively, we observe in the snapshots of Figure 5.5 that the ordering at the beginning of the second branch ($\chi_o N = 0$, $\lambda N = 20$) is slightly more pronounced than in the final stage ($\chi_{\text{final}} N = 20$, $\lambda N = 0$). We also note that the integrand along the second branch seems to develop a singular curvature upon approaching its end point, ($\chi N = 20$, $\lambda N = 0$). This effect can be partially explained by the observation that the local positions of the AB interfaces of the lamellae are harmonically coupled to their ideally flat position by the external ordering field, λN, and capillary waves gradually build up as one decreases $\lambda N \to 0$ [66, 145]. If the AB interfaces of neighboring lamellae were not coupled but freely fluctuated, one would expect the integrand to contain a term proportional to $\lambda N \ln \lambda N$ due to capillary waves. Importantly, the variation of the integrands along both branches of the path is completely gradual, indicating the absence of a first-order transition.

The absence of a first-order transition is also confirmed by Figure 5.7, where we show the evolution of the strength of the ordering field, λN, during the course of the expanded-ensemble simulation (without replica-exchange Monte Carlo moves). The simulation data presented in Figure 5.7 correspond to a single configuration that samples all different external field strengths, λN, of a branch. Four Intel Xeon dual core processors needed about 10 days to generate these data. The system freely diffuses along the λN-axis and there is no "kinetic barrier" between neighboring λN-states. This observation demonstrates that the regions of configuration space associated with neighboring λN-values overlap.

We also observe that the system visits all λN-states with roughly equal probability (see the inset of Figure 5.7). This demonstrates that the weighting factors are very accurate and the error in the free energy difference is on the order of a few $k_B T$ that, in turn, is much smaller than the total free energy difference, $\mathcal{O}(10^4) k_B T$. This

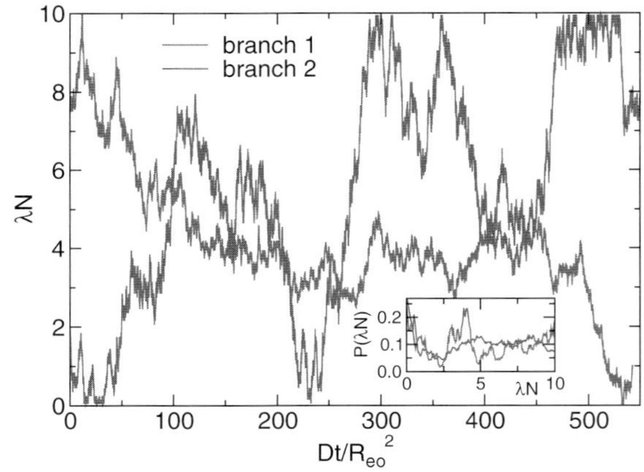

Figure 5.7 Evolution of the ordering field, λN, in the course of the expanded ensemble simulation along both branches. The system parameters are identical to Figure 5.5. Smart Monte Carlo moves are used to update the molecular conformations. The local segment motion gives rise to Rouse-like dynamics for all but the very first Monte Carlo steps. "Time" is measured in units of the Rouse-time of the macromolecules. The inset presents the probability, $P(\lambda N)$, with which the different states, $(\lambda N, \chi_o(\lambda N))$, of the expanded ensemble are visited. No replica-exchange Monte Carlo moves are performed for this simulation run, and the figure presents the data for a single configuration. Reprinted with permission from Ref. [43], © 2008, American Institute of Physics.

observation demonstrates that we can accurately measure free energy differences of the order of $10^{-3}k_BT$ per molecule.

From the graph, we can also obtain a rough idea of the relaxation time of the expanded ensemble. With the ratio of 100 : 1 between SMC moves and attempts to switch between neighboring states of the expanded ensemble, structural and thermodynamic quantities relax on the same order of timescales. Note that the relaxation time of the morphology is significantly larger than the single-chain relaxation time, R_{eo}^2/D.

In Figure 5.8, the variation of the free energy along the two branches is presented. The dependence of the free energy at the beginning of the first branch can be described by the random-phase approximation. The structure factor of composition fluctuations, $S_{coll}(\mathbf{k})$, quantifies the wavevector-dependent susceptibility of the disordered melt (at $\chi_{init}N = 0$) with respect to an external ordering field. Within the random-phase approximation, the average composition variation that is induced by the external field, $\tilde{h}(\mathbf{k}) = \lambda N \tilde{f}_{ext}(\mathbf{k})$ in the disordered state is given by

$$\langle \phi_A(\mathbf{k}) - \phi_B(\mathbf{k}) \rangle = 2\frac{S_{coll}(\mathbf{k})}{N}\tilde{h}(\mathbf{k}). \tag{5.81}$$

The Fourier transform of the external ordered field and its inverse are defined by $\tilde{h}(\mathbf{k}) = \int d^3\mathbf{r}\, h(\mathbf{r}) \exp(-i\mathbf{kr})$ and $h(\mathbf{r}) = \frac{1}{V}\sum_{\mathbf{k}} h(\mathbf{k}) \exp(i\mathbf{kr})$, respectively. To linear order in λN, the integrand of Eq. (5.72) is given by

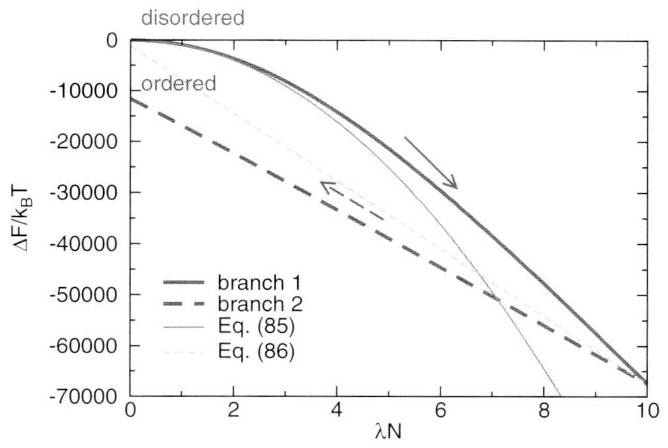

Figure 5.8 Changes of the free energy along the two branches of the transformation path (thick lines). The approximate expressions based on the random phase approximation (cf. Eqs. (5.85) and (5.86)) are also shown. Reprinted with permission from Ref. [43]. © 2008, American Institute of Physics.

$$\frac{\langle \mathcal{I}_{\text{ext}}\rangle R_{\text{eo}}^3}{\sqrt{\mathcal{N}}} \approx \int \mathrm{d}^3\mathbf{r}\tilde{f}_{\text{ext}}(\mathbf{r})\langle \phi_{\text{A}}(\mathbf{r})-\phi_{\text{B}}(\mathbf{r})\rangle \tag{5.82}$$

$$= \frac{1}{V}\sum_{\mathbf{k}} \tilde{f}_{\text{ext}}(-\mathbf{k})\langle \phi_{\text{A}}(\mathbf{k})-\phi_{\text{B}}(\mathbf{k})\rangle \tag{5.83}$$

$$= \frac{2\lambda N}{V}\sum_{\mathbf{k}} \frac{S_{\text{coll}}(\mathbf{k})}{N}\left|\tilde{f}_{\text{ext}}(\mathbf{k})\right|^2. \tag{5.84}$$

The second integrand in Eq. (5.72) does not contribute because $\chi_{\text{o}}N = \chi_{\text{init}}N = 0$ does not change. Thus, we obtain for the free energy change $\Delta F^{(1)}$ in the vicinity of the starting point of the first branch:

$$\frac{\Delta F^{(1)}}{k_{\text{B}}T} \approx -(\lambda N)^2 \frac{\sqrt{\mathcal{N}}V}{R_{\text{eo}}^3}\sum_{\mathbf{k}} \frac{S_{\text{coll}}(\mathbf{k})}{N}\left|\frac{\tilde{f}_{\text{ext}}(\mathbf{k})}{V}\right|^2, \tag{5.85}$$

which depends quadratically on λN. Within the random-phase approximation [5], the collective structure factor for a symmetric diblock copolymer $\chi_{\text{o}}N = 0$ takes the particularly simple form $\frac{S_{\text{coll}}(\mathbf{k})}{N} = \frac{S_{\text{AA}}-S_{\text{AB}}}{N}$, where the structure factors of a single block and between different blocks are given by $S_{\text{AA}}/N = 2/x^2[\mathrm{e}^{-x/2}-1+x/2]$ and $S_{\text{AB}}/N = ([1-\mathrm{e}^{-x/2}]/x)^2$ with $x = (kR_{\text{eo}})^2/6$, respectively. For the specific case, $L_{\text{o}} = 1.686R_{\text{eo}}$, we evaluate $\sum_{\mathbf{k}} \frac{S_{\text{coll}}(\mathbf{k})}{N}|h(\mathbf{k})/V|^2 = 0.0760684$ and the prediction of Eq. (5.85) is compared with the simulation data in Figure 5.8. Excellent agreement is found for $\lambda N \to 0$, but the induced composition fluctuations and the concomitant free energy change is smaller than estimated by Eq. (5.85) for larger values of λN.

Although the structural changes along the second branch of the reversible path are small, the variation of the free energy, $\Delta F^{(2)}$, is substantial. Much of the free energy change is related to the difference how pairwise intermolecular interactions and external ordering fields enter ΔF. To illustrate this effect, let us assume an idealized structure, which was completely segregated, where AB-interfaces were perfectly sharp, and the external ordering field, $h(\mathbf{r})$, perfectly matched this structure. Then, the integrands in Eqs. (5.73) and (5.74) take the simple values $\mathcal{I}_{\text{ord}} \approx \frac{\sqrt{\bar{\mathcal{N}}}V}{4R_{\text{eo}}^3}$ and $\mathcal{I}_{\text{ext}} \approx \frac{\sqrt{\bar{\mathcal{N}}}V}{R_{\text{eo}}^3}$ independent of λN. Inserting this result in Eq. (5.72), one obtains a bound for the free energy change along the second branch of the transformation path

$$\frac{\Delta F^{(2)}}{k_\text{B}T} \approx -\lambda N \frac{\sqrt{\bar{\mathcal{N}}}V}{2R_{\text{eo}}^3}, \tag{5.86}$$

which depends linearly on λN. This prediction is also shown in Figure 5.8.

Since we calculate free energy differences in Eq. (5.72), we arbitrarily set the free energy of the initial, disordered state ($\chi_\text{o} N = 0$) to zero, and we have matched the free energy at the end of the first branch with that of the beginning of the second. From the data, we obtain a free energy difference of $\Delta F/k_\text{B}T = 11\,607(10)$ or $\Delta F/nk_\text{B}T = -0.87659(75)$ for a lamellar spacing, $L_\text{o} = 1.686R_{\text{eo}}$. The error estimate refers to the statistical error of the result but does not include a possible systematic overestimation of the free energy because of the deviation in the lamellar spacing from its equilibrium value due to the finite size of the simulation cell.

This simulation technique can also be applied to calculate the free energy of grain boundaries and T-junctions (see Figure 5.9) or the free energy difference of diblock copolymer morphologies on chemically patterned surfaces (cf. Figure 5.10) [146].

5.4.4
Simultaneous Calculation of Pressure and Chemical Potential in Soft, Off-Lattice Models

A third, alternative method for computing the free energy of self-assembling systems and for calculating phase equilibria between different morphologies consists in simultaneously calculating the pressure, p, and the chemical potential, μ. At coexistence, the different structures are characterized by the same pressure p, temperature T, and thermodynamic potential $\sum_i \mu_i n_i$, where the sum is taken over the different species of which the system is comprised. The Helmholtz free energy of the canonical ensemble can be obtained via $F = -pV + \sum_i \mu_i n_i$.

An advantage of a soft, coarse-grained, off-lattice model is the ability to simultaneously and accurately calculate the pressure, p, and the chemical potential, μ. Abandoning the lattice-description allows a precise calculation of the pressure, p, and simulations at constant pressure or tension. This is also possible in off-lattice models with harsh excluded volume interactions (e.g., a Lennard-Jones bead-spring model). The accurate calculation of the chemical potential by particle insertion methods,

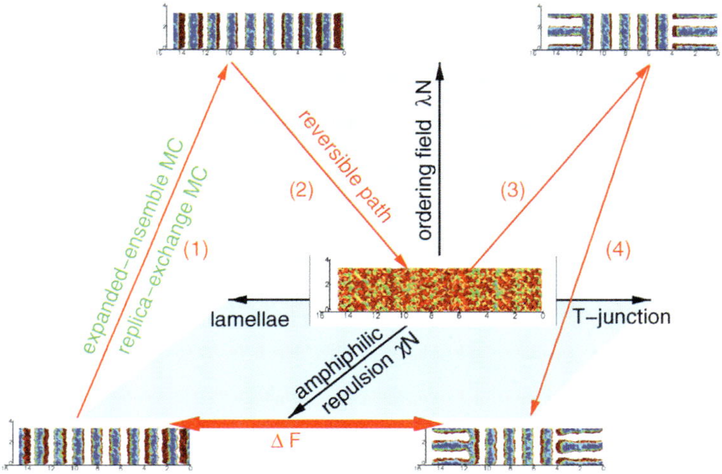

Figure 5.9 Sketch of the integration path to obtain the excess free energy of a T-junction. The snapshots illustrate the system configurations at different stages along the path. For clarity, the simulation cell and one periodic replica are depicted. The B-component has been removed, and the *A*-component is shaded in blue. The interface between A-rich and B-rich regions is shown as red surface. The thermodynamic integration scheme is applied to a melt of symmetric diblock copolymers with $\bar{\mathcal{N}} = 16\,384$, $\chi_o N = 20$, $\varkappa_o N = 50$, $N = 128$, and $\Delta L/R_{eo} \approx 0.09 R_{eo}$ and yields $\frac{\Delta F R_{eo}^2}{k_B T A \sqrt{\bar{\mathcal{N}}}} = 0.19(2)$, where *A* denotes the area of the T-junction. This result is in agreement with self-consistent field calculations of Duque *et al.* [159], predicting $\frac{\Delta F R_{eo}^2}{k_B T A \sqrt{\bar{\mathcal{N}}}} = 0.21$. Adapted from Ref. [146]. Reproduced by permission of the PCCP Owner Society.

however, is greatly facilitated by the soft interactions and accurate methods to estimate its error exist [147].

This method has been recently utilized to estimate the location of the fluctuation-induced, first-order transition between the disordered and the lamellar phase of a symmetric diblock copolymer melt [144] and this application is illustrated in Figure 5.11.

5.5 Outlook

In this chapter, we have discussed computational approaches for describing the equilibrium properties of multicomponent polymer melts. The universal behavior of dense multicomponent systems permits the use of minimal, coarse-grained models that parameterize the relevant interactions via a small number of invariants or coarse-grained parameters, R_e, $\varkappa N$, χN, and $\bar{\mathcal{N}}$. Both particle-based models and field-theoretic representations have been devised to describe these universal properties.

Often, field-theoretic models are considered within the mean field approximation. Provided that the coarse-grained parameters have been identified and describe local

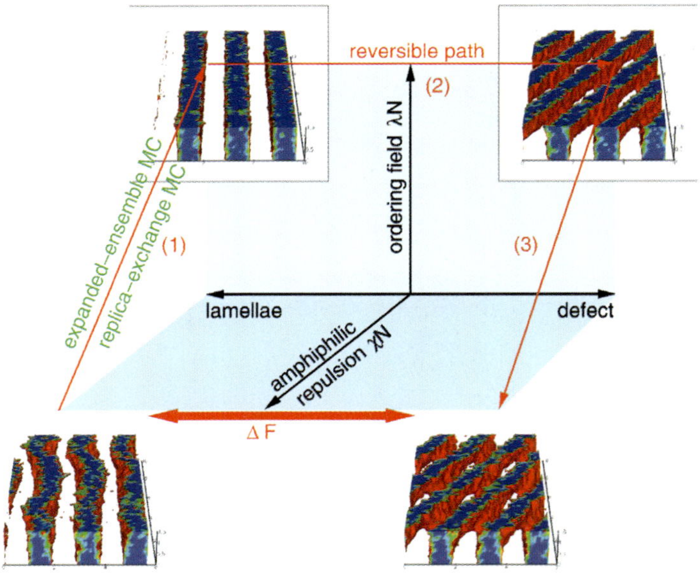

Figure 5.10 Sketch of the integration path to obtain the free energy difference between a stretched lamellar phase on a pattern substrate and a morphology that contains an interface between stretched lamellae at the substrate, which perfectly register with the substrate pattern, and a bulk-like lamellar morphology, which is rotated with respect to the morphology at the bottom substrate. The snapshots illustrate the system configurations at different stages along the reversible transformation path. The B-component has been removed, and the A-component is shaded in blue. The interface between A-rich and B-rich regions is shown as red surface. The thermodynamic integration scheme is applied to a supported thin film of symmetric diblock copolymers with $\bar{\mathcal{N}} = 16\,384$, $\chi_o N = 20$, $\varkappa_o N = 50$, $N = 128$, and $\Delta L/R_{eo} \approx 0.09 R_{eo}$, which assembles on a substrate pattern consisting of stripes. The film thickness is $1.5R_{eo}$ and the lateral dimensions are $5.7R_{eo} \times 9.87269R_{eo}$. The stripes at the supporting substrate attract the different components of the diblock copolymer. The top surface of the film is modeled as hard and nonpreferential. The periodicity of the stripe pattern, L, is 19.5% greater than the equilibrium lamellar spacing of the diblock melt in the bulk. On the left, lower panel, the diblock self-assembles into standing lamellae that register with the stripes. On the right, lower panel, the morphology reconstructs at the patterned substrate: at the bottom, the lamellae register with the stripe pattern, on the top they tilt and adopt a smaller lamellar spacing that is closer to the equilibrium periodicity in the bulk. The thermodynamic integration scheme predicts the free energy difference between these two morphologies to be very close, $\frac{\Delta F R_{eo}^2}{k_B T A \sqrt{\bar{\mathcal{N}}}} = 0.01(3)$, indicating that a mismatch of 20% is about the maximal tolerable mismatch for obtaining defect-free ordering and registration. Adapted from Ref. [146]. Reproduced by permission of the PCCP Owner Society.

correlation effects on the scale of a segment, the mean field theory becomes accurate in the limit $\bar{\mathcal{N}} \to \infty$. On the other hand, particle-based models, which include harsh repulsions or excluded volume interactions, are limited to modest values of $\bar{\mathcal{N}}$. Therefore, Lennard-Jones bead-spring models or the bond fluctuation models exhibit

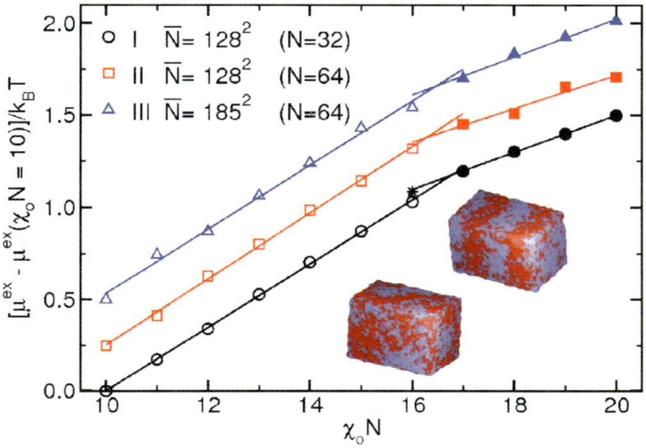

Figure 5.11 Order–disorder transition (ODT) of a symmetric diblock copolymer studied by a soft, coarse-grained, off-lattice model. Monte Carlo simulations are performed in the *npT* ensemble and the pressure is kept constant at $pb^3/k_BT = 18$ (with $b = R_{eo}/\sqrt{N-1}$). $\varkappa_o = 1.5625$. The invariant degree of polymerization, $\bar{\mathcal{N}}$, and the chain discretization are indicated in the key. The figure presents the excess thermodynamic potential, μ_{ex}, per particle as a function of $\chi_o N$. The curves have been shifted for clarity. Empty and filled symbols denote the disordered and lamellar phases, respectively. Lines are linear fits to the data. The crossing points identify the location of the order–disorder transition. Errors are comparable to the symbol size. Snapshots illustrate the structure before and after the ODT (i.e., at $\chi_o N = 16$ and 17). Adapted from Ref. [44].

pronounced fluctuation effects. The experimentally relevant range of large but finite invariant degrees of polymerization, $\bar{\mathcal{N}} \sim 10^4$, can only be addressed with soft coarse-grained models because increasing the density, ϱ_o, of coarse-grained segments is significantly more efficient in increasing $\bar{\mathcal{N}}$ than increasing the number, N, of segment per molecule. These particle-based models correspond to discretized versions of the standard field-theoretic description with a finite compressibility term. The soft, interparticle potential avoids crystallization or vitrification, which occur in models with harsh excluded volume; it should also be chosen as to suppress the formation of cluster crystals [107–109] that may form in dense systems of soft, purely repulsive particles.

These soft, particle-based models can be employed in conjunction with a variety of simulation techniques including dissipative particle dynamics [100], Brownian dynamics [35], Monte Carlo simulations [39, 41, 105], and single-chain-in-mean-field simulations [40–42]. These simulation techniques permit the investigation of the structure of large three-dimensional polymer systems with experimentally relevant $\bar{\mathcal{N}}$.

Special simulation techniques have recently been devised to calculate free energies of these structure-forming fluids [43–45]. We have discussed several methods, which have been inspired by related approaches for calculating free energy of crystals in hard-condensed matter systems [43], rely on a field-theoretic representation via lattice-based fields [45] or exploit the possibility of simultaneously and accurately measuring the pressure and chemical potential due to the softness of the off-lattice potentials [44].

The combination of efficient simulation models and computational methods for calculating free energies opens a wide range of applications for soft coarse-grained models not only limited to polymeric systems but also encompassing lipid membranes and colloid-polymer mixtures. Given the small free energies (per molecule) that distinguish different morphologies and the experimentally observed metastability and protracted relaxation of the structure formation process, it remains an important and challenging task to devise efficient numerical models for studying the kinetics of structure formation. To this end, both entanglement effects and a description of collective flow have to be incorporated. Progress on these issues has recently begun [35, 36, 87, 91, 148–157].

Acknowledgments

It is a great pleasure to thank K.Ch. Daoulas, F.A. Detcheverry, M. Hömberg, Y. Norizoe, J.J. de Pablo, and D.Q. Pike for stimulating discussions and fruitful collaborations. Financial support has been provided by the Volkswagen Foundation and the DFG Mu1674. Ample computing time at the JSC Jülich, the HLRN-II Hannover, and the GWDG Göttingen is greatfully acknowledged.

References

1 Cahn, R., Haasen, P., and Kramer, E. (1993) *Materials Science and Technology: A Comprehensive Treatment*, vol. 12, VCH, Weinheim.
2 Garbassi, F., Morra, M., and Occhiello, E. (2000) *Polymer Surface: From Physics to Technology*, John Wiley & Sons, Inc., Chichester.
3 Leibler, L. (1988) *Makromol. Chem.-Macromol. Symp.*, **16**, 1.
4 Milner, S.T. and Xi, H.W. (1996) *J. Rheol.*, **40**, 663.
5 Leibler, L. (1980) *Macromolecules*, **13**, 1602.
6 Matsen, M.W. and Schick, M. (1994) *Phys. Rev. Lett.*, **72**, 2660.
7 Tyler, C.A. and Morse, D.C. (2005) *Phys. Rev. Lett.*, **94**, 208302.
8 Thurn-Albrecht, T., Schotter, J., Kastle, C.A., Emley, N., Shibauchi, T., Krusin-Elbaum, L., Guarini, K., Black, C.T., Tuominen, M.T., and Russell, T.P. (2000) *Science*, **290**, 2126.
9 Park, C., Yoon, J., and Thomas, E.L. (2003) *Polymer*, **44**, 6725.
10 Stoykovich, M.P., Müller, M., Kim, S.O., Solak, H.H., Edwards, E.W., de Pablo, J.J., and Nealey, P.F. (2005) *Science*, **308**, 1442.
11 Baschnagel, J., Binder, K., Doruker, P., Gusev, A.A., Hahn, O., Kremer, K., Mattice, W.L., Müller-Plathe, F., Murat, M., Paul, W., Santos, S., Suter, U.W., and Tries, V. (2000) *Adv. Polym. Sci.*, **152**, 41.
12 Doi, M. and Edwards, S.F. (1986) *The Theory of Polymer Dynamics*, Oxford University Press.
13 des Cloizeaux, J. and Jannink, G. (1990) *Polymers in Solution: Their Modeling and Structure*, Oxford Science Publications, Oxford.
14 Müller-Plathe, F. (2002) *Chem. Phys. Chem.*, **3**, 754.
15 Faller, R. (2004) *Polymer*, **45**, 3869.
16 Müller, M. (2006) *Soft Matter*, vol. 1 (eds G. Gompper and M. Schick), Wiley-VCH Verlag, Weinheim p. 179.
17 de Gennes, P.G. (1972) *Phys. Lett. A*, **38**, 339.

18 Freed, K.F. (1987) *Renormalization Group Theory of Macromolecules*, Wiley-Interscience.

19 Schäfer, L. (1999) *Excluded Volume Effects in Polymer Solutions*, Springer, Berlin.

20 Müller, M. and Binder, K. (1995) *Macromolecules*, **28**, 1825.

21 Müller, M. (1995) *Macromolecules*, **28**, 6556.

22 Müller, M. (1999) *Macromol. Theory Simul.*, **8**, 343.

23 Wang, Z.-G. (2002) *J. Chem. Phys.*, **117**, 481.

24 Morse, D.C. (2006) *Ann. Phys.*, **321**, 2318.

25 Grzywacz, P., Qin, J., and Morse, D.C. (2007) *Phys. Rev. E*, **76**, 061802.

26 Hohenberg, P.C. and Halperin, B.I. (1977) *Rev. Mod. Phys.*, **49**, 435.

27 Müller, M. and Schmid, F. (2005) *Adv. Polym. Sci*, **185**, 1.

28 Fredrickson, G.H., Ganesan, V., and Drolet, F. (2002) *Macromolecules*, **35**, 16.

29 Matsen, M.W. (2006) *Soft Matter*, vol. 1 (eds G. Gompper and M. Schick), Wiley-VCH Verlag, Weinheim, p. 87.

30 Müller, M. and Werner, A. (1997) *J. Chem. Phys.*, **107**, 10764.

31 Marcelja, S. (1973) *Nature*, **241**, 451.

32 Ben-Shaul, A. and Szleifer, I. (1985) *J. Chem. Phys.*, **83**, 3597.

33 Szleifer, I. and Carignano, M.A. (1996) *Adv. Chem. Phys.*, **94**, 165.

34 Müller, M. and Schick, M. (1998) *Phys. Rev. E*, **57**, 6973.

35 Saphiannikova, M.G., Pryamitsyn, V.A., and Cosgrove, T. (1998) *Macromolecules*, **31**, 6662.

36 Narayanan, B., Pryamitsyn, V.A., and Ganesan, V. (2004) *Macromolecules*, **37**, 10180.

37 Avalos, J.B., Mackie, A.D., and Diez-Orrite, S. (2005) *Macromolecules*, **37**, 1124.

38 Avalos, J.B., Mackie, A.D., and Diez-Orrite, S. (2004) *Macromolecules*, **37**, 1143.

39 Detcheverry, F.A., Daoulas, K.C., Müller, M., and de Pablo, J.J. (2008) *Macromolecules*, **41**, 4989.

40 Müller, M. and Smith, G.D. (2005) *J. Polym. Sci. B: Polymer Physics*, **43**, 934.

41 Daoulas, K.C. and Müller, M. (2006) *J. Chem. Phys.*, **125**, 184904.

42 Daoulas, K.C., Müller, M., de Pablo, J.J., Nealey, P.F., and Smith, G.D. (2006) *Soft Matter*, **2**, 573.

43 Müller, M. and Daoulas, K.C. (2008) *J. Chem. Phys.*, **128**, 024903.

44 Detcheverry, F.A., Pike, D.Q., Nealey, P.F., Müller, M., and de Pablo, J.J. (2009) *Phys. Rev. Lett.*, **102**, 197801.

45 Lennon, E.M., Katsov, K., and Fredrickson, G.H. (2008) *Phys. Rev. Lett.*, **101**, 138302.

46 Kremer, K. and Binder, K. (1988) *Comput. Phys. Rep.*, **7**, 259.

47 Grest, G.S. and Kremer, K. (1986) *Phys. Rev. A*, **33**, 3628.

48 Theodorou, D.N. (1988) *Macromolecules*, **21**, 1391.

49 Deutsch, H.P. and Binder, K. (1991) *J. Chem. Phys.*, **94**, 2294.

50 de Gennes, P.G. (1971) *J. Chem. Phys.*, **55**, 572.

51 Bennemann, C., Paul, W., Binder, K., and Dünweg, B. (1998) *Phys. Rev. E*, **57**, 843.

52 Müller, M. and MacDowell, L.G. (2000) *Macromolecules*, **33**, 3902.

53 Carmesin, I. and Kremer, K. (1988) *Macromolecules*, **21**, 2819.

54 Müller, M., Binder, K., and Oed, W. (1995) *J. Chem. Soc. Faraday Trans.*, **91**, 2369.

55 Müller, M. and Binder, K. (1998) *Macromolecules*, **31**, 8323.

56 Müller, M. and Binder, K. (2001) *Phys. Rev. E*, **63**, 021602.

57 Deutsch, H.P. and Binder, K. (1992) *Macromolecules*, **25**, 6214.

58 Grest, G.S., Lacasse, M.D., Kremer, K., and Gupta, A.M. (1996) *J. Chem. Phys.*, **105**, 10583.

59 Murat, M., Grest, G.S., and Kremer, K. (1999) *Macromolecules*, **32**, 595.

60 Helfand, E. (1975) *J. Chem. Phys.*, **62**, 999.

61 Scheutjens, J.M.H.M. and Fleer, G.J. (1979) *J. Phys. Chem.*, **83**, 1619.

62 Hong, K.M. and Noolandi, J. (1981) *Macromolecules*, **14**, 727.

63 Reister, E., Müller, M., and Binder, K. (2001) *Phys. Rev. E*, **64**, 041804.

64 Maurits, N.M. and Fraaije, J.G.E.M. (1997) *J. Chem. Phys.*, **107**, 5879.

65 Duchs, D., Ganesan, V., Fredrickson, G.H., and Schmid, F. (2003) *Macromolecules*, **36**, 9237.

66 Geisinger, T., Müller, M., and Binder, K. (1999) *J. Chem. Phys.*, **111**, 5241.
67 Werner, A., Schmid, F., Müller, M., and Binder, K. (1997) *J. Chem. Phys.*, **107**, 8175.
68 Werner, A., Schmid, F., Müller, M., and Binder, K. (1999) *Phys. Rev. E*, **59**, 728.
69 Yethiraj, A. and Woodward, C.E. (1995) *J. Chem. Phys.*, **102**, 5499.
70 Yethiraj, A. (1998) *J. Chem. Phys.*, **109**, 3269.
71 Müller, M. and MacDowell, L.G. (2001) *Europhys. Lett.*, **55**, 221.
72 Yu, Y.X. and Wu, J.Z. (2002) *J. Chem. Phys.*, **116**, 7094.
73 Yu, Y.X. and Wu, J.Z. (2002) *J. Chem. Phys.*, **117**, 2368.
74 Müller, M., MacDowell, L.G., and Yethiraj, A. (2003) *J. Chem. Phys.*, **118**, 2929.
75 Bryk, P. and Sokolowsky, S. (2004) *J. Chem. Phys.*, **121**, 11314.
76 Wu, J.Z. (2006) *AIChE J.*, **52**, 1169.
77 Wu, J.Z. and Li, Z.D. (2007) *Annu. Rev. Phys. Chem.*, **58**, 85.
78 Rasmussen, K. and Kalosakas, G. (2002) *J. Polym. Sci. B: Polym. Phys.*, **40**, 1777.
79 Ceniceros, H.D. and Fredrickson, G.H. (2004) *Multiscale Model. Simul.*, **2**, 452.
80 Cochran, E.W., Garcia-Cervera, C.J., and Fredrickson, G.H. (2006) *Macromolecules*, **39**, 2449.
81 Morse, D.C. and Fredrickson, G.H. (1994) *Phys. Rev. Lett.*, **73**, 3235.
82 Schmid, F. and Müller, M. (1995) *Macromolecules*, **28**, 8639.
83 Daoulas, K.C., Theodorou, D.N., Harmandaris, V.A., Karayiannis, N.C., and Mavrantzas, V.G. (2005) *Macromolecules*, **38**, 7134.
84 Ferrenberg, A.M. and Swendsen, R.H. (1988) *Phys. Rev. Lett.*, **61**, 2635.
85 Putz, M., Curro, J.G., and Grest, G.S. (2001) *J. Chem. Phys.*, **114**, 2847.
86 Daoulas, K.C., Müller, M., Stoykovich, M.P., Park, S.M., Papakonstantopoulos, Y.J., de Pablo, J.J., Nealey, P.F., and Solak, H.H. (2006) *Phys. Rev. Lett.*, **96**, 036104.
87 Müller, M. and Pastorino, C. (2008) *Europhys. Lett.*, **81**, 28002.
88 Dawson, J.M. (1983) *Rev. Mod. Phys.*, **55**, 403.
89 Eastwood, J.W., Hockney, R.W., and Lawrence, D.N. (1980) *Comput. Phys. Commun.*, **19**, 215.
90 Deserno, M. and Holm, C. (1998) *J. Chem. Phys.*, **109**, 7678.
91 Müller, M. and Daoulas, K.C. (2008) *J. Chem. Phys.*, **129**, 164906.
92 Mavrantzas, V.G., Boone, T.D., Zervopoulou, E., and Theodorou, D.N. (1999) *Macromolecules*, **32**, 5072.
93 Daoulas, K.C. and Müller, M. (2009) *Adv. Polym. Sci.*, **224**, 187.
94 Müller, M., Binder, K., and Schäfer, L. (2000) *Macromolecules*, **33**, 4568.
95 Wittmer, J.P., Meyer, H., Baschnagel, J., Johner, A., Obukhov, S., Mattioni, L., Müller, M., and Semenov, A.N. (2004) *Phys. Rev. Lett.*, **93**, 147801.
96 Wittmer, J.P., Beckrich, P., Meyer, H., Cavallo, A., Johner, A., and Baschnagel, J. (2007) *Phys. Rev. E*, **76**, 011803.
97 Hoogerbrugge, P.J. and Koelman, J.M.V.A. (1992) *Europhys. Lett.*, **19**, 155.
98 Koelman, J.M.V.A. and Hoogerbrugge, P.J. (1993) *Europhys. Lett.*, **21**, 363.
99 Warren, P. and Espanol, P. (1995) *Europhys. Lett.*, **30**, 191196.
100 Groot, R.D. and Warren, P.B. (1997) *J. Chem. Phys.*, **107**, 4423.
101 Groot, R.D. and Madden, T.J. (1998) *J. Chem. Phys.*, **108**, 8713.
102 Groot, R.D., Madden, T.J., and Tildesley, D.J. (1999) *J. Chem. Phys.*, **110**, 9739.
103 Soddemann, T., Dünweg, B., and Kremer, K. (2003) *Phys. Rev. E*, **68**, 46702.
104 Ihle, T. and Kroll, D.M. (2003) *Phys. Rev. E*, **67**, 066705.
105 Laradji, M., Guo, H., and Zuckermann, M.J. (1994) *Phys. Rev. E*, **49**, 3199.
106 Likos, C.N., Lang, A., Watzlawek, M., and Löwen, H. (2001) *Phys. Rev. E*, **63**, 031206.
107 Mladek, B.M., Gottwald, D., Kahl, G., Neumann, M., and Likos, C.N. (2007) *J. Phys. Chem. B*, **111**, 12799.
108 Likos, C.N., Mladek, B.M., Gottwald, D., and Kahl, G. (2007) *J. Chem. Phys.*, **126**, 224502.

109 Mladek, B.M., Fernaud, M.J., Kahl, G., and Neumann, M. (2005) *J. Phys. Condens. Matter*, **8**, 135.

110 Mladek, B.M., Gottwald, D., Kahl, G., Neumann, M., and Likos, C.N. (2006) *Phys. Rev. Lett.*, **96**, 045701.

111 Mladek, B.M., Charbonneau, P., and Frenkel, D. (2007) *Phys. Rev. Lett.*, **99**, 235702.

112 Frenkel, D. and Ladd, A.J.C. (1984) *J. Chem. Phys.*, **81**, 3188.

113 Alexander-Katz, A., Moreira, A.G., Sides, S.W., and Fredrickson, G.H. (2005) *J. Chem. Phys.*, **122**, 014904.

114 Sheu, S.Y., Mou, C.Y., and Lovett, R. (1995) *Phys. Rev. E*, **51**, R3795.

115 Grochola, G. (2004) *J. Chem. Phys.*, **120**, 2122.

116 Eike, D.M., Brennecke, J.F., and Maginn, E.J. (2005) *J. Chem. Phys.*, **122**, 014115.

117 Müller, M. and de Pablo, J.J. (2006) *Lect. Notes Phys.*, **703**, 67.

118 Yan, Q. and de Pablo, J.J. (2000) *J. Chem. Phys.*, **113**, 1276.

119 Rossky, P.J., Doll, J.D., and Friedman, H.L. (1978) *J. Chem. Phys.*, **69**, 4628.

120 Rouse, P.E. (1953) *J. Chem. Phys.*, **21**, 1272.

121 Lyubartsev, A.P., Martsinovski, A.A., Shevkunov, S.V., and Vorontsov-Velyaminov, P.N. (1992) *J. Chem. Phys.*, **96**, 1776.

122 Hukushima, K. and Nemoto, K. (1996) *J. Phys. Soc. Jpn.*, **65**, 1604.

123 Hansmann, U.H.E. (1997) *Chem. Phys. Lett.*, **281**, 140.

124 Sugita, Y. and Okamoto, Y. (1999) *Chem. Phys. Lett.*, **314**, 141.

125 Wang, F.G. and Landau, D.P. (2001) *Phys. Rev. Lett.*, **86**, 2050.

126 Smith, G.R. and Bruce, A.D. (1995) *J. Phys. A*, **28**, 6623.

127 Wang, J.S., Tay, T.K., and Swendsen, R.H. (1999) *Phys. Rev. Lett.*, **82**, 476.

128 Fitzgerald, M., Picard, R.R., and Silver, R.N. (1999) *Europhys. Lett.*, **46**, 282.

129 Wang, J.S. and Swendsen, R.H. (2002) *J. Stat. Phys.*, **106**, 245.

130 Berg, B.A. (1996) *J. Stat. Phys.*, **82**, 323.

131 Virnau, P. and Müller, M. (2004) *J. Chem. Phys.*, **120**, 10925.

132 Yan, Q.L. and de Pablo, J.J. (2003) *Phys. Rev. Lett.*, **90**, 035701.

133 Müller, M., Katsov, K., and Schick, M. (2003) *J. Polym. Sci. B: Polym. Phys.*, **41**, 1441.

134 Müller, M., Katsov, K., and Schick, M. (2006) *Phys. Rep.*, **434**, 113.

135 Shillcock, J.C. and Lipowsky, R. (2002) *J. Chem. Phys.*, **117**, 5048.

136 Kranenburg, M., Venturoli, M., and Smit, B. (2003) *J. Phys. Chem. B*, **107**, 11491.

137 Kranenburg, M. and Smit, B. (2005) *J. Phys. Chem. B*, **109**, 6553.

138 Shelley, J., Shelley, M., Reeder, R., Bandyopadhyay, S., Moore, P., and Klein, M. (2001) *J. Phys. Chem. B*, **105**, 9785.

139 Loison, C., Mareschal, M., Kremer, K., and Schmid, F. (2003) *J. Chem. Phys.*, **119**, 13138.

140 Brannigan, G. and Brown, F.L.H. (2004) *J. Chem. Phys.*, **120**, 1059.

141 Cooke, I., Kremer, K., and Deserno, M. (2005) *Phys. Rev. E*, **72**, 011506.

142 Brannigan, G., Lin, L.C.L., and Brown, F.L.H. (2006) *Eur. Biophys. J.*, **35**, 104.

143 Fraaije, J.G.E.M. (1993) *J. Chem. Phys.*, **99**, 9202.

144 Fraaije, J.G.E.M., van Vlimmeren, B.A.C., Maurits, N.M., Postma, M., Evers, O.A., Hoffmann, C., Altevogt, P., and Goldbeckwood, G. (1997) *J. Chem. Phys.*, **106**, 4260.

145 Geisinger, T., Müller, M., and Binder, K. (1999) *J. Chem. Phys.*, **111**, 5251.

146 Müller, M., Daoulas, K.C., and Norizoe, Y. (2009) *Phys. Chem. Chem. Phys.*, **11**, 2087.

147 Shing, K.S. and Gubbins, K.E. (1982) *Mol. Phys.*, **46**, 1109.

148 Miao, L., Guo, H., and Zuckermann, M.J. (1996) *Macromolecules*, **29**, 2289.

149 Hua, C.C. and Schieber, J.D. (1998) *J. Chem. Phys.*, **109**, 10018.

150 Padding, J.T. and Briels, W.J. (2001) *J. Chem. Phys.*, **115**, 2846.

151 Padding, J.T. and Briels, W.J. (2002) *J. Chem. Phys.*, **117**, 925.

152 Schieber, J.D., Neergaard, J., and Gupta, S. (2003) *J. Rheol.*, **47**, 213.

153 Likhtman, A.E. (2005) *Macromolecules*, **38**, 6128.

154 Nair, D.M. and Schieber, J.D. (2006) *Macromolecules*, **39**, 3386.

155 Narayanan, B. and Ganesan, V. (2006) *Phys. Fluids*, **18**, 042109.

156 Schieber, J.D., Nair, D.M., and Kitkrailard, T. (2007) *J. Rheol.*, **51**, 1111.
157 Ramirez, J., Sukumaran, S.K., and Likhtman, A.E. (2007) *J. Chem. Phys.*, **126**, 244904.
158 Helfand, E. and Sapse, A.M. (1975) *J. Chem. Phys.*, **62**, 1327.
159 Duque, D., Katsov, K., and Schick, M. (2002) *J. Chem. Phys.*, **117**, 10315.

6
Simulations and Theories of Single Polyelectrolyte Chains

Arindam Kundagrami, Rajeev Kumar, and Murugappan Muthukumar

6.1
Introduction

Understanding charged polymers in aqueous solutions continues to be a major challenge. Even the very first question about the net charge of an isolated flexible polyelectrolyte in dilute solutions is elusive from a full understanding. The key contributor to the challenge is the topological correlation arising from the chain connectivity, which is embedded on a collection of charges, which on their own are long-range correlated. At the same time, the solution is electrically neutral, whereby the polyelectrolyte chains are present in a neutralizing plasma constituted by the counterions and other dissociated salt ions. As can be expected intuitively, there is an interplay between the large polyelectrolyte macromolecules and small electrolyte ions that are coupled electrostatically. Furthermore, there has been a resurge in experimental activities in order to determine the effective charge of polyelectrolyte molecules in solutions and their structure. The interpretation of experimental observations depends on availability of reliable theories. An accurate theoretical treatment of such a complex problem is a challenge. In view of this, only approximate theories have been developed. In a parallel track, lots of computer simulation activities have been pursued by many groups, in order to gain insight. While computer simulation results can provide explanations of some trends observed in experiments, the value of closed form analytical formulas cannot be underestimated in providing explanation of already existing data, both simulation and experiments, and in making predictions.

In the light of the above arguments, we present a brief review on the theoretical methods and simulation results. We restrict ourselves to isolated flexible polyelectrolyte chains in solutions. In terms of simulations, we only consider the Langevin dynamics method. For the theory part, we present a variational theory and self-consistent field theory (SCFT), both of which start from an effective Hamiltonian appropriate for flexible charged chains. The background plasma is treated at different levels of sophistication, ranging from the Debye–Hückel (DH) to the Poisson–Boltzmann descriptions. In this chapter, we have stayed away from less

Modeling and Simulation in Polymers. Edited by P.D. Gujrati and A.I. Leonov

ISBN: 978-3-527-32415-6

diluted solutions, semiflexible polyelectrolytes, and phase behaviors of polyelectrolyte solutions.

For an uncharged polymer chain, there are three length measures: contour length of the chain, L; the Kuhn segment length l, which parameterizes the local chemical details, which in turn determines the number of Kuhn steps N in the chain; and the radius of gyration of the chain. With respect to uncharged systems, there are two additional length scales traditionally considered in theoretical treatments of salty polyelectrolyte solutions. One is the Bjerrum length l_B, which stipulates the strength of the Coulomb interaction at a particular temperature T in a specific solvent of dielectric constant ε, given by

$$l_\mathrm{B} = \frac{e^2}{4\pi\varepsilon_0\varepsilon k_\mathrm{B} T}, \tag{6.1}$$

where e is the electron charge, ε_0 is the vacuum permittivity, and k_B is the Boltzmann constant. The other one is the Debye screening length $\varkappa^{-1}$, which sets the length scale for the screening of electrostatic interactions due to the dissociated ions, and is given by

$$\varkappa^2 = 4\pi l_\mathrm{B} \sum_i Z_i^2 c_i, \tag{6.2}$$

where the sum is over all species (i) of mobile ions of valency Z_i and number concentration c_i. In addition to the above two, a third length scale also arises [1] related to the dielectric heterogeneity, which accounts for the difference in the dielectric constant in the vicinity of the polymer chains and in the bulk solvent. In this chapter, we consider these three length scales for the theoretical treatment of the effective charge and conformation of a single isolated polyelectrolyte chain in both salt-free and salty solutions, the latter in the presence of either monovalent or divalent ions. Simulation models, however, ignore the dielectric inhomogeneity and consider the bulk dielectric constant of the implicit solvent.

In experiments, the major variables are the molecular weight of the polymer, its chemical identity in terms of the backbone structure, valency and size of counterions, amount and valencies of dissolved electrolyte ions, dielectric constant of the solvent, dielectric mismatch between the material constituted by the chain backbone and the bulk solvent, and temperature. Many phenomena related to the effects from these experimental variables have been documented in the literature. The typical non-monotonic dependence of the average size of flexible polyelectrolyte chains on temperature is now well known from computer simulations and experiments. At very high temperatures, the chains in dilute salt-free solutions are in their athermal states with self-avoiding walk (SAW) statistics. Although the chains are fully charged at these conditions, the electrostatic repulsion among the monomers remains negligible compared to the thermal fluctuations. With decreasing temperature, electrostatics becomes progressively important and the chains expand even beyond the excluded volume swelling due to intermonomer repulsion. At even lower temperatures, counterions condense on the chains reducing the net polymer charge, and consequently the chains tend to contract again. Small-molecular monovalent salts have long been known to enhance this condensation effect, and the resulting contraction occurs at higher temperatures. Monovalent salts, however, gradually reduce the size of flexible polyelectrolyte chains at temperatures typically way below their modest values (e.g., room temperature for aqueous solutions). Addition of divalent (or multivalent) salts, however, leads to a drastic qualitative change in polyelectrolyte behaviors. A modest number of divalent counterions in water can effectively neutralize and contract polyelectrolytes close to its Gaussian size at room temperatures. Furthermore, additional divalent salt can even reverse the charge on the polymer (the phenomenon is known as *overcharging* or *charge inversion* or *charge reversal*) at certain physical conditions. A typical simulation model or theory must address and successfully predict these well-known properties of charged chains in salty environments.

The Manning model [2], originally designed for infinitesimally thin and infinitely long rod-like molecules, has traditionally explained condensation of counterions on flexible polyelectrolyte chains. However, Manning's argument remains inadequate [1] for flexible polyelectrolytes. A flexible chain is allowed to bend significantly due to its charge compensation at lower temperatures, which renders substantial changes in its conformational entropy. Furthermore, Manning's assumption that the discrete nature of the charged groups has a secondary effect becomes entirely invalid for multivalent ions. This discreteness plays the key role in complete charge compensation (and resulting contraction of polyelectrolytes) and subsequent overcharging at modest temperatures by multivalent salt counterions. This overcharging behavior is unexplainable within the Poisson–Boltzmann formalism that considers a continuum description of the charge density. Initial theories [3], in order to address the precipitation of chains for high counterion valence, considered the translational free energy of the polyions and salt ions along with the screened electrostatic interaction between charges. With prefixed values of the excluded volume exponent ν (that means prefixed radius of gyration, R_g), the free energy was minimized in terms of counterion species, and the correlated multivalent ions were shown [4] to induce attraction between monomers (through ion-"bridging") leading to the collapse of a chain. Redissolution of chains was also observed at higher (multivalent) salt concentrations, but it was explained by a reduced bridging force due to electrostatic screening (as opposed to overcharging). Later, unscreened Coulomb interaction within condensed ion-pairs was first addressed [5] without considering the chain entropy, and the theory predicted dependencies of the degree of ionization, f, (which is the total effective charge density of polyelectrolytes after accounting for the condensed ions) on temperature and salt concentrations similar to Manning's argument. A two-state (rod-like and collapsed) model for condensation predicted [6] that the chain collapse occurs when the total charge of the multivalent cations equals to that of the ionizable groups of the polymer, implying the condensation of almost *all* added multivalent ions at modest temperatures. The two-state theory [6, 7] treats the collapsed state at low temperatures as an amorphous ionic solid similar to simple electrolytes (say, NaCl) and, therefore, still ignores the chain entropy and bending-related reorganization of condensed charges at low temperatures.

A newly developed theory [1] successfully addresses the chain entropy of flexible polyelectrolytes and its role in counterion distribution. This continuum theory considers the condensation of counterions as an adsorption process. Condensation in this argument is facilitated at lower temperatures but, unlike in previous theories, is coupled with the configurational free energy of the polymer. This adsorption theory [1] considers nondiscrete variable values for the size (R_g) of a single chain that is treated as a continuous curve, and which facilitates an appropriate description of entropy of a flexible chain. The theory treats both salt-free and salty conditions with monovalent counterions in dilute solutions of a flexible polyelectrolyte. The competition between the translational entropy of counterions and the electrostatic energy gain of condensed ions is analyzed parametrically reproducing all classical results including the chain-collapse (below Gaussian dimensions) due to short-ranged dipole interactions at low temperatures [8, 9]. As mentioned before, in addition to the length scales l_B (Bjerrum length) and κ^{-1} (Debye length) in the charged system the adsorption theory uses the concept of a dielectric mismatch parameter, δ, which captures the fact that the dielectric constant has much lower values near the chain backbone of a polyelectrolyte or protein than in the bulk [10–12] solvent. In a typical polyelectrolyte solution, δ is the ratio of the bulk to local dielectric constants, and the distance from the backbone in which ε assumes its bulk value sets a new length scale. This mismatch in ε, if substantial, creates higher potential gradients that can electrostatically guide counterions toward oppositely charged monomers. The theory confirmed that this may significantly increase counterion condensation at modest temperatures leading to a lower effective charge and smaller size of the polyelectrolyte. Monovalent counterion was, however, shown to contract a chain to its Gaussian conformation at modest temperatures (say, room temperature at which $l_B \sim 7$ Å for sodium polystyrene sulfonate (NaPSS) in water) only at higher concentrations of the added salt. The recent extension of the adsorption theory to divalent counterions [13], however, predicts the expected charge neutralization and overcharging under moderate conditions.

One of the merits of the variational theory is its utility in terms of simple formulas for the effective degree of ionization and radius of gyration. Therefore, it is necessary to validate the predictions of this theory with simulation and experimental results. Furthermore, the variational theory has used the severe approximation of uniform electrostatic expansion of the chain. In order to rely on the predictions of the adsorption (variational) theory, it is necessary to assess the severity of the approximations by comparing with more advanced field theoretic calculations [14, 15]. We have organized this chapter along the following line of thought. First, we provide simulation technique and key results. This is then followed by a detailed description of Muthukumar's adsorption theory by using a variational formalism. The results of this theory are discussed along with simulation and experimental results. After the variational formalism, the SCFT is presented with adequate details to enable a beginner to master this technique for this class of problems. In the formalism of SCFT, we present both the saddle-point approximation leading to the self-consistent mean field theory

(SCMFT) and the role of fluctuations. Finally, we offer a comparison between different theoretical tools.

6.2 Simulation

6.2.1 Simulation Method

In the simulation, the polyelectrolyte solution containing added salt is modeled as a collection of n chains each with N spherical beads of point unit electric charge $-e$, nN/Z_c counterions (Z_c being the valency of the counterion), n_+ cations of added salt with valency Z_+, and n_+Z_+/Z_- anions of added salt with valency Z_-, all placed in a cubic medium of dielectric constant ε and volume L^3. We have systematically investigated the role of n, N, Z_c, $Z_+ = 1$ and 2, and $Z_- = 1$ in determining the various structural properties of the polyelectrolyte and counterion distribution. In this model, the key parameters of electrostatic interaction between two ions of valencies Z_i and Z_j are the Bjerrum length l_B, given by Eq. (6.1) and Coulomb strength parameter Γ defined by

$$\Gamma = \frac{|Z_i Z_j| l_B}{l_0}, \tag{6.3}$$

with $k_B T$ being the Boltzmann constant times the absolute temperature T, ε_0 is the permittivity of vacuum, and l_0 is the equilibrium bond length connecting two successive beads of the chain. We first systematically explore the effect of the Coulomb strength, Γ, which dictates the relative importance of thermal fluctuations to Coulomb interactions. The interesting range of Γ is $3.2 > \Gamma > 2.4$ for aqueous solutions ($0 < T < 100\,^\circ\text{C}$) of polyelectrolytes with chemical charge separation along chain backbone of about 0.25 nm and $Z_c = 1$. Since the vast majority of the experiments on polyelectrolytes uses water as the solvent, we consider $\Gamma = 3$ as a special case in the simulations and the following theory. Both our simulations and the theory are motivated by experimental systems such as solutions of sodium polystyrene-sulfonate in water at room temperature containing salts of NaCl type (monovalent) or $BaCl_2$ type (divalent). The major issues addressed are the following. The counterion distribution around a flexible polyelectrolyte molecule, being qualitatively different from the Manning condensation, reduces the net charge of the polyelectrolyte from its highest value pertaining to the maximum ionization. Simulations find that the net charge decreases with an increase in polymer concentration (C_p) or an increase in the concentration of added salt (C_s). We have further added divalent counterions from an additional salt to the solution containing the polyelectrolyte chains and their monovalent counterions and monitored the redistribution of counterion clouds. The divalent counterions competitively replace and dominate over the monovalent counterions in shaping the counterion cloud around the polymer. The extent of this competitive replacement depends on the concentration of the added divalent salt. For the simulated conditions, we have not observed any overcharging of the polymer. The theory, however, predicts charge reversal for certain ranges of parameters. Simultaneous to the computation of counterion distribution and its correlation with polymer configuration, we have monitored the radius of gyration R_g and static structure factor $S(\mathbf{k})$ ($\mathbf{k}$ being the scattering wave vector) of the polymer as functions of N, C_p, C_s, and Z_+.

We model each of the polyelectrolyte molecule as a freely jointed chain of N spherical beads, each carrying a point unit charge of an electron, $-e$. All n chains and various ions are placed in a medium of uniform dielectric constant ε and the medium is taken to be a cubic box of volume L^3. Periodic boundary conditions are used in our simulation. One must note that the dielectric heterogeneity of the solvent and the medium has been treated in the theory but is not addressed in our simulations.

The total potential energy of the system consists of the following three parts:

1) **Excluded volume**: The nonelectrostatic part of potential interaction between nonbonded beads of the chain is taken as a purely repulsive Lennard-Jones (LJ) potential,

$$\begin{aligned} U_{\mathrm{LJ}} &= \varepsilon_{\mathrm{LJ}}\left[\left(\frac{\sigma}{r}\right)^{12} - 2\left(\frac{\sigma}{r}\right)^{6} + 1\right] \quad r \leq \sigma \\ &= 0 \quad r > \sigma, \end{aligned} \tag{6.4}$$

where $\varepsilon_{\mathrm{LJ}}$ is the strength, σ is the hard-core distance, and r is the distance between two beads. $\varepsilon_{\mathrm{LJ}}$ is used as the unit of energy. By using the repulsive LJ potential, we are considering only a hydrophilic system. We have not addressed any hydrophobic interaction that requires an additional parameter. We have used the same form as in Eq. (6.4) to capture the nonelectrostatic excluded volume interactions among the polymer beads and counterions. The values of σ (hard-core distance) are $0.8l_0$, $0.6l_0$, and $0.4l_0$ for the pairs of bead–bead, bead–counterion, and counterion–counterion, respectively, where l_0 is the equilibrium bond length.

2) **Bond stretch**: The potential energy associated with bond stretching of each bond of the chain is taken to be

$$U_{\mathrm{bond}} = k_s(l - l_0)^2, \tag{6.5}$$

where l is the bond length and l_0 is the equilibrium bond length. The spring constant k_s is taken to be high enough ($5000\varepsilon_{\mathrm{LJ}}/l_0^2$) to allow fluctuations of the bond length only within 10% of l_0.

3) **Electrostatic interaction**: The electrostatic interaction among the charged beads and counterions is taken to be the full Coulomb energy,

$$U_{\mathrm{C}}(r_{ij}) = \frac{Z_i Z_j e^2}{4\pi\varepsilon_0\varepsilon r_{ij}} = \frac{Z_i Z_j l_B k_B T}{r_{ij}}, \tag{6.6}$$

where r_{ij} is the distance between the ions i and j, and Z_k is the valency of the kth ion. As already mentioned in the preceding section, we take $l_B/l_0 = 3$ to correspond to a typical experimental case (say, sodium polystyrene sulfonate in water at room temperature). In fact, $l_B/l_0 = 3.0$ defines the room temperature for our model system. After performing the systematic analysis as a function of Γ, all representative simulations reported in this chapter are carried out at room temperature. The Coulomb interaction is calculated with the standard Ewald summation [16] technique. We have taken the Ewald parameters k, r_c, and K_{max} to be $5/L$, $L/2$, and 4, respectively.

The solvent is modeled as a uniform dielectric medium. The dynamics of the ith particle (either a labeled bead or a labeled ion) is taken as

$$m\frac{d^2\mathbf{r}_i}{dt^2} = -\varsigma\mathbf{v}_i - \nabla_{\mathbf{r}_i} U + \mathbf{F}_i(t), \qquad (6.7)$$

where m and ς are the mass and friction coefficient, respectively, of the ith particle. U is the net potential energy (described above) acting on the ith particle. $\mathbf{F}_i(t)$ is the noise from the bath acting on the ith particle and is stipulated to satisfy the fluctuation dissipation theorem,

$$\langle \mathbf{F}_i(t) \cdot \mathbf{F}_j(t') \rangle = \delta_{ij} 6 k_B T \varsigma \delta(t-t'). \qquad (6.8)$$

In the present simulations, the mass of beads is taken as unit mass, and $m = 0.5$ for mass of all other ions. Friction coefficient is chosen as constant $1.0\tau^{-1}$, where τ is the time unit of the system. The velocity Verlet finite-differencing scheme is chosen for the integration of Eq. (6.7). By fixing ε, s, and $k_B T$ the salt concentration is varied from salt-free to high salt. The monomer density of the solution is varied from $1 \times 10^{-5} l_0^{-3}$ to $3 \times 10^{-2} l_0^{-3}$. The chain lengths considered are of $N = 20$, 40, 60, and 100. For $N = 100$, only one monomer density of $8 \times 10^{-4} l_0^{-3}$ is considered, but at many values of salt concentration. The Brownian time step δt is adjusted between 0.007τ and 0.01τ accordingly. The total duration of each simulation run takes from 5×10^5 to 4×10^6 time steps depending on the specific case. The largest system simulated in this work consists of about 5000 charges. For the case of about 1000 charges, computation time required for 2 million steps on a single Alpha 533 MHZ processor is 5 days. The computational time increases as a 3/2 power law of the total number of charges [17].

Typical simulation protocol is as follows. First, the initial state consisting of n polyelectrolyte chains each of N beads and exactly neutralizing the number of counterions and certain known number of salt ions is randomly generated inside the simulation box. Then, the Langevin dynamics simulation is performed. Data on position and velocity of all particles, as well as the energies of the system are gathered at every 1000 steps during the whole simulation course. Computations of physical quantities discussed below are carried out separately from the stored data on simulations.

6.2.2 Degree of Ionization

To obtain a quantitative measure of the counterion distribution around a polyelectrolyte molecule, the following procedure is undertaken. For a given chain configuration, a tube is constructed around the chain backbone. The tube is a nonoverlapping superposition of spheres of fixed radius r_c centered on each bead position. r_c is a cutoff parameter. All ions other than the chain beads inside this tube constitute the counterion cloud (worm) of this given polymer configuration. Knowing the number and charges of all ions inside this worm and adding the charges of all beads of the chain, one gets the net charge of the polymer in this configuration. Averaging over many configurations in equilibrium gives the average net charge (eQ_{eff}) of the polymer, and the degree of ionization (α) of the polymer is defined by $\alpha = -Q_{eff}/N$. The dependence of Q_{eff} on the cutoff parameter r_c is given in Figure 6.1 for a representative situation ($N = 100$, $C_p = 8 \times 10^{-4} l_0^{-3}$) for both monovalent and divalent counterions. It is seen that the net polymer charge approaches an asymptotic value for sufficiently large cutoff length. The separation distance r_0 at which the electrostatic energy of a pair of monovalent ions ($l_B k_B T/r_0$) is comparable to the kinetic energy ($3k_B T/2$) of an ion is $2l_0$ for our choice of temperature ($l_B = 3l_0$). This distance is indicated in Figure 6.1 by the vertical dotted line. At this value of r_c the net polymer charge is close to the asymptotic value within the error bars for the case of monovalent counterions. For the case of divalent counterions, the net polymer charge has already reached the asymptotic value for $r_c = r_0$. In view of these observations, the cutoff parameter r_c is taken to be $2l_0$ in the simulations discussed in the remainder of the chapter.

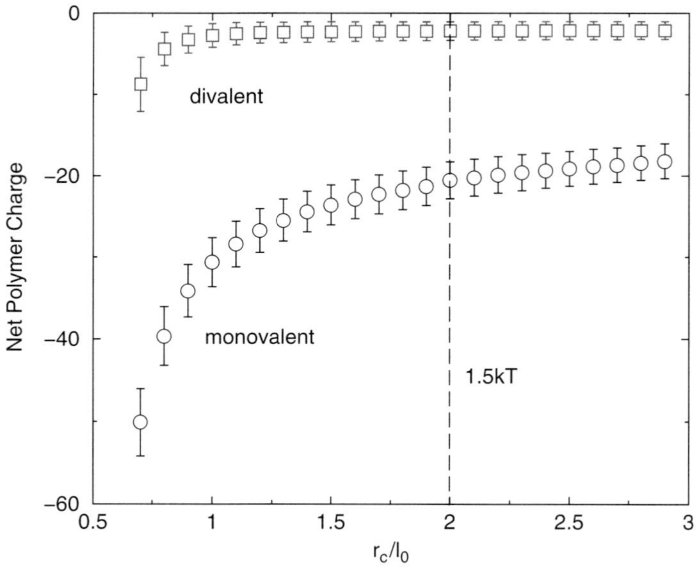

Figure 6.1 Net polymer charge as a function of the cutoff radius r_c of the counterion worm, for $N = 100$, $C_p l_0^3 = 8 \times 10^{-4}$, and the salt-free case. The vertical dashed line denotes the choice of r_c in this chapter. Circles: monovalent counterions; squares: divalent counterions.

6.2.3
Size and Shape of the Polyelectrolyte

The average radius of gyration R_g, defined by

$$R_g^2 = \frac{1}{N}\sum_i \langle \mathbf{r}_i^2 \rangle \tag{6.9}$$

is calculated for different values of N, L, and Γ, for three different valencies of Z_c(=1, 2, and 3 corresponding to monovalent, divalent, and trivalent, respectively). In Eq. (6.9), $\mathbf{r}_i$ is the distance of the ith bead from the center-of-mass of the chain and angular brackets indicate the averaging over chain configurations. The typical results are presented for $N = 100$ and $L = 50l_0$ (monomer density of $8 \times 10^{-4} l_0^{-3}$) in Figure 6.2, where R_g^2/N is plotted against Γ. It must be pointed out that for water as the solvent, the interesting temperature range of 0–100 °C corresponds roughly to the narrow range $3.2 > \Gamma > 2.4$ for monovalent counterions. For multivalent counterions, this range is expanded by the multiple of Z_c. We assume that values of Γ outside these ranges represent solvents different from water. In Figure 6.2, we have also included the data for R_g when the chain is neutral and has only repulsive Lennard-Jones interaction (that means $U_c = 0$, to illustrate the role of electrostatics

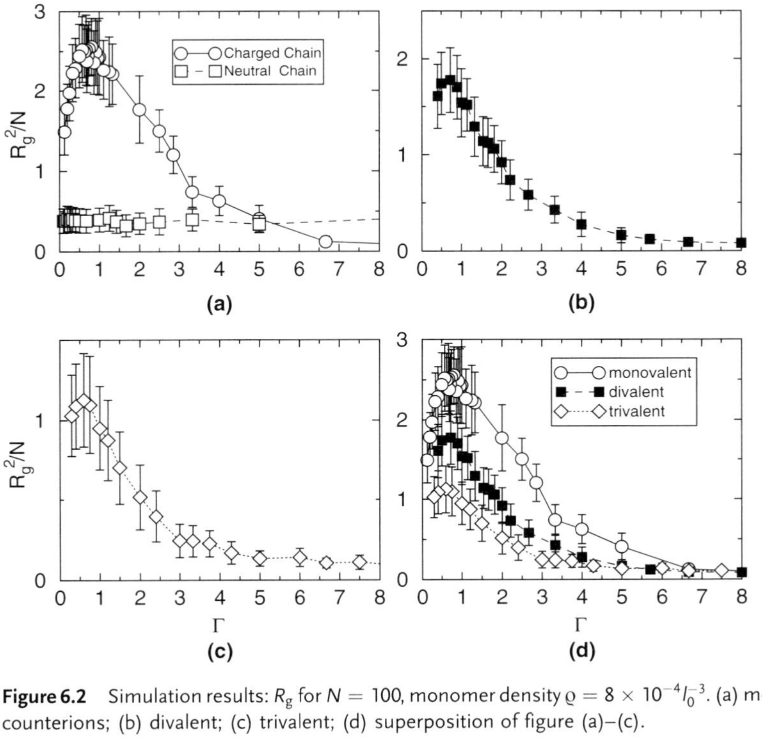

Figure 6.2 Simulation results: R_g for $N = 100$, monomer density $\varrho = 8 \times 10^{-4} l_0^{-3}$. (a) monovalent counterions; (b) divalent; (c) trivalent; (d) superposition of figure (a)–(c).

on R_g. Also, the data for the Lennard-Jones chain are collected in the same temperature range as for the polyelectrolyte chain.

First, we consider Figure 6.2a, corresponding to monovalent counterions. At very high temperatures ($\Gamma \rightarrow 0$), weak electrostatic repulsion is present. Consequently, R_g is higher than that for the LJ chain. As the temperature is lowered (that means Γ is increased), the electrostatic repulsion between beads becomes even stronger and consequently R_g begins to increase with Γ.

As the temperature is decreased even further (that means Γ is close to almost 0.5), the intrachain electrostatic repulsion begins to be mitigated by electrostatic attraction between beads and counterions. The rate of chain swelling with decrease in T begins to decrease. As the temperature is lowered even more (that means Γ goes beyond roughly 1), there are significant number of counterions close to the chain backbone, as described below in detail, creating many dipoles. The interaction between these dipoles leads to intrachain attraction, working against the intrachain swelling arising from the uncompensated charges on the chain backbone. The net result is that R_g decreases as Γ increases (that means Tdecreases). Yet, until $\Gamma \approx 5.0$ (that means for $T > 175$ K if a solution with dielectric constant of 80 can be realized at these low temperatures), R_g of the polyelectrolyte chain is bigger than the value expected for a neutral chain in good solvents. Of course, as Γ becomes much higher than 5, the chain begins to collapse into a compact structure, as theoretically expected. Typical configurations of the chain are given in Figure 6.3 for $N = 100$, $Z_c = 1$, and $L = 50l_0$ ($\varrho = 8 \times 10^{-4} l_0^{-3}$) at different values of Γ. The equivalent theoretical result is presented in Figure 6.11b.

The analogous results of Figure 6.2a for divalent and trivalent counterions are presented in Figure 6.2b and c, respectively. The results are compounded in Figure 6.2d. It is clear from these figures that the extent of chain expansion is weaker at lower values of Γ (that means higher temperatures) for higher valences of counterions. This result agrees with recent experiments [3, 18, 19].

6.2.4
Effect of Salt Concentration on Degree of Ionization

We now consider the effect of salt concentration on α. In the simulation, first a salt-free solution of polyelectrolyte chains and their monovalent counterions are equilibrated. For specificity, we denote the counterion as X, and there are N counterions ($Z_c = 1$), and take the polymer to be uniformly negatively charged. Next, we add a fixed quantity of salt of either the type XY ($Z_+ = 1 = Z_-$) or the type AY_2 ($Z_+ = 2$, $Z_- = 1$) to the system and again equilibrate the system. We then collect the statistics and compute the average number of various ions (X, Y, and A) in the counterion worm surrounding the polymer defined as above with $r_c = 2l_0$. We illustrate the key results only for the case of $N = 100$ and $C_p = 8 \times 10^{-4} l_0^{-3}$.

For the case of XY-type salt (say NaCl added to a solution of sodium polystyrene-sulfonate), the number of X^+ and Y^- ions inside the counterion worm are plotted in Figure 6.4a, as C_s is increased from 0 to 1.36 M. (For example, $C_s = 0.01$ M corresponds to $n_+ = 187$, $L = 126$, and $l_0 = 0.25$ nm.) In Figure 6.4a, $N = 100$ and

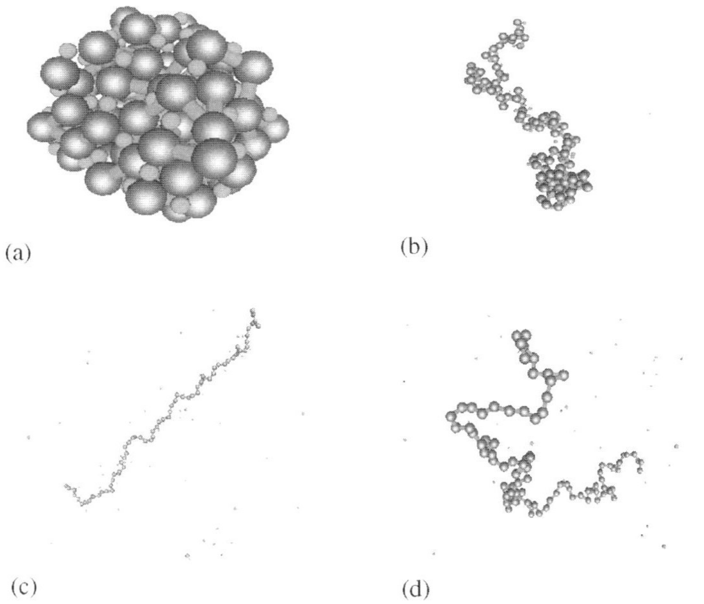

Figure 6.3 Snapshots from the simulations. Structure and ionization rate of the polyelectrolyte change as parameter Γ changes. (a) $\Gamma = 20$; (b) $\Gamma = 7$; (c) $\Gamma = 1$; (d) $\Gamma = 0.13$. Please note that pictures shown here are not in the same scale.

$C_p l_0^3 = 8 \times 10^{-4}$. As noted in Figure 6.1 for the salt-free solution, about 80% of counterions (X^+) are condensed (that means inside the counterion worm). As the salt concentration is increased, the number of condensed counterions around the polymer backbone also increases. At even higher salt concentrations, as the counterion concentration inside the counterion worm increases, increasing number of coions (Y_-) are also brought inside the worm in an effort to maintain electroneutrality. By adding the numbers of X^+ and Y_- ions and their charges and combining with the bare charge of the polymer chain, the effective degree of ionization of the polymer is obtained as given in Figure 6.4b. The degree of ionization decreases monotonically with salt concentration.

In case of adding AY_2-type salt to a solution of polyelectrolyte with X^+ being the counterion (say $BaCl_2$ added to a solution of sodium polystyrene-sulfonate), the number of X^+, A^{2+}, and Y_- ions inside the counterion worm around the polymer is monitored as a function of concentration of AY_2 salt. The results for $N = 100$ and $C_p l_0^3 = 8 \times 10^{-4}$ are given in Figure 6.4c. As soon as small amount of divalent cations are present, these effectively replace the condensed monovalent counterions. Depending on C_s, there are equilibrium concentrations of X^+ and A^{2+} inside the counterion worm. For example, as C_s changes from the salt-free case to 0.05 M, the number of the monovalent X^+ ions decreases from 80 to 5, compensated by an increase in the number of A^{2+} ions from 0 to about 45. In this low C_s regime,

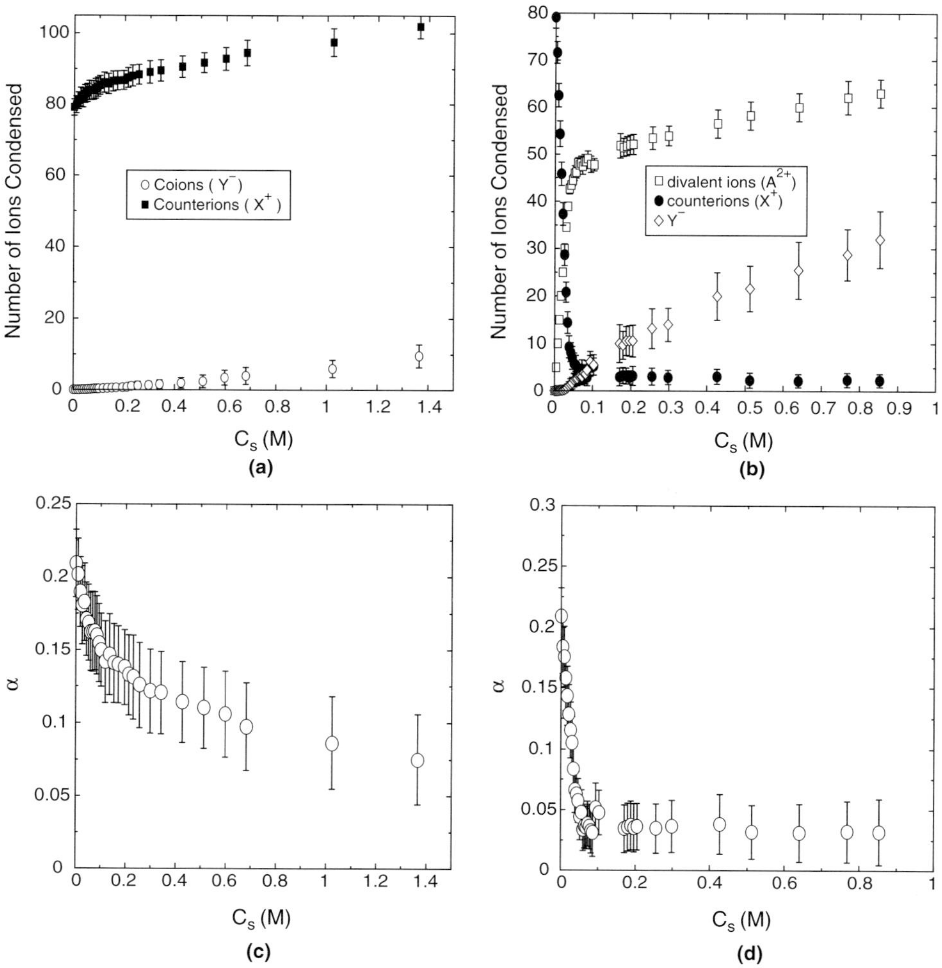

Figure 6.4 Distribution of counterions and coions around the polymer as a function of concentration of salt for $N = 100$ and $C_p l_0^3 = 8 \times 10^{-4}$. ($X^+$, Y^-)-type salt (the counterion of the polymer is X^+): (a) Numbers of condensed X^+ and Y^- ions. (b) Net degree of ionization of the polymer. (A^{2+}, $2Y^-$)-type salt (the counterion of the polymer is X^+): (c) Numbers of condensed X^+, A^{2+}, and Y^- ions. (d) Net degree of ionization of the polymer.

effectively all added divalent ions are condensed by replacing the monovalent counterions. As more AY_2 salt is added, the concentration of X^+ remains at a low value close to zero. But now the concentrations of A^{2+} and Y^- increase monotonically with an increase in C_s. By counting all ions inside the counterion worm, α is obtained and its dependence on C_s is presented in Figure 6.4d. As expected, a comparison of Figure 6.4b and d shows that α is reduced sharply by the divalent ions of the salt in comparison with the case of monovalent ions of the salt. Further, it is to be noted that for both monovalent and divalent counterions from the salt, there is no overcharging of the polymer at the temperature studied in the present simulations (that means α is not negative). The equivalent theoretical result for the effective charge and size as a function of salt concentration is given in Figure 6.12. The degree of condensation for various types of ions is presented in Figure 6.13. We will later notice that theory predicts overcharging in the presence of dielectric inhomogoneity of the solvent. This variation of the dielectric constant is not addressed in present simulations.

6.2.5
Radial Distribution Functions

Following the standard practice, we define the radial distribution function for the beads, $g_{pp}(r)$, as the ratio of number of pairs of beads with distance of separation in the interval between r and $r+\delta r$ to the volume of the shell with inner radius and outer radius of r and $r+\delta r$. In the present binning, δr is taken as roughly $0.1l_0$. Similar to g_{pp}, the radial distribution functions for counterions, $g_{cc}(r)$, and for pairs of bead and counterions, $g_{pc}(r)$, are calculated. These distribution functions are plotted in Figure 6.5a–c, for $\Gamma = 3$, $N = 100$, and $L = 50l_0$. The effect of counterion valency is also included in these figures. As pointed out earlier, $\Gamma = 3$ corresponds roughly to the system of sodium polystyrene sulfonate in water at room temperature. At this $\Gamma = 3$ value, significant number of counterions are close to the beads, as revealed in Figure 6.5b, where the first peak is located at $r = 0.55l_0$ corresponding to the hard core distance between a bead and a counterion. In Figure 6.5a, the first peak at $r = l_0$ reflects the chain connectivity as the equilibrium bond length l_0 has been chosen as unit length. Weak second peaks in Figure 6.5a and b reflect the relatively poor order of second shells in the expanded coil. However, it is worth noting that these peaks are stronger as the counterion valency is higher. Figure 6.5c shows that the packing of higher valency counterions is looser than that for monovalent counterions at this Coulomb strength.

6.2.6
Dependence of Degree of Ionization on Polymer Density

In an effort to understand the role of translational entropy of counterions in countering the attraction between the polymer and the counterions, the volume is varied systematically and one calculates α. The representative results are given in Figure 6.6a for $\varrho = 8 \times 10^{-4} l_0^{-3}$ and $8 \times 10^{-4} l_0^{-3}$, $N = 100$, $r_c = 1.0l_0$, and $Z_c = 1$. As expected, the degree of ionization is higher if the volume of the system is larger.

The value of the effective degree of ionization α of a chain is determined as described above for different values of N and C_p in the salt-free case ($n_+ = 0$). Eight different values of $C_p l_0^3$ in the range of 10^{-5}–10^{-2} have been considered for $N = 20$, 40, and 60. The results are presented in Figure 6.6b where the single data point corresponds to $N = 100$. By estimating the overlap concentration $C_p^* l_0^3$ by

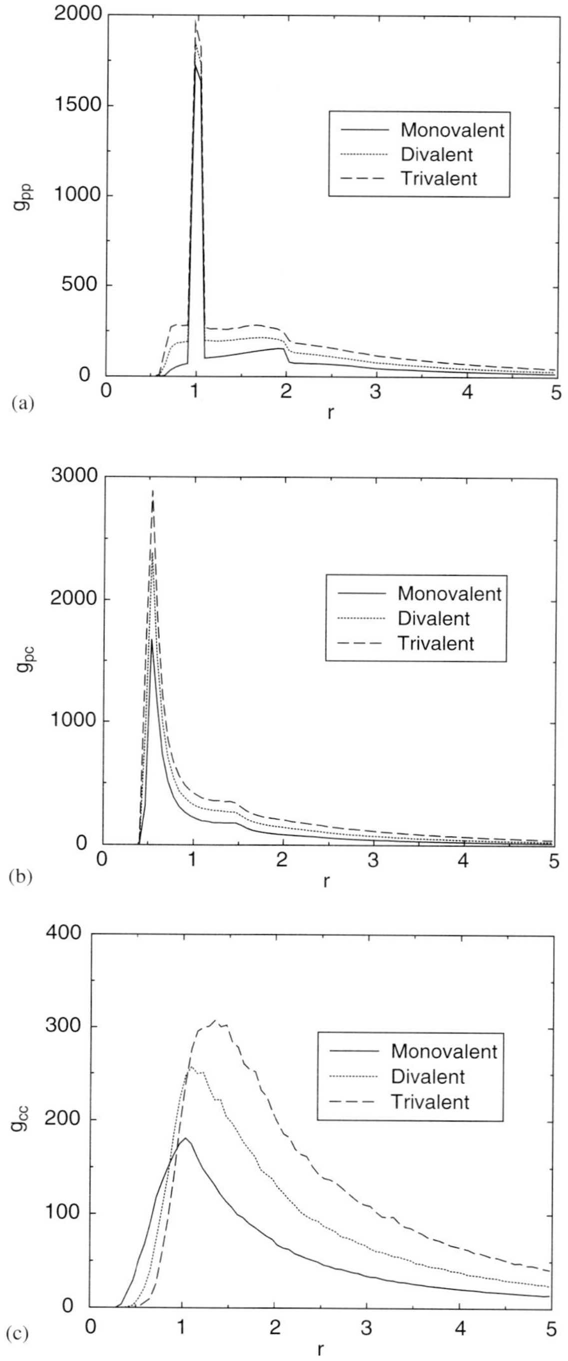

Figure 6.5 Simulation result: radial distribution functions for $N = 100$, $\varrho = 8 \times 10^{-4} l_0^{-3}$, with valency $= 1, 2, 3$ at $\Gamma = 3.0$: (a) g_{pp}; (b) g_{pc}; and (c) g_{cc}.

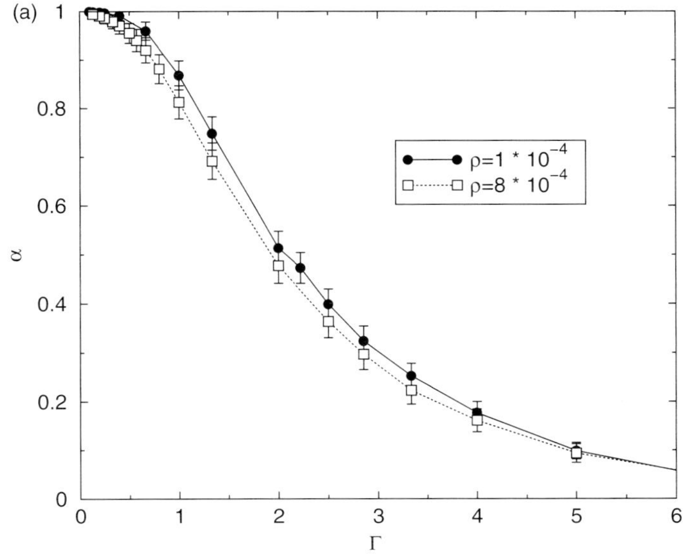

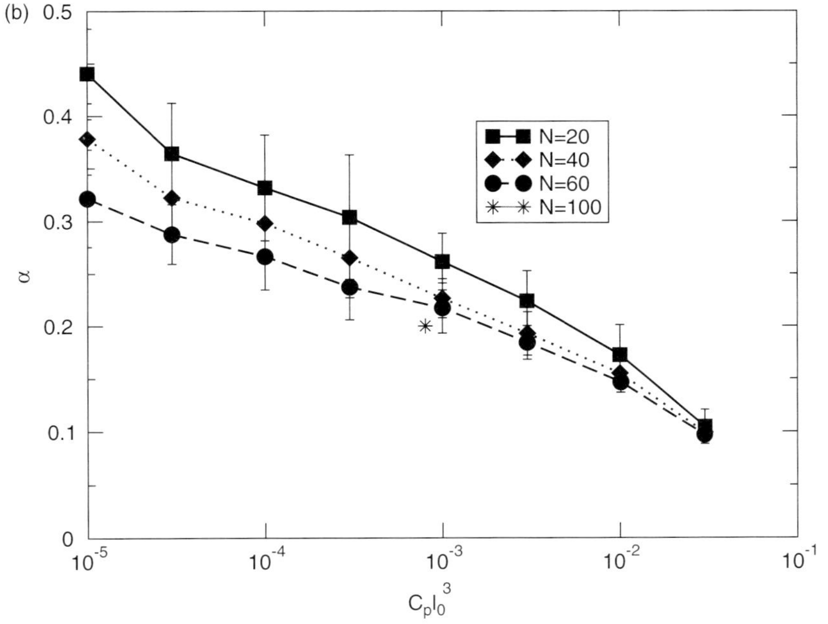

Figure 6.6 (a) Degree of ionization as a function of monomer density ϱ ($N = 100$, $r_c = 1.0 l_0$) and (b) dependence of degree of ionization on polyelectrolyte concentration C_p for different values of N in salt-free solutions.

$3Nl_0^3/4\pi R_g^3$, $C_p^* l_0^3$ are 0.3, 0.1, and 0.03, respectively, for $N = 20, 40$, and 60. Therefore, polymer concentrations investigated are either below or comparable to the overlap concentration. We have carried out simulations at higher concentrations, where we observe a substantial slowing down of dynamics. Reliable averages cannot be constructed at these concentrations, based on our present simulations. Therefore, we do not present our simulation data for such higher concentrations. It is seen from Figure 6.6b that the degree of ionization (proportional to the net polymer charge) decreases with C_p. In other words, the number of counterions in the counterion worm increases with polyelectrolyte concentration. This result is in stark contrast to some recent claims [20]. However, our simulation results are intuitively obvious as the counterions are required by electrostatics to be in the vicinity of polymer segments. Furthermore, there appears to be a systematic N dependence of α at lower values of C_p, where α depends on N for fixed values of C_p. However, as C_p approaches C_p^*, α becomes independent of N. The equivalent theoretical results are presented in Figure 6.15.

6.2.7
Size and Structure of the Polyelectrolyte

The average radius of gyration R_g, defined by

$$R_g^2 = \frac{1}{N}\sum_i \langle \mathbf{r}_i^2 \rangle \tag{6.10}$$

is calculated for different values of N, C_p, C_s, and Z_+. In Eq. (6.10), r is the distance of the ith bead from the center of mass of a labeled chain and angular brackets indicate the averaging over chain configurations. The dependence of R_g on N, C_p, C_s, and the valency of the counterion is given below. Naturally, this dependence is controlled by the electrostatic interaction between the polymer and the counterions resulting in α, electrostatic repulsion between polymer beads, excluded volume interaction between polymer beads, configurational entropy of polymer chains, and the translational entropy of counter ions and salt ions. While there are many scaling arguments discussed extensively in the literature, there are only a few works [21–23] providing formula for R_g with explicit numerical factors, capable of comparison with simulations and experiments. Later in the section discussing our theory, we explain the self-consistent double minimization with respect to the effective charge and size of the chain. In this section, the effective charge is taken as an input from simulations and used in a variational theory [21] in which the free energy is minimized as a function of the size. The formula for R_g in solutions of isolated chains and many chains obtained by the earlier work of Muthukumar are summarized here to facilitate a comparison between theory and simulations.

6.2.7.1 Theoretical Background

In infinitely dilute solutions, the polyelectrolyte chains are essentially isolated. In Muthukumar's treatment [21] of this limit, the degrees of freedom of counterions and salt ions are integrated out within the Debye–Hückel approximation of small electrolytes to obtain an effective electrostatic interaction between chain segments. This effective potential interaction (in units of $k_B T$) between two beads of valency αZ_p separated by distance r is the Debye–Hückel interaction

$$\frac{w_c}{4\pi}\frac{e^{-\varkappa r}}{r}, \tag{6.11}$$

where

$$w_c = l_B Z_p^2 \alpha^2, \tag{6.12}$$

and the inverse Debye length $\varkappa$ is given by

$$\varkappa^2 = 4\pi l_B \left(Z_c^2 \varrho_c + \sum_\gamma Z_\gamma^2 \varrho_\gamma \right). \tag{6.13}$$

ϱ_c and ϱ_γ are the number densities of the counterion of the polymer and the γth salt ions, respectively. Z_c and Z_γ are the valencies of the counterion and the γth salt ion. In addition, the usual nonelectrostatic excluded volume interaction between two segments i and j is modeled as $w l_0^3 \delta(r_i - r_j)$. Using a variational procedure and assuming uniform expansion of the polymer by excluded volume and screened electrostatic interaction, the result for R_g is given by [21]

$$R_g^2 = \frac{L l_1}{6}, \tag{6.14}$$

where $L = N l_0$ is the contour length and l_1/l_0 is the square of the expansion factor of the root mean square end-to-end distance of the polymer. l_1, which may be called the renormalized Kuhn length, is given by the formula

$$l_1^{3/2}\left(\frac{1}{l_0} - \frac{1}{l_1}\right) = \frac{4}{3}\left(\frac{3}{2\pi}\right)^{3/2} \frac{w l_0 L^{1/2}}{l_1} + \frac{4}{45}\left(\frac{6}{\pi}\right)^{1/2} \times \left[\frac{15\sqrt{\pi}\,e^a}{2a^{5/2}}(a^2 - 4a + 6)\,\mathrm{erfc}\left(\sqrt{a}\right) + \frac{15}{\sqrt{\pi}}\left(-\frac{3\pi}{a^{5/2}} - \frac{\pi}{a^{3/2}} + \frac{6\sqrt{\pi}}{a^2}\right)\right], \tag{6.15}$$

where $a = \varkappa^2 L l_1/6$. We call Eqs. (6.12)–(6.15) as Muthukumar's single screening theory, labeled as ssM, for the convenience of references below. There are two parameters, namely, α and w, in the theory. For a solution containing many polyelectrolyte chains, the segment–segment interaction (both nonelectrostatic and electrostatic) is progressively screened as C_p increases, in addition to the usual Debye–Hückel screening by randomly distributed small ions. The coupled double screening was approximately treated in Refs [22, 23], and the final result for R_g can be written in the form of Eq. (6.14). Now l_1 is dependent on C_p also, and its explicit form is available in Ref. [23]. We call this theoretical result Muthukumar's double screening theory, labeled as dsM. This theory is strictly valid for $C_p > C_p^{**}$. Again, there are two input parameters w and α.

Both parameters α and w are known in our simulations. α is already discussed above. For the repulsive LJ potential used in our simulations, w is given by the binary cluster integral between beads,

$$w = \frac{1}{l_0^3}\int_0^\infty \left(1 - e^{-U_{LJ}(r)/k_B T}\right) dr. \tag{6.16}$$

Substituting Eq. (6.4) in Eq. (6.16), w is found to be 2.0. The knowledge of w and α allows a comparison between simulations and theory (ssM and dsM) without any adjustable parameters.

Both ssM and dsM theories are based on the assumption that condensed counterions are uniformly distributed along the chain backbone so that the degree of ionization is uniform everywhere along the chain. This approximation fails drastically when multivalent counterions condense on the chain. The role of bridging between nonbonded beads mediated by multivalent counterions has already been emphasized [4, 24] by the groups of Joanny and Olvera de la Cruz. In the case of divalent counterions considered here, each bridging contact is equivalent to a cross-link junction of functionality four. As shown in Ref. [22], such a bridging results in an attractive two-body interaction. Therefore, when bridging effects are present with divalent counterions, w is replaced by

$$w + \frac{E_{br}}{k_B T}\alpha_{2+}(1 - \alpha_{2+} - \alpha_+), \tag{6.17}$$

where E_{br} is the attractive energy associated with the formation of one bridge, α_{2+} and α_+ are the ratios of the numbers of condensed divalent and monovalent counterions, respectively, within the counterion worm of a chain, to N. In making this mean field estimate of bridging contribution, allowance is not made for any correlation such as the interaction between the counterions and coions inside the counterion worm.

6.2.7.2 Dependence of Radius of Gyration on Salt with Monovalent Counterions

The dependence of R_g on the concentration of XY-type salt (X^+ is the counterion) is illustrated in Figure 6.7a for $N = 100$ and $C_p l_0^3 = 8 \times 10^{-4}$. Again, the decrease in R_g as C_s increases is expected due to the increased electrostatic screening, and R_g eventually reaching the value of a neutral chain. The prediction of ssM theory is included in Figure 6.7a as the continuous curve. In calculating this curve, α from Figure 6.3b is used in Eq. (6.15). Although the theoretical curve is within the error bars of simulation data, it underestimates R_g due to the various approximations employed in the analytical derivation of Eqs. (6.12)–(6.15). The same trend was noticed in comparing [25] with experimental data. Nevertheless, the agreement between the ssM theory and the simulation data is good, given the complexity of the problem and that there are no adjustable parameters. According to Eqs. (6.12)–(6.15), for high enough salt concentrations, R_g is expected to approach the asymptotic limit

$$R_g \sim \left(w l_0^3 + \frac{w_c}{\kappa^2}\right)^{1/5} \sim C_s^{-1/5}. \tag{6.18}$$

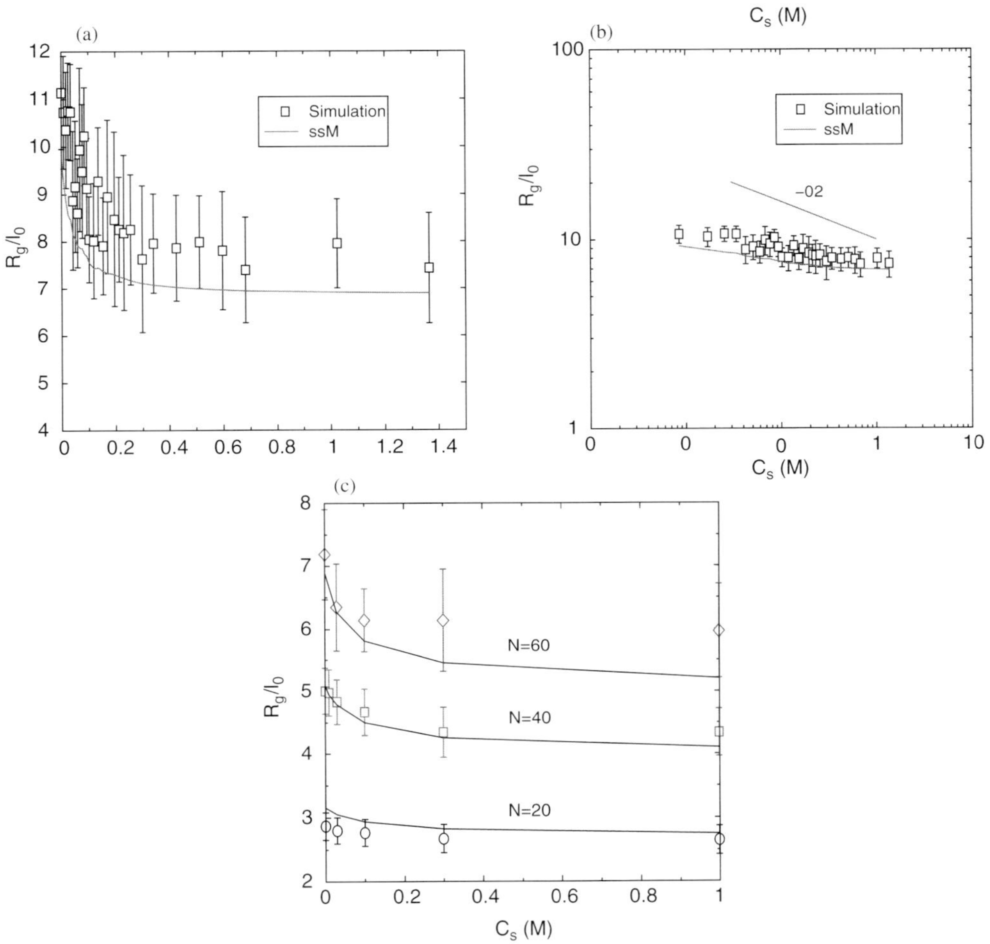

Figure 6.7 Dependence of R_g on salt concentration C_s for (X^+, Y^-)-type salt ($N = 100$, $C_p l_0^3 = 8 \times 10^{-4}$). Open symbols are simulation data and the curve is ssM theory without any adjustable parameters. (b) is the double logarithmic plot of (a).

In view of this, the simulation data of Figure 6.7a are given in Figure 6.7b as a double logarithmic plot of R_g versus C_s. A crossover from a constant value of R_g at low C_s to the scaling form of Eq. (6.18) at high C_s is only barely visible. The dependence of R_g on C_s for different values of $N(20, 40, 60)$ is given in Figure 6.7c for $C_p l_0^3 = 10^{-3}$. The simulation data are given by open symbols and the predictions of ssM are given by filled symbols. The agreement between the theory and the simulation is excellent.

6.2.7.3 **Bridging Effect by Divalent Counterions**

The C_s dependence of R_g for the case of divalent counterions (AY_2-type salt) is presented in Figure 6.8a for $N = 100$ and $C_p l_0^3 = 8 \times 10^{-4}$. As is seen in this figure, the divalent counterions are very effective in shrinking the polyelectrolyte chain. The prediction of ssM theory, where the bridging effect is completely ignored (i.e., $E_{\text{br}} = 0$ in Eq. (6.17)) is given by the solid curve. In obtaining this curve, α from Figure 6.3d is used in Eq. (6.12). There is a significant discrepancy between the simulation data and the ssM theory, and we attribute this discrepancy to the bridging effect caused by the divalent counterions. In an effort to account for the bridging effect, we estimate E_{br} in our simulations as follows. When the divalent counterion is involved in a bridge between two beads, the beads can be as far away as $2r_c$ and as close as $0.8l_0$. The counterion can be directly in between these beads or can be at one vertex of a triangle (and the other two vertices being occupied by the two beads). Instead of performing a detailed calculation of E_{br} corresponding to all shapes of this triangle, we estimate E_{br} to be an average corresponding to the beads being separated by $2.6l_0$ and the counterion placed right in the middle. For such a geometry, E_{br} turns out to be $-9.23k_BT$. By scanning various configurations of the polymer, we find the number of condensed counterions inside the counterion worm. The ratios of the average numbers of ions (n_{2+} for A^{2+}, n_+ for X^+, and n_- for Y^-) to $N = 100$ are given in Table 6.1 for different representative values of C_s (at $C_p l_0^3 = 8 \times 10^{-4}$). Substituting the values of α_{2+} and α_+ from Table 6.1 in Eq. (6.17) and combining with ssM (Eqs. (6.12)–(6.15)), we get the dashed curve, ssM′, displayed in Figure 6.8a. The agreement between simulation data and ssM′ demonstrates the significant role played by divalent ions in forming bridges between segments.

The data of Figure 6.8a are given as a double logarithmic plot in Figure 6.8b. It is clear that R_g shrinks from a swollen state at low C_s to a compact state at high C_s through a crossover. The scaling prediction of Eq. (6.18) is not adequate to describe the bridging effect. The predictions of ssM and ssM′ are also included in the figure for comparison. The N dependence of R_g as the concentration of AY_2-type salt varies is given in Figure 6.8c. The discrepancy between the simulation data and the ssM theory and a better agreement with ssM′ clearly demonstrate the effect of bridging by the divalent counterions. The theoretical double minimization predicts a first-order coil to globule transition with increasing divalent salt in the presence of a dielectric inhomogeneity. The absence of the dielectric inhomogeneity does not allow to have a similar transition in the calculations.

6.3 The Variational Theory

We consider [1, 13] a linear flexible polyelectrolyte chain of N monomers in a solution of volume Ω, the center of mass of the chain being at the origin of the coordinate system. Each monomer is monovalently charged (negative) and of length l. The solution (e.g., water) contains either no salt, only monovalent salt (Figure 6.9a), or both mono- and divalent salts (Figure 6.9b). As we shall see later, all the results for the case when only the monovalent salt is present can be readily obtained as a specific

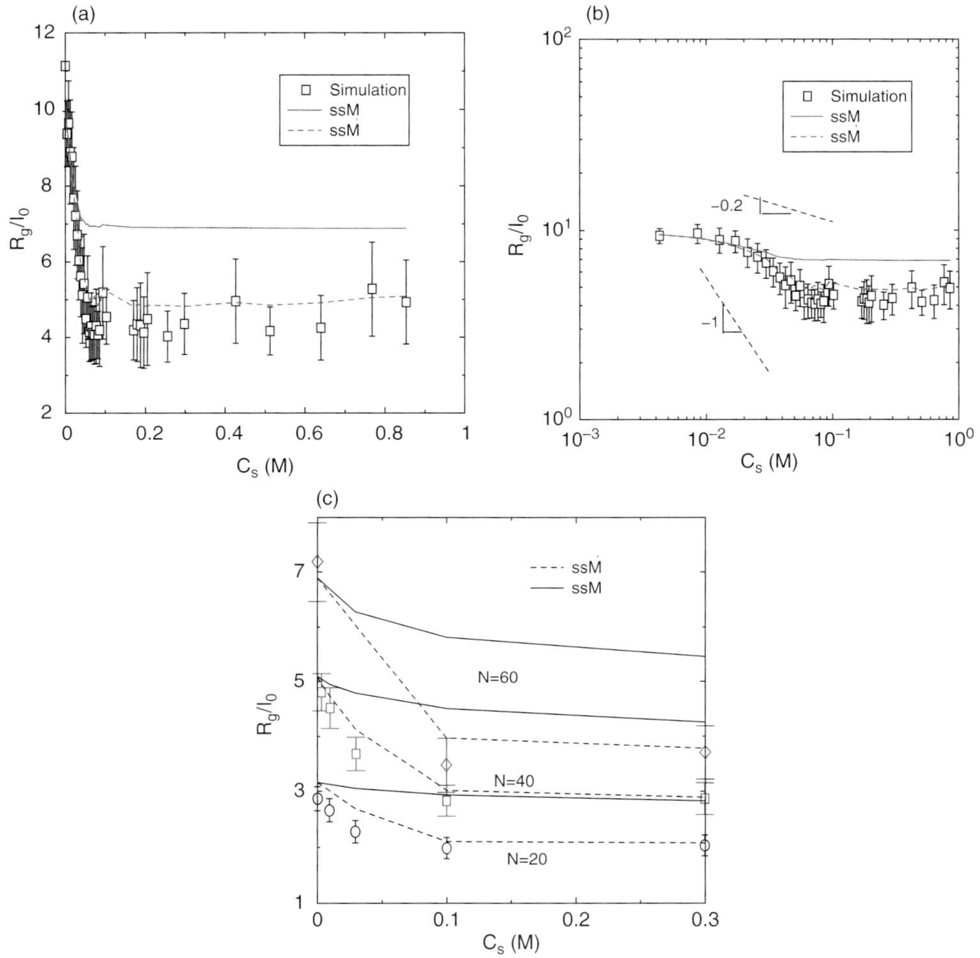

Figure 6.8 Dependence of R_g on salt concentration C_s for $(A^{2+}, 2Y^-)$-type salt ($N = 100$, $C_p l_0^3 = 8 \times 10^{-4}$). Open symbols: simulation data; solid curve: ssM theory; dashed curve: ssM theory accounting for bridging. (b) is the double logarithmic plot of (a).

example of the general case in which both salts are present. The system being electroneutral for all cases will have a maximum of N monovalent counterions in addition to the salt ions. We assume that the counterion from the monovalent salt (e.g., Na^+ from NaCl) is chemically identical to the counterion from the polymer (e.g., Na^+ from NaPSS). Similarly, the coions from both types of salts, if present, are of the same species (e.g., Cl^- from NaCl and $BaCl_2$). At any time, both monovalent and divalent counterions (e.g., Ba^{2+} from $BaCl_2$ as divalent counterions) can adsorb on separate monomers. In addition, in the general scenario with the presence of both types of salts, the Ba^{2+}-monomer ion-pair is viewed as a positive monovalent ion, and the negative coions (Cl^-) will adsorb onto some of these pairs as counterions

Table 6.1 Simulation data of average numbers of condensed ions ($n_{2+} = f_{2+}N$ for A^{2+} ions, $n_+ = f_+N$ for X^+ ions, and $n_- = f_-N$ for Y^- ions) at different concentrations of (A^{2+}, $2Y^-$)-type salt.

C_s	f_-	f_+	f_{2+}
0.0000	0.00	0.79	0.00
0.0085	0.00	0.62	0.10
0.0170	0.00	0.46	0.20
0.0256	0.00	0.29	0.30
0.0341	0.01	0.14	0.39
0.0426	0.01	0.08	0.43
0.0512	0.02	0.06	0.46
0.0597	0.03	0.03	0.48
0.0682	0.03	0.03	0.48
0.0768	0.04	0.03	0.48
0.0853	0.04	0.03	0.49
0.1024	0.05	0.05	0.48
0.1792	0.10	0.03	0.51
0.1962	0.11	0.03	0.52
0.2560	0.13	0.03	0.53
0.4266	0.20	0.03	0.56
0.6400	0.25	0.02	0.60
0.8533	0.32	0.02	0.63

$N = 100$ and $C_p/\ell_0^3 = 8 \times 10^{-4}$.

(Figure 6.9). Therefore, if M_1 monovalent counterions and M_2 divalent counterions get adsorbed onto the chain ($M_1 + M_2 \leq N$), and M_3 (negative) coions condense on to the Ba^{2+}-monomer ion-pairs ($M_3 \leq M_2$), the effective (or average) degree of ionization of the entire chain will be $f = (N - M_1 - 2M_2 + M_3)/N$. We define our variables further as follows: R_g is the radius of gyration of the chain. c_{s1} and c_{s2} are, respectively, the number concentrations of the added monovalent and divalent salts. Both types of salts are fully dissociated into n_1 monovalent counterions (Na^+), n_2 divalent counterions (Ba^{2+}), and $n_1 + 2n_2$ coions (Cl^-). Therefore, $c_{s1} = n_{1+}/\Omega$ and $c_{s2} = n_{2+}/\Omega$. The free energy of the system, consisting of the chain, condensed and mobile counterions, and the solution, would depend on four independent variables: M_1, M_2, M_3, and R_g. The theory aims to evaluate M_1, M_2, M_3, and R_g self-consistently by expressing the free energy F of the system as a function of all these variables and electrostatic parameters, and further obtain their equilibrium values minimizing F simultaneously with respect to these variables. As discussed later, this four-dimensional minimization has to be extended with one more variable when the electrostatic bridging by divalent cations is considered.

The free energy F has six contributions [1, 13] F_1, F_2, F_3, F_4, F_5, and F_6 related, respectively, to (i) entropy of mobility of the adsorbed counterions and coions along the polymer backbone, (ii) translational entropy of the unadsorbed counterions and coions (including salt ions) that are free to move within the volume Ω, (iii)

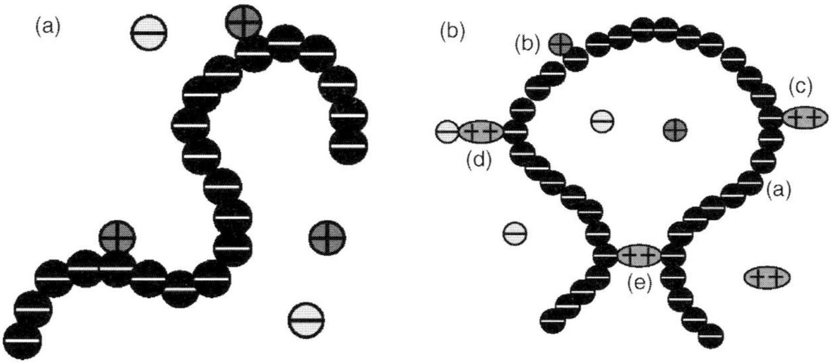

Figure 6.9 Schematic diagram of the system consisting of the isolated polymer chain, condensed counterions, dissociated mobile ions, and the solution as the background interacting only through the dielectric constant ε. (a) Only monovalent salt. (b) Both mono- and divalent salt: possible charge complexes for each monomer: (a) monomer (−1), (b) monomer-monovalent (−1, +1), (c) monomer-divalent (−1, +2), (d) monomer-divalent-monocoion (−1, +2, −1), and (e) a bridging configuration of monomer-divalent-monomer. The dielectric constant ε_l in the vicinity of the chain is much lower than the bulk value. To reach equilibrium, the major competition is between the translational entropy of the dissociated ions and the Coulomb energy gain of the adsorbed ions.

fluctuations in densities of all these mobile ions (in the Debye–Hückel form) except the polymer, (iv) the unscreened electrostatic (Coulomb) energy of the monomer counterion bound pairs (both monovalent and divalent counterions) and the monomer counterion–coion triplets, (v) the entropic, electrostatic, and excluded volume interactions between monomers of the polyelectrolyte with an average degree of ionization f, and (vi) electrostatic correlation involving the charged monomers, the neutral ion-pairs, and the ion-triplets along the backbone of the polymer (also known as dipole–dipole interactions).

6.3.1 Free Energy

The entropic contribution arising from the various distributions of the adsorbed counterions and coions is determined as follows. We note that for the general case of both mono- and divalent salts being present, there are N monomers, M_1 adsorbed monovalent counterions (Na^+), $M_2 - M_3$ adsorbed divalent counterions (Ba^{2+}) with no coion (Cl^-') condensation, and M_3 ion-triplets ("monomer-Ba^{2+}-Cl^-") in the system. Therefore, $N - M_1 - M_2$ monomers remain with their bare charge uncompensated. Consequently, the partition function is

$$Z_1 = \frac{N!}{(N-M_1-M_2)!M_1!(M_2-M_3)!M_3!}. \tag{6.19}$$

Let us define

$$\alpha_1 = \frac{M_1}{N}; \quad \alpha_2 = \frac{M_2}{N}; \quad \alpha_3 = \frac{M_3}{N}. \tag{6.20}$$

Then, using $F_1 = -k_B T \ln Z_1$, we have

$$\begin{aligned} \frac{F_1}{Nk_BT} &= (1-\alpha_1-\alpha_2)\log(1-\alpha_1-\alpha_2) + \alpha_1 \log \alpha_1 \\ &+ (\alpha_2-\alpha_3)\log(\alpha_2-\alpha_3) + \alpha_3 \log\alpha_3. \end{aligned} \tag{6.21}$$

The above expression implies two obvious limits:

$$\alpha_1 + \alpha_2 \leq 1 \quad \text{and} \quad \alpha_3 \leq \alpha_2. \tag{6.22}$$

To determine the translational entropy of the unadsorbed ions that are distributed in the bulk volume Ω, we count mobile ions of various species as $N-M_1+n_{1+}$ monovalent counterions (Na^+), $n_{2+}-M_2$ divalent counterions (Ba^{2+}), and $n_{1+}+2n_{2+}-M_3$ monovalent coions (Cl^-). Therefore, the partition function related to the translational entropy in volume Ω is

$$\begin{aligned} Z_2 &= \frac{\Omega^{N-M_1+n_{1+}+n_{2+}-M_2+n_{1+}+2n_{2+}-M_3}}{(N-M_1+n_{1+})!(n_{2+}-M_2)!(n_{1+}+2n_{2+}-M_3)!} \\ &= \frac{\Omega^{N-M_1-M_2-M_3+2n_{1+}+3n_{2+}}}{(N-M_1+n_{1+})!(n_{2+}-M_2)!(n_{1+}+2n_{2+}-M_3)!}. \end{aligned} \tag{6.23}$$

Let us further define

$$N = \varrho\Omega; \quad n_{1+} = \frac{c_{s1}N}{\varrho}; \quad n_{2+} = \frac{c_{s2}N}{\varrho}. \tag{6.24}$$

Using $F_2 = -k_B T \ln Z_2$ and after some calculations we arrive at

$$\begin{aligned} \frac{F_2}{Nk_BT} &= \left(1-\alpha_1+\frac{c_{s1}}{\varrho}\right)\log(\varrho(1-\alpha_1)+c_{s1}) + \left(\frac{c_{s2}}{\varrho}-\alpha_2\right)\log(c_{s2}-\varrho\alpha_2) \\ &+ \left(\frac{c_{s1}}{\varrho}+2\frac{c_{s2}}{\varrho}-\alpha_3\right)\log(c_{s1}+2c_{s2}-\varrho\alpha_3) \\ &- \left\{(1-\alpha_1-\alpha_2-\alpha_3)+2\frac{c_{s1}}{\varrho}+3\frac{c_{s2}}{\varrho}\right\}, \end{aligned} \tag{6.25}$$

with the constraint

$$M_2 \leq n_{2+}. \tag{6.26}$$

The free energy contribution from the correlations of all dissociated ions is given by the Debye–Hückel electrostatic free energy,

$$F_3 = -k_B T \frac{\Omega\varkappa^3}{12\pi}, \tag{6.27}$$

where the inverse Debye length $\varkappa$ is given by

$$\varkappa^2 = 4\pi l_B \sum_i Z_i^2 n_i/\Omega. \tag{6.28}$$

This result is obtained from the Debye–Hückel theory, which is valid only in specific regimes of salt concentration and temperature. One must be careful in identifying regimes where the DH theory ceases to be valid (e.g., for very high salt concentrations). Here, Z_i is the valency of the dissociated ion of the ith species. For our present case,

$$\begin{aligned}\varkappa^2 &= 4\pi l_B\{N-M_1+n_{1+}+4(n_{2+}-M_2)+n_{1+}+2n_{2+}-M_3\}/\Omega \\ &= 4\pi l_B(N-M_1-4M_2-M_3+2n_{1+}+6n_{2+})/\Omega \\ &= 4\pi l_B\{\varrho(1-\alpha_1-4\alpha_2-\alpha_3)+2c_{s1}+6c_{s2}\}.\end{aligned} \tag{6.29}$$

Using the definitions from Eq. (6.24), we reach

$$\frac{F_3}{Nk_BT} = -\frac{1}{3}\sqrt{4\pi}l_B^{3/2}\frac{1}{\varrho}\{\varrho(1-\alpha_1-4\alpha_2-\alpha_3)+2c_{s1}+6c_{s2}\}^{3/2}. \tag{6.30}$$

To determine the electrostatic energy gain due to the adsorption of all sorts of ions (Na^+, Ba^{2+}, and Cl^-), we recount different numbers of ion-pairs and triplets that form due to counterion adsorption. On the polymer chain, there are $\alpha_1 N$ pairs of "monomer(−1) and Na^+ ion," $(\alpha_2-\alpha_3)N$ pairs of "monomer(−1) and Ba^{2+} ion," and $\alpha_3 N$ triplets of "monomer(−1), Ba^{2+}, and Cl^- ions". In addition, $(1-\alpha_1-\alpha_2)N$ monomers(−1) remain without being charge compensated. The dielectric mismatch parameter, δ, describes [1] a local dielectric constant, ε_l, in the vicinity of the chain backbone. Experiments have shown [10] that ε_l can be as small as one-tenth of the bulk value ε (around 78 in water) near the polyelectrolyte or protein backbone. The dielectric constant increases exponentially [11, 12] from the local value (corresponding to the material made from the chain backbone) to its full bulk value over a distance of 1–10 Å from the chain monomers. The parameter $\delta = (\varepsilon l/\varepsilon_l d)$ is, therefore, introduced [1], where d is the dipole length of the bound ion-pair. The value of d is probably comparable to the Bjerrum length l_B. The schematic of Figure 6.9 reveals that δ in the above form applies only to the monomer-monovalent (Na^+) and monomer-divalent (Ba^{2+}) ion-pairs and not to the divalent counterion–monovalent coion (Ba^{2+}-Cl^-) ion-pair in the monomer-divalent–monovalent triplet. In an ion-pair, there are two ions involved with a fixed distance between them. For the triplet, however, there are three lengths involved (in this case, Ba^{2+}-monomer, Ba^{2+}-Cl^- and Cl^--monomer), and the interpretation of δ is not simple. We introduce a parameter δ_2 for the "monomer-Ba^{2+}-Cl^-" triplet. δ_2 is expected to be less than 4δ (the value it would have assumed if there were two point charges, $+2e$ and $-2e$, respectively), but the determination of its actual value would require a microscopic treatment. In principle, δ_2 would be a function of δ. For simplicity, we assume all ions and monomers to be of the same size, and estimate δ_2 as follows. First, we write the electrostatic energy of counterion adsorption in terms of δ and δ_2:

$$\frac{F_4}{Nk_BT} = -\alpha_1\delta\tilde{l}_B - 2(\alpha_2 - \alpha_3)\delta\tilde{l}_B - \alpha_3\delta_2\tilde{l}_B, \tag{6.31}$$

where $\tilde{l}_B = l_B/l$, and the terms are written as Coulomb energies of bound ion-pairs. To determine the limit of δ_2 in terms of δ, one may note that if one assumes the local dielectric constant ε_l to apply only to the Ba^{2+}-monomer pair energy, but not to the Ba^{2+}-Cl^- pair ($\varepsilon = \varepsilon_{water}$ in that case) energy, then δ_2 turns out to be

$$\delta_2 = \left(2 + \frac{2}{\delta}\right)\delta. \tag{6.32}$$

This should be the lowest value of δ_2. On the other hand, if ε_l applies to both Ba^{2+}-monomer and Ba^{2+}-Cl^- pairs, then

$$\delta_2 = 4\delta. \tag{6.33}$$

This should be the highest value of δ_2, as already noticed before. In reality, δ_2 is likely to be somewhere between these two limiting values. We choose the dielectric constant to be ε_l (the local value pertaining to the chain backbone) for the Ba^{2+}-monomer pair and $(\varepsilon_l + \varepsilon_{bulk})/2$ for the Ba^{2+}-Cl^- pair. Then, δ_2 turns out to be

$$\delta_2 = \left(2 + \frac{4}{\delta + 1}\right)\delta. \tag{6.34}$$

In the above analysis, the repulsion between monomer and Cl^- ions has been ignored; it would bring a very small correction in all above three cases. Although the counterion distribution and chain conformations sensitively depend on δ_2, it is verified that the basic qualitative results are not affected regardless of δ_2 if it is assigned any value in the range mentioned above. Therefore, we use Eq. (6.34) in all of our representative calculations, unless mentioned otherwise.

The free energy of the isolated and flexible polyelectrolyte chain is obtained by the variational method [1, 21] starting from the Edwards Hamiltonian,

$$\begin{aligned} H = {} & \frac{3}{2l}\int_0^L ds\left(\frac{\partial\mathcal{R}(s)}{\partial s}\right)^2 + \frac{w}{2}\int_0^L ds\int_0^L ds'\,\delta(\mathcal{R}(s) - \mathcal{R}(s')) \\ & + \frac{l_B}{2}\int_0^L ds\int_0^L ds'\,\frac{1}{|\mathcal{R}(s) - \mathcal{R}(s')|}\exp[-\varkappa|\mathcal{R}(s) - \mathcal{R}(s')|], \end{aligned} \tag{6.35}$$

where $L = Nl$, $\mathcal{R}(s)$ is the position vector of the chain at arc length s, and w is the strength parameter for all short-ranged hydrophobic or excluded volume effects. An effective expansion factor l_1 is defined as follows:

$$\langle R^2 \rangle = Nll_1 \equiv Nl^2\tilde{l}_1 = 6R_g^2, \tag{6.36}$$

where $\tilde{l}_1 = l_1/l$. Here, $\langle R^2 \rangle$ is the mean square end-to-end distance, and l_1 is the effective expansion factor that measures the swelling of the chain compared to a

Gaussian chain. Assuming uniform, spherically symmetric expansion or contraction of the chain, one extremizes the free energy to obtain

$$\begin{aligned}\frac{F_5}{Nk_BT} &= \frac{3}{2N}(\tilde{l}_1-1-\log\tilde{l}_1)+\frac{4}{3}\left(\frac{3}{2\pi}\right)^{3/2}\frac{w}{\sqrt{N}}\frac{1}{\tilde{l}_1^{3/2}}\\ &\quad+2\sqrt{\frac{6}{\pi}}f^2\tilde{l}_B\frac{N^{1/2}}{\tilde{l}_1^{1/2}}\Theta_0(a),\end{aligned} \tag{6.37}$$

where

$$\Theta_0(a)=\frac{\sqrt{\pi}}{2}\left(\frac{2}{a^{5/2}}-\frac{1}{a^{3/2}}\right)\exp(a)\,\mathrm{erfc}(\sqrt{a})+\frac{1}{3a}+\frac{2}{a^2}-\frac{\sqrt{\pi}}{a^{5/2}}-\frac{\sqrt{\pi}}{2a^{3/2}}, \tag{6.38}$$

where

$$a\equiv\tilde{\kappa}^2N\tilde{l}_1^2/6. \tag{6.39}$$

Here, $\tilde{\kappa}=\kappa l$. We further define two more dimensionless variables, $\tilde{\varrho}=\varrho l^3$ and $\tilde{c}_{si}=c_{si}l^3$, where i stands for the ion species. The important factor f is our previously defined average degree of ionization given by

$$f=1-\alpha_1-2\alpha_2+\alpha_3. \tag{6.40}$$

In obtaining the free energy of the polymer, only a variational result has been used. This method is equivalent to ignoring the vertex terms discussed in Ref. [26]. Inclusion of vertex terms is very complicated but leads to only minor corrections in quantitative details for the case of an excluded volume chain. Similar minor numerical corrections are expected for a chain with screened Coulomb interaction because the Debye screening length is realistically finite. Another key assumption is that the flexible polyelectrolyte chain swells uniformly with spherical symmetry in order to facilitate an analytical expression for the free energy of the chain. In spite of these approximations for polymer chain swelling, the predictions of $\tilde{l}_1$, with f taken as an input from simulations, have been found to be in reasonable agreement with simulation results as seen in Figures 6.7 and 6.8 of Ref. [27]. These arguments are used to justify the use of Eq. (6.37) for the free energy of the chain. An alternative expression for F_5, as might emerge in the future, can be used in the present theory of counterion adsorption without any loss of generality.

The electrostatic interaction only between monomers with nonzero monopole charges, with or without condensed ions, is considered till now in this work (the third-term in the polymer free energy F_5). Further, F_5 considers only the monopole contribution of each ion-pair or ion-triplet. For example, a monomer-Na^+ pair and a monomer-Ba^{2+}-Cl^- triplet would contribute identically to F_5, although they have different electrostatic effects. Similarly, a monomer-Ba^{2+} pair would be simply treated as a $+1$ charge, although the pair has additional dipole effects. These additional dipole or higher order multipole effects are critical when the average charge of the chain is close to zero. We have already observed in simulations [1, 8, 9] that these ion-pair effects play a key role to collapse a chain below its Gaussian size in the presence of monovalent counterions at very low temperatures (that means when the degree of ionization is negligible). These correlations among neutral ion-pairs and between neutral ion-pairs and charged monomers are addressed theoretically by short-ranged, δ-function potentials [1] that lead to a free energy contribution of the form

$$\frac{F_6}{Nk_BT} \sim \frac{4}{3}\left(\frac{3}{2\pi}\right)^{3/2} w_i\delta^2\tilde{l}_B{}^2 \frac{1}{\sqrt{N}} \frac{1}{\tilde{l}_1{}^{3/2}}, \tag{6.41}$$

where w_i (< 0) are temperature dependent parameters, and are different for dipole–dipole and dipole–monopole interactions. These contributions are attractive and would modify the excluded volume interaction (the second term in F_5, Eq. (6.37)). They can significantly reduce the size of the chain only around the isoelectric point ($f \sim 0$), and the type of collapse is generally continuous or second order.

In addition to the short-ranged dipole correlations, there can be long-ranged attraction between monomers mediated by multivalent counterions [28, 29]. This attraction may compensate the residual Coulomb repulsion of the chain even at higher degrees of ionization [30], and the extended conformation of the chain may become unstable. This can as well be treated with the concept of ion "bridging" [27, 31]. It is still not conclusively known what kind of collapse this correlation-induced long-ranged attraction may induce. We briefly mention the short-ranged correlation effects between the dipoles near the isoelectric point in this chapter. In addition, we present an ion-bridging theory based on our model predicting the global instability of a polyelectrolyte chain. One should notice here that the bridging interaction reduces [27] the effective value of the excluded volume parameter w. Therefore, for higher values of w, only very high Coulomb strength or divalent salt concentration will allow the bridging effect to take place. In most of our analysis, we assume w to be high enough to render the bridging effect negligible. Although we assume w to be zero except in Section 6.3.11, choosing nonzero positive values of w only brings minor quantitative changes to our results (and that even only near the isoelectric point) in the "no bridging" scenario. When bridging is included (Section 6.3.11), however, w is a very important parameter affecting the salt concentration or the Coulomb strength required for the first-order transition to the collapsed state.

The total free energy $F = F_1 + F_2 + F_3 + F_4 + F_5$ is given in terms of the fraction of adsorbed counterions and coions (α's), the size of the polymer (l_1), temperature and the bulk dielectric constant (l_B), the degree of polymerization (N), the monomer density (ϱ), monovalent and divalent salt concentrations (c_s), and local dielectric mismatch parameters (δ and δ_2). The goal is to self-consistently determine the fractions of the adsorbed ions (α_1, α_2, and α_3) and the size ($R_g = \sqrt{(Nll_1/6)}$) that minimize the free energy. Note that, for the salt-free case or if only the monovalent salt is present, there are only two variables, α_1 and R_g, which require self-consistent determination. Further, δ_2 does not play any role for monovalent salts. Therefore, it is a simultaneous minimization with respect to two variables

(α_1, R_g) if there are only monovalent counterions, but to four variables ($\alpha_1, \alpha_2, \alpha_3, \tilde{l}_1$) if there are additional divalent counterions. Compared to a neutral system, there are two additional length scales in a charged system. These are the Bjerrum length (l_B) related to the strength of the Coulomb interaction and the Debye length (κ^{-1}) related to the screening of electrostatic forces. In addition, this formalism invokes a third length scale due to the dielectric mismatch parameter δ. In summary, the important parameters that are varied in our analysis are l_B, the salt concentrations (c_{s1} and c_{s2}), and δ.

6.3.2 **Effect of Coulomb Strength on Degree of Ionization and Size**

6.3.2.1 Salt-Free Solutions

We start with an isolated polyelectrolyte chain at low concentrations and at modest temperatures. Our results are first compared directly with the Manning result in Figure 6.10 where f is plotted against $1/\tilde{l}_B\delta = 3, 3.5$, and 4. According to the Manning argument, f is linear with $1/\varepsilon$, until it saturates at unity for $\tilde{l}_B \geq 1$. In contrast, our prediction is that f is a sigmoidal function of $1/\varepsilon$ and ultimately saturates asymptotically at $\tilde{l}_B \to 0$. Our results are in qualitative agreement with experimental deductions [25] on the dependence of f on the dielectric constant of the solvent. The shape of the curves in Figure 6.11a is representative of an adsorption process and is different from the Manning postulate. The dependence of f on $\tilde{l}_B$ is given in Figure 6.11a for $N = 1000$, $\tilde{\varrho} = 0.0005$, $w = 0$, $w_1 = 0$, $w_2 = 0$, at different values of δ. For all values of δ, f decreases monotonically with $\tilde{l}_B$. For a given $\tilde{l}_B$, counterion adsorption is more (that means lower f) if the dielectric mismatch parameter is larger. While the dielectric constant of pure water is about 80, the local dielectric constant in

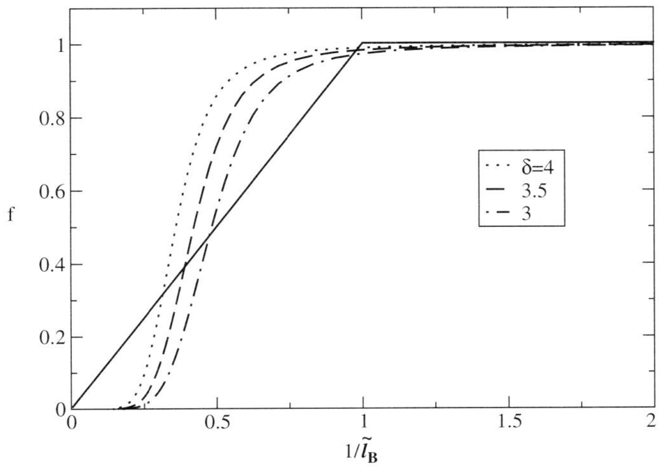

Figure 6.10 Comparison with Manning argument. f versus $1/\tilde{l}_B$ for $\delta = 3$ (dot-dashed), 3.5 (dashed), and 4 (dotted). The solid line is the Manning argument.

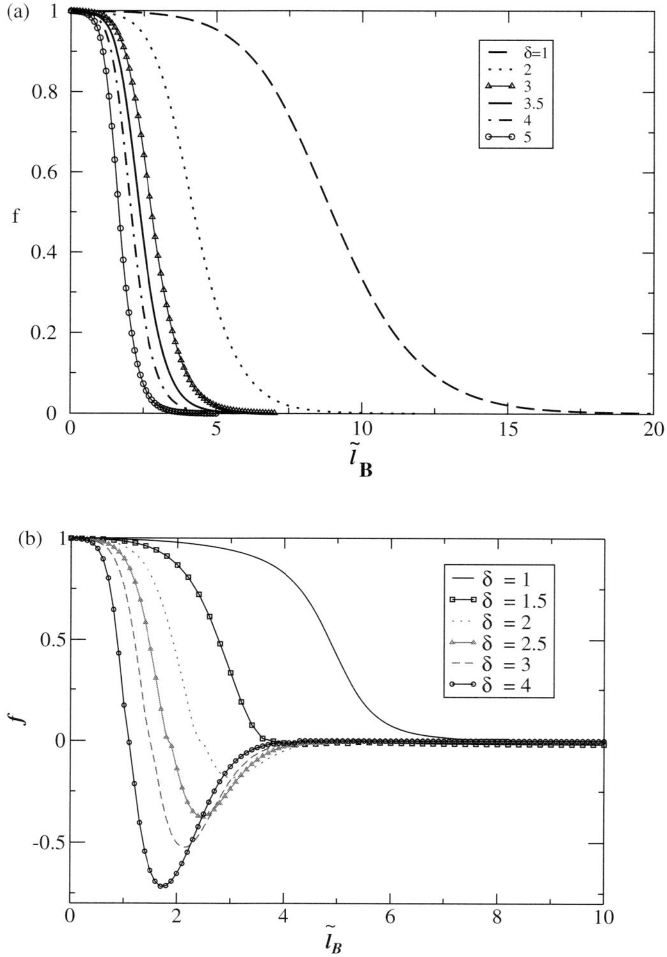

Figure 6.11 (a) Effective charge f and (b) chain expansion factor $\tilde{l}_1$ versus $\tilde{l}_B$ at salt-free conditions with only monovalent counterions for $N = 1000$, $\tilde{\varrho} = 0.0005$, $w = 0$, $w_1 = 0$, and $w_2 = 0$, $\tilde{c}_{s1} = 0 = \tilde{c}_{s2}$. $\delta = 1$ (dashed), 2 (dotted), 3 (triangle), 3.5 (solid), 4 (dot-dashed), and 5 (circle). (c), (d) Divalent salt and overcharging: dependency on $\tilde{l}_B$: the degree of ionization (f) in (c), and the size expansion factor ($\tilde{l}_1$) in (d), of the polyelectrolyte chain plotted against $\tilde{l}_B$ for different values of δ. Parameters are $N = 1000$, $\tilde{\varrho} = 0.0005$, $\tilde{c}_{s1} = 0$, and $\tilde{c}_{s2} = 0.0005$. Collapse and subsequent overcharging occur for higher values of δ. The isoelectric point is reached at lower values of l_B for higher values of δ. The absolute value of overcharging for $\delta = 1.5$-1.7 is less than 2%. The difference in effects of mono- and divalent salt is evident.

the region of ion-pair formation can be smaller [10] by an order of magnitude. Also, the distance d between the ions constituting an ion-pair is comparable to the Bjerrum length, which can be larger than l, the chemical distance between consecutive charges on chain backbone. Consequently, δ can be around 3 for synthetic polyelectrolytes in

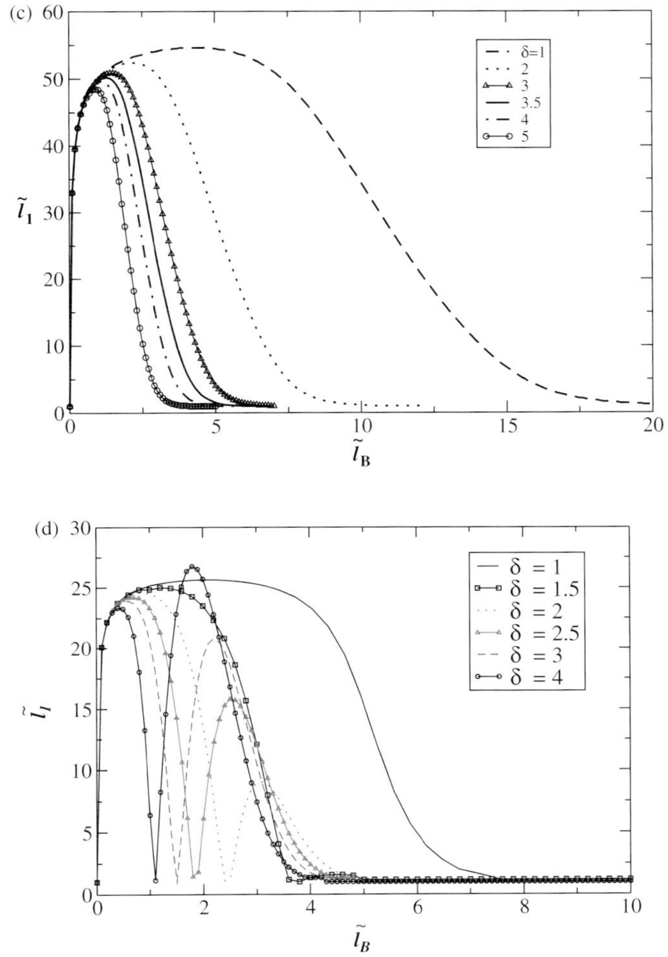

Figure 6.11 *(Continued)*.

aqueous solutions. Although continuum concepts are invoked here, the parameter δ arises from local details and demands a development of a more microscopic theory. In our theory, local dielectric mismatch is accounted for by taking δ as only a parameter. The results of Figure 6.11a clearly show the highly sensitive dependence of f on δ. For the commonly studied synthetic systems such as aqueous solutions of sodium polystyrene sulfonates and polyvinylpyridinium salts, δ is probably around 3.5 and $\tilde{l}_B$ is around 3. Under such assumptions, f is around 0.3. The theoretical profiles have remarkable similarity to the simulation results (Figure 6.6a).

The dependence of the expansion factor $\tilde{l}_1$ for the mean square end-to-end distance on the Coulomb strength $\tilde{l}_B$ given in Figure 6.11b for the same values of N, $\tilde{\varrho}$, w, w_1, w_2, and δ as in Figure 6.11a. The nonmonotonic dependence of $\tilde{l}_1$ on $\tilde{l}_B$ is in agreement with simulation results [8, 9, 27, 32]. The chain is Gaussian ($\tilde{l}_1 = 1$) for $\tilde{l}_B = 0$. As $\tilde{l}_B$ increases, the intrachain electrostatic interaction is manifest and consequently the chain expands. In this regime, the extent of counterion adsorption is negligible due to the dominance of translational entropy of unadsorbed counterions at higher temperatures. When $\tilde{l}_B$ increases further (that means temperature is reduced), more counterions adsorb onto the chain with a reduction of f and a consequent less chain expansion. For higher values of $\tilde{l}_B$, $\tilde{l}_1$ approaches unity (that means Gaussian statistics) in the present case where the short-ranged excluded volume effects are ignored. Comparison to simulation results (Figure 6.2) suggests excellent agreement. As discussed later in the article, the chain can shrink to sizes smaller than the Gaussian size when correlations among the dipoles constituting the ion-pairs are taken into account. It is also evident from Figure 6.11b that the extent of chain expansion depends sensitively on the dielectric mismatch parameter, as in the case of the effective charge of the polymer.

6.3.2.2 **Divalent Salt and Overcharging**

The ion distribution changes qualitatively once a divalent salt is added to the solution. Our theory predicts that charge neutralization and subsequent charge reversal would occur to an isolated flexible polyelectrolyte in aqueous solutions at room temperature and at a modest presence of a divalent salt. The parameter δ in our theory plays an important role in charge reversal induced by counterion adsorption. Temperature is also an important factor regulating the relative weight of the electrostatic energy gain of ion adsorption. To show these effects, we plot the degree of ionization f and the expansion factor $\tilde{l}_1$ of the chain as functions of Bjerrum length l_B (inverse temperature and bulk dielectric constant) for various δ values in Figure 6.11c–d. The other parameters are $N = 1000$, $\tilde{\varrho} = 0.0005$, $\tilde{c}_{s1} = 0$, and $\tilde{c}_{s2} = 0.0005 = \tilde{\varrho}$. The concentrations of the divalent salt and the polymer are chosen to be equal to ensure the availability of enough divalent ions to adsorb on every monomer if physical conditions permit. In Figure 6.11c, we notice that there is a negligible adsorption for $\delta = 1$ (which is the comparable value of δ in simulations [27, 33] (Figure 6.3c–d). Similar to the monovalent case (Figure 6.11a and b), the chain is neutralized only at very low temperatures (there is a factor of two in l_B because the Coulomb energy gain for each ion-pair is twofold for divalent ions). At no temperature there is overcharging for $\delta = 1$. However, for $\delta = 1.5$, we start to see nominal overcharging that is only manifest through a little reswelling of the chain. The absolute value of overcharging is less than 1% at these δ values and we can extend our prediction of the lower threshold value of the dielectric heterogeneity to be 1.5 for the whole range of temperature (threshold $\delta \simeq 1.7$ for $\tilde{l}_B = 3.0$). There is, however, a drastic qualitative change in the dependencies of f and R_g on l_B for δ values 2 and above. At a particular temperature T_0, the chain is neutralized and if T is further reduced, overcharging occurs (and the chain swells). T_0 is higher for higher values of δ as expected (T^{-1} and δ, both favor higher degree of adsorption). The absolute value of the maximum overcharge and the reswelled size increase with δ as well. In particular, the reswelled size is larger than the original swelling for $\delta = 4.0$. This is despite the absolute effective charge being lower at the maximal *reswelling* because, at this point, the Coulomb strength for this large δ value is high enough (as we increase $\tilde{l}_B$) to have repulsion between monomers stronger than that at the point of maximal *swelling*. Another point of note is that for higher temperatures, just as for monovalent counterions, only a fraction of available divalent ions adsorb. The optimal temperature at which the chain re-expansion is maximum shifts to a higher value with higher values of δ. For very low temperatures, sufficient number of negative coions (Cl^-) adsorb to gradually reneutralize the chain for all δ values.

6.3.3 Chain Contraction: Contrasting Effects of Mono- and Divalent Salts

In Figure 6.12, we plot the polyelectrolyte net charge versus increasing divalent salt concentration. The net charge due to the adsorption of divalent counterions becomes negligible as soon as the salt concentration reaches half the polymer concentration ($c_{s2} \sim \varrho/2$). Consequently, the chain shrinks to its Gaussian size at around this isoelectric point. To compare with the case of monovalent counterions, we plot both the degree of ionization f and the expansion factor $\tilde{l}_1$ at $\tilde{l}_B = 3.0$ in Figure 6.12. The other parameters are $N = 1000$, $\delta = 2.5$, and $\tilde{\varrho} = 0.0005$. We notice that for the monovalent salt, degree of ionization f of the polyelectrolyte decreases moderately and monotonically, and never changes sign. Consequently, the size (l_1 or R_g) also decreases monotonically with the Gaussian statistics being obtained only at very high salt concentrations (or at very low temperatures). For the divalent salt, however, the isoelectric point is achieved as soon as there are sufficient number of divalent counterions available to neutralize the chain. That happens at a very low c_{s2}. As a result, the polyelectrolyte shrinks to Gaussian statistics near this isoelectric point.

This continuous shrinkage of a generic polyelectrolyte (NaPSS) chain in water occurs for modest values of δ, at a modest presence of the divalent salt, and at room temperature. This phenomenon has been noticed theoretically [6] and in experiments [34–37] and simulations [27, 33].

6.3.4 Competitive Adsorption of Divalent Salts

To see the effect of gradual overcharging, the concentration of the divalent salt (c_{s2}) is increased, while keeping the concentration of the monovalent salt at zero ($c_{s1} = 0$). We have chosen a higher and a lower value of $\delta = 2.5$ and 1.5, respectively, to illustrate the role of the parameter δ. Generally, at higher values of δ, fractions of adsorbed ion species are expected to increase. We notice that (Figure 6.13) both divalent counterions and negative monovalent coions (α_2 and α_3, respectively) adsorb progressively in higher numbers with increasing divalent salt concentration. The number of monovalent counterions (α_1), however, decreases with increasing c_{s2}. This implies that in this competitive adsorption process, condensed monovalent counterions, when challenged by a divalent salt, are replaced by

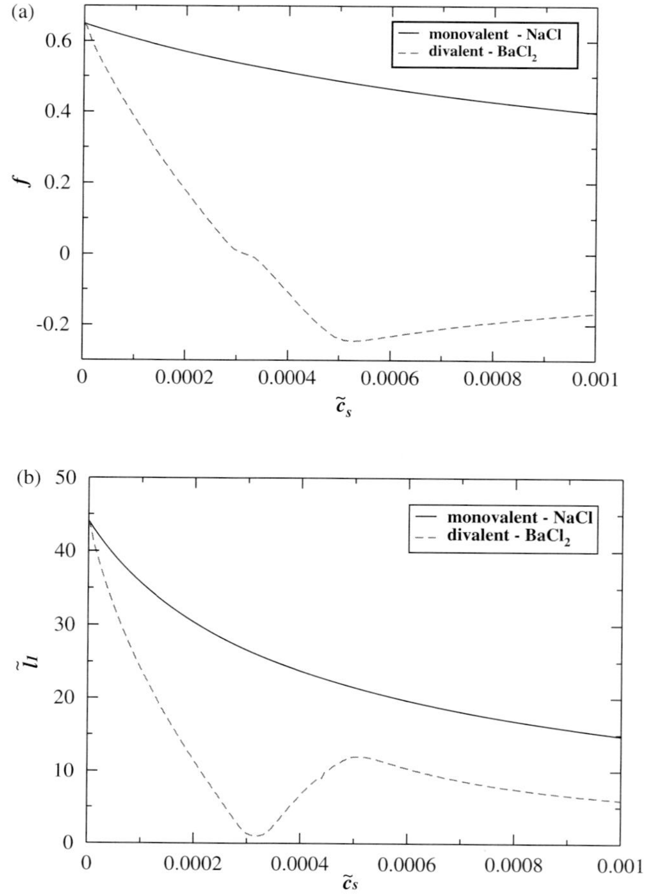

Figure 6.12 Effect of valency of counterions: comparison of the degree of ionization (f) in (a), and the size expansion factor ($\tilde{l}_1$) in (b), of the polyelectrolyte in the presence of either monovalent or divalent salt. $\delta = 2.5$ and other parameters are $N = 1000$, $\tilde{\varrho} = 0.0005$, $\tilde{l}_B = 3.0$, $w = 0$, and $\tilde{c}_{s1} = 0$. $\tilde{c}_{s2} = 0$ is zero when $\tilde{c}_{s1}$ is added and vise versa. Divalent counterions can neutralize and consequently contract the polymer (to a Gaussian chain) under moderate conditions of $\tilde{l}_B = 3.0$ for NaPSS. For monovalent counterions, similar contraction is possible only at very low temperatures. If $\tilde{c}_{s2}$ is increased beyond the isoelectric point, the chain expands due to overcharging.

divalent counterions. This happens for the entire physical range of the dielectric mismatch parameter (as shown below in the diagrams of charged states). The variables chosen in this specific calculation are the degree of polymerization $N = 1000$ and the monomer density $\tilde{\varrho} = \varrho l^3 = 0.0005$ at $\tilde{l}_B = 3.0$ (value related to flexible polymers of the sodium polystyrene sulfonate type in water at room temperature). δ_2 is given by Eq. (6.34) throughout the chapter, unless noted

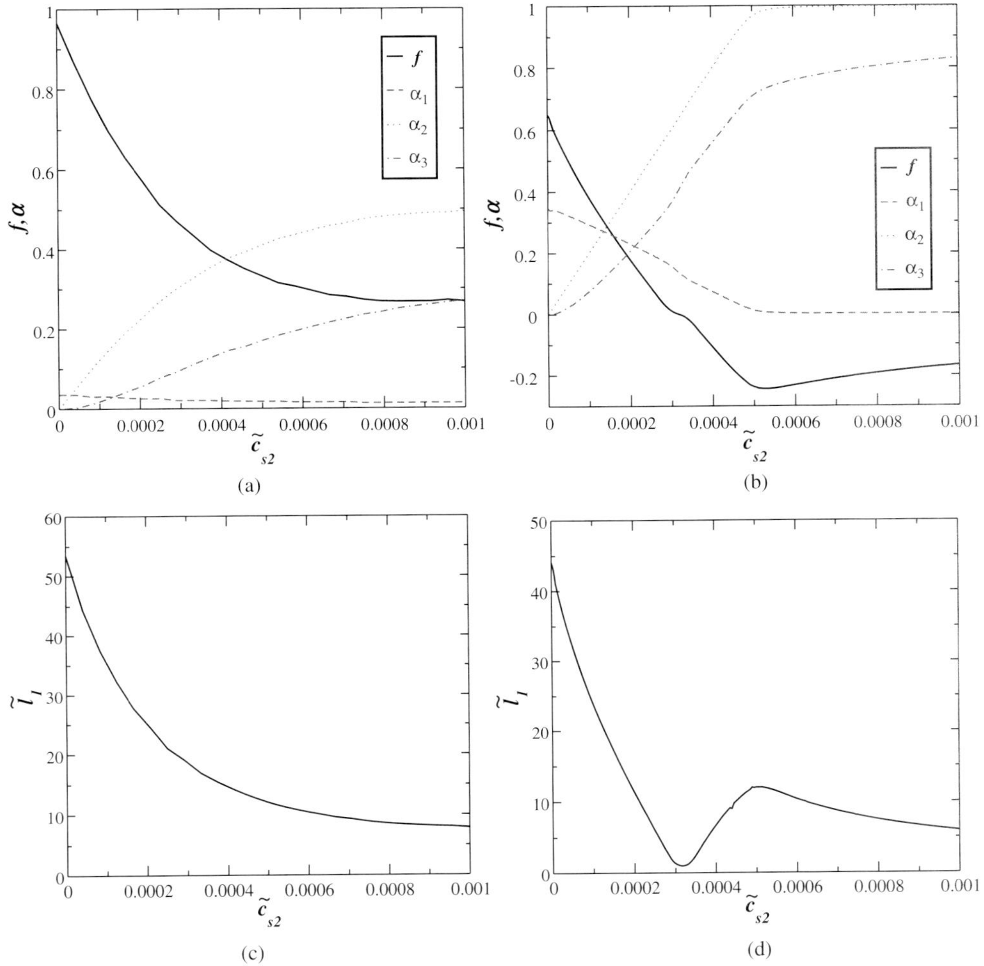

Figure 6.13 Competitive displacement of monovalent counterions by divalent counterions: fractions of adsorbed ions $(\alpha_1, \alpha_2, \alpha_3)$, the degree of ionization $(f = 1-\alpha_1-2\alpha_2+\alpha_3)$ in (a) and (b) for $\delta = 1.5$ and 2.5, respectively, and the expansion factor $(\tilde{l}_1 = 6R_g^2/Nl^2)$ in (c) and (d) for the same δ values plotted against the divalent salt concentration. For a Gaussian chain, $\tilde{l}_1 = 1$. Other parameters are the same as in Figure 6.12. Note, for lower δ, there is no overcharging. For higher δ, almost all divalent counterions adsorb on the chain replacing the monovalent ones. The number of adsorbed coions (α_3) closely follows α_2 for this particular value of δ. The sign of f is reversed (*overcharging*) at some concentration of $BaCl_2$. Near the isoelectric point $(f \sim 0)$, the chain is Gaussian due to minimal electrostatic repulsion. It swells due to overcharging if $\tilde{c}_{s2}$ is further increased.

otherwise. For $\delta = 1.5$, only 5% of the monomers are neutralized by monovalent counterions for the salt-free situation ($c_{s1}, c_{s2} = 0$) (Figure 6.13a) whereas the number increases to 35% for $\delta = 2.5$ (Figure 6.13b). At this higher δ value, almost all available divalent counterions adsorb displacing the monovalent counterions with increasing c_{s2}. Negative monovalent coions (Cl^-) also adsorb on the monomer-divalent ion-pair substantially. f decreases monotonically and *reverses sign* when c_{s2} is still well below the monomer concentration ϱ. At $c_{s2} \sim 80\%$ of ϱ, α_1 drops below 5% and α_2 is about 80% implying that almost all available divalent counterions have adsorbed. The size of the chain decreases steeply (Figure 6.13d) due to rapid neutralization but increases beyond the isoelectric point ($f = 0$) due to repulsion among divalent cations that overcharge the chain. The size is dictated by the third term in Eq. (6.37) at these salt concentrations. For higher values of c_{s2}, more negative coions condense on the chain to marginally reduce the (over)charge of the chain. The number of condensed monovalent ions, however, decreases to zero monotonically.

For the lower value of δ, the original sign of the polyelectrolyte charge (f) is preserved even at higher divalent salts (Figure 6.13a) with the minimum absolute degree of ionization being around 0.27. Consequently, the size of the chain (Figure 6.13c) remains substantially bigger than the Gaussian value for the entire range of salt concentration.

6.3.5 Effect of Dielectric Mismatch Parameter

Both in salt-free and salty cases, the polymer charge depends sensitively on the dielectric mismatch parameter. This is illustrated in the case of only monovalent counterions in Figure 6.14a for $N = 1000$, $\tilde{l}_B = 3$, $\tilde{\varrho} = 0.0005$, $\tilde{c}_{s1} = 0.001$, $\tilde{c}_{s2} = 0$, $w = 0$, $w_1 = 0$, and $w_2 = 0$. In the presence of divalent salts, we further explore the issue of overcharging by plotting f and $\tilde{l}_1$ against δ for $\tilde{l}_B = 3$ in Figure 6.14b and c. The other parameters are the same as in Figure 6.11. At room temperature in aqueous solutions, there can be no overcharging unless δ is higher than a threshold value that is around 1.7 for NaPSS. Only for δ as high as 1.7, the dielectric heterogeneity would be strong enough to electrostatically guide enough divalent cations to be adsorbed and consequently reverse the charge of the chain. The strong sensitivity of the total charge and conformation of the polymer to δ is manifest in Figure 6.14, in which f decreases from about 93% to zero (and subsequently $\tilde{l}_1$ decreases from about 25 to 1, the Gaussian value) for δ changing only from 1 to 1.7. For very high values of δ, Cl^- ions progressively condense at higher numbers to reduce overcharging.

6.3.6 Effect of Monomer Concentration and Chain Length

For a fixed N, as the volume of the system is reduced, the translational entropy of dissociated ions is reduced with a consequent effect of enhanced counterion

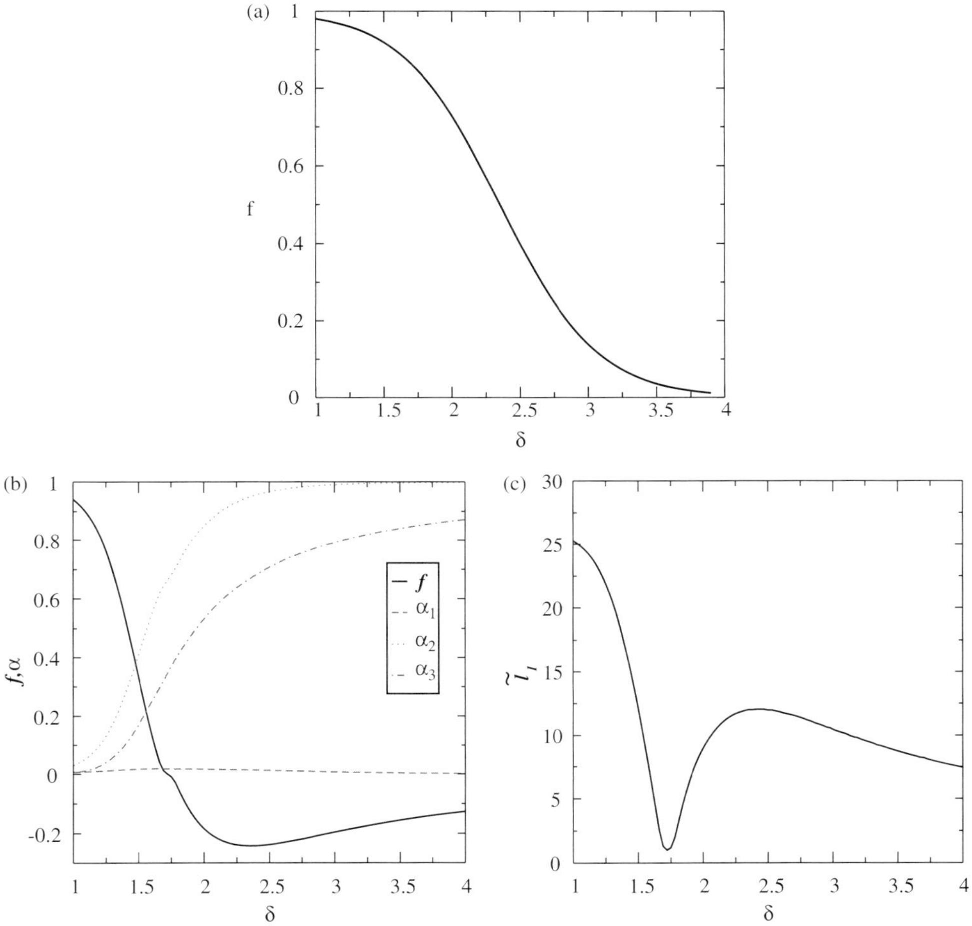

Figure 6.14 Dependency of f on the local dielectric mismatch parameter δ for (a) monovalent salt with $N = 1000$, $\tilde{\varrho} = 0.0005$, $\tilde{c}_{s1} = 0.001$, $w = w_1 = w_2 = 0$. (b and c) Dependency of overcharging on δ: fractions of adsorbed ions ($\alpha_1, \alpha_2, \alpha_3$), the degree of ionization (f) in (b), and the expansion factor ($\tilde{l}_1$) in (c), plotted against the δ-parameter for $\tilde{l}_B = 3.0$. Other parameters are the same as in Figure 6.11c and d. Charge neutralization occurs for $\delta \sim 1.7$; overcharging is possible at this l_B value only if $\delta > 1.7$. Cl^- adsorption (α_3) is higher for higher values of δ, and that reduces overcharging.

adsorption on polymer. Therefore, f and $\tilde{l}_1$ are expected to decrease as $\tilde{\varrho}$ is increased by keeping all other parameters fixed. This result is illustrated in Figure 6.15a and b, where f and $\tilde{l}_1$, respectively, are plotted against $\tilde{l}_B$ for $N = 1000$, $\tilde{c}_{s1} = 0 = \tilde{c}_{s2}$, $\delta = 3.5$, $w = 0$, $w_1 = 0$, and $w_2 = 0$ (simulation results are shown in Figure 6.6a). The solid and dashed curves correspond to $\tilde{\varrho} = 0.0005$ and 0.001, respectively. As the chain length is decreased at a fixed monomer concentration, f and $\tilde{l}_1$ decrease as illustrated in Figure 6.16a and b, where $\tilde{\varrho} = 0005$, $w = 0$, $w_1 = 0$, $w_2 = 0$, and the solid and

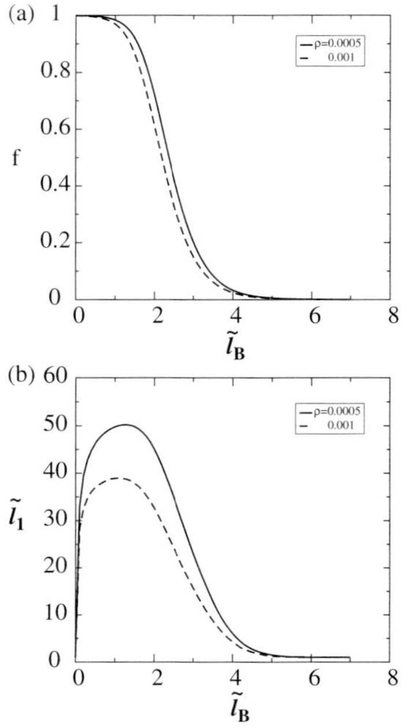

Figure 6.15 Effect of monomer concentration. $\tilde{\varrho} = 0.001$ (dashed) and 0.005 (solid). (a) f versus $\tilde{l}_B$. (b) $\tilde{l}_1$ versus $\tilde{l}_B$. These are under salt-free conditions with only monovalent counterions.

dashed curves correspond, respectively, to $N = 1000$ and 500 (simulation results are in Figure 6.6b). In the present theory, the effects of N and $\tilde{\varrho}$ are nontrivially coupled. This is evident in Figure 6.16 by a comparison between the solid curves ($N = 1000$ and $\tilde{\varrho} = 0.0005$) and dot–dashed curves ($N = 500$ and $\tilde{\varrho} = 0.001$), where $N\tilde{\varrho}$ is fixed at 0.5.

6.3.7
Free energy Profile

One of the advantages of our equilibrium adsorption theory is that it is possible to compare the contributions of different factors in the total free energy (F_1 to F_5) as functions of the critical parameters. The major conclusion of the theory [1] has been that the equilibrium distribution of the adsorbed counterions and the size of the polyelectrolyte are determined essentially by a competition between the translational entropy of dissociated ions and the Coulomb energy gain of adsorbed ions. This is indeed borne out by our calculation in the presence of divalent salts too, as shown in Figure 6.17. In Figure 6.17a, the separate parts of the free energy are plotted against the Bjerrum length $\tilde{l}_B$ for a fixed divalent salt concentration ($\tilde{c}_{s2} = 0.0005$, equal to the monomer density) and for a specific strength of dielectric mismatch ($\delta = 2.5$). The major contributions to the total free energy come from the translational entropy F_2 (Eq. (6.25)) and the adsorption energy F_4 (Eq. (6.31)). For higher temperatures

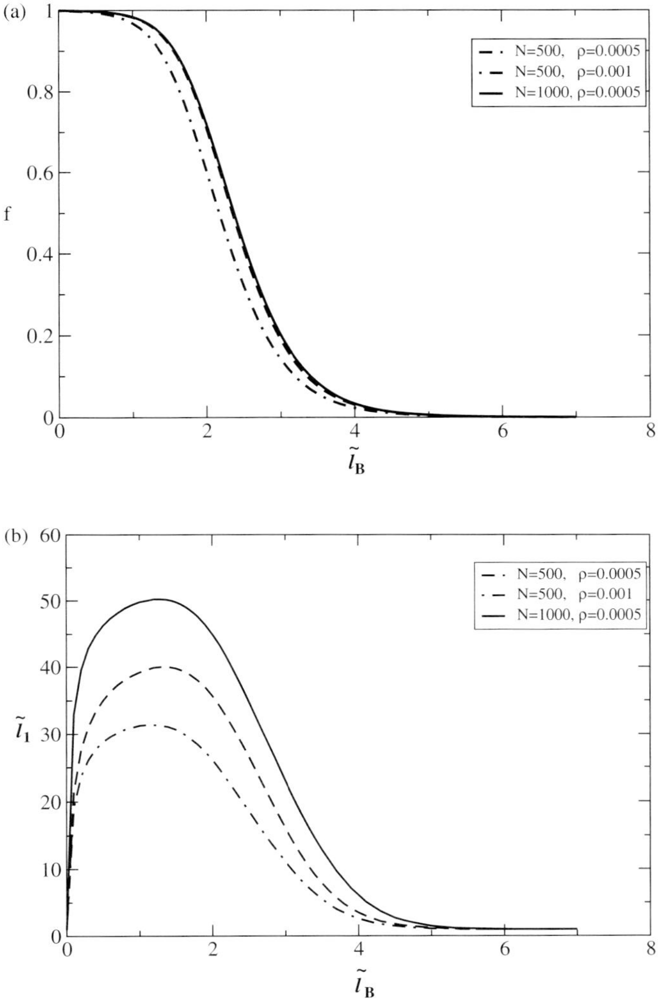

Figure 6.16 Effect of chain length in (a) f versus $\tilde{l}_B$, (b) $\tilde{l}_1$ versus $\tilde{l}_B$ (dashed, $N = 500$, $\tilde{\rho} = 0.0005$), (dot-dashed, $N = 500$, $\tilde{\rho} = 0.001$), and (solid, $N = 1000$, $\tilde{\rho} = 0.0005$). These are under salt-free conditions with only monovalent counterions.

(lower l_B), the entropic term is favored as electrostatics remains negligible compared to thermal fluctuations. For lower temperatures, electrostatics becomes progressively relevant, and many ions condense reflecting substantial gain in F_4. The entropic contribution F_1 (Eq. (6.21)) related to the mobility of condensed ions along the backbone has negligible effect, and so does the Debye–Hückel contribution F_3 (Eq. (6.30)) at these salt concentrations.

In Figure 6.17b, similar free energy components are plotted against δ at the same salt concentration and for $\tilde{l}_B = 3.0$. The curves in (a) and (b) are remarkably similar

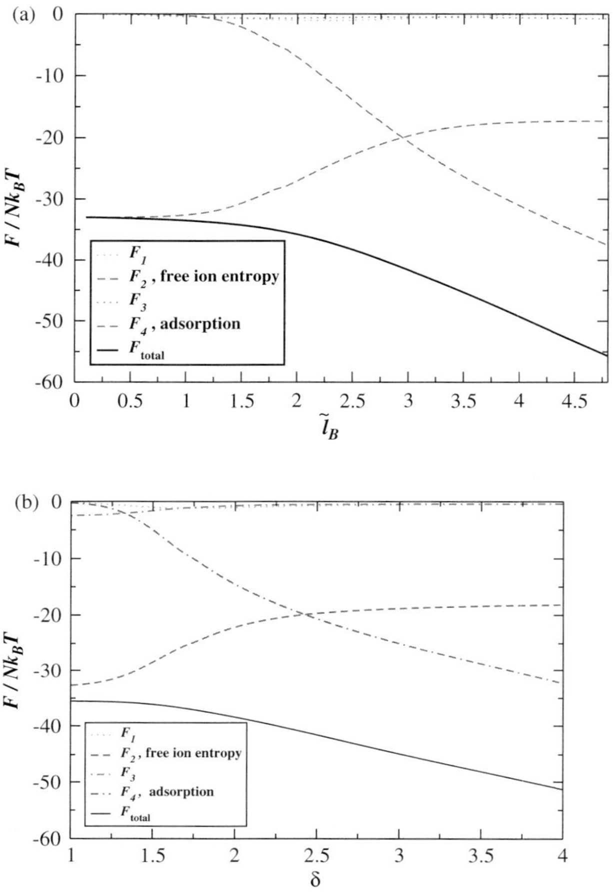

Figure 6.17 Contributions to free energy: separate parts of the free energy for a fixed divalent salt concentration ($\tilde{c}_{s2} = 0.0005$) as functions of (a) the Bjerrum length $\tilde{l}_B$ and (b) the dielectric mismatch strength (δ). For (a), $\delta = 2.5$ and (b), $\tilde{l}_B = 3.0$. Parameters are $N = 1000, \tilde{\varrho} = 0.0005, \tilde{c}_{s1} = 0$. Energies are F_1 (dotted) = entropy of mobility along the chain, F_2 (dashed) = translational entropy of mobile ions, F_3 (dot-dashed) = Debye–Hückel correlation between mobile ions, F_4 (dot-dashed-dashed) = Coulomb attraction between adsorbed ions, and F_{tot} (solid) = total free energy. For a fixed salt concentration, the major competition is between the translational entropy (increases with temperature) and the Coulomb attraction (increases with both $\tilde{l}_B$ and δ). The similarity in the roles of $\tilde{l}_B$ and δ is evident.

demonstrating the equivalence of the parameters l_B and δ. According to this adsorption theory, reduction of any of temperature, the bulk dielectric constant ε, or the local dielectric constant ε_l (near the hydrophobic regions of the chain backbone) by a similar factor would induce very similar effects to polyelectrolyte behaviors. This is especially valid for modest values of l_B and δ.

6.3.8
Diagram of Charged States: Divalent Salt

A typical state diagram for polyelectrolyte charge is predicted (Figure 6.18a) in which below a critical Coulomb strength (proportional to $1/T$, $1/\varepsilon$ or δ) there is no overcharging with increasing divalent salt. Above this critical strength, electrostatics is strong enough to adsorb sufficient number of counterions inducing overcharging. For higher salt concentrations, we expect recharging of the polyelectrolyte due to either screening of electrostatics [38] (for low Coulomb strengths) or coion (Cl^-) condensation (for high Coulomb strengths). The dashed line indicates that in this regime of the salt concentration, the theory only predicts qualitative results. Our theory shows that overcharging is an outcome of both correlation-induced adsorption related to the discreteness of divalent cations and dielectric heterogeneity related to the local chemical structure of polyelectrolytes.

Following this proposed, tentative state diagram of the total charge f, we continue to present the actual state diagrams calculated from our theory as functions of three major variables – the Bjerrum length (l_B), the divalent salt concentration (c_{s2}), and the dielectric mismatch parameter (δ). In what follows, one of these variables is fixed and the diagram of states (regions of negative and positive degrees of ionization) is calculated numerically as functions of the other two. Figure 6.18b–d describes the complete limiting charged states, parts of which have already been discussed in detail in the preceding subsections.

In Figure 6.18b, the calculated state diagram at $\tilde{l}_B = 3.0$ is presented as a function of the divalent salt concentration $\tilde{c}_{s2}$ and the dielectric mismatch parameter δ. The state diagram is qualitatively similar to the proposed one (Figure 6.18a), with the strength of the Coulomb interaction being represented by δ (Eq. (6.31)), $\tilde{l}_B$ being fixed. To explain the diagram, we first choose a specific value of $\delta = 2.5$ (see Figures 6.12 and 6.13b) and monitor the charged state with increasing divalent salt concentration. For low salt, there are not enough divalent counterions (Ba^{2+}) to neutralize the chain and the polyelectrolyte preserves its sign of charge (state A) of salt-free conditions. At around $c_{s2} \sim \varrho/2$, which is half the monomer concentration, the charge of the polymer becomes zero (on the locus of first isoelectric points – the solid line). If c_{s2} is increased further, the polymer charge is reversed (state B), and at around $c_{s2} \sim \varrho$, almost all monomers are neutralized by divalent counterions. The charge reversal is maximum at around this point (on the locus of maximum overcharging points – the dotted line). With c_{s2} increasing even further, more coions (Cl^-) are available in the solution and some of them adsorb on the monomer-Ba^{2+} ion-pairs to reduce the degree of overcharging (state C). The first isoelectric points between states A and B are reached at a higher c_{s2} for a lower δ because a higher fraction of divalent counterions would remain dissociated in the solution due to a lower Coulomb energy gain. For values of δ higher than $\simeq 3$, a substantial fraction of monovalent counterions of the polymer (Na^+) too remain adsorbed on the chain and the chain charge is neutralized with fewer divalent counterions. If δ is less than $\simeq 1.7$, the state of overcharging (state B) is never reached and with increasing salt concentration, the polymer charge goes through a minimum (on the locus of points

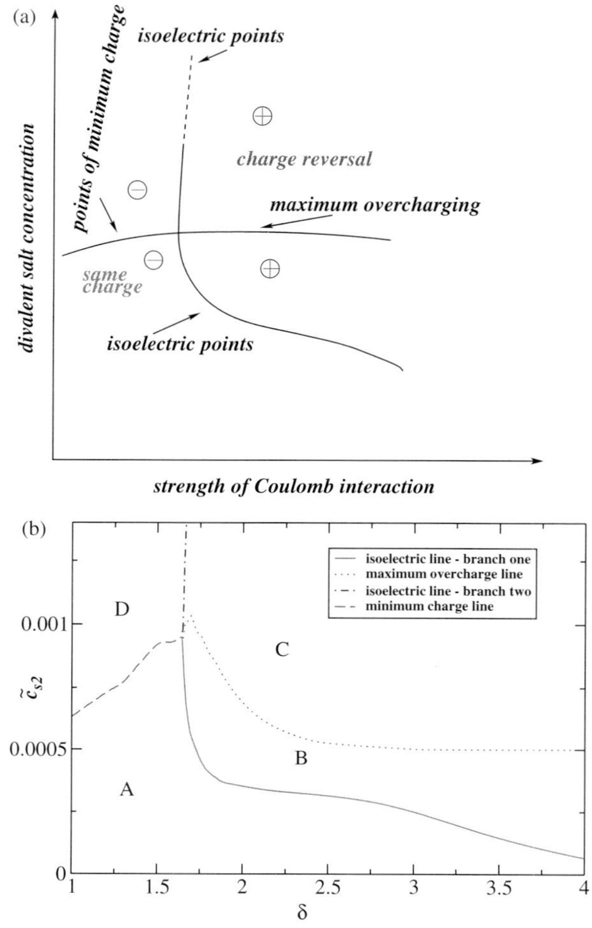

Figure 6.18 (a) Sketch of the charged states for an isolated polyelectrolyte chain (of NaPSS type) in dilute solutions (water) in the presence of a divalent salt (of $BaCl_2$ type) as functions of the Coulomb strength ($l_B \sim (\varepsilon T)^{-1}$) and the salt concentration. Points left to the isoelectric line (on which the net effective charge (degree of ionization) of the polymer is zero) correspond to states in which the sign of polymer charge is unchanged (negative). However, there is a locus of points for intermediate values of the salt concentration at which the net charge is a minimum. Right to the isoelectric line, the effective polymer charge is reversed (positive). If the isoelectric point is crossed along the line of minimum charge from left to right, it becomes the line of maximum overcharging. Dashed part of the isoelectric line is beyond the validity of the Debye–Hückel theory employed here. (b) The state diagram of the effective charge of the polymer (f) for $\tilde{l}_B = 3.0$ as functions of the dielectric mismatch δ and the divalent salt concentration $\tilde{c}_{s2}$. Parameters are $N = 1000$, $\tilde{\varrho} = 0.0005$, $\tilde{c}_{s1} = 0$. Charged states are A, D – negative; B, C – positive (note: original polymer charge is negative). Lines are isoelectric branch one (solid), maximum overcharging (dotted), isoelectric branch two (dot-dashed), and minimum charge (dashed); (c) the state diagram of f at a fixed dielectric mismatch strength ($\delta = 2.5$) as functions of the Bjerrum length $\tilde{l}_B$ and the divalent salt concentration $\tilde{c}_{s2}$. Other parameters, states, and lines are the same as in (c). (d) The state diagram of the total charge density on the polymer (f) at a fixed divalent salt concentration ($\tilde{c}_{s2} = 0.0005$) as functions of the dielectric mismatch strength δ and the Bjerrum length $\tilde{l}_B$. Other parameters are the same as in (b), (c). Charged states are A – negative; B, C – positive; and C1 – zero. Lines are isoelectric (solid), maximum overcharging (dotted), and zero charge (dashed and dot-dashed).

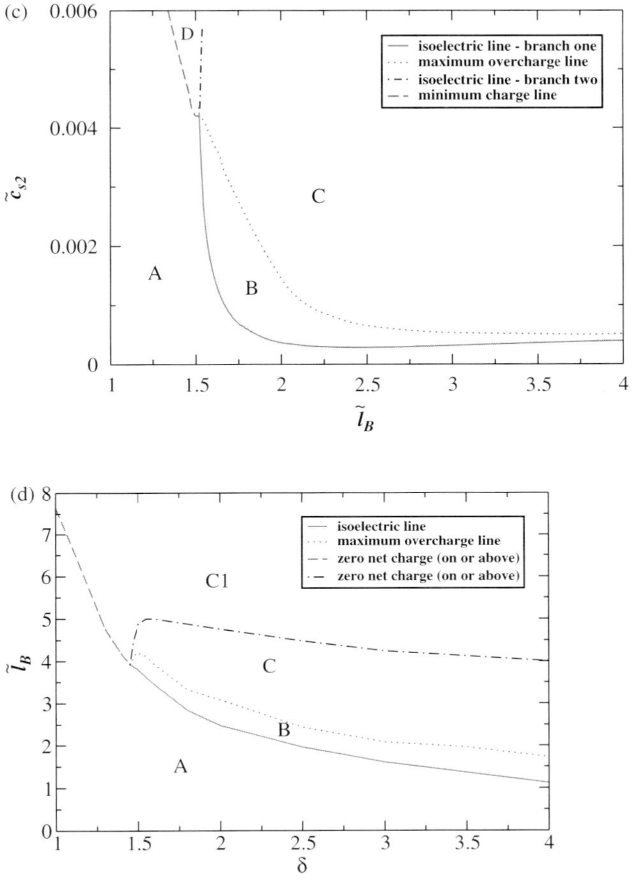

Figure 6.18 *(Continued)*.

of minimum charge – the dashed line) before increasing again due to Cl^- ion adsorption. The line of minimum charge (for δ less than $\simeq 1.7$) expectedly continues to be the line of maximum charge reversal (for δ greater than $\simeq 1.7$). For very high salt concentrations, the Coulomb interaction is progressively screened and all adsorbed ions begin to rejoin the solution (not included in the state diagram). We must, however, be cautious that the DH approximation (and consequently F_3 in Eq. (6.30)) is not valid at this high salt regime. A salt concentration for which the Debye length is equal to the Bjerrum length ($\varkappa^{-1} \geq l_B$) can be tentatively set as the highest limit of validity of the DH theory. For divalent salts, it turns out to be (see Eq. (6.29))

$$c_{s2-\max} \simeq \left(24\pi l_B^3\right)^{-1}. \tag{6.42}$$

The steepness of the state boundary (the locus of second isoelectric points, the dot-dashed line) implies that the polymer charge becomes zero again (only applicable for $\delta > 1.7$) due to redissolution of condensed ions at least an order of magnitude higher in salt concentrations.

The state diagram as a function of l_B and c_{s2} for a fixed value of $\delta = 2.5$ is presented in Figure 6.18c. The diagram is qualitatively similar to the previous one, although a much higher salt concentration (note the difference in the scale of the coordinate) is needed to reach the line of minimum charge (dashed) and the line of maximum overcharging (dotted) at low values of $\tilde{l}_B$ (higher temperatures). In this regime, electrostatics becomes progressively weaker with increasing temperature, and consequently lower fractions of available divalent ions condense. Regarding this diagram too, the degree of ionization and overcharging (absolute value of f) can be obtained for the particular value of $\tilde{l}_B = 3.0$ from Figure 6.13b. The slopes of the isoelectric lines (branch one) in Figure 6.18b and c are opposite at high Coulomb strengths because in the latter case a higher degree of Cl^- adsorption at high l_B requires more divalent salt to neutralize the chain.

The state diagram as a function of $\tilde{l}_B$ and δ for a fixed salt concentration $\tilde{c}_{s2} = 0.0005$ (equal to the monomer concentration) is presented in Figure 6.18d. One essential characteristic is that the degree of ionization remains steadily zero (state C1) above a certain value of $\tilde{l}_B$ (that means below a certain temperature) because the Coulomb attraction is strong enough to form the monomer-Ba^{2+}-Cl^- ion-triplet on every monomer location. This critical value of $\tilde{l}_B$ (on the lower boundary of the zero charge state, the dot-dashed line) generally decreases with higher values of δ (higher electrostatic energy gain). The magnitude of the degree of ionization as a function of $\tilde{l}_B$ for fixed values of δ can be obtained in Figure 6.11c and d, and as a function of δ for a fixed value of $\tilde{l}_B$ in Figure 6.14b and c. Both figures can be analyzed in conjunction with this state diagram. It is worth noting that there will be no overcharging regardless of how high l_B is if δ is less than a critical value ~ 1.50 (see Figure 6.11). We previously had established the critical δ to be around 1.7 for $\tilde{l}_B = 3.0$. It is to be noted that although we see overcharging for δ between 1.5 and 1.7 for $\tilde{l}_B > 3.0$, the actual absolute value of excess charge is negligible (less than 2%) in this range.

6.3.9
Effect of Ion-Pair Correlations

The general consequence of interactions among ion-pairs and those between ion-pairs and charged monomers is to introduce a negative short-ranged excluded volume effect with its magnitude varying as $\delta^2\tilde{l}_B^2$. Therefore, this contribution becomes progressively more important at higher $\tilde{l}_B$ (that means lower temperature). The chain size becomes smaller than the Gaussian size (that means $\tilde{l}_1 < 1$) at higher $\tilde{l}_B$ values, as illustrated in Figure 6.19, for $N = 1000$, $\tilde{\varrho} = 0.0005$, $\tilde{c}_{s1} = 0 = \tilde{c}_{s2}$, $\delta = 3.5$, $w = 0$, $w_1 = -0.01$, and $w_2 = 0$. Similarly, chain shrinkage occurs for $w < 0$ and $w_2 < 0$. In obtaining the curve with chain collapse, a term with repulsive three-body interactions

$$F_{5,\text{three-body}} = \frac{w_3}{N\tilde{l}_1^3} \qquad (6.43)$$

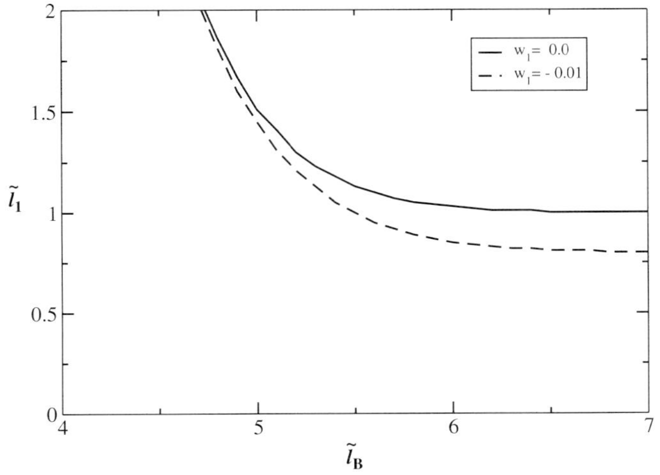

Figure 6.19 Ion-pair correlations shrink polymer size. $\tilde{l}_1$ versus $\tilde{l}_B$ for $N = 1000$, $\tilde{\varrho} = 0.0005$, $\tilde{c}_{s1} = 0 = \tilde{c}_{s2}$, $\delta = 3.5$, $w = 0$, and $w_2 = 0$, $w_1 = -0.01$ (dashed), and $w_1 = 0$ (solid).

is included in the free energy (Eq. (6.37)) of the chain. The solid curve in Figure 6.19 is the same as in Figure 6.11b, without ion-pair correlations.

6.3.10 Collapse in a Poor Solvent

It has already been remarked that the free energy described in Eqs. (6.21)–(6.41) is for a single polyelectrolyte chain in a dilute solution, and it is valid for concentrations of salt not too high (so that $\kappa^{-1} \geq l_B$, i.e., $c_s \leq (8\pi l_B^3)^{-1}$ for a monovalent salt). It is valid, however, for all temperatures, and can easily be modified for any degree of ionization (or ionizability) [39] of the polymer. In addition, the free energy is equally applicable for multichain systems in infinitely dilute solutions in which the chains have negligible interchain interaction (either excluded volume or electrostatic). Qualitative analysis of the free energy shows that the size and charge of the polyelectrolyte chain are primarily determined by the energy gain of ion-pairs (which is linearly proportional to an effective Coulomb strength $(\tilde{l}_B\delta)$) relative to the translational entropy of the mobile ions in the expanded state, and by the relative strength of w to w_3 in the collapsed state. We further use this theoretical model to explain the experimental data [39] for the collapse of polyelectrolyte chains in a poor solvent. The parameter $(\varepsilon_l d)^{-1}$ in Eq. (6.31) is the only adjustable parameter taken to fit the experimental data. One notes that calibration of the excluded volume parameter w by the respective uncharged chain is necessary to eliminate the uncertainty in determining the nonelectrostatic interactions in charged polymers and that can be performed by setting $w_3 = 0 = f$ in Eq. (6.37) (with Eq. (6.43)). Minimizing F_5 with respect to $\tilde{l}_1$, in this case, yields the familiar formula for chain expansion

$$\alpha^5 - \alpha^3 = \frac{4}{3}\left(\frac{3}{2\pi}\right)^{3/2} w\sqrt{N}, \tag{6.44}$$

where $\alpha^2 = \tilde{l}_1$. The functional dependence of the nonelectrostatic parameter w on the solvent composition is established first by using Eq. (6.44) to determine w from the expansion factor of the uncharged polymer chain. Later, by performing the double minimization of the free energy of the charged polymer, α and f were determined with $(\varepsilon_1 d)^{-1}$ as a parameter and then compared with the experimental data.

In order to quantify the solvent quality of the uncharged polyvinylpyridine (PVP) the expansion factor $\alpha = R_g/R_{g,\theta}$ and the second virial coefficient A_2 were measured as functions of w_{ns}, the nonsolvent volume fraction, and compared to the theory (Eq. (6.44)), yielding the solvent quality parameter w as a function of w_{ns}. Note that Eq. (6.44) is valid only for $w \geq 0$. For $w < 0$, we used values of w linearly extrapolated from its value at the θ-condition. θ-Dimensions ($A_2 = 0$, $R_{g,\theta} = 29$ nm) were observed at $w_{ns} = 0.91$. Below theta dimensions ternary interactions were included (Eq. (6.43)). As mentioned in the theory, a nonzero positive value of w_3 was required to stabilize the chain collapse below θ-dimensions. A fixed value of $w_3 = 0.00165$ was chosen for both the uncharged and charged polymers used in our experiments.

In Figure 6.20, the expansion factor, $\alpha = R_g^{app}/R_{g,\theta}$, of the uncharged PVP chains (a) and of the polyions (b) is plotted versus w_{ns} along with the theoretical prediction. Quantitative agreement is observed except close to the phase transition where the experimental data show a broader phase transition regime as discussed in some detail, below.

When the same dependency of w is used on w_{ns}, the expansion factor α for the charged chain is fitted by the theoretical curve, with only one adjustable parameter, $(\varepsilon_1 d)^{-1} = 0.183\ \text{nm}^{-1}$, which reflects the local dielectric constant in the vicinity of the polyion backbone, ε_1. Because the ion-pair energy ($\tilde{l}_B\delta$) and the temperature remain constant for the entire range of the experiment, the effective charge f has negligible variation in the expanded state. The chain free energy (Eq. (6.37) with Eq. (6.43)) in this state is dominated by the electrostatic term and, consequently, the theoretical chain dimension (corresponding to an approximately constant charge density) varies little. Nevertheless, one notes that with increasing proportions of w_{ns} (the nonsolvent fraction) there is a slight increase in chain size due to a small increase in the value of Bjerrum length (with decreasing dielectric constant) that marginally enhances the intrachain monomeric repulsion captured in the third term of the free energy (Eq. (6.37)). This small increase in the dimensions predicted by theory is smaller than the experimental uncertainty for the R_g determination in the regime $0 < w_{ns} < 0.8$. However, the size and shape of the chain undergoes a drastic change at a threshold poorness of the solvent. Beyond the threshold poorness the chain collapses, and that leads it to collect its counterions.

It must be noted that the theory presented above predicts a first-order coil-globule transition for the chains if the excluded volume parameter w is smaller than a certain threshold value and further provided that the three-body interaction parameter w_3 is also smaller than a critical value. The strength of the three-body interaction parameter pertinent to our analysis is substantially lower than the critical value (contrast this

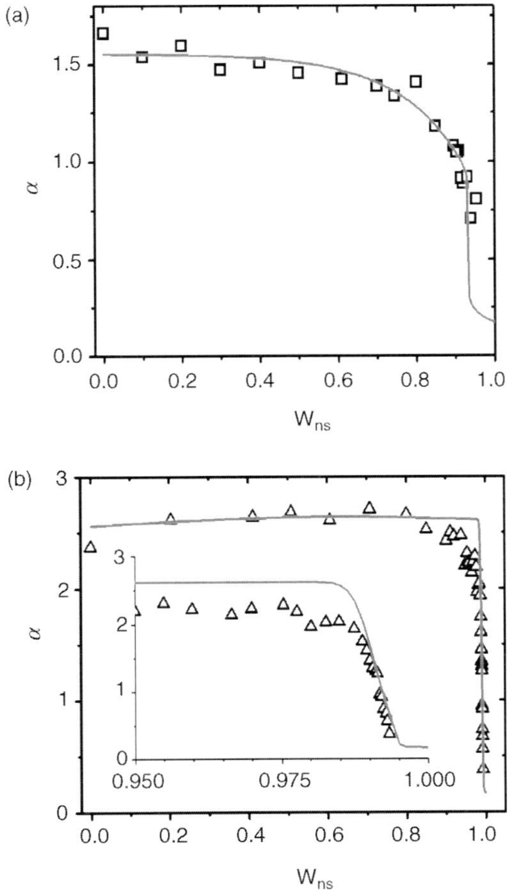

Figure 6.20 Expansion factor α is plotted against the volume fraction of the nonsolvent, w_{ns}, for the neutral polyvinylpyridine (a) and for the charged QPVP4.3 (4.3% quarternized) sample (b). The lines represent the fit according to Eq. (6.44), with $f = f_m = 0$ and $w_3 = 0.00165$ (a) and by minimizing the five contributions to the free energy, as described in the theoretical part, with $(\varepsilon_l d)^{-1} = 0.183\ \text{nm}^{-1}$ as the only fit parameter.

with Figure 6.19) and, hence, the theory predicts a first-order phase transition sharper than the relatively broad transition regime observed in the experiments. One should note that polydispersity in the chain length (N) and in the maximum degree of ionization ($f_m = N_c/N$, where N_c is the number of ionizable monomers) could broaden the transition due to a distribution of the threshold value of w. While the chain length distribution has little effect (data not shown), the variation of the number of charges per chain at constant chain length assuming a Gaussian distribution was utilized for the fit shown in Figure 6.20b. So far, no explanation can be given for the experimentally observed small decrease in α in the regime $0.8 < w_{ns} < 0.985$. The value of $0.183\ \text{nm}^{-1}$ for $(\varepsilon_l d)^{-1}$ is equivalent to $\varepsilon_l = 10.9$) if

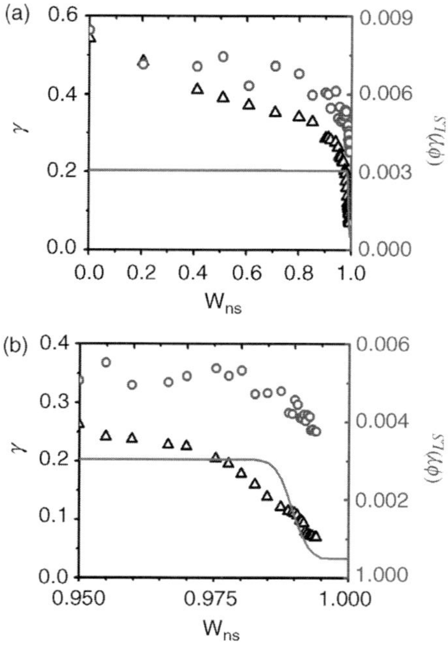

Figure 6.21 (a) Effective charge density $\gamma = f/f_m$ (triangles, left scale) and the osmotic coefficient (circles, right scale) as a function of the nonsolvent fraction, w_{ns}. The solid curve shows the theoretical charge density $\gamma = f/f_m$. (b) Magnification, symbols as in (a).

the dipole length is assumed to be 0.5 nm. This value of ε_l is in between the dielectric constant ($\varepsilon_l = 8.33$) of 2-ethylpyridine, which is chemically close to the chain backbone and that of the solvent ($16 < \varepsilon < 21$). However, given the uncertainty in the dipole length d, that is, the mean distance of the bound counterions from the respective polyion charges (which can vary from 0.3 nm to a few nm), the value for ε_l given above should not be overinterpreted. Nevertheless, we notice that the value of $(\varepsilon_l d)^{-1}$ is remarkably close to the value 0.175 nm^{-1} estimated for polymers of sodium polystyrene sulfonate type in the original theory [1].

In Figure 6.21, the fraction of the effective charges $\gamma = f/f_m$ derived from the experimental conductivity data is compared with the theoretically predicted charge density obtained through the double minimization of the free energy (Eqs. (6.21)–(6.37) and (6.43)). The observed qualitative agreement was to be expected in view of the perfect match of the expansion factor shown above. As mentioned before, the theoretical charge density in the expanded state is found to be virtually constant due to the absence of variation in the effective Coulomb strength $\tilde{l}_B\delta$. Note that this happens despite the somewhat decreasing value of the bulk dielectric constant with increasing w_{ns} because the Coulomb strength relevant to the ion-pair energy depends only on the local (not the bulk) dielectric constant related to the material of the polymer backbone. Again the experimental data show a broader phase transition regime, but the location of the phase boundary where no free counterions exist is well reproduced. The theory

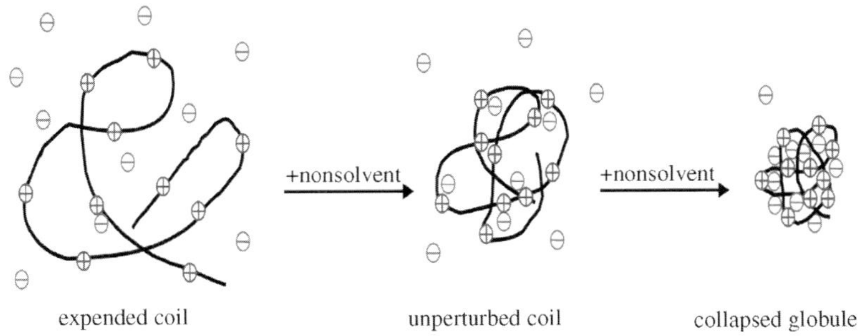

Figure 6.22 Schematic of the collapse of a polyelectrolyte chain in a poor solvent. The chain collects all its counterions as it collapses.

predicts a first-order coil-globule transition for both, size and effective charge of the polymer chain.

Approaching the phase transition the polyion chain starts to collect and bind its counterions as the chain dimensions become successively smaller (Figure 6.22). Eventually, the collapsed polyion chain preserves a few charges only, most probably some surface charges known from colloids. This experimental observation is in remarkable qualitative agreement with the results of explicit solvent simulations [40]. Interestingly, the polyion mobility is already significantly reduced well before the unperturbed θ-dimension is reached. The obvious strong charge reduction in a regime where the Bjerrum length changes only by 5% questions the applicability of the Manning condensation concept [2, 41] to flexible polyelectrolyte chains at least for poor solvent conditions.

The combination of conductivity and light scattering measurements is well suited to investigate cooperative effects of counterion binding and chain collapse mediated by solvent quality and electrostatic interaction. Because the dielectric constant of the solvent remains virtually constant during the chain collapse, the counterion binding is entirely caused by the reduction in the polyion chain dimension. Remarkably, the counterion binding occurs already well above the theta dimension of the polyion that was also reported for the Sr^{2+}-induced collapse of sodium polyacrylate (NaPA) in aqueous sodium chloride solution [42]. The theory of uniform collapse induced by concomitant counterion binding agrees quantitatively with the location of the phase boundary, but does not properly reproduce the width of the transition as mentioned above. Besides, possible anisotropic chain conformation-specific ion-solvation effects could also be the origin of the observed discrepancy.

6.3.11
Bridging Effect: Divalent Salt

We now consider the bridging configuration of nonbonded monovalent monomers mediated by divalent counterions. Let us assume that some of the adsorbed divalent ions participate in bridging. When bridging is included, the minimization of the free energy needs to be performed with respect to five variables. In our theory, we will address the bridging interaction as an effective two-body interaction between monomers with uncompensated charge. A pronounced shrinking of individual polyelectrolyte coils (such as polyacrylate chains [34]) in the presence of divalent ions has long been observed in light-scattering experiments and viscosity measurements. The shrinkage has been interpreted as intramolecular coil collapse [34, 43–46] induced by ion-bridging that has further been addressed [43, 47] as a hydrophobic intramolecular interaction. In response to an increasing salt to polymer ratio, a gradual shrinking of the polymer chains was observed [44] to be followed by a sharp collapse to a compact spherical shape (R_g/R_h going down from 1.5 to 0.8). To theoretically formulate this phenomenon, we note that a bridge formed by a divalent counterion is tantamount to a cross-link junction of functionality four [22, 27], which in turn can be treated as an attractive two-body interaction of local nature (like two-body excluded volume interaction). Therefore, in the presence of the bridging effects induced by divalent counterions, w in Eq. (6.37) is replaced by,

$$w' = w + \frac{E_{br}}{k_B T}\alpha_{2b}, \tag{6.45}$$

where α_{2b} is the ratio of the number of divalent ions that participate in bridging (M_{2b}) to the number of monomers (that means $\alpha_{2b} = M_{2b}/N$). E_{br} is the attractive energy associated with one bridge, and hence is negative. To calculate E_{br} the relevant dielectric constant should be the local one (ε_l) since the divalent cation in the monomer cation–monomer charge complex sits between and in the vicinity of both monomers (Figure 6.9b). With the definition [1] $\delta = \varepsilon l/\varepsilon d$, we have

$$\begin{aligned} E_{br} &= -\frac{4e^2}{4\pi\varepsilon_l d} + \frac{e^2}{4\pi\varepsilon_l \cdot 2d} \\ &= -\frac{7}{2}\tilde{l}_B \delta k_B T. \end{aligned} \tag{6.46}$$

We must add the third virial term in the chain free energy (F_5, Eq. (6.37)) to maintain stability in the system in the case of a negative w'. Combining Eqs. (6.45) and (6.46), therefore, the chain free energy takes the form,

$$\begin{aligned} \frac{F_5}{Nk_BT} &= \frac{3}{2N}(\tilde{l}_1 - 1 - \log\tilde{l}_1) + \frac{4}{3}\left(\frac{3}{2\pi}\right)^{3/2}\left(w - \frac{7}{2}\tilde{l}_B\delta\alpha_{2b}\right)\frac{1}{\sqrt{N}}\frac{1}{\tilde{l}_1^{3/2}} \\ &+ \frac{1}{N}\frac{w_3}{\tilde{l}_1^{3}} + 2\sqrt{\frac{6}{\pi}}f^2\tilde{l}_B\frac{N^{1/2}}{\tilde{l}_1^{1/2}}\Theta_0(a), \end{aligned} \tag{6.47}$$

where w_3 is the third-virial coefficient that is necessarily positive. Furthermore, we note that if a fraction α_{2b} of the condensed divalent counterions participates in bridging, a fraction $\alpha_{2a} = \alpha_2 - \alpha_{2b}$ does not. Therefore, the electrostatic energy related to the formation of monomer-cation monocomplexes (F_4, Eq. (6.31)) is modified after the inclusion of the bridging interaction as,

$$\frac{F_4}{Nk_BT} = -\alpha_1\delta\tilde{l}_B - 2(\alpha_{2a}-\alpha_3)\delta\tilde{l}_B - \alpha_3\delta_2\tilde{l}_B, \tag{6.48}$$

where $\alpha_3 \leq \alpha_{2a}$. In addition, the entropic contribution to the free energy coming from the number of ways the adsorbed ions can be distributed (F_1, Eq. (6.19)) will also be modified. Following the same arguments as in Section 6.3.1, we calculate the number of ways to distribute M_1 monovalent counterions (Na^+), $M_{2a}-M_3$ divalent counterions (Ba^{2+}) (with no coion (Cl^-') adsorption), M_3 ion-triplets ("monomer-Ba^{2+}-Cl^-"), and M_{2b} pairs of ion-bridges among N monomers (ignoring the permutations among the ion-bridges but taking into account of the double counting of them) to be

$$Z_1 = \frac{N!}{(N-M_1-M_{2a}-2M_{2b})!M_1!(M_{2a}-M_3)!M_3!(M_{2b}!)^2 2^{M_{2b}}}, \tag{6.49}$$

where $\alpha_{2a} \equiv M_{2a}/N$ and $\alpha_{2b} \equiv M_{2b}/N$. Therefore,

$$\begin{aligned}\frac{F_1}{Nk_BT} &= \alpha_{2b}\log 2 + (1-\alpha_1-\alpha_{2a}-2\alpha_{2b})\log(1-\alpha_1-\alpha_{2a}-2\alpha_{2b}) + 2\alpha_{2b}\log\alpha_{2b} \\ &+ \alpha_1\log\alpha_1 + (\alpha_{2a}-\alpha_3)\log(\alpha_{2a}-\alpha_3) + \alpha_3\log\alpha_3.\end{aligned} \tag{6.50}$$

The other parts of the free energy depend only on the numbers and valencies of different species of free ions and therefore remain unaltered. They are given as F_2 in Eq. (6.25) and F_3 in Eq. (6.30). In all these cases, $\alpha_2 = \alpha_{2a} + \alpha_{2b}$.

The total free energy $F = \sum_i F_i$ is minimized now for a new set of five variational variables, $\alpha_1, \alpha_{2a}, \alpha_{2b}, \alpha_3$, and $\tilde{l}_1$, and the polymer and counterions are free to explore every possible degree of freedom. The representative result is given in Figure 6.23. The parameters chosen are $N = 100, \tilde{\varrho} = 0.0008$, $\tilde{l}_B = 3.0, \delta = 1.9$, $\tilde{c}_{s1} = 0$, $w = 2.0$, $w_3 = 0.25$. For very low divalent salt concentrations, the conformations are very similar to the case in which bridging is absent. At modest temperatures ($\tilde{l}_B = 3.0$ in water for NaPSS) and for low salt ($\tilde{c}_{s2} < 0.00027$), almost all added divalent counterions adsorb on the chain backbone, but they form monocomplexes (no bridging, Figure 6.23a). At a particular $\tilde{c}^*_{s2}$, which depends on the prevalent physical conditions, all divalent ions suddenly form dicomplexes (bridging) accompanied by a collapse of the chain (Figure 6.23b, in which $\tilde{l}_1 \ll 1$ for $\tilde{c}_{s2} > 0.00027$) and a huge gain in the electrostatic bridging free energy (Figure 6.23c). The Cl^- ions adsorb onto the monomer-divalent ion-pair, as they do in the no-bridging scenario, only if the divalent salt concentration is lower than the collapse concentration $\tilde{c}^*_{s2}$. Above that, Cl^- ions become free as every condensed divalent cation is attached to two monomers. The effect of the excluded volume parameter w is evident in Eqs. (6.45) and (6.47) as they show that a higher w will require a higher Coulomb strength or divalent salt concentration to effect the bridging collapse. Before the collapse, the distribution of the adsorbed counterions and the polymer conformations are quite similar to those of the "no-bridging" cases (see Figure 6.13 for example). This explains our choice of $w = 0$ for the rest of the chapter (except for this subsection). In the "no-bridging" scenario, different positive values of w would only render minor quantitative changes to our results.

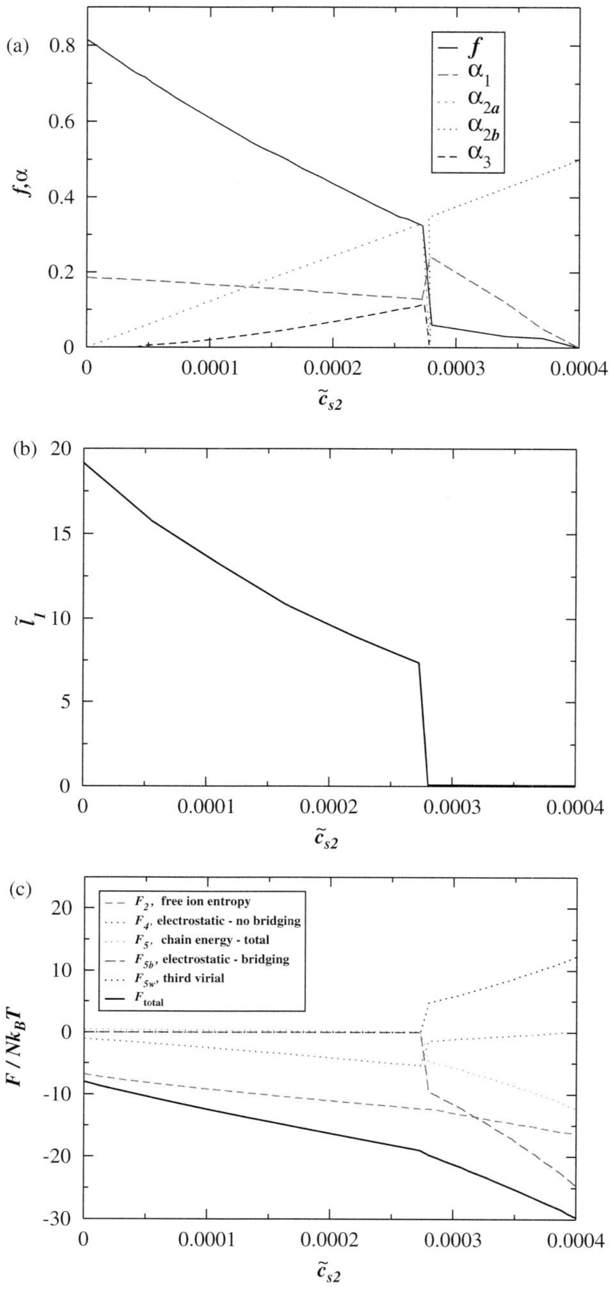

Figure 6.23 Effect of bridging: the degree of ionization (f), fractions of adsorbed ions (α) (a), the size expansion factor ($\tilde{l}_1$) (b), and separate parts of the free energy (F) (c) as functions of the divalent salt concentration (c_{s2}) when monomer-bridging by divalent cations is taken

The sharp collapse of polyelectrolyte chains in solutions from a coil to a compact spherical globule with increasing number of divalent cations is strongly supported by experiments [45]. Regardless of the nature of the intermediate phases (e.g., the pearl-necklace phase that is beyond the scope of our theory limited to uniform spherical expansion), the compact globular phase is found to be a certainty [45, 46] before the system reaches the precipitation phase boundary at which the cation-polyelectrolyte salt is formed. The value of the expansion factor we get ($\tilde{l}_1 \sim 0.08$) does agree quite well to the experimental values of neutral polymers in the globule phase. We notice that, in the collapsed state, the monovalent counterions adsorb on almost all remaining monomers (Figure 6.23a) (also noticed in recent experiments [46]) and the effective charge almost vanishes. This indicates that the polyelectrolyte in the collapsed state virtually behaves as a neutral polymer in poor solvent. In this case, the sign inversion of the excluded volume parameter (w) is effected by the attractive bridging interaction (Eq. (6.45)).

We further notice that the salt concentration, $\tilde{c}^*_{s2}$, at which the polymer collapses, decreases with increasing Coulomb strength (Figure 6.24a and b) further confirming that the first-order collapse induced by the ion-bridging by divalent (or multivalent) cations is an electrostatic phenomenon. In addition, Figure 6.24c shows that $\tilde{c}^*_{s2}$ varies approximately inversely with both forms of the Coulomb strength, $\tilde{l}_B$ and δ. We find $\tilde{l}_B \tilde{c}^*_{s2} \simeq 0.0006$ (for $\delta = 2.5$) and $\delta \tilde{c}^*_{s2} \simeq 0.0005$ (for $\tilde{l}_B = 3.0$).

In conclusion, our model predicts a coil-globule transition, mediated by bridging due to divalent ions, which depends sensitively on temperature and the dielectric heterogeneity, as well as on the availability of divalent counterions.

6.3.12
Role of Chain Stiffness: The Rodlike Chain Limit

The rod-like limit of the chain corresponds to ignoring F_5 given in Eq. (6.37). The consequence of the F_5 term in f is given in Figure 6.25 for $N = 1000$, $\tilde{\varrho} = 0.0005$, $\delta = 3.5$, $\tilde{c}_{s1} = 0 = \tilde{c}_{s2}$, $w = 0$, $w_1 = 0$, and $w_2 = 0$. The solid and dot-dashed curves correspond, respectively, to the flexible and rod-like chains. In the calculation for the rod-like chain, the orientational entropy of the chain has not been taken into account. Our results in the rod-like limit are qualitatively similar to the results of [5] for finite cylinders where equilibration of counterions is allowed as in the present case. It is seen in Figure 6.25 that chain flexibility results in slightly higher counterion adsorption with a lower effective polymer charge.

into account. Parameters are $N = 100$, $\tilde{\varrho} = 0.0008$, $\tilde{l}_B = 3.0$, $\delta = 1.9$, $\tilde{c}_{s1} = 0$, $w = 2.0$, $w_3 = 0.25$. Bridging induces a first-order collapse transition with a sudden gain in the electrostatic ion-bridging energy. At the transition salt concentration, all monomer-divalent cation ion-pairs ("monocomplexes") give way to monomer cation–monomer ion bridges ("dicomplexes").

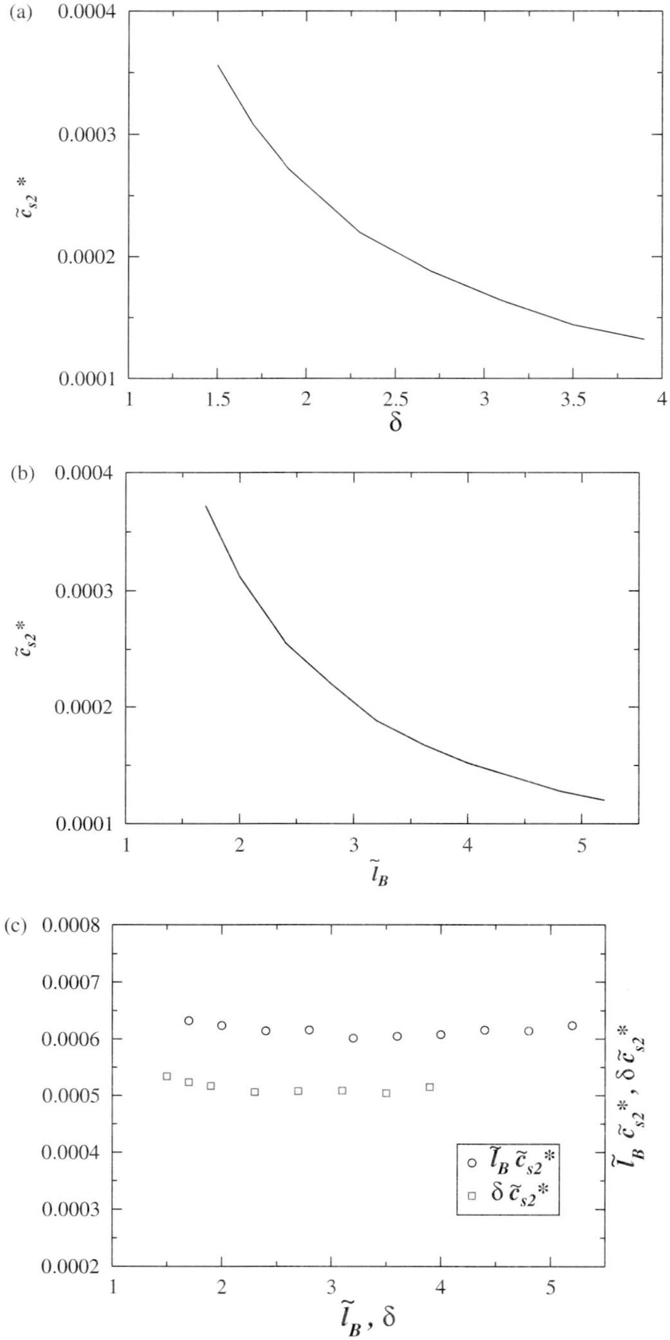

Figure 6.24 Effect of Coulomb strength on bridging collapse: the salt concentration $\tilde{c}_{s2}^*$, at which the first-order collapse occurs, as functions of $\tilde{l}_B$ (a), and δ (b). In (a), $\tilde{l}_B = 3.0$. In (b), $\delta = 2.5$. All other parameters are the same as in Figure 6.23 Lowering of $\tilde{c}_{s2}^*$ with the Coulomb strength indicates that the collapse is due to electrostatic interactions. $\tilde{l}_B$ and δ play similar roles, as expected. In (c), $\tilde{l}_B\tilde{c}_{s2}^* \simeq 0.0006$ (for $\delta = 2.5$) and $\delta\tilde{c}_{s2}^* \simeq 0.0005$ (for $\tilde{l}_B = 3.0$).

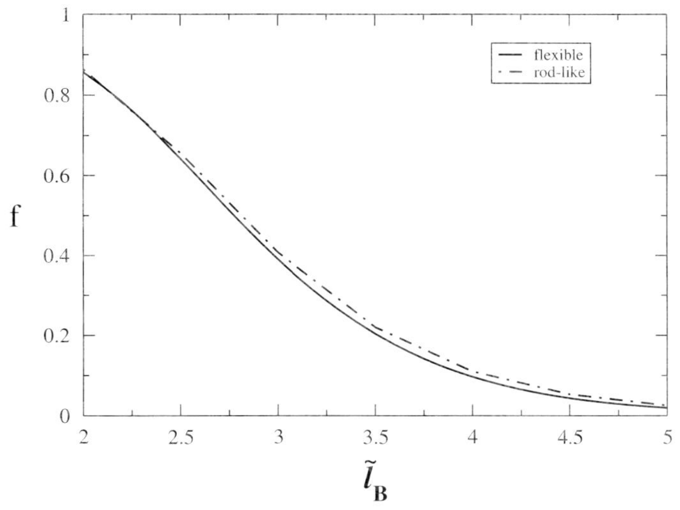

Figure 6.25 Effect of chain stiffness: rod-like (dot-dashed); flexible (solid). $N = 1000$, $\tilde{\varrho} = 0.0005$, $\delta = 3.5$, $\tilde{c}_{s1} = 0 = \tilde{c}_{s2}$, $w = 0$, $w_1 = 0 = w_2$.

6.4 The Self-Consistent Field Theory

Owing to the enormous computational cost of simulating polyelectrolytes with full intricate details, coarse-grained models provide an enticing computational framework for understanding what to expect for a vast parameter space. One of the well-developed coarse-grained models for neutral polymers is known as the self-consistent field theory, which is based on the field theory first proposed by Edwards [48–50]. The theory captures the essential thermodynamics and presents opportunities for carrying out systematic studies on many complicated systems such as polyelectrolytes. A number of different extensions of the original field theory proposed by Edwards for a single neutral chain have been carried out to study neutral polymers [51–55]. Extensive use of the theory to get an insight into problems of different kinds arises due to the general nature of the field theoretical formalism and a reasonable computational demand of the theory in comparison with simulations. Besides the general nature of the theory, it presents an avenue for a systematic study to include or remove certain effects while analyzing the problems.

The theory has its foundation laid on the path integral representation [53, 56] of the distribution function for realizing all possible conformations of a phantom chain (a chain whose connected segments do not interact with each other), whose ends are fixed in space. The distribution function is the sum over all the possible conformations for given locations of the ends and the ingenuity of Edwards was to write the sum as a path integral over all the possible paths between two points representing the ends. The limits and the integrand of the path integral represent the physical positions of the ends of the chain and the probability distribution function for realizing a particular path, respectively. For a phantom chain, all the paths are equally probable and the probability distribution function for realizing a particular path turns out to be the Wiener measure [50].

For computational purposes, the path integral representation of the distribution function can be written in the form of a diffusion equation [50, 52–55]. Physically, this means that the problem of describing the different conformations of the phantom chain can be mapped on to a simple diffusion of a particle between the two points representing the ends. Due to the Gaussian nature of the solution of the diffusion equation, this particular description for the phantom chain is also known as the "standard Gaussian model" in the polymer literature [57, 58]. The model has been found to be quite amenable to study polymers, whose monomers interact with each other by an arbitrary interaction potential. Originally, the effect of short range excluded volume interactions on the conformations of a single neutral chain was studied by Edwards in dilute [48] and concentrated regimes [49]. Later on, the model was extended by Helfand [59] to study other multichain problems including polymer blends, diblock copolymer melts, and so on. There are a number of review articles [57, 58] describing this particular field theoretical model for neutral polymers, and only an extension of the model to describe polyelectrolytes will be presented here.

In the presence of interactions between the connected segments of a single chain, aforementioned simple diffusion or random walks get affected and the walks are no more random. However, the intricate coupling of the different components such as monomers, solvent, or small ions in the case of polyelectrolytes via the interaction potentials complicates the theoretical analysis. In order to decouple different components, the conformations of the chain can be envisioned as the walks in the presence of fields, which arise solely due to the fact that there are interactions present in the system. This physical argument is the basis of the use of certain field theoretical transformations such as Hubbard–Stratonovich [60] transformation, which is well known in the field theory. So, the conformational characteristics of a polymer chain in the presence of different kinds of intrachain interactions can be described once the fields are known. In general, an exact computation of these fields is almost an impossible task. That is the reason theoretical developments resort to certain approximations for computing these fields, which work well for most of the practical purposes. Once these fields are known, the physical properties can be described in terms of these fields. It was shown by Edwards [50] that the similar analysis can be carried out for systems with many chains, where interchain interactions also affect the properties in addition to intrachain interactions.

Recently, the field theory developed for neutral polymers has been extended to describe various polyelectrolytic systems in the absence/presence of externally added salt ions. The theory has been used to investigate the micro- and macrophase separation in polyelectrolyte systems [61–63], adsorption of polyelectrolytes on to the charged surfaces [64, 65], polyelectrolyte brushes [66, 67], confinement effects [14], counterion adsorption [15], translocation of polyelectrolytes (R. Kumar and M. Muthukumar, unpublished), and the assembly of single stranded RNA viruses (J. Wang, R. Kumar, and M. Muthukumar, unpublished). In this chapter, we review the general methodology behind the SCFT for polyelectrolytes.

A general background on the field theoretical formalism for polyelectrolytes is presented in Section 6.4.1. Details of the commonly used transformations in order to switch from a particle to the field description are presented in Section 6.4.2. Different kinds of charge distributions along the polyelectrolyte chain and the well-known saddle-point approximation for computing the free energy are described in Sections 6.4.3 and 6.4.4, respectively. Numerical techniques to solve the nonlinear set of equations and one-loop expansions to go beyond the well-known saddle-point approximation are presented in Sections 6.4.5 and 6.4.6, respectively.

6.4.1 Extension of Edward's Formulation

Here, we present a general outline of the self-consistent field theory for polyelectrolyte solutions containing externally added salt ions. The theory is a generalization of the field theoretical formalism developed by Edwards [48–50] for neutral polymers to polyelectrolytes. We start from the path integral representation of a polymer chain and readers interested in the derivation of the path integral representation are referred to Ref. [56].

We consider computation of the free energy of solutions containing n_p monodisperse polyelectrolyte chains, each containing a total of N segments. In addition to this, there are small ions due to the added salt (in total volume Ω) along with the counterions coming from the polyelectrolyte chains so that the whole system is globally electroneutral. Let Z_j and n_j be the valency and number of the jth charged species. Subscripts p, s, c, $+$, and $-$ are used to represent monomers from the polymer, solvent molecules, counterions from the polyelectrolyte, positive and negative salt ions, respectively. Using the path integral formulation [48, 56], we represent a polyelectrolyte chain as a continuous curve of length Nl, where l is the Kuhn segment length [50]. For the treatment shown below, we assume that the volume occupied by each monomer is the same ($= l^3 \equiv 1/\varrho_o, \varrho_o$ being the bulk density) and that the system is incompressible. We use an arc length variable t_β to represent any segment along the backbone of βth chain. Also, the position vector of a particular segment (t_β) on βth chain is represented by $R_\beta(t_\beta)$. For this system, the partition function Z can be written after carrying out a straightforward extension of the Edward's Hamiltonian by

$$Z = \frac{1}{n_\mathrm{p}!}\int \mathrm{d}\mathbf{r} \int \mathrm{d}\mathbf{r}' \int_r^{r'} \prod_{\beta=1}^{n_\mathrm{p}} D[\mathbf{R}_\beta(t_\beta)] \sum_{\left\{g_{t_\beta}\right\}} \int \prod_j \frac{1}{n_j!} \prod_{\lambda=1}^{n_j} \mathrm{d}\mathbf{r}_\lambda \exp(-H[\mathbf{R}_\beta(t_\beta), \mathbf{r}_\lambda])$$
$$\prod_{\mathbf{r}} \delta\left(\sum_\gamma \hat{\varrho}_\gamma(\mathbf{r}) - \varrho_0\right), \tag{6.51}$$

where $\gamma = \mathrm{p}, \mathrm{s}$ represents monomers and solvent molecules and $j = \mathrm{s}, \mathrm{c}, +, -$ depicts all the small molecules, namely, solvent, counterions from polyelectrolytes, and positive and negative salt ions, respectively.

In writing the partition function, we have summed over all the possible locations of the small molecules in the volume under investigation (represented by volume integrals over $\mathbf{r}_\lambda$) and the sum over all possible conformations of an individual chain is represented by functional integrals over all the possible paths originating from one end and ending at the other (written as functional integral over $\mathbf{R}_\beta(t_\beta)$). Depending on the problem of interest, the ends of the chains ($\mathbf{r}$ and $\mathbf{r}'$ in Eq. (6.51)) can be fixed (e.g., in the case of polymer brushes) or free to enjoy the translational degrees of freedom. In the latter case, the integrals over the positions of the ends need to be carried out to compute the partition function. In Eq. (6.51), it is assumed that the ends can be anywhere in space and hence, the integrals over the possible locations of the ends of chains have been carried out. Due to the fact that two molecules of same species are indistinguishable from each other, the partition function has to be divided by $n_j!$ and $n_\mathrm{p}!$ to avoid double counting in the configurational states of the system.

Polyelectrolytes can have different kinds of charge distributions along the chain backbone depending on the chemistry, and the charge distribution may have significant effect on the properties of polyelectrolyte solutions. For example, in the case of polyacids or polybases, the effective charge on a monomer depends on the pH of the solution, that is, the charge distribution along the chain backbone depends on the pH. Different kinds of charge distributions have been considered in the literature [68] to describe different situations. The simplest kind of charge distribution is called the "smeared" charge distribution, where total charge on a single chain is distributed (or smeared) uniformly along the backbone. Other kinds of charge distributions, which are used quite frequently, include "annealed" and "permuted" charge distributions. In the former, charge on a segment along the chain is associated with a probability of finding it. Latter kind of charge distribution represents the situation, where counterions can move along the chain backbone.

In order to take into account the effect of different kinds of charge distributions along the chain, a summation over charge parameter g_{t_β}, which represents the charge distribution on the segment t_β, have been carried out in Eq. (6.51). The sum over this variable in Eq. (6.51) is defined by

$$\sum_{\{g_{t_\beta}\}}[\cdots] = \int \prod_{\beta=1}^{n_\mathrm{p}} \mathrm{d}g_{t_\beta}[\cdots] \prod_{\beta=1}^{n_\mathrm{p}} P(g_{t_\beta}), \tag{6.52}$$

where $P(g_{t_\beta})$ is the probability distribution function for g_{t_β} and defines different kinds of charge distributions on the chain. For example, $P(g_{t_\beta}) = \prod_m \delta(g_{t_\beta} - \alpha)$ and $P(g_{t_\beta}) = \prod_m \{\alpha\delta(g_{t_\beta} - 1) + (1-\alpha)\delta(g_{t_\beta})\}$ for the "smeared" and "annealed" charge distributions [68] on all the chains, respectively, and m is the index for different monomers along the chains. Also, α is the probability of finding a charge at the mth monomer along a polyelectrolyte chain. Physically, this means that in smeared charge distribution, each monomer on the chain has a charge equal to $Z_\mathrm{p}\alpha e$, where e is the electronic charge and in annealed charge distribution, charged and uncharged sites on the chain are randomly distributed with the probability of finding a charged site as α and uncharged site as $(1-\alpha)$.

In order to simplify the calculation, incompressibility condition (represented by the delta function in the above equation) is used by assuming small ions to be point charges. The effect of the finite size of the small ions can be included partially by taking into account the volume fraction of small ions also while writing the incompressibility condition (let us call it modified incompressibility condition). However, in the case of monovalent ions, the results for the monomer densities obtained after using the modified incompressibility condition are almost indistinguishable from that obtained after modeling them as point charges. The origin of this agreement between the finite size and point-like ions lies in the fact that typical radii of monovalent ions [69] are very small in comparison with the size of monomers (in terms of Kuhn step length). On the other hand, the use of the modified incompressibility condition becomes almost indispensable in the case of multivalent ions. It should be noted that the effect of finite size of the ions [70] on the electrostatic potential distribution is nontrivial and cannot be captured only through the use of modified incompressibility condition. For the theoretical description presented here, we consider the case of monovalent ions to demonstrate the methodology and model them as point charges.

Hamiltonian, H, in Eq. (6.51) can be split into the connectivity and interaction terms, so that

$$\begin{aligned} H[\mathbf{R}_\beta(t_\beta), \mathbf{r}_\lambda] = & \sum_{\beta=1}^{n_p} H_0[\mathbf{R}_\beta(t_\beta)] + \sum_{\beta=1}^{n_p}\sum_{\lambda=1}^{n_p} H_{pp}[\mathbf{R}_\beta(t_\beta), \mathbf{R}_\lambda(t_\lambda)] \\ & + \sum_j \sum_{\beta=1}^{n_p}\sum_{\lambda=1}^{n_j} H_{pj}[\mathbf{R}_\beta(t_\beta), r_\lambda] + \sum_j \sum_a \sum_{\beta=1}^{n_j}\sum_{\lambda=1}^{n_a} H_{ja}[\mathbf{r}_\beta, \mathbf{r}_\lambda], \end{aligned} \tag{6.53}$$

where $j, a = \text{s}, \text{c}, +, -$ and represent all the small molecular species in the system. Also, $H_0[\mathbf{R}_\beta]$ is the chain connectivity part, which comes from the fact that in the absence of interactions, the probability distribution function for the chains must be a Wiener measure. In the continuum representation [50], this term is written explicitly as

$$H_0\left[\mathbf{R}_\beta(t_\beta)\right] = \frac{3}{2l}\int_0^{Nl} \mathrm{d}t_\beta \left(\frac{\partial R_\beta(t_\beta)}{\partial t_\beta}\right)^2. \tag{6.54}$$

We must stress here that the connectivity part represented by the functional integrals over the possible paths in Eq. (6.51) is not properly normalized. This leads to some unknown constants in the computation of absolute free energy of the system. However, for most of the practical purposes, either the relative free energy or the derivatives of the free energy such as osmotic pressure and so on are the required quantities. So, the normalization factor can be taken care of by choosing an appropriate reference system. In the polymeric problems, where one is interested in studying the effect of different kinds of interactions, the reference system for each chain can be taken as the chain of the same number of segments *without* any interactions (that means a phantom chain of the same length) in free space (or vacuum). For a single phantom chain in free space, the partition function, Z_0, can be written as

$$Z_0 = \int D[\mathbf{R}] \exp[-H_0[\mathbf{R}]], \tag{6.55}$$

which is divergent.

Second term, $H_{\text{pp}}[\mathbf{R}_\beta(t_\beta), \mathbf{R}_\lambda(t_\lambda)]$, in Eq. (6.53) is the polymer–polymer interaction term, which includes both inter- and intrachain monomer–monomer interactions arising from the excluded volume and electrostatic effects. To compute the polymer–polymer interaction energy, we have to include the interactions among all monomeric species. If the excluded volume interaction terms are written by using delta functional form for the potential as used by Edwards [48, 50] for neutral polymers, then

$$\begin{aligned} H_{\text{pp}}[\mathbf{R}_\beta(t_\beta), \mathbf{R}_\lambda(t_\lambda)] = {} & \frac{1}{2l^2}\int_0^{Nl} \mathrm{d}t_\beta \int_0^{Nl} \mathrm{d}t_\lambda \left[w_{\text{pp}}\delta[\mathbf{R}_\beta(t_\beta) - \mathbf{R}_\lambda(t_\lambda)] \right. \\ & \left. + \frac{Z_{\text{p}}^2 e^2 g_{t_\beta} g_{t_\lambda}}{k_{\text{B}}T} \varepsilon^{-1}[\mathbf{R}_\beta(t_\beta), \mathbf{R}_\lambda(t_\lambda)] \left\{ \frac{1}{|\mathbf{R}_\beta(t_\beta) - \mathbf{R}_\lambda(t_\lambda)|} \right\} \right], \end{aligned} \tag{6.56}$$

where $\delta(x)$ represents the three-dimensional Dirac delta function and e is the charge of an electron.

Some comments regarding the delta functional form for the excluded volume interaction energy are in order here. In Eq. (6.56), w_{pp} is a parameter to assess the strength of monomer–monomer excluded volume interactions and has the dimensions of volume. This form for the excluded volume term was suggested by Edwards by realizing that for large length scales, properties of the system should not depend upon the specific details of interactions, which may be due to steric effects, van der Waals interactions [69], and so on. As far as the interaction term is written in terms of a short-range function, the predictions of the theory should not change. However, we must point out here that the delta functional form for the interaction potential leads to divergences when $\mathbf{R}_\beta(t_\beta) = \mathbf{R}_\lambda(t_\lambda)$. Although these short-range divergences do not affect any physically measurable quantity, these divergences cause the absolute free energy of the system to diverge. However, the free energy differences, which are experimentally important, remain well behaved. Also, it should be noted that Eq. (6.56) is written after splitting second virial coefficient term into short- and long-range parts through the relation

$$\begin{aligned} V(\mathbf{r}) &= \left[1 - \exp\left(-\frac{V_{\text{pp}}(\mathbf{r}) + V_{\text{cc}}(\mathbf{r})}{k_{\text{B}}T}\right)\right] \\ &\simeq \left[1 - \frac{V_{\text{cc}}(\mathbf{r})}{k_{\text{B}}T}\right]\left[1 - \exp\left(-\frac{V_{\text{pp}}(\mathbf{r})}{k_{\text{B}}T}\right)\right] + \frac{V_{\text{cc}}(\mathbf{r})}{k_{\text{B}}T} \end{aligned} \tag{6.57}$$

$$\equiv w_{\text{pp}}\delta[\mathbf{r}] + \frac{V_{\text{cc}}(\mathbf{r})}{k_{\text{B}}T}, \tag{6.58}$$

where $V_{pp}(\mathbf{r})$ is the potential energy of interaction coming from van der Waals forces between two monomers separated from each other by distance $r = |\mathbf{r}|$. $V_{cc}(\mathbf{r})$ is the electrostatic charge–charge interaction energy, given by Coulomb's law. The second equation is written after expanding the exponential in powers of V_{cc} and retaining up to linear terms. Also, the second virial coefficient (B) is related to $V(\mathbf{r})$ by $B = \int \mathrm{d}\mathbf{r}\, V(\mathbf{r})$ and an analogue of excluded volume parameter in the case of neutral polymers [50] can be defined for polyelectrolytes by

$$w_{pp} = \int \mathrm{d}\mathbf{r}\left[1-\frac{V_{cc}(\mathbf{r})}{k_B T}\right]\left[1-\exp\left(-\frac{V_{pp}(\mathbf{r})}{k_B T}\right)\right]. \tag{6.59}$$

The functional form for w_{pp} reveals the complicated dependence of excluded volume parameter on temperature and expansion of the exponential up to linear terms in V_{cc} (cf. Eq. (6.57)) signifies the validity of the theory for weakly charged polyelectrolytes so that V_{cc} is small. The rightmost term in Eq. (6.56) is the electrostatic interaction energy (V_{cc}), which is written after describing the response of the inhomogeneous systems to an applied electric field by a nonlocal response function (also known as the inverse dielectric function [71–73]), $\varepsilon^{-1}(\mathbf{r}, \mathbf{r}')$) defined by

$$\int \mathrm{d}\mathbf{r}'\, \varepsilon^{-1}(\mathbf{r}, \mathbf{r}')\varepsilon(\mathbf{r}', \mathbf{r}'') = \delta(\mathbf{r}-\mathbf{r}''), \tag{6.60}$$

where $\varepsilon(\mathbf{r}, \mathbf{r}')$ is the dielectric function in real space. In principle, the inverse dielectric function can be written in terms of molecular polarizabilities of the charged species. Computations of the inverse dielectric function add another set of complexity in assessing the effect of long-range electrostatic interactions. For the length scales relevant to the coarse-grained models, the dielectric function is either taken to be constant ($= \varepsilon$) or local in nature so that $\varepsilon(\mathbf{r}, \mathbf{r}')$ is replaced by $\varepsilon(\mathbf{r})$ and $\varepsilon^{-1}(\mathbf{r}, \mathbf{r}')$ by $1/\varepsilon(\mathbf{r})$ in Eq. (6.56). Also, note the similarity in divergences arising from this term when $\mathbf{R}_\beta(t_\beta) = \mathbf{R}_\lambda(t_\lambda)$ and the divergences in excluded volume interaction terms. Both of these divergences set the length scale below which this coarse-grained model fails to properly describe the system.

Third term, $H_{pj}[\mathbf{R}_\beta(t_\beta), \mathbf{r}_\lambda]$, in Eq. (6.53) is the monomer–small molecule interaction term, which depends on the small molecular species. For polymer–solvent interactions, polarization effects are ignored and the interactions are modeled by delta functional form for the excluded volume interactions so that

$$H_{ps}\left[\mathbf{R}_\beta(t_\beta), \mathbf{r}_\lambda\right] = \frac{w_{ps}}{l}\int_0^{Nl} \mathrm{d}t_\beta \delta\left[\mathbf{R}_\beta(t_\beta)-\mathbf{r}_\lambda\right], \tag{6.61}$$

w_{ps} being the monomer–solvent excluded volume parameter. As the small ions (counterions and coions) are taken to be point like in this study, so their interactions with monomers are taken to be purely electrostatic in nature, written by

$$H_{pj}\left[\mathbf{R}_\beta, \mathbf{r}_\lambda\right] = \frac{1}{l}\int_0^{Nl} \mathrm{d}t_\beta\left[\frac{Z_p Z_j e^2 g_{t_\beta}}{k_B T}\varepsilon^{-1}(\mathbf{R}_\beta(t_\beta), \mathbf{r}_\lambda)\left\{\frac{1}{|\mathbf{R}_\beta(t_\beta)-\mathbf{r}_\lambda|}\right\}\right], \tag{6.62}$$

where $j = \mathrm{c}, +, -$.

Last term on the right-hand side in Eq. (6.53) takes care of interactions among small molecules. Similar to monomer–monomer excluded volume interactions, we model solvent–solvent interaction energy as

$$H_{ss}\left[\mathbf{r}_\beta, \mathbf{r}_\lambda\right] = \frac{w_{ss}}{2}\delta(\mathbf{r}_\beta - \mathbf{r}_\lambda), \tag{6.63}$$

w_{ss} being the solvent–solvent excluded volume parameter. Like monomer–solvent electrostatic interactions, we ignore the solvent–ion electrostatic interactions so that $H_{sj} = 0$ for $j = \mathrm{c}, +, -$ due to the point-like sizes of the ions, which exhibit zero excluded volume. Also, taking ion–ion interactions to be purely electrostatic in nature, we can write

$$H_{ja}\left[\mathbf{r}_\beta, \mathbf{r}_\lambda\right] = h_{ja}\frac{Z_j Z_a e^2}{k_B T}\varepsilon^{-1}(\mathbf{r}_\beta, \mathbf{r}_\lambda)\left\{\frac{1}{|\mathbf{r}_\beta - \mathbf{r}_\lambda|}\right\}, \tag{6.64}$$

where $j, a = \mathrm{c}, +, -$, and $h_{ja} = 1/2$ for $j = a$ and 1 otherwise.

The complicated Hamiltonian as presented in Eq. (6.53) can be written in a simplified form using microscopic densities for different species in the system, defined as

$$\hat{\varrho}_p(\mathbf{r}) = \frac{1}{l}\sum_{\beta=1}^{n_p}\int_0^{Nl} dt_\beta \delta(\mathbf{r} - R_\beta(t_\beta)) \tag{6.65}$$

and

$$\hat{\varrho}_j(\mathbf{r}) = \sum_{\beta=1}^{n_j}\delta(\mathbf{r} - \mathbf{r}_\beta) \tag{6.66}$$

for monomers and small molecular species, respectively. Using these definitions of number densities and using identity

$$\delta(\mathbf{R} - \mathbf{R}') = \int d\mathbf{r}\delta(\mathbf{r} - \mathbf{R})\delta(\mathbf{r} - \mathbf{R}'). \tag{6.67}$$

Eq. (6.53) can be written as

$$H[\mathbf{R}_\beta(t_\beta), \mathbf{r}_\lambda] = \sum_{\beta=1}^{n_p} H_0[\mathbf{R}_\beta(t_\beta)] + H_w + H_e, \tag{6.68}$$

where H_w and H_e are the contributions coming from the excluded volume and the electrostatic interactions among different components, respectively. Explicitly,

$$H_w = \chi_{ps} l^3 \int d\mathbf{r}\hat{\varrho}_p(\mathbf{r})\hat{\varrho}_s(\mathbf{r}) + \frac{\varrho_0}{2}\sum_{\gamma=p,s} w_{\gamma\gamma} n_\gamma \tag{6.69}$$

and

$$H_e = \frac{1}{2}\int d\mathbf{r}\int d\mathbf{r}' \frac{\hat{\varrho}_e(\mathbf{r})\varepsilon^{-1}(\mathbf{r}, \mathbf{r}')}{k_B T}\left\{\frac{\hat{\varrho}_e(\mathbf{r}')}{|\mathbf{r} - \mathbf{r}'|}\right\}. \tag{6.70}$$

In these equations, χ_{ps} is the dimensionless Flory's chi parameter [74] defined as

$$\frac{w_{pp}+w_{ss}}{2}-w_{ps}=-\chi_{ps}l^3 \tag{6.71}$$

and $\hat{\varrho}_e(\mathbf{r})$ is the microscopic charge density at $\mathbf{r}$ defined as

$$\hat{\varrho}_e(\mathbf{r})=e\left[\frac{Z_p}{l}\sum_{\beta=1}^{n_p}\int_0^{Nl}\mathrm{d}t_\beta g_{t_\beta}\delta(\mathbf{r}-\mathbf{R}_\beta(t_\beta))+\sum_{j=c,+,-}Z_j\hat{\varrho}_j(\mathbf{r})\right]. \tag{6.72}$$

6.4.2
Transformation from Particles to Fields

So far, the partition function is written in terms of the microscopic density variables starting from a microscopic description of the interactions among different components in the system. However, the level of the complexity of the problem is still the same due to intricate coupling of these densities. In order to carry out further calculations, these couplings need to be decoupled. This can be done using two different transformation schemes, both of which give the same results within normalization constants. The first scheme is based on some functional integral identities [52, 55] and the second scheme is based on a well-known identity for Gaussian functional integrals known as Hubbard–Stratonovich [60] transformation. Here, we present the details of these two methods for the sake of completeness.

6.4.2.1 Transformation Using Functional Integral Identities

For any arbitrary functional, f, of microscopic variables $\hat{\varrho}$

$$f[\hat{\varrho}]=\int D[\varrho]\delta(\varrho-\hat{\varrho})f[\varrho] \tag{6.73}$$

and

$$\delta(\varrho-\hat{\varrho})=\mu\int D[w]\exp\left[i\int \mathrm{d}\mathbf{r}w(\mathbf{r})(\varrho(\mathbf{r})-\hat{\varrho}(\mathbf{r}))\right] \tag{6.74}$$

$$\Rightarrow f[\hat{\varrho}]=\mu\int D[\varrho]\int D[w]\exp\left[i\int \mathrm{d}\mathbf{r}w(\mathbf{r})(\varrho(\mathbf{r})-\hat{\varrho}(\mathbf{r}))\right]f[\varrho], \tag{6.75}$$

where μ is the appropriate normalization factor. Using this transformation, any functional of microscopic variable $\hat{\varrho}(\mathbf{r})$ can be written as functional integral over a collective density variable $\varrho(\mathbf{r})$ and a field variable $w(\mathbf{r})$. By introducing a density and field variable for the charge density variable, $\hat{\varrho}_e(\mathbf{r})$, each microscopic number density variable involved in the incompressibility constraint, $\hat{\varrho}_\gamma(\mathbf{r})$, and by replacing the incompressibility constraint on the microscopic densities using the functional integral identity for delta function as in Eq. (6.74), the partition function in Eq. (6.51) can be written as

$$Z = \Lambda \int \prod_{\gamma} D[\varrho_\gamma] \int D[w_\gamma] \int D[\eta] \int D[\varrho_e] \int D[\psi] \sum_{\{g_{t_\beta}\}} \exp[-H], \tag{6.76}$$

where $\gamma = \mathrm{p, s}$ and $\eta(\mathbf{r})$ is the field variable corresponding to the incompressibility constraint. $\varrho_e(\mathbf{r})$ and $\psi(\mathbf{r})$ are the collective charge density and field variables. In this equation, Λ is the normalization factor and functional dependence of H on $\varrho_j, \varrho_\gamma, w_j, w_\gamma, \eta$ and g_{t_β} has been suppressed for notational convenience. Note here that all the collective field and density variables are real. It should also be kept in mind that the functional integrals over density variables (ϱ's) can be carried out exactly for the two-body interaction model used in this work, and only functional integrals over field variables need to be approximated by some appropriate approximation scheme to make a progress. An alternative way is to introduce two collective variables for *each* microscopic density variable along with collective charge variables $\varrho_e(\mathbf{r}), \psi(\mathbf{r})$, and $\eta(\mathbf{r})$. We do not follow that approach in this chapter due to redundant collective variables used in the formulation. However, the approach [63] is particularly useful in making connections with the density functional theories [75–77] in the context of diblock copolymers.

6.4.2.2 Hubbard–Stratonovich Transformation

This transformation is a generalization of a result for multivariate Gaussian integrals to functionals so that for any real, symmetric, positive-definite operator $A(\mathbf{r}, \mathbf{r}')$,

$$\begin{aligned} &\exp\left[-\frac{1}{2}\int \mathrm{d}\mathbf{r} \int \mathrm{d}\mathbf{r}' J(\mathbf{r}) A^{-1}(\mathbf{r}, \mathbf{r}') J(\mathbf{r}')\right] \\ &= \frac{\int D[f] \exp\left[-(1/2)\int \mathrm{d}\mathbf{r} \int \mathrm{d}\mathbf{r}' f(\mathbf{r}) A(\mathbf{r}, \mathbf{r}') f(\mathbf{r}') + i\int \mathrm{d}\mathbf{r} J(\mathbf{r}) f(\mathbf{r})\right]}{\int D[f] \exp\left[-(1/2)\int \mathrm{d}\mathbf{r} \int \mathrm{d}\mathbf{r}' f(\mathbf{r}) A(\mathbf{r}, \mathbf{r}') f(\mathbf{r}')\right]}, \end{aligned} \tag{6.77}$$

where $J(\mathbf{r})$ and $f(\mathbf{r})$ are arbitrary functions and $i = \sqrt{-1}$. Similarly, another functional integral identity can be written for exponents of Gaussian quantities with positive sign as

$$\begin{aligned} &\exp\left[\frac{1}{2}\int \mathrm{d}\mathbf{r} \int \mathrm{d}\mathbf{r}' J(\mathbf{r}) A^{-1}(\mathbf{r}, \mathbf{r}') J(\mathbf{r}')\right] \\ &= \frac{\int D[f] \exp\left[-(1/2)\int \mathrm{d}\mathbf{r} \int \mathrm{d}\mathbf{r}' f(\mathbf{r}) A(\mathbf{r}, \mathbf{r}') f(\mathbf{r}') + \int \mathrm{d}\mathbf{r} J(\mathbf{r}) f(\mathbf{r})\right]}{\int D[f] \exp\left[-(1/2)\int \mathrm{d}\mathbf{r} \int \mathrm{d}\mathbf{r}' f(\mathbf{r}) A(\mathbf{r}, \mathbf{r}') f(\mathbf{r}')\right]}. \end{aligned} \tag{6.78}$$

Note that both of these identities are also valid when positive sign in front of linear $J(\mathbf{r})$ term is replaced by a negative sign. This is a generalization of the fact that for simple Gaussian integrals

$$\int_{-\infty}^{\infty} \mathrm{d}x \quad \exp\left[-\frac{ax^2}{2} \pm iJx\right] = \sqrt{\frac{2\pi}{a}} \quad \exp\left[-\frac{J^2}{2a}\right]. \tag{6.79}$$

Furthermore, in these equations, inverse operator $A^{-1}(\mathbf{r},\mathbf{r}')$ is defined through the relation

$$\int d\mathbf{r}'\, A^{-1}(\mathbf{r},\mathbf{r}')A(\mathbf{r}',\mathbf{r}'') = \delta(\mathbf{r}-\mathbf{r}''). \tag{6.80}$$

For simple operators such as $A(\mathbf{r},\mathbf{r}') = 1/|\mathbf{r}-\mathbf{r}'|$, it can be shown that inverse operator is $A^{-1}(\mathbf{r},\mathbf{r}') = -\delta(\mathbf{r}-\mathbf{r}')\nabla_{\mathbf{r}}^2/4\pi$ using the property [78], $\nabla_{\mathbf{r}}^2\left[\frac{1}{4\pi|\mathbf{r}-\mathbf{r}'|}\right] = -\delta(\mathbf{r}-\mathbf{r}')$. Similarly, taking into account the fact that the Poisson's equation must be satisfied even for position-dependent dielectric constant [78], it can be shown that inverse operator for $A(\mathbf{r},\mathbf{r}') = 1/\varepsilon(\mathbf{r})|\mathbf{r}-\mathbf{r}'|$ is $A^{-1}(\mathbf{r},\mathbf{r}') = -\delta(\mathbf{r}-\mathbf{r}')\nabla_{\mathbf{r}'}\varepsilon(\mathbf{r}')\nabla_{\mathbf{r}'}/4\pi$.

In order to use this transformation for the Hamiltonian as represented by Eq. (6.68), microscopic density terms that are quadratic in nature need to be written in the form given on the left-hand side in Eqs. (6.77) and (6.78). Electrostatic terms in H_e are already in the appropriate form. It is only the terms in H_w that needs to be rewritten. This can be achieved by rewriting H_w in terms of order parameters and total density. For an n component system, all microscopic densities can be described by $n-1$ independent order parameters (due to the incompressibility constraint serving as the nth relation among the densities). There are many different ways of defining these order parameters. One convenient definition, which makes mathematics simple, is the deviation of densities of solutes from the solvent density, that is, defining $\phi_j(\mathbf{r}) = \hat{\varrho}_j(\mathbf{r})-\hat{\varrho}_s(\mathbf{r})$ for $j = 1, 2, \ldots (n-1)$, where j is the index for different solutes (monomers, counterions, and the salt ions). Using the transformation for each quadratic term in the Hamiltonian (cf. Eq. (6.68)), the partition function becomes

$$Z = \frac{1}{\Lambda_\eta \Lambda_\psi \prod_j \Lambda_j}\int \prod_j D[\varsigma_j]\int D[\eta]\int D[\psi]\sum_{\{g_{t_\beta}\}}\exp[-H'], \tag{6.81}$$

where ς_j is the field variable introduced for the quadratic term involving $\phi_j(\mathbf{r})$ and Λ_j is the corresponding normalization factor. Similarly, ψ is the field variable introduced for the quadratic electrostatic energy term and Λ_ψ is the normalization factor arising as a result. As mentioned earlier, η is the field variable introduced for the incompressibility constraint and Λ_η is the *unknown* normalization constant.

Sometimes, it is advantageous to use this technique rather than the method using functional integral identities as presented in previous section due to the fact that the functional integrals over density variables do not appear in the formulation, and one has to deal with only the functional integrals over the fields with appropriate normalization factors. On the other hand, this technique is plagued with two main shortcomings. One is the fact that in general it is not easy to find the inverse operator A^{-1} for any given A. Second, the technique can only be used for quadratic terms in Hamiltonian. In case, there are higher order terms such as in the problems considering polymers in poor solvent conditions, the method presented in the previous section should come handy. For quadratic functionals, after carrying out Gaussian integrals over collective density variables such as in Eq. (6.76), it can be shown that the method presented in the previous section is the same as the one using Hubbard–Stratonovich transformation. An equivalence of the two techniques for the transformation from particle to field description is demonstrated recently for a single polyelectrolyte chain in spherical cavities [14]. For the discussion in this chapter, let us use the transformation using functional integral identities (cf. Eq. (6.76)).

6.4.3
Sum Over Charge Distributions

Explicitly, the sum over charge distributions in Eq. (6.76) is given by

$$S = \sum_{\{g_{t_\beta}\}} \exp\left[-\sum_{\beta=1}^{n_p} \frac{eZ_p}{l} \int_0^{Nl} dt_\beta g_{t_\beta} \psi(\mathbf{R}_\beta)\right]. \tag{6.82}$$

For "smeared" charge distributions on all polyelectrolyte chains, the sum becomes [62, 68] (using Eq. (6.52))

$$S = \exp\left[-\sum_{\beta=1}^{n_p} \frac{eZ_p\alpha}{l} \int_0^{Nl} dt_\beta\, \psi(\mathbf{R}_\beta)\right] \tag{6.83}$$

and for the "annealed" distribution [62, 68]

$$S = \exp\left[\ln\left\{\alpha \exp\left(-\sum_{\beta=1}^{n_p} \frac{eZ_p}{l} \int_0^{Nl} dt_\beta\, \psi(\mathbf{R}_\beta)\right) + (1-\alpha)\right\}\right]. \tag{6.84}$$

6.4.4
Saddle-Point Approximation

For the sake of discussion, let us consider the "smeared" charge distribution so that Eq. (6.76) can be written as

$$Z = \Lambda \int \prod_\gamma D[\varrho_\gamma] \int D[w_\gamma] \int D[\eta] \int D[\varrho_e] \int D[\psi] \exp[-f], \tag{6.85}$$

where f is given by

$$\begin{aligned} \exp(-f) = {} & \frac{Q_p^{n_p} Q_s^{n_s} \prod_j Q_j^{n_j}}{n_p! n_s! \prod_j n_j!} \exp\left[-\frac{\varrho_0}{2} \sum_\gamma w_{\gamma\gamma} n_\gamma - \chi_{ps} l^3 \int d\mathbf{r} \varrho_p(\mathbf{r}) \varrho_s(\mathbf{r})\right. \\ & - \frac{1}{2} \int d\mathbf{r} \int d\mathbf{r}' \frac{\varrho_e(\mathbf{r})}{k_B T} \varepsilon^{-1}(\mathbf{r}, \mathbf{r}') \left\{\frac{\varrho_e(\mathbf{r}')}{|\mathbf{r}-\mathbf{r}'|}\right\} + i \int d\mathbf{r} \varrho_e(\mathbf{r}) \psi(\mathbf{r}) \\ & \left. + i \int d\mathbf{r} \sum_\gamma w_\gamma(\mathbf{r}) \varrho_\gamma(\mathbf{r}) + i \int d\mathbf{r}\, \eta(\mathbf{r}) \left\{\sum_\gamma \varrho_\gamma(\mathbf{r}) - \varrho_0\right\}\right]. \end{aligned} \tag{6.86}$$

In Eq. (6.86), the Q's are the partition functions of individual components in the presence of a field. Explicitly, for the polyelectrolyte chain with smeared charge distribution along the backbone, single chain partition function is given by

$$Q_{\mathrm{p}} = \int D[\mathbf{R}]\exp\left[-\frac{3}{2l}\int_0^{Nl} \mathrm{d}t\left(\frac{\partial \mathbf{R}}{\partial t}\right)^2 - \frac{i}{l}\int_0^{Nl} \mathrm{d}t\,\{eZ_{\mathrm{p}}\alpha\psi(\mathbf{R}) + w_{\mathrm{p}}(\mathbf{R})\}\right]. \quad (6.87)$$

Similarly, the partition function for a solvent molecule is written as

$$Q_{\mathrm{s}} = \int \mathrm{d}\mathbf{r}\exp[-iw_{\mathrm{s}}(\mathbf{r})] \quad (6.88)$$

and the partition function for the small ions of type $j = \mathrm{c}, +, -$ is given by

$$Q_j = \int \mathrm{d}\mathbf{r}\exp[-ieZ_j\psi(\mathbf{r})]. \quad (6.89)$$

As mentioned earlier, all the functional integrals over collective variables cannot be carried out exactly. One of the approximations used extensively in the literature to evaluate these functional integrals is called the saddle-point approximation [52, 55, 57, 58]. In this approximation, functional integrals over collective variables are approximated by the value of the integrand at the saddle point, that is, free energy is approximated to be

$$\frac{F}{k_{\mathrm{B}}T} = -\ln Z \simeq f\left\{\varrho_{\gamma}^*, w_{\gamma}^*, \eta^*, \varrho_e^*, \psi^*\right\}, \quad (6.90)$$

where $\varrho_{\gamma}^*, w_{\gamma}^*, \eta^*, \varrho_e^*$, and ψ^* are to be obtained by solving the set of equations

$$\left.\frac{\delta f}{\delta \varrho_{\gamma}}\right|_{\varrho_{\gamma}=\varrho_{\gamma}^*} = 0, \quad \left.\frac{\delta f}{\delta w_{\gamma}}\right|_{w_{\gamma}=w_{\gamma}^*} = 0, \quad \left.\frac{\delta f}{\delta \eta}\right|_{\eta=\eta^*} = 0, \quad \left.\frac{\delta f}{\delta \varrho_e}\right|_{\varrho_e=\varrho_e^*} = 0, \quad \left.\frac{\delta f}{\delta \psi}\right|_{\psi=\psi^*} = 0. \quad (6.91)$$

Details of carrying out the functional derivatives [79] are presented elsewhere [53, 55]. The equations obtained after taking functional derivatives are presented here in the order presented in Eq. (73).

$$iw_{\mathrm{p}}^*(\mathbf{r}) = \chi_{\mathrm{ps}}l^3\varrho_{\mathrm{s}}^*(\mathbf{r}) + i\eta^*(\mathbf{r}), \quad (6.92)$$

$$iw_{\mathrm{s}}^*(\mathbf{r}) = \chi_{\mathrm{ps}}l^3\varrho_{\mathrm{p}}^*(\mathbf{r}) + i\eta^*(\mathbf{r}), \quad (6.93)$$

$$\varrho_{\mathrm{s}}^*(\mathbf{r}) = \frac{n_{\mathrm{s}}\exp[-iw_{\mathrm{s}}^*(\mathbf{r})]}{\int \mathrm{d}\mathbf{r}\exp[-iw_{\mathrm{s}}^*(\mathbf{r})]}, \quad (6.94)$$

$$\varrho_{\mathrm{p}}^*(\mathbf{r}) = \frac{n_{\mathrm{p}}\int_0^N \mathrm{d}t q(\mathbf{r},t)q^*(\mathbf{r},N-t)}{\int \mathrm{d}\mathbf{r} q(\mathbf{r},N)}, \quad (6.95)$$

$$\varrho_p^*(\mathbf{r}) + \varrho_s^*(\mathbf{r}) = \varrho_0, \tag{6.96}$$

$$i\psi^*(\mathbf{r}) = \frac{1}{k_B T}\int d\mathbf{r}' \varepsilon^{-1}(\mathbf{r},\mathbf{r}')\frac{\varrho_e^*(\mathbf{r}')}{|\mathbf{r}-\mathbf{r}'|}, \tag{6.97}$$

$$\varrho_e^*(\mathbf{r}) = e\left[Z_p\alpha\varrho_p^*(\mathbf{r}) + \sum_{j=+,-,c} \frac{n_j Z_j \exp[-ieZ_j\psi^*(\mathbf{r})]}{\int d\mathbf{r} \exp[-ieZ_j\psi^*(\mathbf{r})]}\right]. \tag{6.98}$$

In these equations, $q(\mathbf{r}, t)$ satisfies the modified diffusion equation [50, 52, 53, 55]

$$\frac{\partial q(\mathbf{r},t)}{\partial t} = \left[\frac{l^2}{6}\nabla_{\mathbf{r}}^2 - i\left\{eZ_p\alpha\psi^*(\mathbf{r}) + w_p^*(\mathbf{r})\right\}\right]q(\mathbf{r},t) \tag{6.99}$$

for $t \in [0, N]$ with the initial condition $q(\mathbf{r}, 0) = 1$. Similarly, $q^*(\mathbf{r}, N-t)$ satisfies

$$\frac{\partial q^*(\mathbf{r},N-t)}{\partial t} = -\left[\frac{l^2}{6}\nabla_{\mathbf{r}}^2 - i\left\{eZ_p\alpha\psi^*(\mathbf{r}) + w_p^*(\mathbf{r})\right\}\right]q^*(\mathbf{r},N-t) \tag{6.100}$$

with initial condition $q^*(\mathbf{r}, 0) = 1$. Also, note that on approximating the nonlocal dielectric function by a local function so that $\varepsilon(\mathbf{r}, \mathbf{r}') \rightarrow \varepsilon(\mathbf{r})$, Eq. (6.97) can be written in the differential form as

$$\nabla_{\mathbf{r}}\varepsilon(\mathbf{r})\nabla_{\mathbf{r}} ie\psi^*(\mathbf{r}) = -\frac{4\pi e^2}{k_B T}\left[Z_p\alpha\varrho_p^*(\mathbf{r}) + \sum_{j=+,-,c} \frac{n_j Z_j \exp[-ieZ_j\psi^*(\mathbf{r})]}{\int d\mathbf{r} \exp[-ieZ_j\psi^*(\mathbf{r})]}\right], \tag{6.101}$$

which is the familiar Poisson–Boltzmann equation for inhomogeneous dielectric media and has been used to incorporate the effect of inhomogeneous dielectric constant using SCFT [62]. Different numerical techniques, which are useful for solving the nonlinear set of equations, are presented in the next section.

6.4.5
Numerical Techniques

To compute the free energy within saddle-point approximation, the coupled nonlinear equations, which include second-degree partial differential equations such as the modified diffusion and Poisson–Boltzmann equations are to be solved. General strategy to solve these equations is to start from an initial guesses for the fields and compute the densities. Using the computed densities and the old guesses for the fields, compute the new guesses for the fields. The guessing process is iterated till the converged solutions for the densities and fields are obtained. Different numerical techniques are used for computing the new guesses using the old guesses and the recently computed densities. An excellent reference to look into the currently available techniques for the guessing process is Ref. [55].

In order to compute the densities for a given guess for the fields, the modified diffusion and the Poisson–Boltzmann equations need to be solved. A number of numerical techniques for solving the modified diffusion equation in the context of neutral polymers already exist in the literature [55, 80]. Similar techniques can be used for solving the modified diffusion equation encountered in the case of polyelectrolytes. Furthermore, these techniques can be readily generalized to solve the Poisson–Boltzmann equation.

In general, these numerical techniques can be classified into three classes: finite difference, spectral and pseudospectral methods. Finite difference methods [81] are straightforward to implement after approximating partial derivatives by finite difference approximation schemes. However, depending on the problem at hand, the accuracy and convergence attained using these methods may not be what is desired for computations. In certain problems, accuracy is the main issue, sometimes even at the cost of memory. That is the reason more accurate (but difficult to implement) numerical techniques such as spectral and pseudospectral methods have been developed. These techniques provide higher accuracy and better convergence. However, these techniques have a common drawback that they are very specialized and work for problems with specific boundary conditions. The choice of a numerical technique for a certain problem depends on the accuracy, convergence, and memory issues, which appear during the implementation of different schemes. Here, we present only a brief description of the relevant details of these techniques for solving Poisson–Boltzmann equation along with the modified diffusion equation.

Consider a general Poisson–Boltzmann equation (cf. Eq. (6.101)) of the form

$$\nabla_{\mathbf{r}}\varepsilon(\mathbf{r})\nabla_{\mathbf{r}}ie\psi(\mathbf{r}) = -f\{\psi(\mathbf{r})\}, \tag{6.102}$$

when $f\{\psi(\mathbf{r})\}$ is known. Also, consider another equation of the form

$$\frac{\partial h(\mathbf{r},t)}{\partial t} = \left[\frac{l^2}{6}\nabla_{\mathbf{r}}^2 - w(\mathbf{r})\right]h(\mathbf{r},t), \tag{6.103}$$

where $w(\mathbf{r})$ is known a priori along with the initial condition $h(\mathbf{r},0)$. This equation corresponds to the time-dependent modified diffusion equation. Three types of commonly used numerical techniques to solve these two equations are presented in the next section.

6.4.5.1 Finite Difference Methods

Traditional finite difference methods [55, 81] for solving time-dependent second-degree partial differential equations (such as modified diffusion equation) include forward time-centered space (FTCS), Crank–Nicholson, and so on. For time-independent second-degree partial differential equations such as Poisson–Boltzmann equation, finite difference equations can be written after discretizing the space and approximating derivatives by their finite difference approximations. For *space-independent* dielectric constant, that is, $\varepsilon(\mathbf{r}) = \varepsilon$, a tridiagonal matrix inversion needs to be carried out in order to obtain a solution for ψ for a given f.

As matrix inversion is computationally very costly, so this particular technique is limited to one-dimensional problems. Also, the generalization of Eq. (6.102) to space-dependent dielectric constant creates extra numerical difficulties while solving Poisson–Boltzmann equation.

In order to deal with the case of space-dependent dielectric constant in multi-dimensional space, alternating direction implicit techniques [82] are developed after rewriting Eq. (6.102) as

$$\varepsilon(\mathbf{r})\frac{\partial\psi(\mathbf{r})}{\partial t} = \nabla_{\mathbf{r}}\varepsilon(\mathbf{r})\nabla_{\mathbf{r}}ie\psi(\mathbf{r}) + f\{\psi(\mathbf{r})\}, \tag{6.104}$$

and the steady-state solution of Eq. (6.104) corresponds to the solution of Eq. (6.102).

6.4.5.2 **Spectral Method: Method of Basis Functions**

This particular technique was introduced to the polymer literature by Matsen and Schick [83] in the context of diblock copolymer morphologies. The technique is based on the series expansion of any unknown function in terms of suitable basis functions and the numerical work is carried out to compute the coefficients of different terms in the series. For example, to solve Eq. (6.103), let us approximate space-dependent quantities such as $h(\mathbf{r},t)$ and $w(\mathbf{r})$ by a *finite* series in terms of orthonormal basis functions, that is, $h(\mathbf{r},t) \simeq \sum_{j=1}^{n} h_j(t)g_j(\mathbf{r})$ and $w(\mathbf{r}) \simeq \sum_{j=1}^{n} w_j g_j(\mathbf{r})$, where $g_j(\mathbf{r})$ represents the appropriate basis function of order j and n is the number of such basis functions required to correctly represent the functions $h(\mathbf{r},t)$ and $w(\mathbf{r})$. The choice of n depends on the desired accuracy for computations. In order to solve Eq. (6.103), the basis functions must have the following properties:

(a) Basis functions, $g_j(\mathbf{r})$, must be the eigenfunctions of the Laplacian operator, that is,

$$\nabla_{\mathbf{r}}^2 g_j(\mathbf{r}) = -\frac{\lambda_j}{L^2} g_j(\mathbf{r}), \tag{6.105}$$

where $j = 2, 3, \ldots, n$, and L is the length scale describing the volume of the system, and λ_j's are the eigenvalues of the Laplacian. The first basis function, $g_j(\mathbf{r})$, is chosen to be a constant and normally unity, that is, $g_1(\mathbf{r}) = 1$.

(b) These basis functions must be the orthonormal basis set, that is, for a given volume of the system $(= \Omega \sim L^3)$, they must satisfy $\frac{1}{\Omega}\int d\mathbf{r} g_j(\mathbf{r})g_k(\mathbf{r}) = \delta_{jk}$.

(c) These basis functions must satisfy the boundary conditions.

In this technique, the basis functions are ordered starting with $g_1(\mathbf{r}) = 1$ such that λ_j is a nondecreasing series. Also, the constraints on the fields and densities are taken care of by fixing the first term in the series, which is independent of $\mathbf{r}$. Now, the goal is to compute the coefficients $h_j(t)$ in the finite series expansion for $h(\mathbf{r},t)$ using the initial values of these coefficients $h_j(0)$ (which comes from the known initial condition) and the known values for w_j's (due to the known values for $w(\mathbf{r})$). Using the finite series expansion and orthonormal property of the basis functions, Eq. (6.103) can be transformed into a set of equations for coefficients $h_j(t)$ as shown below.

$$\frac{\partial h_k(t)}{\partial t} = \sum_{j=1}^{n} A_{kj} h_j(t), \tag{6.106}$$

where

$$A_{kj} = \frac{-\lambda_k l^2}{6L^2}\delta_{kj} - \sum_{i=1}^{n} w_i \Gamma_{ijk}, \tag{6.107}$$

where the function Γ_{ijk} is defined as $\Gamma_{ijk} = \frac{1}{\Omega}\int \mathrm{d}\mathbf{r} g_i(\mathbf{r}) g_j(\mathbf{r}) g_k(\mathbf{r})$. Now, the equations for $h_k(t)$ in the matrix form become

$$\frac{\partial h(t)}{\partial t} = Ah(t), \tag{6.108}$$

where $h(t)$ is a column vector and A is a square matrix given by

$$h(t) = \begin{bmatrix} h_1(t) \\ h_2(t) \\ h_3(t) \\ \vdots \\ h_n(t) \end{bmatrix}, \quad A = \begin{bmatrix} A_{11} & A_{12} & \dots & A_{1n} \\ A_{21} & A_{22} & \dots & A_{2n} \\ A_{31} & A_{32} & \dots & A_{3n} \\ \vdots & \vdots & \vdots & \vdots \\ A_{n1} & A_{n2} & \dots & A_{nn} \end{bmatrix}$$

Formal solution of Eq. (6.108) is given by $h(t) = \mathrm{e}^{At} h(0)$, where $h(0)$ is given by the initial condition for $h(\mathbf{r}, t)$. e^{At} is the exponential of the matrix At and can be calculated by diagonalizing the matrix At. Note here that t is a scalar.

Computation of exponential of a matrix is a numerically intensive job and very difficult in general. However, for real, symmetric square matrices such as A here, it can be calculated by diagonalizing the matrix after computing its eigenvalues and eigenvectors. An approach to calculate the exponential of a matrix is to compute the exponential of a matrix times a scalar, we need to solve the matrix problem $\frac{\mathrm{d}X(t)}{\mathrm{d}t} = AX(t)$ for a given $X(0)$, when the matrix A is diagonalizable into a diagonal matrix d. Owing to the fact that A can be diagonalized, there exists a nonsingular matrix P such that $P^{-1}AP = d$. Now, to solve the matrix equation for X, we change variables as

$$X(t) = PY(t), \tag{6.109}$$

$$\Rightarrow \frac{\mathrm{d}X(t)}{\mathrm{d}t} = AX(t) \Rightarrow \frac{\mathrm{d}Y(t)}{\mathrm{d}t} = \mathrm{d}Y(t), \quad Y(0) = P^{-1}X(0), \tag{6.110}$$

$$\Rightarrow X(t) = P\,\mathrm{diag}\left[\mathrm{e}^{d_1 t}, \mathrm{e}^{d_2 t}, \dots, \mathrm{e}^{d_n t}\right] P^{-1} X(0), \tag{6.111}$$

where diag stands for a diagonal matrix. If the matrix P is made up of eigenvectors of matrix A as its columns, then for symmetric, real matrices it is orthogonal, that is,

$P^{\mathrm{T}} = P^{-1}$, where superscripts T and -1 represent the transpose and inverse of the matrix, respectively. This means, for real symmetric square matrices such as A here, exponential of the matrix At is given by

$$\mathrm{e}^{At} = P\,\mathrm{diag}\left[\mathrm{e}^{d_1 t}, \mathrm{e}^{d_2 t}, \ldots, \mathrm{e}^{d_n t}\right] P^{\mathrm{T}}, \tag{6.112}$$

where d_j is the jth eigenvalue of matrix A and P is a matrix whose columns are made up of the eigenvectors of the matrix A.

The computation of exponential of matrix is in fact the most expensive part in the numerical solution of modified diffusion equations. Also, as we can see that as the order of the matrix (n) increases, which in turn determines the accuracy of the method, the computational cost substantially increases. That is the reason this method should be used in problems where low values of n meet the accuracy requirements. From our experience, problems requiring $n \leq 30$ can be readily solved using this technique. The second issue with this technique is the availability of an appropriate basis set. There are only a few known orthonormal basis sets and that too depends on certain specialized geometries and specific boundary conditions. These issues make this technique very specialized.

Using this technique, the solution of Poisson–Boltzmann equation as in Eq. (6.102) for the case $\varepsilon(\mathbf{r}) = \varepsilon$ becomes trivial. The solution is given by $\psi_j = L^2 f_j/\lambda_j$ for $j > 1$, where f_j's are computed from the given values of $f\{\psi(\mathbf{r})\}$. The first component, ψ_1, is generally set to zero, which also fixes the unknown constant in ψ by assuring that $\int \mathrm{d}\mathbf{r}\psi(\mathbf{r}) = 0$ in these computations. Spectral method has recently been used to study microphase separation in charged-neutral diblock copolymer melts [63]. The Poisson–Boltzmann for the inhomogeneous dielectric media has not been solved using the spectral method so far.

6.4.5.3 **Pseudospectral Method**

With the increase in the desired number of basis functions in the finite series expansion, the spectral method demands a lot of memory and becomes extremely expensive. The sets of problems, where the required number of basis functions are large can be solved by pseudospectral method [84], which optimize both speed and accuracy.

This method is based on the use of operators in solving Eq. (6.103). Within the operator formalism employed by the technique, formal solution of Eq. (6.103) is given by

$$h(\mathbf{r}, t+\mathrm{d}t) = \exp\left[\mathrm{d}t\left(\frac{l^2}{6}\nabla_{\mathbf{r}}^2 - w(\mathbf{r})\right)\right] h(\mathbf{r}, t). \tag{6.113}$$

In Eq. (6.113), three-dimensional Laplacian and $w(\mathbf{r})$ are treated as operators. Now, the goal is to approximate the exponential of the operator on the right-hand side in Eq. (6.113) by some technique. Approximation scheme must respect the fact that for any two operators A and B, $\mathrm{e}^{A+B} \neq \mathrm{e}^A \mathrm{e}^B$ unless they commute. The approximation scheme employed by the pseudospectral method uses the fact that if the operator on the right-hand side in Eq. (6.113) can be split into two parts, then it is relatively easy to implement numerically. A powerful approximation scheme for computing the exponential of an operator by splitting into a sum of noncommuting operators uses the well-known Baker–Campbell–Hausdorff [85] formula. According to this formula, if $e^A e^B = e^C$, then

$$C = A + B + \frac{1}{2}[A, B] + \frac{1}{12}[[A, B], (B-A)] + \frac{1}{24}\left[B, A^2, B\right] + \cdots, \tag{6.114}$$

where $[x, y] = xy - yx$ is the commutator, $[x, y^2] = [[x, y], y]$ and $[x, y^2, x] = [[x, y^2], x]$.

Splitting of the operators can be quite tricky. In the case of Eq. (6.113), the splitting is motivated by the observation that it is easy to implement $\exp(-w(\mathbf{r}))$ in real space (at all points on the mesh), but it is difficult to take double derivatives at all points on the mesh. This difficulty can be overcome by using the fact that the derivatives can be implemented trivially in Fourier space. So, if somehow the Laplacian part could be split from $w(\mathbf{r})$ part, then the progress can be made. This is the motivation for splitting the operator in Eq. (6.113) into derivative and nonderivative parts by Baker–Campbell–Hausdorff formula. For most practical purposes, a symmetric decomposition is carried out so that within an error of the order dt^3

$$\exp\left[\mathrm{d}t\left(\frac{l^2}{6}\nabla_{\mathbf{r}}^2 - w(\mathbf{r})\right)\right] \simeq \exp\left[-\frac{\mathrm{d}t}{2}w(\mathbf{r})\right]\exp\left[\mathrm{d}t\frac{l^2}{6}\nabla_{\mathbf{r}}^2\right]\exp\left[-\frac{\mathrm{d}t}{2}w(\mathbf{r})\right]. \tag{6.115}$$

So, solution of Eq. (6.103) can be approximated by

$$h(\mathbf{r}, t + \mathrm{d}t) \simeq \exp\left[-\frac{\mathrm{d}t}{2}w(\mathbf{r})\right]\exp\left[\mathrm{d}t\frac{l^2}{6}\nabla_{\mathbf{r}}^2\right]\exp\left[-\frac{\mathrm{d}t}{2}w(\mathbf{r})\right]h(\mathbf{r}, t), \tag{6.116}$$

which is correct within an error of $\mathrm{d}t^3$.

Numerical implementation of Eq. (6.116) uses the fact that the rightmost operator in the exponential can be implemented as it is in real space. Let us say $g(\mathbf{r}, t)$ is the outcome of this operation, which is a function of position $(\mathbf{r})$ at some t, that is,

$$g(\mathbf{r}, t) = \exp\left[-\frac{\mathrm{d}t}{2}w(\mathbf{r})\right]h(\mathbf{r}, t). \tag{6.117}$$

For implementing the exponential of the Laplacian operator for systems with periodic boundary conditions

$$\exp\left[\mathrm{d}t\frac{l^2}{6}\nabla_{\mathbf{r}}^2\right]g(\mathbf{r}, t) = \mathrm{FT}^{-1}\left[\exp\left\{-\mathrm{d}t\frac{k^2 l^2}{6}\right\}\mathrm{FT}\{g(\mathbf{r}, t)\}\right], \tag{6.118}$$

where FT stands for Fourier transform and FT^{-1} stands for inverse Fourier transform, which can be carried out quite efficiently using fast Fourier transforms (FFTs). Hence, the numerical solution of Eq. (6.103) can be obtained by using

$$h(\mathbf{r}, t + \mathrm{d}t) \simeq \exp\left[-\frac{\mathrm{d}t}{2}w(\mathbf{r})\right]\mathrm{FT}^{-1}\left[\exp\left\{-\mathrm{d}t\frac{k^2 l^2}{6}\right\}\mathrm{FT}\left\{\exp\left[-\frac{\mathrm{d}t}{2}w(\mathbf{r})\right]h(\mathbf{r}, t)\right\}\right]. \tag{6.119}$$

So, in pseudospectral method, one has to switch between real space and Fourier space back and forth. Symmetric decomposition leads to the implementation of half time step in real space and the other half in Fourier space. That is the reason this method is also known as split-step method in the literature [84].

Owing to the use of Fourier Transform in the implementation of the exponential of the Laplacian operator, the technique is mainly useful for systems with periodic boundary conditions. Also, note that the technique cannot be used to solve Poisson–Boltzmann equation due to time-independent nature of the equation. However, the Poisson–Boltzmann equation can be solved using finite difference techniques or by a combination of fast Fourier transforms and the finite difference techniques [68, 81].

6.4.6 Fluctuations Around the Saddle Point

A number of attempts have been made to incorporate the effect of composition fluctuations [86–88] in theories involving neutral polymers. Here, we present systematic one-loop expansion to go beyond the saddle-point approximation described in the previous section. In order to carry out the loop expansion, it is advantageous to use Hubbard–Stratonovich transformation to get rid of redundant functional integrals over collective density variables (ϱ in Eq. (6.85)) and use Eq. (6.81) as the starting point for the partition function with the explicitly known normalization constants except Λ_η. Saddle-point approximation within this formalism now requires taking functional derivatives with respect to fields only.

Let us say after summing over charge distributions, the partition function becomes (cf. Eq. (6.81))

$$Z = \frac{1}{\Lambda_\eta \Lambda_\psi \prod_j \Lambda_j} \int \prod_j D[\varsigma_j] \int D[\eta] \int D[\psi] \exp[-f']. \tag{6.120}$$

Within saddle-point approximation, the partition function is approximated by $Z \simeq f'\left\{\varsigma_j^*, \eta^*, \psi^*\right\}$, where saddle-point values for the fields are to be obtained by extremizing f' with respect to the fields. For n component system, there are $n+1$ field variables designated by ς_j, η, and ψ, for $j = 1, 2, \ldots, (n-1)$. For notational convenience, let us write them as ς_j, where $j = 1, 2, \ldots (n+1)$ and $j = n, n+1$ represent η and ψ, respectively. To go beyond the saddle-point approximation, we use the functional Taylor expansion [79] of the integrand and neglect all the terms beyond quadratic terms, that is, we write

$$f'\left\{\varsigma_j\right\} = f'\left\{\varsigma_j^*\right\} + \frac{1}{2}\int d\mathbf{r} \int d\mathbf{r}' \sum_{jk} K_{jk}(\mathbf{r}, \mathbf{r}')(\varsigma_j(\mathbf{r}) - \varsigma_j^*(\mathbf{r}))(\varsigma_k(\mathbf{r}') - \varsigma_k^*(\mathbf{r}')), \tag{6.121}$$

where the linear terms in ς_j do not appear due to the saddle-point conditions and

$$K_{jk}(\mathbf{r}, \mathbf{r}') = \frac{\delta^2 f'\{\varsigma_j\}}{\delta\varsigma_j(\mathbf{r})\delta\varsigma_k(\mathbf{r}')}\Big|_{\varsigma_j = \varsigma_j^*}, \tag{6.122}$$

for $j, k = 1, 2, \ldots, (n+1)$. Plugging the functional Taylor expansion for f' in Eq. (6.120), the functional integrals to be carried out are Gaussian in nature and can be carried out. Formally, the result is the one-loop approximation [14, 53] for the free energy in terms of a ratio of determinants of continuous block matrices, that is,

$$\frac{F}{k_B T} \simeq f'\left\{\varsigma_j^*\right\} + \frac{1}{2}\ln\frac{\det K}{\det K_0}, \qquad (6.123)$$

where K is the square block matrix of order $n+1$ with its block elements represented by K_{jk} and K_0 is the diagonal square block matrix of order $n+1$ (which appear because of the normalization factors and unknown normalization constant Λ_η has been used to define K_0). It turns out that the individual determinants appearing in Eq. (6.123) are divergent due to the presence of divergent terms on the principal diagonal, which in turn highlights the fact that the model is ill defined at very small length scales. This is similar to the familiar ultraviolet divergences appearing in one-loop calculations in field theory. The unknown Λ_η can be estimated by identifying the divergent terms in $\det K$, which should cancel out exactly at one-loop level.

Similar attempts to include fluctuations beyond one-loop in SCFT have also been exercised in the context of neutral polymers using field theoretical simulations [55, 90] or by bridging SCFT with Monte Carlo techniques [91]. However, these techniques have not been applied for the case of polyelectrolytes with counterions and added salt ions due to very high computational cost.

In the next section, we compare the results of the degree of ionization of a flexible polyelectrolyte chain obtained from the self-consistent field theory within saddle-point approximation and compare with the previously developed variational theory [21] by considering the chain in a spherical cavity in the presence of a monovalent salt. Physically, such a situation is realized in extremely dilute polyelectrolyte solutions, where interchain interactions can be ignored safely and a finite volume can be carved out for each chain, and in pores confining polyelectrolyte chains, depending on the ratio of the cavity size to the radius of gyration of the polymer. We use the self-consistent field theory to compute the equilibrium degree of counterion adsorption after taking a "permuted" charge distribution on the chain so that the "adsorbed" counterions are allowed to move along the backbone. SCFT computes the free energy of the system by summing over all possible conformations of the chain, and hence, provides a more accurate description of the system (in fact, it provides the exact free energy at the mean field level) compared to the variational formalism. Also, as the electrostatics in SCFT is treated at full, nonlinear Poisson–Boltzmann level, we can assess the validity of the Debye–Hückel potential to describe the electrostatic energy in the variational formalism. Although SCFT provides an accurate and clear picture, it is computationally expensive to calculate the degree of ionization due to a vast parameter space in the case of polyelectrolytes. On the other hand, the variational theory put forward by Muthukumar [1] is transparent, analytically tractable (to some extent), and very inexpensive in terms of the computational needs. The aim of the comparison presented in the next section is to provide a simple, accurate, and easy-to-use method to compute the degree of ionization and assess the approximations used in the variational formalism.

6.5 Comparison of Theories: SCFT and Variational Formalism

To carry out the field theoretical study for a single chain, the chain is represented as a continuous curve of length Nl, N being the number of Kuhn segments, each of length l. Arc length variable t is used to represent any segment along the backbone so that $t \in [0, Nl]$. Also, we assume that the chain (negatively charged) is surrounded by n_c monovalent counterions (positively charged) released by the chain along with n_γ ions of species $\gamma(= +, -)$ coming from added salt so that the whole system is globally electroneutral. Keeping the notation the same as in the previous section, that is, Z_j being the valency (with sign) of the charged species of type j, and n_s being the number of solvent molecules (satisfying the incompressibility constraint after assuming the monovalent small ions to be point-like), we assume that the volume of a solvent molecule (v_s) is equal to the volume of the monomer (that means $v_s \equiv l^3 = 1/\varrho_0$, ϱ_0 being the total number density of the system). Also, subscripts p, s, c, +, and − are used to represent monomer, solvent, counterion from polyelectrolyte, and positive and negative salt ions, respectively.

In order to study counterion adsorption using the two theoretical frameworks, we use the so-called "two-state" model for the counterions so that there are two populations of counterions in the system. One population of the counterions is free to enjoy the available volume (called the "free" counterions) and the other population is "adsorbed" on the backbone. However, the adsorbed counterions are allowed to enjoy translational degrees of freedom along the backbone, maintaining a total charge of $efNZ_p$ on the chain, where e is the electronic charge and f is the degree of ionization of the chain (that means there are $-(1-f)NZ_p/Z_c$ "adsorbed" counterions on the chain). In the literature, this kind of charge distribution has been referred to as a "permuted" charge distribution [68].

For a particular set of parameters, we compute the free energy of the system comprising the single chain, its counterions, the salt ions, and the solvent as a function of the degree of ionization (f) using two different computational frameworks: SCFT [14, 55] and the variational [1, 21, 92] formalism. In both the formalisms, we ignore the electrostatic interactions between the solvent molecules and the small ions, and model the dielectric constant (ε) of the medium to be independent of temperature (T) to extract energy and entropy of the system. Also, for comparison purposes, we divide the free energy into a mean field part and an additional part, which goes beyond the mean field theory. Mean field part is further divided into the contributions coming from the "adsorbed" counterions (F_a^*) and the "free" ions, and from the chain entropy, and so on (F_f^*). All contributions are properly identified (or subdivided into) as enthalpic or/and entropic parts. In all of what follows, the superscript $*$ represents the mean field part. A brief description of the derivation for the two formalisms is presented in Sections 6.5.1 and 6.5.2, and the original references [1, 14] may be consulted for details.

6.5.1
Self-Consistent Field Theory for Single Chain

The theoretical description presented above for multichain polyelectrolyte solutions can be easily adapted to study a single flexible polyelectrolyte chain confined in a spherical cavity of volume $\Omega = 4\pi R^3/3$. For a single flexible polyelectrolyte chain with a fixed degree of ionization $(=f)$ in the presence of salt ions and solvent molecules, the Hamiltonian can be written as

$$\exp\left(-\frac{F-F_0}{k_B T}\right) = \frac{\exp[-E_a/k_B T]}{\mu\prod_j n_j!}\int D[\mathbf{R}]\int\prod_j\prod_{i=1}^{n_j}\mathrm{d}\mathbf{r}_i\ \left[\exp\left\{-\frac{3}{2l}\int_0^{Nl}\mathrm{d}t\left(\frac{\partial\mathbf{R}(t)}{\partial t}\right)^2\right.\right.$$
$$\left. -\chi_{ps}b^3\int\mathrm{d}\mathbf{r}\,\hat{\varrho}_p(\mathbf{r})\hat{\varrho}_s(\mathbf{r}) - \frac{1}{2}\int\mathrm{d}\mathbf{r}\int\mathrm{d}\mathbf{r}'\frac{\hat{\varrho}_e(\mathbf{r})\hat{\varrho}_e(\mathbf{r}')}{4\pi\varepsilon_0\varepsilon k_B T|\mathbf{r}-\mathbf{r}'|}\right\}$$
$$\left.\prod_{\mathbf{r}}\delta(\hat{\varrho}_p(\mathbf{r})+\hat{\varrho}_s(\mathbf{r})-\varrho_0)\,\delta\left(\frac{1}{l}\int_0^{Nl}\mathrm{d}t\,\gamma(t)-fNe\right)\right]_\gamma, \tag{6.124}$$

where $\mathbf{R}(t)$ represents the position vector for the tth segment and subscripts $j = \mathrm{s,c},+,-$. In Eq. (6.124), $k_B T$ is the Boltzmann constant times absolute temperature. In writing the interaction energies between the polyelectrolyte segments and the small ions, we have taken the small ions to be point charges so that they have zero excluded volume, and hence, interactions are purely electrostatic in nature. As we consider the "permuted" charge distribution, the partition function has an additional sum over all possible locations of the "adsorbed" ions on the backbone, which appears as an average over the parameter γ in Eq. (6.124). We define the average over γ as $[\cdots]_\gamma = \int \mathrm{d}\gamma[\cdots]g(\gamma)$, where $g(\gamma) = f\delta(\gamma(t)-1)+(1-f)\delta(\gamma(t))$. For a detailed description of the different terms in Eq. (6.124), see Section 6.4.1. Also, $\hat{\varrho}_p(\mathbf{r}), \hat{\varrho}_j(\mathbf{r})$ and $\hat{\varrho}_e(\mathbf{r})$ stand for the monomers, small molecules (both ions and solvent molecules), and the local charge density, respectively. These are defined by Eqs. (6.65), (6.66) and (6.72), respectively, after putting $n_p = 1$.

The additional delta function involving γ is a constraint that for all the charge distributions to be considered for one particular value of f, the net charge on the chain must be a constant $(= fNe)$. Taking different charge distributions of the chain for the same net charge $(= fNe)$ and a particular chain conformation to be degenerate, the partition function is divided by the number of ways (μ) the "adsorbed" counterions can be distributed along the chain. If M out of total N sites on the backbone are occupied at any particular instance, then μ is given by $\mu = N!/(M!(N-M)!)$ so that $1-f = M/N$.

Defining the dimensionless Flory parameter for chemical mismatch, χ_{ps} by Eq. (6.71) after using w_{pp}, w_{ss}, and w_{ps} as the excluded volume parameters characterizing the short-range excluded volume interactions of type monomer–monomer, solvent–solvent, and monomer–solvent pairs, respectively, the self-energy (F_0) and ion-pair energy (E_a) contributions are given by

$$\frac{F_0}{k_B T} = N w_{pp} + n_s w_{ss}, \tag{6.125}$$

$$\frac{E_a}{k_B T} = -(1-f) N \delta l_B / l, \tag{6.126}$$

where $\delta = \varepsilon l / \varepsilon_l d$, ε_l, and d being the local dielectric constant and the dipole length, respectively, are used to characterize the formation of an ion-pair on the backbone due to "adsorbed" counterion.

Now, using the methods of collective variables (cf. Section 6.4.2.1) for decoupling all the interactions except the electrostatics and the Hubbard–Stratonovich transformation [14, 55] (cf. Section 6.4.2.2) for the electrostatic part in Eq. (6.124), the partition function can be written as integrals over the collective densities and corresponding fields so that Eq. (6.124) becomes

$$\exp\left(-\frac{F-F_0}{k_B T}\right) = \int D[w_p] D\left[\varrho_p\right] D[\psi] D[\eta] \mathrm{d}u D[w_s] D[\varrho_s] \left[\exp\left\{-\frac{H_{scf}}{k_B T}\right\}\right]_{\gamma}. \tag{6.127}$$

Here, w_p, w_s are the collective fields experienced by the monomers and solvent, respectively, and ϱ_p, ϱ_s represent their respective collective densities. All charged species (excluding the ion-pairs formed due to adsorption of counterions) experience a field ψ (which is equivalent to the electrostatic potential). η and u are Lagrange's multipliers corresponding to, respectively, the incompressibility and net charge constraints in the partition function.

Invoking the saddle-point approximation, extremization of the integrand leads to a number of nonlinear equations for the fields and the densities. The saddle-point approximation with respect to u gives equations similar to a "smeared" charge distribution, where every monomer has a charge ($=fe$). The extremization with respect to $\psi, w_p, \varrho_p, \eta, w_s$, and ϱ_s leads to the saddle point specified by equations similar to Eqs. (6.92)–(6.101). Using these saddle-point equations and employing the Stirling approximation for $\ln n!$, we obtain the approximated free energy, that is, $F - F_0 \simeq H^*_{scf}$ after taking $k_B T = 1$. The superscript $*$ represents the saddle-point estimate of the free energy.

After solving these equations for fields (and, in turn, for densities), the free energy at the saddle point, $F_{SCFT} = F_0 + F^*_a + F^*_f$, is divided into enthalpic and entropic contributions due to different components. To start with, we note that the contributions coming from "adsorbed" counterions can be divided as

$$F^*_a = E_a - T S_a, \tag{6.128}$$

$$E_a = -(1-f) N \delta l_B / l, \tag{6.129}$$

$$-T S_a = N[f \log f + (1-f) \log (1-f)], \tag{6.130}$$

Table 6.2 Comparison of contributions to F_{f}^{*} in SCFT and variational formalism.

Term	SCFT	Variational formalism
$E_{\mathrm{w}}-TS_{\mathrm{s}}$	$\chi_{\mathrm{ps}} l^3 \int \mathrm{d}\mathbf{r}\, \varrho_{\mathrm{p}}(\mathbf{r})\varrho_{\mathrm{s}}(\mathbf{r}) + \varrho_0 \int \mathrm{d}\mathbf{r}\, \eta(\mathbf{r})$ $+ \int \mathrm{d}\mathbf{r}\, \varrho_{\mathrm{s}}(\mathbf{r})\{\ln[\varrho_{\mathrm{s}}(\mathbf{r})]-1\}$	$\frac{4}{3}\left(\frac{3}{2\pi}\right)^{3/2}(1-2\chi_{\mathrm{ps}})\sqrt{N}/\tilde{l}_1^{3/2} + \chi_{\mathrm{ps}} N l^3 - \Omega$
E_{e}	$\frac{1}{2}\int \mathrm{d}\mathbf{r}\, \psi(\mathbf{r})\varrho_{\mathrm{e}}(\mathbf{r})$	$2\sqrt{\frac{6}{\pi}} f^2 \tilde{l}_{\mathrm{B}} N^{3/2} \Theta_0(a)/\tilde{l}_1^{1/2}$
$-TS_{\mathrm{i}}$	$\sum_{j=\mathrm{c},+,-} \int \mathrm{d}\mathbf{r}\, \varrho_j(\mathbf{r})\left\{\ln\left[\varrho_j(\mathbf{r})\right]-1\right\}$	$(fN+n_+)\ln\frac{fN+n_+}{\Omega} + n_- \ln\frac{n_-}{\Omega}$ $-(fN+n_+ + n_-)$
$-TS_{\mathrm{p}}$	$-\ln\left[\int \mathrm{d}\mathbf{r}\, q(\mathbf{r},N)/\int \mathrm{d}\mathbf{r}\, q_0(\mathbf{r},N)\right] - \varrho_0 \int \mathrm{d}\mathbf{r}\, \eta(\mathbf{r})$ $- \int \mathrm{d}\mathbf{r}[\{Z_{\mathrm{p}} f \psi(\mathbf{r}) + w_{\mathrm{p}}(\mathbf{r})\}\varrho_{\mathrm{p}}(\mathbf{r})]$	$\frac{3}{2}\left[\tilde{l}_1 - 1 - \ln \tilde{l}_1\right]$

so that E_{a} is the electrostatic binding energy of the ion-pairs formed on the polymer backbone due to the adsorption of ions and S_{a} is the translational entropy of the "adsorbed" counterions along the backbone.

Similarly, F_{f}^{*} is divided into enthalpic and entropic contributions due to the excluded volume and electrostatic interactions and into entropic contributions because of small ions, solvent molecules, and the polyelectrolyte chain. Denoting these contributions by E_{w}, E_{e}, S_i, S_{s}, and S_{p}, respectively, F_{f}^{*} is given by

$$F_{\mathrm{f}}^{*} = E_{\mathrm{w}} + E_{\mathrm{e}} - T(S_i + S_{\mathrm{s}} + S_{\mathrm{p}}), \tag{6.131}$$

Explicit expressions for different constituents of F_{f}^{*} are presented in Table 6.2 in terms of densities and fields at the saddle point.

In order to compare the free energies obtained from SCFT and the variational formalism for a given N and R, a single Gaussian chain of contour length Nl in the volume Ω is chosen as the reference frame, whose free energy is taken to be zero. This reference free energy of confinement for a single Gaussian chain has been subtracted from the polymer conformational entropy in Table 6.2. The free energy of confining a single Gaussian chain with N Kuhn segments of length l each in a spherical cavity of radius R can be computed exactly and is given by [93]

$$F_{\mathrm{Gaussian}} = -\ln\left[\int \mathrm{d}\mathbf{r} q_0(\mathbf{r},N)\right] = -\ln\left[\frac{6\Omega}{\pi^2}\sum_{k=1}^{\infty}\frac{1}{k^2}\exp\left[-\frac{k^2\pi^2 N l^2}{6R^2}\right]\right]. \tag{6.132}$$

6.5.2
Variational Formalism

For the sake of completeness, we present the procedure to obtain the variational free energy as presented in Ref. [1] in the absence of ion-pair correlations. In Ref. [1], it has been assumed that the counterions from the polyelectrolyte are indistinguishable from the counterions from the salt. So, we start from a partition function similar to Eq. (6.124) with the solvent, counterions (from the polyelectrolyte and the salt), coions and the chain as distinguishable species. After using the method of collective variables (cf. Section 6.4.2.1), the partition function can be written as

$$\exp\left(-\frac{F-F_0}{k_B T}\right) = \int D[w_p]D[\varrho_p]D[\eta]\prod_j D\left[\varrho_j\right]D\left[w_j\right]\exp\left\{-\frac{h}{k_B T}\right\}, \tag{6.133}$$

where $j = \mathrm{s, c, -}$, and where the integral over u has already been evaluated by the saddle-point method so that the functional h corresponds to a single chain with a "smeared" charge distribution. Also, note that we have introduced collective fields and densities for small ions instead of using the Hubbard–Stratonovich transformation for the electrostatic part. This is the analogue of Eq. (6.127) in SCFT. Now, evaluating the path integrals over w_j by the saddle-point method, Eq. (6.133) can be written in terms of the densities ϱ_j. Functional integrals over η and ϱ_s can be carried out in a trivial way. To carry out functional integrals over small ion densities, that is, the $\varrho_j \log \varrho_j$ terms, which emerge after integrations over fields w_j, are expanded up to the quadratic terms after writing $\varrho_j(\mathbf{r}) = n_j/\Omega + \delta\varrho_j(\mathbf{r})$ so that $\int \mathrm{d}\mathbf{r}\delta\varrho_j(\mathbf{r}) = 0$, and the resulting integrals are Gaussian. This procedure also gives one-loop corrections to the free energy coming from the small ions density fluctuations ($\Delta F/k_B T$). Now, expanding the $(1-\varrho_p)\log(1-\varrho_p)$ term up to the quadratic terms in ϱ_p, the problem of carrying out the functional integrals over w_p and ϱ_p is equivalent to a single chain problem whose monomers interact with each other via a renormalized excluded volume parameter and an electrostatic potential. The renormalized excluded volume parameter comes out to be $w = 1-2\chi_{ps}$ and the electrostatic potential comes out to be the Debye–Hückel potential, where the inverse Debye length ($\varkappa$) depends on the "free" ions only. Eventually, Eq. (6.133) becomes

$$\exp\left(-\frac{F-F_0-\delta F}{k_B T}\right) = \frac{1}{\mu}\exp\left\{-\frac{E_a - TS_i}{k_B T}\right\}\int D\left[w_p\right]D\left[\varrho_p\right]\exp\left\{-\frac{H_{var}}{k_B T}\right\}, \tag{6.134}$$

where $-TS_i$ is the translational entropy of the "free" ions as presented in Table 6.2 for the variational theory. Now, writing the Hamiltonian of a single polyelectrolyte chain using an effective excluded parameter (w) and the Debye–Hückel potential [1], the functional integrals over w_p and ϱ_p can be computed using the variational technique developed by Muthukumar [21]. Taking $k_B T = 1$, the variational *ansatz* of the total free energy ($F_{variational}$) is given by $F_{variational} = F_0 + F_a^* + F_f^* + \Delta F$.

Although F_a^* in the variational theory is the same as in SCFT (cf. Eq. (6.128)), other contributions involving the "free " ions, the chain entropy, and so on (that means F_f^*) differ significantly in terms of computational details. In SCFT, F_f^* is computed after solving for fields experienced by different components in the system, which arise as a result of interactions of a particular component with the others. On the other hand, in variational calculations [1], a single polyelectrolyte chain, whose monomers interact with the excluded volume and the electrostatic interactions in the presence of the small ions is approximated by an *effective* Gaussian chain, whose conformational statistics depend on the different kinds of interactions in the system. To compute the equilibrium free energy, its variational *ansatz* is minimized with respect to the variational parameter l_1, which is related to the radius of gyration (R_g) of the chain by $R_g^2 = Nll_1/6$. Physically, this corresponds to the minimization of the free energy of the single chain system with respect to the size of the chain. For the computation of the equilibrium degree of ionization, an additional minimization of the free energy with respect to the degree of ionization has to be carried out. However, due to the intricate coupling between the size of the chain and the degree of ionization, the minimizations have to be carried out self-consistently.

The F_f^* part of the variational *ansatz* [1] is tabulated in Table 6.2. The function $\Theta_0(a)$ in Table 6.2 is a crossover function given by [1, 21]

$$\Theta_0(a) = \frac{\sqrt{\pi}}{2}\left(\frac{2}{a^{5/2}} - \frac{1}{a^{3/2}}\right)\exp(a)\,\mathrm{erfc}(\sqrt{a}) + \frac{1}{3a} + \frac{2}{a^2} - \frac{\sqrt{\pi}}{a^{5/2}} - \frac{\sqrt{\pi}}{2a^{3/2}}, \tag{6.135}$$

where $a \equiv \varkappa^2 Nll_1/6$, and $\varkappa l$ is the dimensionless inverse Debye length. Furthermore, $\tilde{l}_1 = l_1/l, \tilde{l}_B = l_B/l$. The number of salt ions (n_+, n_-) are related to the salt concentration (c_s) by the relation $Z_+n_+ = -Z_-n_- = 0.6023c_s\Omega$, where c_s is in units of moles per liter (molarity). Also, all the terms in the free energies are in units of k_BT.

In this work, we have ignored one-loop corrections to the free energy within SCFT. However, one-loop corrections to the free energy coming from the density fluctuations of small ions, within the variational formalism, is given by

$$\Delta F = -\frac{\Omega\varkappa^3}{12\pi}, \tag{6.136}$$

where $\varkappa^2 = 4\pi l_B(fN + n_+ + n_-)/\Omega$ and $\varkappa$ is the inverse Debye length.

6.5.3
Numerical Techniques

We solve SCFT equations for the single polyelectrolyte chain within spherical symmetry (that means $\mathbf{r} \to r = |\mathbf{r}|$), using the Dirichlet boundary conditions for $q(r,t)$ and all the fields except $\eta(r)$. Also, due to the use of spherical symmetry in these calculations, we use

$$\frac{\partial\psi(r)}{\partial r}\Big|_{r=0} = \frac{\partial q(r,t)}{\partial r}\Big|_{r=0} = 0, \quad \text{for all } t. \tag{6.137}$$

Starting from an initial guess for fields, new fields and densities are computed after solving the modified diffusion and Poisson–Boltzmann equation by finite difference methods [81]. Broyden's method [81] has been used to solve the set of nonlinear equations. The equilibrium value of the degree of ionization (f^*) is obtained after minimizing the free energy with respect to f. We carry out the numerical minimization of free energy over f using Brent's method [81]. The results presented in this paper were obtained by using a grid spacing of $\Delta r = 0.1$ and contour steps of $\Delta t = 0.01$.

On the other hand, the self-consistent minimization of the free energy in the variational method has been carried out by assuming a uniform expansion of the chain within spherical symmetry. In this formalism, the free energy is minimized simultaneously with respect to f and l_1, and both these quantities at equilibrium (f^*, l_1^*) are computed self-consistently. The radius of gyration of the chain (which is confined to a finite volume, $\Omega = 4\pi R^3/3$) is obtained from the equilibrium value of the expansion factor l_1^*. For these calculations, the upper bound for the radius of gyration of the chain is specified to be the radius of the confining volume (that means $R_g \leq R$) to mimic the confinement effects. Also, the Kuhn step length l is taken to be unity in both variational and SCFT calculations.

6.5.4
Degree of Ionization

We have carried out an exhaustive comparison between the SCFT and variational formalisms by calculating the equilibrium degree of ionization (f^*) of a negatively charged single flexible polyelectrolyte chain (that means $Z_p = -1, Z_c = 1$) in the presence of a monovalent salt. The equilibrium degree of ionization is determined as a function of the strength of the electrostatic interaction (or Coulomb strength) that is proportional to the Bjerrum length l_B for a given solvent. For both cases, the effective charge expectedly decreases (Figure 6.26) with higher Coulomb strengths that help a progressively larger degree of adsorption of counterions on the chain backbone. However, f^* obtained from the variational procedure is systematically

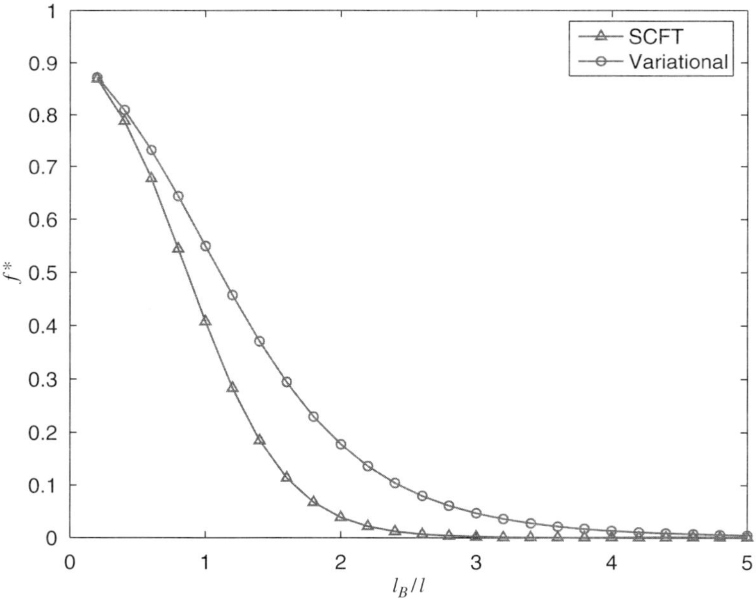

Figure 6.26 Comparison of SCFT and the variational formalism (with one-loop corrections) to illustrate the effect of correlations among small ions on the effective degree of ionization (f^*). $Z_p = -Z_c = -1, R/l = 10, N = 100, c_s = 0.1\,\text{M}, \chi_{ps} = 0.45$, and $\delta = 3$.

higher than that from SCFT. It is to be noted that the degree of ionization is essentially zero in SCFT for experimentally relevant values of l_B/l (around 3 for aqueous solutions), whereas f^* is reasonable in the variational theory. Although both theories use different approximations and computational procedures, there is one major conceptual input that distinguishes these theories. While the variational formalism of Ref. [1] includes the density fluctuations of small ions as one-loop corrections to the free energy, the SCFT does not address these fluctuations. In an effort to quantify the consequences of small ion density fluctuations and then compare the consequences of the rest of the terms in the variational theory against SCFT (which does not contain small ion density fluctuations by construction), we subtract ΔF from $F_{\text{variational}}$ and then compute f^*. The results are given in Figures 6.27, and 6.29 and 6.30.

Remarkably, in different conditions corresponding to widely varying degrees of confinement, the f^* obtained by the minimization of SCFT free energies is indistinguishable from that obtained using variational free energies without one-loop corrections (that means $F_{\text{variational}} - \Delta F$). We demonstrate this in Figure 6.27 where we have plotted f^* as a function of l_B/l for different spherical volumes (that means different R). Thus, we arrive at two conclusions: (a) density fluctuations of small ions included in the full variational formalism significantly contribute in determining the equilibrium degree of ionization and lead to better values of f^* than

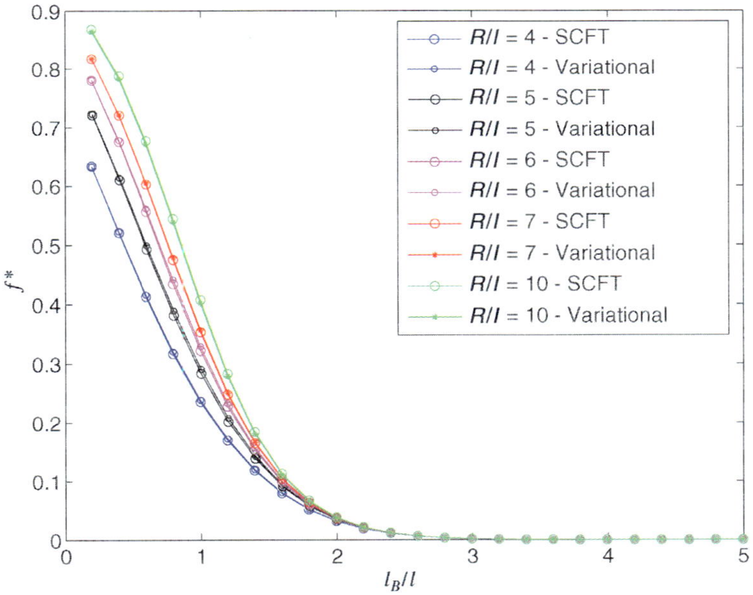

Figure 6.27 Comparison of f^* computed using SCFT and the variational formalism (without one-loop correction) for different values of R and l_B/l. $Z_p = -Z_c = -1$, $N = 100$, $c_s = 0.1$ M, $\chi_{ps} = 0.45$, and $\delta = 3$. Plot for SCFT when $R/l = 10$ is the same as in Figure 6.26.

SCFT; (b) the value of f^* is remarkably indistinguishable between the SCFT and the variational formalism with deliberate suppression of small ion density fluctuations. The first conclusion can be readily rationalized as follows by considering the two curves in Figure 6.26.

The increase in f^* with the inclusion of ΔF can be understood by the fact that the density fluctuations of the small ions lower the free energy, and its contribution to the total free energy increases with the increase in the number of "free" ions (goes like $-n^{3/2}$ in salt-free case, where n is the number of "free" ions – cf. Eq. (6.136)). For higher values of l_B/l (above 4), all counterions are adsorbed on the chain so that the degree of ionization of the chain is zero irrespective of the density fluctuations. In contrast, for lower values of l_B/l (below 0.5), the chain is fully ionized and the effect of the density fluctuations of the small ions on the effective degree of ionization is minimal. However, for the intermediate values of l_B/l, the density fluctuations of the small ions significantly affect the degree of ionization, and nonmonotonic deviations from the SCFT results as a function of l_B/l are observed in this regime. Also, term-by-term comparison of the free energy components reveals that the discrepancy arises solely due to the term accounting for density fluctuations of the small ions. This disagreement highlights the fact that the effect of density fluctuations of the small ions is not included in SCFT within the saddle-point approximation.

The above second conclusion requires further scrutiny. The remarkable agreement between the two formalisms is surprising since these theories use different approximations and different computational procedures. In the variational formalism of Ref. [1], which is used in this chapter, the chain swelling due to electrostatic interaction is assumed to be spherically uniform at all length scales and at the level of Debye–Hückel potential between the segments. However, this scheme is more tractable analytically with different contributing factors (Table 6.2) having explicit physical interpretation. On the other hand, in SCFT, the electrostatic interaction is at the nonlinear Poisson–Boltzmann level, and the chain expansion is addressed at all local length scales through fields generated by intersegment potentials. Although the chain conformations are not readily accessible in the standard version of SCFT used here, the free energy of the system can be calculated and its resolution into entropic and enthalpic parts is possible. In view of such apparently divergent approaches in SCFT and the variational formalisms, we now proceed to make quantitative comparisons between the two in terms of the various contributing factors.

6.5.5
Term-by-Term Comparison of Free Energy: SCFT and Variational Formalism

To assess the approximations used in the variational theory and to find out the origin of the remarkable agreement in terms of the equilibrium effective charge (f^*) obtained from SCFT and variational theory (with deliberate suppression of one-loop corrections for small ion density fluctuations), we have compared individual contributions to the free energies in these two formalisms. Before presenting the numerical results, the role of different contributions in driving the counterion adsorption can be understood qualitatively by considering the following physical picture.

The driving forces for the counterion adsorption are the formation of ion-pairs due to the presence of strong attractive interactions in the process (self-energy of the dipoles) and the decrease in intramolecular electrostatic repulsions (compared with a fully ionized chain, where these repulsions are the strongest). However, an extensive counterion adsorption on the chain backbone is unfavorable due to the loss in translational degrees of freedom of the "free" counterions. Another factor, which plays a role in this competition of the energy and entropy, is the translational entropy of the "adsorbed" counterions. This entropic feature alone favors a state where half the charges on the chain backbone are free and the other half adsorbing the counterions. With an increase in the electrostatic interaction strength (that means Bjerrum length), the driving forces for the counterion adsorption increase and drive more and more counterions to the backbone. On the other hand, an extensive counterion adsorption leads to the chain contraction due to lower intramolecular electrostatic repulsions (a result of the lower number of bare charged sites) even when the electrostatic interaction strength is high. However, we show that the counterion adsorption leading to the lowering of the effective charge (that decreases the electrostatic energy due to the formation of ion-pairs) has a bigger effect than the chain contraction (which affects the chain conformational entropy and polymer–solvent entropy) or the increase in intramolecular electrostatic repulsions among the unadsorbed segments as we gradually increase the Bjerrum length. Of course, in addition, correlations of small ion density fluctuations also contribute to f^*, in the full variational calculation. Numerical results on the relative importance of the various contributions to the total free energy along with their role in driving the counterion adsorption are presented below.

To start with, in Figure 6.28, we have plotted the total free energy calculated in both methods for the following set of parameters: $Z_p = -Z_c = -1$, $R/l = 10$, $N = 100$, $c_s = 0.1$ M, $\delta = 3$, and $\chi_{ps} = 0.45$. It is clear that the total free energies obtained from SCFT and the variational theory are in quantitative agreement with each other.

To analyze this striking agreement between the two methods, we focus on the individual components of the free energy as listed in Table 6.2. In Figures 6.29 and 6.30, we have compared these different constituents of the free energy obtained from both SCFT and the variational formalisms for low monomer densities. It is evident that both theories predict that the major contributions to the free energy are from the ion-pair energy (Figure 6.29a), the "adsorbed" counterion translational entropy (Figure 6.29b), the polymer–solvent interaction energy and the solvent entropy (Figure 6.29c), and the "free" ions translational entropy (Figure 6.29d). Contributions due to the chain conformational entropy (Figure 6.30a) and the electrostatic energy (Figure 6.30b) are almost negligible (less than 0.1% in the total free energy) compared to others. For low monomer densities (monomer volume fractions lower than 0.1), the dominant contributions to the total free energies come from the polymer–solvent interaction energy and the solvent entropy. For the particular single chain dilute system investigated here, these contributions account for more than 50% of the total free energy. Although large, these contributions are

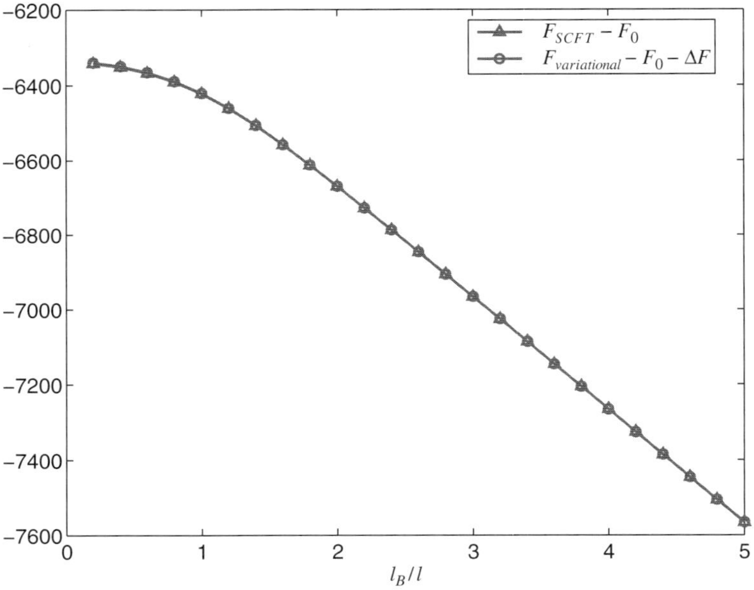

Figure 6.28 Comparison of total free energies (at equilibrium, for $f = f^*$) obtained from SCFT and the variational calculations (without one-loop correction). $Z_p = -Z_c = -1$, $R/l = 10$, $N = 100$, $c_s = 0.1$ M, $\chi_{ps} = 0.45$, and $\delta = 3$.

found to be almost insensitive to f. For example, polymer–solvent interaction energy changes only by less than $0.5k_BT$ when l_B/l is varied from 0.2 to 5.0. On the other hand, f^* changes from 1 to 0 in the same range of l_B/l. In fact, the f-dependent terms contributing significantly to the total free energy are the ion-pair energy and the "free" ions translational entropy. At lower electrostatic interaction strengths (that means low l_B/l), the translational entropy of the "free" ions dominates, and at higher electrostatic strengths, the ion-pair energy term contributes significantly to the free energy. Together, these two contributions account for as high as 99% of the f-dependent part in the total free energy (cf. Figures 6.28 and 6.29a and d). Relatively very small contributions ($\sim 1\%$) to the free energies come from the translational entropy of the "adsorbed" ions. We will see below, however, that the relative importance of a particular contribution in determining the equilibrium degree of ionization is not necessarily related to its actual contribution to the total free energy.

We now discuss the various trends seen in Figures 6.29 and 6.30, based on conceptual arguments aided by the different terms in Table 6.2. Intuitively, stronger ion-pair energy should promote counterion adsorption. As l_B/l is increased, the energy due to counterion adsorption should decrease monotonically, as seen in Figure 6.29a. From Eq. (6.129), it is clear that the ion-pair energy (note the negative contribution) favors counterion adsorption with a linear dependence on f and the Coulomb strength l_B, hence a progressive gain in adsorption energy with increasing Coulomb strength (Figure 6.29a). On the other hand, the counterion adsorption is opposed by the translational entropy of the "free" ions (see the expression for TS_i in

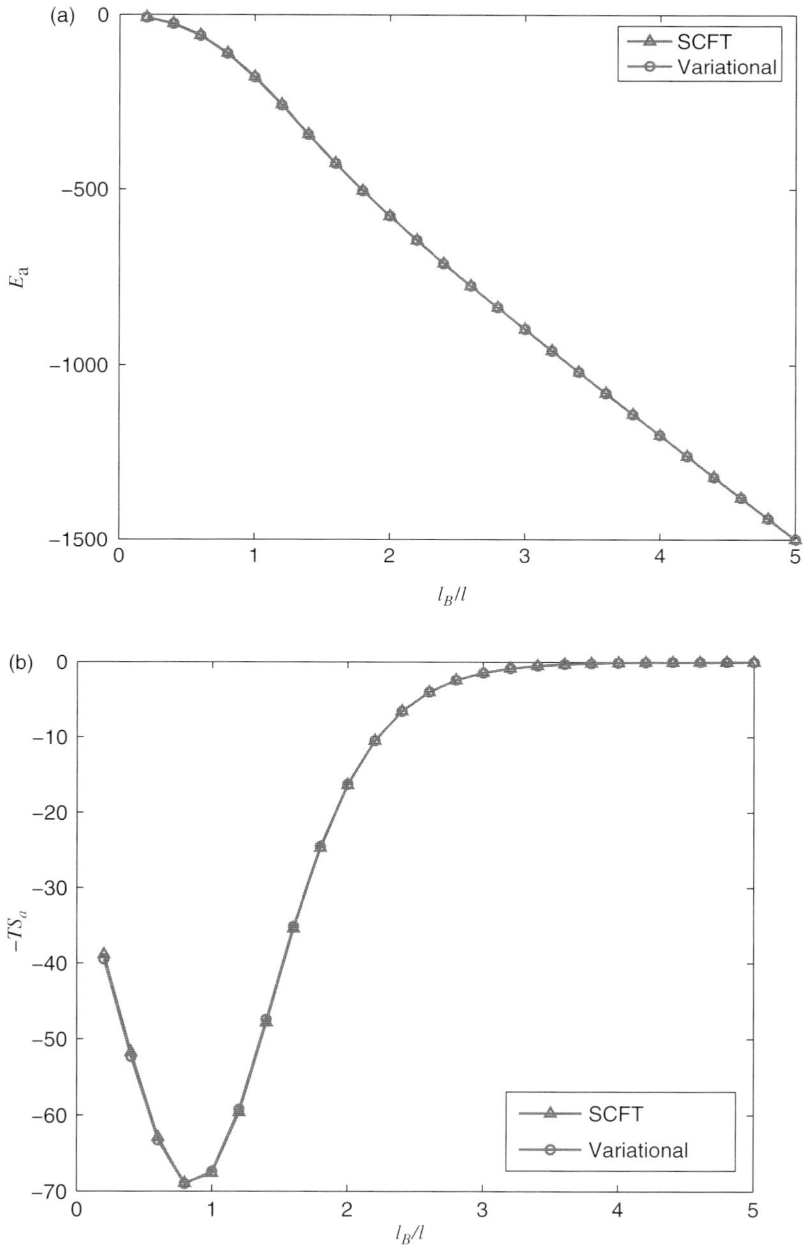

Figure 6.29 Comparison of major contributions to the free energies (presented in Figure 6.28) obtained from SCFT and the variational formalism. (a) Ion-pair energy contributions (E_a), (b) translational entropy of the "adsorbed" counterions ($-TS_a$), (c) polymer–solvent interaction energy and solvent entropy ($E_w - TS_s$); and (d) translational entropy of the "free" ions ($-TS_i$). The variational theory captures SCFT results quantitatively.

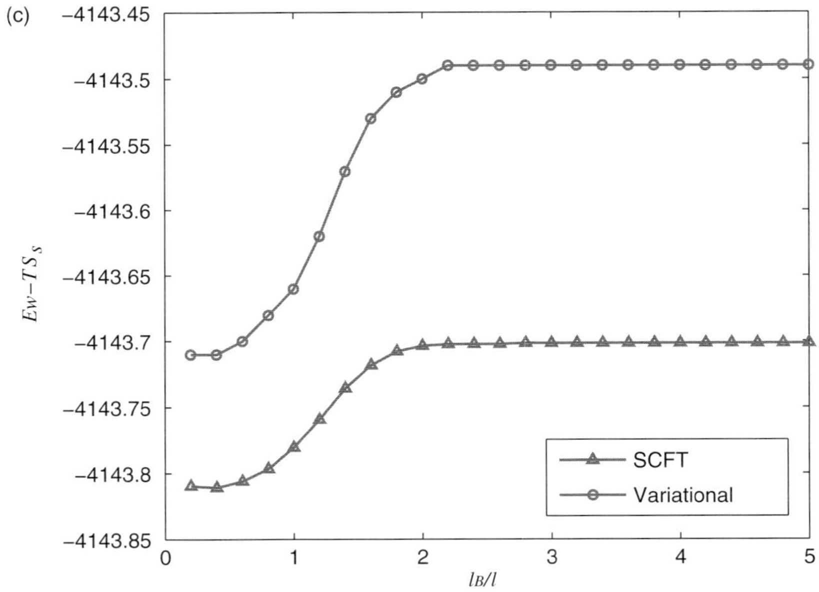

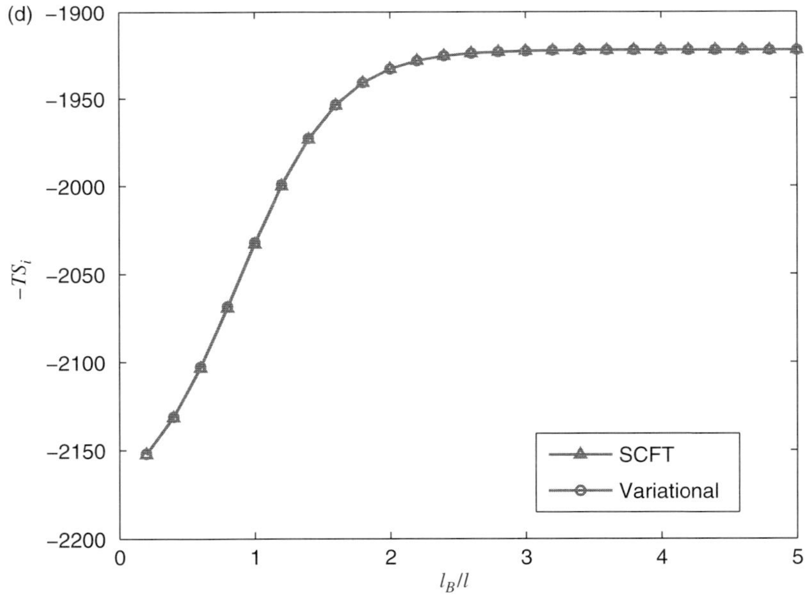

Figure 6.29 *(Continued)*.

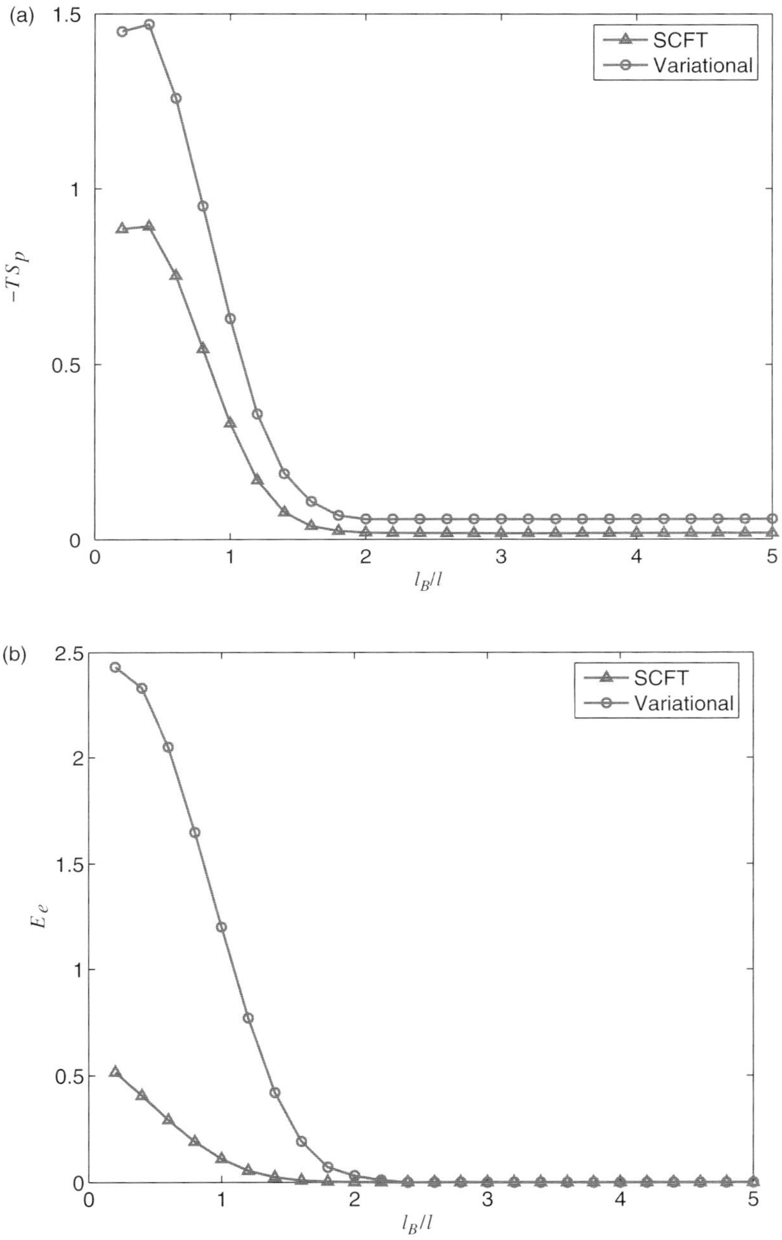

Figure 6.30 Comparison of minor contributions to the free energies (presented in Figure 6.28). (a) Conformational entropy of the chain ($-TS_p$) and (b) electrostatic energy (E_e).

Table 6.2), and hence a progressive loss of the part of the free energy related to the translational entropy (Figure 6.29d). The plateau in Figure 6.29d arises due to the completion of the adsorption of all counterions that limits the loss in the number of free counterions (which is the number of salt ions) with increasing Coulomb strength. However, there is no plateau in Figure 6.29a because even after all counterions are adsorbed, the ion-pair energy continually decreases due to an increasing Coulomb strength. In addition, the translational entropy of the "adsorbed" counterions ($-TS_a$) drives the adsorption toward $f^* = 0.5$ to optimize this part of the entropy (cf. Eq. (6.130)). Physically, it can be understood from the fact that the complete adsorption of the counterions leads to the lowering of the translational entropy of the "adsorbed" counterions due to the unavailability of sites. Similarly, a complete desorption of the counterions also leads to the lowering of translational entropy of the "adsorbed" counterions due to the unavailability of the "adsorbed" counterions on the chain backbone. For a given N, the translational entropy of the "adsorbed" counterions is optimum at $f^* = 0.5$. We note, however, that the other two contributions might overwhelm $-TS_a$ so that at equilibrium it is not necessarily at its minimum (Figure 6.29b). With varying Coulomb strength, $-TS_a$ is minimum at around $l_B/l = 0.8$ at which $f \simeq 0.5$, which is prevalently determined by the first two components mentioned above. The role of other contributions, that is, the polymer–solvent interaction energies and the solvent entropy (that means $E_w - TS_s$), the electrostatic energy involving the "free" ions and the monomers (E_e) and the conformational entropy of the chain ($-TS_p$) is minuscule in driving the counterion adsorption in a particular direction. However, these three contributions dictate the effective size of the chain (through l_1) at the equilibrium (note the dependence of these terms on l_1 in Table 6.2). Further, the equilibrium counterion distribution specified by the first three contributions stipulates the actual contributions of the last three parts of the free energy at equilibrium. We have noticed before that with increasing Coulomb strength (l_B/l) the number of free counterions (and, therefore, the effective charge of the chain) decreases. Owing to a decreasing electrostatic repulsion between the monomers, the polymer chain progressively contracts until it reaches its Gaussian size at zero effective charge. Consequently, there is less mixing between the polymer and the solvent at higher Coulomb strengths leading to a gradual loss of polymer–solvent interaction energy (Figure 6.29c) that reaches a plateau when all counterions adsorb, the physical condition that creates plateaus in all these curves. Also accompanying the decreasing size of the chain, there is a gain in conformational entropy (which is maximum at the Gaussian size) observed in Figure 6.30a. In addition, a gradual decrease in the effective charge of the chain progressively reduces the electrostatic energy penalty observed in Figure 6.30b. However, this effect is very small compared to the lowering of electrostatic energy due to the formation of ion-pairs as mentioned earlier.

The quantitative agreement between the first three contributions to the free energy in two formalisms explains the observed agreement in the results obtained for f^* (Figure 6.27). Despite E_e having a negligible contribution to the total free energy (less than 0.1%), the comparison reveals that the Debye–Hückel estimate for the electrostatic energy (E_e) used in the variational formalism is an overestimation (as large as five times the full, nonlinear Poisson–Boltzmann at low l_B/l). In other words, the

Debye–Hückel approximation underestimates the degree of screening, which is in agreement with other theoretical [94] and simulation results [95]. Note that the electrostatic energy in Figure 6.30b includes all the charged species in the system except the ion-pairs formed on the chain by the adsorbing counterions. Nevertheless, contributions due to the electrostatic energy to total free energy are almost negligible and hence do not affect f^* significantly.

We have also carried out the same comparison between the two formalisms at higher monomer densities (above monomer volume fractions of 0.1). It is found that the discrepancy in the polymer–solvent interaction energy and the solvent entropy between the two schemes is significant (see Figure 6.31). All other contributing factors are essentially the same between the two theories. The origin of this discrepancy lies in the expansion of the $(1-\varrho_p)\log(1-\varrho_p)$ term, which is carried up to only terms quadratic in polymer density in the variational calculations . The higher order terms in the expansion are ignored in the variational calculations to carry out the analysis analytically, which limits the applicability of the variational theory to sufficiently low monomer concentrations. The discrepancy clearly highlights the breakdown of the variational procedure at high densities and questions the use of an effective excluded volume parameter in variational calculations. However, we have not attempted to compute the boundary of the disagreement between the theories because the main focus of this chapter is on f^* that is insensitive to this discrepancy. Also, the variational formalism predicts the polymer–solvent interaction

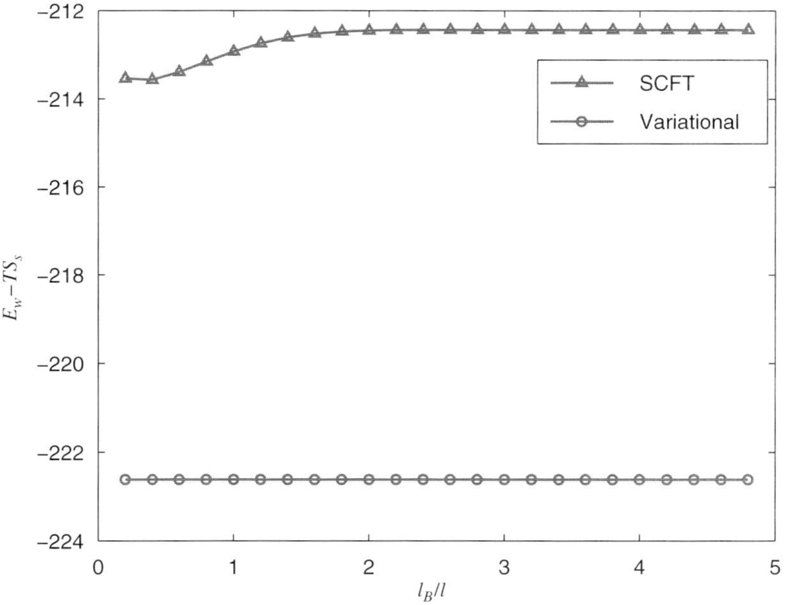

Figure 6.31 A discrepancy between SCFT and variational theory for the polymer–solvent interaction energy and solvent entropy contributions arise at high monomer densities. $Z_p = -Z_c = -1$, $R/l = 4$, $N = 100$, $c_s = 0.1$ M, $\chi_{ps} = 0.45$, and $\delta = 3$.

energy and the solvent entropy to be completely independent of the electrostatic interaction strength l_B/l in the high-density regime, in contrast to SCFT predictions of a weak dependence on l_B/l (see Figure 6.31). This is a result of the constraint $R_g \leq R$ used in variational calculations for mimicking the confinement effects and shows the inability of the constraint to capture the confinement effects in an appropriate fashion. While the radius of gyration R_g of the chain follows readily from l_1 in the variational formalism, it is nontrivial to compute this quantity in SCFT. In view of this, we have not addressed R_g in the present paper.

Finally, we remark on the experimental relevance of the radius parameter R for the confining cavity. The variational calculation readily gives R_g without any confinement for fixed values of monomer density and other parameters such as l_B, χ_{ps}, and c_s. Knowing this result, we have investigated the role played by the cavity radius R in the above analysis. If R is larger than R_g, then the above conclusions are relevant to unconfined dilute polyelectrolyte solutions. On the other hand, if R is less than R_g, then confinement effect is manifest and now our results are relevant to a polyelectrolyte chain inside a spherical pore. As an example, for $N = 100, c_s = 0.1$ M, $\chi_{ps} = 0.45$, and $\delta = 3$, the calculated value of R_g/l from the variational procedure depends on l_B and attains a maximum value of 7.29, whereas $R/l = 10.0$ in Figures 6.28–6.30. Therefore, the conclusions drawn above based on these figures are generally valid for dilute polyelectrolyte solutions. On the other hand, $R/l = 4.0$ in Figure 6.31, whereas the maximum value of R_g would have been 5.92 if confinement were to be absent. Under these conditions, the conclusions regarding the discrepancy between SCFT and variational theory is pertinent to a polyelectrolyte chain confined inside a spherical cavity.

In summary, we have computed the effective charge of a single flexible polyelectrolyte chain using SCFT and compared it with the results obtained from a variational theory. It is found that for all sets of parameters, the effective degree of ionization (f^*) computed from SCFT and the variational theory is in quantitative agreement if one-loop fluctuation corrections are deliberately suppressed in the latter. The origin of this agreement lies in the fact that f^* is determined as an interplay of the ion-pair energy and the translational entropy of the "adsorbed" counterions as well as of all "free" ions. The conformational entropy of the chain, the electrostatic energy involving the "free" ions and the chain, the polymer–solvent interaction energy and the solvent entropy do not play significant roles in affecting f^*.

The comparison of different components in free energy reveals that the Debye–Hückel approximation underestimates screening effects as compared to the Poisson–Boltzmann theory. Despite the fact that there are small discrepancies in the different contributing factors to the total free energy, the effective degree of ionization (f^*) comes out to be the same in SCFT and the truncated variational theory. Furthermore, the density fluctuations of the "free" ions, which are included in the full variational theory, are predicted to increase the equilibrium degree of ionization. As this latter effect is not captured by SCFT calculations within the saddle-point approximation and due to the close agreement between SCFT and the variational theory for all other contributing factors, the variational theory appears to be a very useful tool for a quick, easy, and transparent estimation of f^*.

6.6 Conclusions

We have given an overview of the present status of theoretical treatments for isolated flexible polyelectrolyte chains in polar solvents under experimentally relevant conditions. The importance of nonuniversal nature of the polymer backbone, ionic radii of counterions and coions, and the dielectric mismatch around the chain backbone are addressed in addition to the entropic effects of chain connectivity and long-ranged electrostatic interactions. By comparing with the Langevin dynamics simulations and numerically solved self-consistent field theories, we find the variational theory to provide an adequate and reliable insight into the charge and size of flexible polyelectrolyte chains. In addition to the usual Flory–Huggins parameter for the immiscibility of a hydrophobic polymer in a polar solvent, a new dielectric mismatch parameter has been introduced to collect most of the nonuniversal aspects of electrostatic interactions between chain segments and solvated ions. Further comparison with experimental data on a variety of systems will help establish the utility of computational schemes reviewed in this chapter.

Acknowledgments

Financial support for this work was provided by the NIH Grant No. 5R01HG002776, National Science Foundation (NSF) Grant No. 0605833, and the MRSEC at the University of the Massachusetts, Amherst.

References

1 Muthukumar, M. (2004) *J. Chem. Phys.*, **120**, 9343.
2 Manning, G.S. (1969) *J. Chem. Phys.*, **51**, 924.
3 Gonzalez-Mozuelos, P. and Olvera de la Cruz, M. (1995) *J. Chem. Phys.*, **103**, 3145.
4 Olvera de la Cruz, M., Belloni, L., Delsanti, M., Dalbiez, J.P., Spalla, O., and Drifford, M. (1995) *J. Chem. Phys.*, **103**, 5781.
5 Kuhn, P.S., Levin, Y., and Barbosa, M.C. (1998) *Macromolecules*, **31**, 8347.
6 Solis, F.J. and Olvera de la Cruz, M. (2000) *J. Chem. Phys.*, **112**, 2030.
7 Solis, F.J. (2002) *J. Chem. Phys.*, **117**, 9009.
8 Winkler, R.G., Gold, M., and Reineker, P. (1998) *Phys. Rev. Lett.*, **80**, 3731.
9 Liu, S. and Muthukumar, M. (2002) *J. Chem. Phys.*, **116**, 9975.
10 Mehler, E.L. and Eichele, G. (1984) *Biochemistry*, **23**, 3887.
11 Lamm, G. and Pack, G.R. (1997) *J. Phys. Chem. B*, **101**, 959.
12 Rouzina, I. and Bloomfield, V.A. (1998) *Biophys. J.*, **74**, 3152.
13 Kundagrami, A. and Muthukumar, M. (2008) *J. Chem. Phys.*, **128**, 244901.
14 Kumar, R. and Muthukumar, M. (2008) *J. Chem. Phys.*, **128**, 184902.
15 Kumar, R., Kundagrami, A., and Muthukumar, M. (2009) *Macromolecules*, **42**, 1370.
16 Ewald, P. (1921) *Ann. Phys.*, **64**, 253.
17 Fincham, D. (1994) *Mol. Simul.*, **13**, 1.
18 Zhang, Y.B., Douglas, J.F., Ermi, B.D., and Amis, E.J. (2001) *J. Chem. Phys.*, **114**, 3299.
19 Prabhu, V.M., Muthukumar, M., Wignall, G.D., and Melnichenko, Y.B. (2001) *Polymer*, **42**, 8935.

20 Bordi, F., Cametti, C., Tan, J.S., Boris, D.C., Krause, W.E., Plucktaveesak, N., and Colby, R.H. (2002) *Macromolecules*, **35**, 7031.

21 Muthukumar, M. (1987) *J. Chem. Phys.*, **86**, 7230.

22 Muthukumar, M. (1989) *Molecular Basis of Polymer Networks* (eds A. Baumgärtner and C.E. Picot), Springer Proceedings in Physics, vol. 42, Springer, New York, p. 28.

23 Muthukumar, M. (1996) *J. Chem. Phys.*, **105**, 5183.

24 Wittmer, J., Johner, A., and Joanny, J.F. (1995) *J. Phys. II France*, **5**, 635.

25 Beer, M., Schmidt, M., and Muthukumar, M. (1997) *Macromolecules*, **30**, 8375.

26 Muthukumar, M. and Nickel, B.G. (1987) *J. Chem. Phys.*, **86**, 460.

27 Liu, S., Ghosh, K., and Muthukumar, M. (2003) *J. Chem. Phys.*, **119**, 1813.

28 Rouzina, I. and Bloomfield, V.A. (1996) *J. Phys. Chem.*, **100**, 9977.

29 Ha, B.-Y. and Liu, A.J. (1997) *Phys. Rev. Lett.*, **79**, 1289.

30 Golestanian, R., Kardar, M., and Liverpool, T. (1999) *Phys. Rev. Lett.*, **82**, 4456.

31 Grosberg, A.Yu., Nguyen, T.T., and Shklovskii, B.I. (2002) *Rev. Mod. Phys.*, **74**, 329.

32 Micka, U. and Kremer, K. (2000) *Europhys. Lett.*, **49**, 189.

33 Hsiao, P.-Y. and Luijten, E. (2006) *Phys. Rev. Lett.*, **97**, 148301.

34 Huber, K. (1993) *J. Phys. Chem.*, **97**, 9825.

35 Zhang, Y., Douglas, J.F., Ermi, B.D., and Amis, E.J. (2001) *J. Chem. Phys.*, **115**, 4367.

36 Murayama, Y., Sakamaki, Y., and Sano, M. (2003) *Phys. Rev. Lett.*, **90**, 018102.

37 Besteman, K., Van Eijk, K., and Lemay, S.G. (2007) *Nat. Phys.*, **3**, 641.

38 Prabhu, V.M., Amis, E.J., Bossev, D.P., and Rosov, N. (2004) *J. Chem. Phys.*, **115**, 4367.

39 Loh, P., Deen, G.R., Vollmer, D., Fischer, K., Schmidt, M., Kundagrami, A., and Muthukumar, M. (2008) *Macromolecules*, **41**, 9352.

40 Reddy, G. and Yethiraj, A. (2006) *Macromolecules*, **39**, 8536.

41 Manning, G.S. (1981) *J. Phys. Chem.*, **85**, 1506.

42 Goerigk, G., Huber, K., and Schweins, R. (2007) *J. Chem. Phys.*, **127**, 154908.

43 Ikeda, Y., Beer, M., Schmidt, M., and Huber, K. (1998) *Macromolecules*, **31**, 728.

44 Schweins, R. and Huber, K. (2001) *Eur. Phys. J. E*, **5**, 117.

45 Goerigk, G., Schweins, R., Huber, K., and Ballauff, M. (2004) *Europhys. Lett.*, **66**, 331.

46 Schweins, R., Goerigk, G., and Huber, K. (2006) *Eur. Phys. J. E*, **21**, 99.

47 Narh, K.A. and Keller, A. (1993) *J. Polym. Sci., Part B: Polym. Phys.*, **31**, 231.

48 Edwards, S.F. (1965) *Proc. Phys. Soc. London*, **85**, 613.

49 Edwards, S.F. (1975) *J. Phys. A: Math. Gen.*, **8**, 1670.

50 Doi, M. and Edwards, S.F. (1986) *The Theory of Polymer Dynamics*, Clarendon Press, Oxford.

51 Fleer, G.J., Cohen Stuart, M.A., Scheutjens, J.M.H.M., Cosgrove, T., and Vincent, B. (1993) *Polymers at Interfaces*, Chapman & Hall, London.

52 Helfand, E. (1975) *J. Chem. Phys.*, **62**, 999.

53 Freed, K.F. (1987) *Renormalization Group Theory of Macromolecules*, John Wiley & Sons, Inc., New York.

54 de Gennes, P.G. (1979) *Scaling Concepts in Polymer Physics*, Cornell University Press, Ithaca and London.

55 Fredrickson, G.H. (2006) *The Equilibrium Theory of Inhomogeneous Polymers*, Oxford University Press, New York.

56 Freed, K.F. (1972) *Adv. Chem. Phys.*, **22**, 1.

57 Schmid, F. (1998) *J. Phys.: Condens. Matter*, **10**, 8105.

58 Matsen, M.W. (2002) *J. Phys.: Condens. Matter*, **14**, R21.

59 Helfand, E. (1975) *Macromolecules*, **8**, 552; Helfand, E. (1976) *Macromolecules*, **9**, 879; Helfand, E. (1978) *Macromolecules*, **11**, 960;Helfand, E. (1980) *Macromolecules*, **13**, 994.

60 Hubbard, J. (1959) *Phys. Rev. Lett.*, **3**, 77.

61 Shi, A. and Noolandi, J. (1999) *Macromol. Theory Simul.*, **8**, 214.

62 Wang, Q., Taniguchi, T., and Fredrickson, G.H. (2004) *J. Phys. Chem. B*, **108**, 6733;Wang, Q., Taniguchi, T., and Fredrickson, G.H. (2005) *J. Phys. Chem. B*, **109**, 9855.

63 Kumar, R. and Muthukumar, M. (2007) *J. Chem. Phys.*, **126**, 214902.

64 Wang, Q. (2005) *Macromolecules*, **38**, 8911.

65 Wang, Q. (2006) *J. Phys. Chem. B*, **110**, 5825.

66 Witte, K.N. and Won, Y.Y. (2006) *Macromolecules*, **39**, 7757.

67 Seki, H., Suzuki, Y.Y., and Orland, H. (2007) *J. Phys. Soc. Jpn.*, **76**, 104601.

68 Borukhov, I., Andelman, D., and Orland, H. (1998) *Eur Phys. J. B*, **5**, 869.

69 Israelachvili, J.N. (1991) *Intermolecular and Surface Forces*, Academic Press, London.

70 McQuarie, D.A. (2000) *Statistical Mechanics*, University Science Books, Sausalito, CA.

71 Falk, D.S. (1960) *Phys. Rev.*, **118**, 105.

72 Mahan, G.D. (2000) *Many-Particle Physics*, 3rd edn, Springer.

73 Böttcher, C.J.F. (1973) *Theory of Electric Polarization*, Elsevier, Amsterdam.

74 Flory, P.J. (1953) *Principles of Polymer Chemistry*, Cornell University Press, New York.

75 Leibler, L. (1980) *Macromolecules*, **13**, 1602.

76 Ohta, T. and Kawasaki, K. (1986) *Macromolecules*, **19**, 2621; Ohta, T. and Kawasaki, K. (1988) *Macromolecules*, **21**, 2972; Ohta, T. and Kawasaki, K. (1990) *Macromolecules*, **23**, 2413.

77 Marko, J.F. and Rabin, Y. (1991) *Macromolecules*, **24**, 2134; Marko, J.F. and Rabin, Y. (1992) *Macromolecules*, **25**, 1503.

78 Jackson, J.D. (1962) *Classical Electrodynamics*, John Wiley & Sons, Inc., New York.

79 Hansen, J.P. and McDonald, I.R. (1996) *Theory of Simple Liquids*, Elsevier Academic Press, San Diego, CA.

80 Ceniceros, H.D. and Fredrickson, G.H. (2004) *Multiscale Model. Simul.*, **2**, 452.

81 Press, W.H., Teukolsky, S.A., Vetterling, W.T., and Flannery, B.P. (1992) *Numerical Recipes in C*, Cambridge University Press, New York.

82 Sayyed-Ahmad, A., Tuncay, K., and Ortoleva, P.J. (2004) *J. Comp. Chem.*, **25**, 1068.

83 Matsen, M.W. and Schick, M. (1994) *Phys. Rev. Lett.*, **72**, 2660; Matsen, M.W. and Bates, F.S. (1996) *Macromolecules*, **29**, 1091.

84 Hermann, M.R. and Fleck, J.A. (1988) *Phys. Rev. A*, **38**, 6000; Tzeremes, G., Rasmussen, K.O., Lookman, T., and Saxena, A. (2002) *Phys. Rev. E*, **65**, 041806.

85 Weiss, G.H. and Maradudin, A.A. (1962) *J. Math. Phys.*, **3**, 771.

86 Fredrickson, G.H. and Helfand, E. (1987) *J. Chem. Phys.*, **87**, 697.

87 Olvera de la Cruz, M. (1991) *Phys. Rev. Lett.*, **67**, 85.

88 Muthukumar, M. (1993) *Macromolecules*, **26**, 5259.

89 Shi, A.C., Noolandi, J., and Desai, R.C. (1996) *Macromolecules*, **29**, 6487; Laradji, M., Shi, A.C., Noolandi, J., and Desai, R.C. (1997) *Phys. Rev. Lett.*, **78**, 2577; Laradji, M., Shi, A.C., Noolandi, J., and Desai, R.C. (1997) *Macromolecules*, **30**, 3242.

90 Fredrickson, G.H., Ganesan, V., and Drolet, F. (2002) *Macromolecules*, **35**, 16.

91 Muller, M. and Schmid, F. (2005) *Adv. Polym. Sci.*, **185**, 1.

92 Muthukumar, M. and Edwards, S.F. (1982) *J. Chem. Phys.*, **76**, 2720.

93 Muthukumar, M. (2003) *J. Chem. Phys.*, **118**, 5174.

94 Stigter, D. (1995) *Biophys. J.*, **69**, 380.

95 Stevens, M.J. and Kremer, K. (1996) *J. Phys. II*, **6**, 1607.

7
Multiscale Modeling and Coarse Graining of Polymer Dynamics: Simulations Guided by Statistical Beyond-Equilibrium Thermodynamics

Patrick Ilg, Vlasis Mavrantzas, and Hans Christian Öttinger

7.1
Polymer Dynamics and Flow Properties We Want to Understand: Motivation and Goals

7.1.1
Challenges in Polymer Dynamics Under Flow

Polymer molecules differ from simple fluids in several aspects: they are extremely diverse in structure (they can have a linear, branched, ring-like, or block copolymer structure), they can be characterized by a molecular weight distribution, and they are capable of exhibiting a huge number of configurations implying that a large number of degrees of freedom should be accounted for in any molecular modeling approach. As a result, polymers exhibit properties that are totally distinct from those of the simpler Newtonian liquids. The drag reduction phenomenon (the substantial reduction in pressure drop during the turbulent flow of a Newtonian liquid when a very small amount of a flexible polymer is added), their unique rheological properties (shear thinning and normal stress differences in simple shear, strain hardening in elongation, complex viscosity, anisotropy in thermal conductivity and diffusivity), and a plethora of other phenomena associated with their elastic character are only a few manifestations of the departure of their behavior from the Newtonian one [1, 2]. Of particular importance from a mechanical or fluid dynamics point of view is their viscoelasticity quantifying the irreversible conversion of the work needed for their deformation to heat loss but also their capability to store part of this work as elastic energy. It is a property closely related to the multiplicity of time- and length scales characterizing the dynamics and structure in these fluids. Thus, even in the viscous regime ($Wi \ll 1$, where Wi is the Weissenberg number empirically defined as $Wi = \tau_p \dot{\gamma}$ with τ_p being the longest relaxation time and $\dot{\gamma}$ the flow rate), the flow can still be strong enough for several degrees of freedom not to be close to equilibrium giving rise to interesting rheological properties also there [2], especially for high molecular weight polymers.

Understanding relaxation processes and structure development occurring over these multiple scales is a prerequisite for deriving reliable constitutive equations

Modeling and Simulation in Polymers. Edited by P.D. Gujrati and A.I. Leonov

ISBN: 978-3-527-32415-6

connecting the stresses developing in these materials in terms of the imposed flow kinematics and certain molecular parameters or functions, and for computing polymer flows [3–10]. It is only through a comprehensive understanding across scales that one can hope to build the relationship between polymer molecular structure, conformation and architecture and macroscopic rheological response. In addition to experiments and theories, molecular simulations can play a significant role by providing high-resolution calculations especially on the crossover from small to intermediate scales. This chapter is devoted to a brief discussion of some of the emerging multiscale simulation approaches in the recent research literature on nonequilibrium systems, with emphasis on those based on well-founded theoretical frameworks. Our goal is to demonstrate that with the help of and guidance from recent advances in the field of nonequilibrium statistical mechanics and thermodynamics, this highly demanding endeavor (modeling across scales) can lead to simulation methodologies that have been elevated from simple, brute-force computational experiments to systematic tools for extracting complete, redundancy-free, and consistent coarse-grained information for the flow dynamics of polymeric systems [11].

7.1.2 Modeling Polymer Dynamics Beyond Equilibrium

Describing macromolecular configurations under nonequilibrium conditions is an extremely difficult problem, which usually requires simplified models for analytical or numerical studies [2]; such simple models have contributed enormously to our understanding of polymer rheology and mechanics. For a review on proposed models, see, for example, [12–16], while for a review on available simulation tools addressing different time- and length scales, see Refs [17–23]. Figure 7.1 shows a schematic of the pertinent models for polymer solutions and melts depending on the length- and timescale of interest.

In general, examples of macroscopic constitutive equations employed to calculate polymer flow behavior include typically conformation tensor models such as the Giesekus, Maxwell, and FENE models as well as the more recently proposed pom-pom, CCR, and Rolie-Poly models [15, 16, 24–27]. For an overview, see Refs [3, 4] as well as the contributions by A.N. Beris and E. Mitsoulis in this volume. Most of these constitutive equations have been derived (or inspired) by simple mechanical models of polymer motion. They are mesoscale, kinetic theory models (based, for example, on the dumbbell, FENE, bead-spring chain, and bead-rod chain analogues) capable of accounting for some important aspects of polymer dynamics, such as chain stretching, nonaffine deformation, diffusion, and hydrodynamic drag forces, either separately or altogether [28]. In polymer solutions, in particular, one should account for the solvent-mediated effect between beads (known as hydrodynamic interaction), which is usually modeled with the hydrodynamic resistance matrix [2, 29]. Recently, an efficient simulation of solvent dynamics with the help of either the lattice Boltzmann [30, 31] or the stochastic rotation dynamics [32, 33] methods has been proposed with a suitable coupling of the bead dynamics. Besides their efficiency,

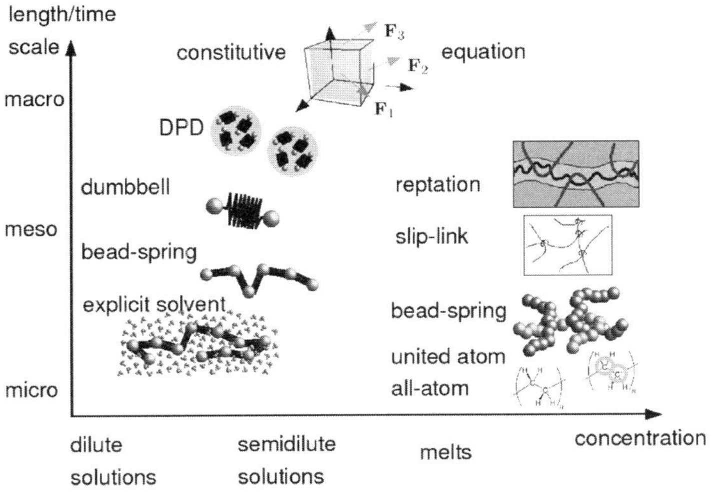

Figure 7.1 Different models of polymer dynamics are schematically shown for solutions and melts.

these combined methods also incorporate nonuniformities and fluctuations in internal flow fields.

Usually, the mesoscopic, kinetic models are considered to be well suited for predicting dynamic properties of polymer solutions on macroscopic scales. Details of the fast solvent dynamics are in most cases irrelevant for macroscopic properties. Exceptions are polyelectrolytes, where the motion of counterions in the solvent can have a major influence on polymer conformation. Therefore, more microscopic models of polyelectrolytes with explicit counterions are sometimes employed [34] (see also the contribution by M. Muthukumar in this volume). Another exception is the dynamics of individual biopolymers, for example, protein folding, which is modeled with an all atomistic model including an explicit treatment of the (water) solvent molecules [35].

For typical polymer melts, dumbbell models are inappropriate since they fail to account for the essential role of entanglements on the long-time dynamics. Successful mesoscopic, mean field descriptions of entangled melts are offered by reptation models [14]. Several modifications of the original Doi–Edwards–de Gennes reptation theory have been proposed over the years in order to improve comparison to experiments in the linear and nonlinear flow regimes [36–43]. Recently, slip-link models have also been proposed [44–50], providing a slightly more detailed description of entanglements, and which agree well with available experimental results.

At the microscopic level, polymer melts are modeled as multichain systems, see Refs [2, 12, 13, 51]. For example, all-atom or united-atom force fields, accounting explicitly for bond angle bending and torsion angle contributions (in addition to bond stretching and intermolecular interactions) [52, 53], are available. Different united-atom force fields are reviewed and compared, for example, in Refs [51, 54, 55]. From such detailed atomistic molecular dynamics (MD) simulations, the linear viscoelastic properties can be computed by Green–Kubo relationships [56–58]. Also, less detailed bead-spring models are available; a prototypical model is that of Kremer and Grest [59] (or variants thereof [60]), which neglects chemical details and instead focuses on the interplay between chain connectivity and excluded volume effects. The nonlinear regime can be studied by nonequilibrium molecular dynamics (NEMD) simulations, which directly address flow effects on polymer structure and conformation, both in shear and planar elongation (see, for example, [12, 60–67] and references therein). They are based on flow-adapted boundary conditions, such as those proposed by Lees and Edwards for planar shear [68] and by Kraynik and Reinelt for planar elongation [69].

7.1.3
Challenges in Standard Simulations of Polymers in Flow

Despite the enormous advances in the field of molecular simulations [4, 18, 21, 22, 70–73], predicting the macroscopic flow properties of polymers from their underlying microstructure presents still major challenges [74]. Available MD and NEMD algorithms can address timescales only on the order of a few microseconds at most, implying that only moderately entangled polymers can be studied in full atomistic detail in a brute-force manner. Extending the simulations to longer, truly entangled polymers is a first big task. Extending these flows to mixed or inhomogeneous flows is another big challenge. Among others, such a development would help understand the origin of interfacial slip and its mode (localized slip versus global slip) in the flow past a solid substrate. On the other hand, with the introduction of the revolutionary set of chain connectivity altering moves, extremely powerful Monte Carlo (MC) algorithms have been developed that have helped overcome the issue of the thermal equilibration of long polymers even at beyond-equilibrium conditions [75–79]. With the help of the end-bridging and double bridging moves, for example, truly long polyethylene (PE) (linear and branched) and polybutadiene systems have been equilibrated over the years, which also opened the way to their topological analysis for the identification of entanglements [80–86].

Arguably the biggest challenge in polymer simulation under nonequilibrium conditions is to build well-founded multiscale tools that can bridge the gap between microscopic information and macroscopically manifested viscoelastic properties, preferably through a constitutive equation founded on the microscopic model [11, 70]. Simulations of metals face the same problem where again the objective is concurrent length scale simulations [87, 88]; to some extent, it is also relevant to simple fluids [89]. For polymers, additional motivation stems from the increased interest in polymer mixtures and interfaces [90] resulting in morphology development at nanoscale.

Here, we aim to briefly review some recent coarse-graining and multiscale methods (see [23, 58, 79, 91–95] and also [96–99] for recent reviews of such methods for polymers and in a more general context, respectively), and also put forward some new ideas for addressing such issues, which could eventually allow to model the macroscale quantities of interest by a suitable coarse-graining procedure. We will see that, if one is guided by nonequilibrium statistical mechanics and thermodynamics, it is possible to design well-founded multiscale modeling tools that can link microscopic models with macroscopic constitutive equations. Such multiscale modeling tools benefit from recently proposed approaches for static coarse graining that are mainly built on potentials of mean force [79, 100, 101]. For dynamic properties, however, coarse-grained models need to account for dissipative effects that arise due to fast degrees of freedom that are eliminated (e.g., via projection operators) in favor of the remaining slowly varying ones [56, 58, 102, 103]. In a flow situation, these slow dynamic variables depart from their values in the quiescent fluid, while all other (faster) degrees of freedom track the evolution of the structural parameters; that is, they are assumed to be in local equilibrium subject to the constraints imposed by the values of the structural parameters at all times. A proper definition of the set of state variables, effectively representing the nonequilibrium states, is the key to the success of such an approach. Linking the microscopic model with a macroscopic model built on these slowly relaxing variables is the second key; as we will discuss in the next sections of this chapter, this is best addressed by getting guidance from a nonequilibrium statistical thermodynamic framework proposing a fundamental evolution equation for the macroscopic model in terms of the chosen structural (dynamic) variables.

7.2
Coarse-Grained Variables and Models

We start with a microscopic polymer model, whose state is specified by a point in $6N$-dimensional phase space, $\mathbf{z} \in \Gamma$ with $\mathbf{z} = (\mathbf{r}_1, \ldots, \mathbf{r}_N; \mathbf{p}_1, \ldots, \mathbf{p}_N)$, a short notation for the positions and momenta of all N particles. The model is described by the microscopic Hamiltonian $H(\mathbf{z})$ with inter- and intramolecular interactions. The coarse-grained model eliminates some of the (huge number of) microscopic degrees of freedom. The level of detail that is retained is specified by the choice of coarse-grained variables $\mathbf{x} = (x_1, \ldots, x_{n_{CG}})$ with

$$x_k = \langle \Pi_k \rangle \equiv \int_\Gamma d\mathbf{z} \Pi_k(\mathbf{z}) \varrho(\mathbf{z}), \quad k = 1, \ldots, n_{CG}, \tag{7.1}$$

where $\varrho(\mathbf{z})$ denotes the probability distribution on Γ and the phase space functions $\Pi_k(\mathbf{z})$ are the instantaneous values of the coarse-grained variables in the microstate $\mathbf{z}$.

Instead of the full, microscopic distribution $\varrho(\mathbf{z})$, the coarse-grained model is already specified by the reduced probability distribution $p(\mathbf{x}) \equiv \langle \delta(\mathbf{x} - \Pi(\mathbf{z})) \rangle$. Knowledge of $p(\mathbf{x})$ allows to calculate averages of quantities $a(\Pi(\mathbf{z}))$ via $\langle a(\Pi) \rangle = \int_\Gamma d\mathbf{z} a(\Pi(\mathbf{z})) \varrho(\mathbf{z}) = \int d\mathbf{x} a(\mathbf{x}) p(\mathbf{x})$. Instead of $p(\mathbf{x})$, coarse-grained models are often described by the so-called potentials of mean force $U_{\mathrm{mf}}(x) \equiv -k_B T \ln p(x)$ that formally replace the Hamiltonian in the calculation of equilibrium, canonical averages. However, U_{mf} is an effective free energy difference and therefore depends on the thermodynamic state of the system. It contains in general effective many-body interactions that arise by partially integrating out microscopic degrees of freedom. For recent reviews, see Refs [91, 100, 104, 105].

Different sets of coarse-grained variables have been suggested in the literature and are briefly reviewed here. Usually, one is interested in a drastic reduction of microscopic complexity, so $n_{CG} \ll 6N$. However, sometimes, for equilibrating atomistic systems, a coarse-grained model with a modest reduction (n_{CG} only a factor 5–10 smaller than $6N$) might be useful. We emphasize that both the mapping (7.1) and the corresponding probability $p(\mathbf{x})$ is not restricted to equilibrium situations. We proceed with a short review of coarse-grained variables and models that capture different levels of detail and briefly discuss their static and dynamic properties.

7.2.1
Beads and Superatoms

Coarse graining to the level of superatoms [106–120] is a method followed when one wishes to reduce chemical complexity in a polymer chain without losing the chemical identity of the molecule. According to this, a certain number of atoms or repeat units along the chain are grouped together into "superatoms" connected by effective bonds and governed by softer or smoother effective nonbonded interactions. The resulting (usually linear) chain sequence is simpler and amenable to fast thermal equilibration through application of the state-of-the-art Monte Carlo algorithms slightly modified to account for the presence of the few different species along the chain. The method has been widely applied to reduce complexity and permit the molecular simulation of a number of polymers. Typical examples include polystyrene, poly(ethylene terepthalate), polycarbonates, polyphenylene dendrimers, even DNA. It involves computing the effective intra- and intermolecular potentials among superatoms such that the coarse-grained model reproduces as faithfully as possible the structural, configurational, and thermodynamic properties of the original atomistic polymer model. For vinyl chains presenting sequences of methylene (CH_2) and pseudoasymmetric methyne ($-CHR$) groups along their backbone, the method should also account for the isotactic, syndiotactric, or atactic stereochemistry of the polymer, based on the succession of meso (m) and racemo (r) dyads (see Figure 7.2).

According to Zwicker and Lovett [121], if all interactions with potentials $V^{(n)}(\mathbf{r}_1, \mathbf{r}_2, \ldots, \mathbf{r}_N)$, with $\mathbf{r}_i$ denoting the position vector of the ith particle, in an N-atom molecular system consist of n-body and lower terms, then the system can be completely described by knowing all n-order correlation functions $g^{(n)}(\mathbf{r}_1, \mathbf{r}_2, \ldots, \mathbf{r}_N)$ and lower. Since, in practice, complete determination of the n-point correlation functions is a huge task for $N > 2$ and $n > 4$, calculations are usually limited to correlation functions that depend only on a single coordinate. For polymers where potential functions are usually separated into intra- and intermolecular ones, examples of such correlation functions include typically the radial distribution function, the distribution of bond lengths, the distribution of bending angles, and the distribution of dihedral angles. The coarse-grained potential then should be chosen such that it matches the distributions of all possible bond lengths, bond

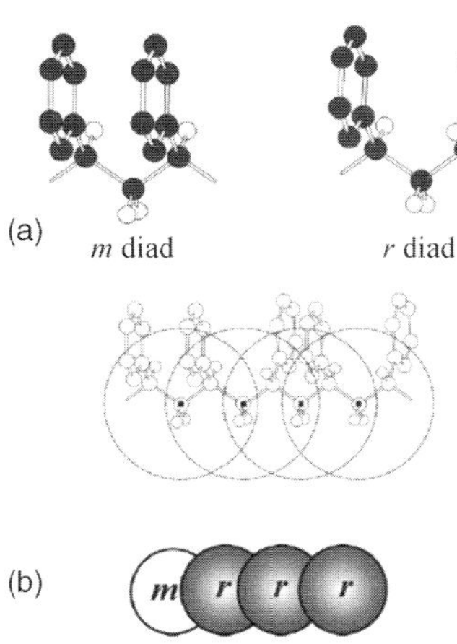

Figure 7.2 (a) Polystyrene *m* and *r* dyads in transplanar conformation (for clarity, hydrogen atoms on phenyl rings have been omitted). (b) Illustration of a mapping scheme from the atomistic to a coarse-grained structure for PS wherein one bead corresponds to an *m* or *r* dyad. (Reproduced with permission from [113].)

angles, and torsional angles, and of all intermolecular pair distribution functions, as extracted from simulations with the corresponding atomistic model.

For a distribution function that depends on a single coordinate, the corresponding effective potential can be computed through the iterative Boltzmann inversion method [106, 108, 110, 122], aimed at matching the distribution of the relevant degrees of freedom (called target distribution) between the chosen coarse-grained model and the initial atomistic model for the polymer; the latter is usually extracted from accurate, brute-force MD simulations on short homologues. The method uses the differences in the potentials of mean force between the distribution function generated from a guessed potential and the true distribution function to improve the effective potential iteratively. Qualitative arguments for the conditions under which convergence should be expected have been discussed by Soper [122].

The naive use of the coarse-grained potentials in standard molecular dynamics simulations leads to wrong predictions of diffusion and relaxation processes [104, 123, 124]. A simple, empirical method for relating the dynamics of superatoms at the coarse-grained level with the dynamics of true atomistic units at the atomic level uses a time rescaling factor [125, 126]. Within this method, effective potentials are used in reversible equations of motion of classical mechanics to perform standard molecular dynamics simulations and then the mean square displacements of the relevant structural units in the atomistic and coarse-grained models are matched, both in amplitude and slope. Noid *et al.* [127, 128] formulate consistency criteria that should be obeyed when using coarse-grained potentials in equilibrium dynamic simulations.

Coarse graining to the level of superatoms has drawn a lot of attention over the years mainly because of the capability to account for the correct stereochemical sequence of the repeat units. Despite the success of effective pair potentials in reproducing many of the structural properties of the corresponding atomistic system, their use in actual simulations is accompanied by a number of thermodynamic inconsistencies:

(a) They perform well for the particular physical properties they were developed for. For example, the value of pressure as computed by using the virial theorem from the effective potentials optimized with the iterative Boltzmann method is higher than what is observed in the atomistic system, unless an attractive perturbation potential is added (ramp correction to the pressure) and the potential is reoptimized. Given that in integral equation methods the mechanical properties of a system (such as pressure, energy, and compressibility) are fixed by the singlet and pair number densities along with proper closures [121, 129, 130], such a pressure inconsistency should be related to the degree of sensitivity of the site–site pair correlation function to the effective pair potential [122]. This is in line with the simulations of Jain *et al.* [131] who showed that although there is a one-to-one correspondence between the structure of a liquid (i.e., the pair correlation function) and its pairwise additive intermolecular potential (Henderson's theorem [129]), the convergence of potentials obtained by standard inversion procedures is extremely slow: although the repulsive part of the potential converges rapidly, its attractive part (to which, for example, the internal energy and pressure are primarily sensitive) converges slowly.
(b) Effective potentials are in general not transferable; they are state-point dependent (e.g., temperature and pressure). In same cases, it has been noticed that temperature changes at about the same density do not drastically affect their parameterization [132, 133]. A newly developed effective force coarse-graining seems to improve transferability to other state points [134]. Developing fully self-consistent and transferable potentials at any arbitrary level of coarse graining remains still a challenge.
(c) The coarse-grained system is considerably more compressible than the corresponding atomistic one.
(d) Despite recent efforts, the proper use of coarse-grained potentials for dynamic simulations remains unclear. In particular, the emergence of dissipation due to the coarse-graining step is mostly ignored or, at best, included phenomenologically via some stochastic thermostat as done, for example, in Refs [123, 135]. For some notable exceptions, see Ref. [58]. Considerable work is definitely needed in order to arrive at a thermodynamically consistent description of a model system at the two levels of analysis (atomistic and coarse-grained), which will eliminate all these undesired symptoms and errors.

7.2.2
Uncrossable Chains of Blobs

Briels and collaborators [136–139] proposed a coarse-graining scheme wherein chains are subdivided into a number of subchains of equal length; the center-of-mass of each subchain is taken as the position of a corresponding mesoscopic particle called the blob. The blobs are connected by springs so that chain connectivity is preserved. Similar to the coarse-graining procedure at the level of superatoms, the method uses a potential of mean force $U_{\mathrm{mf}} = V(\mathbf{R}^{(n)})$ for the position vectors of n blobs, which ensures that blob distributions in atomistic and coarse-grained systems are the same.

In order to describe shear flow effects in a velocity field of the from $v_x(\mathbf{r}) = \dot{\gamma} r_y$, Kindt and Briels [136] proposed the use of the SLLOD algorithm [140]. Starting with a Langevin equation, such a method results in the following expression for the blob dynamics:

$$\begin{aligned} M_i \frac{\mathrm{d}^2\mathbf{R}_i}{\mathrm{d}t^2} &= \mathbf{F}_i^{\mathrm{S}} - \varsigma^{\mathrm{eff}}\left(\frac{\mathrm{d}\mathbf{R}_i}{\mathrm{d}t} - \dot{\gamma} P_{iy}\hat{\mathbf{e}}_x\right) + \mathbf{F}_i^R \\ \varsigma^{\mathrm{eff}} &= \frac{\varsigma + \left[\sum_i (\mathbf{F}_i^{\mathrm{S}} \cdot \mathbf{P}_i - \dot{\gamma} P_{ix} P_{iy})\right],}{\sum_i \mathbf{P}_i^2} \end{aligned} \tag{7.2}$$

where $\hat{\mathbf{e}}_x$ denotes the unit vector along the flow (x) direction, M_i is the mass of the ith blob particle, $\mathbf{R}_i$ its position vector, $\mathbf{F}_i^{\mathrm{S}} = -\partial V/\partial \mathbf{R}_i$ the systematic force on particle i, ς the friction coefficient, and $\mathbf{F}_i^{\mathrm{R}}$ the random force on particle i. Since the coarse-grained bonded and nonbonded interactions are so soft that unphysical crossing of two bonds would not be prohibited, Eq. (7.2) is supplemented with an uncrossability constraint of the blob chains. Padding and Briels [137, 141] realized this constraint by a method that explicitly detects entanglements and prevents chain crossings through a geometric procedure. The procedure detects possible chain crossings and defines an entanglement point $\mathbf{X}$ at the prospective crossing site. Padding and Briels [136, 137] also proposed some nontrivial order-altering moves that lead to creation-removal of entanglements; these are important for the best possible realistic treatment of uncrossability constraints in simulations with the blob model. The Langevin equation of motion (7.2) contains the blob friction coefficient ς, whose calculation is not a straightforward issue even under equilibrium conditions. Despite this and its simplicity, the blob method has been found to correctly capture the viscoelastic properties of polymer melts with molecular length several times their entanglement length. From a numerical point of view, the method suffers from large requirements in CPU time, associated with the minimization algorithm for the location of entanglements that eventually limits simulations to chains made up of a finite number of blobs.

7.2.3 Primitive Paths

A method to project atomistically detailed chains to smoother paths was proposed by Kröger *et al.* [84] through a projection operation that maps a set $\{\mathbf{r}_i\}$, $i = 1, 2, \ldots, N$, of N atomistic coordinates of a linear discrete chain to a new set $\{\mathbf{R}_i\}$ of N coarse-grained ones defining a smoother path for the chain that avoids the kinks of the original chain but preserves somewhat its topology (the main chain contour).

The projection involves only a single parameter, ξ, whose value was obtained by Öttinger [142] by mapping the Porod–Kratky model (an atomistic model for a polymer chain) to a smoother chain with a Kuhn length equal to the entanglement length. In the limit of infinitely long chains, such a mapping suggests that the Kuhn length of the coarse-grained chain (the length of a segment between two entanglements) is equal to twice the tube diameter. The ξ-based method maps a particular chain onto a smoother path; however, the reduction of an ensemble of atomistic polymer chains to a mesh of primitive paths (PPs) as defined by the Doi–Edwards theory [14] requires that the projection satisfy not only chain continuity but also chain uncrossability. This subtle problem has been addressed only very recently through the seminal works of Everaers *et al.* [82], Kröger [80], and Tzoumanekas and Theodorou [143]. Nevertheless, the simple ξ-based mapping has been very helpful in many respects; for example, it has allowed [85, 144] to successfully calculate the zero shear rate viscosity of model polymer melts in the crossover regime from Rouse to entangled.

The topological analysis of Everaers *et al.* [82] is based on the idea that PPs can be simultaneously identified for all polymer molecules in a bulk system by keeping chain ends fixed in space, disabling intrachain excluded volume interactions and retaining the interchain ones, and minimizing the energy of the system by slowly cooling down toward the zero Kelvin temperature. This causes bond springs to reduce their length to zero, pulls chains taut, and results in a mesh of PPs consisting of straight segments of strongly fluctuating length and more or less sharp turns at the entanglement points. The method can be modified [81, 145, 146] to preserve self-entanglements or to distinguish between local self-knots and entanglements between different sections of the same chain.

Kröger [80] also presented an algorithm that returns a shortest path and the related number of entanglements for a given configuration of a polymeric system in 2D and 3D space, based on geometric operations designed to minimize the contour length of the multiple disconnected path (i.e., the contour length summed over all individual PPs) simultaneously for all chains in the simulation cell. The number of entanglements is simply obtained from the shortest path, as the number of interior kinks, or from the average length of a line segment. Application of the algorithm to united atom models of linear polyethylene [147] allowed the calculation of a number of important statistical properties characterizing its PP network at equilibrium and helped make the connection with an analytic expression for the PP length of entangled polymers by Khaliullin and Schieber [148] following earlier works [149, 150]. A representative snapshot of the entanglement network computed for a linear *trans*-1,4-PB polymer (40 chains of C_{500} at $T = 450$ K and $P = 1$ atm) with Kröger's method is shown in Figure 7.3.

The third methodology for reducing chains to shortest paths has been presented by Tzoumanekas and Theodorou [143] where topological (chain uncrossability) constraints are defined as the nodes of an entanglement network. Through their contour reduction topological analysis (CReTA) algorithm, an atomistic configuration of a model polymer sample is reduced to a network of corresponding PPs defined by a set of rectilinear segments (entanglement strands) coming together at nodal points

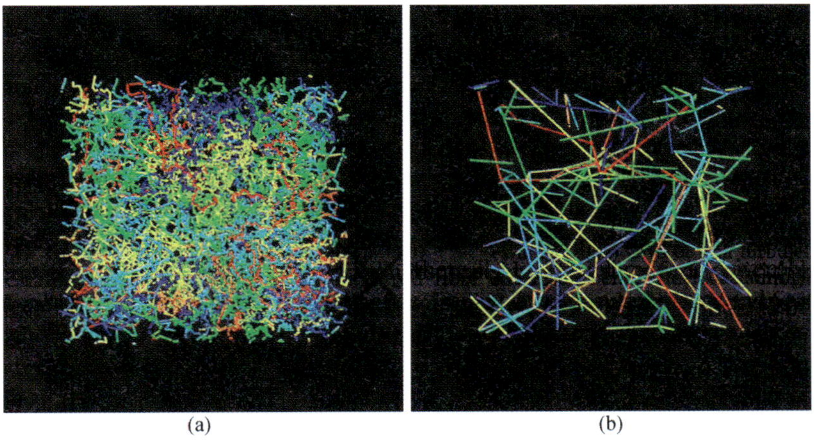

Figure 7.3 (a) A snapshot of a fully equilibrated atomistic configuration of a 40-chain C_{500}*trans*-1,4-PB melt at 500 K and 1 atm. (b) The corresponding entanglement network mesh computed with Kröger's method [80].

(entanglements) by implementing random aligning string moves to polymer chains and hard-core interactions. In addition to obtaining topological measures for a number of entangled polymers, Tzoumanekas and Theodorou [143] found that data for the normalized distribution of the reduced number of monomers (united atoms or beads) in an entanglement strand are well described by a master curve suggesting a universal character for linear polymers. As analyzed by Tzoumanekas and Theodorou [143], the master curve is also obtainable in terms of a renewal process generating entanglement events stochastically along the chain.

Apart from some algorithmic details, these three methods lead to practically similar conclusions as far as the topological state of many entangled linear polymer melts (PE, PB, PET, and PS) is concerned. A significant result of all of them (see Ref. [143]) is that the ensemble average of the number of monomers per entanglement strand $\bar{N}_{ES}$, as computed directly from the topological analysis, is significantly smaller than the corresponding quantity N_e measured indirectly through $N_e = N\langle R^2\rangle/\langle L\rangle^2$ by assuming that PP conformations are random walks. This is due to directional correlations between entanglement strands along the same PP, which decay exponentially with entanglement strand separation. Therefore, PPs are not random walks at the length scale of the network mesh size.

7.2.4 Other Single-Chain Simulation Approaches to Polymer Melts: Slip-Link and Dual Slip-Link Models

Doi and Edwards [151] and Doi and Takimoto [152] have proposed a description of an entangled polymer in terms of a slip-link model that can cumulatively account for chain confinement and constraint release in a consistent way. Slip-links do not represent an entanglement junction in real space; they are rather virtual links representing effective constraints whose statistical character is determined by other polymers. In the dual slip-link version of the model, the slip-link confines a pair of chains (and not a single chain). Masubuchi and collaborators [49, 50, 153, 154] generalized the idea by regarding a slip-link as an actual link in real space. In their formulation, each polymer chain is represented as a linear sequence of entanglement strands considered as segments (phantom entropic springs) joining consecutive entanglement points (the beads) along the chain. These Rouse-like chains are all interconnected by slip-links at the entanglement points to form a 3D primitive chain network. The system is described by the number Z of segments in each chain, the number n of monomers in each chain segment, and the position vectors $\mathbf{R}_i$ of the slip-links or entanglement points. These state variables are postulated to obey certain Langevin-type governing equations in which the single relevant parameter of the primitive chain network is the average value $\langle Z \rangle$ of entanglements per chain. Defining the model functions and parameters on the basis of the results obtained from one of the three topological analysis methods discussed above leads to rheological predictions that follow quite satisfactorily experimental data for many polymer melts in shear, but deviations are observed when the model is used to describe the elongational rheology of these systems. A generalization of the slip-link idea by Schieber and collaborators [44, 46] to a full-chain slip-link model with a mean field implementation of constraint release and constant chain friction (as opposed to constant entanglement friction) has been shown to provide accurate predictions of the G' and G'' spectra for many polymer melts (such as PS, PB, and PIB).

7.2.5
Entire Molecules

In dilute polymer solutions, coarse graining a polymer coil or star polymers to a system of interacting soft particles has been explored in Refs [155–157]. Kindt and Briels [141, 158] proposed such a type of coarse graining also for polymer melts where an entire chain is represented as a single particle. To account for the presence of entanglements that are considered to be responsible for the distinct viscoelastic properties exhibited by polymers, Kindt and Briels [141, 158] introduced a second set of variables, the number n_{ij} of entanglements between chains i and j. This governs the degree of interpenetration or overlapping of two chains whose centers of mass are fixed at a given distance. The state of the system is thus fully determined by the position vectors $\mathbf{R}_i$ of the centers of mass of the N_{ch} chains and the $N_{\mathrm{en}} = N_{\mathrm{ch}}(N_{\mathrm{ch}}-1)/2$ entanglement numbers n_{ij}. The equilibrium density distribution function Ψ for such a system is of the following form:

$$\Psi(\mathbf{R}^{(N_{\mathrm{ch}})}, n^{(N_{\mathrm{en}})}) \sim \exp\left\{ -\frac{1}{k_{\mathrm{B}}T} \left[U_{\mathrm{mf}}(\mathbf{R}^{(N_{\mathrm{ch}})}) + \sum_{i,j} \frac{1}{2}\alpha(n_{ij}-n_0(r_{ij}))^2 \right] \right\}, \tag{7.3}$$

where k_{B} is Boltzmann's constant, T the temperature, U_{mf} the potential of mean force, the double summation is over all interacting particle pairs, and α is a constant determining the strength of the fluctuations around a mean number $n_0(r_{ij})$ of entanglements between chains i and j. $n_0(r_{ij})$ is like an order parameter governing the "friction" felt by each chain and generating restoring elastic forces. According to Eq. (7.3), integration over n_{ij} results in a Boltzmann distribution for the N_{ch} coordinates $\mathbf{R}_i$, thus the equilibrium statistics of the system is not altered by the introduction of entanglements. Typical expressions for U_{mf} have been discussed by Padding and Briels [136, 137] and also by Pagonabarraga and Frenkel [159, 160] in their derivation of the "multiparticle dissipative particle dynamics" method. The method is capable of providing structural information only about the radial distribution function $g(r)$ of the centers of mass of the chains. Representative results for a number of linear PE melts revealed a small correlation hole effect at the level of entire chains, which is consistent with data reported by Mavrantzas and Theodorou [161] through atomistic Monte Carlo simulations. No other signals of local structure could be discerned. Clearly, accounting for entanglements, which is necessary in order to produce the correct viscoelastic properties, seems to have a negligible effect on the structural properties at the level of entire chains.

The single-particle model has been proposed to describe systems where memory effects are dominant. This is the case, for example, of complex fluids involving colloidal particles floating in a solvent in which a small amount of polymer is also dissolved. In this model, dynamics is described [158, 162] by generating (according to standard expressions for Smoluchowski-type equations) at every time step $\mathrm{d}t$ not only a displacement $\mathrm{d}\mathbf{R}_i$ in the position of each particle i but also a change $\mathrm{d}n_{ij}$ in the number of entanglements. Memory effects are taken into account through transient forces: when two particles come together such that temporarily $n_{ij} < n_0$, their coronas are pushed apart causing a repulsion between the two particles. On the other hand, if the coronas are separated enough such that temporarily $n_{ij} > n_0$, the particles experience attractive forces. These phenomena cannot be studied by traditional Brownian dynamics simulations where delta-correlated random displacements are assumed. Figure 7.4 shows the viscosity obtained from such a method referring to a typical resin with particles having a hard-core diameter of 100 nm. The very same model has also been successful in describing shear banding and the chaining of dissolved colloids in viscoelastic systems [163].

7.2.6
Conformation Tensor

Based on the idea that in a flow situation certain structural variables depart from their values in the quiescent fluid while all other (faster) degrees of freedom are at equilibrium subject to the constraints imposed by the values of the slow variables at all times, coarse-grained models have also been developed where chains are described at the level of the conformation tensor $\mathbf{c}$ [2, 12, 15, 164, 165]. The latter might be defined via the average gyration tensor or via the tensorial product of the end-to-end vectors. The choice of $\mathbf{c}$ among the structural parameters marks a description in terms of a *tensorial* variable. In addition to a single conformation tensor one can envision a description in terms of many (higher mode) conformation tensors,

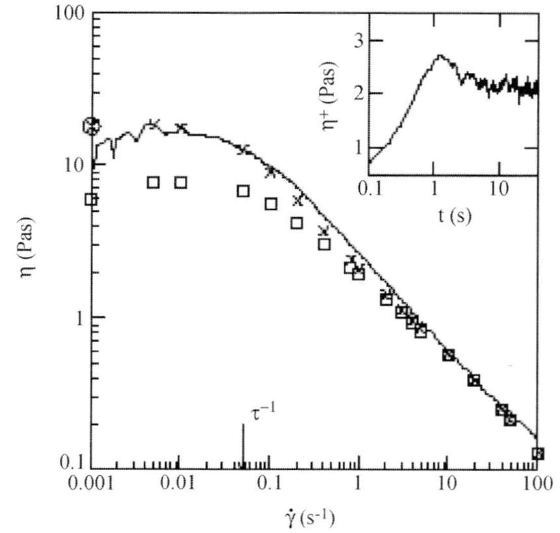

Figure 7.4 Viscosity versus shear rate for a typical resin, as obtained from the particle model of Kindt and Briels [158] and van der Noort *et al.* [162]. The solid line represents experimental data while the crosses are simulation results including all forces in the stress tensor. The squares represent viscosities based on the conservative forces only. The circle at the vertical axis gives the zero shear viscosity according to the Green–Kubo formula. In the inset, we show the overshoot in the instantaneous value of the viscosity at $\dot{\gamma} = 1\,\text{s}^{-1}$. (Reproduced with permission from [162].)

corresponding to the Rouse or bead-spring chain model. In general [164], from an N-mer chain, $N(N-1)/2$ different conformation tensors $\mathbf{c}_{ij}$, $i, j = 1, 2, \ldots, N-1$ can be constructed, each one being identified as a properly dimensionalized average dyadic $\langle \mathbf{Q}_i \mathbf{Q}_j \rangle$ with $\mathbf{Q}_i$ denoting the connector vector between mers $i+1$ and i along the chain [2].

For entangled polymer chain systems, the reptation theory suggests the distribution function of primitive path orientations as a structural variable, see previous section. If chains contain long-chain branches that can significantly affect the rheological response of the system, their contribution to overall system dynamics should also be accounted. This is the case, for example, with H-shaped polymers for which a model built on two structural parameters, the tube orientation tensor $\mathbf{S}$ from one branch to the other branch along the chain, and a scalar quantity Λ describing the length of the tube divided by the backbone tube length at equilibrium, has been proposed [24, 166]. This marks a description at the level of *a tensorial and a scalar.*

The description in terms of a few, well-defined structural parameters (such as the conformation tensor, the configurational probability function, and the orientation tensor and a scalar) is very appealing because of the existence of well-founded models developed under the GENERIC framework of nonequilibrium thermodynamics [167]. In Section 7.4, we will see, among other things, that this allows to build thermodynamically guided multiscale approaches by expanding the equilibrium statistical ensemble to incorporate terms involving conjugate variable(s) driving the corresponding structural parameters away from equilibrium. The GENERIC formalism here is an extremely useful tool since it can guide us in linking the conjugate variable(s) to the applied flow field. It is not surprising, therefore, that this class of coarse-graining procedures constitutes one of the most understood and best-founded today.

7.2.7
Mesoscopic Fluid Volumes

For the numerical simulation of flowing polymers, several mesoscopic models have been proposed in the last few years that describe polymer (hydro-)dynamics on a mesoscopic scale of several micrometers, typically. Among these methods, we like to mention dissipative particle dynamics (DPD) [168], stochastic rotation dynamics (sometimes also called multiparticle collision dynamics) [33], and lattice Boltzmann algorithms [30]. Hybrid simulation schemes for polymer solutions have been developed recently, combining these methods for solvent dynamics with standard particle simulations of polymer beads (see Refs [32, 169, 170]). Extending the mesoscopic fluid models to nonideal fluids including polymer melts is currently in progress [30, 159, 160, 171].

7.3
Systematic and Thermodynamically Consistent Approach to Coarse Graining: General Formulation

7.3.1
The Need for and Benefits of Consistent Coarse-Graining Schemes

Under equilibrium conditions, statistical thermodynamics forms a bridge between thermodynamics (whose goal is to understand and predict macroscopic phenomena) and molecular physics (which focuses on intermolecular interactions between atoms making up the system). It provides therefore an interpretation of thermodynamic quantities from a molecular point of view. For a number of complex fluids (such as colloids, liquid crystals, and polymers), information at a mesoscopic level of description (intermediate to molecular and macroscopic ones) is often extremely useful in understanding and predicting material behavior (see Section 7.2 and Ref. [12]). At equilibrium, statistical thermodynamics also provides the framework for understanding system properties at these intermediate scales. For example, as mentioned in Section 7.2.1, effective potentials describing interactions between coarse-grained (pseudo-)particles can systematically be derived by integrating out irrelevant degrees of freedom.

Although extensions to capture dynamics at a mesoscopic level are in progress (see Section 7.2), most descriptions in terms of coarse-grained particles are so far largely restricted to equilibrium situations. For example, the recently proposed time rescaling approach (see [114, 126]) for coarse-grained models does not seem appropriate to fully account for the increase in dissipation inherent in any meaningful coarse graining technique. We mention the case of hydrodynamic interactions in polymer solution that necessitate a description not in terms of a scalar frictional variable but in terms of a tensorial friction matrix.

Dissipation and friction are more properly accounted for in Refs [58, 136, 138, 139, 172]. These authors, however, arrive at a daunting assessment: "We therefore conclude that coarse-grained models lack thermodynamic consistency" [138]. As we will demonstrate below, for appropriately defined coarse-grained models, there is a way to restore thermodynamic consistency. So, contrary to the authors of [138], we believe that the recently introduced GENERIC formalism of nonequilibrium thermodynamics offers a framework for the development of true and complete coarse-graining strategies [11], in the sense that (a) the resulting model is well-behaved and thermodynamically consistent, (b) it can be parameterized on the basis of the information provided by a lower resolution model, and (c) it can be improved with microscopic simulations targeted to address the relevant structural variables and their dynamic evolution. Of course, a word of caution is in place here: coarse-grained models based on the GENERIC framework will rely on a number of strong assumptions (inherent to most projection operator-based methods) implying a description in terms of a set of carefully chosen, slowly evolving state variables. The underlying assumption behind such a description is that of the existence of a clear timescale separation between the evolution of these (slow) variables and that of the (eliminated) fast or irrelevant ones. The coarse-grained model of Ref. [138], for example, involves lumping 10 beads along a chain into 1 or 2 blobs, for which the timescale separation argument is questionable. Their negative conclusion about the thermodynamic consistency of the model is therefore not surprising.

7.3.2
Different Levels of Description and the Choice of Relevant Variables

Coarse-graining connects (at least) two descriptions of the same system at two different levels of detail: a low-resolution level and a high-resolution level. We focus attention here to the case where the high-resolution level is the atomistic one, although this is not necessary [173].

We consider a point in phase space $\mathbf{z} \in \Gamma$, where $\mathbf{z} = (\mathbf{r}_1, \ldots, \mathbf{r}_N;\ \mathbf{p}_1, \ldots, \mathbf{p}_N)$ is a shorthand notation for the positions $\mathbf{r}_i$ and momenta $\mathbf{p}_i$ of all N particles, at the microscopic level; this, for example, could be an all-atom or a united atom model or even the simpler and computationally more convenient FENE bead-spring model [12]. All these three models are classified here as microscopic models due to the absence of dissipation and irreversibility. Dynamics at the microscopic model is governed by Hamilton's equation of motion

$$\dot{\mathbf{z}} = \mathbf{J} \cdot \frac{\partial H}{\partial \mathbf{z}}, \tag{7.4}$$

where $H(\mathbf{z})$ is the microscopic Hamiltonian and $\mathbf{J}$ the symplectic matrix. Equivalently, Hamilton dynamics can be formulated by $\dot{A} = \{A, H\}$, where $\{A, B\}$ denotes the microscopic Poisson bracket between arbitrary functions $A(\mathbf{z})$ and $B(\mathbf{z})$. We recall that the basic properties of Poisson brackets are their antisymmetry $\{A, B\} = -\{B, A\}$, the Leibniz rule $\{AB, C\} = A\{B, C\} + \{A, C\}B$, and the Jacobi identity, $\{A, \{B, C\}\} + \{B, \{C, A\}\} + \{C, \{A, B\}\} = 0$ [167].

Owing to their very long relaxation times, there is a clear gap between the timescales that can be addressed in microscopic simulations of polymer melts and the relevant timescales in experimental studies. Although this prevents the direct applicability of brute-force microscopic simulations, it renders them ideal systems for a comprehensive understanding over multiple time- and length scales by embodying the concept of multiscale modeling. The first and more important step in this context is the proper choice of relevant variables at the coarser level. For simple fluids, densities of conserved quantities (mass, momentum, and energy) are the proper variables to consider if one is interested in hydrodynamic properties. For systems with broken symmetries, the corresponding order parameters constitute additional candidates for slow variables [174]. In the case of complex fluids, however, no general rules are available how the appropriate relevant variables should be chosen, and this emphasizes the importance of physical intuition [56, 103] for the choice of variables beyond equilibrium.

For polymers, one can be guided by available theoretical models. For example, orientational ordering in liquid crystals and liquid crystalline polymers can be described by the alignment tensor within the Landau–de Gennes theory [14]. Birefringence and viscous properties in the case of unentangled polymer melts can be addressed by models based on the concept of a conformation tensor, see Section 7.2.6. For branched polymers, a scalar variable is added to the conformation tensor in order to capture additional contributions to the stress tensor [24] due to long arm relaxations. For entangled polymer melts, the reptation theory provides a description in terms of a probability distribution function for the orientation of segments along the primitive path [2, 14].

These theories are examples of mesoscopic or macroscopic models that lead to closed-form constitutive equations. Furthermore, they can all be described in the context of the single-generator bracket [175] or the GENERIC [167] formalisms of nonequilibrium thermodynamics,

$$\dot{\mathbf{x}} = \mathbf{L} \cdot \frac{\delta E}{\delta \mathbf{x}} + \mathbf{M} \cdot \frac{\delta S}{\delta \mathbf{x}}. \tag{7.5}$$

In Eq. (7.5), $E(\mathbf{x})$ and $S(\mathbf{x})$ are the coarse-grained energy and entropy functions, respectively. The antisymmetric operator $\mathbf{L}$ defines a generalized Poisson bracket $\{A, B\} = \frac{\delta A}{\delta \mathbf{x}} \cdot \mathbf{L} \cdot \frac{\delta B}{\delta \mathbf{x}}$ that possesses the same properties as the classical Poisson bracket described above. The last term in Eq. (7.5) is new compared to Hamiltonian dynamics (7.4) and describes dissipative, irreversible phenomena. The friction matrix $\mathbf{M}$ is symmetric[1)] and positive, semidefinite. Together with the degeneracy

1) A more detailed discussion of the Onsager–Casimir symmetry is given in Section 7.3.2.1 of Ref. [167].

requirements $\mathbf{L}\cdot(\delta S/\delta\mathbf{x}) = \mathbf{M}\cdot(\delta E/\delta\mathbf{x}) = 0$, these properties ensure that the total energy E is preserved and S is not decreasing in time [167].

The mesoscopic and macroscopic models come with a number of parameters, for example, mean field potentials, friction coefficients, effective relaxation times, and so on, whose connection with molecular terms is not straightforward. It is the purpose of thermodynamically guided, systematic coarse-graining methods to address this issue.

7.3.3 GENERIC Framework of Coarse Graining

Coarse graining implies a description in terms of a few, carefully chosen variables after the elimination of all irrelevant degrees of freedom. Inevitably, this comes together with entropy generation (irreversibility) and dissipation. In the GENERIC formalism [167, 173], the emphasis is then shifted from the fundamental time evolution equation itself to the individual building blocks of that theory (7.5), and paves the way for the development of consistent coarse-graining strategies. The interested reader may refer to Ref. [11]. More details on the statistical mechanics of coarse graining can be found in Refs [56, 167, 176]; for coarse graining of simple fluids within the GENERIC framework, see Refs [177, 178].

7.3.3.1 Mapping to Relevant Variables and Reversible Dynamics

For every microstate $\mathbf{z}$ of the system, the instantaneous values of the relevant variables are defined by a set of phase space functions $\Pi(\mathbf{z})$. The functions $\Pi(\mathbf{z})$ cannot generally be identified with $\mathbf{x}$; they are rather connected with $\mathbf{x}$ through $\mathbf{x} = \langle\Pi(\mathbf{z})\rangle$, that is, as averages based on a suitable probability density $\varrho_{\mathbf{x}}(\mathbf{z})$ at the microscopic phase space Γ. Thus, the coarse-grained energy $E(\mathbf{x})$ is obtained from the microscopic Hamiltonian $H(\mathbf{z})$ by straightforward averaging,

$$E(\mathbf{x}) = \langle H(\mathbf{z})\rangle_x, \qquad (7.6)$$

and, similarly, the coarse-grained Poisson bracket is obtained from the average of the classical Poisson bracket,

$$\{A, B\} = \frac{\delta A}{\delta x_k}\cdot L_{kl}\cdot\frac{\delta B}{\delta x_l}; \quad L_{kl}(\mathbf{x}) = \langle\{\Pi_k, \Pi_l\}\rangle_x. \qquad (7.7)$$

Equations (7.6) and (7.7) define the reversible part of GENERIC (7.5) in terms of a coarse-grained Poisson bracket [174, 179]. The additional terms related to dissipation and increase in entropy have to be accounted for by the irreversible contribution to GENERIC (7.5) and are described in the following section.

7.3.3.2 Irreversibility and Dissipation Through Coarse Graining

The fact that we do not account explicitly for the irrelevant variables at the level of the GENERIC framework leads to entropy increase and additional dissipation at the coarser level of description [11].

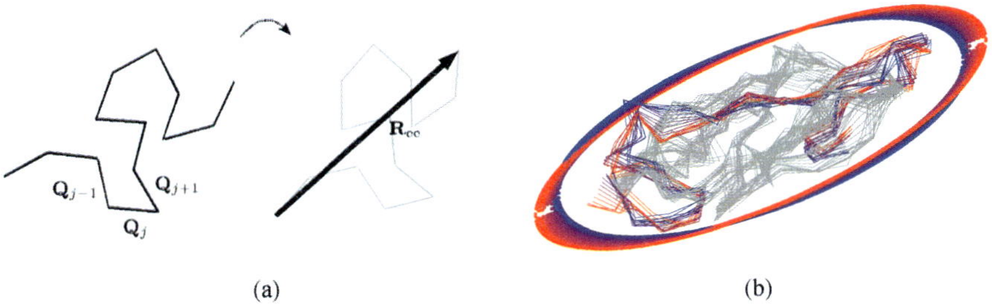

Figure 7.5 (a) Schematic illustration of freely jointed chain with end-to-end vector $\mathbf{R}_{ee}$. (b) Fluctuations of polymer chain in shear flow around stationary state. Ellipses indicate the eigenvalues and orientation of eigenvector of $\langle \mathbf{R}_{ee}\mathbf{R}_{ee}\rangle$. (Figure courtesy of M. Kröger, ETH Zürich.)

To illustrate this, let us consider the simple example of a freely jointed chain, shown schematically in Figure 7.5a. We can describe such a chain by the set of all connector vectors $\mathbf{Q}_j \equiv \mathbf{r}_{j+1} - \mathbf{r}_j$, $j = 1, \ldots, N$. All admissible configurations with $|\mathbf{Q}_j| = b$ have equal probability. If we decide to choose the end-to-end vector $\mathbf{R}_{ee}$ as the only relevant variable, then there are in general many configurations $\{\mathbf{Q}_j\}$ that are compatible with $\mathbf{R}_{ee}$. The coarse-grained entropy $S(\mathbf{R}_{ee})$ is a measure of the number of these configurations. One finds that the probability of $\mathbf{R}_{ee}$ is Gaussian around $\mathbf{R}_{ee} = 0$, which implies that there are many more coiled configurations compared to stretched ones. The associated entropy $S(\mathbf{R}_{ee}) = S(0) - 3\mathbf{R}_{ee}^2/2Nb^2$ decreases for chains undergoing stretching and leads to a restoring force that is known as "entropic spring" in coarse-grained polymer models [2]. This illustrates the emergence of (additional) entropy through coarse graining.

We turn now to the discussion of the probability distribution $\varrho_{\mathbf{x}}(\mathbf{z})$. In sharp contrast to equilibrium statistical mechanics, there are unfortunately no general results for the probability distribution of nonequilibrium states. Even for nonequilibrium stationary states there are at present only a few results for very special model systems available (see Ref. [180]). Systems, however, where the timescale separation assumption holds are well described within the quasiequilibrium approximation that treats the nonequilibrium system as an equilibrium one for the present values of relevant variables [181–183]. In the generalized microcanonical ensemble, all microstates $\mathbf{z}$ compatible with the given values of relevant variables $\Pi(\mathbf{z})$ have equal probability. The corresponding entropy is a measure of the number of such microstates $\mathbf{z}$ that are compatible with a given coarse-grained state. For practical calculations, it is more convenient to pass to the generalized canonical distribution. In an analogy to equilibrium statistical mechanics, the average values $x_k = \langle \Pi_k(\mathbf{z})\rangle_x$ are constrained to prescribed values with the help of Lagrange multipliers Λ_k. The generalized canonical distribution can then be obtained from the maximum entropy principle and reads

$$\varrho_{\mathbf{x}}(\mathbf{z}) = \frac{e^{-\sum_k \Lambda_k \Pi_k(\mathbf{z})}}{\int_\Gamma d\mathbf{z}\, e^{-\sum_l \Lambda_l \Pi_l(\mathbf{z})}}, \tag{7.8}$$

where the Λ_k have to be chosen so as to satisfy $x_k = \int \mathrm{d}\mathbf{z}\Pi_k(\mathbf{z})\varrho_{\mathbf{x}}(\mathbf{z})$. The quasi-equilibrium entropy associated with Eq. (7.8) is

$$S(\mathbf{x}) = k_\mathrm{B} \sum_k \Lambda_k x_k + k_\mathrm{B} \ln \int_\Gamma \mathrm{d}\mathbf{z}\, \mathrm{e}^{-\sum_l \Lambda_l \Pi_l(\mathbf{z})}. \qquad (7.9)$$

The coarse-grained entropy plays the role of an effective potential for the relevant variables. Determining the functional form of $S(\mathbf{x})$ from (7.9) presents a challenge since the explicit expression for the Lagrange multipliers $\Lambda_k(\mathbf{x})$ is in general unknown. A successful method for extracting at least partial information on $S(\mathbf{x})$ has been explored in Ref. [164] from atomistic simulations of a polymer melt in elongational flow. We discuss this issue further in Sections 7.4.2 and 7.4.3.

After specifying the nonequilibrium ensemble and coarse-grained entropy, we finally like to discuss the increase of dissipation through coarse graining in more general terms. We have seen above that many microstates (values of connector vectors) are in general compatible with a given coarse-grained state (defined by the value of the end-to-end vector in the above example). Conversely, this implies that a coarse-grained state does not uniquely determine the microstate. The dynamics on the coarse-grained level has necessarily a stochastic character known as fluctuations (see Figure 7.5b). If these fluctuations are correlated in time, they have to be accompanied by dissipation, as required by the fluctuation-dissipation theorem [56]. These qualitative observations are put into a solid theoretical framework by the projection operator formalism [103]. It should be emphasized that projection operators provide exact relations for any set of variables in terms of complicated integro-differential equations. Simpler, closed form equations without a memory integral, however, result only in cases when the timescale of the chosen variables is well separated from those of the irrelevant ones [103, 184]. A prominent example where this assumption seems not to be met is the dynamics of glassy polymers, where usually mode-coupling approximations for the memory kernel are employed [185, 186]. We here insist on the timescale separation, which severely restricts possible choices of relevant variables where such a separation can hold. For glassy polymers or glasses, in general, an appropriate set of relevant variables is not known at present, although some promising first steps have recently been taken [187, 188]. These restrictions are the price to pay for a proper coarse-grained description with a well-defined entropy and without accounting for memory effects. In this case, the dissipation matrix $\mathbf{M}$ as derived from the projection operator formalism reads

$$M_{kl} = \frac{1}{k_\mathrm{B}} \int_0^{\tau_\mathrm{s}} \mathrm{d}t\, \langle \dot{\Pi}_k^\mathrm{f}(t) \dot{\Pi}_l^\mathrm{f}(0) \rangle, \qquad (7.10)$$

where $\dot{\Pi}_k^\mathrm{f}$ is the fast part of the time derivative of the macroscopic variables [167, 173, 182]. The separating timescale τ_s should be chosen large enough to comprise all the fast fluctuations that are not captured on the coarse-grained level [167]. Thus, the friction matrix $\mathbf{M}$ arises due to fast fluctuations that are not resolved at the coarser level. The numerical evaluation of the dissipation matrix (7.10) for a polymer melt is described in Section 7.4.3.

7.4 Thermodynamically Guided Coarse-Grained Polymer Simulations Beyond Equilibrium

7.4.1 GENERIC Coarse-Graining Applied to Unentangled Melts: Foundations

Unentangled polymer melts are usually described in terms of the conformation tensor that provides an overall picture of the entire polymer chain. From the point of view of the GENERIC formalism, this implies a description where, in addition to the hydrodynamic fields mass ϱ, momentum $\mathbf{g}$, and energy density ε, the conformation tensor $\mathbf{c}$ is also included in the vector of state variables,

$$\mathbf{x} = (\varrho, \mathbf{g}, \varepsilon, \mathbf{c}). \tag{7.11}$$

Let $\mathbf{z} = (\mathbf{r}_1, \ldots, \mathbf{r}_N; \mathbf{p}_1, \ldots, \mathbf{p}_N)$ denote the microstate defined by the positions and momenta of all particles. The chosen macroscopic variables $\mathbf{x} = \langle \Pi(\mathbf{z}) \rangle$ are defined as follows. The mass density is defined by

$$\varrho(\mathbf{r}; t) = \left\langle \sum_j m_j \delta(\mathbf{r} - \mathbf{r}_j(t)) \right\rangle \equiv \langle \Pi_\varrho \rangle, \tag{7.12}$$

where m_j denotes the mass of particle j. Similarly, the momentum density is obtained by

$$\mathbf{g}(\mathbf{r}; t) = \left\langle \sum_j \mathbf{p}_j \delta(\mathbf{r} - \mathbf{r}_j(t)) \right\rangle \equiv \langle \Pi_g \rangle. \tag{7.13}$$

From ϱ and $\mathbf{g}$, the macroscopic velocity field $\mathbf{v}(\mathbf{r})$ is defined by $\mathbf{v}(\mathbf{r}) = \mathbf{g}(\mathbf{r})/\varrho(\mathbf{r})$. The total energy density can be expressed as

$$\varepsilon(\mathbf{r}; t) = \left\langle \sum_j \hat{e}_j \delta(\mathbf{r} - \mathbf{r}_j(t)) \right\rangle \equiv \langle \Pi_\varepsilon \rangle, \tag{7.14}$$

where $\hat{e}_j = (1/2) m_j \mathbf{u}_j^2 + \Phi_j$ with $\mathbf{u}_j = \mathbf{p}_j / m_j - \mathbf{v}(\mathbf{r}_j)$ the peculiar velocity of particle j and Φ_j the potential energy of particle j. Finally, the additional, internal variable $\mathbf{c}$ is a symmetric, second-rank tensor that is defined by

$$\mathbf{c}(\mathbf{r}; t) = \frac{1}{N_{\text{ch}}} \sum_{a=1}^{N_{\text{ch}}} \langle \hat{\Pi}^a \delta(\mathbf{r} - \mathbf{r}_c^a(t)) \rangle \equiv \langle \Pi_c \rangle, \tag{7.15}$$

where N_{ch} is the number of chains in the system and $N_a = \sum_{j \in I_a}$ the number of particles in chain a. The center of mass of polymer a is denoted by $\mathbf{r}_c^a = N_a^{-1} \sum_{j \in I_a} \mathbf{r}_j$. The tensor $\hat{\Pi}^a$ is a conformation tensor of a single chain and quantifies the instantaneous, internal structure of polymer a. Examples are the gyration tensor $\hat{\Pi}^a = N_a^{-1} \sum_{j \in I_a} (\mathbf{r}_j - \mathbf{r}_c^a)(\mathbf{r}_j - \mathbf{r}_c^a)$ or the tensor product formed either by the end-to-end vector or by the first Rouse mode.

The macroscopic energy E is obtained by straightforward averaging of the microscopic Hamiltonian, see Eq. (7.6).

The resulting expression for the distribution function of the generalized canonical ensemble reads

$$\varrho_x(\mathbf{z}) = Z^{-1}\exp[-\beta p\mathcal{V} - \beta\Phi - \lambda : \Pi_c], \tag{7.16}$$

where $\beta = (k_B T)^{-1}$, $\mathcal{V}$ the volume occupied by the N particles, Φ the total potential energy, Z the normalization integral and p the pressure (see Refs [164, 189] and Section 8.2.3 in Ref. [167] where Eq. (7.16) is used in nonequilibrium situations). The macroscopic entropy associated with the generalized canonical distribution is given by Eq. (7.9), which here reads

$$S(\mathbf{x}) = S_0(T, V, N) + k_B\left[\ln\left(\frac{Z}{V^N}\right) + \beta p V + \beta\langle\Phi\rangle_x + \lambda : \mathbf{c}\right], \tag{7.17}$$

where $V = \langle\mathcal{V}\rangle_x$ is the average volume and S_0 the entropy of an ideal gas of N particles. In addition to the usual Lagrange multipliers β and βp that are associated with total energy and volume (for homogeneous density), respectively, the additional Lagrange multiplier λ is identified as

$$\lambda = k_B^{-1}\frac{\partial S}{\partial \mathbf{c}}. \tag{7.18}$$

For a numerical calculation of the Lagrange multiplier λ for a model polymer melt, see Sections 7.4.2 and 7.4.3.

The matrix $\boldsymbol{L}$ defining the coarse-grained Poisson bracket (7.7) is obtained by inserting the definitions (7.12)–(7.15) of the coarse-grained variables into Eq. (7.7). Details of the straightforward calculations are presented in Ref. [190].

From the degeneracy requirement on the Poisson bracket $\{S, E\} = 0$ mentioned in Section 7.3.2, one finds that the entropic part of the macroscopic stress tensor has to be of the form

$$\sigma = -p_{\text{eff}}\mathbf{1} - 2T\mathbf{c}\cdot\frac{\partial s}{\partial \mathbf{c}}, \tag{7.19}$$

where p_{eff} is the effective scalar pressure and s the local entropy density. The same form (7.19) has been previously found in Refs [167, 191].

As far as the dissipative bracket and the associated friction matrix $\mathbf{M}$ are concerned, a direct calculation of the fast time evolution $\dot{\Pi}^f$ appearing in Eq. (7.10) shows that $\dot{\Pi}^f_\varrho = 0$, $\dot{\Pi}^f_g = \nabla\cdot\hat{\sigma}^{\text{tot}}$, where $\hat{\sigma}^{\text{tot}}$ is the instantaneous value of the total stress tensor. The expression for $\dot{\Pi}^f_\varepsilon$ containing the heat flux and viscous heating can be found in Ref. [190]. The integral of the time correlation function of these fast fluctuations that appears in Eq. (7.10) can in most cases be determined only numerically. How these quantities can be extracted from molecular dynamics simulations for a model polymer melt is described in Section 7.4.3.

The resulting GENERIC equations (7.5) for the present choice of relevant variables are [167, 190, 191]

$$\begin{aligned}
\frac{\partial}{\partial t}\varrho &= -\nabla_\beta(v_\beta \varrho) \\
\frac{\partial}{\partial t}g_\alpha &= -\nabla_\beta(v_\beta g_\alpha) + \nabla_\beta \sigma^{\text{tot}}_{\beta\alpha} \\
c_{\alpha\beta,[1]} &= -\frac{1}{T} M_{c_{\alpha\beta}c_{\mu\nu}} \left[\frac{\partial \varepsilon}{\partial c_{\mu\nu}} - T \frac{\partial s}{\partial c_{\mu\nu}} \right],
\end{aligned} \tag{7.20}$$

where $\mathbf{c}_{[1]}$ denotes the upper convected derivative of $\mathbf{c}$, $\mathbf{c}_{[1]} \equiv \partial_t \mathbf{c} + \mathbf{v} \cdot \mathbf{c} - \mathbf{c} \cdot \varkappa^{\text{T}} - \varkappa \cdot \mathbf{c}$, $\varkappa = (\nabla \mathbf{v})^{\text{T}}$ the transpose of the velocity gradient, and ε and s the energy and entropy density, respectively. Since we consider in the following only isothermal conditions, the reader may refer to Refs [190, 191] for the rather lengthy expression of the internal energy balance. Furthermore, additional second-order dissipative processes appearing in Eq. (7.20) are discussed in Ref. [190].

The macroscopic stress tensor appearing in the momentum balance Eq. (7.20) is given by

$$\sigma^{\text{tot}} = -p_{\text{eff}}\,\mathbf{1} + 2\mathbf{c} \cdot \left[\frac{\partial \varepsilon}{\partial \mathbf{c}} - T \frac{\partial s}{\partial \mathbf{c}} \right] - \frac{1}{T} \mathbf{C}^{(\sigma\sigma)} : \varkappa, \tag{7.21}$$

where $\mathbf{C}^{(\sigma\sigma)} = \int_0^{\tau_s} \text{d}t \langle \sigma^{\text{f}}(t) \sigma^{f}(0) \rangle_x$ is a Green–Kubo formula for the viscosity contribution of fast (on timescale shorter than τ_{s}) stress fluctuations. This finding is in agreement with previous simulation studies on bead-spring chain polymer melts that found it necessary to include a simple fluid background viscosity in their analysis [57, 192].

7.4.2
Thermodynamically Guided Atomistic Monte Carlo Methodology for Generating Realistic Shear Flows

In this section, we discuss how one, guided by the principles of nonequilibrium thermodynamics, can use the Monte Carlo technique to drive an ensemble of system configurations to sample statistically appropriate steady-state nonequilibrium phase-space points corresponding to an imposed external field [161, 164, 193–195]. For simplicity, we limit our discussion to the case of an unentangled polymer melt. The starting point is the probability density function ϱ_x of the generalized canonical GENERIC ensemble (7.16) for the same set of slow variables (7.11) as in Section 7.4.1. Then, following Mavrantzas and Theodorou [161], we extend the Helmholtz free energy, A, of equilibrium systems to nonequilibrium systems as

$$\text{d}\left(\frac{A}{V}\right) = -\frac{S}{V}\text{d}T + \mu\,\text{d}\left(\frac{N_{\text{ch}}}{V}\right) - k_{\text{B}} T \lambda : \text{d}\mathbf{c}, \tag{7.22}$$

where μ is the chemical potential. The last term accommodates the effect of the external field (e.g., a flow) for which λ represents a nonequilibrium force variable conjugate to $\mathbf{c}$. According to Eqs. (7.16) and (7.22), one can carry out Monte Carlo simulations in the expanded $\{N_{ch}NPT\lambda\}$ ensemble exactly as in the corresponding $\{N_{ch}NPT\}$ equilibrium ensemble (with N denoting the total number of atoms in the system) by assigning nonzero values to the field λ. This is the key point of the new method opening up the way toward sampling steady-state nonequilibrium phase points of the system corresponding to a given flow field with Monte Carlo simulation by suitably choosing the components of λ. For the case of a simple shear flow, for example, from the symmetry property of $\mathbf{c}$, we recognize that λ is to have only four independent nonzero components: $\lambda_{xx}, \lambda_{xy}, \lambda_{yy}$, and λ_{zz}. In order to specify their numerical values for a given shear rate $\dot{\gamma}$ (these are needed to be used as input in the GENERIC MC simulations), one can resort to the fundamental GENERIC evolution law for the set of state variables $\mathbf{x}$. Based on this and Eq. (7.16) for the definition of λ, we see that we can, indeed, assign a kinematic interpretation to the Lagrange multiplier λ since for a nonequilibrium system that has reached a steady state, Eq. (7.5) simplifies to

$$\lambda = -\frac{1}{k_B}\mathbf{M}^{-1}\cdot\mathbf{L}(\mathbf{x})\cdot\frac{\delta E(\mathbf{x})}{\delta\mathbf{x}}. \tag{7.23}$$

For example, for all known conformation tensor viscoelastic models, the corresponding evolution equation for the conformation tensor reads

$$\hat{c}_{\alpha\beta,[1]} = -\Lambda_{\alpha\beta\gamma\varepsilon}\frac{\delta A(\mathbf{c})}{\delta c_{\gamma\varepsilon}} = -nk_B T\Lambda_{\alpha\beta\gamma\varepsilon}\alpha_{\gamma\varepsilon}; \quad \alpha_{\alpha\beta} = \frac{1}{nk_B T}\frac{\delta A(\mathbf{c})}{\delta c_{\alpha\beta}}, \tag{7.24}$$

where (for simplicity) we have replaced the tensor λ with the tensor α defined through $\lambda = -(N_{ch}/V)\alpha$. In Eq. (7.24), $\hat{c}_{\alpha\beta}$ denotes the upper convected derivative of $c_{\alpha\beta}$ and n the chain number density, and the Einstein summation convention has been employed for repeated indices. Note also that in the case considered here, the element M_{44} of the $\mathbf{M}$ matrix in Eq. (7.23) has the form of $T\Lambda_{\alpha\beta\gamma\varepsilon}$ (see Refs [164, 193] for details) where the fourth-order relaxation matrix Λ for most single-conformation tensor models can be cast into the following general form:

$$\Lambda_{\alpha\beta\gamma\varepsilon}(\mathbf{c}) = f_1(I_1)(c_{\alpha\gamma}\delta_{\beta\varepsilon} + c_{\alpha\varepsilon}\delta_{\beta\gamma} + c_{\beta\gamma}\delta_{\alpha\varepsilon} + c_{\beta\varepsilon}\delta_{\alpha\gamma}) + 2f_2(I_1)(c_{\alpha\gamma}c_{\beta\varepsilon} + c_{\alpha\varepsilon}c_{\beta\gamma}), \tag{7.25}$$

where I_1 is the first invariant of $\mathbf{c}$ (i.e., the trace of $\mathbf{c}$), δ the unit tensor, and f_1 and f_2 arbitrary functions of I_1. With the help of Eqs. (7.24) and (7.25), for the case of a steady-state flow described by the kinematics

$$\nabla\mathbf{v} = \begin{pmatrix} 0 & 0 & 0 \\ \dot{\gamma} & 0 & 0 \\ 0 & 0 & 0 \end{pmatrix} \tag{7.26}$$

we find that the form of α that generates shear is

$$\alpha = \begin{pmatrix} \alpha_{xx} & \alpha_{xy} & 0 \\ \alpha_{xy} & \alpha_{yy} & 0 \\ 0 & 0 & 0 \end{pmatrix}. \tag{7.27}$$

Although nonequilibrium thermodynamics has helped us define the functional form of α, the exact relationship between its three nonzero components (α_{xx}, α_{xy}, and α_{yy}) on the applied shear rate $\dot{\gamma}$ remains still undetermined. One way to come around this problem is to explicitly use specific expressions for the matrix **M** proposed by GENERIC for a viscoelastic model. In such a case, however, the results will be model dependent and not representative of the true structure developing in the system in response to the applied shear rate $\dot{\gamma}$. Baig and Mavrantzas [193] proposed overcoming this by computing α iteratively so that, for a given value of $\dot{\gamma}$, the resulting average conformation of the simulated melt is the same as that predicted by a brute-force application of the NEMD method.

Baig and Mavrantzas [193] demonstrated the applicability of such a hybrid GENERIC MC-NEMD approach for a relatively short unentangled PE system, $C_{50}H_{102}$, for different nonequilibrium states corresponding to different values of the Werssenberg number (*Wi*). *Wi* is defined as the product of the imposed shear rate $\dot{\gamma}$ and the longest relaxation (Rouse) time of the system, τ_R, at simulation temperature and pressure. If x is the flow direction and y and z the velocity gradient and neutral directions, respectively, then the three nonzero components of α can be computed iteratively so that the values of the conformation tensor **c** for the system (at the given value of *Wi*) from the GENERIC MC and the NEMD methods coincide. Representative results are shown in Figures 7.6 and 7.7. Figure 7.6 presents the values of the nonzero components of α that were found to accurately reproduce the corresponding nonequilibrium state for the simulated $C_{50}H_{102}$ system as a function of the imposed *Wi*. Figure 7.7, on the other hand, presents comparisons of the conformation tensor between the GENERIC MC simulations (corresponding to the α-values shown in Figure 7.6) and the direct NEMD simulations, confirming that c_{xx}, c_{xy}, and c_{yy} from the GENERIC MC and the NEMD simulations, respectively, superimpose. It is only for the c_{zz} component of the conformation tensor that

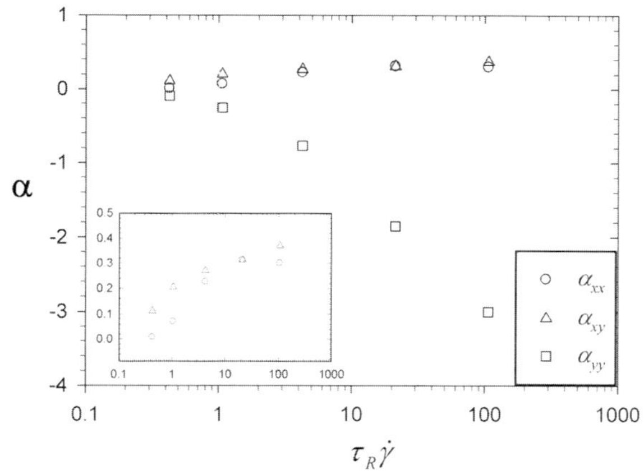

Figure 7.6 Plot of the thermodynamic force field, α, versus *Wi* number for the $C_{50}H_{102}$ PE melt ($T = 450$ K, $P = 1$ atm). (Reproduced with permission from Ref. [193].)

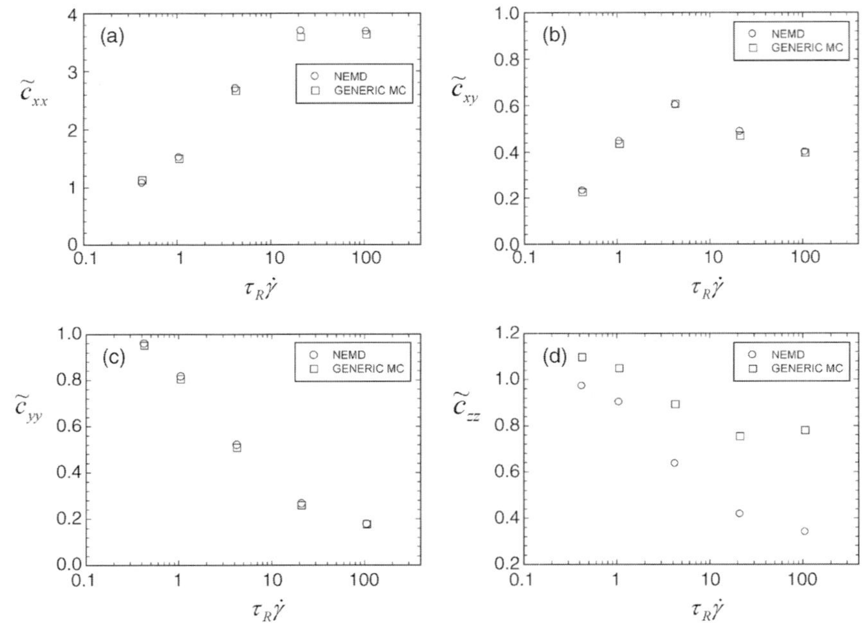

Figure 7.7 Comparison of the conformation tensor **c** components between NEMD and GENERIC MC simulations, as a function of *Wi* number: (a) c_{xx}, (b) c_{xy}, (c) c_{yy}, and (d) c_{zz}. The error bars are smaller than the size of the symbols. (Reproduced with permission from Ref. [193].)

Figure 7.7 reveals an inconsistency between the two methods. As argued by Baig and Mavrantzas [193], this is related to the selection of a zero value for the α_{zz} component of α, as suggested by the general expression, Eq. (7.25). Also, to exactly reproduce the zz-component of **c**, a nonzero α_{zz} component should be incorporated in the GENERIC MC simulations. This is a significant accomplishment of the new methodology since it suggests that the rather general form of the friction matrix, Eq. (7.25), for this conformation tensor family of models is not complete. As demonstrated by Baig and Mavrantzas in a recent publication [196], this can be achieved by including in the relaxation matrix α terms beyond the symmetries implied by Eq. (7.25), without violating the Onsager–Casimir reciprocity relationships or the second law of thermodynamics.

The information provided by the GENERIC MC simulations is important in many aspects:

- The dependence of the components of the tensor α on *Wi* is directly related to the (nonequilibrium) free energy of the system – see Eq. (7.22). Therefore, with the proposed methodology one can accurately calculate the free energy of the simulated system by requiring a series of simulations, according to the thermodynamic state points, by varying one component of α and fixing the rest and then using thermodynamic integration. This can serve as a starting point for developing more accurate viscoelastic models.

- The new thermodynamically guided method can help overcome the problem of long relaxation times (and of statistical noise in the Newtonian plateau) faced in brute-force NEMD simulations by providing the initial configuration at the relevant nonequilibrium state for a given *Wi*.
- The new method can also be combined with recently proposed coarse-graining simulation strategies for long polymer melts to enable the simulation of the viscoelastic properties of high molecular weight polymers, comparable to those encountered in practical polymer processing.

7.4.3
Systematic Timescale Bridging Molecular Dynamics for Flowing Polymer Melts

We consider again a description of the polymer melt coarse-grained to the level of the conformation tensor. The corresponding Poisson bracket is known analytically, see Section 7.4.1. Same as in Section 7.4.2, we investigate the nonequilibrium stationary state of the polymer melt in a given flow situation, and therefore face the same problem of solving the stationary GENERIC equation(7.23) self-consistently. Here, we complete the studies reported in Section 7.4.2 and consistently determine the friction matrix **M** from microscopic fluctuations according to general formula (7.10). Our presentation mainly follows Ref. [197].

7.4.3.1 Systematic Timescale Bridging Algorithm

The coarse-grained energy and entropy, as well as the Poisson bracket, require only static information and can therefore be determined very efficiently by Monte Carlo simulation methods, see Section 7.4.2 and Refs [18, 198]. Only the friction matrix **M** depends on dynamic properties, thus its numerical evaluation requires dynamical simulation, in our case molecular dynamics. The GENERIC coarse-graining approach therefore suggests to combine the strengths of MC and MD simulations in a well-defined way to break the timescale gap between microscopic and macroscopic scales. In order to implement these ideas consistently, we propose a hybrid algorithm [197] schematically illustrated in Figure 7.8 as a general strategy for timescale bridging simulations based on GENERIC. For the special case of flowing, unentangled polymer melts, the algorithm was implemented and tested in Ref. [197].

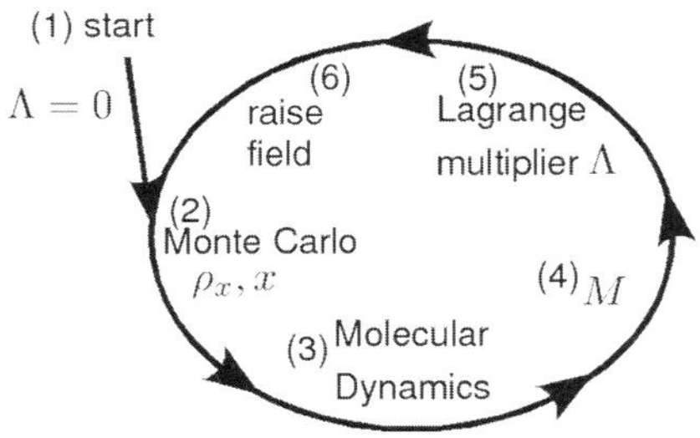

Figure 7.8 Schematic illustration of systematic timescale bridging algorithm that consistently combines Monte Carlo and molecular dynamics simulations.

For simplicity and speed of calculations, the classical FENE bead-spring model introduced in Ref. [60] was used in these studies, although using an atomistic model does not pose any extra difficulty.

For this model system subject to a stationary flow with fixed velocity gradient $\varkappa$, the algorithm illustrated in Figure 7.8 can be implemented as follows:

1) Start with an equilibrium system with $\varkappa = \lambda = 0$.
2) Use a Monte Carlo scheme in order to generate an ensemble of n_s (typically $n_s = 500$) independent configurations that are distributed according to the generalized canonical distribution (7.16) with the current value of λ. Calculate the value of the relevant variables in this ensemble, $\mathbf{x} = \langle \Pi \rangle_x$. In order to efficiently generate an ensemble of such n_s configurations, a slight modification of the Monte Carlo algorithm proposed in [199] was used in [197]. The numerical values of the coarse-grained variables $\mathbf{x} = \langle \Pi \rangle_x$ can then be estimated as the ensemble average $\mathbf{x} = (1/n_s) \sum_{k=1}^{n_s} \Pi(\mathbf{z}_k)$ of the n_s configurations $\{\mathbf{z}_k\}$. In the last step, Maxwellian distributed velocities are assigned to the particles, realizing equilibrium in momentum space for the present choice of (velocity-independent) relevant variables.
3) The Monte Carlo generated ensemble is used as an initial condition for MD simulations of Hamilton's microscopic dynamics. We use a standard velocity-Verlet algorithm that preserves the symplectic structure to simulate trajectories $\mathbf{z}_k(t)$, $k = 1, \ldots, n_s$, during a "short" time interval $0 \leq t \leq \tau_s$. The separating timescale τ_s is short enough, such that the relevant variables $\mathbf{x}$ do not change significantly during the MD simulation. For this reason, the MD part of the simulation does not need any constraints such as thermo- or barostats or flow-adapted boundary conditions. Performing short time, unconstrained, microcanonical molecular dynamics simulations is one of the great benefits of the present approach as it makes the scheme both highly efficient and applicable to arbitrary flow situations that – due to the lack of corresponding boundary conditions – could not be simulated so far.
4) From the particle trajectories $\mathbf{z}_k(t)$, we evaluate the friction matrix $\mathbf{M}$ from Eq. (7.10). We make use of time-translational invariance to equivalently rewrite Eq. (7.10) as

$$\mathbf{M} = \langle \mathcal{M}(z) \rangle_x, \quad \mathcal{M}(\mathbf{z}) = \frac{1}{2k_B\tau_s} \Delta_{\tau_s}\Pi(\mathbf{z})\Delta_{\tau_s}\Pi(\mathbf{z}), \tag{7.28}$$

   where $\Delta_{\tau_s}\Pi(\mathbf{z}) \equiv \Pi(\mathbf{z}(\tau_s)) - \Pi(\mathbf{z}(0))$ denotes fast fluctuations of Π (on the timescale τ_s). Equation (7.28) is more convenient for numerical evaluation than (7.10).
5) Updated values of the Lagrange multiplier λ are calculated from the stationary GENERIC equation(7.23) by inverting the symmetric, positive semidefinite matrix $\mathbf{M}$.
6) The procedure is now repeated until consistent values $\mathbf{x}, \mathbf{M}, \lambda$ for given $\varkappa$ are obtained. Alternatively, one may use an efficient reweighting scheme if $\varkappa$ is changed only slightly and λ is already close to the true value $\lambda \rightarrow \lambda + \delta\lambda$. Then, the

explicit form of the generalized canonical distribution can be exploited to solve the nonlinear system of equations

$$\mathbf{0} = \sum_{k=1}^{n_s} [\mathbf{R}_k + k_B \mathcal{M}(\mathbf{z}_k) : \delta\lambda] w_k, \quad w_k \equiv \frac{e^{-\delta\lambda:\Pi(\mathbf{z}_k)}}{\sum_{k'} e^{-\delta\lambda:\Pi(\mathbf{z}_{k'})}} \tag{7.29}$$

for $\delta\lambda$. This first-order scheme solves the stationary GENERIC equation(7.23), where $\mathbf{R}_k \equiv \varkappa \cdot \Pi(\mathbf{z}_k) + \Pi(\mathbf{z}_k) \cdot \varkappa^T + k_B \mathcal{M}(\mathbf{z}_k) : \lambda$ is the error in the previous value of λ. In a shear flow, for example, Eq. (7.29) represents six equations and six unknowns. The solution $\delta\lambda$ of (7.29) allows one to calculate the reweighted slow variables and friction matrix, $\mathbf{x} = \sum_k w_k \Pi(\mathbf{z}_k)$, $\mathbf{M} = \sum_k w_k \mathcal{M}(\mathbf{z}_k)$, as well as updated Lagrange multipliers, $\lambda \to \lambda + \delta\lambda$. Finally, the flow rate $\varkappa$ is increased, and the procedure is started again, until the control parameter space has been swept through.

With such a scheme, we establish the coarse-grained model along one-dimensional paths in the parameter space. Choosing, for example, viscometric flows of varying strength $\varkappa$ is analogous to the situation encountered in experiments.

7.4.3.2 **Fluctuations, Separating Timescale, and Friction Matrix**

We have already emphasized several times that "fast" but correlated fluctuations give rise to dissipation on the coarse-grained level of description, which is described here by the friction matrix **M**, Eq. (7.10) or (7.28). The notion "fast" is defined here by times t smaller than the timescale τ_s, which separates the evolution of the relevant variables **x** from rapid dynamics of the remaining degrees of freedom. The existence of such a timescale (which is equivalent to the crucial assumption of timescale separation discussed in Section 7.3) is not obvious. Here, we observe that the correlation functions $C_{kl}(t) = \langle \dot{\Pi}_k^f(t) \dot{\Pi}_l^f(0) \rangle_x$ decay monotonically over a few molecular (Lennard-Jones) time units τ. This shows that those fast fluctuations are indeed correlated only over short times compared to typical polymer relaxation times (which are huge relative to τ). Therefore, we find that the friction matrix, which is proportional to the integral over $C(t)$, rapidly converges toward a value that is approximately independent of τ_s in a broad range $5 \leq \tau/\tau_s \leq 50$ (see Figure 7.9).

7.4.3.3 **Results**

Before discussing the results obtained with the proposed timescale bridging algorithm, we like to mention several consistency checks that can be performed in order to test the range of applicability of the coarse-grained model. First, we compare two expressions for the macroscopic stress tensor. One is the standard virial expression $\sigma = -V^{-1}\langle \mathbf{rF} \rangle_x$, where **r** and **F** are the relative position and forces between particles. The kinetic contribution is found to be negligible in dense systems such as polymer melts as long as the flow rates are not too high [12]. Evaluating the expression for the stress tensor σ in the generalized canonical ensemble leads to the expression $\sigma^p = -2V^{-1}k_B T\mathbf{x} \cdot \lambda$ for the (entropic part of the) polymer contribution to σ, see Eq. (7.19). From Section 7.4.1, we know that σ and σ^p differ by a simple fluid contribution. Accounting for this contribution via the nonbonded short-range

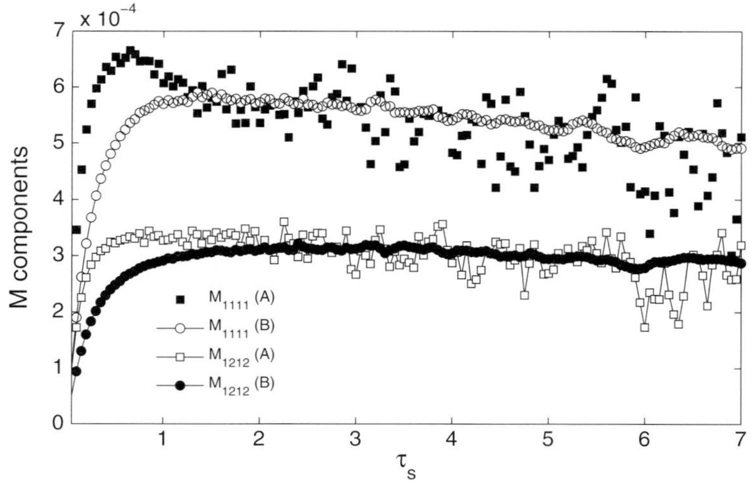

Figure 7.9 The friction matrix calculated from (a): Eq. (7.10) compared to (b): the values obtained from Eq. (7.28), as a function of the separating timescale τ_S. (Reprinted with permission from Ref. [197].)

repulsive interactions, we have verified that the two expressions for the stress tensor agree with each other for the flow rates studied. Next, by its definition and the symmetry of Π, the matrix **M** possesses some basic symmetries that can be used to test the statistical accuracy of the ensemble averages. Finally, for the case of simple shear, $\varkappa = \dot{\gamma}\mathbf{e}_x\mathbf{e}_y$, we have used the identity $(x_{11}-x_{22})x_{12}^{-1} = (\lambda_{11}-\lambda_{22})\lambda_{12}^{-1}$, which can be derived from the stationary GENERIC equations [193], in order to check the consistency of our results. In our studies, this identity holds within error margins for the flow rates considered. We observed that the breakdown of this relation at high flow rates signaled problems with the coarse-grained model as it can no longer capture the relevant dynamic processes at these elevated rates.

For the case of simple shear flow, we validated the algorithm by reproducing the chain-length dependence of the zero shear rate viscosity and of the first normal stress coefficient, which are known in the literature [60] (see Figure 7.10). Also, the shear rate dependence of the viscosity obtained with the timescale bridging algorithm is in very good agreement with standard NEMD results [60, 200, 201], as shown in Figure 7.10. More results can be found in Ref. [197]. As mentioned above, the flexibility of our timescale bridging simulations allows us to study arbitrary flow fields. We therefore could perform the first steady-state equibiaxial simulation for polymer melts. Results for this and other elongational flows can also be found in Ref. [197].

7.5 Conclusions and Perspectives

The tremendous multiplicity of length- and timescales in polymeric systems clearly calls for systematic, multiscale modeling approaches in which a higher resolution

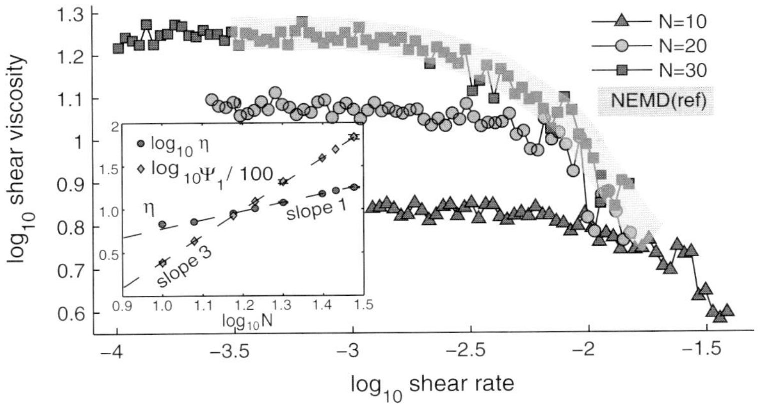

Figure 7.10 Polymer contribution to the shear viscosity as a function of shear rate for different molecular weights. Reference results [60] obtained with standard NEMD simulations for $N = 30$ are indicated. Inset: zero shear rate viscosity η_0 and first normal stress coefficient Ψ_1 as a function of chain length. The expected scaling $\eta_0 \propto N$ and $\Psi_1 \propto N^3$ is observed, as shown by the dashed lines. (Reprinted with permission from Ref. [197].)

model is consistently coupled with a lower resolution one. In particular, if one is interested in describing relaxation processes and structure development under nonequilibrium conditions, most present-day coarse-graining strategies based on the use of effective potentials are of limited use since they do not account for the additional dissipation and irreversibility accompanying inevitably the elimination of fast degrees of freedom in favor of a smaller set of slowly-relaxing structural variables. Therefore, *thermodynamically guided simulations* are very important and useful, since one takes full advantage of the underlying principles of nonequilibrium thermodynamics and statistical mechanics. There, the emphasis is shifted from the time evolution equations (which respect important physical laws such as the Onsager reciprocity relationships for the transport coefficients and the second law of thermodynamics) to its four building blocks, the energy E, the entropy S, the Poisson matrix $\mathbf{L}$, and the friction matrix $\mathbf{M}$, describing the reversible and dissipative contributions to the dynamics.

We have outlined such a methodology for the case of unentangled polymer melts for which, guided by network theory approaches to polymer elasticity, the appropriate coarse-grained variable $\mathbf{x}$ is the conformation tensor. The underlying, microscopic model is simulated by the nonequilibrium molecular dynamics method. The relevant nonequilibrium state is assumed to be given by a generalized canonical distribution incorporating a conjugate variable (the Lagrange multiplier) λ to the conformation tensor. Monte Carlo simulations in this ensemble can then be employed in order to calculate the values of the slow variables $\mathbf{x}$ and the static building blocks E and $\mathbf{L}$. For a given value of imposed flow rates, the Lagrange multiplier can be determined iteratively so that the solutions of the micro- and macrosolvers for the coarse-grained structural variables coincide. Through this one can compute *model-independent* values of the Lagrange multiplier, which for a wide range of strain rates (covering both the linear and the nonlinear viscoelastic regimes) bring results for the overall polymer conformation from the two models (microscale and macroscale) on top of each other. We presented two approaches with which one can obtain the missing blocks of the macroscopic model. In Section 7.4.2, the computed values of the Lagrange multiplier were compared with those corresponding to specific choices of the friction or relaxation matrix **M** in the macroscopic GENERIC model (addressing the chosen structural variable; here the conformation tensor); based on this, one can identify shortcomings and suggest improvements. And this is the biggest advantage of the new framework since the multiscale model proceeds without a priori knowledge of the exact form of the macroscopic model. Being built on the GENERIC framework of nonequilibrium thermodynamics, what is only needed is just to rely on the nature of the chosen structural variables at the coarse level. This, further, emphasizes the significance of the choice of variables in the method. In the second approach, Section 7.4.3, we introduced a novel, low-noise, timescale bridging strategy for the same system (low molecular weight, unentangled polymers) subjected to homogeneous flow fields. Through an alternating Monte Carlo-molecular dynamics iteration scheme, we were able to obtain the model equations for the slow variables. For a chosen flow (including elongational ones), the method predicts both structural and material functions beyond the regime of linear response. The method is simple to implement and allows the calculation of time-dependent behavior through quantities readily available from the nonequilibrium steady states. In the end, it is only when all three different methodologies (macromodel, micromodel, and the macroscopic viscoelastic GENERIC equation bridging them) come together to complement each other that the entire multiscale strategy can be considered as successful. Then, simulation techniques are elevated from brute-force computational tools to sophisticated techniques capable of mapping the detailed description of the system to a handful of carefully chosen variables whose dynamics (time evolution) is also faithfully described by an accurate analytical model.

Future efforts will address other systems such as entangled (linear and branched) polymers where, inspired by the corresponding GENERIC formalism, one should resort to a description in terms of the orientational distribution function of an entanglement segment along the primitive path of the chain.

References

1 Dealy, J. and Larson, R.G. (2006) *Structure and Rheology of Molten Polymers*, Hanser Publications.
2 Bird, R.B., Curtiss, C.F., Armstrong, R.C., and Hassager, O. (1987) *Dynamics of Polymeric Liquids*, 2nd edn, John Wiley & Sons, Inc., New York.
3 Owens, R.G. and Phillips, T.N. (2002) *Computational Rheology*, World Scientific.
4 Keunings, R. (2001) Advances in the computer modeling of the flow of polymeric liquids. *Comput. Fluid Dyn. J.*, 9, 449–458.
5 Malkin, A.Ya. (2009) The state of the art in the rheology of polymers: achievements and challenges. *Polym. Sci., Ser. A*, **51** (1), 80–102.
6 Bent, J., Hutchings, L.R., Richards, R.W., Gough, T., Spares, R., Coates, P.D., Grillo, I., Harlen, O.G., Read, D.J., Graham, R.S., Likhtman, A.E., Groves, D.J., Nicholson, T.M., and McLeish,

T.C.B. (2003) Neutron-mapping polymer flow: scattering, flow visualization, and molecular theory. *Science*, **301** (5640), 1691–1695.

7 Soulages, J., Schweizer, T., Venerus, D.C., Kröger, M., and Öttinger, H.C. (2008) Lubricated cross-slot flow of a low density polyethylene melt. *J. Non-Newtonian Fluid Mech.*, **154**, 52–64.

8 Peters, G.W.M., Schoonen, J.F.M., Baaijens, F.P.T., and Meijer, H.E.H. (1999) On the performance of enhanced constitutive models for polymer melts in a cross-slot flow. *J. Non-Newtonian Fluid Mech.*, **82** (2–3), 387–427.

9 Verbeeten, W.M.H., Peters, G.W.M., and Baaijens, F.P.T. (2002) Viscoelastic analysis of complex polymer melt flows using the extended Pom-Pom model. *J. Non-Newtonian Fluid Mech.*, **108** (1–3), 301–326.

10 Hassell, D.G., Auhl, D., McLeish, T.C.B., and Mackley, M.R. (2008) The effect of viscoelasticity on stress fields within polyethylene melt flow for a cross-slot and contraction–expansion slit geometry. *Rheol. Acta*, **47** (7), 821–834.

11 Öttinger, H.C. (2007) Systematic coarse graining: "Four lessons and a caveat" from nonequilibrium statistical mechanics. *MRS Bull.*, **32**, 936–940.

12 Kröger, M. (2005) Models for polymeric and anisotropic liquids, in *Lecture Notes in Physics*, vol. **675**, Springer, Berlin.

13 Kremer, K. (2005) Modeling soft matter, in *Handbook of Materials Modeling*, Springer, pp. 2675–2686.

14 Doi, M. and Edwards, S.F. (1986) *The Theory of Polymer Dynamics*, Oxford University Press, Oxford.

15 Bird, R.B. and Wiest, J.M. (1995) Constitutive equations for polymeric liquids. *Ann. Rev. Fluid Mech.*, **27**, 169–193.

16 Bird, R.B. and Öttinger, H.C. (1992) Transport properties of polymeric liquids. *Ann. Rev. Phys. Chem.*, **43**, 371–406.

17 Colbourn, E.A. (ed.) (1994) *Computer Simulations of Polymers*, Longman Scientific & Technical, Harlow.

18 Binder, K. (ed.) (1995) *Monte Carlo and Molecular Dynamics Simulations in Polymer Science*, Oxford University Press, New York.

19 Glotzer, S.C. and Paul, W. (2002) Molecular and mesoscale simulation methods for polymer materials. *Annu. Rev. Mater. Res.*, **32**, 401–436.

20 Kremer, K. (2004) Entangled polymers: from universal aspects to structure property relations, in *Computational Soft Matter: From Synthetic Polymers to Proteins, Lecture Notes* (eds N. Attig, K. Binder, H. Grubmüller, and K. Kremer), NIC Series, vol. 23, John von Neumann Institute for Computing, Jülich, pp. 141–168.

21 Kremer, K. (2003) Computer simulations for macromolecular science. *Macromol. Chem. Phys.*, **204** (2), 257–264.

22 Kotelyanskii, M. and Theodorou, D.N. (eds) (2004) *Simulation Methods for Polymers*, Marcel Dekker, New York.

23 Girard, S. and Müller-Plathe, F. (2004) Coarse-graining in polymer simulations, in *Novel Methods in Soft Matter Simulations*, Lecture Notes in Physics, vol. 640, Springer, Berlin, pp. 327–356.

24 McLeish, T.C.B. and Larson, R.G. (1998) Molecular constitutive equations for a class of branched polymers: the pom-pom polymer. *J. Rheol.*, **42** (1), 81–110.

25 Wapperom, P. and Keunings, R. (2001) Numerical simulation of branched polymer melts in transient complex flow using pom-pom models. *J. Non-Newtonian Fluid Mech.*, **97** (2–3), 267–281.

26 Likhtman, A.E. and Graham, R.S. (2003) Simple constitutive equation for linear polymer melts derived from molecular theory: Rolie-Poly equation. *J. Non-Newtonian Fluid Mech.*, **114** (1), 1–12.

27 Ianniruberto, G. and Marrucci, G. (2001) A simple consitutive equation for entangled polymers with chain stretch. *J. Rheol.*, **45** (6), 1305–1318.

28 Stephanou, P.S., Baig, C., and Mavrantzas, V.G. (2009) A generalized differential constitutive equation for polymer melts based on principles of non-equilibrium thermodynamics. *J. Rheol.*, **53**, 309–337.

29 Kim, S. and Karrila, S.J. (2005) *Microhydrodynamics: Principles and Selected Applications*, Dover Publications Inc.

30 Succi, S. (2001) *The Lattice Boltzmann Equation for Fluid Dynamics and Beyond. Numerical Mathematics and Scientific Computation*, Oxford University Press, New York.

31 Dünweg, B. and Ladd, A.J.C. (2009) Lattice Boltzmann simulations of soft matter systems. *Adv. Polym. Sci.*, **221**, 89.

32 Winkler, R.G., Mussawisade, K., Ripoll, M., and Gompper, G. (2004) Rod-like colloids and polymers in shear flow: a multi-particle-collision dynamics study. *J. Phys.: Condens. Matter*, **16**, S3941–S3954.

33 Gompper, G., Ihle, T., Kroll, D.M., and Winkler, R.G. (2009) Multi-particle collision dynamics: a particle-based mesoscale simulation approach to the hydrodynamics of complex fluids. *Adv. Polym. Sci*, **221**, 1.

34 Holm, C., Joanny, J.F., Kremer, K., Netz, R.R., Reineker, P., Seidel, C., Vilgis, T.A., and Winkler, R.G. (2004) Polyelectrolyte theory, in *Polyelectrolytes with Defined Molecular Architecture II, Advances in Polymer Science*, vol. 166, Springer, pp. 67–111.

35 Shea, J.-E. and Brooks, C.L. III (2001) From folding theories to folding proteins: a review and assessment of simulation studies of protein folding and unfolding. *Ann. Rev. Phys. Chem.*, **52**, 499–535.

36 Graham, R.S., Likhtman, A.E., McLeish, T.C.B., and Milner, S.T. (2003) Microscopic theory of linear, entangled polymer chains under rapid deformation including chain stretch and convective constraint release. *J. Rheol.*, **47** (5), 1171–1200.

37 Auhl, D., Ramirez, J., Likhtman, A.E., Chambon, P., and Fernyhough, C. (2008) Linear and nonlinear shear flow behavior of monodisperse polyisoprene melts with a large range of molecular weights. *J. Rheol.*, **52** (3), 801–835.

38 Ianniruberto, G. and Marrucci, G. (2002) A multi-mode CCR model for entangled polymers with chain stretch. *J. Non-Newtonian Fluid Mech.*, **102** (2), 383–395.

39 Öttinger, H.C. and Beris, A.N. (1999) Thermodynamically consistent reptation model without independent alignment. *J. Chem. Phys.*, **110**, 6593–6596.

40 Öttinger, H.C. (2000) Thermodynamically admissible reptation models with anisotropic tube cross sections and convective constraint release. *J. Non-Newtonian Fluid Mech.*, **89**, 165–185.

41 Öttinger, H.C. (1999) A thermodynamically admissible reptation model for fast flows of entangled polymers. *J. Rheol.*, **43**, 1461–1493.

42 Fang, J., Kröger, M., and Öttinger, H.C. (2000) A thermodynamically admissible reptation model for fast flows of entangled polymers. II. Model predictions for shear and elongational flows. *J. Rheol.*, **44**, 1293–1317.

43 Mead, D.W., Larson, R.G., and Doi, M. (1998) A molecular theory for fast flows of entangled polymers. *Macromolecules*, **31** (22), 7895–7914.

44 Schieber, J.D., Neergaard, J., and Gupta, S. (2003) A full-chain, temporary network model with slip-links, chain-length fluctuations, chain connectivity and chain stretching. *J. Rheol.*, **47** (1), 213–233.

45 Shanbhag, S. and Larson, R.G. (2004) A slip-link model of branch-point motion in entangled polymers. *Macromolecules*, **37** (21), 8160–8166.

46 Nair, D.M. and Schieber, J.D. (2006) Linear viscoelastic predictions of a consistently unconstrained Brownian slip-link model. *Macromolecules*, **39** (9), 3386–3397.

47 Likhtman, A.E. (2005) Single-chain slip-link model of entangled polymers: simultaneous description of neutron spin-echo, rheology, and diffusion. *Macromolecules*, **38** (14), 6128–6139.

48 Doi, M. and Takimoto, J. (2003) Molecular modelling of entanglement. *Philos. Trans. R. Soc. Lond. A*, **361** (1805), 641–652.

49 Masubuchi, Y., Takimoto, J.-I., Koyama, K., Ianniruberto, G., Marrucci, G., and Greco, F. (2001) Brownian simulations of a network of reptating primitive chains. *J. Chem. Phys.*, **115** (9), 4387–4395.

50 Masubuchi, Y., Ianniruberto, G., Greco, F., and Marrucci, G. (2003) Entanglement molecular weight and frequency response of sliplink networks. *J. Chem. Phys.*, **119** (13), 6925–6930.

51 Rigby, D. and Eichinger, B.E. (2001) Polymer modeling. *Curr. Opin. Solid State Mater. Sci.*, **5** (5), 445–450.

52 Siepmann, J.I., Karaborni, S., and Smit, B. (1993) Simulating the critical behaviour of complex fluids. *Nature*, **365**, 330–332.

53 Padilla, P. and Toxvaerd, S. (1991) Structure and dynamical behavior of fluid *n*-alkanes. *J. Chem. Phys.*, **95** (1), 509–519.

54 Dysthe, D.K., Fuchs, A.H., and Rousseau, B. (2000) Fluid transport properties by equilibrium molecular dynamics. III. Evaluation of united atom interaction potential models for pure alkanes. *J. Chem. Phys.*, **112**, 7581–7590.

55 Rehan, M., Mattics, W.L., and Suter, U.W. (1997) Rotational isomeric state models in macromolecular systems. *Adv. Polym. Sci., volumes* 131/132, Springer, Berlin.

56 Kubo, R., Toda, M., and Hashitsume, N. (1991) *Statistical Physics II. Nonequilibrium Statistical Mechanics,* 2nd edn, Springer, Berlin.

57 Vladkov, M. and Barrat, J.-L. (2006) Linear and nonlinear viscoelasticity of unentangled polymer melts: molecular dynamics and Rouse mode analysis. *Macromol. Theory Simul.*, **15**, 252–262.

58 Padding, J.T. and Briels, W.J. (2007) Ab-initio coarse-graining of entangled polymer systems, in *Nanostructured Soft Matter, NanoScience and Technology*, Springer, Netherlands, pp. 437–460.

59 Kremer, K. and Grest, G.S. (1990) Dynamics of entangled linear polymer melts: a molecular-dynamics simulation. *J. Chem. Phys.*, **92**, 5057–5086.

60 Kröger, M., Loose, W., and Hess, S. (1993) Structural changes and rheology of polymer melts via nonequilibrium molecular dynamics. *J. Rheol.*, **37**, 1057.

61 Todd, B.D. (2001) Computer simulation of simple and complex atomistic fluids by nonequilibrium molecular dynamics techniques. *Comput. Phys. Commun.*, **142**, 14–21.

62 Todd, B.D. and Daivis, P.J. (2007) Homogeneous non-equilibrium molecular dynamics simulations of viscous flow: techniques and applications. *Mol. Simul.*, **33**, 189.

63 Kröger, M. and Hess, S. (2000) Rheological evidence for a dynamical crossover in polymer melts via nonequilibrium molecular dynamics. *Phys. Rev. Lett.*, **85**, 1128–1131.

64 Edwards, B.J., Baig, C., and Keffer, D.J. (2005) An examination of the validity of nonequilibrium molecular-dynamics simulation algorithms for arbitrary steady-state flows. *J. Chem. Phys.*, **123**, 114106.

65 Edwards, B.J., Baig, C., and Keffer, D.J. (2006) A validation of the p-SLLOD equations of motion for homogeneous steady-state flows. *J. Chem. Phys.*, **124**, 194104.

66 Baig, C., Edwards, B.J., Keffer, D.J., and Cochran, H.D. (2005) A proper approach for nonequilibrium molecular dynamics simulations of planar elongational flow. *J. Chem. Phys.*, **112**, 114103.

67 Baig, C., Edwards, B.J., Keffer, D.J., Cochran, H.D., and Harmandaris, V.A. (2006) Rheological and structural studies of linear polyethylene melts under planar elongational flow using nonequilibrium molecular dynamics simulations. *J. Chem. Phys.*, **124**, 084902.

68 Lees, A.W. and Edwards, S.F. (1972) The computer study of transport processes under extreme conditions. *J. Phys. C*, **5** (15), 1921–1930.

69 Kraynik, A.M. and Reinelt, D.A. (1992) Extensional motions of spatially periodic lattices. *Int. J. Multiphase Flow*, **18** (6), 1045–1059.

70 Doi, M. (2003) Challenge in polymer physics. *Pure Appl. Chem.*, **75** (10), 1395–1402.

71 Paul, W., Smith, G.D., Yoon, D.Y., Farago, B., Rathgeber, S., Zirkel, A., Willner, L., and Richter, D. (1998) Chain motion in an unentangled polyethylene melt: a critical test of the Rouse model by molecular dynamics simulations and neutron spin echo spectroscopy. *Phys. Rev. Lett.*, **80** (11), 2346–2349.

72 Bair, S., McCabe, C., and Cummings, P.T. (2002) Comparison of nonequilibrium molecular dynamics with experimental measurements in the nonlinear shear-thinning regime. *Phys. Rev. Lett.*, **88**, 058302.

73 Tuzun, R.E., Noid, D.W., Sumpter, B.G., and Wozny, C.E. (2003) Recent advances in polymer molecular dynamics

simulation and data analysis. *Macromol. Theor. Simul.*, **6** (5), 855–880.

74 Ober, C.K., Cheng, S.Z.D., Hammond, P.T., Muthukumar, M., Reichmanis, E., Wooley, K.L., and Lodge, T.P. (2009) Research in macromolecular science: challenges and opportunities for the next decade. *Macromolecules*, **42**, 465–471.

75 Mavrantzas, V.G., Boone, T., Zervopoulou, E., and Theodorou, D.N. (1999) End-bridging Monte Carlo: an ultra-fast algorithm for the equilibration of the condensed phases of long polymers. *Macromolecules*, **32**, 5072–5096.

76 Mavrantzas, V.G. (2005) Monte Carlo simulation of chain molecules, in *Handbook of Materials Modeling* (ed. S. Yip), Springer, Netherlands, pp. 2583–2597.

77 Karayiannis, N.C., Mavrantzas, V.G., and Theodorou, D.N. (2002) A novel Monte Carlo scheme for the rapid equilibration of atomistic model polymer systems of precisely defined molecular architecture. *Phys. Rev. Lett.*, **88**, 105503.

78 Auhl, R., Everaers, R., Grest, G.S., and Kremer, K. (2003) Equilibration of long chain polymer melts in computer simulations. *J. Chem. Phys.*, 12718.

79 Theodorou, D.N. (2006) Equilibration and coarse-graining methods for polymers, in *Computer Simulations in Condensed Matter Systems: From Materials to Chemical Biology, Volume 2, Lecture Notes in Physics*, vol. 704, Springer, Berlin, pp. 419–448.

80 Kröger, M. (2005) Shortest multiple disconnected path for the analysis of entanglements in two- and three-dimensional polymeric systems. *Comput. Phys. Commun.*, **168** (3), 209–232.

81 Sukumaran, S.K., Grest, G.S., Kremer, K., and Everaers, R. (2005) Identifying the primitive path mesh in entangled polymer liquids. *J. Polym. Sci. B: Polym. Phys.*, **43** (8), 917–933.

82 Everaers, R., Sukumaran, S.K., Grest, G.S., Svaneborg, C., Sivasubramanian, A., and Kremer, K. (2004) Rheology and microscopic topology of entangled polymeric liquids. *Science*, **303**, 823–826.

83 Shanbhag, S., Park, S.J., Zhou, W., and Larson, R.G. (2007) Implications of microscopic simulations of polymer melts for mean-field tube theories. *Mol. Simul.*, **105** (2–3), 249–260.

84 Kröger, M., Ramrez, J., and Öttinger, H.C. (2002) Projection from an atomistic chain contour to its primitive path. *Polymer*, **43**, 477–487.

85 Harmandaris, V.A., Mavrantzas, V.G., Theodorou, D.N., Kröger, M., Ramrez, J., Öttinger, H.C., and Vlassopoulos, D. (2003) Crossover from the Rouse to the entangled polymer melt regime: signals from long, detailed atomistic molecular dynamics simulations, supported by rheological experiments. *Macromolecules*, **36**, 1376–1387.

86 Kim, J.M., Keffer, D.J., Kröger, M., and Edwards, B.J. (2008) Rheological and entanglement characteristics of linear-chain polyethylene liquids in planar Couette and planar elongational flows. *J. Non-Newtonian Fluid Mech.*, **152**, 168–183.

87 Miller, R.E. and Tadmor, E.B. (2007) Hybrid continuum mechanics and atomistic methods for simulating materials deformation and failure. *MRS Bull.*, **32**, 920–926.

88 Lu, G., Tadmor, E.B., and Kaxiras, E. (2008) From electrons to finite elements: a concurrent multiscale approach for metals. *Phys. Rev. B*, **73**, 024108.

89 De Fabritiis, G., Delgado-Buscalioni, R., and Coveney, P.V. (2006) Multiscale modeling of liquids with molecular specificity. *Phys. Rev. Lett.*, **97**, 134501.

90 Milner, S.T. (1991) Polymer brushes. *Science*, **251** (4996), 905–914.

91 Guenza, M.G. (2008) Theoretical models for bridging timescales in polymer dynamics. *J. Phys.: Condens. Matter*, **20**, 033101.

92 Praprotnik, M., Junghans, C., Site, L.D., and Kremer, K. (2008) Simulation approaches to soft matter: generic statistical properties vs. chemical details. *Comput. Phys. Commun.*, **179** (1–3), 51–60.

93 Praprotnik, M., Site, L.D., and Kremer, K. (2008) Multiscale simulation of soft matter: from scale bridging to adaptive resolution. *Ann. Rev. Phys. Chem.*, **59**, 545–571.

94 Kremer, K. and Müller-Plathe, F. (2002) Multiscale simulation in polymer science. *Mol. Simul.*, **28**, 729–750.

95 Baschnagel, J., Binder, K., Doruker, P., Gusev, A.A., Hahn, O., Kremer, K., Mattice, W.L., Müller-Plathe, F., Murat, M., Paul, W., Santos, S., Suter, U.W., and Tries, V. (2000) Bridging the gap between atomistic and coarse-grained models of polymers: status and perspectives. *Adv. Polym. Sci.*, **152**, 41–156.

96 Nielaba, P., Mareschal, M., and Cicotti, G. (eds) (2002) *Bridging Time Scales: Molecular Simulations for the Next Decade, Lecture Notes in Physics*, vol. **606**, Springer, Berlin.

97 Weinan, E., Li, X., and Vanden-Eijnden, E. (2004) Some recent progress in multiscale modeling, in *Multiscale Modelling and Simulation, Lecture Notes in Computational Science and Engineering*, vol. 39 (eds S. Attinger and P. Koumoutsakos), Springer, pp. 3–23.

98 Voth, G.A. (ed.) (2009) *Coarse-Graining of Condensed Phases and Biomolecular Systems*, CRC Press, Boca Raton.

99 Murtola, T., Bunker, A., Vattulainen, I., Deserno, M., and Karttunen, M. (2009) Multiscale modeling of emergent materials: biological and soft matter. *Phys. Chem. Chem. Phys.*, **11** (12), 1869–1892.

100 Denton, A.R. (2007) Effective interactions in soft materials, in *Nanostructured Soft Matter, NanoScience and Technology*, Springer, Netherlands, pp. 395–433.

101 Faller, R. (2007) Coarse-grain modeling of polymers, in *Reviews in Computational Chemistry*, vol. 23 (eds T.R. Cundari and K.B. Lipkowitz), John Wiley & Sons, Inc., New York, pp. 233–262.

102 Gorban, A.N., Kazantzis, N.K., Kevrekidis, I.V., Öttinger, H.C. and Theodoropoulos, C. (eds) (2006) *Model Reduction and Coarse-Graining Approaches for Multiscale Phenomena*, Springer, Berlin.

103 Grabert, H. (1982) *Projection Operator Techniques in Nonequilibrium Statistical Mechanics*, Springer, Berlin.

104 Nielsen, S.O., Lopez, C.F., Srinivas, G., and Klein, M.L. (2004) Coarse grain models and the computer simulation of soft materials. *J. Phys.: Condens. Matter*, **16**, R481.

105 Hansen, J.P. and Löwen, H. (2002) Effective interactions for large-scale simulations of complex fluids, in *Bridging Time Scales: Molecular Simulations for the Next Decade, Lecture Notes in Physics*, vol. 606 (eds P. Nielaba, M. Mareschal, and G. Cicotti), Springer, Berlin, pp. 167–196.

106 Sun, Q. and Faller, R. (2005) Systematic coarse-graining of atomistic models for simulation of polymeric systems. *Comput. Chem. Eng.*, **29** (11–12), 2380–2385.

107 Faller, R. (2004) Automatic coarse graining of polymers. *Polymer*, **45** (11), 3869–3876.

108 Spyriouni, T., Tzoumanekas, C., Theodorou, D.N., Müller-Plathe, F., and Milano, G. (2007) Coarse-grained and reverse-mapped united-atom simulations of long-chain atactic polystyrene melts: structure, thermodynamic properties, chain conformation, and entanglements. *Macromolecules*, **40**, 3876–3885.

109 Kamio, K., Moorthi, K., and Theodorou, D.N. (2007) Coarse grained end bridging Monte Carlo simulations of poly(ethylene terephthalate) melt. *Macromolecules*, **40**, 710–722.

110 Reith, D., Pütz, M., and Müller-Plathe, F. (2003) Deriving effective mesoscale potentials from atomistic simulations. *J. Comput. Chem.*, **24** (13), 1624–1636.

111 Reith, D., Meyer, H., and Müller-Plathe, F. (2001) Mapping atomistic to coarse-grained polymer models using automatic simplex optimization to fit structural properties. *Macromolecules*, **34** (7), 2335–2345.

112 Milano, G. and Müller-Plathe, F. (2005) Mapping atomistic simulations to mesoscopic models: a systematic coarse-graining procedure for vinyl polymer chains. *J. Phys. Chem. B*, **109**, 18609–18619.

113 Santangelo, G., Di Matteo, A., Müller-Plathe, F., and Milano, G. (2007) From mesoscale back to atomistic models: a fast reverse-mapping procedure for vinyl polymer chains. *J. Phys. Chem. B*, **111**, 2765–2773.

114 Harmandaris, V.A., Reith, D., van der Vegt, N.F.A., and Kremer, K. (2007) Comparison between coarse-graining models for polymer systems: two mapping schemes for polystyrene. *Macromol. Chem. Phys.*, **208** (19–20), 2109–2120.

115 Harmandaris, V.A., Adhikari, N.P., van der Vegt, N.F.A., and Kremer, K. (2006) Hierarchical modeling of polystyrene: from atomistic to coarse-grained simulations. *Macromolecules*, **39**, 6708–6719.

116 Abrams, C.F. and Kremer, K. (2003) Combined coarse-grained and atomistic simulation of liquid bisphenol A-polycarbonate: liquid packing and intramolecular structure. *Macromolecules*, **36** (1), 260–267.

117 Leo, S., van der Vegt, N., Delle Site, L., and Kremer, K. (2005) Bisphenol A polycarbonate: entanglement analysis from coarse-grained MD simulations. *Macromolecules*, **38**, 8078–8092.

118 Carbone, P., Negri, F., and Müller-Plathe, F. (2007) Coarse-grained model for polyphenylene dendrimers: switching and backfolding of planar three-fold core dendrimers. *Macromolecules*, **40** (19), 7044–7055.

119 Knotts, T.A., I.V. Rathore, N., Schwartz, D.C., de Pablo, J.J. (2007) A coarse grain model for DNA. *J. Chem. Phys.*, **126**, 084901.

120 Peter, C., Site, L.D., and Kremer, K. (2008) Classical simulations from the atomistic to the mesoscale and back: coarse graining an azobenzene liquid crystal. *Soft Matter*, **4**, 859–869.

121 Zwicker, J. and Lovett, R.J. (1990) When does a pair correlation function fix the state of an equilibrium system? *J. Chem. Phys.*, **93**, 6752–6755.

122 Soper, A.K. (1996) Empirical potential Monte Carlo simulation of fluid structure. *Chem. Phys.*, **202** (2), 295–306.

123 Qian, H.-J., Liewb, C.C., and Müller-Plathe, F. (2009) Effective control of the transport coefficients of coarse-grained liquid and polymer models using the dissipative particle dynamics and Lowe-Andersen equations of motion. *Phys. Chem. Chem. Phys.*, **11** (12), 1962.

124 Chen, X., Carbone, P., Cavalcanti, W.L., Milano, G., and Müller-Plathe, F. (2007) Viscosity and structural alteration of a coarse-grained model of polystyrene under steady shear flow studied by reverse nonequilibrium molecular dynamics. *Macromolecules*, **40**, 8087–8095.

125 Depa, P.K. and Maranas, J.K. (2005) Speed up of dynamic observables in coarse-grained molecular-dynamics simulations of unentangled polymers. *J. Chem. Phys.*, **123**, 094901.

126 Harmandaris, V.A. and Kremer, K. (2009) Predicting polymer dynamics at multiple length and time scales. *Soft Matter*, **5**, 3920–3926.

127 Noid, W.G., Chu, J.-W., Ayton, G.S., Krishna, V., Izvekov, S., Voth, G.A., Das, A., and Andersen, H.C. (2008) The multiscale coarse-graining method. I. A rigorous bridge between atomistic and coarse-grained models. *J. Chem. Phys.*, **128**, 244114.

128 Noid, W.G., Liu, P., Wang, Y., Chu, J.-W., Ayton, G.S., Izvekov, S., Andersen, H.C., and Voth, G.A. (2008) The multiscale coarse-graining method. II. Numerical implementation for coarse-grained molecular models. *J. Chem. Phys.*, **128**, 244115.

129 Henderson, R.L. (1974) Uniqueness theorem for fluid pair correlation-functions. *Phys. Lett. A*, **49** (3), 197–198.

130 Gray, C.G. and Gubbins, K.E. (1984) *Theory of Molecular Fluids*, Oxford University Press, New York.

131 Jain, S., Garde, S., and Kumar, S.K. (2006) Do inverse Monte Carlo algorithms yield thermodynamically consistent interaction potentials? *Ind. Eng. Chem. Res.*, **45**, 5614–5618.

132 Carbone, P., Varzaneh, H.A.K., Chen, X., and Müller-Plathe, F. (2008) Transferability of coarse-grained force fields: the polymer case. *J. Chem. Phys.*, **128**, 064904.

133 Krishna, V., Noid, W.G., and Voth, G.A. (2009) The multiscale coarse-graining method. IV. Transferring coarse-grained potentials between temperatures. *J. Chem. Phys.*, **131** (2), 024103.

134 Wang, Y., Noid, W.G., Liu, P., and Voth, G.A. (2009) Effective force coarse-

graining. *Phys. Chem. Chem. Phys*, **11** (12), 2002–2015.

135 Guerrault, X., Rousseau, B., and Farago, J. (2004) Dissipative particle dynamics simulations of polymer melts. I. Building potential of mean force for polyethylene and *cis*-polybutadiene. *J. Chem. Phys.*, **121**, 6538–6546.

136 Padding, J.T. and Briels, W.J. (2003) Coarse-grained molecular dynamics simulations of polymer melts in transient and steady shear flow. *J. Chem. Phys.*, **118** (22), 10276–10286.

137 Padding, J.T. and Briels, W.J. (2001) Uncrossability constraints in mesoscopic polymer melt simulations: non-Rouse behavior of $C_{120}H_{242}$. *J. Chem. Phys.*, **115** (6), 2846–2859.

138 Akkermans, R.L.C. and Briels, W.J. (2001) A structure-based coarse-grained model for polymer melts. *J. Chem. Phys.*, **114** (2), 1020–1031.

139 Akkermans, R.L.C. and Briels, W.J. (2000) Coarse-grained dynamics of one chain in a polymer melt. *J. Chem. Phys.*, **113** (15), 6409–6422.

140 Evans, D.J. and Morriss, G.P. (1984) Nonlinear-response theory for steady planar Couette flow. *Phys. Rev. A*, **30**, 1528–1530.

141 Kindt, P. and Briels, W.J. (2005) Scaling of mesoscale simulations of polymer melts with the bare friction coefficient. *J. Chem. Phys.*, **123**, 224903.

142 Öttinger, H.C. (2004) Coarse-graining of wormlike polymer chains for substantiating reptation. *J. Non-Newtonian Fluid Mech.*, **120**, 207–213.

143 Tzoumanekas, C. and Theodorou, D.N. (2006) Topological analysis of linear polymer melts: a statistical approach. *Macromolecules*, **39**, 4592–4604.

144 Tsolou, G., Mavrantzas, V.G., and Theodorou, D.N. (2005) Detailed atomistic molecular dynamics simulation of *cis*-1,4-poly(butadiene). *Macromolecules*, **38**, 1478–1492.

145 Kremer, K., Sukumaran, S.K., Everaers, R., and Grest, G.S. (2005) Entangled polymer systems. *Comput. Phys. Commun.*, **169**, 75–81.

146 Uchida, N., Grest, G.S., and Everaers, R. (2008) Viscoelasticity and primitive path analysis of entangled polymer liquids: from F-actin to polyethylene. *J. Chem. Phys.*, **128**, 044902.

147 Foteinopoulou, K., Karayiannis, N.C., Mavrantzas, V.G., and Kröger, M. (2006) Primitive path identification and entanglement statistics in polymer melts: results from direct topological analysis on atomistic polyethylene models. *Macromolecules*, **39**, 4207–4216.

148 Khaliullin, R.N. and Schieber, J.D. (2008) Analytic expressions for the statistics of the primitive-path length in entangled polymers. *Phys. Rev. Lett.*, **100**, 188302.

149 Doi, M. and Kuzuu, N.Y. (1980) Rheology of star polymers in concentrated solutions and melts. *J. Polym. Sci., Polym. Lett. Ed.*, **18**, 775–780.

150 Shanbhag, S. and Larson, R.G. (2005) Chain retraction potential in a fixed entanglement network. *Phys. Rev. Lett.*, **94**, 076001.

151 Doi, M. and Edwards, S.F. (1978) Dynamics of concentrated polymer systems. Part 2. Molecular motion under flow. *J. Chem. Soc. Faraday Trans. 2*, **74**, 1802–1817.

152 Doi, M. and Takimoto, J.-I. (2003) Molecular modelling of entanglement. *Phil. Trans. R. Soc. Lond. A*, **361**, 641–652.

153 Yaoita, T., Isaki, T., Masubuchi, Y., Watanabe, H., Ianniruberto, G., Greco, F., and Marrucci, G. (2004) Highly entangled polymer primitive chain network simulations based on dynamic tube dilation. *J. Chem. Phys.*, **121** (24), 12650–12654.

154 Masubuchi, Y., Ianniruberto, G., Greco, F., and Marrucci, G. (2008) Quantitative comparison of primitive chain network simulations with literature data of linear viscoelasticity for polymer melts. *J. Non-Newtonian Fluid Mech.*, **149**, 87–92.

155 Louis, A.A., Bolhuis, P.G., Finken, R., Krakoviack, V., Meijer, E.J., and Hansen, J.P. (2002) Coarse-graining polymers as soft colloids. *Physica A*, **306**, 251–261.

156 Louis, A.A., Bolhuis, P.G., Hansen, J.P., and Meijer, E.J. (2000) Can polymer coils be modeled as "soft colloids"? *Phys. Rev. Lett.*, **85**, 2522.

157 Bolhuis, P.G., Louis, A.A., Hansen, J.P., and Meijer, E.J. (2001) Accurate effective

pair potentials for polymer solutions. *J. Chem. Phys.*, **114**, 4296.

158 Kindt, P. and Briels, W.J. (2007) A single particle model to simulate the dynamics of entangled polymer melts. *J. Chem. Phys.*, **127** (13), 134901.

159 Pagonabarraga, I. and Frenkel, D. (2000) Non-ideal DPD fluids. *Mol. Simul.*, **25**, 167–175.

160 Pagonabarraga, I. and Frenkel, D. (2001) Dissipative particle dynamics for interacting systems. *J. Chem. Phys.*, **115** (11), 5015–5026.

161 Mavrantzas, V.G. and Theodorou, D.N. (1998) Atomistic simulation of polymer melt elasticity: calculation of the free energy of an oriented polymer melt. *Macromolecules*, **31** (18), 6310–6332.

162 van den Noort, A., den Otter, W.K., and Briels, W.J. (2007) Coarse-graining of slow variables in dynamic simulations of soft matter. *Europhys. Lett.*, **80**, 28003.

163 van den Noort, A. and Briels, W.J. (2008) Coarse grained simulations of elongational viscosities, superposition rheology and shear banding in model core-shell systems. *Macromol. Theory Simul.*, **16** (8), 742–754.

164 Mavrantzas, V.G. and Öttinger, H.C. (2002) Atomistic Monte Carlo simulations of polymer melt elasticity: their nonequilibrium thermodynamics GENERIC formulation in a generalized canonical ensemble. *Macromolecules*, **35**, 960–975.

165 Bernardin, F.E. III and Rutledge, G.C. (2008) Estimation of macromolecular configurational properties from atomistic simulations of oligomers under nonequilibrium conditions. *Macromol. Theor. Simul.*, **17**, 23–31.

166 Öttinger, H.C. (2001) Thermodynamic admissibility of the pom-pom model for branched polymers. *Rheol. Acta*, **40**, 317–321.

167 Öttinger, H.C. (2005) *Beyond Equilibrium Thermodynamics*, John Wiley & Sons, Inc., Hoboken.

168 Hoogerbrugge, P.J. and Koelman, J.M.V.A. (1992) Simulating microscopic hydrodynamic phenomena with dissipative particle dynamics. *Europhys. Lett.*, **19** (3), 155–160.

169 Fyta, M., Sircar, J., Kaxiras, E., Melchionna, S., Bernaschi, M., and Succi, S. (2008) Parallel multiscale modeling of biopolymer dynamics with hydrodynamic correlations. *Int. J. Multiscale Comp. Eng.*, **6** (1), 25.

170 Ripoll, M., Mussawisade, K., Winkler, R.G., and Gompper, G. (2004) Low-Reynolds-number hydrodynamics of complex fluids by multi-particle-collision dynamics. *Europhys. Lett.*, **68**, 106.

171 Ihle, T., Tüzel, E., and Kroll, D.M. (2006) Consistent particle-based algorithm with a non-ideal equation of state. *Europhys. Lett.*, **73**, 664.

172 Padding, J.T. and Briels, W.J. (2002) Time and length scales of polymer melts studied by coarse-grained molecular dynamics simulations. *J. Chem. Phys.*, **117** (2), 925–943.

173 Öttinger, H.C. (1998) General projection operator formalism for the dynamics and thermodynamics of complex fluids. *Phys. Rev. E*, **57**, 1416.

174 Chaikin, P.M. and Lubensky, T.C. (2000) *Principles of Condensed Matter Physics*, Cambridge University Press.

175 Beris, A.N. and Edwards, B.J. (1994) *Thermodynamics of Flowing Systems with Internal Microstructure*, Oxford University Press, New York.

176 Español, P. (2004) Statistical mechanics of coarse-graining, in *Novel Methods in Soft Matter Simulations, Lecture Notes in Physics*, vol. 640, Springer, pp. 69–115.

177 de Pablo, J.J. and Öttinger, H.C. (2001) An atomistic approach to general equation for the nonequilibrium reversible-irreversible coupling. *J. Non-Newtonian Fluid Mech.*, **96**, 137–162.

178 Öttinger, H.C. and Struchtrup, H. (2007) The mathematical procedure of coarse graining: from Grad's ten-moment equations to hydrodynamics. *Multiscale Model. Simul.*, **6**, 53–69.

179 Dzyaloshinskii, I.E. and Volovick, G.E. (1980) Poisson brackets in condensed matter physics. *Ann. Phys. (N.Y.)*, **125** (1), 67–97.

180 Zia, R.K.P. and Schmittmann, B. (2006) A possible classification of nonequilibrium

steady states. *J. Phys. A: Math. Gen.*, **39** (24), L407–L413.

181 Jaynes, E.T. (1957) Information theory and statistical mechanics. *Phys. Rev. A*, **106** (4), 620–630.

182 Zubarev, D., Morozov, V., and Röpke, G. (1997) *Statistical Mechanics of Nonequilibrium Processes*, Akademie Verlag, Berlin.

183 Gorban, A.N. and Karlin, I.V. (2005) *Invariant Manifolds for Physical and Chemical Kinetics, Lecture Notes in Phys.*, vol. 660, Springer, Berlin.

184 Hijón, C., Serrano, M., and Español, P. (2006) Markovian approximation in a coarse-grained description of atomic systems. *J. Chem. Phys.*, **125**, 204101.

185 Schweizer, K.S. (1989) Microscopic theory of the dynamics of polymeric liquids: general formulation of a mode-mode coupling approach. *J. Chem. Phys.*, **91** (9), 5802–5821.

186 Chong, S.-H. and Fuchs, M. (2002) Mode-coupling theory for structural and conformational dynamics of polymer melts. *Phys. Rev. Lett.*, **88**, 185702.

187 Öttinger, H.C. (2006) Nonequilibrium thermodynamics of glasses. *Phys. Rev. E*, **74**, 011113.

188 Del Gado, E., Ilg, P., Kröger, M., and Öttinger, H.C. (2008) Non-affine deformations of inherent structure as signature of cooperativity in supercooled liquids. *Phys. Rev. Lett.*, **101**, 095501.

189 Bernardin, F.E. III and Rutledge, G.C. (2007) Simulation of mechanical properties of oriented glassy polystyrene. *Polymer*, **48**, 7211–7220.

190 Ilg, P. (2008) Macroscopic thermodynamics of flowing polymers derived from systematic coarse-graining procedure. *Physica A*, **387**, 6484–6496.

191 Dressler, M., Edwards, B.J., and Öttinger, H.C. (1999) Macroscopic thermodynamics of flowing polymeric liquids. *Rheol. Acta*, **38**, 117–136.

192 Kröger, M., Luap, C., and Muller, R. (1997) Polymer melts under uniaxial elongational flow: stress-optical behavior from experiments and NEMD computer simulations. *Macromolecules*, **30**, 526–539.

193 Baig, C. and Mavrantzas, V.G. (2007) Thermodynamically guided nonequilibrium Monte Carlo method for generating realistic shear flows in polymeric systems. *Phys. Rev. Lett.*, **99**, 257801.

194 Ionescu, T.C., Edwards, B.J., Keffer, D.J., and Mavrantzas, V.G. (2008) Energetic and entropic elasticity of nonisothermal flowing polymers: experiment, theory, and simulation. *J. Rheol.*, **52** (1), 105–140.

195 Ionescu, T.C., Edwards, B.J., Keffer, D.J., and Mavrantzas, V.G. (2008) Atomistic simulation of energetic and entropic elasticity in short-chain polyethylenes. *J. Rheol.*, **52** (2), 567–602.

196 Baig, C. and Mavrantzas, V.G. (2009) Multiscale simulation of polymer melt viscoelasticity guided from nonequilibrium statistical thermodynamics: atomistic non-equilibrium molecular dynamics coupled with Monte Carlo in an expanded statistical ensemble. *Phys. Rev. B*, **79**, 144302.

197 Ilg, P., Öttinger, H.C., and Kröger, M. (2009) Systematic time-scale-bridging molecular dynamics applied to flowing polymer melts. *Phys. Rev. E*, **79**, 011802.

198 Landau, D.P. and Binder, K. (2000) *A Guide to Monte Carlo Simulations in Statistical Physics*, Cambridge University Press.

199 Kröger, M. (1999) Efficient hybrid algorithm for the dynamic creation of semiflexible polymer solutions, brushes, melts and glasses. *Comput. Phys. Commun.*, **118** (278), 298.

200 Daivis, P.J., Matin, M.L., and Todd, B.D. (2007) Nonlinear shear and elongational rheology of model polymer melts at low strain rates. *J. Non-Newtonian Fluid Mech.*, **147** (1–2), 35–44.

201 Daivis, P.J., Matin, M.L., and Todd, B.D. (2003) Nonlinear shear and elongational rheology of model polymer melts by non-equilibrium molecular dynamics. *J. Non-Newtonian Fluid Mech.*, **111** (1), 1–18.

8
Computational Mechanics of Rubber and Tires

Michael J. Poldneff and Martin W. Heinstein

8.1
Introduction

Computational modeling of rubber involves issues that have been traditionally difficult in computational mechanics. Rubber exhibits relatively low stiffness for loads that are typically applied. This leads to large displacements and high strains implying nonlinearities in strain–displacement relationships. In addition, when the strains are high, the constitutive relationships become nonlinear, too. As a result, the discrete equations of the finite element method that incorporate both geometric and constitutive nonlinearities are highly nonlinear and require a large number of iterations in order to follow the loading path.

Some other complications in the course of numerical simulation of rubber and tires are rubber near-incompressibility and reinforcement in tires. The rubber near-incompressibility makes the system of finite element equations ill conditioned since in this case the volumetric stiffness greatly exceeds the shear stiffness. The near-incompressibility and incompressibility conditions are essentially constraints imposed on the solution, and depending on the ratio of the number of discrete equations and discrete number of constraints solution may or may not exist. Therefore, the design of specific finite elements to satisfy these conditions becomes very important.

Contact conditions add even more difficulty and complexity to an already very complex and difficult analysis of rubber products and tires. Contact conditions are unilateral and need to be constantly checked during the incremental nonlinear analysis. In addition, they are not smooth, thus degrading the performance of nonlinear solvers. A number of numerical regularization parameters need to be introduced to prevent chattering and ensure robustness of a finite element analysis (FEA) with frictional contact.

In this chapter, we will both discuss the above issues and deal with special topics of the finite element analysis of tires.

Modeling and Simulation in Polymers. Edited by P.D. Gujrati and A.I. Leonov

ISBN: 978-3-527-32415-6

8.2
Nonlinear Finite Element Analysis

As we have mentioned in the introduction, rubber parts typically experience large displacements and strains during their deformation history and, therefore, linearization based on the theory infinitesimal strains and small displacements that is traditionally employed for steel, reinforced concrete, and so on will produce inaccurate results. In order to retain the accuracy and realistic description of the deformation process in rubber, a fully nonlinear description of the deformation process should be considered. In the following discussion, we will obtain discretized finite element equations and outline their solution methods.

To outline a finite element analysis approach, we will formulate a boundary value problem, transform it into a weak or variational form, and obtain discretized finite element equations. We begin with the equations of equilibrium that are written in the deformed configuration [1]:

$$\sigma_{ij,j} + \varrho f_i = 0, \tag{8.1}$$

where σ_{ij} are the components of the Cauchy stress tensor, f_i are the components of the body force vector, and ϱ is the density. All the above quantities are defined in the deformed configuration.

We add to these set of equations the constitutive equations that relate stresses to strains. One form of constitutive equations for isotropic highly deformable materials is of the generalized Mooney–Rivlin type [2] in which the strain energy density W is expressed in terms of the strain invariants:

$$W = W(I_1, I_2, I_3), \tag{8.2}$$

where I_j are the strain invariants of the right Cauchy–Green deformation tensor C_{ji} that is expressed through the deformation gradient tensor F_{jk} as follows:

$$C_{ij} = F_{ki} F_{kj}. \tag{8.3}$$

In Eq. (8.3), $F_{ij} = \partial x_i / \partial X_j$ and x_i, X_j are respectively the coordinates in the deformed and the undeformed configurations. The stress–strain relationships are derived from (8.2) by differentiation of the strain energy density with respect to the strain measures obtaining as results the conjugate measure of strain. For incompressible materials such as tire rubber, the incompressibility condition at every point of continuum is expressed as $J = \det F = 1$ and has to be enforced separately in addition to the stress–strain equations. This constraint presents difficulties in the course of the analysis and will be addressed later.

Finally, the addition of boundary conditions (BCs) are needed for a unique solution of (8.1). Dirichlet or displacement BCs are prescribed on the part of the boundary where displacements are known:

$$\Gamma_u : u_i = u_{ib} \tag{8.1a}$$

and Neumann or traction BCs on the part of the boundary where tractions are known:

$$\Gamma_t : \sigma_{ij} n_i = t_j. \tag{8.1b}$$

This completes the set of equations in the boundary value problem expressed in the so-called strong form.

In order to develop equations suitable for a finite element discretization, we cast Eq. (8.1) in a weak form by selecting test functions w_i and integrate over the deformed configuration v:

$$\int_v (\sigma_{ij,j} + \varrho f_i) w_i = 0, \tag{8.4}$$

where the test functions are continuous in the deformed configuration v and vanish on the boundary Γ. Green's theorem is used to transform the volume integral into a surface integral to obtain

$$\int_v \sigma_{ij} w_{i,j}\, \mathrm{d}v + \int_{\Gamma_t} t_i w_i\, \mathrm{d}v + \int_v \varrho f_i w_i\, \mathrm{d}v = 0. \tag{8.5}$$

Thus, we have arrived at an alternative form of equilibrium equations where differentiability requirements are relaxed for the stress. Therefore, the stresses can be related through the constitutive and strain–displacement relationships to the primary variable, displacements, in the discretized equations.

To simplify Eq. (8.5) further, we take advantage of the symmetry of the stress tensor by splitting $w_{i,j}$ into the sum of symmetric and skew symmetric parts, that is, $(1/2)(w_{i,j} + w_{j,i})$ and keep only the symmetric part of the product $\sigma_{ij} w_{i,j}$. Thus, Eq. (8.5) becomes the equation of virtual work (also called virtual power) that can be found in textbooks on continuum mechanics, for example [1]:

$$\int_v \frac{1}{2} \sigma_{ij} (w_{i,j} + w_{j,i})\, \mathrm{d}v + \int_{\Gamma_t} t_i w_i\, \mathrm{d}v + \int_v \varrho f_i w_i\, \mathrm{d}v = 0. \tag{8.6}$$

Equation (8.6) is now suitable to obtain a set of discrete finite element equations. Specifically, the undeformed domain is discretized by subdividing it into a collection of nodes and elements, and after making an assumption of how nodal variables vary locally within each element, Eq. (8.6) represents a set of discrete equations consisting of a set of primary unknown nodal velocities.

Using the standard Galerkin finite element method, within any element the spatial distributions of the undeformed coordinates, velocities, and test functions are approximated using the same shape functions and nodal values as follows:

$$X_i = N_K(X) X_{Ki}, \quad u_i(X) = N_K(X) u_{Ki}, \quad w_i(X) = N_K(X) w_{Ki}, \tag{8.7}$$

where N_K are the local shape functions within the element and they are equal to 1 at node K and 0 at all other nodes of an element. In the above equations, $X_{Ki,} u_{Ki}, w_{Ki}$

are the nodal values of the undeformed coordinates, velocity, and test functions, respectively, at node K. The second subscript i denotes the coordinate direction.

In order to meet the continuity requirements imposed by the weak form of the equilibrium Eq. (8.6), the shape functions need to be continuous across the element boundaries. Physically, this also means that strains or strain rates that are first derivatives of the velocities could be discontinuous, but they must remain finite.

Substituting discretizations (8.7) in (8.6), we obtain the following equations with the nodal variables w_{iK}:

$$w_{iK}\left(\left(\int_{v} \sigma_{ij} N_{K,j}\, dv + \int_{\Gamma_t} t_i N_K\, dv + \int_{v} \varrho f_i N_K\, dv\right)\right) = 0. \tag{8.8}$$

Since the nodal test functions w_{iK} are arbitrary, we obtain a set of nonlinear finite element equations:

$$\int_{v} \sigma_{ij} N_{K,j}\, dv + \int_{\Gamma_t} t_i N_K\, dv + \int_{v} \varrho f_i N_K\, dv = 0. \tag{8.9}$$

The number of equations in (8.9) is equal to the number of nodes times the number of degrees of freedom (dof) at a node. We note that the number of elements and nodes can be of the order of millions to achieve an accurate solution for tire problems. All material and geometric nonlinearities in (8.9) are in the first term where the stress σ_{ij} is a nonlinear function of the velocities through the constitutive Eq. (8.2) and the strain displacement Eq. (8.3).

The integration in (8.9) is over the deformed domain while the shape functions are expressed in terms of the undeformed coordinates. However, since the body is subdivided into a collection of finite elements, all the integrals in (8.9) become a sum of element integrals. The element shape functions are written in terms of the reference coordinates where an element is mapped into a standard unit element and both the deformed and the undeformed coordinates are expressed in terms of the reference coordinates $\eta : X_i = N_K(\eta)X_{Ki}$, $x_i = N_K(\eta)x_{Ki}$. Then, the integrals in Eq. (8.9) are calculated over the reference domain with the appropriate domain and boundary Jacobians J and J_Γ:

$$\sum_{e} \int_{V_R} J\sigma_{ij} N_{K,j} dv + \sum_{\Gamma e} \int_{\Gamma_{tR}} J_\Gamma t_i N_K dv + \sum_{e} \int_{V_R} J\varrho f_i N_K\, dv = 0. \tag{8.10}$$

The summation in (8.10) is over all elements and element boundaries. In addition, the integrals over elements in (8.10) in practice are calculated using the Gauss numerical integration that means the integrands are calculated at a number of well-chosen integration points in the reference configuration. Even though the integration in (8.10) is over the reference configuration, the gradients of the shape functions are still with respect to the deformed coordinates. However, they can be easily calculated at the Gauss integration points using the chain rule $\frac{\partial N_k}{\partial x_i} = \frac{\partial N_k}{\partial \eta_j}\frac{\partial \eta_j}{\partial x_i}$, and the matrix $\frac{\partial \eta_j}{\partial x_i}$ (3×3 for 3D computations) is obtained by inversion of the matrix $\frac{\partial x_j}{\partial \eta_i}$, which is readily available.

The method of obtaining the nonlinear finite element Eqs. (8.9) and (8.10) using the deformed configuration is called the Updated Lagrangian approach. Equations can also be derived by considering equilibrium equations in the undeformed configuration utilizing the Second Piola–Kirchoff stress and the Green–Lagrange strain in the methods that is called Total Lagrangian. Then, the integrals in (8.9) will be over the undeformed configuration, but the discretized equations are equivalent.

As was mentioned, the integrations in the first and last terms in (8.10) are over the whole collection of finite elements. Therefore, when it comes to obtaining a nonlinear equation at a specific degree of freedom of a specific node, only the integrals from the elements adjacent to this node contribute to the equation. The process of accounting for contributions from adjacent elements is called assembly and is one of the basic procedures in the finite element method. The same is true for the second term in (8.10) with the only difference that the element integrals are computed over the boundary, but the process is the same. The above-mentioned treatment can be found in a number of references, for example [3–5].

8.3 Incompressibility Conditions

We now turn our attention to an important aspect of rubber materials in tires, which is its nearly incompressible response. Incompressibility conditions may cause considerable complications in computational mechanics if they are not treated correctly. Initially, people who used the displacement-based finite element methods tried to approximate incompressible solutions in linear elasticity by using the Poisson ratio close to 0.5. Clearly, this is not a viable approach because some stiffness terms approach infinity. The next step was to introduce mixed finite element methods [6] in which the incompressibility conditions were treated as constraints that were enforced by Lagrange multipliers. The Lagrange multipliers are introduced as additional variables and have their own interpolation shape functions that for performance reasons are different from those used for displacements. Near-incompressibility constraints can be handled by penalty methods where the bulk modulus is assigned a very large, perhaps even realistic value. For example, in case of rubber compounds the bulk modulus can be thousands of times larger than the shear modulus. The penalty approach, however, causes problems since the problem becomes ill conditioned unless the numerical integration is carried out differently for the shear and the bulk modes of deformation. The incompressibility problems reveal themselves in a more or less the same fashion in both linear and nonlinear problems, and, therefore, for simplicity we will deal with the incompressibility for linear problems.

To develop equations in a weak form for incompressible problems, we start with the same equations of equilibrium (8.1) and boundary conditions (8.1a) and (8.1b). But we split the stresses and strains into the bulk and the deviatoric parts:

$$\sigma_{ij} = -p\delta_{ij} + 2\mu\varepsilon_{ij}, \quad \varepsilon_{ij} = \frac{1}{2}(u_{i,j} + u_{j,i}). \tag{8.11}$$

And also add the incompressibility constraint,

$$\varepsilon_{ii} = 0. \tag{8.12}$$

Essentially, we have divided the Hook's law into two parts and introduced a new variable, the hydrostatic pressure. We have also eliminated the bulk modulus that becomes infinite in our case.

Augmenting the weak form Eq. (8.6) with a product of a Lagrange multiplier λ and the constraint (8.12) yields

$$\int_v \frac{1}{2}\sigma_{ij}(w_{i,j} + w_{j,i})\,\mathrm{d}v + \int_v \lambda u_{i,i}\,\mathrm{d}v + \int_{\Gamma_t} t_i w_i\,\mathrm{d}v + \int_v \varrho f_i w_i\,\mathrm{d}v = 0, \tag{8.13}$$

where the second term in (8.13) is the weak form of the incompressibility condition. This term, as we will see, has important implications for the element formulation used.

The discretization follows the same route as before with the only exception that the pressure p and Lagrange multiplier λ are an extra unknown and test function in additions to u and w.

We also use shape functions for p and λ: $p = M_N(\eta)p_N$ and $\lambda = M_N(\eta)\lambda_N$. Substituting the approximations together with the displacement approximations in (8.13), we obtain linear equations in matrix form:

$$\begin{bmatrix} \mathbf{K} & \mathbf{C} \\ \mathbf{C}^{\mathrm{T}} & \mathbf{0} \end{bmatrix} \begin{bmatrix} \mathbf{U} \\ \mathbf{P} \end{bmatrix} = \begin{bmatrix} \mathbf{F} \\ \mathbf{0} \end{bmatrix}, \tag{8.14}$$

where **K** is the standard linear elasticity stiffness, but its elasticity matrix does not have the volumetric terms, and

$$\mathbf{C} = \left[\int_v M_K N_{J,1}\,\mathrm{d}v \quad \int_v M_K N_{J,2}\,\mathrm{d}v \quad \int_v M_K N_{J,3}\,\mathrm{d}v \right]. \tag{8.15}$$

In Eq. (8.14), **F** denotes the nodal forces derived from the body forces and surface tractions as before, and **U** and **P** are the nodal displacements and pressures.

The pressure nodes and approximating functions are different from those used for the displacements, and not every combination of the displacement and pressure shape functions will perform numerically well. For many combinations of the shape functions such incompressible elements will display a phenomenon that is called locking. This means that regardless of the applied loads and BC, the only displacements that can be calculated are zero. This is related to the number of kinematic degrees of freedom in the model and the number of constraints.

Let us consider the 2D mesh of Figure 8.1 that comprises triangular elements and assume that nodes 1, 2, 3, 4, and 7 are fixed. The elements are incompressible and we have one incompressibility constraint per element that means the element area remains constant during the deformation process.

Considering element 1, we see that nodes 1 and 2 are fixed, and, since the area must remain constant, node 5 can move only horizontally. However, when we consider element 2, we can conclude that node 5 can move only vertically. Therefore, node 5 cannot move at all. That is, the only possible displacements at the end for node

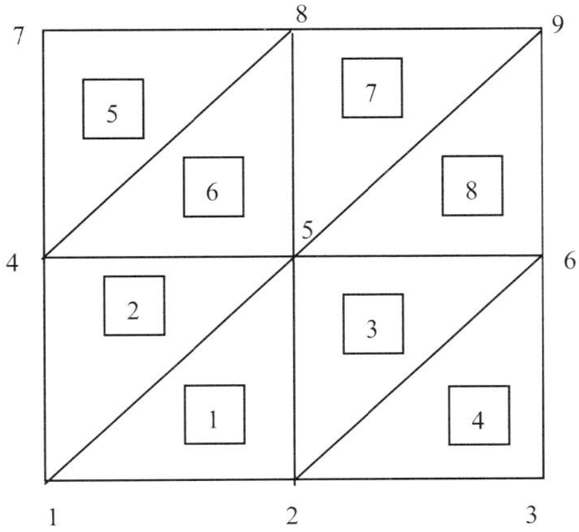

Figure 8.1 Simplified 2D mesh of incompressible triangular elements.

5 are zero. Then, we continue with elements 3 and 4 and also conclude that node 6 cannot move at all. Continuing with the same kind of reasoning, the only possible displacement for all the nodes is zero, meaning that the nodes of the mesh cannot move – they are locked.

It can be easily seen from this simple example that the problem is that the number of displacement boundary constraints (or BCs) and incompressibility constraints is equal to the number of kinematic dof, and therefore, the system cannot be solved for any deformation mode. The mathematical theory behind this phenomenon is based on the Babuska–Brezzi condition [7] that numerically is related to the solvability of Eq. (8.14).

Different designs of elements can be evaluated on the basis of the number of kinematic dof and constraints [8]. For that, the element is fixed along two sides (three faces in 3D) and the ratio of dof to constraints is calculated. The rule of thumb is that this ratio should mimic the continuum situation where we have three PDEs for displacements and one for incompressibility. The optimal ratio, therefore, is 3. It is 2 in 2D case. Some examples of "good" elements are given in Figure 8.2.

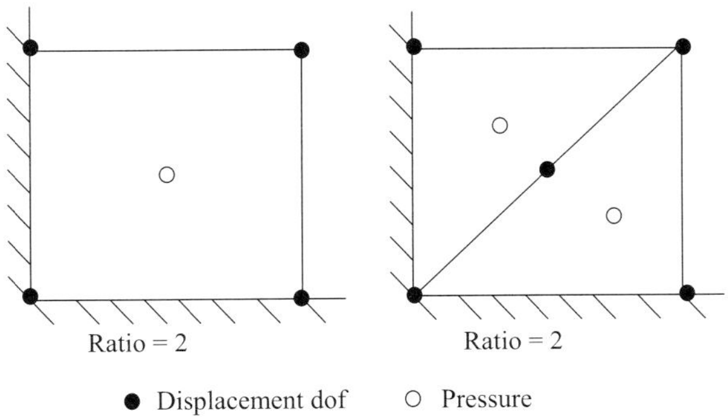

Figure 8.2 Elements with reasonable incompressible behavior.

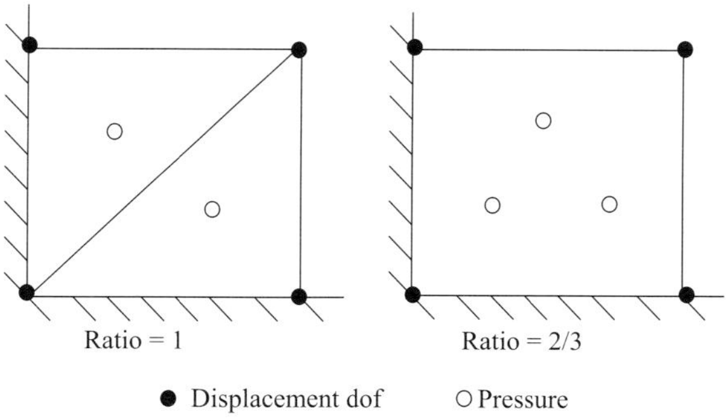

Figure 8.3 Elements with poor incompressible response.

In Figure 8.3, examples of "not so good" elements are shown. These elements are overconstrained and display poor numerical performance. The first element in Figure 8.3 is the same one that was used in Figure 8.1, and, as we have seen, it locks.

If the medium is slightly compressible that is typically the case for real materials, one might be tempted to employ the penalty method. In such a situation, the value of the bulk modulus is considerably higher than that of the shear modulus – exactly the case for rubber as was noted in the introduction. Then, the actual value of the bulk modulus is specified in the Hook's law. However, such displacement methods display the same problems as the improperly designed mixed finite element methods, and the problem comes down again to the improper ratio of the dof and the constraints.

When the bulk modulus is sufficiently high, that implies imposing a constraint similar to one with the Lagrange multiplier method since the dilatation is close to zero. This constraint is enforced at the element integration points because the elements are numerically integrated. Therefore, the number of integration points defines the number of the incompressibility constraints provided the constraints possess enough independent parameters. For example, the incompressibility constraint for the triangular element we have used in Figure 8.1 has only one independent parameter because displacements are linearly approximated in the element, and $u_{i,i}$ is constant. Thus, no matter how many integration points are used for incompressibility in such an element, there is still going to be only one constraint per element.

It is clear that the situation in the penalty method is very similar to the one based on the Lagrange multiplier method. The technique to reduce the number of constraints in the penalty formulation is to split the constitutive equations into the deviatoric and the dilatational parts, and then use a lower order integration scheme for the dilatational part (thus reducing the number of the incompressibility constraints) while keeping the normal order of numerical integration for the deviatoric part.

Then, the Hook's law becomes

$$\sigma_{ij}^{\mathrm{vol}} = B\varepsilon_{kk}\delta_{ij}, \tag{8.16a}$$

$$\sigma_{ij}^{\text{dev}} = 2\mu\varepsilon_{ij}^{\text{dev}}, \tag{8.16b}$$

$$B \gg \mu, \tag{8.16c}$$

where B is the bulk modulus.

Correspondingly, the stiffness matrix is split into parts, the volumetric K_{vol} and deviatoric K_{dev} parts. To reduce the number of the incompressibility constraints, a lower order of integration is used for the K_{vol} part of the stiffness matrix. Such a numerical integration procedure is called selectively reduced integration and is very effective for near-incompressible problems. One can see a definite analogy between the Lagrange multiplier methods and the penalty methods with the selectively reduced integration. In fact, for elements shown in Figures 8.2 and 8.3, one can use the same reduced integration schemes as for Lagrange multiplier nodes and arrive at similar conclusions [8].

In the large strain situation, we can split the deviatoric and volumetric terms [9] by redefining the deformation gradient tensor as $\bar{F} = J^{1/3}F$. Then, the right Cauchy–Green deformation tensor invariants become

$$\bar{I}_1 = J^{-\frac{2}{3}}I_1, \tag{8.17a}$$

$$\bar{I}_2 = J^{-\frac{4}{3}}I_2. \tag{8.17b}$$

By doing so, $\bar{I}_1, \bar{I}_2$ define a volume preserving deformation mode. Then, the strain energy function can be written as follows:

$$W = W(\bar{I}_1, \bar{I}_2) + \frac{1}{2}K(J-1)^2 \tag{8.18}$$

with the term $\frac{1}{2}K(J-1)^2$ representing the volumetric part of the strain energy. The same procedure of selectively reduced integration can be applied.

8.4 Solution Strategy

Equation (8.10) can be expressed in a compact matrix vector form suitable for programming. The first term in (8.10) is often called the vector of internal forces, $\mathbf{F}^{\text{int}}$ because it is derived from the internal stresses arising in the body. This vector contains the left-hand side of the equations with unknown velocities **v**. The second term and third term together are called the right-hand side, or vector forces external forces, $\mathbf{F}^{\text{ext}}$, with contributions from the surface tractions applied to the deformed body from the body forces distributed in the domain. In addition, to solve Eq. (8.10), the displacement boundary conditions have to be imposed at the boundary nodes.

Equation (8.10) is a nonlinear algebraic equation and is difficult to solve in general. Depending on the load magnitude and the stiffness of the structure, the nonlinearities may be too significant for any iterative solution method. That is why the load or external excitations like applied displacements are usually applied in steps. When the solution is obtained, say, in the first step that may contain only a certain percentage of the external load, then the load is incremented, and the solution is obtained in the second step, and so on. This way, nonlinearities of the problem are dealt with incrementally, and there is a better chance of arriving at the solution.

Within a load step, equations are most commonly solved by the Newton method. There is a plethora of modern iterative methods for solution of algebraic equations, but the Newton method is so far the most robust and popular and the modern nonlinear finite element software [3–5]. Let us write (8.10) in the following matrix form:

$$\mathbf{r} = \mathbf{F}^{\text{ext}}(\mathbf{d}(t-\Delta t) + \Delta\mathbf{d}(t)) - \mathbf{F}^{\text{int}}(\mathbf{d}(t-\Delta t) + \Delta\mathbf{d}(t)) = 0, \tag{8.19}$$

where $\mathbf{r}$is called the residual of (8.10). The internal force vector, $\mathbf{F}^{\text{int}}$, depends on the primary displacement variables in a nonlinear manner. The velocities $\mathbf{v}(t)$ at time t have been written to reflect the load stepping scheme, that is, $\mathbf{d}(t-\Delta t)$ denotes the known (solved) displacements at the beginning of the load step and $\Delta\mathbf{d}(t) = \Delta t\mathbf{v}(t)$ denotes the unknown (to be solved) displacement increments at time (t).

In order to apply Newton's method to solve the nonlinear Eq. (8.19), they need to be linearized via the "tangent" or matrix of derivatives:

$$\mathbf{K} = [\mathbf{F}^{\text{int}}_{i,j}]. \tag{8.20}$$

The derivatives here are of the vector of internal forces with respect to the unknown displacements, $\mathbf{d}$. Iterations are performed by solving the system of linear equations to get displacement increments: $\mathbf{K}\,\Delta\mathbf{d}^k = \mathbf{F}^{\text{ext}} - \mathbf{F}^{\text{int}}$. Then, the vector of displacements, $\mathbf{d}$, are updated: $\mathbf{d}^k = \mathbf{d}^{k-1} + \Delta\mathbf{d}^k$ and are used to calculate the new residual $\mathbf{r} = \mathbf{F}^{\text{ext}} - \mathbf{F}^{\text{int}}$, and the iterations continue.

We note that while the Newton method is the most robust and most widely used in nonlinear finite element software, it is also computationally expensive primarily due to the necessity to solve a system of linear equations. It also imposes considerable computer memory requirements since a global system matrix is used. This method also is not as easily parallelized as some other iterative methods. In order to achieve the optimal performance of the Newton method, it is crucial to calculate the tangent stiffness matrix that is indeed "tangent" or, in other words, is the derivative with respect to unknowns that are calculated very accurately.

8.5 Treatment of Contact Constraints

We now turn our attention to an important aspect of solving tire problems – the treatment of contact constraints. The contact problem to be solved is one of the two bodies contacting across their respective surfaces. The impenetrability of the two bodies (normal contact) will manifest itself as a set of unilateral constraints

$$g_{\text{N}} \geq 0, \tag{8.21a}$$

$$t_N \geq 0, \tag{8.21b}$$

$$t_N g_N = 0, \tag{8.21c}$$

$$t_N \dot{g}_N = 0, \tag{8.21d}$$

where g_N is the normal gap between the two bodies and t_N the normal traction, $t_N = \mathbf{n} \cdot \mathbf{t}$.

The frictional behavior of the two contacting bodies (we consider Coulomb friction here) is represented by the tangential contact constraints:

$$\Phi := \|\mathbf{t}_T\| - \mu t_N \leq 0, \tag{8.22a}$$

$$L_v g_T - \varsigma \frac{\mathbf{t}_T}{\|\mathbf{t}_T\| = 0,} \tag{8.22b}$$

$$\varsigma \geq 0, \tag{8.22c}$$

$$\Phi\varsigma = 0, \tag{8.22d}$$

where g_T is the relative slip vector between two bodies, ς is the (scalar) slip rate in the direction of the tangential traction given by the projection of the traction, **t**, onto the slip surface, $t_T = -(\mathbf{I} - \mathbf{n} \otimes \mathbf{n}) \cdot \mathbf{t}$. Note that the form of tangential constraint set is similar to the normal constraint set in that it represents a discontinuous behavior between sticking $\varsigma = 0$ and sliding $\varsigma > 0$.

Discretization of the contact problem posed leads the definition of the constraint operator, **G**. Referring to Figure 8.4, we define the constraint operator at slave node, s as

$$\mathbf{G}|_s = \lfloor \ldots 1 \ \ldots -N_{m_1} - N_{m_2} - N_{m_3} - N_{m_4} \rfloor. \tag{8.23}$$

Thus, the discrete representation of the contact constraints expressed in Eqs. (21d) and (22d) are

$$\mathbf{g}_N = \mathbf{G}_N \cdot \mathbf{d}(t) = 0, \tag{8.24a}$$

$$\mathbf{g}_T = \mathbf{G}_T \cdot \mathbf{d}(t) = 0, \tag{8.24b}$$

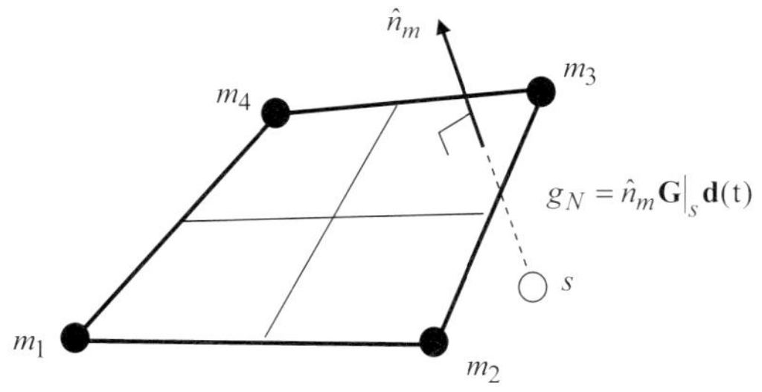

Figure 8.4 Definition of the constraint operator at a surface slave node s against a master surface m via a closest point projection along the surface normal $\hat{n}_m$.

where a projection of the nodal displacement onto the constraint normal and its tangent plane is used to construct the normal and tangential constraint operator $\mathbf{G}_{\mathrm{N}}$ and $\mathbf{G}_{\mathrm{T}}$, respectively. It is a common practice to use, and we do so here, the so-called "master–slave" algorithm where one side of the interface is the slave surface on which the constraints are defined and enforced and the other side of the interface is the master surface that defines the interface geometry. In that way, the interface is unambiguously defined. For tire problems, this is the preferred approach anyway since the tire is in contact with a comparably rigid rim and road.

Focusing now on the enforcement of these constraints, we consider the similarity of the discrete contact constraint Eqs. (24a) and (24b) and the incompressibility constraint (8.12).

When considering the usual approaches for constraint enforcement, that is, penalty, Lagrange multiplier, or Augmented Lagrangian, we consider the class of problems being solved. Tires by nature of their design are not intended to slip significantly relative to the rim or the road. Thus, an approach that strongly enforces the no-slip condition is favored. In addition, there is an observed strong coupling between the enforcement of the contact stick-slip condition and the near incompressibility of the material response. For these reasons, we choose to enforce the contact rate constraints during the Newton iterations – that is, slave nodes are assumed to be either in contact and sticking or not in contact at all.

With this choice, two important omissions in the contact treatment are evident. First, our choice to enforce rate conditions rather than impenetrability and frictional stick-slip constraints during the solution of the Newton iteration leaves us vulnerable to inaccurate choices of active constraint sets, over which the rate constraint is enforced. Second, our argument for linearity of the constraint enforcement scheme significantly depends on the matrices $\mathbf{G}_{\mathrm{N}}$ and $\mathbf{G}_{\mathrm{T}}$ being invariant during the Newton iterations, so that both the set of active constraints and the geometrical variations on the surfaces should be independent of the deformation. Of course, this invariance will not hold in general unless we enforce it algorithmically.

So, we introduce a second iteration loop outside the Newton iterations, where the normal gaps are kinematically removed and some amount of frictional slip is allowed if required. Also in this loop, which we will take to be indexed by k, the contact kinematic matrices $\mathbf{G}_{\mathrm{N}}$ and $\mathbf{G}_{\mathrm{T}}$ are recomputed based on current geometry that are then fixed during the subsequent Newton iteration loop. The proposed algorithm is therefore somewhat similar in spirit to some recent proposed augmented Lagrangian treatments of contact problems, in which Uzawa's method for multiplier updates is applied (see Ref. [10]). Both the proposed algorithm and the augmented Lagrangian treatments require nested iteration strategies, where the inner iteration loop is primarily concerned with equilibrium iterations and where the outer loop iterates on the quality of constraint satisfaction. As is the case with augmented Lagrangian strategies, it should also be noted that the convergence rate associated with this outer loop is expected to be linear only.

Associating an index k with this outer constraint loop, an incremental kinematic prescription of the gap removal is utilized via

$$\Delta\mathbf{d}_{k+1} = \Delta\mathbf{d}_k + (\Delta\mathbf{d}_{j*})_k + \beta_{\mathrm{N}}\mathbf{G}_{\mathrm{N}_k}^{\mathrm{T}}\mathbf{G}_{\mathrm{N}_k}(\mathbf{d}(t-\Delta t) + \Delta\mathbf{d}_k + (\Delta\mathbf{d}_{j*})_k) \qquad (8.25)$$

for the normal gap constraints. In Eq. (8.25), $\Delta\mathbf{d}_k$ is the accumulated displacement increment of the kth intermediate solution, $(\Delta\mathbf{d}_{j*})_k$ is the displacement increment for the current (k) intermediate solution, andβ_N is a pushback factor, that is, $0 < \beta_N \leq 1$ on the normal gap. Equation (8.25), in the case where $\beta_N = 1$, simply gives the displacement increment $\Delta\mathbf{d}_{k+1}$ necessary to remove violations of the impenetrability constraint, as would occur, for example, if a new node came into contact so that a change in the active constraint set was necessitated. In practice, β_N is generally set to a value less than 1, as too large a value will excessively upset the global equilibrium of the system and require needless extra Newton iterations to restore it.

The tangential gap constraints are treated in a similar manner, resulting in the following displacement update for the frictional case:

$$\begin{aligned}\Delta\mathbf{d}_{k+1} &= \Delta\mathbf{d}_k + (\Delta\mathbf{d}_{j*})_k + \alpha\mathbf{M}^{-1}(\mathbf{G}_T^s)^T(\beta_T\mathbf{r}_k^{tan}) \\ &\quad + \beta_N\mathbf{G}_{N_k}^T\mathbf{G}_{N_k}(\mathbf{d}(t-\Delta t) + \Delta\mathbf{d}_k + (\Delta\mathbf{d}_{j*})_k),\end{aligned} \tag{8.26}$$

where α is the line search parameter, β_T is an allowable slip factor, that is, $0 < \beta_T \leq 1$, and $\mathbf{r}_k^{tan}$ is the residual tangential force unbalance, that is,

$$\mathbf{r}_k^{tan} = \min[0, (\mathbf{G}_T^s - \mu\mathbf{G}_N^s)(\mathbf{F}^{ext} - (\mathbf{F}_{j*}^{int})_k)]. \tag{8.27}$$

As with the case for the impenetrability constraint, the tangential (stick-slip) constraints are also gradually enforced. The frictional slip is determined from a line search along the steepest descent direction $\mathbf{M}^{-1}(\mathbf{G}_T^s)^T(\beta_T\mathbf{r}_k^{tan})$, and the rate constraints are active during the Newton iterations while assuming sticking conditions, that is,

$$\mathbf{t}_{T_j} = (\mathbf{t}_{T_j})_{stick} = \mathbf{G}_T^s(\mathbf{F}^{ext} - (\mathbf{F}_j^{int}). \tag{8.28}$$

Thus, although the stick-slip decision could be built into the Newton iteration strategy, our experience shows that the approach where complete stick is assumed during the Newton iterations is more robust and cost-effective for tire applications.

8.6 Tire Modeling

Tires are complex structural systems that present significant challenges to finite element analysis. In order to simulate tire performance, FE modeling must account for both geometrical and material nonlinearities; it must deal effectively with soft composites and with rubber near-incompressibility; and it must accommodate the complicated contact conditions existing at the tire/rim and tire/road interfaces. In addition, a large number of degrees of freedom (on the order of millions) are required to accurately model the structural details of a realistic tread design. Moreover, an FE rolling analysis of such a design is extremely CPU intensive. For all these reasons, numerical modeling of tires is extremely difficult, and in addition to numerical treatment described in previous sections, parallel computing is highly desirable and is becoming a key component in the tire FEA.

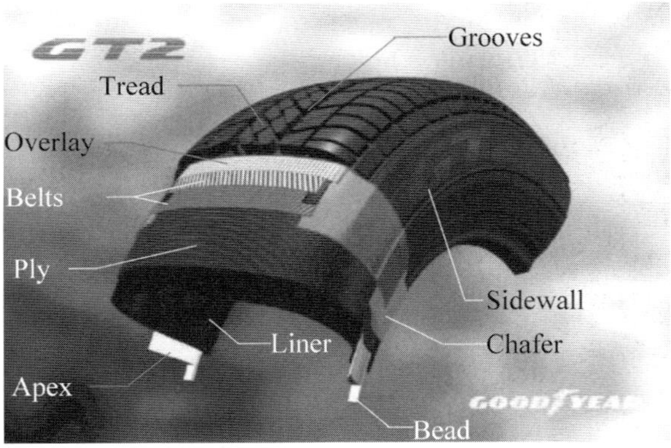

Figure 8.5 Schematic of a pneumatic radial tire.

Tires consist of many components that include rubber compounds as well as steel and synthetic reinforcement, as shown in Figure 8.5. Reinforcement components such as belts, plies, beads, and overlays shown in the figure are considerably stiffer than the rubber compounds surrounding them.

Therefore, in addition to difficulties arising from nonlinearities due to large deformations, incompressibility, and contact conditions, a tire is a structure with components possessing a wide variation in stiffness, which makes the overall assembled stiffness matrix ill conditioned. This ultimately can affect the accuracy of the analysis without careful consideration of the overall solution strategy. Finally, we note that owing to a large number of components and their elaborate shapes, for example, the tread with many grooves and sipes, a tire model requires a considerable number of degrees of freedom to satisfy the accuracy requirements.

A realistic tire analysis involves several stages: inflation and rim seating, deflection or loading with a vertical force, and rolling. During the inflation and rim seating stage, a tire is considered to be loaded with a uniformly distributed pressure **P** on the inner surface of the liner and contacting the rim, Figure 8.6. The pressure remains normal to the inner surface of the liner during the formation. This has to be accounted for in both the pressure direction change and in what is called the load-stiffness matrix [11] since the pressure in effect depends on displacements and, therefore, a derivative of the external load – the pressure – with respect to the primary unknowns – the displacements – does not vanish. The most numerical difficulties in tire modeling during the inflation and rim seating stage occurs due to the contact of the tire with the rim and the frictional sliding of the chafer rubber over the rim curved surface (Figure 8.5). This involves severe deformations of the chafer and many contact iterations in both the normal and the tangential directions.

After a tire model has been inflated, it is loaded with a vertical force to simulate the weight of the vehicle, Figure 8.7. In this stage, another contact condition takes effect – contact of the tire tread with the road. The inflation **P** is still normal to the surface and the contact with the rim continues to be engaged.

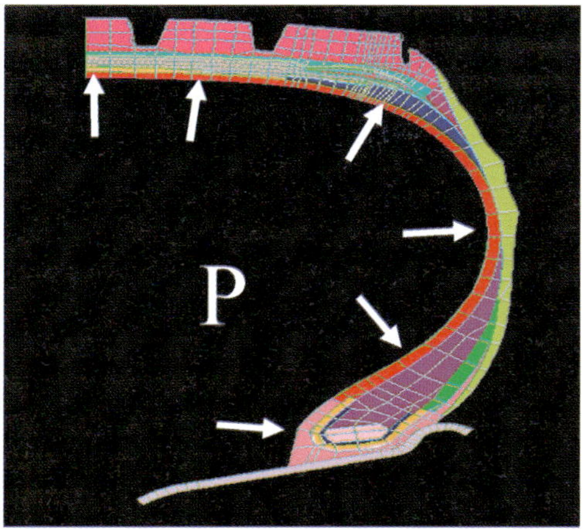

Figure 8.6 Tire inflation.

Next, the tire model is rolled on the road (Figure 8.8), where the tire model first cambered, that is, tilted about the *X*-axis and then rotated about the *Y*-axis. The road meanwhile can move freely in the direction that is in the *X*–*Y* plane but at an angle with respect to the *X*-axis. Thus, we can model tires rolling with a slip angle or simulate turning of the vehicle.

Model generation for this analysis is also a multistage process. Since the tire cross-section geometry (without tread) is symmetric about the centerline, one half of the cross section is meshed using 2D tools. The mesh is reflected to create a full cross-section mesh and is then swept about the tire axis of symmetry to produce an axisymmetric 360°, 3D mesh.

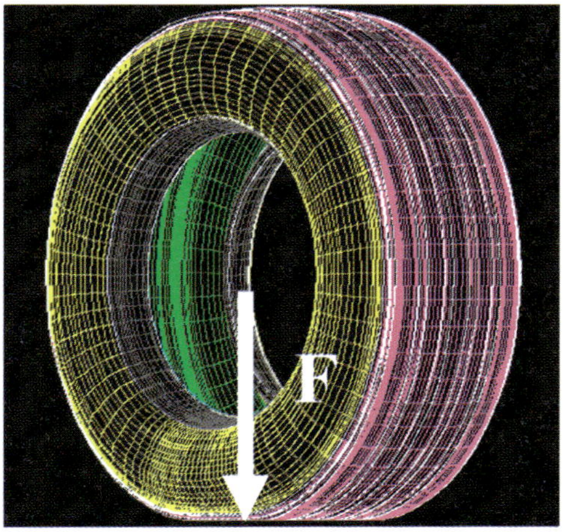

Figure 8.7 Vertical loading on a tire.

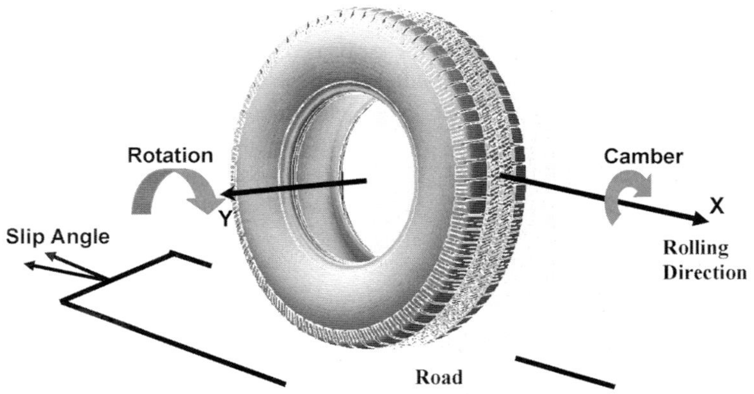

Figure 8.8 Definition of camber and slip angle and rolling direction in the global coordinate system.

A realistic tread design is meshed in a separate process and is snapped onto the carcass mesh. Because of the complexity of the tread geometry, it is difficult to achieve mesh contiguity between carcass and tread. Mesh incompatibility is handled by imposing multipoint constraints (MPCs) between contacting surfaces.

8.6.1
Example: Bump Envelopment [12]

A bump envelopment analysis is a similar sequence of FE stages in which a tire model is mounted against a rigid rim, inflated, deflected against a rigid road, and then rolled at constant road deflection over an imperfection in the road. Although such an analysis is used for tire comfort and harshness assessment it can answer several other important questions:

1) What loads will the axle see?
2) What is the nature of the carcass deformations?
3) How severe is the loss of contact at the tread/road interface?
4) Will the tire dismount from the rim?

This sort of analysis run on a highly refined, realistically treaded model demands parallel execution to obtain timely results. Table 8.1 shows relevant data for the FE model.

Table 8.1 Model data for bump impact analysis.

Total elements	Total nodes	Degrees of freedom	Circumferential sectors in mesh
938,843	910,468	3,670,247	180

Figure 8.9 Treaded tire model.

Figure 8.9 shows the fully treaded FE mesh and Figure 8.10 shows the tire as it envelops the road bump after approximately 60° of quasistatic rolling.

Figure 8.11 shows footprint contact pressures after mounting, inflation, and application of the nominal vertical axle load of 6230 N. All pressures are in MPa.

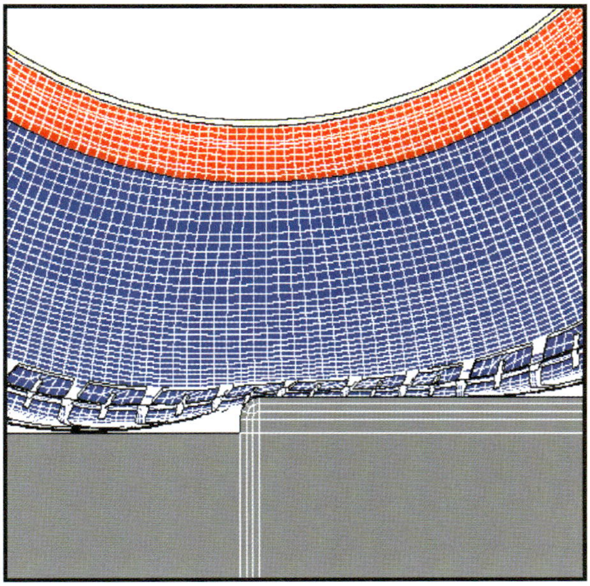

Figure 8.10 Tire running over an obstacle.

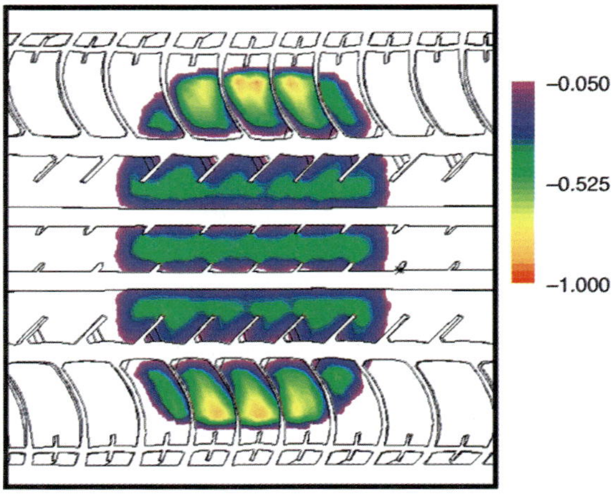

Figure 8.11 Tire-road pressure distribution.

Figure 8.12 shows this footprint as the tire envelops the bump. Note the band of high contact stresses that coincides with the leading edge of the bump. There are many other quantities of interest that were calculated but are not presented here: for example, global forces and moments, slip distributions in the tread material, heat build-up, and so on. For a model of the size and complexity presented here, parallel computing was the only viable means for obtaining these quantities.

In conclusion, a very elaborate analysis system that includes a robust and parallel aware finite element code capable of effectively handling nonlinearities, incompressibility, contact and a wide range of stiffness variations must be in place in order to

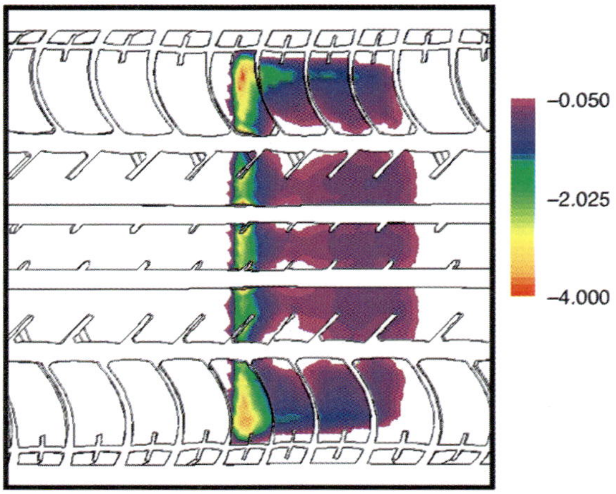

Figure 8.12 Tire-bump pressure distribution.

handle tire stress analysis applications. In addition, for practical applications, the code should be equipped with both an effective meshing preprocessor and a finite element variable (such as stresses, strains, etc.) visualizer.

References

1 Malvern, L.E. (1969) *Introduction to the Mechanics of a Continuous Medium*, Prentice Hall.
2 Ogden, R.W. (1984) *Non-Linear Elastic Deformations*, John Wiley & Sons, Inc., New York.
3 Ziekiewicz, O.C. and Taylor, R.L. (1989) *The Finite Element Method*, McGraw-Hill.
4 Bathe, K.-J. (1996) *Finite Element Procedures*, Prentice Hall.
5 Belytschko, T., Liu, W.K., and Moran, B. (2004) *Nonlinear Finite Elements for Continua and Structures*, John Wiley & Sons, Inc., New York.
6 Hermann, L.R. (1965) Elasticity equations for nearly incompressible materials by a variational theorem. *AIAA J.*, **3**, 1896–1900.
7 Oden, J.T. and Carey, G.F. (1984) *Finite Elements: Mathematical Aspects*, vol. 4, Prentice Hall.
8 Hughes, T.J.R. (1987) *The Finite Element Method. Linear Static and Dynamic Finite Element Analysis*, Prentice Hall.
9 Crisfield, M.A. (1997) *Nonlinear Finite Element Analysis of Solids and Structures, Volume 2: Advanced Topics*, John Wiley & Sons, Inc., New York.
10 Laursen, T.A. *Computational Contact and Impact Mechanics*, Springer Verlag.
11 Hibbitt, H.D. (1979) Some follower forces and load stiffness. *Int. J. Numer. Meth. Eng.*, **14**, 937–941.
12 Blanford, M.L., Heinstein, M.W., Assaker, D., Assaker, R., Davis, T.A., Jeusette, J.-P., Poldneff, M.J., and Quoirin, D. (2002) Parallel computations in tire analysis. WCCM V: Fifth World Congress on Computational Mechanics, July 7–12, Vienna, Austria.

9
Modeling the Hydrodynamics of Elastic Filaments and its Application to a Biomimetic Flagellum

Holger Stark

9.1
Introduction

In nature, microorganisms employ beating elastic filaments called flagella or cilia to propel themselves forward in their highly viscous environment [1–3]. In addition, the cilia are used to transport fluid. This strategy can be viewed as nature's answer to the fact that on the micron scale water is highly viscous and thus in the regime of low Reynolds numbers, where inertia does not play any role. Since microorganisms cannot rely on drifting by inertia, as we do when we swim in water, they immediately come to a halt when they stop their beating motion. The chapter gives a condensed overview of how one can model an elastic filament moving in a viscous fluid. It then introduces a superparamagnetic elastic filament that mimics a flagellum and that was used to create an artificial swimmer [4]. Finally, we demonstrate how this filament is capable of transporting fluid. The latter is especially attractive for the field of microfluidics [5], where fluids have to be transported along microchannels or thoroughly mixed without having turbulent flow fields available.

9.1.1
Lessons from Nature

In 1977, Purcell pointed out in his famous article "Life at Low Reynolds Number" that microorganisms have to perform a nonreciprocal periodic motion to be able to move forward [6] (see also [7]). Nonreciprocal means that the time-reversed motion is not the same as the original one (e.g., see Refs [8–13]). The reason lies in the Stokes equations [14] governing the fluid flow around microorganisms for negligible inertia: they allow for a time-inverted flow pattern when all the external and pressure forces are inverted. Especially, a scallop that just opens and closes its two shells would fail to proceed in the microscopic world [6]. At least two hinges are necessary to perform a nonreciprocal motion as realized in the Purcell swimmer [6, 10, 15].

Bacteria such as *Escherichia coli* and *Salmonella typhimurium* employ marvelous rotary motors to crank a bundle of relatively stiff filaments of helical shape [16, 17]. The rotating helices create a thrust on the bacterium since each local piece of one

Modeling and Simulation in Polymers. Edited by P.D. Gujrati and A.I. Leonov

ISBN: 978-3-527-32415-6

helix has an anisotropic friction coefficient. This means that the two friction coefficients for moving, respectively, along or perpendicular to the local axis are different. Sperm cells as one example for eukaryotic cells propel themselves by creating planar bending waves that move along their elastic flagella from the head to the tail [2, 18–20]. These waves are generated by the collective motion of internal molecular motors and the action of frictional forces from the surrounding fluid [21, 22]. Some unicellular microorganisms, such as the protozoan paramecium or opalina, are covered by a carpet of shorter flagella called cilia. Each cilium performs a characteristic three-dimensional stroke that is coordinated with the strokes of neighboring cilia. As a result, the so-called metachronal waves run along the surfaces of the protozoans and propel them. Hydrodynamic interactions between the beating cilia are discussed as the major source of their synchronization [23–32] (J. Elgeti and G. Gompper, in preparation).

Nature also uses arrays of collectively beating cilia to transport mucus in the respiratory tract or fluid in the brain [33]. During the early stage of a developing embryo, arrays of cilia that rotate about their anchoring points and thereby produce a vortex in the surrounding fluid are responsible for establishing the left-right asymmetry in the placement of organs [34]. Genetic defects that produce defective cilia then lead to the so-called *situs invertus* where organs are placed on the wrong side of the animal or human body. Finally, fluid transport and mixing on the microscopic level is a fascinating problem that is at the center of a successful lab-on-chip technology [5, 35]. Therefore, experimental efforts have been initiated to copy nature's successful concept by developing biomimetic or artificial cilia that are actuated by external fields [4, 35, 36] or to move fluid with the help of bacterial carpets [37].

9.1.2
A Historical Overview

For modeling a beating flagellum correctly, one important ingredient is the correct description of its hydrodynamic friction with the surrounding fluid. An account of this problem is given by Brennen and Winet in Ref. [1]. We first consider a long thin rod of length L and with circular cross section of diameter $2a$. It has two characteristic friction coefficients per unit length, $\gamma_{\parallel}$ and $\gamma_{\perp}$, for moving, respectively, along and perpendicular to its axis in a fluid with shear viscosity η [38]:

$$\gamma_{\parallel} = \frac{2\pi\eta}{\ln(L/a) + C_1} \quad \text{and} \quad \gamma_{\perp} = \frac{4\pi\eta}{\ln(L/a) + C_2}, \tag{9.1}$$

where $C_1 = -3/2 + \ln 2 = -0.807$ and $C_2 = C_1 + 1 = 0.193$. In general, these constants depend on the axial variation of the cross-sectional radius [38]. For finite values L/a corrections of the order $[\ln(L/a)]^{-3}$ occur and for $L/a \to \infty$ one obtains $\gamma_{\perp}/\gamma_{\parallel} = 2$. In what is called resistive force theory [1], one tries to apply these friction coefficients to a long slender body that is allowed to bend as, for example, a beating flagellum. However, in such cases the meaning of L is not obvious. Clearly, resistive force theory is only an approximation since it does not appropriately treat the hydrodynamic interactions between different parts of a filament. Interactions are mediated by flow fields that are initiated when these parts move. In slender body theory [39, 40], a more accurate treatment of the problem is achieved. The boundary value problem for a long filament is solved by distributing the so-called stokeslets and their spatial derivatives (Stokes doublet, quadrupole, stresslet, etc.) along the filament's centerline. A stokeslet gives the velocity field initiated by a point force and is therefore proportional to the Green function of the Stokes equations, called Oseen tensor. On the basis of such concepts, Gray and Hancock [19] suggested to use $L = 2\lambda$ in the local friction coefficients per unit length of Eq. (9.1), where λ is the wavelength of the wave traveling along a flagellum, whereas Lighthill suggested for a "suboptimal representation" $L = 0.18\lambda$ using a careful analysis of slender body theory [40]. However, in general, this theory is somehow cumbersome.

In the physics community, the growing interest in the topic developed a decade ago. Motivated by Purcells work on swimmers and by experiments on driven microfilaments such as actin [41, 42], Wiggins and Goldstein explored the *elastohydrodynamics* of long elastic filaments. They combined the bending elasticity of an elastic rod with the resistive force theory to arrive at a hyperdiffusion equation for a filament in the limit of small bending [43] (see also Section 9.2.2). In this limit, only $\gamma_\perp$ is relevant. Going beyond pure planar undulations, rotating filaments were also considered, where twist deformations besides bending become important [44, 45]. In parallel, Camalet, Jülicher, and Prost studied the self-organized beating of a flagellum, for instance, of a sperm cell, based on elastohydrodynamics and a modeling of the collective motion of internal motors that drive the flagellum [21, 22]. More recent work contains a detailed investigation of swimming with actuated filaments [46, 47]. Finally, elastohydrodynamics has been extended to viscoelastic fluids [48] and its implications for propulsion are addressed in Ref. [48] also using an elastic sheet instead of a filament [49].

In parallel to the works just reported that use the continuum description of an elastic filament, an alternative approach was developed. It approximates the filament by rigidly connected spheres and writes down dynamic equations for the spheres by balancing the bending force derived from a discrete version of the bending free energy with the anisotropic friction force from resistive force theory [50–52]. The authors investigated the dynamics of driven filaments. They compared it with the shape of a sperm cell's flagellum [52] and also determined the velocity and efficiency of a one-armed swimmer [51]. The authors extended their approach by including hydroynamic interactions between the spheres using the Oseen tensor and studied the shapes of single and hydrodynamically interacting filaments during sedimentation [53, 54]. An experimental realization with electrophoretically driven microtubules has appeared recently [55].

A modified and extended approach uses a bead-spring model with bending elasticity and takes into account hydrodynamic interactions via Rotne–Prager mobilities for the spheres [56–58]. Configurations of a rotating nanorod were investigated [59], later confirmed by experiments [60, 61], and then beating grafted filaments for pumping fluids were simulated [24]. Implementing the full elasticity theory for a rotating rod including twist deformations revealed a discontinuous shape transition [62]. The elasticity theory of a helical rod was used for studying kink-pair propagation in the propulsion of the Spiroplasma bacterium [63] and for investigating stretching induced transitions between polymorphs of a bacterial flagellum [64].

Finally, we mention a new development in treating the hydrodynamic properties of elastic filaments. A recently introduced particle-based mesoscopic simulation method, called multiparticle collision dynamics [65, 66], was used to study the hydrodynamic interaction and cooperation of sperm cells embedded in a two-dimensional fluid [67].

9.1.3
A Biomimetic Flagellum

In an effort to mimic nature's successful strategies for propulsion and fluid transport, an artificial cilium or flagellum actuated by an external magnetic field has been constructed recently on the basis of a superparamagnetic elastic filament. The filament is made of superparamagnetic colloidal particles of micron size. A static external magnetic field induces dipoles in the colloids so that they form a chain. In the gaps between the charged colloids, chemical linkers such as double-stranded DNA are attached to the particles and an elastic filament resisting bending and stretching is formed [68–70] (for similar systems, see Refs [71, 72]). Dreyfus *et al.* attached the superparamagnetic filament to a red blood cell and thereby introduced the first artificial microswimmer [4]. While the bending waves moving along the flagellum of a sperm cell are generated by the collective motion of internal molecular motors, an oscillating external magnetic field induces a nonreciprocal beating motion of the superparamagnetic filament. It is therefore able to move the attached red blood cell forward.

The modeling of the dynamics of the superparamagnetic filament followed two strategies. On the one hand, the elastohydrodynamics of an elastic rod was used supplemented by a continuum version for the interaction of the magnetic field-induced dipoles [4, 73, 74] or a simpler description for the interaction with the magnetic field [75–78]. The authors are able to describe the dynamics of the filament [73, 75–78] and the velocity curve of the artificial swimmer [4]. On the other hand, an alternative modeling based on the bead-spring configuration with bending elasticity was employed [57]. It fully takes into account the hydrodynamic and dipole–dipole interactions. A detailed investigation of the artificial swimmer was performed [57], and recent studies explore how the artificial cilium can be employed for fluid transport by attaching it to a surface [79, 92]. We will summarize the results of these studies in Section 9.3.

9.2
Elastohydrodynamics of a Filament

9.2.1
Theory of Elasticity of an Elastic Rod

The theory of elasticity of an elastic rod [80] also called worm-like chain model [81] introduces a bending free energy for an elastic filament, which is described by the space curve $\boldsymbol{r}(s)$. The arc length s ranges from zero to the filament length L. From the tangent vector $\hat{\boldsymbol{t}} = \mathrm{d}\boldsymbol{r}(s)/\mathrm{d}s$ with $|\hat{\boldsymbol{t}}| = 1$, one derives the local curvature $1/R = |\mathrm{d}\hat{\boldsymbol{t}}/\mathrm{d}s|$ of the space curve. Its inverse R denotes the curvature radius. The bending free energy of an elastic filament in harmonic approximation then reads

$$H^{\mathrm{B}} = \frac{1}{2} k_{\mathrm{B}} T l_{\mathrm{p}} \int_0^L \left(\frac{\mathrm{d}\hat{\boldsymbol{t}}}{\mathrm{d}s}\right)^2 \mathrm{d}s. \tag{9.2}$$

The bending constant $k_{\mathrm{B}} T l_{\mathrm{p}}$ contains the thermal energy $k_{\mathrm{B}} T$ and the so-called persistence length l_{p} meaning that thermal energy is able to bend the filament with a curvature radius $R = l_{\mathrm{p}}$. In addition, in thermal equilibrium, correlations of the tangent vectors along the filament decay exponentially

$$\langle \hat{\boldsymbol{t}}(0) \cdot \hat{\boldsymbol{t}}(s) \rangle = \mathrm{e}^{-s/l_{\mathrm{p}}}, \tag{9.3}$$

where "·" denotes the scalar product. If $L \ll l_{\mathrm{p}}$, the tangent vectors are all parallel to each other. Therefore, the filament hardly bends and is rigid. On the other hand, in the case of $L \gg l_{\mathrm{p}}$ correlations between the tangent vectors are lost beyond a distance of the order of the persistence length. So, the filament is flexible. Finally, for $L \approx l_{\mathrm{p}}$ the filament is considered semiflexible.

Bending forces within the filament are derived from the variation δF^{B} of the free energy that occurs when the space curve $\boldsymbol{r}(s)$ is varied by small displacements $\delta\boldsymbol{r}(s)$ along the filament. With the help of two partial integration, one arrives at

$$\delta H^{\mathrm{B}} = k_{\mathrm{B}} T l_{\mathrm{p}} \int_0^L \frac{\mathrm{d}^4 \boldsymbol{r}}{\mathrm{d}s^4} \cdot \delta\boldsymbol{r}\, \mathrm{d}s + k_{\mathrm{B}} T l_{\mathrm{p}} \left(\frac{\mathrm{d}^2 \boldsymbol{r}}{\mathrm{d}s^2} \cdot \frac{\mathrm{d}}{\mathrm{d}s} \delta\boldsymbol{r} - \frac{\mathrm{d}^3 \boldsymbol{r}}{\mathrm{d}s^3} \cdot \delta\boldsymbol{r} \right) \Bigg|_0^L. \tag{9.4}$$

The first integrand on the right-hand side is the functional derivative of H^{B} and its negative is interpreted as bending force:

$$\boldsymbol{F}^{\mathrm{B}}(s) = -\frac{\delta H^{\mathrm{B}}}{\delta \boldsymbol{r}(s)} = -k_{\mathrm{B}} T l_{\mathrm{p}} \frac{\mathrm{d}^4 \boldsymbol{r}(s)}{\mathrm{d}s^4}. \tag{9.5}$$

The "surface term" will be used in Section 9.2.2.

Normally, the filament is considered as inextensible. So, during the temporal evolution of the filament described by the space curve $\boldsymbol{r}(s, t)$, one always has to fulfill $|\mathrm{d}\boldsymbol{r}/\mathrm{d}s| = 1$. Here, s always indicates the same material point of the filament and always ranges from zero to L. This corresponds to the Lagrangian formulation of the elasticity theory for solid bodies. The constraint $|\mathrm{d}\boldsymbol{r}/\mathrm{d}s| = 1$ is formally included into the variation of the free energy by adding the term

$$H^{\mathrm{C}} = \frac{1}{2} \int_0^L \lambda(s) \left(\frac{\mathrm{d}\boldsymbol{r}}{\mathrm{d}s}\right)^2 \mathrm{d}s \tag{9.6}$$

to the bending free energy, where $\lambda(s)$ is the local Lagrange parameter. This term can be interpreted as a stretching free energy that describes the stress

$$\boldsymbol{\tau}(s) = \lambda(s) \frac{\mathrm{d}\boldsymbol{r}}{\mathrm{d}s} \tag{9.7}$$

necessary to guarantee the constraint $|\mathrm{d}\boldsymbol{r}/\mathrm{d}s| = 1$. The variation of H^{C} gives

$$\delta H^{C} = -\int_0^L \frac{d}{ds}\left(\lambda(s)\frac{d\boldsymbol{r}}{ds}\right)\cdot\delta\boldsymbol{r}\,ds + \lambda(s)\frac{d\boldsymbol{r}}{ds}\cdot\delta\boldsymbol{r}\Big|_0^L, \tag{9.8}$$

from which we read the functional derivative or stretching force

$$F^{C}(s) = -\frac{\delta H^{S}}{\delta\boldsymbol{r}(s)} = \frac{d}{ds}\boldsymbol{\tau}(s), \tag{9.9}$$

where $\boldsymbol{\tau}(s)$ is defined in Eq. (9.7).

In numerical simulations, one needs a discretized version of the free energy. We describe the space curve of the filament by $N+1$ space points $\boldsymbol{r}_i$ $(i = 0, \ldots, N)$ with an equilibrium distance l_0 and introduce the normalized tangent vectors via $\hat{\boldsymbol{t}}_i = (\boldsymbol{r}_i - \boldsymbol{r}_{i-1})/l_i$, where $l_i = |\boldsymbol{r}_i - \boldsymbol{r}_{i-1}|$ is the momentary distance between the space points. Instead of implementing the inextensibility constraint, we introduce a concrete stretching free energy in harmonic approximation,

$$H^{S} = \frac{1}{2}k\sum_{i=1}^{N}(l_i - l_0)^2, \tag{9.10}$$

where k is the spring constant of the springs connecting the space points. Furthermore, replacing $d\hat{\boldsymbol{t}}/ds$ in Eq. (9.2) by $(\hat{\boldsymbol{t}}_{i+1} - \hat{\boldsymbol{t}}_i)/l_0$, one arrives at the discretized version of the bending free energy:

$$H^{B} = \frac{k_B T l_p}{l_0}\sum_{i=1}^{N-1}(1 - \hat{\boldsymbol{t}}_{i+1}\cdot\hat{\boldsymbol{t}}_i). \tag{9.11}$$

Finally, bending and stretching forces on point $\boldsymbol{r}_i$ of the filament follow from

$$\boldsymbol{F}_i^{S} = -\nabla_{\boldsymbol{r}_i}H^{S} \quad \text{and} \quad \boldsymbol{F}_i^{B} = -\nabla_{\boldsymbol{r}_i}H^{B}, \tag{9.12}$$

where $\nabla_{\boldsymbol{r}_i}$ is the nabla operator with respect to $\boldsymbol{r}_i$. Explicit expressions for $\boldsymbol{F}_i^{S}$ and $\boldsymbol{F}_i^{B}$ are given in Ref. [57].

9.2.2 Hydrodynamic Friction of a Filament: Resistive Force Theory

Dynamic equations for the elastic filament need a proper account of the frictional forces with the surrounding fluid. Within resistive force theory, they are proportional to the local velocities of the filaments. Furthermore, in the low Reynolds number limit they have to balance the bending and stretching forces introduced in Eqs. (9.5) and (9.9). Using the two friction coefficients per unit length, $\gamma_{\parallel}$ and $\gamma_{\perp}$, one thus arrives at the highly nonlinear dynamic equations:

$$\begin{aligned}\left[\gamma_{\parallel}\hat{\boldsymbol{t}}\otimes\hat{\boldsymbol{t}} + \gamma_{\perp}(\mathbf{1} - \hat{\boldsymbol{t}}\otimes\hat{\boldsymbol{t}})\right]\frac{d\boldsymbol{r}}{dt} &= -\frac{\delta H^{B}}{\delta\boldsymbol{r}(s)} + \frac{d}{ds}\boldsymbol{\tau}(s)\\ &= -k_B T l_p\frac{d^4\boldsymbol{r}(s)}{ds^4} + \frac{d}{ds}\left(\lambda(s)\frac{d\boldsymbol{r}}{ds}\right).\end{aligned} \tag{9.13}$$

The dyadic product $\hat{\boldsymbol{t}} \otimes \hat{\boldsymbol{t}}$ projects the velocity $\mathrm{d}\boldsymbol{r}/\mathrm{d}t$ on its component parallel to $\hat{\boldsymbol{t}}$ $[(\hat{\boldsymbol{t}} \otimes \hat{\boldsymbol{t}})\boldsymbol{a} = \hat{\boldsymbol{t}}\hat{\boldsymbol{t}} \cdot \boldsymbol{a}]$ and the projector $\mathbf{1} - \hat{\boldsymbol{t}} \otimes \hat{\boldsymbol{t}}$ extracts the component perpendicular to $\hat{\boldsymbol{t}}$. The dynamic Eqs. (9.13) have to be supplemented by the inextensibility constraint $|\mathrm{d}\boldsymbol{r}/\mathrm{d}s| = 1$. Assuming free boundary conditions for the ends of the filament, which means that in the "surface terms" of Eqs. (9.4) and (9.8) the variations $\delta\boldsymbol{r}$ and $\delta\mathrm{d}\boldsymbol{r}/\mathrm{d}s$ are arbitrary, results in the conditions

$$-k_B T l_p \frac{\mathrm{d}^3\boldsymbol{r}}{\mathrm{d}s^3} + \lambda(s)\frac{\mathrm{d}\boldsymbol{r}}{\mathrm{d}s} \quad \text{and} \quad \frac{\mathrm{d}^2\boldsymbol{r}}{\mathrm{d}s^2} = 0. \tag{9.14}$$

The first and second conditions mean that the respective external forces and torques at the free ends of the filament vanish.

A linearized version of Eq. (9.13) follows with the help of the Monge representation. The undistorted ground state of the filament is represented by $\boldsymbol{r}_0(x) = x\boldsymbol{e}_x$, where x ranges from zero to L and $\boldsymbol{e}_x$ is a unit vector along the x-axis. For small deviations from the straight filament, the inextensibility constraint is negligible and the space curve of the filament can be described by $\boldsymbol{r}(x,t) = x\boldsymbol{e}_x + y(x,t)\boldsymbol{e}_y$. Inserting this parameterization into Eq. (9.13) and linearizing in $y(x,t)$ gives the so-called hyperdiffusion equation:

$$\frac{\mathrm{d}y}{\mathrm{d}t} = -\frac{k_B T l_p}{\gamma_\perp}\frac{\mathrm{d}^4 y}{\mathrm{d}x^4}. \tag{9.15}$$

We solve it with the ansatz $y(x,t) = y_0 \exp[i(kx - \omega t)]$ and the boundary conditions $y(x \to \infty, t) = 0$ and $\mathrm{d}^2 y(0,t)/\mathrm{d}x^2 = 0$; here the latter means vanishing torque. The solution reads

$$y(x,t) = \frac{y_0}{2}\left[\mathrm{e}^{-c_2 x/\xi}\mathrm{e}^{i(c_1 x/\xi - \omega t)} + \mathrm{e}^{-c_1 x/\xi}\mathrm{e}^{-i(c_2 x/\xi + \omega t)}\right], \tag{9.16}$$

where we introduced the penetration length

$$\xi = \left(\frac{k_B T l_p}{\gamma_\perp \omega}\right)^{1/4} \tag{9.17}$$

and the constants $c_1 = \cos(\pi/8)$ and $c_2 = \sin(\pi/8)$. Equation (9.16) is a superposition of two damped waves propagating, respectively, in positive and negative x-direction and the point at $x = 0$ oscillates in time, $y(0,t) = y_0 \exp(-i\omega t)$. So, solution (9.16) represents an infinitely long filament whose one end is oscillated in time and the penetration length gives the distance along which this oscillation penetrates into the filament.

To characterize the elastohydrodynamic properties of an elastic filament of length L, we define the characteristic number

$$S_p = \frac{L}{\xi} = \left(\frac{\gamma_\perp \omega L^4}{k_B T l_p}\right)^{1/4}, \tag{9.18}$$

also termed *sperm number* [51, 52]. It compares the frictional force acting on the filament with the bending force. Introducing reduced spatial and temporal coordinates, $\bar{y} = y/L$, $\bar{x} = x/L$, and $\bar{t} = \omega t$, the hyperdiffusion equation (9.15) becomes

$$\frac{d\bar{y}}{d\bar{t}} = -Sp^{-4}\frac{d^4\bar{y}}{d\bar{x}^4}. \tag{9.19}$$

The reduced version demonstrates that the sperm number Sp completely determines the behavior of an elastic filament. For $Sp \ll 1$, for example, for sufficiently low frequencies ω, it behaves as a rigid rod meaning that frictional forces are not large enough to bend the filament. For $Sp \approx 1$ hydrodynamic friction can bend the whole filament and for $Sp \gg 1$ the forced oscillation is visible only along the length ξ.

9.2.3
Hydrodynamic Friction of a Filament: Method of Hydrodynamic Interaction

Here, the filament is modeled by a bead-spring configuration that additionally resists bending like a worm-like chain [81]. Thus, each bead in the filament experiences a force caused by stretching and bending as described in Eq. (9.12). This offers an approach to treat hydrodynamic friction of the filament with the surrounding fluid beyond resistive force theory. Each bead moving under the influence of a force initiates a flow field that influences the motion of other beads and vice versa, so a complicated many-body problem arises. At low Reynolds number the flow field $\boldsymbol{u}(\boldsymbol{r}, t)$ around the spheres is described by the Stokes equations and the incompressibility condition:

$$\mathbf{0} = -\nabla p + \eta\nabla^2\boldsymbol{u} \quad \text{and} \quad \text{div}\, \upsilon = \mathbf{0}, \tag{9.20}$$

where p is pressure. In addition, the no-slip boundary condition on bead surfaces is assumed. The Stokes equations are linear in the flow field; hence, the velocities υ_i of the beads are proportional to the forces $\boldsymbol{F}_j$ acting on them and the beads obey the following equations of motion [14]:

$$\upsilon_i = \sum_j \boldsymbol{\mu}_{ij}\boldsymbol{F}_j, \tag{9.21}$$

where $\boldsymbol{F}_j = \boldsymbol{F}_j^{\mathrm{S}} + \boldsymbol{F}_j^{\mathrm{B}}$ is the sum of stretching and bending forces introduced in Eq. (9.12) for the discretized filament. The important quantities are the mobilities $\boldsymbol{\mu}_{ij}$. In general, they depend on all the coordinates $\boldsymbol{r}_i$ of the beads. If the mobilities are known, Eq. (9.21) can numerically be integrated, for example, by the simple Euler method.

The Green function of the Stokes equations in an unbounded fluid is the Oseen tensor

$$\boldsymbol{O}(\boldsymbol{r}) = \frac{1}{8\pi\eta r}\left(\mathbf{1} + \frac{\boldsymbol{r}\otimes\boldsymbol{r}}{r^2}\right). \tag{9.22}$$

It provides the flow field $\boldsymbol{u}(\boldsymbol{r}) = \int \boldsymbol{O}(\boldsymbol{r}-\boldsymbol{r}')\boldsymbol{b}(\boldsymbol{r}')\, d^3\boldsymbol{r}'$ for body forces $\boldsymbol{b}(\boldsymbol{r}')$ acting on the fluid. In particular, the flow field of a point force $\boldsymbol{F}_0$ located at $\boldsymbol{r}'$ is $\boldsymbol{O}(\boldsymbol{r}-\boldsymbol{r}')\boldsymbol{F}_0$. We already introduced it as stokeslet in Section 9.1.2. It is realized in the far field of a spherical particle dragged through the fluid by the force $\boldsymbol{F}_0$. The complete flow field initiated by the particle with radius a at position $\boldsymbol{r}_\mathrm{p}$ reads

$$\boldsymbol{u}(\boldsymbol{r}) = \left(1 + \frac{1}{6}a^2\nabla_\mathrm{p}^2\right)\mathbf{O}(\boldsymbol{r}-\boldsymbol{r}_\mathrm{p})\boldsymbol{F}_0, \tag{9.23}$$

where $\boldsymbol{v}_\mathrm{p} = \mu_0\boldsymbol{F}_0$ is the particle velocity and $\mu_0 = 1/(6\pi\eta a)$ the Stokes mobility of an isolated spherical particle. For large distances, the particles can be treated as pointlike. So, the force $\boldsymbol{F}_j$ acting on particle j initiates a stokeslet, the velocity of which at space point $\boldsymbol{r}_i$ is taken over by particle i, $\boldsymbol{v}_i = \boldsymbol{O}(\boldsymbol{r}_i-\boldsymbol{r}_j)\boldsymbol{F}_j$. The cross-mobility for pointlike particles therefore is $\boldsymbol{\mu}_{ij} = \boldsymbol{O}(\boldsymbol{r}_i-\boldsymbol{r}_j)$. Taking into account the finite size of the moving particles leads to corrections. First of all, the flow field of particle j (see Eq. (9.23)) includes corrections from fulfilling the no-slip boundary condition at its surface. Second, according to Faxén's theorem, particle i placed into the flow field $\boldsymbol{u}_j(\boldsymbol{r})$ of particle j possesses the velocity [14]

$$\boldsymbol{v}_i = \mu_0\boldsymbol{F}_i + \left(1 + \frac{1}{6}a^2\nabla_i^2\right)\boldsymbol{u}_j(\boldsymbol{r}_i). \tag{9.24}$$

Note that Faxén's theorem is valid for any flow field $\boldsymbol{u}_j(\boldsymbol{r})$ satisfying the Stokes equations. Using Eq. (9.23) in Eq. (9.24) and comparing with Eq. (9.21), one obtains the Rotne–Prager mobilities as an expansion up to terms $1/r_{ij}^3$,

$$\boldsymbol{\mu}_{ii} = \mu_0\mathbf{1}, \quad \mu_0 = 1/(6\pi\eta a) \tag{9.25}$$

$$\begin{aligned}\boldsymbol{\mu}_{ij} &= \left(1 + \frac{1}{6}a^2\nabla_i^2\right)\left(1 + \frac{1}{6}a^2\nabla_j^2\right)\boldsymbol{O}(\boldsymbol{r}_i-\boldsymbol{r}_j)\\ &= \mu_0\left[\frac{3}{4}\frac{a}{r_{ij}}(\mathbf{1} + \hat{\boldsymbol{r}}_{ij}\otimes\hat{\boldsymbol{r}}_{ij}) + \frac{1}{2}\left(\frac{a}{r_{ij}}\right)^3(\mathbf{1}-3\hat{\boldsymbol{r}}_{ij}\otimes\hat{\boldsymbol{r}}_{ij})\right], \quad i \neq j,\end{aligned} \tag{9.26}$$

where $\boldsymbol{r}_{ij} = \boldsymbol{r}_i-\boldsymbol{r}_j$ and $\hat{\boldsymbol{r}}_{ij} = \boldsymbol{r}_{ij}/r_{ij}$. Higher-order corrections to the Rotne–Prager mobilities arise since the flow field initiated by particle i acts back on particle j. In addition, many-body interactions due to the presence of additional particles have to be taken into account. The method of induced forces provides a systematic expansion of the mobilities in $1/r_{ij}$ [82]. For particles in close contact, lubrication theory has to be used for determining the mobilities [83, 84]. A program was developed that incorporates all these effects and calculates mobilities for a given cluster of spherical particles [85]. We checked that the Rotne–Prager approximation is in good agreement with the more exact values of the mobilities down to distances of $3a$. Finally, we mention that Eq. (9.26) is generalized to particles with different radii a_i and a_j when $\mu_0 a$ and a^2 are replaced, respectively, by $1/(6\pi\eta)$ and $(a_i^2 + a_j^2)/2$.

Close to a planar surface with no-slip boundary condition, the traditional Rotne–Prager mobilities can no longer be employed. The velocity and pressure fields of a point force for this boundary condition were first derived by Lorentz more than 100 years ago [86]. Blake put these results into a modern form replacing the Oseen tensor by the appropriate Green function, now called Blake tensor [87]. The condition of a vanishing fluid velocity field on an infinitely extended plane is satisfied with the help of appropriate mirror images, similar to the image charge approach used in electrostatics. However, in contrast to electrostatics, where it suffices to simply mirror the charge distribution, the hydrodynamic image system is more complicated due to the vectorial argument of the Stokes equations and the incompressibility condition compared to the Poisson equation. Therefore, the so-called stresslet and source-dipole contributions are needed in addition to the stokeslet of the mirrored point disturbance (also called anti-stokeslet). This yields Blake's tensor,

$$\begin{aligned} \boldsymbol{G}^{\mathrm{Blake}}(\boldsymbol{r},\boldsymbol{r}') &= \boldsymbol{O}(\boldsymbol{r}-\boldsymbol{r}') + \boldsymbol{G}^{\mathrm{im}}(\boldsymbol{r},\bar{\boldsymbol{r}}') \\ &= \boldsymbol{O}(\boldsymbol{r}-\boldsymbol{r}') - \boldsymbol{O}(\boldsymbol{r}-\bar{\boldsymbol{r}}') + \delta\boldsymbol{G}^{\mathrm{im}}(\boldsymbol{r},\bar{\boldsymbol{r}}')\,, \end{aligned} \tag{9.27}$$

where $\bar{\boldsymbol{r}}'$ is the position of the anti-stokeslet source, that is, the stokeslet source at $\boldsymbol{r}'$ mirrored at the bounding plane, and $\delta\boldsymbol{G}^{\mathrm{im}}(\boldsymbol{r},\bar{\boldsymbol{r}}')$ denotes the source-dipole and stresslet contributions. Equation (9.23) for the flow field initiated by a spherical particle close to the bounding plane remains approximately valid when the Oseen tensor is replaced by the Blake tensor $\boldsymbol{G}^{\mathrm{Blake}}$. A non-uniform contribution to the force distribution on the particle surface is neglected. Faxén's theorem stated in Eq. (9.24) also applies if the flow field initiated by the image system $\boldsymbol{G}^{\mathrm{im}}$ of particle i is added to $\boldsymbol{u}_j(\boldsymbol{r}_i)$. Therefore, the cross mobilities $\boldsymbol{\mu}_{ij}$ in Rotne–Prager approximation are calculated as in Eq. (9.26) but with $\boldsymbol{O}(\boldsymbol{r}-\boldsymbol{r}')$ replaced by $\boldsymbol{G}^{\mathrm{Blake}}(\boldsymbol{r}-\boldsymbol{r}')$. A correction of the standard form has to be added to the self-mobilities $\boldsymbol{\mu}_{ii}$:

$$\boldsymbol{\mu}_{ii} = \mu_0\mathbf{1} + \left(1 + \frac{1}{6}a^2\nabla_{\bar{i}}^2\right)\left(1 + \frac{1}{6}a^2\nabla_i^2\right)\boldsymbol{G}^{\mathrm{im}}(\boldsymbol{r}_i,\bar{\boldsymbol{r}}_i), \tag{9.28}$$

where $\nabla_{\bar{c}i}$ means gradient with respect to the image coordinate $\bar{\boldsymbol{r}}_i$. Concrete formulas for the mobilities are given, for example, in Ref. [57].

9.3 A Biomimetic Flagellum and Cilium

In Section 9.1.3, we introduced a filament made of superparamagnetic colloidal particles that are linked to each other by double-stranded DNA. We have modeled this filament as a bead-spring configuration with bending elasticity as described in Section 9.2.3. In addition, we have to include now the forces on the beads due to the dipole–dipole interaction induced by the external magnetic field. Our model very well describes the constructed artificial swimmer [4] and allows to explore the filament's capacity for transporting fluid.

9.3.1 Details of the Modeling

All superparamagnetic beads with radius a and magnetic susceptibility χ subject to a homogeneous external magnetic field $\boldsymbol{B}$ develop a dipole moment with identical orientation and strength,

$$\boldsymbol{p} = \frac{4\pi a^3}{3\mu_0}\chi \boldsymbol{B}, \tag{9.29}$$

where $\mu_0 = 4\pi \times 10^{-7}\,\text{N/A}^2$ is the permeability of free space. In Eq. (9.29), we neglect that the local magnetic field determining $\boldsymbol{p}$ differs from the applied field due to the dipolar field from neighboring particles in the chain. The dipoles of the beads then give rise to the total dipole–dipole interaction energy

$$H^{\mathrm{D}} = \frac{4\pi a^6}{9\mu_0}(\chi B)^2 \sum_{i<j} \frac{1-3(\hat{\boldsymbol{p}}\cdot\hat{\boldsymbol{r}}_{ij})}{r_{ij}^3}, \tag{9.30}$$

where $r_{ij} = |\boldsymbol{r}_j - \boldsymbol{r}_i|$ and $\hat{\boldsymbol{r}}_{ij} = (\boldsymbol{r}_j - \boldsymbol{r}_i)/r_{ij}$. From this energy, a dipolar force acting on bead i is calculated as in Eq. (9.12), $\boldsymbol{F}_i^{\mathrm{D}} = -\nabla_{\boldsymbol{r}_i} H^{\mathrm{D}}$, and the total force in the dynamic Eq. (9.21) for particle velocity $\boldsymbol{v}_i$ reads $\boldsymbol{F}_j = \boldsymbol{F}_j^{\mathrm{S}} + \boldsymbol{F}_j^{\mathrm{B}} + \boldsymbol{F}_j^{\mathrm{D}}$.

The dynamic Eq. (9.21) for particle velocity $\boldsymbol{v}_i$ can be written in reduced form [57]. It shows that the dynamics of the superparamagnetic filament depends on three characteristic numbers. One of them is the sperm number

$$Sp = \left(\frac{6\pi\eta \frac{a}{l_0} \omega L^4}{k_{\mathrm{B}} T l_{\mathrm{p}}}\right)^{1/4}, \tag{9.31}$$

already introduced in Eq. (9.18). Note, when hydrodynamic interactions are neglected, the local friction coefficient per unit length of the bead-spring chain reads $\gamma_\perp = 6\pi\eta a/l_0$. A second characteristic number is the reduced magnetic field strength,

$$B_{\mathrm{s}} = \frac{2\pi^{1/2} a^3 \chi N}{3\mu_0^{1/2} l_0 (k_{\mathrm{B}} T l_{\mathrm{p}})^{1/2}} B. \tag{9.32}$$

It determines the influence of the external magnetic field on the superparamagnetic filament. The number B_{s}^2 compares dipolar to bending forces and it is proportional to the magnetoelastic number introduced in Refs [4, 73, 74]. An alternative dimensionless number for characterizing the influence of the magnetic field is the Mason number introduced in the literature on magnetorheological suspensions [75, 88],

$$Ma = Sp^4/B_{\mathrm{s}}^2. \tag{9.33}$$

It is the ratio of frictional to magnetic forces and determines the behavior of the superparamagnetic filament when magnetic forces dominate over bending forces. Finally, a reduced spring constant

$$k_{\mathrm{s}} = \frac{N^2 l_0^3}{k_{\mathrm{B}} T l_{\mathrm{p}}} k \tag{9.34}$$

appears. The superparamagnetic filament is not strictly inextensible, so k_{s} is another material parameter. In our modeling, we only introduced it for numerical reasons and always chose a sufficiently large k_{s} to keep overall length fluctuations of the filament well below 10%. All simulation results presented in the following two sections were obtained with realistic parameter values close to experiments [57, 79].

9.3.2 Microscopic Artificial Swimmer

In order to model the microscopic artificial swimmer of Ref. [4], we attach a spherical load particle of radius a_0, typically several bead radii a large, to the bead-spring chain. The load particle also experiences hydrodynamic interactions with the filament's beads. In our simulations, the filament is actuated with a magnetic field $\boldsymbol{B}(t)$ whose strength is constant but whose direction oscillates about the z-axis with an angle $\varphi(t) = \varphi_{\max}\sin(\omega t)$ (see Figure 9.1). Note that this time protocol differs from the one used in Ref. [4]). The swimmer moves with an average velocity $\bar{\upsilon}$ along the z-axis. However, in contrast to spermatozoa, where the head is pushed forward by damped waves traveling from the head to the tail [2, 18–20], the superparamagnetic filament drags the passive load behind itself by performing a sort of paddle motion with its free end as indicated in Figure 9.1.

We discuss the performance of the swimmer by studying two quantities. First, we determine its average speed $\bar{\upsilon}$ by averaging the velocity υ_0 of the load particle over one actuation cycle:

$$\bar{\upsilon} = \frac{1}{T}\int_{\tau}^{\tau+T} \upsilon_0(t)\,\mathrm{d}t, \qquad (9.35)$$

where $T = 2\pi/\omega$. Second, we introduce the efficiency of the swimmer in transporting a load by comparing the energy dissipated by the load particle, when moved uniformly with velocity $\bar{\upsilon}$, with the total energy dissipated by the swimmer:

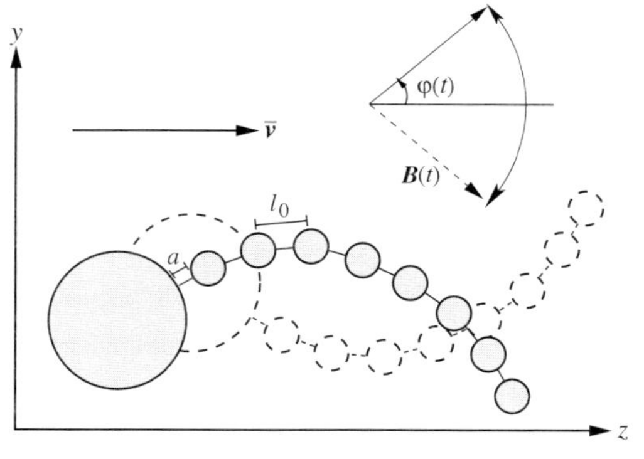

Figure 9.1 To model the artificial swimmer, a larger load particle is attached to the filament. It is actuated by a magnetic field whose direction oscillates about the z-axis with $\varphi(t) = \varphi_{\max}\sin(\omega t)$. The two configurations of the swimmer are schematic drawings for positive and negative $\varphi(t)$, respectively. (Reprinted with permission from Ref. [57]. Copyright (2006) by the American Physical Society.)

$$\xi = \frac{6\pi\eta a_0 \bar{v}^2}{\overline{\sum_{i=0}^{N} \boldsymbol{F}_i \cdot \boldsymbol{v}_i}}. \tag{9.36}$$

Here, the bar in the denominator means average over one actuation cycle. The efficiency ξ indicates how much energy from the total energy used to actuate the swimmer is employed to move the load particle forward with velocity $\bar{\boldsymbol{v}} = |\bar{\boldsymbol{v}}|$.

Figure 9.2 demonstrates mean velocity $\bar{\boldsymbol{v}} = |\bar{\boldsymbol{v}}|$ and efficiency ξ as a function of the sperm number Sp. The reduced velocity $\bar{v}/(L\omega)$ in Figure 9.2a exhibits the same behavior as reported in Refs [51, 52], where the elastic filament is driven by an oscillating torque acting on one of its ends. For small sperm numbers around $Sp = 3$,

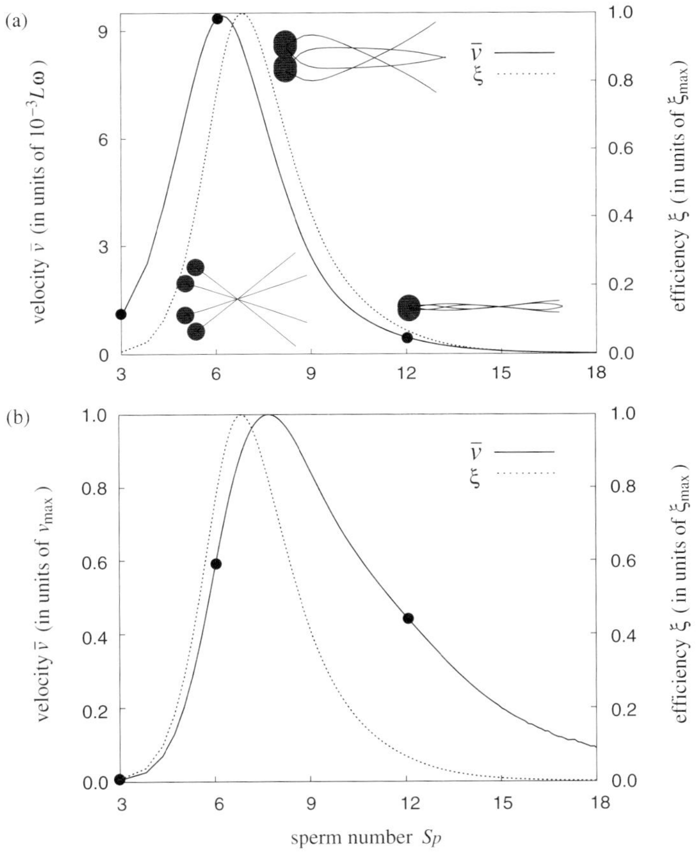

Figure 9.2 Swimming velocity $\bar{v}$ and efficiency ξ in units of $\xi_{max} = 1.58 \cdot 10^{-3}$ as a function of sperm number Sp for reduced magnetic-field strength $B_s = 5.76$. (a) Reduced velocity $\bar{v}/(L\omega)$, (b) absolute velocity $\bar{v}$ in units of $v_{max} = 5.56 \cdot 10^{-5}$m/s. The insets show several snapshots of the filament's configuration at $Sp = 3, 6$, and 12, respectively, indicated by the dots. (Reprinted with permission from Ref. [57]. Copyright (2006) by the American Physical Society.)

the reduced velocity is small since the superparamagnetic filament behaves nearly like a rigid rod, as illustrated by the snapshots of the filament on the lower left-hand side of the figure. The oscillating motion of a rigid rod is reciprocal and therefore does not produce a directed motion of the swimmer. Increasing the actuating frequency ω increases the sperm number and speeds up the artificial swimmer to the maximum value of $\bar{v}/(L\omega)$ at around $Sp = 6$. Here, the enhanced friction with the surrounding fluid is able to bend the whole filament (see upper snapshots) and obviously promotes a high swimming velocity. Further increase in Sp leads to a decrease in $\bar{v}/(L\omega)$; due to the strong friction with the fluid the whole filament cannot follow the magnetic field and only a small wiggling of its free end remains as expected for a penetration depth $\xi \ll L$ (see Eq. (9.17)). The efficiency as a function of Sp exhibits a similar behavior as $\bar{v}/(L\omega)$: oscillating a rigid rod (small Sp) or fast wiggling of the filament (large Sp) dissipates energy but does not produce an effective motion. So, one expects a maximum of ξ close to the maximum of $\bar{v}/(L\omega)$ since ξ is determined by $\bar{v}^2$. Note that the swimmer possesses only a small absolute efficiency of around 10^{-3} since a large amount of energy is dissipated by the motion of the filament. The shape of the velocity curve changes when absolute velocities are plotted (see Figure 9.2b). At $Sp = 3$, the absolute velocity is nearly zero and the maximum is shifted to a larger value around $Sp = 7.5$. Interestingly, the absolute velocities of the oscillating filaments at $Sp = 6$ and 12 are similar in contrast to the rescaled velocities in Figure 9.2a. One therefore concludes that the absolute swimming velocity is determined by two factors: (1) the shape of the oscillating filament, where bending the whole filament favors large velocities, and (2) the oscillation frequency. Owing to the influence of the oscillation frequency, the absolute velocity $\bar{v}$ only slowly decreases with increasing Sp. A comparison with the narrow maximum of ξ, however, shows that at large Sp most of the energy is dissipated in the small wiggling motion of the filament. So, operating the artificial swimmer at around $Sp = 7$ between the two close maxima ensures highest swimming velocities with very efficient energy consumption. Finally, we note that with increasing B_s the maxima of both the velocity and the efficiency curves move to larger sperm numbers or frequencies. We understand this since larger magnetic fields mean stronger alignment of the dipoles and, therefore, larger resistance to bending.

We expect the swimming velocity $\bar{v}$ to depend on the size or the radius a_0 of the load particle and therefore plot its absolute value as a function of sperm number Sp and radius a_0 for constant B_s (see Figure 9.3a). A pronounced maximum at $Sp = 8$ and $a_0 \approx 3a$ exists, indicated by a filled circle. The velocity $\bar{v}$ decreases for large a_0 since the friction coefficient of the load particle increases and therefore resists efficient transport by the oscillating filament. At $a_0 = a$, we expect zero velocity due to symmetry. However, the load particle is not superparamagnetic and a small asymmetry remains, as observed in Figure 9.3a. Figure 9.3b shows efficiency ξ as a function of Sp and a_0. The absolute maximum, indicated by an open circle, is at $Sp = 6.6$ and $a_0 \approx 5a$. For comparison, the location of the maximum of the swimming velocity is shown by the filled circle. So, to operate the swimmer one has to choose a compromise between the largest swimming velocity and the best efficiency. Clearly, our analysis shows that operating the swimmer optimally also

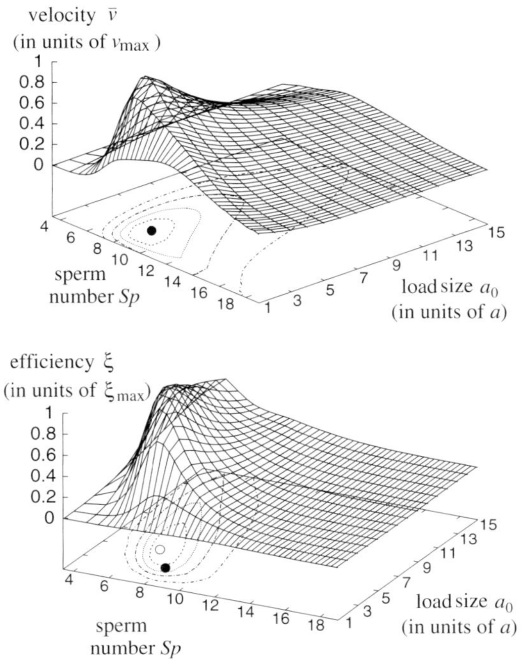

Figure 9.3 Swimming velocity $\bar{v}$ (in units of $v_{max} = 7.31 \cdot 10^{-5}$ m/s) and efficiency ξ (in units of $\xi_{max} = 1.54 \cdot 10^{-3}$) as a function of sperm number *Sp* and load size a_0 in units of *a*. The filled and open circles indicate, respectively, the absolute maxima of $\bar{v}$ and ξ. $B_s = 5.76$. (Reprinted with permission from Ref. [57]. Copyright (2006) by the American Physical Society.)

needs the right choice of the size of the load particle, which we also expect to increase with B_s.

The studies presented so far were performed for angular amplitudes φ_{max} of the oscillating field between 40° and 60°. If this angle is increased further and for sufficiently large *Sp*, a symmetry breaking transition occurs and the swimmer does not move any longer along the *z*-axis [57].

We also applied the time protocol of Ref. [4] (i.e., a constant *z*-component and an oscillating *y*-component of the magnetic field) to actuate the one-armed swimmer. For one set of experimental data points of Ref. [4], a nearly quantitative agreement is documented in Figure 9.4. To achieve this, we had to rescale the actuating magnetic field by a numerical factor to account for the larger distance of the beads in our modeling and therefore to compensate for the weaker dipole interaction compared to the swimmer in Ref. [4]. Deviations between our simulations and the experimental results might be due to the fact that we use a spherical load particle compared to the oblate shape of the red blood cell in Ref. [4], that our modeling is performed in the bulk whereas in experiments the swimmer moved close to a surface, and that we neglect corrections to the actuating external field due to the induced dipole fields.

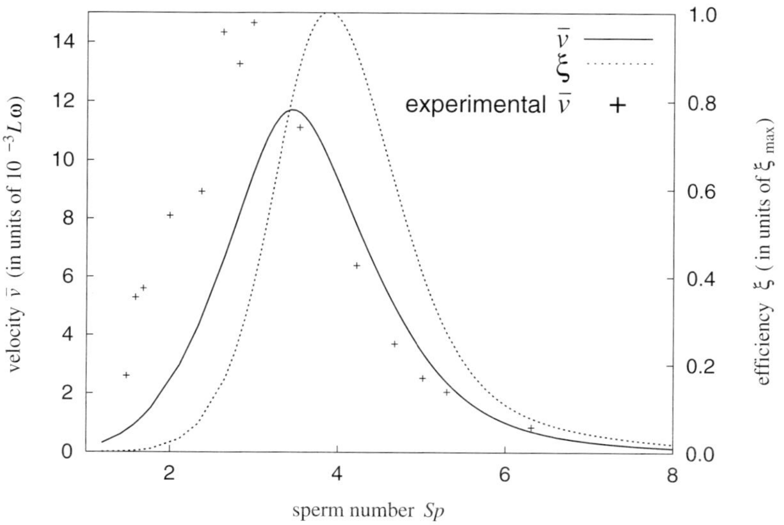

Figure 9.4 Swimming velocity $\bar{v}$ and efficiency ξ in units of $\xi_{max} = 1.8 \cdot 10^{-3}$ as a function of sperm number *Sp*. Parameters are the same as the one used for the red experimental data points of Figure 9.4 in Ref. [4]. The experimental points are included in this figure with the symbol +. (Reprinted with permission from Ref. [57]. Copyright (2006) by the American Physical Society.)

9.3.3
Fluid Transport

The superparamagnetic elastic filament when attached to a planar surface is an ideal system for investigating strategies for generating fluid transport at low Reynolds number. We have studied such strategies using either a two-dimensional [79] or a three-dimensional beating pattern [92]. They consist of a transport stroke where fluid is pumped and a recovery stroke where the filament is returned to its initial position with the goal to keep the amount of fluid moved opposite to the pumping direction as small as possible. An ideal transport stroke keeps the filament away from the surface and moves it perpendicular to its axis where friction with the surrounding fluid is large and as a result the amount of pumped fluid. On the other hand, the recovery stroke occurs preferentially close to the surface where fluid cannot be moved due to the no-slip boundary condition and/or is performed along the filament axis where hydrodynamic friction is smaller by an approximate factor of 2.

In order to compare different strategies with regard to their ability of pumping fluid, we introduce a measure that we call pumping performance. In principle, it integrates the time-averaged fluid flow (initiated by the beating filament) over a whole plane parallel to the bounding surface and situated above the beating filament. We assume that the bounding surface is defined by $z = 0$ and study the integrated fluid flow along the y-axis. It is determined by the laterally averaged Blake tensor [89],

$$\bar{G}(z, z') = \int dx\, dy G_{yy}^{\text{Blake}}(x, y, z, z') = \frac{\min(z, z')}{\eta}. \tag{9.37}$$

Introducing the distance z_i between the wall and bead i of the filament and the force F_{yi} acting on it in y-direction, the integrated flow $\mathcal{F}$ generated by all the beads is approximated by summing over all stokeslets:

$$\mathcal{F} = \frac{1}{\eta} \sum_i z_i F_{yi}. \tag{9.38}$$

Here, we already considered the case $z > z_i$ for all beads so that $\bar{G}(z, z_i) = z_i/\eta$. Contributions to $\mathcal{F}$ from the forces F_{xi} and F_{zi} vanish by symmetry. Note also that $\mathcal{F}$ does not depend on the position z of the integrated flow. The filament is actuated periodically in time and the time-averaged fluid flow amounts to

$$\bar{\mathcal{F}} = \frac{1}{T} \int_t^{t+T} \mathcal{F} dt', \tag{9.39}$$

where T is the period of one actuation cycle. Finally, we introduce the pumping performance as the unitless quantity

$$\xi = \bar{\mathcal{F}} / \bar{\mathcal{F}}^{\text{ref}}, \tag{9.40}$$

where $\bar{\mathcal{F}}^{\text{ref}}$ is a reference value typical for a filament of length L. By dimensional analysis one defines $\bar{\mathcal{F}}^{\text{ref}} = L^3/T$. In Ref. [79], we have introduced an idealized stroke pattern for which we expect optimum fluid transport with a filament of length L and calculated an alternative value for $\bar{\mathcal{F}}^{\text{ref}}$. In the transport stroke, the filament is oriented perpendicular to the bounding surface and it is dragged parallel to the surface along a distance L. Now, the filament is rotated by 90° and then in the recovery stroke it is dragged along its long axis to its original position keeping it always close to the surface. Again, one finds $\bar{\mathcal{F}}^{\text{ref}} \approx L^3/T$. Finally, we note that $\mathcal{F}/\pi$ agrees with the volume flow rate initiated by the filament through a plane perpendicular to the flow direction [90, 91].

9.3.3.1 Two-Dimensional Stroke

The two-dimensional or planar stroke is initiated by a magnetic field of strength B, whose direction oscillates in the yz plane about the normal of the bounding surface that we identify with the z-axis,

$$\boldsymbol{B}(t) = (0, B \sin \varphi(t), B \cos \varphi(t)), \tag{9.41}$$

where $\varphi(t)$ is the angle the field encloses with the z-axis. In our modeling, the angular amplitude $\varphi_{\max}$ was always 60°.

In order to accomplish net fluid transport along the y-direction, not only a nonreversible motion of the filament is required [6] but also the motion of the filament in positive and negative y-direction has to be asymmetric. We define the transport stroke by slowly rotating the filament about its anchoring point. Owing to the small frictional forces, the filament hardly bends. The beating cycle is

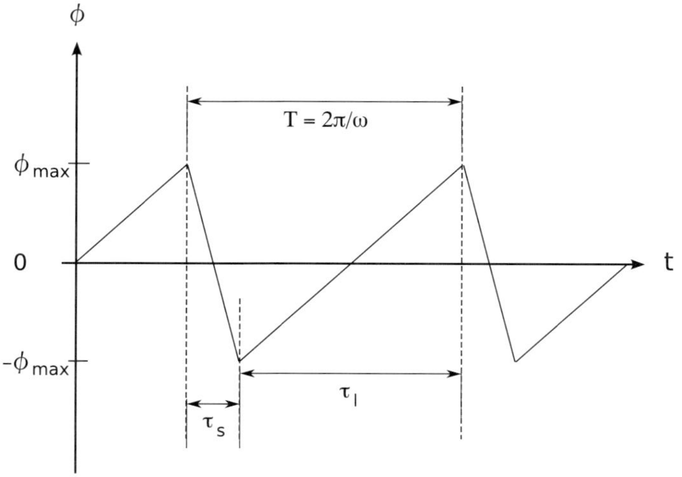

Figure 9.5 The angle φ enclosed by the magnetic field $\boldsymbol{B}(t)$ and the z-axis is shown as a function of time. φ has different velocities when decreasing and increasing. (Reprinted with permission from Ref. [79]. Copyright (2009) by EDP Sciences.)

complemented by a fast recovery stroke in the reversed direction where the filament bends due to increased hydrodynamic friction. Note that for real cilia in nature the speeds are just reversed: the transport stroke is fast and the recovery stroke is slow. For the magnetically actuated filament, the time protocol of the angle $\varphi(t)$ is illustrated in Figure 9.5. To quantify the asymmetry in the actuation cycle and therefore in the beating pattern of the filament, we define the asymmetry parameter

$$\varepsilon = \frac{\tau_l - \tau_s}{\tau_s + \tau_l}, \tag{9.42}$$

where τ_l and τ_s are the respective durations of the transport and recovery strokes indicated in Figure 9.5. The asymmetry parameter is zero for $\tau_s = \tau_l$ and tends to one in the limit of $\tau_l \gg \tau_s$.

Figure 9.6 illustrates the pumping performance ξ of a single filament as a function of the Sperm number Sp and the asymmetry parameter ε at a fixed magnetic field strength. The most striking feature is the pronounced peak for ε close to 1 and at $Sp \approx 3$. It is similar to the peak of the swimming velocity of the artificial micro-swimmer reported in Figure 9.2 [4, 57, 74]. The corresponding stroke pattern for $Sp \approx 3$ is illustrated in the middle picture of Figure 9.7a. In the slow transport stroke, the filament rotates clockwise being nearly straight, whereas in the fast recovery stroke the filament bends due to large hydrodynamic friction forces and then relaxes back to the initial configuration. As the inset in Figure 9.6 demonstrates, fluid transport is also noticeable in the recovery stroke ($\mathcal{F} < 0$). So, the pumping performance, even for the most efficient stroke pattern, is the result of a small asymmetry in the amount of fluid transported to the right and left. In the example of the inset, which is close to the optimum stroke pattern, only 4.3% of the total amount

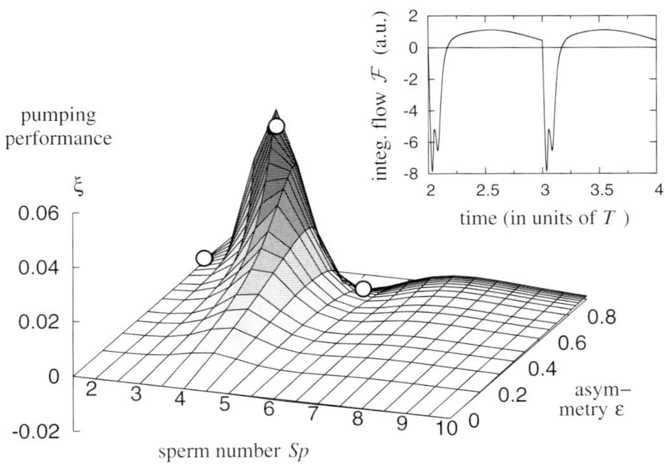

Figure 9.6 Pumping performance ξ for a single filament as a function of sperm number *Sp* and asymmetry parameter ε. The reduced magnetic field strength is $B_s = 2.5$. The white dots mark parameters for which stroboscopic snapshots of the filament are shown in Figure 9.7a. Inset: Integrated flow $\mathcal{F}$ in arbitrary units as a function of time in units of *T* for the parameters $Sp = 3$, $B_s = 3$, and $\varepsilon = 0.9$. (Reprinted with permission from Ref. [79]. Copyright (2009) by EDP Sciences.)

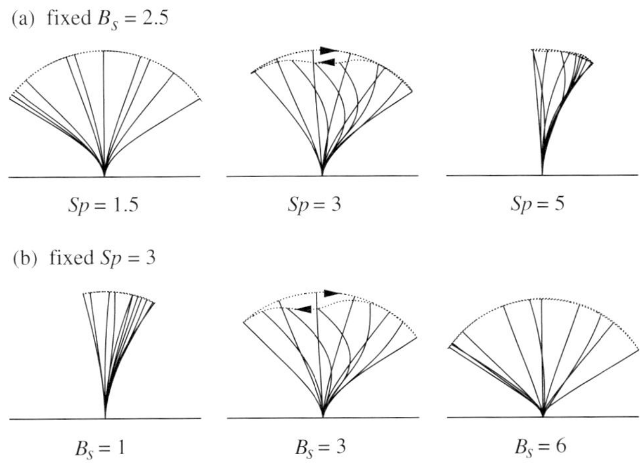

Figure 9.7 Stroboscopic snapshots of the filament at different times during the beating cycle for $\varepsilon = 0.9$. The trajectory of the top bead during one beating cycle is also indicated. In the slow transport stroke, the filament rotates clockwise, the fast recovery stroke occurs to the left, as indicated by the arrows. A pronounced bending of the filament occurs only at intermediate sperm number *Sp* and magnetic field strength B_s. (Reprinted with permission from Ref. [79]. Copyright (2009) by EDP Sciences.)

of moved fluid are effectively transported in positive y-direction. As a result, the maximum pumping performance in Figure 9.6 is only 6% of the reference stroke described above. As expected, the pumping performance vanishes for symmetric beating ($\varepsilon = 0$) about the z-axis. The same is true for $Sp \to 0$ where the filament follows the actuating magnetic field instantaneously and therefore remains straight leading to a reciprocal stroke as illustrated by the graph on the left-hand side of Figure 9.7a. Even a reversal of the pumping direction ($\xi < 0$) is observed at $Sp \approx 5.5$, albeit only with a rather weak performance. Finally, the pumping performance goes to zero for increasing Sp or frequency since the filament can no longer follow the actuating field as shown by the graph on the right-hand side of Figure 9.7a. Hence, optimal pumping performance is achieved only for intermediate values of Sp.

The pumping performance ξ exhibits a pronounced dependence on the strength B_s of the actuating magnetic field. Figure 9.8a clearly demonstrates this behavior for different values of Sp. When the magnetic field increases from zero, the pumping performance first remains close to zero and then, beyond a threshold value, it grows until a maximum is reached. Finally, ξ decreases and even becomes negative. The

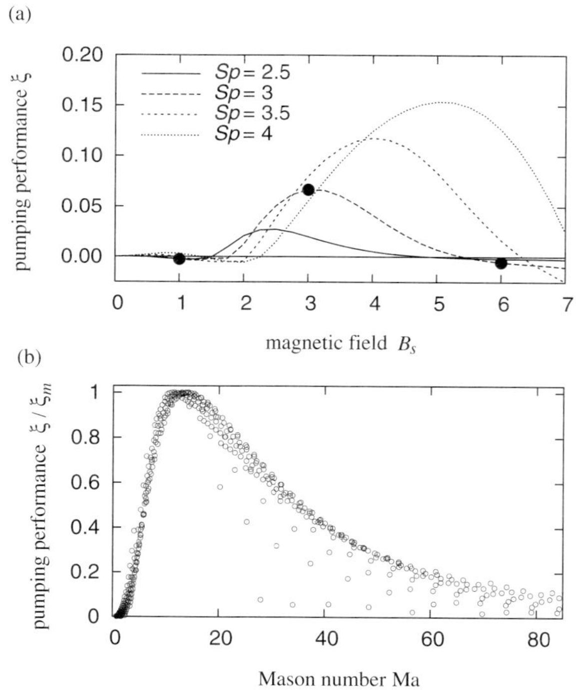

Figure 9.8 (a) Pumping performance ξ versus B_s for different Sp and $\varepsilon = 0.9$. The black dots mark parameters for which stroboscopic snapshots of the filament are shown in Figure 9.7b. (b) Pumping performance ξ as a function of the Mason number $Ma = Sp^4/B_s^2$, where ξ is given in units of the maximum values ξ_m when B_s is kept constant. (Reprinted with permission from Ref. [79]. Copyright (2009) by EDP Sciences.)

snapshots in Figure 9.7b again help to clarify this behavior. Small field strengths B_s are too small to overcome hydrodynamic friction forces. Therefore, the motion of the filament is very limited. On the other hand, at large strengths B_s the filament is always straight and thus performs a reciprocal motion. So, the optimal stroke exists in an intermediate regime for the strength B_s. Clearly, the optimal performance shifts with increasing Sp to larger values of B_s since a larger field is needed to move the filament through the fluid. In other words, for larger B_s the filament is stiffer and the optimum stroke is realized at higher frequencies $\omega \propto Sp^4$. This means larger frictional forces and therefore a larger pumping performance, which Figure 9.8a demonstrates. When the magnetic forces on the filament exceed the bending forces, one expects the dynamics of the filament to be determined by the ratio of the hydrodynamic friction to magnetic forces, which we introduced in Eq. (9.33) as Mason number Ma. We rescaled curves $\xi(Sp)$ for different $B_s \geq 2$ by their respective maximum values ξ_m. When plotted as a function of the Mason number Ma, the data points indeed fall on a master curve as illustrated in Figure 9.8b. Deviations occur for data points with B_s close to 2.

We close with two remarks. First, we also investigated the influence of defects on the lower part of the filament, where the bending stiffness $k_B T l_p$ is strongly reduced, and found that they significantly increase the pumping performance [92]. Second, we studied the pumping performance of several filaments placed along the y-axis [79]. Each filament was actuated separately so that phase shifts between neighboring beating filaments could be adjusted. Our studies revealed that the pumping performance is very sensitive to the imposed phase lag, which means to the details of the initiated metachronal wave. In particular, the pumping performance per filament increases relative to a single filament within a range of nonzero phase lag when the metachronal wave propagates opposite to the transport stroke. These waves are then termed antiplectic [1]. Creating them for a field of superparamagnetic filaments is certainly a challenge to experimenters. Nevertheless, due to our results we expect that metachronal waves in real cilia systems also increase the pumping performance for fluid transport.

9.3.3.2 **Three-Dimensional Stroke**

As reviewed in Section 9.1.1, in nature also three-dimensional stroke patterns exist, for example, of cilia that cover the surfaces of the protozoan paramecium or opalina. In the case of the superparamagnetic filament, the actuating magnetic field can be used to initiate three-dimensional stroke patterns [92]. In the "cone stroke," the magnetic field vector follows the surface of a cone with opening angle ϕ tilted at an angle θ to the surface normal (see Figure 9.9a). The idea is that through the tilt of the cone toward the x-direction a clear asymmetry between fluid transport in positive and negative y-direction is produced. This is also the stroke pattern of the cilia that generate the nodal flow for establishing the left-right asymmetry in mammals already discussed in Section 9.1.1 [91, 93]. The transport stroke in the alternative "hybrid stroke" is induced by the field vector rotating in the yz plane followed by a rotation around the z-axis again on the surface of a cone with opening angle ϕ (see Figure 9.9b). Both parts of the cone stroke are executed in equal time.

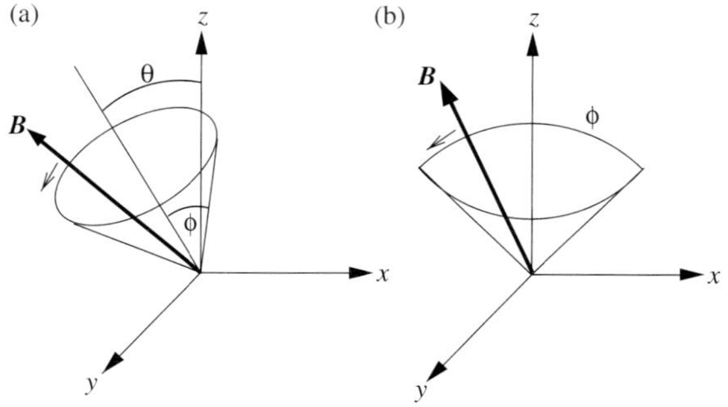

Figure 9.9 Kinematics of the actuating magnetic field for three-dimensional stroke patterns. (a) Cone stroke and (b) hybrid stroke.

When the filament is driven slowly, it will remain straight. In this case, approximations for the pumping performance can be calculated analytically by assuming that the strengths of the forces in Eq. (9.38) driving the single beads are proportional to the beads' velocities. For the cone stroke, for example, these velocities are determined by the geometry and therefore one assumes that the fluid flow is produced by forces whose strengths increase linearly along the length of the filament starting from the anchoring point. Then from Eq. (9.38) one finds the pumping performance scales as $\xi \propto \sin^2(\phi)\sin(\theta)$. In the parameter space of the cone stroke, $\theta + \phi \le \pi/2$, where the equal sign means that the filament just touches the surface at its lowest point, the pumping performance is maximized for $\phi = \arccos\left(1/\sqrt{(3)}\right) \approx 54^\circ$ and $\theta = \pi/2 - \phi$. We have checked the approximation for ξ by comparing it with numerical results and found that the agreement is excellent [92].

Figure 9.10 illustrates the pumping performance of cone (a) and hybrid (b) stroke for several opening angles ϕ as a function of *Sp*. It is immediately obvious that for the largest ϕ values the pumping performance is more than a factor 10 larger compared

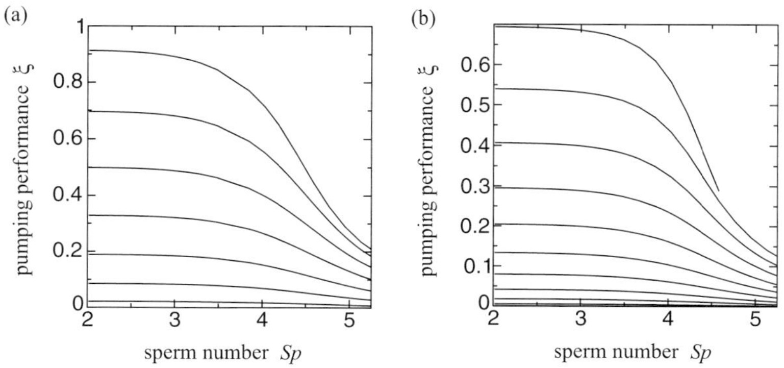

Figure 9.10 Pumping performance ξ as a function of *Sp*. (a) Tilted cone ($\theta = \pi/4$): each curve corresponds to a different opening angle ϕ with changes of $\Delta\phi = 0.1$ (in radians) between curves up to a maximum $\phi = 0.7$ (upper curve). (b) Hybrid stroke: each curve corresponds to a different opening angle ϕ with spacing $\Delta\phi = 0.1$ up to a maximum $\phi = 1.1$ (upper curve).

to the planar case in Figure 9.6. Therefore, the three-dimensional strokes are much more efficient in pumping fluid. Increasing *Sp* or the frequency, the pumping performance decreases since the filament starts to bend due to the increased hydrodynamic friction. For example, in the cone stroke the cone defined by the rotating bent filament is narrower than the magnetic-field cone and therefore fluid transport is reduced [92].

9.4 Conclusions

In this chapter, we have thoroughly reviewed how the dynamics of an elastic filament moving in a strongly viscous environment can be modeled. We especially introduced a bead-spring model with bending free energy whose constituent beads interact via hydrodynamic interactions. This enabled us to model a recently introduced superparamagnetic elastic filament. The appealing system was used to construct a first artificial microswimmer whose properties we presented with regard to swimming velocity and efficiency of energy consumption. The one-armed swimmer now offers the interesting vision for propelling micromachines that perform their work in the microscopic world. One example would be a device that moves through blood vessels [94].

We have also attached the filament to a surface and studied different stroke patterns for transporting fluid that are realized by the actuating magnetic field. Our studies clearly show that three-dimensional strokes are much more efficient for pumping fluid. In addition, we briefly explained that in a line of beating artificial cilia the pumping performance increases for an appropriate nonzero phase lag reminiscent of metachronal waves. Thus, the superparamagnetic elastic filament not only helps elucidate biological features of beating flagella and cilia but also offers possible exciting applications with regard to the transport and mixing of fluids in the field of microfluidics.

References

1 Brennen, C. and Winet, H. (1977) Fluid mechanics of propulsion by cilia and flagella. *Ann. Rev. Fluid Mech.*, **9**, 339–398.

2 Linck, R.W. (2001) Cilia and flagella, in *Encyclopedia of Life Sciences*, John Wiley & Sons, Ltd., Chichester, www.els.net.

3 Bray, D. (2001) *Cell Movements: From Molecules to Motiliy*, 2nd edn, Garland Publishing, New York.

4 Dreyfus, R., Baudry, J., Roper, M.L., Fermigier, M., Stone, H.A., and Bibette, J. (2005) Microscopic artificial swimmers. *Nature*, **437**, 862–865.

5 Squires, T.M. and Quake, S.R. (2005) Microfluidics: fluid physics at the nanoliter scale. *Rev. Mod. Phys.*, **77**, 977–1026.

6 Purcell, E.M. (1977) Life at low Reynolds number. *Am. J. Phys.*, **45**, 3–11.

7 Berg, H.C. (1996) Symmetries in bacterial motility. *Proc. Natl. Acad. Sci. USA*, **93**, 14225–14228.

8 Shapere, A. and Wilczek, F. (1987) Self-propulsion at low Reynolds number. *Phys. Rev. Lett.*, **58**, 2051–2054.

9 Stone, H.A. and Samuel, A.D.T. (1996) Propulsion of microorganisms by

surface distortions. *Phys. Rev. Lett.*, **77**, 4102–4104.

10 Becker, L.E., Koehler, S.A., and Stone, H.A. (2003) On self-propulsion of micro-machines at low Reynolds number: Purcell's three-link swimmer. *J. Fluid Mech.*, **490**, 15–35.

11 Avron, J.E., Gat, O., and Kenneth, O. (2004) Optimal swimming at low Reynolds numbers. *Phys. Rev. Lett.*, **93**, 186001–186004.

12 Najafi, A. and Golestanian, R. (2004) Simple swimmer at low Reynolds number: three linked spheres. *Phys. Rev. E*, **69**, 062901–062904.

13 Dreyfus, R., Baudry, J., and Stone, H.A. (2005) Purcell's "rotator": mechanical rotation at low Reynolds number. *Eur. Phys. J. B*, **47**, 161–164.

14 Dhont, J.K.G. (1996) *An Introduction to Dynamics of Colloids*, Elsevier, Amsterdam.

15 Tam, D. and Hosoi, A.E. (2007) Optimal stroke patterns for Purcell's three-link swimmer. *Phys. Rev. Lett.*, **98**, 068105-1–068105-4.

16 Berg, H.C. and Anderson, R.A. (1973) Bacteria swim by rotating their flagellar filaments. *Nature*, **245**, 380–382.

17 Berg, H.C. (2004) *E. coli in Motion*, Springer Verlag, New York.

18 Taylor, G. (1951) Analysis of the swimming of microscopic organisms. *Proc. R. Soc. A*, **209**, 447–461.

19 Gray, J. and Hancock, G.J. (1955) The propulsion of sea-urchin spermatozoa. *J. Exp. Biol.*, **32**, 802–814.

20 Satir, P. (1968) Studies on cilia. III. Further studies on the cilium tip and a "sliding filament" model of ciliary motility. *J. Cell Biol.*, **39**, 77–94.

21 Camalet, S., Jülicher, F., and Prost, J. (1999) Self-organized beating and swimming of internally driven filaments. *Phys. Rev. Lett.*, **82**, 1590–1593.

22 Camalet, S. and Jülicher, F. (2000) Generic aspects of axonemal beating. *New J. Phys.*, **2**, 24.1–24.23.

23 Gueron, S. and Levit-Gurevich, K. (1998) Computation of the internal forces in cilia: application to ciliary motion, the effects of viscosity, and cilia interactions. *Biophys. J.*, **74**, 1658–1676.

24 Kim, Y. and Netz, R. (2006) Pumping fluids with periodically beating grafted elastic filaments. *Phys. Rev. Lett.*, **96**, 158101-1–158101-4.

25 Gueron, S., Levit-Gurevich, K., Liron, N., and Blum, J.J. (1997) Cilia internal mechanism and metachronal coordination as the result of hydrodynamical coupling. *Proc. Natl. Acad. Sci. USA*, **94**, 6001–6006.

26 Gueron, S. and Levit-Gurevich, K. (1999) Energetic considerations of ciliary beating and the advantage of metachronal coordination. *Proc. Natl. Acad. Sci. USA*, **96**, 12240–12245.

27 Lagomarsino, M.C., Bassetti, B., and Jona, P. (2002) Rowers coupled hydrodynamically. Modeling possible mechanisms for the cooperation of cilia. *Eur. Phys. J. E*, **26**, 81–88.

28 Lagomarsino, M.C., Jona, P., and Bassetti, B. (2003) Metachronal waves for deterministic switching two-state oscillators with hydrodynamic interaction. *Phys. Rev. E*, **68**, 021908-1–021908-9.

29 Vilfan, A. and Jülicher, F. (2006) Hydrodynamic flow patterns and synchronization of beating cilia. *Phys. Rev. Lett.*, **96**, 058102-1–058102-4.

30 Lenz, P. and Ryskin, A. (2006) Collective effects in ciliar arrays. *Phys. Biol.*, **3**, 285–294.

31 Guirao, B. and Joanny, J.-F. (2007) Spontaneous creation of macroscopic flow and metachronal waves in an array of cilia. *Biophys. J.*, **92**, 1900–1917.

32 Niedermayer, T., Eckhardt, B., and Lenz, P. Synchronization, phase locking, and metachronal wave formation in ciliary chains. *CHAOS*, **18**, 037128-1–037128-10.

33 Ibañez-Tallon, I., Pagenstecher, A., Fliegauf, M., Olbrich, H., Kispert, A., Ketelsen, U.-P., North, A., Heintz, N., and Omran, H. (2004) Dysfunction of axonemal dynein heavy chain Mdnah5 inhibits ependymal flow and reveals a novel mechanism for hydrocephalus formation. *Hum. Mol. Genet.*, **13**, 2133–2141.

34 Stern, C.D. (2002) Fluid flow and broken symmetry. *Nature*, **418**, 29–30; Essner, J.J., Vogan, K.J., Wagner, M.K., Tabin, C.J., Jost Yost, H.J., and Brueckner,

M. (2002) Conserved function for embryonic nodal cilia. *Nature*, **418**, 37–38; Nonaka, S., Shiratori, H., Saijoh, Y., and Hamada, H. (2002) Determination of left-right patterning of the mouse embryo by artificial nodal flow. *Nature*, **418**, 96–99.

35 den Toonder, J., Bos, F., Broer, D., Filippini, L., Gillies, M., de Groede, J., Mol, T., Reijme, M., Talen, W., Wilderbeek, H., Khatavkar, V., and Anderson, P. (2008) Artificial cilia for active micro-fluidic mixing. *Lab Chip*, **8**, 533–541.

36 Evans, B.A., Shields, A.R., Lloyd Carroll, R., Washburn, S., Falvo, M.R., and Superfine, R. (2007) Magnetically actuated nanorod arrays as biomimetic cilia. *Nano Lett.*, **7**, 1428–1434.

37 Darnton, N., Turner, L., Breuer, K., and Berg, H.C. (2004) Moving fluid with bacterial carpets. *Biophys. J.*, **86**, 1863–1870.

38 Cox, R.G. (1970) The motion of long slender bodies in a viscous fluid. Part 1. General theory. *J. Fluid Mech.*, **44**, 791–810.

39 Lighthill, M.J. (1975) *Mathematical Biofluiddynamics*, SIAM, Philadelphia.

40 Lighthill, M.J. (1976) Flagellar hydrodynamics. *SIAM Rev.*, **18**, 161–230.

41 Riveline, D., Wiggins, C.H., Goldstein, R.E., and Ott, A. (1997) Elastohydrodynamic study of actin filaments using fluorescence microscopy. *Phys. Rev. E*, **56**, R1330–R1333.

42 Wiggins, C.H., Riveline, D., Ott, A., and Goldstein, R.E. (1998) Trapping and wiggling: elastohydrodynamics of driven microfilaments. *Biophys. J.*, **74**, 1043–1060.

43 Wiggins, C.H. and Goldstein, R.E. (1998) Flexive and propulsive dynamics of elastica at low Reynolds number. *Phys. Rev. Lett.*, **80**, 3879–3882.

44 Wolgemuth, C.W., Powers, T.R., and Goldstein, R.E. (1998) Twirling and whirling: viscous dynamics and rotating elastic filaments. *Phys. Rev. Lett.*, **84**, 1623–1226.

45 Koehler, S.A. and Powers, T.R. (1998) Twirling elastica: kinks, viscous drag, and torsional stress. *Phys. Rev. Lett.*, **85**, 4827–4830.

46 Yu, T.S., Lauga, E., and Hosoi, A.E. Experimental investigations of elastic tail propulsion at low Reynolds number. *Phys. Fluids*, **18**, 091701-1–091701-4.

47 Lauga, E. (2007) Floppy swimming: viscous locomotion of actuated elastica. *Phys. Rev. E*, **75**, 041916-1–041916-16.

48 Fu, H.C., Powers, T.R., and Wolgemuth, C.W. (2007) Theory of swimming filaments in viscoelastic media. *Phys. Rev. Lett.*, **99**, 258101-1–258101-4.

49 Lauga, E. (2007) Propulsion in a viscoelastic fluid. *Phys. Fluids*, **19**, 083104-1–083104-13.

50 Lowe, C.P. (2001) A hybrid particle/continuum model for microorganism motility. *Future Generat. Comput. Syst.*, **17**, 853–862.

51 Lagomarsino, M.C., Capuani, F., and Lowe, C.P. (2003) A simulation study of the dynamics of a driven filament in an Aristotelian fluid. *J. Theor. Biol.*, **224**, 215–224.

52 Lowe, C.P. (2003) Dynamics of filaments: modelling the dynamics of driven microfilaments. *Phil. Trans. R. Soc. Lond. B*, **358**, 1543–1550.

53 Lagomarsino, M.C., Pagonabarraga, I., and Lowe, C.P. (2005) Hydrodynamic induced deformation and orientation of a microscopic elastic filament. *Phys. Rev. Lett.*, **94**, 148104-1–148104-4.

54 Llopis, I., Pagonabarraga, I., Lagomarsino, M.C., and Lowe, C.P. (2007) Sedimentation of pairs of hydrodynamically interacting semiflexible filaments. *Phys. Rev. E*, **76**, 061901-1–061901-10.

55 van den Heuvel, M.G.L., Bondesan, R., Lagomarsino, M.C., and Dekker, C. (2008) Single-molecule observation of anomalous electrohydrodynamic orientation of microtubules. *Phys. Rev. Lett.*, **101**, 118301-1–118301-4.

56 Schlagberger, X. and Netz, R.R. (2005) Orientation of elastic rods in homogeneous Stokes flow. *Europhys. Lett.*, **70**, 129–135.

57 Gauger, E. and Stark, H. (2006) Numerical study of a microscopic artificial swimmer. *Phys. Rev. E*, **74**, 021907-1–021907-10.

58 Manghi, M., Schlagberger, X., Kim, Y.-W., and Netz, R.R. (2006) Hydrodynamic

effects in driven soft matter. *Soft Matter*, **2**, 653–668.

59 Manghi, M., Schlagberger, X., and Netz, R. (2006) Propulsion with a rotating elastic nanorod. *Phys. Rev. Lett.*, **96**, 068101-1–068101-4.

60 Qian, B., Powers, T.R., and Breuer, K.S. (2008) Shape transition and propulsive force of an elastic rod rotating in a viscous fluid. *Phys. Rev. Lett.*, **100**, 078101-1–078101-4.

61 Coq, N., du Roure, O., Martelot, J., Bartolo, D., and Fermigier, M. (2008) Rotational dynamics of a soft filament: wrapping transition and propulsive forces. *Phys. Fluids*, **20**, 051703-1–051703-4.

62 Wada, H. and Netz, R.R. (2006) Non-equilibrium hydrodynamics of a rotating filament. *Europhys. Lett.*, **75**, 645–651.

63 Wada, H. and Netz, R.R. (2007) Model for self-propulsive helical filaments: kink-pair propagation. *Phys. Rev. Lett.*, **99**, 108102-1–108102-4.

64 Wada, H. and Netz, R.R. (2008) Discrete elastic model for stretching-induced flagellar polymorphs. *Europhys. Lett.*, **82**, 28001-1–28001-6.

65 Malevanets, A. and Kapral, R. (1999) Mesoscopic model for solvent dynamics. *J. Chem. Phys.*, **110**, 8605–8613.

66 Malevanets, A. and Kapral, R. (2000) Solute molecular dynamics in a mesoscale solvent. *J. Chem. Phys.*, **112**, 7260–7269.

67 Yang, Y., Elgeti, J., and Gompper, G. (2008) Cooperation of sperm in two dimensions: synchronization, attraction and aggregation through hydrodynamic interactions. *Phys. Rev. E*, **78**, 061903-1–061903-9.

68 Goubault, C., Jop, P., Fermigier, M., Baudry, J., Bertrand, E., and Bibette, J. (2003) Flexible magnetic filaments as micromechanical sensors. *Phys. Rev. Lett.*, **91**, 260802-1–260802-4.

69 Cohen-Tannoudji, L., Bertrand, E., Bressy, L., Goubault, C., Baudry, J., Klein, J., Joanny, J.-F., and Bibette, J. (2005) Polymer bridging probed by magnetic colloids. *Phys. Rev. Lett.*, **94**, 038301-1–038301-4.

70 Koenig, A., Hébraud, P., Gosse, C., Dreyfus, R., Baudry, J., Bertrand, E., and Bibette, J. (2005) Magnetic force probe for nanoscale biomolecules. *Phys. Rev. Lett.*, **95**, 128301-1–128301-4.

71 Biswal, S.L. and Gast, A.P. (2003) Mechanics of semiflexible chains formed by poly(ethylene glycol)-linked paramagnetic particles. *Phys. Rev. E*, **68**, 021402-1–021402-9.

72 Biswal, S.L. and Gast, A.P. (2004) Rotational dynamics of semiflexible paramagnetic particle chains. *Phys. Rev. E*, **69**, 041406-1–041406-9.

73 Roper, M.L., Dreyfus, R., Baudry, J., Fermigier, M., Bibette, J., and Stone, H.A. (2006) On the dynamics of magnetically driven elastic filaments. *J. Fluid Mech.*, **554**, 167–190.

74 Roper, M.L., Dreyfus, R., Baudry, J., Fermigier, M., Bibette, J., and Stone, H.A. (2008) Do magnetic micro-swimmers move like eukaryotic cells? *Proc. Royal Soc. A*, **464**, 877–904.

75 Cebers, A. (2003) Dynamics of a chain of magnetic particles connected with elastic linkers. *J. Phys. Condens. Matter*, **15**, S1335–S1344.

76 Cebers, A. (2005) Flexible magnetic filaments. *Curr. Opin. Coll. Interface Sci.*, **10**, 167–175.

77 Cebers, A. (2005) Flexible magnetic swimmer. *Magnetohydrodynamics*, **41**, 63–72.

78 Belovs, M. and Cebers, A. (2006) Nonlinear dynamics of semiflexible magnetic filaments in an AC magnetic field. *Phys. Rev. E*, **71**, 051503-1–051503-11.

79 Gauger, E.M., Downton, M.T., and Stark, H. (2009) Fluid transport at low Reynolds number with magnetically actuated artificial cilia. *Eur. Phys. J. E*, **28**, 231–242.

80 Landau, L.D. and Lifschitz, E.M. (1991) *Lehrbuch der Theoretischen Physik Band VII: Elastizitätstheorie*, Akademie Verlag, Berlin.

81 Kratky, O. and Porod, G. (1949) Röntgenuntersuchung gelöster Fadenmoleküle. *Recl. Trav. Chim. Pays-Bas*, **68**, 1106.

82 Cichocki, B., Felderhof, B.U., and Hinsen, K. (1994) Friction and mobility of many spheres in Stokes flow. *J. Chem. Phys.*, **100**, 3780–3790.

83 Jeffrey, D.J. and Gnishi, Y. (1984) Calculation of the resistance and mobility functions for two unequal rigid spheres in low-Reynolds-number flow. *J. Fluid Mech.*, **139**, 261–290.

84 Kim, S. and Kanila, S.J. (1991) *Microhydrodynamics: Principles and Selected Applications*, Butterworth-Heinemann, Boston.

85 Hinsen, K. (1995) HYDROLIB: a library for the evaluation of hydrodynamic interactions in colloidal suspensions. *Comput. Phys. Commun.*, **88**, 327–340.

86 Lorentz, H.A. (1896) A general theorem concerning the motion of a viscous fluid and a few consequences derived from it. *Zittingsverlag Akad. v. Wet.*, **5**, 168–175.

87 Blake, J.R. (1971) A note on the image system for a stokeslet in a no-slip boundary. *Proc. Camb. Phil. Soc.*, **70**, 303.

88 Melle, S., Calderón, O.G., Rubio, M.A., and Fuller, G.G. (2003) Microstructure evolution in magnetorheological suspensions governed by Mason number. *Phys. Rev. E*, **68**, 041503-1–041503-11.

89 Blake, J. (1972) A model for the micro-structure in ciliated organisms. *J. Fluid Mech.*, **55**, 1–23.

90 Liron, N. (1978) Fluid transport by cilia between parallel plates. *J. Fluid Mech.*, **86**, 705–726.

91 Smith, D.J., Blake, J.R., and Gaffney, E.A. (2008) Fluid mechanics of nodal flow due to embryonic primary cilia. *J. R. Soc. Interface*, **5**, 567–573.

92 Downton, M.T. and Stark, H. (2009) Beating kinematics of magnetically actuated cilia. *Europhys. Lett.*, **85**, 44002-1–44002-6.

93 Hirokawa, N., Tanaka, Y., Okada, Y., and Takeda, S. (2006) Nodal flow and the generation of left-right asymmetry. *Cell*, **125**, 33–45.

94 Ishiyama, K., Sendoh, M., Yamazaki, A., Inoue, M., and Arai, K.I. (2001) Swimming of magnetic micro-machines under a very wide range of Reynolds number conditions. *IEEE Trans. Mag.*, **37**, 2868.

10
Energy Gap Model of Glass Formers: Lessons Learned from Polymers

Puru D. Gujrati

10.1
Introduction

10.1.1
Equilibrium and Metastable States: Supercooled Liquids

Supercooled (SCL) and superheated states are ubiquitous in Nature and are usually treated as metastable states (MSs) such as those associated with the van der Waals loop in the celebrated van der Waals equation. The states have higher free energies than the corresponding equilibrium state (EQS) and violate the fundamental thermodynamic property that the equilibrium free energy be *minimized* or the equilibrium partition function (PF) be maximized. Therefore, they *cannot* be rigorously derived from *equilibrium* statistical mechanics [1]. We need to go beyond it to explain their existence. As a consequence, many standard results of equilibrium thermodynamics will not hold for metastable states [2–4],[1)] even when they are manipulated to exist for an abnormally long time, a situation that occurs for glasses, which is the subject matter of this chapter [5–10]. Being metastable, glasses have higher energies compared to their crystalline form at absolute zero. This difference in their energies is what we call the *energy gap*. The idea of the energy gap is properly introduced in Section 10.1.5.1, and the energy gap model of glass formers is introduced and elaborated in Section 10.2 by carefully analyzing an established polymer model. We do not expect the reader to be an expert in the field of glass transition and the general phenomenology of glass formers. We, therefore, provide a brief introduction to glass phenomenology in Section 10.3, localization and confinement of glasses in Section 10.4, and some current important theories in Section 10.5.

In this section, we discuss general properties of MSs. Stable and abnormally long-lasting MSs can be easily prepared in the laboratory; we only have to recall the stability

1) For example, the singularity in the equilibrium free energy at the melting transition may not occur when we extend the liquid state into its supercooled state. This point is elaborated in Section 10.8. This should be contrasted with how an essential singularity appears in the droplet model [2, 3], where one does not restrict the microstates in the partition function. Limitations of the droplet model are discussed by Domb (see Ref. [4], pp. 217–218).

Modeling and Simulation in Polymers. Edited by P.D. Gujrati and A.I. Leonov

ISBN: 978-3-527-32415-6

of medieval glasses. They also appear in many mean field theories, including the van der Walls equation such as the Bragg–Williams theory [11] after we abandon minimizing the free energy. This has led to the argument that MSs are the results of approximate calculations. This is not true, as they also emerge in exact calculations [12]; we report two such calculations in Sections 10.10 and 10.11. They emerge solely because we abandon the *free energy minimization principle* in calculation. (In experiments, they emerge by the very nature of preparing the system such as a fast quench.) The violation of this principle can still lead to a stable solution, except that it is metastable. The *stability* only requires the specific heat, compressibility, and so on to be *nonnegative* (Section 10.7). Such mathematically stable solutions in theoretical models imply that they will never decay. We therefore call them *stationary metastable states* (SMSs) in this chapter to distinguish them from MSs that are encountered in experiments and that usually change with time. The state associated with the stable solution with the lowest free energy represents the equilibrium state such as the equilibrium liquid (EL) and is time independent. All other states represent non-equilibrium states (NESs), whether time dependent or not.

10.1.2 Common Folklore

The Bragg–Williams [11] approximation or most mean field approximations are valid in the limit of infinite lattice coordination number q ($q \to \infty$) and vanishing interaction strength indicated by J ($J \to 0$), keeping their product (qJ) fixed and finite; see Ref. [13] and references therin for a more recent discussion of this approximation. The approximation is equivalent to solving the models in an *infinite-dimensional* space with vanishing interaction. This has given rise to the common folklore that SMSs occur only in an infinite-dimensional space. This is incorrect as we demonstrate by an exact calculation in Section 10.10, where we deal with a model of branched polymers. It is a *nonmean field calculation* carried out in a one-dimensional model and captures SMS without any singularity.[1] The model also exhibits the famous *entropy crisis* in SMS, which was first discussed by Kauzmann [5] in a very forceful and convincing manner. The crisis is elaborated in Sections 10.3.5 and 10.3.6. An alternative interpretation of the above approximations is to allow long-range interactions. Thus, another folklore is that SMSs do not exist for short-range models. Even frustration is considered in the folklore to be necessary for the glassy behavior; see for example Ref. [14]. To overcome this folklore, we consider an Ising model of a binary mixture with *short-range interactions and no frustration* in Section 10.11, which shows a glass transition. The model is solved exactly on a special recursive lattice, the Husimi lattice. From all the experience we have accumulated, models on recursive lattices provide a much better description of the regular lattice models than the conventional mean field approximation, as shown elsewhere [15]. Because of the exactness of both calculations, the thermodynamics is proper in that all solutions are stable. We supplement the exact results by general proofs that are based on our energy gap model described in Section 10.2. The model itself is based on the experimentally supported observation (10.3) and the universally accepted second law of thermodynamics encapsulated in (10.12) and Nernst–Planck postulate (the third law) [16],[2)] so the results are general. Furthermore, our conclusions are not limited to polymers alone, where the entropy crisis has been first discussed by Gibbs [17] and later justified analytically by Gibbs and Di Marzio [18]. They apply to all systems that form glasses via supercooling. They are also consistent with the formal random energy model [19] exhibiting a transition similar to a glass transition.

Our goal, and our hope, in this chapter is to convince the reader that it is the presence of an energy gap that invariably leads to the experimental glass transition at a nonzero temperature in most of the glass formers.

10.1.3 Systems Being Considered

Materials, natural or man-made, can be broadly classified into two classes [7, 8]:

(A) **Crystallizable Materials**: Their equilibrium stable state is crystalline at low temperatures with certain symmetries. Examples are rock or quartz crystals, various forms of ice, metals, and salt. The MSs in these systems are defined with respect to the ordered crystalline phase, an EQS.

(B) **Noncrystallizable Materials**: They remain *amorphous* or *noncrystalline* even at absolute zero due to their structural randomness. Examples are spin glasses (not to be confused with ordinary glasses, which are the focus of this chapter), materials with quenched or frozen impurities, atactic polymers, and so on. Supercooling is not an issue for these materials.

Materials in class A can also be prepared in a state that is random and disordered, very similar to their liquid state. We will refer to these states as *random* or *amorphous states* (also called *glassy states* or *glasses*) to distinguish them from amorphous materials in class B; the latter are *always* random (no regularity) even when they are in equilibrium. The glassy states do not have their free energy at its minimum. We will use glass formers as a general term for materials in class A. It should also be mentioned here that there exist materials known as quasi crystals, which are ordered but not periodic. An example is a metallic solid (Al-14 at.%-Mn) with long-range orientational order, but with icosahedral point group symmetry first discovered by Shechtman and coworkers [20]. In this chapter, we are only interested in materials in class A.

We will mostly consider the canonical ensemble with fixed number of particles N and volume V and its appropriate extension to describe time-independent metastable states (SMSs). Therefore, the temperature T will play an important role. We will usually not show N explicitly unless clarity is needed. The central quantity of interest will be the configurational multiplicity $W(E, V) \geq 1$, the number of configurations

2) We will assume that the entropy $S(T)$ of stationary states (EQS's or SMSs) satisfy $TS(T) \to 0$ as the temperature $T \to 0$. This is a much weaker condition than the conventional Nernst–Planck postulate $S(T) \to 0$ for EQS's, but is consistent with all the consequences of the latter. Our version is also applicable to SMSs, for which the entropy need not vanish at absolute zero (see Ref. [16], Section 64).

or microstates of *potential energy* E in a volume V. In terms of the multiplicity, the entropy (we define entropy as a dimensionless quantity) is defined by the Boltzmann relation [16]:

$$S(E, V) \equiv \ln W(E, V) \geq 0, \tag{10.1}$$

which assumes that all microstates counted in $W(E, V)$ are *equally probable,* which happens only in equilibrium. Away from equilibrium, we use the Gibbs definition [16] for the entropy

$$S(E, V, t) \equiv -\sum p_i(E, V, t) \ln p_i(E, V, t) \geq 0 \tag{10.2}$$

to obtain the time-dependent entropy, where $p_i(E, V, t)$ is the probability of the ith microstate at time t. This entropy will, in time, increase and approach $S(E, V)$ from below as the system equilibrates ($t \to \infty$). This is consistent with the second law of thermodynamics: the entropy of a closed system at fixed E, and V *cannot* decrease in time. In equilibrium, $p_i(E, V, t) \to 1/W(E, V)$ for each microstate as $t \to \infty$, and (10.2) reduces to (10.1). The entropy is a *continuous* and *concave*[3] function of its arguments E and V.

10.1.4 Long-Time Stability

There are usually two different mechanisms operative in MSs. The "fast" mechanism (timescale τ_f) creates a metastable state in the system, followed by a "slow" mechanism (timescale τ_s) for nucleation of the stable phase and the eventual decay of the metastable state. For an MS to exist for a while, we need to require $\tau_s > \tau_f$. The time-dependent NESs include not only states that will eventually turn into equilibrium states (such as crystals) but also states that will eventually turn into SMSs (such as glasses) as we wait infinitely long (in principle), depending on how they are prepared. To study glass dynamics, we need to compare the two timescales with the longest feasible experimental observation time τ_{exp}. From the experimental point of view, the inequality $\tau_s \gg \tau_{exp}$ for supercooled liquids is almost equivalent to the long time, that is, the *stationary* limit $\tau_s \to \infty$ of metastable states; we again appeal to the stability of medieval glasses. Thus, to a first approximation, we can treat real glasses and SCLs as SMSs, states that never decay. (According to Maxwell [21], this can, in principle, be achieved by ensuring that the equilibrium state nuclei are absent in MSs). This approximation is quite reasonable near the experimental glass transition temperature T_g, the region of interest in this chapter. Thus, we will assume the existence of these SMSs, as they will play a central role in our modeling and understanding of glass formers.

Metastable states (SCLs and glasses) at low temperatures can remain thermodynamically stable with no hint of any decay for a long period of times [5, 22, 23]

3) A concave function $f(x)$ is a function that always lies above the line connecting $f(x_1)$ and $f(x_2)$ over any of its interval $[x_1, x_2]$.

comparable to the longest feasible experimental observation time τ_{exp} due to the high viscosity, so in this case $\tau_s > \tau_{exp} \gg \tau_f$. It is found that supercooled glycerol does not readily crystallize over experimental timescales even if one disperses some glycerol crystallites in it. The large viscosity in glycerol near freezing also makes crystallization extremely slow, thereby enhancing the stability of the supercooled glycerol [24]. Supercooled viscous liquids also do not usually exhibit spinodals. Rather, they usually undergo a so-called (experimental) *glass transition* at T_g, about two-thirds of their melting temperature T_M, provided the liquid is cooled in such a way that crystallization does *not* intervene [5–9, 25]. Care must be exercised to suppress the nuclei of the stable equilibrium phase, the crystal phase (CR), from forming while cooling the viscous liquid to form a glass. Glasses constitute nonequilibrium states that are not unique; they depend on external controls such as the rate of cooling. They remain disordered even at absolute zero [16]. It is usually much easier to accomplish suppressing stable phase nuclei in SCLs than in supercooled vapors and superheated liquids, presumably because of the low temperatures (and high viscosity) and the stable phases (crystals) that have *distinct symmetries* from supercooled liquids[4)] in the former case. This also makes the decay of the metastable state even less probable, and strengthens the inequalities to $\tau_s \gg \tau_{exp} \gg \tau_f$. The idea enunciated above is consistent with Maxwell's idea [21] that to observe long-lasting metastable states, stable phase nuclei must *not* be present.

10.1.5 High Barriers, Confinement, and the Cell Model

10.1.5.1 Cell Model

Another, and probably the most important, property of SCLs at low temperatures is that glasses and CRs have very similar vibrational heat capacities below T_g, except that glasses have higher potential energies than the corresponding CRs [5, 6, 16]. Let E_{NES} denote the lowest possible energy of a NES such as a glass and E_0 the energy of the ideal crystal at $T = 0$. Then, we empirically have

$$E_{NES} > E_0. \tag{10.3}$$

The difference

$$\Delta_G \equiv E_{NES} - E_0 > 0 \tag{10.4}$$

is called the *energy gap*. Otherwise, glasses and crystals are confined to execute quite similar vibrations (not necessarily harmonic) within their *potential wells* or *basins*, although their minima are at different energies, notwithstanding the clear evidence to the contrary [23] and the presence of boson peaks observed in many glasses; see, for example, Refs [26–29]. All that is important is that both glasses and crystals exhibit localized motion. This property has led to the enormous popularity of the *potential*

4) The two coexisting phases in liquid–gas transitions have identical symmetry in that they are simply related by a symmetry operation like the up–down symmetry in the Ising model. No symmetry operation can transform crystals into liquids, and vice versa.

energy landscape picture, originally proposed by Goldstein [22], to investigate glass transition. The potential energy $E(\{\mathbf{r}\})$ as a function of the set $\{\mathbf{r}\}$ of particles' positions uniquely determines this landscape. A glass is confined to one of the myriads of potential wells, while the crystal is most commonly believed to be confined to just one potential well (see Refs [30], in particular, Figure 5, and [31]). The number of these basins determines the *basin entropy*, which can be derived either from (10.1) or from (10.2) as the case may be. Goldstein identifies this entropy as the *residual entropy* [23]; we will comment on this deep connection in Section 10.12.3.

A glass is SCL trapped in one of the many basins at T_g, and executes vibrations within this potential well. The resulting glass can be characterized by the properties of this basin [5, 6], which determine the *average* configuration of the glass in that basin. The high barriers of these potential wells also provide high stability to very slowly cooled SCLs, at least near T_g, and are responsible for the strong inequality $\tau_s \gg \tau_{exp}$. This also implies that barriers to the formation of stable nuclei must also be extremely high in SCLs. This was first argued by Goldstein in his seminal work on viscous liquids [22]. In other words, the metastability of SCLs is controlled by the almost *complete absence of stable nuclei*, and will be central in our analysis of the low-temperature metastability of SCLs. The high-temperature metastability in the liquid-gas transition, in which both phases have the same *symmetry*,[4)] has been extensively studied by various workers [32, 33], to which we refer the reader for more information. In our opinion, the supercooled metastability, where one is confronted with the distinct *symmetry* of the stable phase and its *absence* in SCL,[4)] has not received the same critical attention so far.

The above picture of slowly cooled SCLs allows considering the liquid cell model of Lennard-Jones and Devonshire [34] (Figure 10.1) and its various elaborations [35]. In the figure, we show a cell representation of a dense liquid in (a) and of a crystal in (b). Each cell is occupied by a particle in which the particle vibrates. A defect in the cell representation corresponds to some empty cells. The regular lattice in (b) is in accordance with Einstein's model of a crystal. In the liquid state, this regularity is absent. We consider the *configurational partition function* $Z(T, V)$ (Appendix 10.A),

$$Z(T, V) \equiv \frac{1}{v_0^N N!} \int e^{-\beta E} d^N\{\mathbf{r}\} \equiv \int W(E, V) e^{-\beta E} dE/\varepsilon_0, \qquad (10.5)$$

in the canonical ensemble at temperature T (measured in the units of the Boltzmann constant k_B; this amounts to effectively setting $k_B = 1$); $\beta \equiv 1/T$ is the inverse temperature and $W(E, V)dE/\varepsilon_0$ represents the number of distinct configurations with energy in the range E and $E + dE$; v_0 and ε_0 respectively represent the small-scale constant volume, such as the cell volume, and the energy constant, such as the average spacing between vibrational energy levels of a single particle in its cell in Figure 10.1b at $T = 0$. We set $v_0 = 1$ and $\varepsilon_0 = 1$ in this chapter.

From now on, we will make a distinction between a quantity and its configurational counterpart by adding a subscript "T" to the quantity. For example, $S_T(T, V)$ will denote the (total) entropy, while $S(T, V)$ the configurational part of $S_T(T, V)$. Most often we will simply use the name of the quantity Q such as the entropy to refer to its

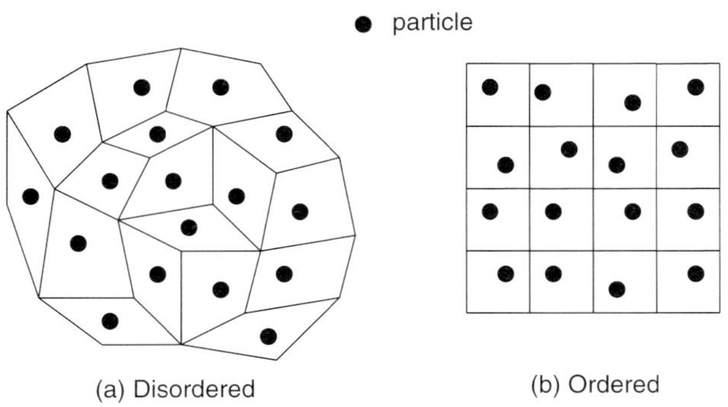

Figure 10.1 Cell representation of a small region of disordered (a) and ordered (b) configurations at full occupation: each cell contains a particle. Each cell representation uniquely defines a potential well or basin in the potential energy landscape. Observe that while each particle is surrounded by four particles in the ordered configuration, this is not the case for the disordered configuration. We have shown a higher volume for the disordered configuration, as found empirically.

configurational part, but will always use total quantity to refer to $Q_{\mathrm{T}}(T, V)$ from now on. The *microcanonical* configurational entropy is given by the Boltzmann relation (10.1), so $S(E, V) \equiv \ln W(E, V)\mathrm{d}E \simeq \ln W(E, V) \geq 0$.[5] The average energy $\bar{E} = E(T, V)$ is used to give the *canonical* configurational entropy $S(T, V) \equiv S[E(T, V), V]$. The slope of the entropy at $\bar{E}$ is related to the inverse temperature by the standard relation valid at equilibrium

$$(\partial S/\partial E)_{\bar{E}} = \beta. \tag{10.6}$$

This entropy differs from the total entropy $S_{\mathrm{T}}(T, V)$, see (10.A7) and (10.A3), which is the entropy associated with the total PF $Z_{\mathrm{T}}(T, V)$ in (10.A1) where the kinetic energy (KE) is also included [36–39]. The entropy contribution $S_{\mathrm{KE}}(T)$ in (10.A3) is the same for all systems as it is independent of the interactions and volume.

Remark

In classical statistical mechanics conventionally used to analyze metastability, the contribution $S_{\mathrm{KE}}(T)$ to $S_{\mathrm{T}}(T)$ from translational degrees of freedom is the same for all phases like SCL or CR at a given T, and is a function only of T.

For this reason, we do not have to include $S_{\mathrm{KE}}(T)$ in any investigation of metastability. We only consider configurational degrees of freedom from now on, unless explicitly mentioned otherwise. For the same reason, there is also no need to explicitly show the dependence on V, unless clarity demands otherwise.

The configurational part of the Helmholtz free energy is given by $F(T, V) \equiv -T \ln Z(T, V)$. The free energy per particle will be denoted by the lower

5) For a macroscopic system, we will always neglect terms that do not grow exponentially fast, as they are not relevant in the thermodynamic limit $N \to \infty$.

case letter $f(T, V) \equiv F(T, V)/N$. The total Helmholtz free energy is obtained by adding to it the contribution $-TS_{KE}(T)$. If we consider a system at constant T, and P, then we need to allow for fluctuating volumes. This requires an additional sum over volumes with a weight $\exp(-\beta PV)$. This defines a *NTP* ensemble with the following partition function:

$$Y(T, P) = \int Z(T, V)e^{-\beta PV} dV/v_0. \tag{10.7}$$

The introduction of the volume element $v_0(\equiv 1)$ is to ensure that $Y(T, P)$ is dimensionless. The extensive free energy corresponding to $Y(T, P)$ is the configurational part of the Gibbs free energy $G(T, P) = -T \ln Y(T, P)$. To obtain the total Gibbs free energy $G_T(T, P)$, we must add to it $-TS_{KE}(T)$. We will mostly consider the canonical ensemble. However, the arguments can be easily extended to the *NTP* ensemble or other ensembles.

10.1.5.2 **Communal Entropy, Free Energy, and Lattice Models**

The communal entropy is defined as the difference between the configurational entropy $S(T)$ of the system and the entropy $S_b(T)$ when the particles are *confined* in their cells [40], that is, the basin (we suppress showing the V dependence):

$$S_{comm}(T) \equiv S(T) - S_b(T) \geq 0; \tag{10.8}$$

see Section 10.4.1 for a precise definition of the basin entropy $S_b(T)$. The communal part of the free energy is given by

$$F_{comm}(T) \equiv F(T) + TS_b(T) = E(T) - TS_{comm}(T). \tag{10.9}$$

The communal entropy is the entropy due to the deconfinement of the system from the basin. Accordingly, it vanishes when the system is confined in a basin such as in Figure 10.1a. Let $T_K > 0$ denote the temperature when this happens for SCL:

$$S_{comm}^{SCL}(T_K) \equiv 0. \tag{10.10}$$

As noted in Section 10.1.5.1, it is a common assumption that CR is also confined to a single basin such as the one shown in Figure 10.1b. We will verify this assumption in this chapter. If the assumption holds, then the confinement will occur at some $T_{CR} > 0$:

$$S_{comm}^{CR}(T_{CR}) \equiv 0. \tag{10.11}$$

In confined or localized states, particles only occupy positions that are within their individual cells. Let us focus on a SCL localized into one such basin; however, the discussion is equally valid for a localized CR. Let the minimum of the basin energy be E_K. It corresponds to the *average state* in the basin in which each particle has its average position. Any deviation from these positions will only raise the potential energy. The relevant average state in the basin at $T = T_K$ remains unchanged below T_K and continues to represent the average state of the system.

In other words, the average state of the system remains frozen in this *inert state* below T_K and is called the *ideal glass* (IG). This localization and freezing of the average state to IG is called the *ideal glass transition* (IGT). Thus, the vanishing of $S_{comm}(T)$ in (10.10) is taken as the condition for the formation of an IG due to its localization or confinement in a single basin, and the temperature T_K at which this occurs is called the ideal glass transition temperature, or the Kauzmann temperature, to honor the pioneering contribution of Kauzmann in the field of supercooled liquid, even though the term is commonly used to denote the temperature where the CR and SCL have the same entropy. This issue has been discussed elsewhere [37–39], to which we refer the reader for additional information. As the above discussion is also applicable to a confined CR, there will be a similar localization transition in CR.

Remark

In a lattice model, particles do not deviate from their fixed lattice positions. Consequently, the total entropy in the lattice model is purely communal in nature.

Thus, a lattice model can be effectively used to analyze the average state of a glass former in real continuum space. Accordingly, we do not feel guilty about using lattice models here for which exact calculations can be carried out. There are other reasons to use *lattice* models to investigate glass formers. This is further discussed in Appendix 10.B.

Since the heat capacity is nonnegative, $S(T)$ and $E(T)$ are monotonic increasing function of T, and must have their minimum values at absolute zero. Assuming CR to be the stable phase at absolute zero, we conclude that it must be in the state with the lowest possible energy $E_0(V)$. This follows from the Nernst–Planck postulate[2)] $TS_{CR}(T, V) \to 0$ as $T \to 0$, so $F(T = 0, V) = E_0(V)$. Thus, $E_0(V)$ *sets the zero of the temperature scale in the system.* As long as the heat capacities of various phases, stable or metastable, remain nonnegative, and we will see that this is true, these higher energies will correspond to temperatures $T > 0$.

10.1.6 Fundamental Postulate: Stationary Limit

There are no general arguments [16, 41, 42] to show that thermodynamically stable states must always be ordered, that is, periodic. The remarkable aperiodic Penrose tilings of the plane, for example, by two differently but suitably shaped tiles are stable. It is found empirically that the volume (or the energy or enthalpy) of a glass, or more generally, a NES at absolute zero is higher than that of the corresponding crystal; see, for example, Ref. [43] for a careful analysis of data. Here, we will focus on the potential energy for which the above observation is in accordance with (10.3) (Figure 10.2). (There will be no energy gap if $E_{NES} = E_0$). According to (10.3), there must be many *defects* in the glass relative to the crystal even at absolute zero to account for this difference in the energy (or enthalpy). The value of E_{NES} depends on the rate of cooling r. As r decreases, this energy falls and approaches a limiting value $E_K \leq E_{NES}$, which is still higher than E_0 (Sections 10.10 and 10.11). As CR is heated, its energy

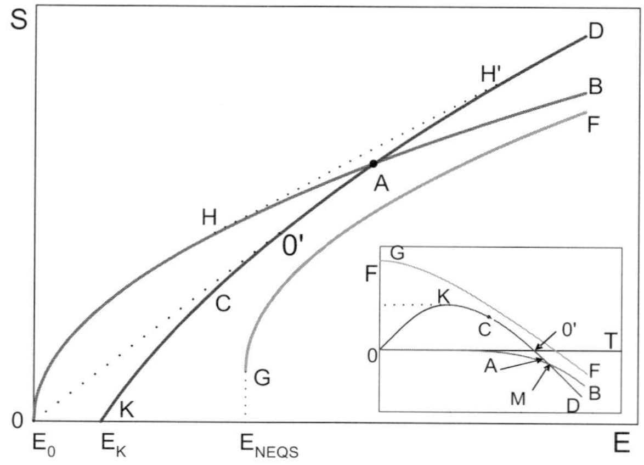

Figure 10.2 Schematic form of generic (total, cofigurational, or lattice communal) entropy functions as continuous and *concave* functions[3)] of E for a fixed volume V for various possible states, and the resulting continuous and concave Helmholtz free energies in the inset. The communal entropy is defined in (10.8). At present. we need not worry about the distinction between the total and configurational entropies except to note that both of these and the communal entropy on a lattice have similar features. Note the presence of an energy gap ($\Delta_G > 0$) in the model. This form, which is shown in an exaggerated fashion to highlight the distinction, will be justified in this chapter. The point O′ represents the point where the free energy DO′CKO of the liquid in the inset is equal to the free energy of the crystal at O (absolute zero), as shown by the gap theorem (Theorem 10.1). The point A on DO′CKO in the inset is slightly below the melting temperature T_M located at M, where it crosses the crystal free energy OMB.

rises due to the defects, but their densities is much smaller in the crystal than in the corresponding glass at the same temperature. As a consequence, one can treat a glass as a highly defective crystal [44].

As we wait longer and longer, NES will approach its two possible limits: EQS or SMS. In the process, the energy E_{NES} will continue to decrease. It will converge to E_0 if NES approaches CR and there will be no energy gap; otherwise, it will converge to some higher energy E_K with an energy gap, if NES converges to IG. The presence of the gap will turn out to imply [45] a nonzero $T_K > 0$. The assumptions leading to this important result are reviewed in Section 10.9.

Since we can never wait infinitely long time ($\tau_{exp} \to \infty$) to actually observe a SMS, its existence can never be verified. This is no different from what is customary in equilibrium statistical mechanics, where the existence of the equilibrium state is taken for granted as a postulate. We quote Huang ([41], p. 127): “Statistical mechanics, however, does not describe how a system approaches equilibrium, nor does it determine whether a system can ever be found to be in equilibrium. It merely states what the equilibrium situation is for a given system.” Ruelle ([42], p. 1) notes that equilibrium states are defined *operationally* by assuming that the state of an isolated system tends to an equilibrium state as time tends to $+\infty$. Whether a real system actually approaches this state cannot be answered. Therefore, we will also assume the *existence* of SMS as a fundamental *postulate*, so that the PF formalism could be applied to study it. The actual dynamics that leads to such a state will not be our focus here, even though it is important in its own right. The hope is that the study of the long time limit of metastable states, which can be carried out using the basic principles of statistical mechanical formalism, though modified by imposing some *restrictions* as described in the chapter later, still has a predictive value.

10.1.7 Thermodynamics of Metastability

To form a glass, *crystallization must be avoided* either by ensuring that the material does not have time to become crystalline or by suppressing the mechanism to form a crystal. As glasses, or more generally NESs, are formed under some sort of constraints (crystallization is forbidden in the case of glasses), their (configurational) multiplicity $W_{\mathrm{NE}}(E, V)$ (which may be a function of time, but we do not show it) must not be greater than $W_{\mathrm{EQ}}(E, V)$ of the corresponding equilibrium state, so that we must always have

$$S_{\mathrm{NE}}(E, V) \leq S_{\mathrm{EQ}}(E, V); \tag{10.12}$$

see Figure 10.2 for fixed V. We use (10.1) for $S_{\mathrm{EQ}}(E, V)$, and (10.2) for $S_{\mathrm{NE}}(E, V)$. The curve OHH′D (with straight segment HH′) represents $S_{\mathrm{EQ}}(E, V)$ and the curve GF represents $S_{\mathrm{NE}}(E, V)$. In time, the curve GF will move upward toward OHH′D: it either converges to it if crystallization occurs, or to KCAD if it is forbidden. This is consistent with the second law of thermodynamics: As the constraints are removed, the entropy of a closed system (fixed E, V) *cannot* decrease in time; it can only increase or remain constant.

The segment H′D represents the entropy of the liquid, the disordered phase, and is determined by the multiplicity $W_{\mathrm{dis}}(E)$ of disordered configurations. The segment OH represents the entropy of the crystal, the ordered phase, and is determined by the multiplicity $W_{\mathrm{ord}}(E)$ of ordered configurations. Because of the straight segment HH′, the equilibrium entropy $S_{\mathrm{EQ}}(E, V)$ is a singular function, which is then reflected in a singular equilibrium free energy at the melting temperature T_{M}; the latter is given by the inverse of the slope of HH′. Let $E_{\mathrm{CR,M}}$ and $E_{\mathrm{EL,M}}$ denote the energies of the coexisting phases CR and EL at T_{M}; see points H and H′ in Figure 10.2. It is hard to imagine that the ordered and disordered configurations terminate at $E_{\mathrm{CR,M}}$ and $E_{\mathrm{EL,M}}$, respectively. Thus, we will assume that the curve KCAH′D represents the entropy $S_{\mathrm{dis}}(E) \equiv \ln W_{\mathrm{dis}}(E)$ of the disordered states in the system even below $E_{\mathrm{EL,M}}$. Similarly, we will assume OHAB to represent the entropy $S_{\mathrm{ord}}(E) \equiv \ln W_{\mathrm{ord}}(E)$ of the ordered states in the system even above $E_{\mathrm{CR,M}}$. The entropy then has two different branches KCAD and OHAB, rather than being a single function given by OHH′D. We only require that the branches be continuous and concave. The existence of the two branches requires that we are able to distinguish between disordered and ordered states. After all, the glassy state is formed by disordered states. So, making such a distinction is not merely an academic curiosity; it is a vital issue for our understanding of how and why glasses are formed.

Assumption

The singularity of the single entropy function OHH′D is absent from either of the two branches or in $S_{\mathrm{NE}}(E, V)$.

At this moment, this is merely an assumption, which needs to be tested. We will show later by exact calculations that it is valid in these calculations. This is also the case in other exact calculations that have been carried out in our group [36, 37, 44, 46–48].

10.1.8 Scope of the Review

While there is no rigorous theory of such SMSs at present, there are some valuable approaches available in the literature. One approach is to use the formalism of Penrose and Lebowitz (PL) [1] using *restricted* ensemble method, which we modify and adapt for our purpose. In particular, the decay of the metastable states will be *completely* suppressed in order to make them stationary. This results in completely banning the nucleation of the stable phase, which is consistent with Maxwell's idea [21] that to observe metastable states, we must ensure that the stable phase is not present. The properties of the SMS are what PL call the static or reversible properties [1]. In their approach, only certain microstates, which we take to be ordered or disordered, out of all are allowed to determine the partition functions. Their multiplicities $W_{\mathrm{ord}}(E)$ and $W_{\mathrm{dis}}(E)$ define the two separate restricted partition functions for each kind of states (Section 10.8). The PFs will not contain any singularity if multiplicities themselves are nonsingular. Despite this, a singularity will appear as we switch over from one PF to another if we insist on only considering equilibrium states; see Sections 10.8 and 10.12 for further details. The second approach requires analytically continuing the eigenvalues of the transfer matrix as presented in Ref. [49]. This accomplishes the same goal as the restricted ensemble but in a somewhat direct fashion. An important issue in both approaches is to unravel the condition or conditions under which extrapolation will represent SMSs that might be observed in Nature.

To develop our approach, we borrow ideas from both approaches. We then derive the consequences of our gap model schematically presented in Figure 10.2. We prove that assuming the existence of the gap in (10.3) and the validity of (10.12), the IG transition must occur at a positive temperature T_{K} (see Section 10.9). This then assures a rapid entropy drop near T_{K} as a precursor of the impending entropy crisis, which is discussed in Section 10.3.6.

In view of the Remark in Section 10.1.5.2, we find it convenient to consider the communal entropy rather than the total or configurational entropy to study the glass transition. The configurational entropy can be broken into a sum of three terms (10.19), each of which must be independently nonnegative. As soon as one of them vanishes at some positive temperature, this component will induce an entropy crisis by becoming negative if we attempt to extrapolate it to lower temperatures. Indeed, we discover that it is the communal entropy in (10.8) and (10.20), whose vanishing describes IG transition (10.10). The ideal glass transition then gives rise to a singularity in the free energy, so the above extrapolation is of no significance. This is established in Section 10.12. The use of the cell model (Section 10.1.5) allows us to use lattice models to provide an unambiguous clue to the entropy crisis (see Remark in Section 10.1.5.2). This gives rise to a tremendous simplification as continuous models are hard to deal with, and also because continuum models suffer from negative entropy (see Appendix 10.B). The negative entropy in a continuum model is unrelated to any glass transition. Therefore, a negative entropy there cannot unambiguously decipher any IG transition.

The inequality (10.12) and the presence of a gap play a central role in our modeling of metastability in terms of the entropy function $S(E)$ with a gap. We are primarily interested in the thermodynamic understanding of the nature of the glass transition, and not in its dynamical aspect. This does not mean that the latter is not important. However, we believe that without a thorough understanding of any SMS, there is no hope to deeply understand its dynamic behavior. Accordingly, we have used the above inequality (10.12) and the gap as the foundation to model the possible form of the generic entropy for any glass former in Figure 10.2. Our hope is that this chapter will make a modest progress in our understanding of glasses, as glass transition (GT) even in a molecular liquid (for example, water and silicate melts) remains a controversial long-standing problem even after many decades of active investigation and presents one of the most challenging problems in theoretical physics [5, 6, 9, 25]. We will justify the gap model on theoretical and numerical grounds. Our approach, which is based on a fundamental principle of thermodynamics, should hopefully provide a solid basis of a new understanding.

Our formulation and its consequences have been tested by us in exact calculations of several lattice models [36, 37, 44, 46–48]. Metastability comes out of our formalism because it allows us to abandon the global free energy minimization principle, not because of any approximation. We discuss two such examples in this chapter. One of them is a one-dimensional lattice model. This model has only nearest-neighbor interactions, and is solved exactly by the use of the transfer matrix. We find that the extrapolation can be carried out without any ambiguity to describe SMSs in this case. Thus, SMS can exist even in nonmean field theories and without long-range interactions, which is in itself an interesting result and disproves the folklore. The other lattice model deals with an Ising model with only short-range interactions and without any frustration. This model represents a binary mixture or a pure component depending on the interpretation and captures SMSs. Both examples show that the extrapolation, although thermodynamically stable to absolute zero, does not represent any observable metastable state at low temperatures because their entropy becomes negative there. The entropy crisis is avoided by the ideal glass transition.

10.2 Modeling Glass Formers by an Energy Gap

10.2.1 Distinct SMSs

It is clear from the presence of the high-energy barriers, at least near T_K, that lend justification to the cell model in Section 10.1.5 that viscous liquids and glasses are *metastable states in which equilibrium (stable) nuclei are almost forbidden to exist*. Thus, if the interest is to only locate T_K and not the dynamics itself, there is no harm in treating very slowly cooled metastable SCLs near T_K and glasses in the mathematical formalism to be developed here as originating from nearby metastable states at higher temperatures in which stable nuclei are also forbidden to exist. In the absence of these nuclei, MSs cannot decay to corresponding crystals, so their long time limit remains distinct from corresponding crystals and represents a *new* and *unique* state, identified here as SMSs. Long-time stability of medieval glasses is a proof in itself about the existence of this new state over that extended long time. We, therefore, accept the existence of this new state associated with SCLs. As SCLs and glasses are random in nature as opposed to the ordered CR, this distinction allows us to consider disordered and ordered microstates separately, so that they will never mix. This will result in two distinct restricted PFs (Section 10.8). This is to be contrasted with the study of high-temperature metastability in which stable nuclei are allowed to be formed; their presence then gives rise to the decay of metastable states [1]. This decay is signaled by the presence of a singularity [32, 33], either an essential singularity or in the form of spinodals, so that they will never mix.

10.2.2 Entropy Extension in the Gap

We consider the energy range over which there is only one transition, the melting transition. We have a disordered phase (the equilibrium liquid EL) above the melting temperature T_M and an ordered phase CR below it. The two phases have distinct symmetries,[4)] allowing us to classify microstates into two distinct sets containing ordered or disordered microstates, as will be shown later. The corresponding entropies $S_{ord}(E) \geq 0$ and $S_{dis}(E) \geq 0$ are schematically shown in Figure 10.2, from which we can obtain the communal and total entropies. The entropy as a function of E must be thought of as the entropy in the microcanonical ensemble [50], which must be at its maximum in the equilibrium state [16], so these curves are given by (10.1). As far as the communal entropy for disordered configurations is concerned, it *vanishes* at K, because of the presence of the gap between E_0 and E_K. This entropy function $S_{comm,\,dis}(E)$ can be *extended* to energies in the gap, though the extension will give negative values and will be unphysical. Despite this, the extension will serve a very useful purpose.

Remark
All we require of the extension of $S_{\mathrm{comm,dis}}(E)$ in the gap is that the resulting entropy be continuous and concave.

In most computations, the analytic form of the communal entropy $S_{\mathrm{comm,dis}}(E)$ will be known over the range $E \geq E_0$, so the extension is unique. Otherwise, the extended portion of the entropy can be arbitrary provided it does maintain continuity and concavity. This arbitrariness is irrelevant as its values over the range $[E_0, E_K]$ will not determine the physics of the system. This will become apparent in Section 10.12.

Since SMS is not the equilibrium state EQS, its entropy *cannot* exceed the entropy of the corresponding equilibrium state at the same E; we keep V fixed. At lower energies, the ordered state must have the highest entropy, while at higher energies the disordered state must have the highest entropy. On the other hand, if a time-dependent metastable state is prepared under the constraint that the stable phase is not allowed, then its entropy will be represented schematically by FG.[6] The free energies corresponding to the above entropy functions are shown in the inset in Figure 10.2. The SMS free energy $F_{\mathrm{dis}}(T)$ *cannot* be lower than the free energy $F_{\mathrm{ord}}(T)$ of CR at the same temperature T, and vice versa. This explains the form of the free energy in the inset. The slope of the tangent line HH′ gives the inverse melting temperature, while the slope of the tangent line OO′ gives the inverse temperature at which the free energy DO′CK in the inset is equal to the free energy of the crystal phase at absolute zero ($T = 0$) (Theorem 10.1).

10.2.3
Gibbs–Di Marzio Theory

Making a distinction between ordered and disordered microstates has been a time-honored practice in theoretical physics. In the context of polymers, this was carried out by Flory [51] in his study of polymer melting, which was later followed by Gibbs and Di Marzio [18] in their highly celebrated work on glass transition in polymers. Flory considered a simple model of semiflexible polymers in which each gauche bond has a penalty $\varepsilon > 0$ each over a trans bond for which the energy is zero. Let g denote the density of gauche bonds, so the energy is simply $E = Ng\varepsilon$, where $N \to \infty$ is the number of lattice sites. Let us consider a single polymer whose monomers cover all the lattice sites of a square lattice. In the crystalline state, the polymer bonds are all trans, so $g = 0$. All bonds are parallel, which can be either in the horizontal or in the vertical directions on a lattice, oriented so that its lattice bonds are either horizontal or vertical. In the disordered state, there are equal number of bonds in both orientations. Let n_{H} and n_{V} respectively denote the density of horizontal and vertical

6) We treat nonstationary metastable state as one of the partial equilibriums and follow Landau and Lifshitz ([16], p. 27) to define its entropy according to (10.2). This entropy will continue to increase with time as different parts of the system move toward the same stationary state. This explains why FG lies below DAK. Using the entropy function given by FG, we can also calculate the corresponding free energy, which is shown in the inset by FG.

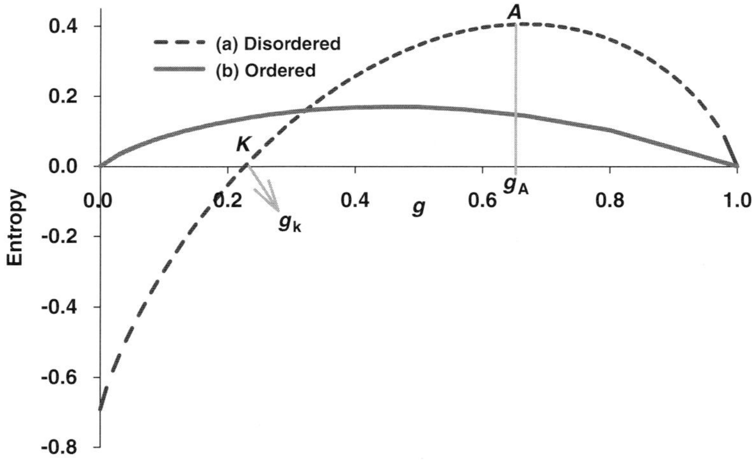

Figure 10.3 Entropy for the disordered (a) and ordered (b) states for the Flory model of semiflexible linear chain. Flory's approximate form for the disordered state gives a negative entropy below g_k, indicating the presence of an energy gap in the energy for the disordered configurations with respect to the crystal energy at absolute zero. We are considering a chain that covers all the N sites of the lattice. The exact calculation by Gujrati and Goldstein provides a lower bound for the ordered state shown by (b). The approximate form of the entropy for the disordered liquid state is given by the Flory–Huggins approximation. Compare (a) with KAD and (b) with OMB in Figure 10.2.

polymer bonds. One can introduce $\varrho \equiv n_H - n_V$ as the order parameter (see Section 10.7.3 for more details), so it takes the value 0 in the disordered state. For the crystalline state, ϱ takes a nonzero value that reaches ± 1 at absolute zero when the crystal is completely ordered. Flory assumed that the entropy of the crystalline phase $S_{ord} = 0$.[5] Thus, the entire curve OHAB reduces to a single point at the origin in Figure 10.2. He also used the crude Flory–Huggins approximation to calculate $S_{dis}(g)$, which per site is shown schematically by the dashed curve (a) in Figure 10.3. The entropy per site is given by

$$s_{FH,dis}(g) = -\ln 2 - g \ln (g/2) - (1-g) \ln (1-g) \tag{10.13}$$

in the Flory–Huggins approximation (see Ref. [52] for details). We show $s_{FH,dis}(g)$ for $g \geq 0$; this entropy is negative for $g < g_K$ [52]. The part $s_{FH,dis}(g) \geq 0$ of this curve is shown in Figure 6 in Ref. [52b]. At $g = 0$, the entropy takes a negative value

$$s_{FH,dis}(g = 0) = -\ln 2. \tag{10.14}$$

We clearly see an energy gap at the lower end of the curve at K; $g_K > 0$. For the square lattice, we find that $g_K \simeq 0.227$ [52], where $s_{FH,dis}(g)$ vanishes with a *finite* (nonzero) slope, as was the case for the communal entropy in Figure 10.2, even though the analytical form of $s_{FH,dis}(g)$ derived by Flory is over the full entire range [0, 1] of g. This entropy is also concave over the entire range, including the gap. As the point H in

Figure 10.2 has moved to the origin in the Flory calculation, the melting transition is obtained by drawing a tangent to the curve (a) and passing through the origin at $g = 0$. The inverse of the slope of this tangent is the melting temperature T_M(10.6). Note that the point A ($g = g_A$) on the curve (a) corresponds to infinite temperatures, so the portion of the curve on the right of A for $g > g_A$ is not relevant.

Gibbs and Di Marzio [18] used the above $s_{FH,dis}(g)$ to demonstrate the entropy crisis in polymers (see Ref. [52] for details). This calculation was the first one of its kind to demonstrate the entropy crisis. Despite its limitation, to be discussed below, the work by Gibbs and Di Marzio has played a pivotal role in elevating the Kauzmann entropy crisis from a mere curious observation to probably the most important mechanism behind the glass transition, even though the demonstration was only for long molecules.

To provide a better description of the crystalline phase than given by Flory, Gujrati and Goldstein [52] considered local excitations in the perfect crystal at $T = 0$ and obtained a lower bound $\bar{s}(g)$ of the entropy, whose optimistic form is numerically given as the solid curve (b) in Figure 10.3 for a square lattice. For small g, this optimistic bound by Gujrati and Goldstein is given by

$$s(g) \geq \bar{s}(g) = (g/8)[2 \ln (4/g-3) - 0.56827 \ldots].$$

Near $g = 0$, this bound should describe the excitation in the perfect crystal quite well so that the bound should be close enough to the exact $s_{ord}(g)$ near $g = 0$. For large g, these and other excitations will disorder the configurations. So, for large g, the configurations counted in the solid bound (b) should not be interpreted as ordered configurations; they most probably represent disordered configurations. However, this point is not relevant for our discussion here.

As $s_{ord}(g) \gtrsim \bar{s}(g)$ in (b) near $g = 0$, the original idea of Flory about the crystalline state turns out to be incorrect. As this idea was also central in the calculation of Gibbs and Di Marzio, their work was severely criticized by Gujrati and Goldstein [52]. Unfortunately, the criticism by Gujrati and Goldstein has been incorrectly interpreted [53] by taking their bounds to be also applicable to the metastable states (SCL) in polymers. To overcome the bounds, Di Marzio and Yang have suggested to replace the crisis condition $S_{SCL} < 0$ by $S_{SCL} < S_{c0}$, where S_{c0} is a small critical value.[7] This is

7) Communal entropy being less than a positive value, no matter how small, cannot be as fundamental an entropy crisis as the requirement $S_{comm} = 0$ to argue for an ideal glass transition. For example, liquid helium shows no glass transition when its entropy becomes equal to such a small positive value. Thus, we will adhere to $S_{comm} = 0$ as the most fundamental requirement for the entropy crisis. This also rules out using the excess entropy $\Delta S_{ex}(T)$ in (10.16) used by Kauzmann and various other authors, to be used as a signal of an entropy crisis when it vanishes at a positive temperature, since thermodynamics itself does not rule out the possibility that the total CR entropy $S_{T,CR}(T)$ can be greater than the total liquid (SCL) entropy as seen recently [36, 46, 47] in exact calculation. Liquid He also has the property that its CR phase can have higher entropy than the liquid phase at low temperatures. Thus, $\Delta S_{ex}(T) < 0$ under extrapolation does not pose any thermodynamic problem of stability or reality, and we will not use it as the signal for any entropy crisis although it is commonly done; rather, we will consider the communal entropy and use its negativity under extrapolation as the criterion for the entropy crisis.

not the correct interpretation of the Gujrati–Goldstein bounds. The bounds near $g = 0$ are only for the equilibrium states, since they are obtained by considering the Gujrati–Goldstein excitations in CR; they are not applicable to SCL. More recently, we have carried out an explicit exact calculation [36, 46, 47] for long semiflexible polymers, which not only satisfies the rigorous Gujrati–Goldstein free energy upper bound [52] for the equilibrium state but also yields a positive Kauzmann temperature $T = T_K$.[7)] These calculations thus provide a very strong support of a glass transition as envisioned by Gibbs [17] and Gibbs and Di Marzio [18]; it has also been tested by simulations [54]. Our results for a single chain covering all the sites of a square lattice are presented in Figures 10.4 and 10.5. Apart from the bending penalty ε between gauche and trans conformations of a bond, there is another parameter $a \equiv \varepsilon'/\varepsilon$ considered in the calculation, which generalizes the Flory model of polymer melting. Here, ε' is the energy of interaction between two parallel bonds in a square cell of the lattice. For $a = 0$, the new model reduces to the Flory model. We set $\varepsilon = 1$ for the results shown in Figures 10.4 and 10.5. We show the free energy $\tilde{F} \equiv F - a$ and the energy $\tilde{E} \equiv E - a$ as a function of T in Figure 10.4 and the entropy S and the specific heat C in Figure 10.5. The model has the following phases. There is an equilibrium liquid EL (dotted, dash–dot, and dash–dot–dot corresponding to different values of a) above the melting temperature T_M. Below T_M, we have a CR (dashed), which exists all the way to $T = 0$. At T_M, EL turns into a metastable liquid SCL (dotted, dash–dot, and dash–dot–dot corresponding to different values of a). There is an unusual feature of the model worth noting. There is another state, called ML (continuous), which exists at *all* temperatures but is quite unusual in its behavior. It is metastable with respect to EL/SCL at temperatures higher than some temperature T_{MC}, where EL/SCL terminates by turning into ML. Below T_{MC}, ML is the only metastable state, whose entropy vanishes at some positive Kauzmann temperature. The free energy of EL/SCL depends on a, but that of ML and CR do not. Therefore, both T_M and T_{MC} depend

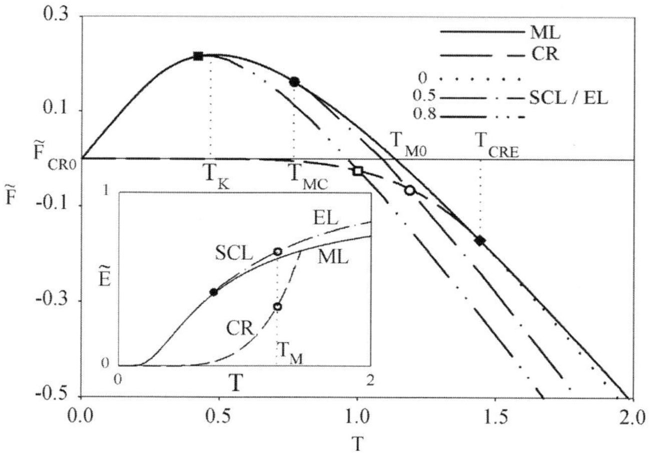

Figure 10.4 $\tilde{F}(T)$ for different states and $a = 0$, 0.5, and 0.8, and (*inset*) for $a = 0.5$. ML (continuous), CR (dashed, and EL/SCL (dotted, dash–dot, and dash–dot–dot).

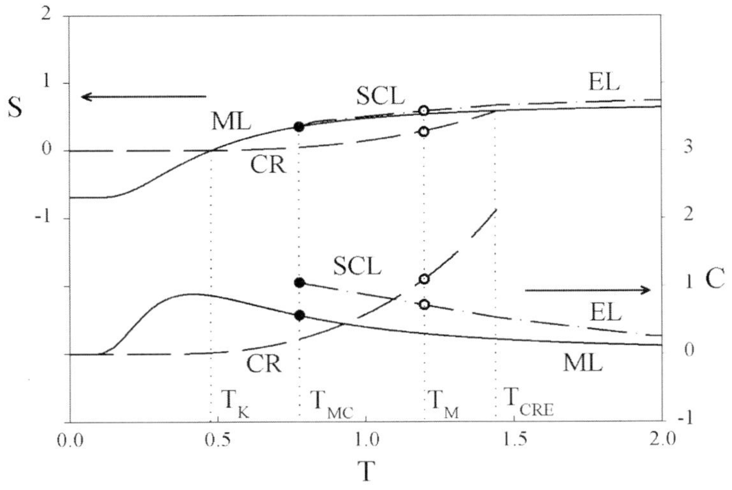

Figure 10.5 $S(T)$ and $C(T)$ for $a = 0.5$; see legend in Figure 10.4.

on a. The melting transition is discontinuous for $a > 0$, but turns into a continuous transition for $a = 0$ (Flory's model). Thus, the melting transition in the Flory model is a critical end point shown by T_{CRE}. We note that free energies of all states are concave to ensure that they remain stable. The calculation gives the state ML over the energy gap. Its entropy at absolute zero is identical to the entropy in (10.13) and is independent of a (see Figure 10.5). These conclusions do not change if the polymers are very long but finite and if we allow a small amount of free volume [36].

10.3 Glass Transition: A Brief Survey

10.3.1 Experimentally Observed Glassy State

Under suitable conditions, many substances can be cooled to a glassy amorphous solid, even though it is not established whether or not every substance can be put into a glass form [55]. Operationally we identify a solid as a fluid whose shear viscosity η exceeds some unusually large cutoff value such as $\eta_{exp} = 10^{14}$–10^{15} poise, which normally corresponds to a relaxation time τ of a day or so. The corresponding cutoff τ_{exp} of the relaxation timescale ranges anywhere between 10^2–10^4 s, with $\tau_{exp} = 10^2$ being the most common choice for experimentalists. This is obviously an arbitrary definition, but will serve to fix our ideas of a solid. The viscosity of most common liquids at room temperature is usually 10^{-2} poise, so by comparison, our cutoff value of the solid viscosity seems quite reasonable. It can be easily seen that such a value will correspond to a deformation of 0.02 mm in a material of size 1 cm^3 over a period of 1 day under a force of 100 N [8].

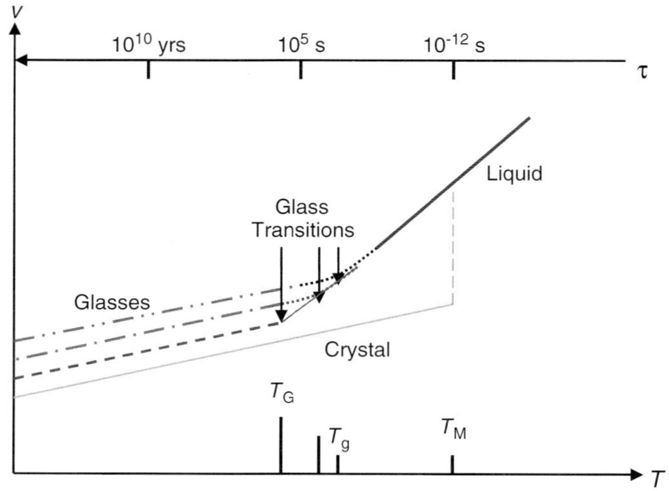

Figure 10.6 Schematic behavior of volume as the liquid is cooled. The freezing transition occurs at T_M to the crystal, which becomes perfectly ordered at absolute zero. If the crystallization is bypassed, we obtain the supercooled liquid, which turns into different glasses at glass transition temperatures (T_g) depending on the rate of cooling. As the cooling rate becomes smaller, T_g of the glass transition decreases (shown by arrows becoming larger), until finally it converges to its limit T_G under infinitely slow cooling rate. This limit is called the *ideal glass transition temperature*, and the corresponding glass shown by the dashed curve is called the *ideal glass*. A similar behavior in the slopes of the densities is also when we increase P at a fixed T; we merely replace T in the figure by $1/P$.

10.3.2 Glass Phenomenology

We plot in Figure 10.6 the volume per particle v as a function of the temperature T, but the following discussion applies equally well to any density function like the energy, enthalpy, or entropy per particle. The equilibrium liquid EL shown by the thick solid curve at higher temperatures $T \geq T_M$ can be forced to bypass crystallization so that it continues as a metastable liquid SCL below T_M, also shown by thick solid curve. The SCL, which exists below T_M, is an example of a NES and is *disordered* with respect to its *ordered* crystalline phase CR, shown by the thin solid curve. There is no abnormal behavior observed in the SCL volume v at T_M; contrary to that, the CR volume shows a discontinuity (see the vertical dashed line at T_M) with respect to the liquid volume at T_M. Depending on the rate of cooling r, the apparent "glass transition" occurs in SCL at temperature T_g (when $\eta = \eta_{exp}$, or $\tau = \tau_{exp}$), which is usually about two-thirds of the melting temperature T_M for the liquid. There is a discontinuity in the slope at T_g, as seen in Figure 10.6. The resulting glassy states are represented by the dashed–dotted curves or the dashed curve. The value of T_g depends on the pressure P at which the glass former is cooled. One can also obtain a glass transition by fixing the temperature and varying the pressure. At the corresponding glass transition pressure $P_g(T)$, which depends on the temperature of the glass former, one will find various densities to have a discontinuity in their slope, this time with respect to P.

It should be emphasized that $T_g(P)$ has no unique value, as it depends on the choice of η_{exp} or τ_{exp}. The relaxation time τ and viscosity η increase by several orders of magnitudes, typically within a range of a few decades of the temperature as it is lowered, and eventually surpass experimental limits τ_{exp} and η_{exp}, respectively; see the upper axis in Figure 10.6 for the relaxation time τ. As it usually happens, τ monotonically increases with lowering the temperature. For EL, $\tau < \tau_{exp}$, so that the system has enough time to come to equilibrium as we lower T. At T_M, we need to carefully ensure the *constraint* that crystal seeds are not allowed to form [21]. Consequently, we obtain a metastable SCL, which can still remain under "constrained" equilibrium as long as $\tau < \tau_{exp}$. The SCL begins to fall out of this constrained equilibrium when $\tau_{exp} \lesssim \tau$. This is shown by the dotted portion of the two curves representing glasses in Figure 10.6. Eventually, at low enough temperature, the system will appear to have no mobility when τ becomes extremely long compared to our experimental timescale τ_{exp} (such as $\tau_{exp} = 10^2$ or $\tau_{exp} \simeq 10^5$ s $\simeq$ one day). The loss of mobility results in "freezing" of the system *without* any anomalous changes in its thermodynamic densities like its specific volume or the entropy density at the glass transition T_g for that particular choice of τ_{exp}, which itself depends on the cooling rate r. The two dashed–dotted curves representing two different experimental glasses will not show any singularity at their respective glass transition temperature T_g. They gradually connect with the thick solid curve describing the liquid. Thus, the experimental glass transition should be thought of as a crossover phenomenon in which the supercooled liquid gradually turns into a glass over a temperature range.

10.3.3 Fragility

A quantity of interest is fragility [56, 57] m, which is expected to increase with τ_{exp},

$$m \equiv \lim_{T \to T_g^+} \frac{\mathrm{d} \ln \tau}{\mathrm{d}(T_g/T)}.$$

For conventional values of τ_{exp}, it takes values between $m \simeq 17$ for "strong" glass formers that show an Arrhenius behavior of η and $m \simeq 150$ for "fragile" glass formers. Empirically, m is found to be strongly correlated with the interaction in the system. Strongly directional bonds such as covalent bonds give rise to "strong" glass formers such as network glasses like silica (SiO_2; $m = 20$). In contrast, isotropic interactions such as van der Walls usually give rise to "fragile" glass formers such as O-terphenyl ($m = 80$). Fragile glasses are those formed by molecular glass formers, including polymers. It is known that the volume (to be precise, the free volume) plays no important role in network glasses. Recent molecular dynamics investigations show that the form of the potential such as its anharmonicity strongly influence the fragility [58].

Dynamics of SCL covers a very broad range from fast dynamics (known as the β relaxation) due to cage effects and slow dynamics (known as the α relaxation) due to structural relaxation that are separated by a plateau region [59]. At high T, SCL has only one kind of relaxation involving the diffusion of particles. As T is lowered, there comes a temperature $T_{MC} > T_g$, where the above two different relaxation processes emerge out of the diffusive motion. The peak of the low-frequency response of the α process vanishes as T_g is approached, while the peak due to the β process persists even below T_g. Johari and Goldstein [59] have emphasized that both relaxations are characteristics of all glass transitions. The time when the α relaxation emerges out of the plateau region increases with decreasing r so that the plateau region gets broader and broader, thus suggesting that the plateau will eventually diverge in width in the hypothetical case when r vanishes. Consequently, any theory of dynamics ($r > 0$) must be able to explain the two separate relaxation phenomena.

The transition at T_{MC} is now known as the mode coupling transition because it has been justified by the mode coupling theory [60]. The relationship between T_{MC} and T_G is not understood at present. We have attempted recently [36, 46, 47] to understand it partially in connection with long polymers. It was discovered that the free volume falls rapidly near T_{MC} for long polymer fluids, with the nature of the drop becoming *singular* in the limit of infinitely long polymers. Thus, even though the dynamics was itself not considered in these works, the vanishing of free volume suggests some deep connection with what one expects near T_{MC}. However, the connection is not very clear and remains speculative at best.

10.3.4
Ideal Glass Transition as $r \to 0$

It is commonly believed that in the theoretical limit $r \to 0$ (or $\tau_{exp} \to \infty$), which can never be accessed in an experimental setup, there will be a precise temperature, the limit

$$T_G (= T_K) \equiv \lim T_g \quad \text{as} \quad r \to 0 \quad \text{or} \quad \tau_{exp} \to \infty, \tag{10.15}$$

where we will observe a singular behavior, as shown in Figure 10.6: There is a sharp break at T_G in the slope, but no discontinuity in the density itself is expected to occur at T_G as is the case at T_M. As $r \to 0$ gives us a SMS, the limiting transition at T_G is the same as the ideal glass transition at T_K introduced in Section 10.1.5.2. This justifies equating T_G with T_K. The glass below T_G shown by the dashed curve in Figure 10.6 represents the *ideal glass*.

Because of the experimental time limitation, no experiment can ever be done at infinitely slow cooling rate. Hence, the limit $T_G = T_K$ can only be inferred by some sort of extrapolation to infinitely slow cooling rate; it cannot be measured in any experiment or directly demonstrated in any simulation as neither can accomplish infinitely slow cooling rate ($r \to 0$) (see Ref. [12] for further details). It is only by extrapolation that the latter two may predict an entropy crisis. The reliability of such extrapolation is debatable, and has been used to argue against an entropy crisis [61, 62]. By analyzing experimental data, they have argued on procedural grounds that an entropy crisis in any experiment must be absent, which is most certainly true. However, such arguments based solely on experimental data or simulation *without extrapolation* can never shed light on the issue of the entropy crisis, which is purely hypothetical. To verify the existence or nonexistence of the crisis, one must resort to theoretical arguments, including exact calculations. Several workers [14, 63] have argued theoretically that it is not possible to have any entropy crisis, not withstanding the explicit demonstration of it in long polymers in Section 10.2.3 and in the abstract random energy model [19]. The argument of Stillinger and Weber [63], in particular, is forceful though not rigorous [9]. While they concede that long polymers may very well have an entropy crisis at T_K, they argue for its absence in viscous liquids of small molecules. From a purely mathematical point of view, it is hard to understand how this scenario could be possible. Using the physical argument of continuity, we expect T_K to be a smooth function of the molecular weight. Thus, it does not seem possible that such a function remains zero over a wide but finite range of the size of the molecules, and abruptly becomes nonzero for very large sizes. A function like this must be a singular function. However, no argument that we can imagine can support a particular large molecular size to play the role of the location of such a singularity. Thus, whether T_K is nonzero or not can only be settled by an exact calculation of the kind reported in Sections 10.10 and 10.11.

Glass formation has been studied numerically by several authors. Abraham [64] carried out Monte Carlo simulation for a system of 108 spherical particles using the Lennard-Jones 6–12 potential and clearly demonstrated the existence of glass transition in simple fluids in a temperature quench or a pressure crush. The density shows a discontinuity in its slope at a temperature that is identified as the glass transition $T_G = T_K$ (see Figure 10.6). Recall that a discontinuity in slope can only occur at T_K, but not at an experimental glass transition at T_g. A nice summary of more recent simulations can be found in Ref. [10].

The uniqueness of SMSs makes investigating the singular behavior at T_K more appealing from a thermodynamic point than the crossover behavior around some T_g. It should be emphasized that the infinite time limit of SCL in (10.15) implies that it can be treated as the "equilibrium" MS as opposed to a NES represented by dashed–dotted curves in Figure 10.6. Because of their time independence, both states can be studied using the (restricted) PF formalism. On the other hand, such a formalism can only be a crude approximation to study nonequilibrium states, where time dependence is present. It is also clear that in the limit $r \to 0$ at T_K, one will only observe the β relaxation; the α relaxation will never emerge because of the broadening of the plateau. This follows immediately since the glass is trapped in a single potential well for an infinitely long time. Therefore, it cannot make any excursion out of this well to another well (see Section 10.1.5).

10.3.5 Kauzmann Paradox and Thermodynamics

A distinctive feature of low-temperature metastability and supercooling is the almost universal rapid drop in the excess entropy:

$$\Delta S_{\text{ex}}(T, V) \equiv S_{\text{T,SCL}}(T, V) - S_{\text{T,CR}}(T, V). \tag{10.16}$$

As discussed in Section 10.1.5, a glass is trapped in a basin or cell representation that does not change with the temperature as it is lowered [5]. Just above T_g, SCL will contain certain basin entropy due to the possible number of basins in which a glass can be trapped. This entropy cannot be negative. As there is no way to uniquely measure this entropy in SCL, Kauzmann [5, 6] proposed to use the excess entropy as a measure of the entropy associated with different basins, the idea being that the vibrational entropy of the glass in any basin or cell representation is almost identical to that of the crystal. The drop in $\Delta S_{\text{ex}}(T, V)$ tends toward zero rapidly near the lowest possible experimental T_g. Indeed, if $\Delta S_{\text{ex}}(T, V)$ is extrapolated to the limit of infinitely slow cooling rate, one invariably finds it to go to zero at a nonzero temperature, as first observed by Kauzmann [5, 6] from thermal data for various systems capable to form a glass. The possibility that the excess entropy drops to zero and becomes negative under extrapolation is known as the *Kauzmann paradox* as it suggests that the number of basins becomes less than one, which is impossible. Just before it could vanish at this temperature, the system undergoes an experimental glass transition at T_g and the system avoids the entropy crisis. Again, whether the excess entropy vanishes can only be answered by an exact calculation, and not by experiments due to time constraints in the latter. This further emphasizes the need to study the long time limit of supercooled states theoretically (see also Ref. [37]).

By a careful analysis, Goldstein [23] has shown that much of the excess entropy arise from nonbasin contributions, such as vibrational differences between glassy and crystalline phases, and is reflected most prominently in boson peaks. Thus, it is obvious that one cannot take the excess entropy as a genuine measure of the basin entropy or the entropy associated with the number of cell representations.

As there are no vibrations on a lattice, the entropy on a lattice is purely due to cell representations. Its rapid drop with lowering T at low temperatures is a thermodynamic requirement due to nonnegative specific heat, as can be seen from an isolated Ising spin in a magnetic field, which is equivalent to noninteracting Ising spins in a magnetic field. The energy of interaction of N mutually noninteracting Ising spins $\sigma_i = \pm 1$, each experiencing an external magnetic field H, is given by $E = -\sum H\sigma_i$, and the resulting entropy and the heat capacity are given by

$$S = N[\ln\{2\cosh(H/T)\} - (H/T)\tanh(H/T)],$$
$$C = N[(H/T)/\cosh(H/T)]^2.$$

The two quantities per spin are plotted in Figure 10.7; compare with Figures 1.11–1.13, in Ref. [65]. The entropy falls off very rapidly at low temperatures with a resulting peak mimicking a discontinuity in the specific heat, similar to what happens in SCLs [6].

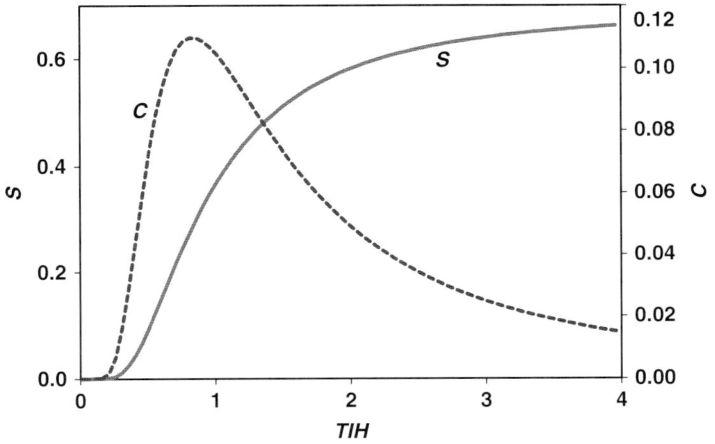

Figure 10.7 Behavior of the entropy S (solid) and the heat capacity C (dashed) for a single spin ($N = 1$) in an external magnetic field H. The entropy rises rapidly around $T/H = 1$, which corresponds to a peak in the heat capacity in the vicinity.

What distinguishes experimentally prepared SCL is that the drop in its extrapolated basin entropy actually becomes negative below $T_K > 0$ [5, 6, 9, 25].

10.3.6 Entropy Crisis and Ideal Glass Transition

For states to exist in Nature, $W(E, V) \geq 1$ in (10.1). For the entropy,

$$S(T, V) \geq 0, \tag{10.17}$$

even in the restricted ensemble, whether it is obtained from (10.1) or (10.2). This condition must also be satisfied whether $S(T, V)$ represents the total, configurational, or communal entropy. Indeed, as it turns out, the entropy can be broken into various parts (10.19). Each part itself must be nonnegative in this partition for the state to occur in Nature.

It should be emphasized that there is no violation of thermodynamics just because $\Delta S_{ex}(T, V)$ has become negative.[7)] Although it is not very common, it is possible for SCL entropy to be less than that of the crystal. On the other hand, a negative entropy is impossible. Thus, it appears more natural to identify the Kauzmann paradox with a component of the entropy becoming negative. Accordingly, as discussed in Section 10.4.1, we interpret the Kauzmann paradox as the following entropy crisis:

Remark
Under extrapolation, a negative component of $S(T, V)$, and not a negative $S_{ex}(T, V)$, signals the entropy crisis. Its implication is simply that such states cannot occur in Nature, and the onset of the crisis is gradually the underlying thermodynamic driving force for GT in molecules of all sizes.

It will be shown in Section 10.9 that the entropy crisis is a consequence of the energy gap. Recent computer simulations have not been able to settle the issue clearly [66–68] in simple fluids. Our previous investigation [37, 39] of a dimer model, which incorporates orientational degrees of freedom, clearly establishes the ideal glass transition in molecular systems. This model calculation should be a reliable representation of glassy plastic crystals that has been investigated by Johari [69]. It has gradually become apparent that the short-ranged orientational order in supercooled liquids plays an important role not only in the formation of glasses but also in giving rise to a liquid–liquid phase transition [70, 71]. Tanaka [72] has proposed a general view in terms of cooperative medium-ranged bond ordering to describe liquid–liquid transitions, based on the original work of Nelson [73].

There are two *independent* aspects of a proper thermodynamics. First, the requirement of *stability* according to which thermodynamic quantities like the heat capacity, the compressibility, and so on must never be negative. Second, the *reality* condition (10.17) independent of the stability criteria, that ensures that such states occur in Nature.[8] Consider our gap model in Figure 10.2 in terms of the communal entropy. The resulting communal free energies of the disordered and ordered states are shown by OKCAD and OMB, respectively, in the inset. Note that both free energies satisfy stability as they are both concave. However, over the portion OK, we obtain a negative entropy, which means that this portion does not represent observable states, even though the states are stable. In other words, the extrapolation of the MS free energy to lower temperatures, while always satisfying the stability criteria everywhere ($T \geq 0$), does not always satisfy the reality condition (10.17).

The situation will be different if we consider real metastable states described by the curve GF, either representing the communal entropy curve or the communal free energy curve in the inset in Figure 10.2. If we consider the long time limit of the two curves, it is clear that since the communal entropy of an experimentally observed SMS can never be negative according to (10.17), its limiting value will also never be negative. Hence, the free energy will also not come down as we lower the temperature. One possible form of the limiting free energy is DACK along with the dotted horizontal line passing through K, as shown in the inset in Figure 10.2; compare the dotted horizontal line through K with dashed line showing the glass free energy F_G in the lower graph in Figure 10.11. This possibility does not match with the calculated free energy DACKO below K. The contradiction is resolved by the result in Section 10.12, where we show that because of the ideal glass transition, the correct free energy will never correspond to any of its entropy component being negative. Below the ideal glass transition, it must look like F_G, and not the unphysical thin line below K in the lower graph in Figure 10.11. The actual form of the unphysical communal branch OK is irrelevant. This justifies our claim earlier in Section 10.2.2

8) The stability criteria such as a nonnegative heat capacity that immediately follow from the PF formulation are independent of the nonnegative entropy requirement. Thus, it is possible for the theoretically generated SMSs to have a negative entropy over some temperature range. These states will not occur in Nature.

that the arbitrary continuation of the entropy in the gap cannot affect the physics of the glass transition.

As long as we describe the ideal glass transition in terms of thermodynamic densities like the volume, energy, and entropy, the ideal glass transition will be continuous (second order) and not discontinuous (first order). This says nothing about the possibility that this transition may turn out to be discontinuous in terms of some other order parameter, which couples to some abstract field unrelated to T and P.

10.4
Localization in Glassy Materials

10.4.1
Communal Entropy, Confinement, and Ideal Glass

The spatial integration in (10.5) can be carried out in the cell model by assuming that each particle is allowed to move about within its cell formed by its neighbors, as shown in Figure 10.1 or in Figure 10.8. In both figures, a disordered *inherent structure* (IS) is shown in (a), while (b) shows an ordered IS. We have explicitly shown voids in Figure 10.8 to contrast with Figure 10.1 that had no voids. These voids represent cells that particles do not visit during their vibration. The discussion below applies equally well to each figure. For each cell representation, the potential energy will have its (local) minimum when each particle has a certain particular position $\bar{\mathbf{r}}$ in its cell. The set $\{\bar{\mathbf{r}}\}$ of these particular positions of all the particles determines a point in the configuration space Γ, where the potential energy has a *local minimum*. This

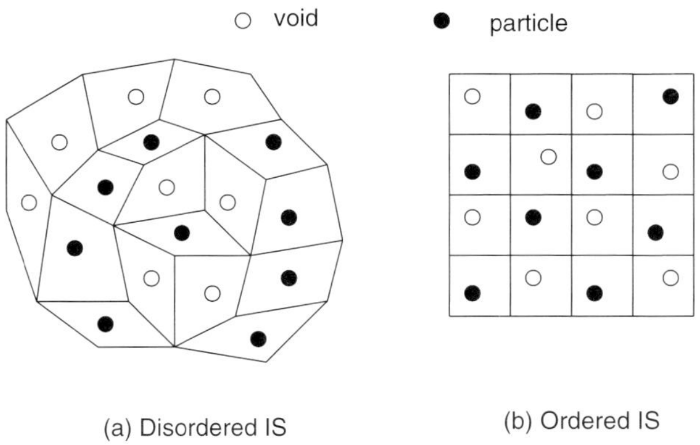

Figure 10.8 Cell representation of a small region of disordered (a) and ordered (b) inherent structures at half occupation: half of the cells contain a particle; other half are empty and are said to contain a void. Observe that while each particle is surrounded by four voids in the ordered configuration, this is not the case for the disordered configuration. We have shown a higher volume for the disordered configuration, as found empirically.

particular point $\{\bar{\mathbf{r}}\}$ is commonly known as the *inherent structure* [9, 74]. In general, there are many possible IS's, which we index by $j = 1, 2, \ldots$. The potential energy $E(\{\mathbf{r}\})$ uniquely determines the potential energy landscape, introduced by Goldstein to describe glasses [22]. Each jth IS uniquely determines a potential well or basin, which we identify as the jth potential well or basin. Indeed, each cell representation in Figure 10.1 represents a basin. The minimum of a basin is at its IS $\{\bar{\mathbf{r}}\}$ and its boundary is formed by points (not necessarily having the same energy) beyond which we can no longer go up in energy as we climb the well [38]. It is commonly believed that at low enough temperatures, the motion of the particles in the system is predominantly small harmonic motions about a *single* IS, whether the system is a CR or a glass (see, for example, Refs [30], in particular, Figure 5 there, and [31] for a recent discussion of the relevance of basins for a CR and SCL).

Confinement Hypothesis
At low enough temperatures, the communal entropy vanishes when SCL or CR is confined to a single basin.

At present, it is merely a hypothesis, which needs to be verified [75]. We first identify the communal entropy in a rigorous manner. We introduce the canonical PF $Z^{(j)}(T)$ of the jth basin obtained by *restricting* the integration in (10.5) over points that belong to the basin. Let $F_j(T) \equiv -T \ln Z^{(j)}(T)$ be the corresponding basin free energy and S_j its entropy. We follow [38] and express $Z(T)$ (we do not show V dependence) as a sum over various disjoint basins:

$$Z(T) \equiv \sum_j Z^{(j)}(T) = \sum_\lambda Z_\lambda(T). \tag{10.18}$$

The quantity $Z_\lambda(T)$ is defined as follows. We introduce IS *classes* $\mathcal{B}_\lambda$, indexed by λ: $\mathcal{B}_\lambda$ contains basins whose IS's have the same energy $\mathcal{E}_\lambda$; otherwise, there is no other restriction such as on their shapes and sizes. Accordingly, basins in each class may have different free energies at a given T. The class PF $Z_\lambda(T)$ is defined as

$$Z_\lambda(T) \equiv \sum_{j \in \mathcal{B}_\lambda} Z^{(j)}(T).$$

Let $N_\lambda(F_\mathrm{b}, T)$ denote the number of basins of free energy F_b (b: basin) in $\mathcal{B}_\lambda$ and $S_\lambda(F_\mathrm{b}, T) \equiv \ln N_\lambda(F_\mathrm{b}, T)$ the corresponding IS class entropy. (Note that basins with the same free energy need not all belong to the same basin class.) At a given T, $Z_{N,\lambda}$ is dominated by those basins in $\mathcal{B}_\lambda$ for which the class free energy $\mathcal{F}_\lambda(F_\mathrm{b}, T) \equiv F_\mathrm{b} - TS_\lambda$ is minimum as a function of F_b. The minimum occurs at $F_\mathrm{b} = \bar{F}_\lambda$, where

$$[\partial S_\lambda / \partial F_\mathrm{b}]_{F_\mathrm{b} = \bar{F}_\lambda} = \beta;$$

compare this with (10.6). Let $\bar{S}_\lambda(T) = S_\lambda(\bar{F}_\lambda, T)$ and $\bar{\mathcal{F}}_\lambda(T) = \bar{F}_\lambda - T\bar{S}_\lambda$ denote the class entropy and free energy, respectively, so $Z_\lambda(T) = \exp[-\beta \bar{\mathcal{F}}_\lambda(T)]$. We now carry out the summation over different classes in (10.18) as follows. Let $\mathcal{F}$ group denote the set of different classes, each having the free energy $\mathcal{F}$, and $\mathcal{N}(\mathcal{F}, T)$ the number of classes in this group with $\mathcal{S}(\mathcal{F}, T) \equiv \ln \mathcal{N}(\mathcal{F}, T)$ the group entropy. The PF Z is dominated by that group for which $F \equiv \mathcal{F} - T\mathcal{S}$ is minimum over $\mathcal{F}$. Let the minimum occur at $\mathcal{F} = \bar{\mathcal{F}}$, where

$$[\partial\mathcal{S}/\partial\mathcal{F}]_{\mathcal{F}=\bar{\mathcal{F}}} = \beta$$

[again, compare this with (10.6)] so that the corresponding group entropy becomes $\bar{\mathcal{S}}(T) \equiv \mathcal{S}(\bar{\mathcal{F}}, T)$. The free energy of the system is $F \equiv \bar{\mathcal{F}} - T\bar{\mathcal{S}}$. Finally, we find that the configurational entropy is given by [38]

$$S(T) \equiv \bar{\mathcal{S}}(T) + \bar{S}_{\text{bc}}(T) + S_{\text{b}}(T) \geq 0, \tag{10.19}$$

where $\bar{S}_{\text{bc}}(T)$ represents the average basin class-entropy ($\lambda = \text{bc}$) and $S_{\text{b}}(T)$ the average basin entropy ($j = \text{b}$). The first two terms in (10.19) represent the *communal entropy*

$$S_{\text{comm}}(T) \equiv \bar{\mathcal{S}}(T) + \bar{S}_{\text{bc}}(T) \geq 0, \tag{10.20}$$

see (10.8). To get an estimate of the communal entropy, we consider an *ideal* gas of N particles in a volume V, for which the configurational entropy is given in (10.A6). The corresponding entropy in the cell model can be calculated by assuming that each cell containing just one particle has a volume V/N, so that the entropy of a particle in a cell is exactly $\ln(V/N)$. The entropy for the cell model is $N\ln(V/N)$, and the communal entropy is their difference:

$$S_{\text{comm}}^{\text{ideal}}(T) = N \ln e \geq 0,$$

which satisfies the nonnegative constraint in (10.20).

At high temperatures, the communal entropy is large. As T is reduced, both components in (10.20) begin to decrease. We first expect the number of (class-) groups contributing to $S(T)$ to decrease. Thus, at some temperature $T = T_{\text{ES}}$, we expect $\bar{\mathcal{S}}(T)$ to vanish, but $\bar{S}_{\text{bc}}(T) > 0$. Thus,

$$S_{\text{comm}}(T_{\text{ES}}) \to \bar{S}_{\text{bc}}(T_{\text{ES}}) > 0$$

at this temperature. The system is now confined to one single class (bc representing the class index λ). Within this class, the system explores only those basins that have the same free energy $\bar{F}_{\text{bc}}$ at T_{ES}. Based on general grounds, we cannot draw any conclusion whether the system belongs to the same class or not at all lower temperatures $T_1 < T_{\text{ES}}$. It is possible that the basins belonging to a different class are explored at T_1 than the class at a higher temperature $T_2 < T_{\text{ES}}$. Thus, T_{ES} cannot be identified as the ideal glass transition. (As said earlier, we can speculate its connection with T_{MC}).

With further reduction of T, we expect the number of participating basins in this class to also fall so that at $T = T_{\text{K}}$, we have

$$\bar{S}_{\text{bc}}(T_{\text{K}}) = 0, \quad S_{\text{comm}}(T_{\text{K}}) = 0; \tag{10.21}$$

see (10.10). At T_{K}, $\bar{S}_{\text{bc}}(T_K)$ also vanishes, which ensures that $S_{\text{comm}}(T_{\text{K}}) = 0$, and the system is finally confined to a *single* basin corresponding to a certain cell representation in Figure 10.8a; we do not distinguish this case with the possibility of nonextensive basins per comment.[5)] Let $j = j_K$ be the particular localizing basin and $E(T_K)$ the potential energy of the system at T_K. As we lower the temperature below T_K, the system will probe microstates of lower energies $E < E(T_K)$. Again, whether these microstates belong to the same basin as the basin at T_K cannot be answered based on general arguments. Depending on their shapes and sizes, it is conceivable that some other basin j'_K has a lower free energy than j_K at a lower temperature. This will suggest that the system will jump to this basin at a lower temperature, thereby giving rise to the α-relaxation. Then, T_K cannot be identified as an ideal glass transition. However, it is commonly believed that the system remains confined to the same single basin j_K for all temperatures below T_K, and that no α-relaxation takes place. If this holds, then the likely situation is that there are no other basins $j \neq j_K$ for energies $E < E(T_K)$ with lower free energies. This then justifies calling the glass at T_K an ideal glass, as its average structure (described by the corresponding IS) remains frozen below T_K. Accordingly, the ideal glass transition is determined by the vanishing of the communal entropy, and not the configurational entropy, which is consistent with the confinement hypothesis above.

The vanishing of $S_{comm}(T_K)$ may leave $S(T)$ positive only if $S_b(T)$ is nonnegative. When $S_{comm}(T_K)$ vanishes, the energy of the system is still higher than that of its IS of the confining basin. This means that the system will probe microstates of higher energies that belong to this basin. The corresponding entropy $S_b(T)$ should be nonnegative. Whether a computation actually give a positive $S_b(T)$ depends on the nature of the calculation. Its sign, however, is irrelevant in locating the Kauzmann temperature T_K, which can be identified by focusing on the communal entropy alone. Accordingly, there is no loss of generality in using lattice models to locate T_K when the confinement occurs; see the Confinement Hypothesis above.

The group-entropy $\bar{\mathbf{S}}(T)$ remains zero below T_{ES}, which will make $\bar{\mathbf{S}}(T)$ a singular function of the temperature. Similarly, the class-entropy $\bar{S}_{bc}(T) = 0$ below T_K, which makes the class-entropy also singular at T_K, the ideal glass transition temperature as said above. It is interesting to investigate if T_{ES} has any relation to T_{MC}. We will not pursue this issue here.

The vanishing of the communal entropy at T_K and the localization of the system to a particular cell or a basin are related to the presence of high-energy barriers in the potential energy landscape. The latter provides another good reason to study the above long time limit of supercooled states under infinitely slow cooling rate. The transition from supercooled liquid to a glass is conceptually a transition between a state in which the system has high mobility and a state in which the mobility is practically nonexistent. Thus, one can view the glass transition as a localization delocalization (confinement ↔ deconfinement) transition [7]. A classic example of such a transition is *percolation* in which an infinite cluster is formed at the percolation threshold [7, 51, 76–80]. The onset of the infinite cluster is similar to the sol–gel transition in systems that undergo gelation such as the formation of silica gel [7, 51]. The formation of the gel can be considered as the formation of a glass from the liquid phase. A gel is a frozen structure similar to a glass. In this respect, the study of gel formation may provide some deep insight into the problem of glass transition. Zallen [7] provides an enlightening discussion of the relevance of percolation for the localization in glass transition, which we highly recommend.

The above discussion then allows us to conclude that (10.10) or (10.21) determines the ideal glass transition. In a lattice model, which will be considered later in this chapter, we bypass the complication due to $S_{KE}(T)$ and $S_b(T)$ all together. Hence, the total entropy in the lattice model is purely configurational, and the condition for an ideal GT reduces to

$$S(T) = 0. \tag{10.22}$$

Let us now turn to the low-temperature behavior of CR. Being an equilibrium state, CR is quite different from the metastable SCL. Therefore, there is no reason a priori to expect the same behavior of the communal entropy in CR. However, according to the above confinement hypothesis, a CR is also confined to a single basin at low enough temperatures. If true, one will also expect $T_{CR} > 0$ in (10.11). Whether an equilibrium state also represents a confined state will be answered only after the exact calculations reported later.

10.4.2
Partitioning of $Z_T(T, V)$

An important comment about the form of the potential energy E in (10.5) and the partitioning in (10.52) is in order. We have implicitly assumed that the potential energy contains all physical interactions, and that no interaction depends on velocities. This restriction means that, for example, we do not consider magnetic interactions. As a consequence, we only deal with conservative forces. This is not a major limitation as it covers the majority of the cases of interest. The potential energy of the system must certainly include the interaction energy that would be responsible for strong chemical bonding, such as in a polymer, if they occur in the system (below some temperature). This approach allows us to describe the vibrational modes associated with chemical bonds also. However, the most important reason to use this approach is to be able to treat all $3N$ Cartesian degrees of freedom (due to $3N$ coordinates $\{\mathbf{r}_i\}$) as independent. If, however, we treat the chemical bonds as fixed in length, this reduces the degrees of freedom due to these holonomic constraints that do not depend on time and temperature. This does not create any complication, as can be seen from the following argument. The theory of Lagrange multiplier allows us to treat the $3N$ degrees of freedom as independent at the expense of adding forces of constraints [81], which we take to be also conservative since the constraints themselves arise from conservative forces. This modifies the potential energy E by an additional potential energy that is independent of time and temperature, and the partitioning in (10.52) continues to remains valid. The issue of constraints due to chemical bonds is also studied by Di Marzio [82], but from a different point of view. The potential energy he considers does not involve interactions responsible for bond formation. Thus, his formulation is quite different from ours. Our approach allows us to avoid the complications noted by Di Marzio.

In case the potential energy depends on velocities, we can still factor out a PF related to the center-of-mass kinetic energy; the remainder PF will play the role of the configurational PF; however, such a case has never been studied within the context of SCL and glasses and will not be pursued anymore in this chapter.

10.5 Some Glass Transition Theories

We briefly review some important theories that have been used to explain glass transitions. None of them at present are able to explain all observed features of glass transition [9, 25, 30] even though some major progress has been made recently [36, 45–47, 56, 83–86]. Thus, we are far from having a complete understanding of the phenomenon of glass transition. It is fair to say that there yet exists no completely satisfying theory of the glass transition. Theoretical investigations mainly utilize two different approaches, which are based either on thermodynamic or on kinetic ideas. The two approaches provide an interesting duality in GT, neither of which seems complete.

10.5.1 Thermodynamic Theory of Adam and Gibbs

The most commonly used thermodynamic theory is due to Adam and Gibbs [87], which attempts to provide a justification of the *entropy crisis* in SCL [5, 6, 9, 25]. The central idea is that the sluggishness observed in a system is a manifestation of the smallness of the entropy, that is, the smallness of the available configurations to the system [88]. According to this theory, the viscosity $\eta(T)$ above the glass transition depends on a quantity that, though not rigorously defined, is also called the configurational entropy. This entropy will be denoted by $S_{\mathrm{conf}}(T)$ so as not to be confused with our rigorous definition of configurational entropy $S(T)$. Another reason to use a different notation is that their entropy seems to be closely related to the communal entropy. Crudely speaking, $S_{\mathrm{conf}}(T)$ represents the entropy of a "rearranging" region containing N_{re} particles able to rearrange themselves collectively in the sense that the entire region participates cooperatively in its rearrangement. Since the rearranging region is a subsystem of the macroscopic system (the latter being described in the canonical ensemble with fixed V and N), its volume will not be fixed but fluctuating. Therefore, the proper description of the region will be given by using the NTP ensemble encoded in (10.7). Let $G_{\mathrm{re}}(T, P, N_{\mathrm{re}})$ denote the (configurational) Gibbs free energy of the N_{re} particles that are required to rearrange cooperatively, and $G(T, P, N_{\mathrm{re}})$ that of the N_{re} particles that need not rearrange cooperatively, that is, of the N_{re} particles of the macroscopic system. The (configurational) Gibbs free energy of the macroscopic system is $G(T, P, N)$. Assuming that we can treat the cooperatively rearranging regions as independent, Adam and Gibbs argue that the average transition probability is given by

$$\bar{p}(T, P) = p_0 \exp(-\Delta G^*/T),$$

where $\Delta G^* \equiv \Delta G(T, P, N_{\text{re}}^*) = G_{\text{re}}(T, P, N_{\text{re}}^*) - G(T, P, N_{\text{re}}^*) \propto N_{\text{re}}^*$, where N_{re}^* is the smallest possible size of the region that can undergo rearrangement. Also, p_0 is taken to be almost a constant. We introduce $\Delta g \equiv \Delta G^*/N_{\text{re}}^*$ and $S_{\text{conf}}^* \equiv S_{\text{conf}}(T, P, N_{\text{re}}^*)$ of the average region by the relation

$$\frac{S_{\text{conf}}^*}{N_{\text{re}}^*} = \frac{S_{\text{conf}}(T, V, N)}{N},$$

in terms of which the average transition probability is given by

$$\bar{p}(T, P) = p_0 \exp(-S_{\text{conf}}^* \Delta g / T s_{\text{conf}}),$$

where we have introduced entropy $s_{\text{conf}} = S_{\text{conf}}(T, V, N)/N$ per particle for the macroscopic system. As N_{re}^* is of small size, S_{conf}^* is finite and can be taken to be almost a constant. Neglecting also the variation of Δg with T and P, we conclude that the numerator in the exponent is almost a constant. From this, we find that the viscosity is given as follows:

$$\ln \eta(T) = A_{\text{AG}} + B_{\text{AG}}/T s_{\text{conf}}, \tag{10.23}$$

where A_{AG} and B_{AG} are system-dependent constants. We refer the reader to Refs [89, 90] for a recent discussion of this theory and its limitations.

An alternative thermodynamic theory for the impending entropy crisis based on spin-glass ideas has also been developed in which proximity to an underlying first-order transition is used to explain the glass transition [83].

10.5.2 Free Volume Theory

The other successful theory that attempts to describe *both* aspects with some respectable success is based on the "free volume" model of Cohen and Turnbull [91]. The concept of free volume has been an intriguing one that pervades throughout physics but its consequences and relevance are not well understood [92], at least in our opinion, especially because there is no consensus on what various workers mean by free volume. Nevertheless, GT in this theory occurs when the free volume becomes sufficiently *small* to impede the mobility of the molecules [93]. The time dependence of the free volume redistribution, determined by the energy barriers encountered during redistribution, provides a *kinetic* view of the transition, and must be properly accounted for. This approach is yet to be completed to satisfaction. Nevertheless, assuming that the change in free volume is proportional to the difference in the temperature $T - T_{\text{V}}$ near the temperature T_{V}, even though there is no thermodynamic requirement for the free volume to drop as T is lowered [36], it is found that the viscosity $\eta(T)$ diverges near T_{V} according to the Vogel–Tammann–Fulcher equation:

$$\ln \eta(T) = A_{\text{VTF}} + B_{\text{VTF}}/(T - T_{\text{V}}), \tag{10.24}$$

where A_{VTF} and B_{VTF} are system-dependent constants. This situation should be contrasted with the fact that there are theoretical models [19, 46, 47] *without* any free volume in which the ideal glass transition occurs due to the entropy crisis at a positive temperature. This occurs because in these models, the communal entropy changes due to internal degrees of freedom (conformational changes) even in the absence of any free volume. Hence, it appears likely that the free volume itself is not the determining cause for the glass transition in all supercooled liquids. In some cases, the entropy crisis persists even when free volume is incorporated. However, too much free volume can destroy the transition [36].

Cohen and Grest [91] extended the free volume theory by introducing the concept of *percolation* for particle diffusion in the liquid by focusing on the random distribution of free volume. The singularity at T_V in (10.24) represents the singularity induced by the percolation threshold. The free volume regions do not percolate below T_V, so the particle diffusion is limited as expected in the glassy region. Above T_V, the percolated network of free volume allows the particle diffusion to occur over the entire volume, which makes the system behave like a fluid.

Drawbacks of the free volume approach should be mentioned. While the decrease of the free volume with temperature drop certainly explains the increase in the viscosity, it is rather difficult to explain in this approach the pressure dependence of the viscosity and negative $\mathrm{d}T_g/\mathrm{d}P$ observed in some SCLs (see, for example, Ref. [94]). Similarly, negative expansion coefficients are also untenable for the theory.

Assuming s_{conf} in (10.23) to be the communal entropy defined in Section 10.1.5, which is known to vanish in theoretical calculations, its rapid decrease to zero should give rise to a diverging viscosity, thus providing a connection between the thermodynamic view and the kinetic view. The suggestion that the rapid rise in the viscosity is due to a sudden drop in $S_{comm}(T)$ seems very enticing, since both phenomena are ubiquitous in glassy states. The experimental data indicate that T_V and T_K are, in fact, very close [95], clearly pointing to a close relationship between the rapid rise in the viscosity and the entropy crisis. This deep connection, if true, provides a very clean reflection of the dual aspects of the glassy behavior mentioned above. Thus, we are driven to the conclusion that we can treat the SCL glass transition within a thermodynamics formalism by demonstrating the existence of the entropy crisis.

It should be noted that what one measures in experiments is the difference in the entropy, and not the absolute entropy. Assuming that the entropy is zero at absolute zero in accordance with the Nernst–Planck postulate, one can determine the absolute entropy experimentally. However, it is well known that SCL is a metastable state, and there is no reason for its entropy to vanish at absolute zero [16]. Indeed, it has been demonstrated some time ago that the residual entropy at absolute zero obtained by extrapolation is a nonzero fraction of the entropy of melting [43], which is not known a priori. Therefore, it is *impossible* to argue from experimental data that the entropy indeed falls to zero, since such a demonstration will certainly require calculating absolute entropy though efforts continue to date [61, 62].

10.5.3 Mode Coupling Theory

The mode coupling theory [60] is an example of theories based on kinetic ideas that deals not with the glass transition but with the transition at T_{MC}. Thus, it is not directly relevant for our chapter. This theory may be regarded as a theory based on first-principle approach, which starts from the static structure factor. In this theory, the ergodicity is lost completely, and structural arrest occurs at a temperature T_{MC}, which lies well above the customary glass transition temperature T_G. Consequently, the correlation time and the viscosity diverge due to the *caging effect*. The diverging viscosity can be related to the vanishing free volume [91, 93], which might suggest that *the MC transition is the same as the glass transition*. This does not seem to be the consensus at present. Thus, it is not clear if the free volume is crucial for the MC transition. Some progress has been made in this direction recently [36], where it has been shown for long polymers that the free volume vanishes at a temperature much higher than the ideal glass transition. It has been speculated that the divergence at T_{MC} is due to the neglect of any activated process in this theory [96, 97]. This, however, has been disputed in a recent study [98]. The mode coupling theory is also not well understood, especially below the glass transition. More recently, it has been argued that this and mean field theories based on an underlying first-order transition may be incapable of explaining dynamic heterogeneities.

10.6 Progigine–Defay Ratio Π and the Significance of Entropy

Let us consider an experimental glass obtained by a nonzero cooling rate such as in an experiment or in a simulation. Because the densities (entropy, volume, etc.) are continuous, but their derivatives such as the expansion coefficient, heat capacity, and so on show a discontinuity at T_g, it appears that there is some similarity between the glass transition and a continuous thermal phase transition. Such a similarity should not be taken too seriously as T_g depends on the history of glass preparation: this temperature can vary over a range of several tens of degrees. (The objection of course is not applicable to the ideal glass transition.) This usually does not happen with thermodynamic transitions such as a melting transition, whose temperature does not vary so drastically solely due to kinetics. Considering the continuity of the entropy S_g and the volume V_g for the two states (supercooled liquid and the glass) along the glass transition curve $T_g(P)$ in the T–P plane, we can easily derive the following relations [99–101]:

$$\frac{dT_g}{dP} = T_g V_g \frac{\Delta\alpha_T}{\Delta C_P}, \quad \frac{dT_g}{dP} = \frac{\Delta\varkappa_T}{\Delta\alpha_T}, \tag{10.25}$$

where ΔQ represents the discontinuity in the quantity Q; here, Q represents the isothermal expansion coefficient α_T, the compressibility $\varkappa_T$, and the isobaric heat capacity C_P. The first equation follows from entropy continuity, while the second equation from volume continuity. As discussed by Goldstein [101], the first equation in (10.25) is found to be generally valid experimentally, but not the second equation. In fact, it is usually found that the right-hand side ratio in the second equation is larger than the slope dT_g/dP. Consequently, the following Progigine–Defay ratio [99, 100]

$$\Pi = \frac{\Delta C_P \Delta\varkappa_T}{T_g V_g (\Delta\alpha_T)^2} \geq 1 \tag{10.26}$$

(see also Refs [101–103]). Due to the general validity of the first equation, Goldstein concludes, though plausibly, that the GT is thermodynamic as hypothesized by Kauzmann [5], and then further elaborated by Gibbs [88]. According to this hypothesis, the GT is associated with the vanishing of the excess entropy or excess enthalpy. For kinetic reasons, this requirement may turn into some fixed value of the excess entropy. Goldstein also concludes that the vanishing of free volume does not seem to be the determining factor in the GT. Di Marzio and Yang [53] come to a similar conclusion about the dominant role of the entropy in glass transitions.

Let us assume that the particles comprising the system have no internal structure. Ideally, we are considering particles to be hard sphere (Figures 10.1 and 10.8). The ability for the center-of-mass of each particle to move about most certainly depends on the volume in the system and contributes to the configurational entropy S. Consequently, S is a function of the volume V, a conventional thermodynamic dependence. For hard spheres, one can equivalently use the volume V_{f}, the excess volume above the smallest possible volume at absolute zero, instead of V. Treating V_{f} as the free volume, we must conclude that S and V_{f} are equivalent, and that S will vanish with V_{f}. In this case, the above conclusion cannot be justified. However, for most glass formers, particles have internal structures that will also contribute to the configurational entropy S. This will most certainly be the case with polymers as glass formers. Accordingly, in this case, the configurational entropy will not vanish with the free volume. Moreover, one can treat S and V_{f} as independent so that the vanishing of the free volume does not necessarily imply absence of configurational change. Thus, free volume does not seem to be the primary cause, though it may be secondary, behind GT as shown recently [36]. This is also consistent with the known fact that there is no thermodynamic requirement for the volume or the free volume to decrease with the lowering of the temperature. Since it is the entropy that plays a role in both cases, this provides some very strong justification for assuming that it is the entropy that plays the central role in dictating the glass transition. At this point, we should also mention an interesting work that provides another argument in terms of temperature being the dominant controlling variable than the volume or pressure [104].

It is found that the slope $\mathrm{d}T_{\mathrm{g}}/\mathrm{d}P$ is nonzero and finite. This will presumably also remain true even in the limit $r \to 0$ at T_{K}. None of the discontinuities ΔC_P, $\Delta \varkappa_T$ and $\Delta \alpha_T$ can vanish at the transition if $\mathrm{d}T_{\mathrm{K}}/\mathrm{d}P$ is to remain nonzero and finite. In other words, any nonzero discontinuity in one quantity implies that all these quantities must have nonzero discontinuities simultaneously. Thus, the actual value of the ideal glass transition temperature T_{K} does not depend on whether we consider the volume V, enthalpy H, or entropy S in Figure 10.6, as a singular behavior (nonzero discontinuity) in any one of these quantities reflects a singular behavior in other thermodynamic quantities. This can be seen directly from the following thermodynamic identities:

$$T\left(\frac{\partial S}{\partial T}\right)_P = \left(\frac{\partial H}{\partial T}\right)_P, \quad \left(\frac{\partial S}{\partial P}\right)_T = \frac{1}{T}\left[\left(\frac{\partial H}{\partial P}\right)_T - V\right] = -\left(\frac{\partial V}{\partial T}\right)_P.$$

Any discontinuity in $(\partial H/\partial T)_P$ or $(\partial H/\partial P)_T$ is reflected in a discontinuity in $(\partial S/\partial T)_P$ or $(\partial V/\partial T)_P$, respectively. This is a very important property of the ideal glass transition.

10.7 Equilibrium Formulation and Order Parameter

Before we introduce our restricted partition function formalism, we need to review some salient aspects of the equilibrium statistical mechanics formalism. This will also help with the clarity and continuity of presentation. We will restrict our discussion to the canonical ensemble, but the extension to other ensembles is straightforward and trivial.

10.7.1 Canonical Partition Function

While studying the lattice models for exact calculations, we will use a discrete version of the configurational partition function (10.5) in the canonical ensemble:

$$Z(T) \equiv \sum W(E)\exp(-\beta E), \quad W(E) \geq 1, \tag{10.27}$$

where the summation is over all possible energies $E \geq E_0$.[9] Since Z is a sum of positive terms, the following two principles of equilibrium statistical mechanics always hold:

- **Minimization Principle**: The free energy $F(T)$ must be minimized as $N \to \infty$.
- **Stability Principle and Concavity**: The heat capacity, which is given by the square of fluctuations in the energy, is nonnegative, and $F(T)$ and $S(E)$ are concave functions of their arguments.

It should be stressed that the nonnegativity of the heat capacity and the maximization principle only require $W(E) \geq 0$ and not $W(E) \geq 1$. Thus, both principles remain valid even if the entropy becomes negative,[8] which is most certainly the case with the ideal gas. The principles refer to two *independent* aspects.[9] What we will see below is that it is the condition $W(E) \geq 1$ that tends to be violated by the low-temperature SMSs, when extrapolated to $T = 0$. There is no such violation for low-temperature equilibrium states (CR). The latter is related to the absence of any energy gap for equilibrium states.

9) It is well known that in classical statistical mechanics, the entropy in continuum space can become negative. This is true of the ideal gas at low temperatures. From the exact solution of the classical Tonks gas of rods in one dimension, one also finds that the entropy becomes negative at high coverage. Thus, $W(E)$ need not always represent the number of configurations, and care is needed to interpret $W(E)$ in all cases.

10.7.2
Free Energy Branches

We focus on a system with a single transition, the melting transition at $T = T_M$. The *disordered* equilibrium liquid EL at and above T_M and the ordered crystal CR at and below T_M correspond to different free energy functions $f_{EL}(T)$ and $f_{CR}(T)$, respectively, from which we can calculate the entropies, and energies per particle. In equilibrium, each of these quantities have two different branches corresponding to EL and CR:

$$f(T) = f_{EL}(T), \quad s(T) = s_{EL}(T), \quad e(T) = e_{EL}(T), \quad T \geq T_M, \tag{10.28}$$

$$f(T) = f_{CR}(T), \quad s(T) = s_{CR}(T), \quad e(T) = e_{CR}(T), \quad T \leq T_M. \tag{10.29}$$

The equilibrium entropy or energy has a discontinuity, but the free energy is continuous at $T_M : f_{EL}(T_M) = f_{CR}(T_M)$.

10.7.3
Order Parameter and Classification of Microstates

The presence of a melting transition at T_M (the inverse of the slope of HH′ in Figure 10.2) means that EL above T_M and CR below T_M correspond to different values of the order parameter ϱ, which is traditionally defined in such a way that $\varrho = 0$ represents the disordered phase and $\varrho \neq 0$ the ordered phase. Therefore, the use of ϱ classifies each configuration or microstate into two *disjoint* classes, to be called ordered and disordered classes. How this can be done in practice will be taken up in Sections 10.10 and 10.11, where the classification is carried out explicitly. Here, we wish to set the formalism for the required classification scheme.

The definition of ϱ depends on the system. For semiflexible polymers, we had identified it as $\varrho \equiv n_H - n_V$. Here, we consider a magnetic system for which the microstates can be divided into the two *disjoint* classes corresponding to the value of the spontaneous magnetization $\varrho = 0$ (ordered) and $\varrho \neq 0$ (disordered). We illustrate this by considering an Ising model in zero magnetic field for its simplicity. Let $\sigma_i = \pm 1$ be the Ising spin at the ith site of the lattice with N sites, and let the angular brackets $\langle\,\rangle$ denote thermodynamic average with respect to the weight

$$\Omega(E) \equiv W(E)\exp(-\beta E)/Z \geq 0. \tag{10.30}$$

We consider a microstate and introduce the average spin over all the sites of the lattice $\bar{\sigma} \equiv (1/N)\sum \sigma_i$, which we use to identify the order parameter ϱ:

$$\varrho \equiv \langle\bar{\sigma}\rangle = \sum_{i=1}^{N} \langle\sigma_i\rangle/N = \sum_{E} \bar{\sigma}\Omega(E), \tag{10.31}$$

in the limit $N \to \infty$. Being a density (an intensive quantity), ϱ should be insensitive to the size. Therefore, we can divide the lattice into several parts, each macroscopically large, so that the order parameter can be calculated for each part. For a macroscopic system, we expect all the parts to have the same value of ϱ.

For any finite N, no matter how large, $\Omega(E) > 0$ for all all energies $E < \infty$. Then, the only way $\varrho = 0$ can occur if $\bar{\sigma} \equiv 0$ for each configuration that contributes to (10.31) not only for the system but also for its various macroscopic parts. This rules out the possibility, for example, that $\bar{\sigma} = 0$ is due to a mixture state of two states $\bar{\sigma} = +1$ and $\bar{\sigma} = -1$ of equal sizes. The rest of the configurations with $\varrho = 0$ are classified as ordered. All remaining configurations are classified as disordered. This prescription allows us to classify configurations into ordered and disordered configurations in an unambiguous manner.

10.8
Restricted Ensemble

10.8.1
Required Extension in the Energy Gap

We now describe the extension of the restricted ensemble formalism developed by Penrose [1] to suit our goal. While $W_{\text{ord}}(E)$ certainly exists for $E \geq E_0$, there is no guarantee that $W_{\text{dis}}(E)$ also exists for $E \geq E_0$. This is abundantly clear from the entropy for linear polymers in Figure 10.3. Most probably, there is an energy gap for $W_{\text{dis}}(E)$. Otherwise, the energy of the disordered phase at absolute zero would also be E_0 (we assume that $TS_{\text{dis}} \to 0$ as $T \to 0$[2)]), the same as that of CR. This would most certainly imply that they would coexist at $T = 0$. While there is no thermodynamic argument against it, it does not seem to be the case normally. Usually, the most stable state at $T = 0$ is that of a crystal. Moreover, it is an experimental fact [5] that all glasses have much higher energies compared to their crystalline forms at low temperatures. Thus, we *assume an energy gap* (Figure 10.2). In other words, $W_{\text{dis}}(E) \geq 1$ for $E \geq E_{\text{K}}$. We find it extremely useful to *extend* $W_{\text{dis}}(E)$ in the gap. All we need to ensure is that the resulting entropy is continuous and concave over the entire range $E \geq E_0$. We denote this extended entropy function by $S^*_{\text{dis}}(E)$ and the multiplicity by $W^*_{\text{dis}}(E) = \exp[S^*_{\text{dis}}(E)]$, $E \geq E_0$. The function $S^*_{\text{dis}}(E)$ is identical to $S_{\text{dis}}(E)$ over $E \geq E_{\text{K}}$.

10.8.2
Restricted and Extended Restricted PF's

We now introduce the following *restricted* ensemble PFs:

$$Z_{\text{ord}}(T) \equiv \sum_{E \geq E_0} W_{\text{ord}}(E)\exp(-\beta E), \quad Z^*_{\text{dis}}(T) \equiv \sum_{E \geq E_0} W^*_{\text{dis}}(E)\exp(-\beta E), \tag{10.32}$$

$$Z_{\text{dis}}(T) \equiv \sum_{E \geq E_{\text{K}}} W_{\text{dis}}(E)\exp(-\beta E), \tag{10.33}$$

and the corresponding free energies per particle $f_{\text{ord}}(T)$, $f^*_{\text{dis}}(T)$, and $f_{\text{dis}}(T)$, which are defined for all temperatures $T \geq 0$.

Remark

The following remark is important to understand the relationship between $Z^*_{\rm dis}(T)$ and $Z_{\rm dis}(T)$. For temperatures so that the average energies $\bar{E}^*_{\rm dis}(T)$ and $\bar{E}_{\rm dis}(T)$ are greater than $E_{\rm K}$, both partition functions are determined by the microstates of energies above $E_{\rm K}$, where $W^*_{\rm dis}(E) = W_{\rm dis}(E)$. Hence, for $T \geq T_{\rm K}$, $f^*_{\rm dis}(T)$ and $f_{\rm dis}(T)$ are the same. They differ only below $T_{\rm K}$. Thus,

$$f_{\rm dis}(T) = f^*_{\rm dis}(T), \quad T \geq T_{\rm K}. \tag{10.34}$$

As long as $W_{\rm dis}(E)$, $W^*_{\rm dis}(E)$, and $W_{\rm ord}(E)$ are nonnegative (≥ 0), the restricted PF's are the sum of positive terms. Therefore, the corresponding free energies satisfy the two principles in Section 10.7.1, so that there will never be any unstable state.

It is clear that the global *free energy minimization* for true equilibrium requires that

$$f(T) = f_{\rm dis}(T) = f^*_{\rm dis}(T), \quad T \geq T_{\rm M}, \tag{10.35}$$

$$f(T) = f_{\rm ord}(T), \quad T \leq T_{\rm M}. \tag{10.36}$$

The switch over from $f_{\rm dis}(T)$ to $f_{\rm ord}(T)$ at $T_{\rm M}$ makes $f(T)$ *singular*. The singularity signals the (melting) transition. But it is not present in $f_{\rm ord}(T)$ or $f_{\rm dis}(T)$ at $T_{\rm M}$ and allows us to continue these free energies across $T_{\rm M}$.

10.8.3
Metastability Prescription

A prescription to describe metastability using the above formalism can now be formulated. We are only interested in SCL and its extension. Accordingly, we *abandon* the above global free energy minimization principle, and use $f_{\rm dis}(T)$ and $f^*_{\rm dis}(T)$ to give the SMS free energy below $T_{\rm M}$. From these free energies, we can obtain the entropy and energy per particle for the supercooled liquid. As said above, SMSs cannot be unstable in the restricted ensembles. Thus, either they terminate in a singularity, or they extrapolate to $T = 0$.

It is easy to calculate the order parameter ϱ for $T < T_{\rm M}$ for the "disordered phase" by using the weights $W_{\rm dis}(E)$ or $W^*_{\rm dis}(E)$ in (10.30) to check if we have properly identified the set of disordered microstates. If properly identified, the phases represented by $f_{\rm dis}(T)$ and $f^*_{\rm dis}(T)$ below $T_{\rm M}$ will correspond to a disordered state ($\varrho = 0$). This is our required description of SMSs to describe SCL below $T_{\rm M}$. We still have to check which one of $f_{\rm dis}(T)$ and $f^*_{\rm dis}(T)$ will describe the correct physics of SCLs and glasses.

10.9
Three Useful Theorems

We now prove some important theorems for our gap model [105, 106]; the proofs themselves are quite general and do not require any detailed knowledge of the nature of the system except that it should have a gap.

Theorem 10.1

Due to the energy gap ($E_K > E_0$), the free energy $F_{dis}(T_{eq})$ at O′ equals the free energy $F_{ord}(T = 0) = E_0$ at O, where $T_{eq} > 0$ is the inverse of the slope of the line OO′ touching the entropy function $S_{dis}(E)$; see point O′ in Figure 10.2 and the inset

The proof is very simple. The slope $1/T_{eq}$ of OO′ is given by

$$1/T_{eq} = S_{dis}(E_{O'})/(E_{O'} - E_0),$$

where $E_{O'}$ is the energy at O′. Thus,

$$E_0 = E_{O'} - T_{eq} S_{dis}(E_{O'}).$$

Since the slope of $S_{dis}(E_{O'})$ at $E_{O'}$ from (10.6) is $1/T_{eq}$, the right side represents the free energy $F_{dis}(T_{eq})$ of the SMS at O′ The left side represents the free energy of CR at $T = 0$. This proves the theorem.

It should be stressed that the proof does not use the vanishing of $S_{dis}(E_K)$. Thus, the equality $F_{dis}(T_{eq}) = E_0$ is also valid if $S_{dis}(E_K) > 0$. The proof also does not depend on the entropy slope at E_K. From the concavity of $S_{dis}(E)$, it should be obvious that slope at E_K is larger than $1/T_{eq}$.

Theorem 10.2

The free energy F^*_{dis} or F_{ord} of all stable phases, mathematically continued or not, are equal at $T = 0$ in that $F^*_{dis}/E_0 \to 1$ and $F_{ord}/E_0 \to 1$, provided they satisfy Nernst–Planck condition $TS^*_{dis} \to 0$ or $TS_{ord} \to 0$ as $T \to 0$. Their entropies, however, may be different.

The proof will require considering finite N and then taking the thermodynamic limit later. This allows us to treat the multiplicity of each microstate as bounded. Let us assume for simplicity that $W_{dis}(E_K) = 1$, which makes $0 < W^*_{dis}(E) < 1$. We factor out the term corresponding to $E = E_0$ from the sum in $Z^*_{dis}(T)$, and express

$$Z^*_{dis}(T) = W^*_{dis}(E_0)e^{-\beta E_0}[1 + Z'_\alpha(T)], \qquad (10.37)$$

where we have introduced a new quantity

$$Z'_{dis}(T) \equiv \sum_{E \neq E_0} [W^*_{dis}(E)/W^*_{dis}(E_0)]e^{-\beta(E-E_0)}. \qquad (10.38)$$

Since $E - E_0 > 0$, we note that $e^{-\beta(E-E_0)} \to 0$ in each summand in $Z'_{dis}(T)$ as $T \to 0$. For finite N, $W^*_{dis}(E)$ is a bounded quantity, and so is the ratio $W^*_{dis}(E)/W^*_{dis}(E_0)$. Hence, each term in the sum in (10.38) vanishes, and so does $Z'_{dis}(T)$ as $T \to 0$. We finally have

$$Z^*_{dis}(T) \to W^*_{dis}(E_0)e^{-\beta E_0} \quad \text{as} \quad T \to 0,$$

so that $F^*_{dis}(T) \to E_0 - TS^*_{dis}(E_0)$ as $T \to 0$, which follows from the boundedness of $W^*_{dis}(E_0) > 0$. Hence,

$$F^*_{dis}/E_0 \to 1 \quad \text{as} \quad T \to 0.$$

As the derivation does not depend on the energy gap, it is just as valid for $Z_{\text{ord}}(T)$ so that

$$F_{\text{ord}}/E_0 \to 1 \quad \text{as} \quad T \to 0.$$

We can now take the limit $N \to \infty$ in both cases without affecting the conclusion. The possibility that the limit is different (indeed, higher) from 1 for F^*_{dis}/E_0 is ruled out because of the Nernst–Planck postulate.[2)] This proves the theorem.

In case $S^*_{\text{dis}}(E)$ has some singular behavior at E_0 in the limit $N \to \infty$, we can limit the extension of $W^*_{\text{dis}}(E)$ to some energy $E^*_0 > E_0$ to avoid the singularity, and define $Z^*_{\text{dis}}(T)$ by restricting the sum in (10.32) to $E \geq E^*_0$. In this case, the temperature T_{eq} in Theorem 10.1 will be defined by the slope of the tangent line drawn not from $E = E_0$, point O, but from $E = E^*_0$. Also, following the same step as above will show that

$$F^*_{\text{dis}}/E^*_0 \to 1 \quad \text{as} \quad T \to 0,$$

with a slight modification in the statement of the theorem. Indeed, one can extend $W^*_{\text{dis}}(E)$ to any energy $E^*_0 > E_0$ in the gap even if there is no singularity encountered at $T = 0$. This is because the actual form of the extension is not relevant as Theorem (10.3) shows.

Theorem 10.3 (Energy Gap Theorem)
It follows from the energy gap that T_{K} is positive.

The proof uses the above results and is quite simple. We apply the above theorems to the communal free energy in (10.9). We will use the same notation as above so that $F^*_{\text{dis}}(T_{\text{eq}})$ denotes the extended communal free energy and so on. This should cause no confusion. From the concavity of $F^*_{\text{dis}}(T)$, see OKCD in the inset in Figure 10.2, it is obvious that since $F^*_{\text{dis}}(T_{\text{eq}}) = E_0$ (or E^*_0) $= F^*_{\text{dis}}(0), F^*_{\text{dis}}(T)$ must have a *maximum* between $T = 0$ and $T = T_{\text{eq}} > 0$. Consequently, the location of the maximum must be at a positive temperature, where the communal entropy vanishes. This proves the theorem.

It follows from this theorem that as the choice of E^*_0 in the gap can be arbitrary (see the comments above), the actual form of the communal free energy $F^*_{\text{dis}}(T)$ below T_{K} is irrelevant. This then proves that the actual form of $S^*_{\text{dis}}(E)$ below E_{K} is irrelevant as long as it remains continuous and concave. But the extension allows us to draw an important conclusion about the presence of a maximum in $F^*_{\text{dis}}(T)$ at $T_{\text{K}} > 0$. From (10.6), it follows that $T_{\text{K}} > 0$ and finite is equivalent to saying that the slope of $S^*_{\text{dis}}(E)$ at E_{K} is finite and positive. Indeed, the possibility of the extension of $S_{\text{dis}}(E)$ in the gap requires the slope at E_{K} to be finite, which we summarize below:

Remark
Possibility of extending $S_{\text{dis}}(E)$ in the gap is equivalent to having its slope at E_{K} finite.

10.10 1D Polymer Model: Exact Calculation

10.10.1 Polymer Model and Classification of Configurations

We relate the high-temperature expansion of a one-dimensional m-component axis spin model to a polymer problem, as has been discussed elsewhere by us [50, 107, 108]. Such a connection between polymers and magnetic systems was first established by deGennes [80, 109], which also provides a motivation to treat m as a real variable rather than an integer. We will be interested in $1 > m > 0$, though at the end we will also consider $m = 0$. For $m = 1$, the axis model reduces to an Ising model, while for $m \to 0$, it reduces to the model of linear chains with no loops [108, 109]. The spin model contains m-component spins $\mathbf{S}_i$ located at the site $i = 1, 2, \ldots, N$ of a one-dimensional lattice of N sites, with periodic boundary condition ($\mathbf{S}_{N+1}$ same as $\mathbf{S}_1$ and $i = N+1$ same as $i = N$). We index the lattice bonds by $k = 1, 2, \ldots, N$; here k denotes the lattice bond between sites k and $k+1$. Each spin can point along or against the axes (labeled $1 \leq \alpha \leq m$) of an m-dimensional spin space and is of length $\sqrt{m}$: $\mathbf{S} = (0, 0, \ldots, \pm\sqrt{m}, 0, \ldots, 0)$. The spins interact via a ferromagnetic nearest-neighbor interaction energy $(-J)$. The energy of interaction is given by

$$E = -J \sum_{i=1,\ldots,N} \mathbf{S}_i \cdot \mathbf{S}_{i+1}.$$

We follow [50, 107, 108] and consider the high-temperature expansion of the PF

$$Z(K, m) \equiv (1/2m)^N \sum \exp(-\beta E) = (1/2m)^N \sum \prod_k \exp(K x_k),$$

where $K = \beta J$ and $x_k \equiv \mathbf{S}_i \cdot \mathbf{S}_{i+1}$. The sum is over all $(2m)^N$ orientations of the N spins. We use the expansion $e^y = 1 + y + y^2/2! + y^3/3! + \cdots$ for each $\exp(K x_k)$ and multiply them for all lattice bonds k. Each term in the product represents a branched polymer diagram [50] whose contribution to the PF is obtained by summing over all spin orientations. Lattice sites in a polymer represent monomers, and those not covered by polymers represent solvent. The details for the model are found in Refs [107, 108]. We quote the result. The following must hold for a diagram to contribute to $Z(K, m)$. A bond originating from y^μ is between two neighboring sites and represents a *multiple* bond of multiplicity $\mu = 1, 2, 3, \ldots$. Each bond of multiplicity μ gives rise to μ chemical bonds and contributes $K^\mu/\mu!$. The power ν of any component S_α of a spin at a site, also called the valence of the site (the number of chemical bonds attached to it), must be even $(= 2, 4, 6, \ldots)$ and contribute $m^{\nu/2}$ to the PF. We set $\nu = 0$ for a site occupied by a solvent. In the following, we only speak of bonds $(\mu \geq 1)$ between monomers, as there are no bonds attached to a solvent. If there is no bond between two neighboring monomers, we treat them as disconnected. There are no interactions between the solvents, and between a solvent and a monomer. As we will see, the model exhibits a first-order transition at $T = T_{\mathrm{M}}$. We take the ordered phase at low temperatures to represent CR (even though there is no real crystalline symmetry), and the disordered phase at high temperatures to represent EL, since the two phases turn out to correspond to distinct symmetries.[4)]

A microstate is determined by a polymer diagram specified uniquely by the set of multiplicity $\{\mu_k\}$ of the bonds. Each diagram can be decomposed into disjoint parts $\mathcal{C}_r, r = 1, 2, \ldots$, in each of which the monomers are sequentially connected. The spin component α must be the same for all spins in a disjoint diagram. Therefore, each disjoint diagram additionally contributes $2m$ due to the sum over α. Let $|\mathcal{C}|$ denote the size of a component $\mathcal{C}$, which is nothing but the number of monomers in it. If each component has a size strictly less than N, then the multiplicity μ_k of each of its bonds must be even. On the other hand, if $|\mathcal{C}| = N$, which represents a *percolating* component, then the multiplicity of each bond in it must be either simultaneously all even ($= 2, 4, 6, \ldots$) or all odd ($= 1, 3, 5, \ldots$).

A bond of multiplicity μ_k gives rise to loops, whose number is given by $l_k \equiv \mu_k - 1$. Let v_i denote the valence at a site. The number of chemical bonds B and the number of loops L are given by

$$B \equiv \sum_k \mu_k = \sum_i (v_i/2), \quad L \equiv \sum_k l_k = B - n, \tag{10.39}$$

where n represents the number of bonds. The PF is given by

$$Z(K, m) = \sum_{\text{microstates}} K^B m^L / (2^n \prod_k \mu_k!) = \sum_{B,L} W(B, L) K^B m^L. \tag{10.40}$$

We see that the activity of each loop is m, and the activity of each chemical bond is K. The multiplicity $W(B, L)$ determines the polymer system entropy. We can treat $(-B)$ as analogous to the energy E so that we can consider the entropy as a function of $E = -B$.

The following will be established. At high temperatures, we have microstates in which each component is finite in size; hence, the multiplicity of each bond is even. This condition then uniquely defines disordered microstates. There are two distinct percolating components ($|\mathcal{C}| = N$). One of them has all bond multiplicities even. We will treat this to represent a disordered microstate for the obvious reason. The other percolating component has all bond multiplicities odd. This is obviously a component with a different "symmetry," and we will see below that it represents the ordered state. The possible odd multiplicities of its bonds uniquely determine ordered microstates. The sets of disordered and ordered microstates are obviously *disjoint*, which ensures that there will be no stable nuclei in the metastable state if it occurs, as discussed in Section 10.1.6, and distinguishes *our approach from the PL approach*. Because of distinct symmetries of the two phases, our model genuinely represents a "melting transition," and not a liquid–gas transition in which the symmetry remains the same in both equilibrium phases.

One can also consider the above model with the free boundary condition. In this case, we define the ordered microstates as the microstates associated with the percolating component ($|\mathcal{C}| = N$), except that the valence of the monomers at each of the two end sites of the lattice is odd, while all interior monomers have even valences. This ensures that each bond has an odd multiplicity, as above. The disordered microstates have even bond multiplicities.

10.10.2 Exact Calculation

We exactly evaluate the PF using the transfer matrix method [110], a standard technique for one-dimensional problems, and use the approach of Newman and Schulman [49] to identify SMSs. In terms of the transfer matrix $\hat{T} \equiv \exp(K\mathbf{S}\cdot\mathbf{S}')$ between two neighboring spins,

$$Z(K,m) \equiv (1/2m)^N Tr\,\hat{T}^N. \tag{10.41}$$

The transfer matrix has the eigenvalues

$$\lambda_{\text{dis}} = u + 2(m-1), \quad \lambda_{\text{ord}} = v, \quad \lambda = u-2, \tag{10.42}$$

that are onefold, m-fold, and $(m-1)$-fold, respectively [110]. Here we have introduced

$$x \equiv \exp(Km), \quad u \equiv x + 1/x, \quad v \equiv x - 1/x.$$

The temperature T of the spin system does not represent the temperature (T_{P}) in the polymer problem but its inverse is related to the polymer activity K [108, 109]. We will see below that small x or K corresponds to high (spin) temperatures where the disordered phase is present, and large x or K corresponds to low (spin) temperatures where the ordered and possible SMS phases are present. Thus, decreasing T amounts to going toward the region where the ordered and metastable disordered phases are present. Therefore, we will continue to use T of the spin system, even though it is not the temperature T_{P} of the polymer system. Because of this, we will study ω, the limiting value $(1/N)\ln Z(K,m)$ as $N\to\infty$, which does not require T_{P}.

The eigenvalue λ_{dis} is dominant at high temperatures for all $m \geq 0$ and describes the disordered phase. Its eigenvector is

$$\left\langle \chi_{\text{dis}} \right| = \sum_i \left\langle i \right| / \sqrt{2m},$$

where $\langle 2k|$ (or $\langle 2k+1|$) denotes the single spin state in which the spin points along the positive (or negative) kth spin axis. It has the correct symmetry to give zero magnetization ($\varrho = 0$). For $m \geq 1, \lambda_{\text{dis}}$ remains the dominant eigenvalue at all temperatures $T \geq 0$. For $0 \leq m < 1$, the situation changes and λ_{ord} becomes dominant at temperatures $T < T_{\text{c}}$, or

$$x \geq x_{\text{c}} = 1/(1-m)$$

where T_{c} is determined by the critical value $x_{\text{c}} \equiv \exp(Jm/T_{\text{c}})$; there is a phase transition at T_{c}. The corresponding eigenvectors are given by the combinations

$$\left\langle \chi_{\text{ord}}^{(k+1)} \right| = \left[\left\langle 2k \right| - \left\langle 2k+1 \right| \right] / \sqrt{2}, \quad k = 0, 2, .., m-1,$$

which are orthogonal to $\langle\chi_{\text{dis}}|$, as can be easily checked. These eigenvectors have the symmetry to ensure $\varrho \neq 0$. The remaining eigenvalue λ is $(m-1)$-fold degenerate with eigenvectors

$$\left\langle\chi^{(k+1)}\right| = \left[\left\langle 2k\right| + \left\langle 2k+1\right| - \left(\langle 2k+2\right| - \langle 2k+3\left|\right)\right]/\sqrt{4}, \quad k = 0, 2, \ldots, m-2.$$

For $m > 0$, this eigenvalue is never dominant. For $m \to 0$, it becomes degenerate with λ_{dis}. Since the degeneracy plays no role in the thermodynamic limit, there is no need to consider this eigenvalue separately for $m \geq 0$.

We now consider the limit $N \to \infty$. In this limit, we only need to consider the dominant eigenvalue for a given x. The eigenvalue of the transfer matrix can be viewed as the contribution from a lattice bond. Let us first consider $x < x_c$, where the dominant eigenvalue is λ_{dis}. We rewrite it as follows:

$$\lambda_{\text{dis}} = 1 + \frac{1}{m}\sum_{k=1}^{\infty}\frac{(mK)^{2k}}{(2k)!}.$$

In this form, it is immediately clear that the first term represents the absence of a polymer bond, while the sum represents the possibility of the presence of a polymer bond of even multiplicity $(2, 4, 6, \ldots)$. Expanding λ_{dis}^N, it is clear that the disordered microstates contain some empty sites, and the rest of the lattice bonds are covered by polymer bonds, each of which must have only even multiplicity. On the other hand, rewriting λ_{ord} as

$$\lambda_{\text{ord}} = \frac{1}{m}\sum_{k=0}^{\infty}\frac{(mK)^{2k+1}}{(2k+1)!},$$

we immediately conclude that it is not possible to have any lattice bond empty. All lattice bonds must be covered and that each polymer bond must have an odd multiplicity $(1, 3, 5, \ldots)$. There are no solvent particles. This is consistent with the earlier discussion. One must still have even valences at all other points connected by polymer bonds, regardless of whether we consider disordered or ordered microstates.

We again consider the limit $N \to \infty$. The free energy per site of the high-temperature equilibrium phase is $f_{\text{dis}}^*(T) \equiv -T_{\text{P}} \ln(\lambda_{\text{dis}}/2m)$. It exists at all temperatures down to $T = 0$, even though the equilibrium osmotic pressure has a singularity at x_c. The singularity at x_c appears only when we consider equilibrium free energy, which requires a switch over from the disordered branch to the ordered branch of the free energy, as discussed in Section 10.8. Similarly, $f_{\text{ord}}(T) \equiv -T_{\text{P}} \ln(\lambda_{\text{ord}}/2m)$ related to the low-temperature equilibrium phase also exists all the way down to $T = 0$. To calculate the entropy density, we proceed as follows. The adimensional free energy ω represents the osmotic pressure of the polymer system [36, 111]. The bond and loop densities are given by

$$\phi_{\text{B}} \equiv \partial\omega/\partial \ln K, \quad \phi_{\text{L}} \equiv \partial\omega/\partial \ln m, \tag{10.43}$$

which are needed to calculate the entropy per site of the polymer system

$$s^{(\mathrm{P})} = \omega - \phi_{\mathrm{B}} \ln K - \phi_{\mathrm{L}} \ln m;$$

the superscript is to indicate that it is the polymer system entropy, and is different from the spin system entropy $s^{(\mathrm{S})} = \partial T\omega/\partial T$.

In the following, we will be only interested in the polymer entropy. The proper stability requirements for the polymer system are

$$(\partial \phi_{\mathrm{B}}/\partial \ln K) \geq 0, \quad (\partial \phi_{\mathrm{L}}/\partial \ln m) \geq 0, \tag{10.44}$$

as can easily be seen from (10.41), and must be satisfied even for SMS. They replace the positivity of the heat capacity of the spin system, which no longer represents a physical spin system for $0 \leq m < 1$ [107]. It is easy to see from the definition of $s_{\mathrm{dis}}^{(\mathrm{P})}$ that $(\partial s_{\mathrm{dis}}^{(\mathrm{P})}/\partial T)_m$ need not be positive, even if the conditions in (10.44) are satisfied. We note from λ_{ord} that $\phi_{\mathrm{B}} - \phi_{\mathrm{L}} = 1$, so the number density $\phi_n \equiv \phi_{\mathrm{B}} - \phi_{\mathrm{L}}$, see (10.39), is $\phi_n = 1$. This means that all lattice sites are covered by the percolating polymer in the ordered state, as discussed above. Let us compute ω as $K \to \infty$ $(T \to 0)$ for λ_{dis} and λ_{ord}. From (10.43), we see that $\phi_{\mathrm{B}} \to mK$ for both states as $T \to 0$. Thus, using $\omega = s^{(\mathrm{P})} + \phi_{\mathrm{B}} \ln K + \phi_{\mathrm{L}} \ln m$, we have

$$\omega_{\mathrm{dis}}(T)/\omega_{\mathrm{ord}}(T) \to 1 \quad \text{as } T \to 0. \tag{10.45}$$

This means that if the eigenvalue λ_{dis} is taken to represent the stationary metastable phase above x_{c}, its osmotic pressure must become equal to that of the equilibrium phase (described by the eigenvalue λ_{ord}) at absolute zero. This is in conformity with Theorem 10.2 [105]. We take $\omega_{\mathrm{dis}}(T)$ to represent the SMS osmotic pressure below T_{c}. We have also checked that $Ts_{\mathrm{dis}}^{(\mathrm{S})} \to 0$, as $T \to 0$.

We will consider $m = 0.7$ and $J = 1$ below for numerical results. With the melting transition at $T_{\mathrm{M}} \cong 0.581$, the EL entropy $s_{\mathrm{EL}}^{(\mathrm{P})} \cong 0.357$ and the entropy discontinuity $\Delta s^{(\mathrm{P})} \cong 0.214$. Thus, there is a latent heat at the transition at x_{c}. Let us consider the SMS osmotic pressure $\omega_{\mathrm{dis}}(T)$ below T_{M}. It is easily checked that the above stability conditions in (10.44) are always satisfied in both phases; see, for example, the behavior of $\phi_{\mathrm{B,dis}}$ in Figure 10.9 for the disordered phase. We also show the number density for the disordered phase. The SCL entropy $s_{\mathrm{dis}}^{(\mathrm{P})}$ becomes negative below $T_{\mathrm{K}} \cong 0.385$, a temperature below $T_{\mathrm{M}} \cong 0.581$. We can use the negative of the bond density as a measure of the energy. At $T_{\mathrm{K}}, \phi_{\mathrm{B,dis}} \simeq 2.716$, while $\phi_{\mathrm{B,ord}} \to \infty$ at absolute zero. This establishes an energy gap. We have also plotted the excess entropy $\Delta s_{\mathrm{ex}}^{(\mathrm{P})} \equiv s_{\mathrm{dis}}^{(\mathrm{P})} - s_{\mathrm{ord}}^{(\mathrm{P})}$ in Figure 10.9. We note that it also becomes negative below $T \cong 0.338$, close to the temperature where $s_{\mathrm{dis}}^{(\mathrm{P})}$ becomes negative. It has a minimum at a lower temperature, and eventually vanishes as $T \to 0$, and the entropies of both phases become the same.

We now make an important observation. As m decreases (below 1), both T_{K} and T_{M} $(T_{\mathrm{K}} < T_{\mathrm{M}})$ move down toward zero simultaneously. As $m \to 0$, the equilibrium ordered phase corresponding to λ_{ord} disappear completely, and the disordered phase corresponding to λ_{dis} becomes the equilibrium phase. There is no transition to any other state. Thus, there is no metastability anymore.

The above calculation for a model with short-range interactions clearly shows that there is no singularity in λ_{dis} or $\omega_{\mathrm{dis}}(T)$ at T_{M}, even though there is a phase transition.

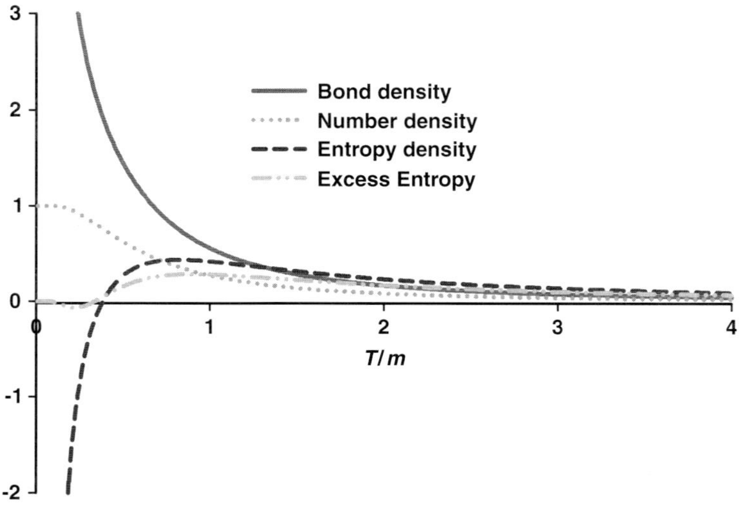

Figure 10.9 The bond, number, and entropy densities for the disordered phase. The bond density is a monotonic function of T, so that the stability is not violated. The excess entropy shows that the entropy of the disordered metastable phase becomes less than that of the ordered phase at low temperatures. The entropy also becomes negative at low temperatures before the excess entropy becomes negative.

Similarly, there is no singularity in λ_{ord} or $\omega_{\text{ord}}(T)$ at T_{M}. Thus, the thermodynamic singularity in the equilibrium free energy does not necessarily create a singularity in $\omega_{\text{dis}}(T)$ or $\omega_{\text{ord}}(T)$ at T_{M}, as was discussed. The existence of a singularity at some other temperature is a different issue with which we are not concerned here.

10.11
Glass Transition in a Binary Mixture

Let us turn to the cell representations in Figure 10.8, where the empty and filled circles represent two kinds of particles A and B in a binary mixture. As we are only interested in the communal entropy, the possibility of vibrations within each cell is not relevant for our consideration. We are only interested in the average positions of particles in each cell. The average position can be used to represent sites of a lattice. Each site is occupied by a particle A or B. No cell is empty. Instead of considering a random and a regular lattice to represent (a) and (b) in Figure 10.8, we further simplify the situation and consider a regular lattice of a fixed coordination number. Despite this, disordered states still occur and are obtained by random placements of particles on the regular lattice. We now consider a simple lattice model of an incompressible (no empty sites) binary mixture of A and B, to be represented by an Ising spin S. The two spin states ($+1$ or up) and (-1 or down) represent A and B, respectively. As we are not interested in their phase separation, but in the possibility of a glass transition, we assume that their mutual interaction is attractive. In

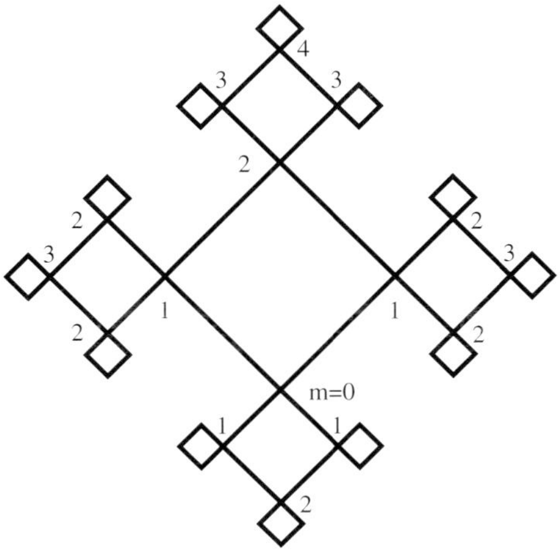

Figure 10.10 A small portion of a recursive Husimi cactus with two squares meeting at each sites. The sites of the squares are labeled as shown, with the site index increasing as we move away from the origin $m = 0$.

addition, we are interested in ordered and disordered phases to have distinct symmetries to give rise to the conventional supercooling. We will, therefore, use an antiferromagnetic (AF) Ising model in zero magnetic field with both two-spin ($J > 0$) and three-spin interactions ($J' \neq 0$) to make the model slightly more complex and interesting. As the model cannot be solved exactly, we consider a Husimi cactus made of squares, on which the model can be solved exactly [15]. We consider the simplest cactus shown in Figure 10.10 in which only two squares meet at a site; they cannot share a lattice bond. The method is easily extended to consider more than two squares meeting at a site. The squares are connected so that there are no closed loops except those formed by the squares. The cactus can be thought as an approximation of a square lattice, so that the exact Husimi cactus solution can be thought of as an approximate solution of the square lattice model. However, we can also think of the solution as the exact solution on a lattice, although artificial. The exactness ensures that stability will always be satisfied. There is a sublattice structure at low temperatures caused by the antiferromagnetic interaction: particles of one species are found on one of the two sublattices. We identify this ordered structure as a crystal. The interaction energy is

$$E = J\sum SS' + J'\sum SS'S''. \tag{10.46}$$

The first sum is over nearest-neighbor spin pairs and the second over neighboring spin triplets within each square. In the absence of the three-spin coupling, the two-spin coupling gives rise to an antiferromagnetic ordering at low temperatures. For $J' > -J$, the AF ordering remains the preferred ordering, while for $J' < -J$, the ferromagnetic ordering is preferred. Therefore, we only consider $J' > -J$ and set $J = 1$ to set the temperature scale.

The model is solved recursively, as has been described elsewhere [15]. Sites are labeled by an index m, which increases sequentially outward from $m = 0$, as shown in Figure 10.10. We introduce partial PFs $Z_m(\uparrow)$ and $Z_m(\downarrow)$, depending on the states of the spin at the mth level. It represents the contribution of the part of the cactus above that level to the total PF. We then introduce the ratio

$$x_m \equiv Z_m(\uparrow)/[Z_m(\uparrow)+Z_m(\downarrow)], \qquad (10.47)$$

which satisfies the recursion relation (RR):

$$x_m \equiv f(x_{m+1}, x_{m+2}, v)/[f(x_{m+1}, x_{m+2}, v) + f(y_{m+1}, y_{m+2}, 1/v)], \qquad (10.48)$$

where

$$f(x, x', v) \equiv x^2 x'/u^4 v^4 + 2xx' y v^2 + x^2 y v^2 + u^4 x' y^2 + 2xyy' + y^2 y'/v^2, \qquad (10.49)$$

with

$$u \equiv e^{\beta}, \quad v \equiv e^{\beta J'}, \quad y \equiv 1-x, \quad y' \equiv 1-x'.$$

There are two kinds of fix point solutions of the RR that describe the bulk behavior [15]. In the 1-cycle solution, the fix point solution becomes independent of the index m as we move toward the origin $m = 0$ on an infinite cactus, and is represented by x^*. For the current problem, it is given by $x^* = 1/2$, as can be checked explicitly by the above RR in (10.48). It is obvious that it exists at all temperatures. There is no singularity in this fix point solution at any temperature. This solution corresponds to the disordered paramagnetic phase at high temperatures and the SMS below the melting transition to be discussed below. The other fix point solution is a 2-cycle solution, which has been found and discussed earlier in the semiflexible polymer problem [36, 37, 44, 46–48], the dimer model [37], and the star and dendrimer solutions [48]. The fix point solution alternates between two values x_1^* and x_2^* on two successive levels. At $T = 0$, this solution is given either by $x_1^* = 1$ and $x_2^* = 0$, or by $x_1^* = 0$ and $x_2^* = 1$. The system picks one of these as the solution. At and near $T = 0$, this solution corresponds to the low-temperature AF ordered phase, which represents the CR and its excitation at equal occupation, and can be obtained numerically. The 1-cycle free energy is calculated by the general method proposed in Ref. [15], and the 2-cycle free energy is calculated by the method given in Refs [36, 37, 44, 46–48].

We now discuss numerical results. We take $J' = 0.01$. The ground-state energy (perfect crystal) per particle is given by $e_0 = -2J$. We shift all free energies and energies by this value so that they have the same common value ($= 0$) at absolute zero. The free energy F_1 $[= f_{\text{dis}}^*(T) - e_0]$ and entropy S_1 $[= s_{\text{dis}}^*(T)]$ associated with the 1-cycle FP solution are shown by the continuous and the long dash curves in Figure 10.11. The free energy F_2 $[= f_{\text{ord}}(T) - e_0]$ and entropy S_2 $[= s_{\text{ord}}(T)]$ associated with the 2-cycle FP solution are shown by the dotted and the dash–dot curves. The energy $E_1(T) = e(T) - e_0$ and $E_2(T) = e(T) - e_0$ as a function of T, and the entropy $S_1(E)$ and $S_2(E)$ as a function of the shifted energy E for the two fix point solutions are shown in Figure 10.12. The shifted F and E represent the contributions of excitations with respect to the ground-state energy e_0, so that they vanish at $T = 0$, as is clearly

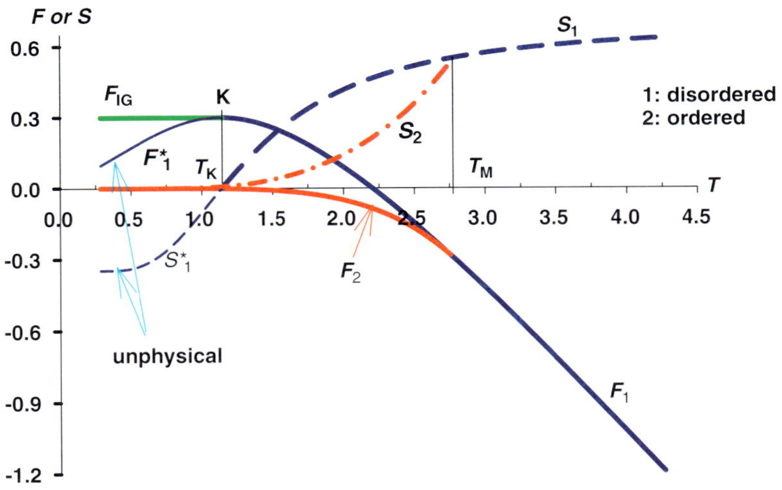

Figure 10.11 The free energies and entropies from the two FP solutions. The entropy crisis occurs at T_k, below which S_1^* (dashed cyan) becomes negative, but the corresponding free energy F_1^* (cyan) remains concave despite an unphysical communal entropy. This explains their labeling as unphysical in the figure. The disordered state is indicated by cyan and ordered state by red; continuous curves show the free energies and the discontinuous curves show the entropies. The ideal glass is represented by the dashed green horizontal line, which has a constant free energy $F_{IG}=E_K$ and replaces F_1^* below T_K.

seen in Figures 10.11 and 10.12. The transition temperature is found to be $T_M \cong 2.7706$. Figure 10.11 shows that F_1 crosses zero and becomes positive below $T = T_{eq} \simeq 2.200$ but again becomes zero (not shown here, but we have checked it) as $T \to 0$. Thus, F_1 possesses a maximum at an intermediate temperature (see point K in Figure 10.11) at $T = T_K \simeq 1.1316$, so that S_1 vanishes there. Below T_K, the *continuation* of F_1 and S_1, shown by their thin portions in Figures 10.11 and 10.12, continue to satisfy the *stability condition* (remaining concave). Despite this, they have to be discarded as *unphysical* due to *negative entropy*. Below T_K, we must extend the metastable state (described by F_1 and S_1 between T_K and T_M) by a glassy phase of a constant free energy $F = F_{IG} = E_K - e_0 \simeq 0.301$ of the ideal glass (E_K denoting the energy per particle at IGT in this section only), see the short dash horizontal line in Figure 10.11, and $S = S_G = 0$. The 1-cycle energy at K is $E_{1K} = F_{IG}$ due to the presence of a large amount of excitations in the IG. The 2-cycle gives the equilibrium crystal at $T \leq T_M$, with a nonnegative entropy S_2. As the CR entropy only vanishes at absolute zero, it is never confined to a single basin except at absolute zero. The energy E_{2K} of the CR at T_K is smaller than E_{1K} of the IG, which is consistent with the previous discussion that the glass is a highly defective CR.

At $T = 0$, $s_{dis}^*(T) \simeq -0.3466$, while the CR entropy is zero, as expected. Thus, both entropies satisfy Nernst–Planck postulate.[2)] We also observe that $f_{dis}^*(T)$ and $f_{ord}(T)$ become identical ($= -2J$) at absolute zero in accordance with the Theorem 10.2. Thus,

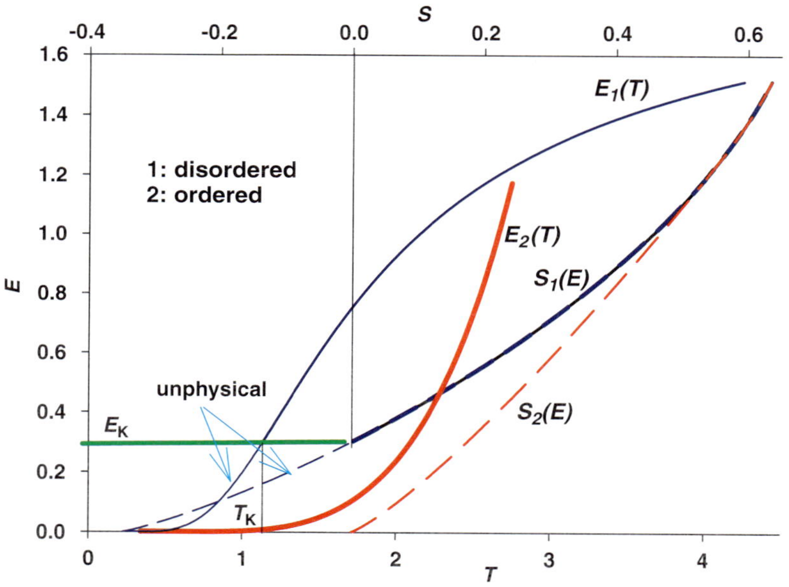

Figure 10.12 $S - E - T$ relationship for the two FP solutions. The excitations near $T = 0$ in them are very different. The excitations in the 1-cycle state near T_k are strongly interacting as opposed to those near $T = 0$.

the free energy diagram we obtain in this case is similar to that in the inset in Figure 10.2.

10.12
Ideal Glass Singularity and the Order Parameter

10.12.1
Singular Free Energy

We will now show that the form of $S_{\text{dis}}(E)$ in the gap has no relevance to the physics of the problem. Therefore, the entropy extension over the gap can be *arbitrary* as long as $S^*_{\text{dis}}(E)$ is continuous and concave. We consider quantities associated with communal entropies in this section, but will not show "comm" for simplicity of notation. The argument can be easily extended to configurational entropies. We compare $Z_{\text{dis}}(T)$ and $Z^*_{\text{dis}}(T)$. They only differ in terms containing $E < E_{\text{K}}$. For $E \geq E_{\text{K}}$, they use the same function $S_{\text{dis}}(E)$. Thus, for $T \geq T_{\text{K}}$ or ($E_{\text{dis}}(T) \geq E_{\text{K}}$), the two PFs are identical. Consider $T = T_{\text{K}}$ and write

$$Z_{\text{dis}}(T_{\text{K}}) = W_{\text{dis}}(E_{\text{K}})\text{e}^{-\beta_{\text{K}}E_{\text{K}}}\left[1 + \sum_{E>E_{\text{K}}} \frac{W_{\text{dis}}(E)}{W_{\text{dis}}(E_{\text{K}})}\text{e}^{-\beta_{\text{K}}\Delta E}\right],$$

where $\Delta E = E - E_K > 0$, and $\beta_K = 1/T_K$. In the thermodynamic limit, the right-hand side approaches $W_{dis}(E_K)e^{-\beta E_K}$, so that

$$\frac{W_{dis}(E)}{W_{dis}(E_K)}e^{-\beta_K \Delta E} \to 0 \text{ for } E > E_K. \tag{10.50}$$

We consider $T < T_K$and

$$Z_{dis}(T) = W_{dis}(E_K)e^{-\beta E_K}\left[1 + \sum_{E>E_K}\frac{W_{dis}(E)}{W_{dis}(E_K)}e^{-\beta\Delta E}\right].$$

Expressing $\beta = \beta_K + \Delta\beta$ in $e^{-\beta\Delta E}$, where $\Delta\beta > 0$, we observe that the summand in the above equation vanishes on account of (10.50). Thus, $Z_{dis}(T)$ reduces to the prefactor in the above equation, and we finally have

$$\begin{aligned} F_{dis}(T) &\equiv -T\ln Z_{dis}(T) = -T\ln W_{dis}(E_K)e^{-\beta E_K} \\ &= (T/T_K)F_{dis}(T_K) + (1 - T/T_K)E_K. \end{aligned}$$

Using the fact that $F_{dis}(T_K) = E_K - T_K S_{dis}(T_K) = E_K$, the above equation reduces to

$$F_{dis}(T) \equiv E_K \quad \text{for } T \leq T_K. \tag{10.51}$$

This is the free energy of the *ideal glass* below T_K shown by F_{IG} in Fig. 10.11, which makes $F_{dis}(T)$ singular at T_K. On the other hand, $F^*_{dis}(T)$ behaves very differently below T_K, as the calculations show. However, the correct physics is described by $F_{dis}(T)$ and not $F^*_{dis}(T)$ below T_K.

10.12.2
Order Parameter

We now proceed to identify an order parameter for IGT. For this, we turn to the melting of crystals into liquids is similar (but not identical) to the "melting" of glasses into SCLs. According to the Lindemann criterion of melting [112], a crystal melts when the mean square displacement $\bar{\mathbf{r}^2}$ becomes so large that the atoms start to get into each other's cells to the point that they begin to diffuse over a large distance, and the melting initiates. If a denotes the interatomic distance in the crystal, then the melting proceeds when $\bar{\mathbf{r}^2} = c_L a^2$, where c_L is Lindemann's constant and is expected to be the same for crystals with similar structure. It should, however, be pointed out that the criterion is expected to be valid only for those systems that have simple crystalline structures. For particles obeying Lennard-Jones potential, Jin *et al.* [113] have tested the validity of the criterion by considering 6912 Lennard-Jones particles. The value of c_L is estimated to be $\simeq 0.12$–0.13 at equilibrium melting, as expected. As the temperature is raised toward the melting temperature, clusters of correlated particles of various sizes are formed. If we now treat a glass as a defective crystal [44] with a disordered cell representation, then it is not hard to imagine that a similar criterion can be applied to the "melting" of a glass into its SCL state. The relevance of $\bar{\mathbf{r}^2}$ for glass transition has been discussed in the literature [114]. We will pursue this analogy a bit further below.

The so-called melting of an ideal glass into a SCL is nothing but a localization–delocalization transition, as we have discussed in Section 10.4. As the temperature is raised, the IG becomes deconfined and begins to probe other basins corresponding to different cell representations, thereby giving rise to the α relaxation. We can make this process somewhat quantitative by considering a particle at some average position $\mathbf{r}_0$ within its cell. We average the square of its displacement. At low temperatures where it is confined to its cell, the particle does not diffuse away from the original point $\mathbf{r}_0$; it merely undergoes vibrations within its cell. Let us introduce a certain length b characterizing the average size of the cell. Then, $\bar{\mathbf{r}^2} \leq c_G b$, with $c_G < 1$ a parameter that depends on the system under investigation, will characterize a "solid-like" atom (S) in that the particles satisfying this constraint are only allowed to vibrate about their equilibrium position $\mathbf{r}_0$ within their cells. This will be the situation in the IG, in which the particles do not diffuse out of their cells. On the other hand, $\bar{\mathbf{r}^2} > c_G b$ will describe a "liquid-like" atom (L) in that the particles escape the neighborhood of $\mathbf{r}_0$ by diffusion and will give rise to a collective motion. Thus, we can classify the atoms as L or S, thereby reducing the system to a binary mixture of L and S particles. This scenario was conjectured by us in Ref. [43] as the possible origin of the β relaxation in SCL. This is the only relaxation in IG in which there are no L-type particles, whereas there are many L-type particles in the SCL state contributing to the α relaxation. Hence, we can use the density n_L of L-type particles as the order parameter ϱ to describe IGT: it is nonzero in SCL and gradually vanishes at T_K as the temperature is reduced; it remains zero in the IG state. We can use the proximity of two L-type particles to determine if they are "connected" to form a cluster. These clusters can be classified as *liquid-like clusters* (formed by L particles). (One can similarly define a solid-like cluster). As the temperature is reduced, the liquid-like clusters not only disappear but their sizes also become smaller until finally they completely disappear at T_K. At all higher temperatures, these clusters themselves will be continuously changing in time, but their average densities will remain constant at fixed T, P, similar to what happens in physical gelation [7, 51, 80]. Above some higher temperature, there will be a percolating L-type cluster. Therefore, we are dealing with the phenomenon of percolation. A picture similar but not identical to this has been developed by Novikov *et al.* [115]. Another picture involving two kinds of regions very similar to the above has been proposed by deGennes [116].

10.12.3
Relevance for Experiments

So far, we have mainly focused on SMSs, which can never be observed in experiments or in simulation because both are conducted under a finite time limit, whereas SMSs require an infinite time limit, at least near the glassy region. Therefore, it is important to understand the experimental relevance of the above singularity and of the gap model. As the singularity appears in SMSs, there is no way to observe it in any experimental setup or in simulation. In particular, there will be no way to ensure that the system is confined to a single cell representation or a potential energy basin. Doing this will require $t \to \infty$. Thus, in reality, the system will always probe many basins in any realistic setup. Accordingly, the communal entropy will never go to zero. In fact, it is well known [43] that there is considerable communal contribution to the entropy at T_g. A careful analysis of calorimetric data showed that, on an average, 70% of the excess entropy at T_g survives at the lowest temperature (where data are available). This fraction can be considered as communal in nature at T_g [43], and is traditionally identified as the *residual entropy* in glasses [23]. The remaining 30% of the excess entropy arises from vibrational and anharmonic differences between glassy and crystalline phases. The implication of the communal entropy is that the system is trapped or confined in a large number of basins at the experimental glass transition temperature. This is because of the experimental time constraint τ_{exp}, which forces the glass to be arrested in one of many basins over this duration. Consequently, this results in smoothing out of the IGT singularity at T_g, which is borne out of our experience. The existence of this huge amount of communal entropy at T_g also explains why no experiment will ever exhibit the Kauzmann entropy crisis. To discover that, one must perform *extrapolation* of experimental data to lower temperatures, as was done by Kauzmann.

Another important consequence of the experimental time constraint τ_{exp} appears in the form of the residual entropy in glasses [23]. This entropy is a measure of the different forms of glasses that can be formed in an experimental glass transition. It is assumed that each basin in which the glass may be trapped represents a different glass form. Let N_f denote the number of different glass forms that can be formed in an experiment conducted with some fixed τ_{exp}. This number will evidently change with τ_{exp}. Then, the residual entropy will also depend on τ_{exp}. It is given by

$$S_R \equiv \ln N_f,$$

if each glass form is equally probable so that the probability of each glass form is $p_f = 1/N_f$. The observation of the residual entropy is very common in Nature ([16], Section 64). It is commonly identified as the amount of entropy that survives at absolute zero, despite the assumption that the glass has been trapped in one of the basins. The mere fact that the glass has been trapped in a single basin does not imply that the communal entropy has vanished, as we are not sure which of the possible basins the glass is trapped in. This ignorance gives rise to the residual entropy. To understand this paradoxical claim, we need to carefully justify it. For this we proceed as follows [16]. Imagine partitioning the system of N particles into a large number of subsystems of equal size with N' particles, each macroscopically large so that surface effects can be neglected. Each subsystem will appear trapped in one of the many possible basins (corresponding to a system of N' particles) or represent a glass form. Each basin j will appear with some probability p_j, given by

$$p_j = A \exp(-\beta F_j),$$

where F_j is the basin free energy introduced in Section 10.4.1. As the basin free energy will be a function of time, this probability will also depend on time. Consequently, we need to apply (10.2) to determine the entropy by using the above probability in (10.2). The sum is over N_f basins or glass forms and determines the entropy associated with different forms of glasses; this was called the communal entropy in Section 10.1.5. Alternatively, one must prepare several glasses under identical conditions. Each glass will be trapped in one of the possible basins with probability p_j, so that the application of (10.2) will give the same communal entropy. Let us consider the situation at T_g. As SCL is a SMS here, all basins will have the same free energy so that the probability p_j is also the same for all basins and equals

$$p_b = 1/N_f.$$

This results in the residual entropy S_R. As the system freezes in one of these basins below T_g, the probaility for each basin remains unchanged and equal p_b. This leaves S_R unchanged, so that the residual entropy will not disappear as the glass temperature is reduced, keeping fixed τ_{exp}; the latter only reduces the basin entropy $S_b(T)$. Thus, the communal entropy trapped at the glass transition persists all the way down to absolute zero and results in, what is customarily called, the residual entropy.

At any temperature below T_g, the system will remain trapped in a particular basin for the duration of τ_{exp}. During this period, it will most probably not reach equilibrium with the surroundings. However, it is usually the case that the choice of τ_{exp} is such that the system has enough time to come to internal equilibrium so that we can define its temperature $T(t)$, which will be different from the temperature T of the surroundings, and all other thermodynamic quantities such as its entropy, energy, and so on, all of which are usually functions of time. During cooling, we expect $T(t) < T$, so that if we wait long enough, $T(t)$ will approach T. During this relaxation, its energy $E(t)$ will also approach the SMS energy E at temperature T from above. This relaxation is not allowed if we cool the system and wait for only τ_{exp} at each step. In this case, the system will never relax past τ_{exp}, and its properties will be dramatically different from the behavior of SMSs that we have investigated in this chapter. If we hold the temperature past τ_{exp}, the system will then relax. Above T_K, we expect the relaxation of the glassy state toward SCL, while the relaxation will be toward the ideal glass below T_K.

10.13 Conclusions

In summary, we have presented an energy gap model of glass formers in class A. These are the materials that form an equilibrium crystal CR at low temperatures. The lowest allowed energy E_K of a SCL is higher than the lowest energy E_0 of the corresponding CR. The difference then gives rise to a gap in SCL. The glass formation occurs in the metastable SCL state after bypassing crystallization at the melting temperature T_M. We have mainly considered equilibrium properties in SCL and glassy state. Accordingly, we have investigated long-time stationary limit of SCLs that we have called stationary metastable states. They and CR are investigated by the use of restricted PF formalism, which we have used to obtain the communal entropy. The configurational entropy can be expressed as a sum of the communal and basin entropies (10.20). The communal entropy vanishes at a positive Kauzmann temperature T_K due to the energy gap as follows from our energy gap theorem, and results in the confinement of the system in a single potential energy basin whose minimum is at E_K. A basin gives rise to a particular cell representation. The cell representation, which is a common mode of description of low-temperature liquids, is equally applicable to SCL and CR and leads quite naturally to the potential landscape description at all temperatures, even though the landscape was originally introduced by Goldstein as useful only at low temperatures. In addition, the cell representation allows us to use lattice models to calculate the communal entropy without any loss of generality.

We hope that we have convinced the reader that the use of the order parameter allows us to classify microstates or configurations as ordered and disordered in a unique manner, though it may be tedious to implement it in practice. The classification is needed for the gap model to make sense. In general, ordered microstates exist for energies $E \geq E_0(N, V)$. According to the Nernst–Planck hypothesis, the corresponding communal entropy $S_{\text{comm,ord}}(E_0) = 0$, its smallest possible value; however, its slope at E_0 is infinite. This implies that the communal entropy of a CR vanishes only at absolute zero ($\beta \to \infty$) so that it can be confined to a single basin *only* at $T = 0$. This then shows that the basin describing the presence of a defect in CR will be a different basin. However, the energy barrier between these basins must be very small as it is very easy to create a f= defect in CR by raising its temperature slightly above absolute zero. The situation for disordered microstates is quite different. Because of the energy gap, disordered microstates exist for energies $E \geq E_K > E_0$. The corresponding entropy $S_{\text{comm,dis}}(E_K) = 0$, but its slope at E_K is finite so that the SCL gets confined to a single basin at $T = T_K > 0$, where it turns into an ideal glass, which remains confined in this basin as the temperature is lowered to $T = 0$. This basin must be separated by other basins by very high energy barriers to account for the stability of the glassy state.

As said above, E_0 sets the zero of the temperature scale. The assumption that $S_{\text{dis}}(E)$ is nonsingular at E_K has allowed us to continue $S_{\text{dis}}(E)$ in the energy gap $[E_0, E_K]$, so that the same zero of the scale is common to both PFs in (10.32). (The fact that both states have the same common temperature T_M at the coexistence where $Z_{\text{dis}}(T_M) = Z_{\text{ord}}(T_M)$ does not depend on this continuation, since $Z_{\text{dis}}(T) = Z^*_{\text{dis}}(T)$ at T_M.) The extension $S^*_{\text{dis}}(E)$ in the energy gap is arbitrary as long as it remains concave and continuous, in consistence with the Remark in Section 10.2.2. In our calculation, we do not need to carry out this extension as the calculation gives $S^*_{\text{dis}}(E)$ over the range $E \geq E_0$, which is found to be nonsingular at E_K and its slope infinitely large at E_0. The latter result is very important in that it shows that SMS in any exact calculation is not divorced from CR; their entropies extend from $E \geq E_0$ and both have a diverging slope ($T = 0$) at the lower end $E = E_0$. In other words, the exact calculations verify the properties of our gap model.

The model is analyzed by using $S_{\text{ord}}(E)$ and $S_{\text{dis}}(E)$ or $S^*_{\text{dis}}(E)$. These entropies allow us to introduce three restricted PFs and restricted free energies per particle $f_{\text{ord}}(T), f_{\text{dis}}(T)$, and $f^*_{\text{dis}}(T)$. These free energies are equal at the melting temperature T_M. The equilibrium free energy $f(T)$ has two branches as shown in (8.2). The switch over from one branch to another gives rise to a singular equilibrium free energy $f(T)$

at T_M, even though none of the three restricted free energies are singular at T_M; the latter property of the free energies basically verifies the Assumption made in Section 10.1.7.

Each of the restricted free energies is a concave function of T and describes a stable state. Thus, we have a thermodynamically valid description of CR and SMS at low temperatures. While $f_{dis}(T)$ and $f^*_{dis}(T)$ are identical for $T \geq T_K$, they are different below T_K. The communal part of $f^*_{dis}(T)$ possesses a maximum at T_K and goes down to E_0 at $T = 0$. The maximum corresponds to $S_{comm,dis}(E_K) = 0$. Otherwise, $f^*_{dis}(T)$ has no singularity anywhere. The only use of $f^*_{dis}(T)$ is to locate T_K, below which $f^*_{dis}(T)$ has no physical significance. On the other hand, $f_{dis}(T)$ has a singularity at T_K: its communal part remains constant equal to $f_{dis}(T_K)$ for any $T \leq T_K$.

Appendix 10.A: Classical Statistical Mechanics

In classical statistical mechanics, the Hamiltonian $\mathcal{H}(\mathbf{p}, \mathbf{q}) \equiv K(\mathbf{p}) + E(\mathbf{r})$ of a system of N particles in a fixed volume V is a sum of the kinetic energy $K(\mathbf{p})$ and the potential energy $E(\mathbf{r})$ of the particles; here $\mathbf{p}$ and $\mathbf{r}$ represent the collective momenta $\{\mathbf{p}_i\}$ and positions $\{\mathbf{r}_i\}$ of the particles, respectively. The fact that the potential energy is taken to be a function of coordinates only is not always true as happens if we have charged particles in a magnetic field. These cases will not be considered here. The dimensionless total canonical PF $Z_T(T, V)$ of the system (we revert back to exhibiting the dependence on V in this section) can be written as a product of two *independent* integrals

$$Z_T(T, V) \equiv Z_{KE}(T) Z(T, V), \tag{10.A1}$$

where

$$Z_{KE}(T) \equiv \frac{v_0^N}{(2\pi\hbar)^{3N}} \int e^{-\beta K} d^N\{\mathbf{p}\}, \tag{10.A2}$$

represent the kinetic partition function due to the translational degrees of freedom of the particles, and the configurational partition function $Z(T, V)$ is given in (10.5). We have kept the constant v_0 (which we earlier set equal to 1) to make the two factors in (10.A1) dimensionless.

The prefactor in terms of $\hbar$ is used to explicitly show the correspondence of Z_T with the corresponding PF in the quantum statistical mechanics in the classical limit $\hbar \to 0$. Despite the classical limit requirement $\hbar \to 0$, we are not allowed to set $\hbar = 0$ in the final result, but keep its actual nonzero value. Accordingly, some problems remain such as Wigner's distribution function not being a classical probability distribution, which we do not discuss any further but refer the reader to the literature [117]. Keeping $\hbar$ at its nonzero value avoids infinities as we will see below but in no way implies that we are dealing with quantum effects. In particular, it does not imply that the entropy is nonnegative, as we have discussed elsewhere [75]. We merely quote the results here. The contribution $S_{KE}(T)$ to the entropy due to the kinetic energy is

$$S_{KE}(T) = N\ln(v_0/\lambda^3), \tag{10.A3}$$

where $\lambda \equiv h/\sqrt{2\pi meT}$, a quantity related to the de Broglie thermal wavelength, and the average kinetic energy $\bar{K}(T)$ is

$$\bar{K}(T) = (3/2)NT, \tag{10.A4}$$

which should come as no surprise. For an ideal gas, the total entropy function $S_T(T)$ is given by

$$S_T(T, V) = N\ln(Ve/\lambda^3 N), \tag{10.A5}$$

which is independent of the choice of v_0 but depends on the volume V, while $S_{KE}(T)$ depends on v_0 but not on the volume. The configurational entropy for an ideal gas, which no longer depends on T, is

$$S(V) = N\ln(Ve/v_0 N). \tag{10.A6}$$

If we set $h = 0$, we encounter an infinity at all temperatures in (10.A3) and in (10.A5). To avoid this, we keep h at its nonzero value.

The contribution $S_{KE}(T)$ is the *same* for *all* systems (that have the same v_0), regardless of their potential energy of interaction and volume. It is most certainly the same for all phases of any system such as SCL and CR at the same temperature, and we do not have to specifically take it into account. Thus, in general, we can focus on the configurational entropy without any loss of generality. It is obtained by subtracting $S_{KE}(T)$ from $S_T(T)$ [36]:

$$S(T, V) \equiv S_T(T, V) - S_{KE}(T). \tag{10.A7}$$

Appendix 10.B: Negative Entropy

At absolute zero, or for $V/v_0 N < 1/e$, $S_T \to -\infty$, a well-known result of classical statistical mechanics. We also note that for $V/v_0 N < 1/e$, the configurational entropy S in (10.A6) and, therefore, S_T diverge to $-\infty$. The problem is due to the continuum nature of the real space. This is easily seen from the exact solution of the 1D Tonks gas, which is a simple model of noninteracting hard rods, each of length l. In one dimension, we will take v_0 to have the dimension of length. The configurational entropy S corresponding to N rods in a line segment of length L is given by [118]

$$S_L = N\ln[(L - Nl)e/v_0 N] \tag{10.B1}$$

in the thermodynamic limit, while S_{KE} is still given by (10.A3), where m now represents the mass of a rod. Comparison with (10.A6) shows that the only difference is that the total volume V in (10.A6) is replaced by the free volume analogue $L - Nl$ in one-dimensional Tonks gas (see (10.B1)). It is clear that the entropy becomes *negative* as soon as $L/v_0N < 1/e + l/v_0$ and eventually diverges to $-\infty$ in the fully packed state. Similarly, the problem with $S_{KE}(T) \to -\infty$ as $T \to 0$ is due to the continuum nature of the momentum space.

This problem disappears as soon as we invoke quantum statistical mechanics to describe the total PF, which no longer can be written as product of two or more PF's as in (10.A1). Here, we consider the number of states $W_T(E_T, V) \geq 1$ as function of the (total) energy eigenvalue E_T. The energy eigenvalue E_T can certainly be broken into the kinetic energy part K and the potential energy part E, but such a partition is not possible for the total entropy $S_T(E_T, V) \equiv \ln W_T(E_T, V)$. Therefore, in general, the notion of the configurational entropy does not make sense in the quantum case. Since we are only concerned with classical statistical mechanics in this work, we will not discuss this point further here, except to note that a negative entropy is impossible to occur on a lattice. Therefore, if the communal entropy becomes negative in a lattice model, it will directly refer to an entropy crisis and will point to an IGT.

Acknowledgments

I thank Andrea Corsi, Sagar Rane, and Fedor Semerianov with whom I have had numerous discussions and have closely collaborated. I also thank Martin Goldstein who inspired me to delve into the field of glass transition and Alexei Sokolov for various interesting discussions.

References

1 Penrose, O. and Lebowitz, J.L. (1979) *Fluctuation Phenomena* (eds E.W. Montroll and J.L. Lebowitz), North-Holland.
2 Fisher, M.E. (1967) *Physics*, **3**, 255.
3 Sütõ, A. (1982) *J. Phys. A*, **15**, L749.
4 Domb, C. (1996) *Critical Point*, Taylor & Francis, London.
5 Kauzmann, W. (1948) *Chem. Rev.*, **43**, 219.
6 Goldstein, M. and Simha, R.(eds) (1976) *Ann. N. Y. Acad. Sci.*, **279**, 117.
7 Zallen, R. (1983) *The Physics of Amorphous Solids*, John Wiley & Sons, Inc., New York.
8 Elliott, S.R. (1983) *Physics of Amorphous Materials*, Longmans, London.
9 Debenedetti, P.G. (1996) *Metastable Liquids: Concepts and Principles*, Princeton University Press, Princeton, NJ.
10 Barrat, J-L., Feigelman, Mi., Kurchan, J., and Dalibard, J. (eds) (2003) For a recent review, see *Slow Relaxations and Nonequilibrium Dynamics in Condensed Matter. Les Houches Session LXXVII*, Springer, Berlin; Lubchenko, V. and Wolynes, P.G. (2007) *Annu. Rev. Phys. Chem.*, **58**, 235.
11 Bragg, W.L. and Williams, E.J. (1934) *Proc. R. Soc. Lond. A*, **A145**, 699.
12 Gujrati, P.D., cond-mat/0412757.
13 Gujrati, P.D. (1996) *Phys. Rev. E*, **54**, 2723.
14 Kievelson, D., Kivelson, S.A., Zhao, X., Nussinov, Z., and Tarjus, G. (1995) *Physica A*, **219**, 27.
15 Gujrati, P.D. (1995) *Phys. Rev. Lett.*, **74**, 809.
16 Landau, L.D. and Lifshitz, E.M. (1986) *Statistical Physics*, vol. 1, 3rd edn, Pergamon Press, Oxford.
17 Gibbs, J.H. (1956) *J. Chem. Phys.*, **25**, 185.
18 Gibbs, J.H. and Di Marzio, E.A. (1958) *J. Chem. Phys.*, **28**, 373.

19 Derrida, B. (1981) *Phys. Rev. B*, **24**, 2613.
20 Shechtman, D., Blech, I., Gratias, D., and Cahn, J.W. (1984) *Phys. Rev. Lett.*, **53**, 1951–1953.
21 Maxwell, J.C. (1965) *Scientific Papers* (ed. W.D. Niven), Dover, New York, p. 425.
22 Goldstein, M. (1969) *J. Chem. Phys.*, **51**, 3728.
23 Goldstein, M. (1976) *J. Chem. Phys.*, **64**, 4767, and references therein.
24 Bermejo, F.J., Criado, A., de Andres, A., Enciso, E., and Schober, H. (1996) *Phys. Rev. B*, **53**, 5259.
25 Ediger, M.D., Angell, C.A., and Nagel, S.R. (1996) *J. Phys. Chem.*, **100**, 13200.
26 Malinovsky, V.K. and Sokolov, A.P. (1986) *Solid State Commun.*, **57**, 757.
27 Malinovsky, V.K., Novikov, V.N., Parshin, P.P., Sokolov, A.P., and Zemlyanov, M.G. (1990) *Europhys. Lett.*, **11**, 43.
28 Sokolov, A.P., Kisliuk, A., Soltwisch, M., and Quitman, D. (1992) *Phys. Rev. Lett.*, **69**, 1540.
29 Gurevich, V.L., Parshin, D.A., and Schober, H.R. (2002) *JETP Lett.*, **76**, 553.
30 Debenedetti, P.G. and Stillinger, F.H. (2001) *Nature*, **410**, 259.
31 Shell, M.S., Debenedetti, P.G., and Stillinger, F.H. (2004) *J. Phys. Chem. B*, **108**, 6772.
32 Andreev, A.F. (1964) *Soviet Physics JETP*, **5**, 1415;Fisher, M.E. (1967) *Physics*, **3**, 255; Langer, J.S. (1967) *Ann. Phys.*, **41**, 108; Landford, O. and Ruelle, D. (1969) *Commun. Math. Phys.*, **13**, 194.
33 Isakov, S.N. (1984) *Commun. Math. Phys.*, **95**, 427; Friedli, S. and Pfister, Ch.-Ed. (2004) *Commun. Math. Phys.*, **245**, 69; (2004) *Phys. Rev. Lett.*, **92**, 015702.
34 Lennard-Jones, J.E. and Devonshire, A.F. (1937) *Proc. R. Soc. Lond. A*, **A163**, 53.
35 Kirkwood, J.G. (1950) *J. Chem. Phys.*, **18**, 380; Janssens, P. and Prigogine, I. (1950) *Physica*, **XVI**, 895; Rowlinson, J.S. and Curtis, C.F. (1951) *J. Chem. Phys.*, **19**, 1519. Leelt, J.M.H. and Hurst, R.P. (1960) *J. Chem. Phys.*, **32**, 96.
36 Gujrati, P.D., Rane, S.S., and Corsi, A. (2003) *Phys. Rev. E*, **67**, 052501.
37 Semerianov, F. and Gujrati, P.D., cond-mat/0401047; Semerianov, F. and Gujrati, P.D. (2005) *Phys. Rev. E*, **72**, 0111102.
38 Gujrati, P.D. and Semerianov, F., cond-mat/0412759.
39 Semerianov, F. (2004) PhD Dissertation. University of Akron.
40 Hill, T.L. (1956) *Statistical Mechanics*, Dover, p. 355.
41 Huang, K. (1963) *Statistical Mechanics*, 2nd edn, John Wiley & Sons, Inc., New York.
42 Ruelle, D. (1982) *Physica (Utrecht)*, **113A**, 619.
43 Gujrati, P.D. and Goldstein, M. (1980) *J. Phys. Chem.*, **84**, 859.
44 Gujrati, P.D., arXiv:0708.2075.
45 Gujrati, P.D., cond-mat/0309143.
46 Gujrati, P.D. and Corsi, A. (2001) *Phys. Rev. Lett.*, **87**, 025701.
47 Corsi, A. and Gujrati, P.D. (2003) *Phys. Rev. E*, **68**, 031502.
48 Corsi, A. and Gujrati, P.D. (2006) *Phys. Rev. E*, **74**, 061121;Corsi, A. and Gujrati, P.D. (2006) *Phys. Rev. E*, **74**, 061122;Corsi, A. and Gujrati, P.D. (2006) *Phys. Rev. E*, **74**, 061123.
49 Newman, C.M. and Schulman, L. (1977) *J. Math. Phys.*, **18**, 23.
50 Gujrati, P.D. (1995) *Phys. Rev. E*, **51**, 957.
51 Flory, P.J. (1953) *The Principles of Polymer Chemistry*, Cornell University Press, Ithaca.
52 (a) Gujrati, P.D. (1980) *J. Phys. A*, **13**, L437; (b) Gujrati, P.D. and Goldstein, M. (1981) *J. Chem. Phys.*, **74**, 2596; (c) Gujrati, P.D. (1982) *J. Stat. Phys.*, **28**, 241.
53 Di Marzio, E.A. and Yang, A.J. (1997) *J. Res. Natl. Inst. Stand. Technol.*, **102**, 135.
54 Baschnagel, J., Wolfgardt, M., Paul, W., and Binder, K. (1997) *J. Res. Natl. Inst. Stand. Technol.*, **102**, 159.
55 Turnbull, D. (1969) *Contemp. Phys.*, **10**, 473.
56 Angell, C.A. (1991) *J. Non-Cryst. Solids*, **131–133**, 13.
57 Böhmer, R., Ngai, K.I., Angell, C.A., and Plazek, D.J. (1993) *J. Chem. Phys.*, **99**, 4201.
58 Bordat, P., Affouard, F., Descamps, M., and Ngai, K.L. (2004) *Phys. Rev. Lett.*, **93**, 105502.
59 Johari, G.P. and Goldstein, M. (1970) *J. Chem. Phys.*, **53**, 2372; Johari, G.P. and Goldstein, M. (1970) *J. Chem. Phys.*, **55**, 4245.

60 Götze, W. (1991) *Freezing and the Glass Transition* (eds J.P. Haasma, D. Levesque, and J. Zinn-Justin Liquids), North-Holland, Amsterdam, p. 287;Götze, W. and Sjögren, L. (1992) *Rep. Prog. Phys.*, **55**, 241.
61 Johari, G.P. (2000) *J. Chem. Phys.*, **113**, 751.
62 Pyda, M. and Wunderlich, B. (2002) *J. Poly. Sci. B*, **40**, 1245.
63 Stillinger, F.H. and Weber, T.A. (1982) *Phys. Rev. A*, **25**, 978.
64 Abraham, F.F. (1980) *J. Chem. Phys.*, **72**, 359.
65 Kubo, R., Ichimura, H., Usui, T., and Hashitsume, N. (1965) *Statistical Mechanics: An Advanced Course with Problems and Solutions*, North-Holland, Amsterdam.
66 Sciortino, F., Kob, W., and Tartaglia, P. (1999) *Phys. Rev. Lett.*, **83**, 3214.
67 Santen, L. and Krauth, W. (2000) *Nature*, **405**, 550.
68 Coluzzi, B., Parisi, G., and Verrocchio, P. (2000) *Nature*, **84**, 306.
69 Johari, G.P. (1976) *Ann. N. Y. Acad. Sci.*, **279**, 117.
70 Angell, C.A. (1995) *Science*, **267**, 1924; Poole, P.H., Grande, T., Angell, C.A., and McMillan, P.F. (1997) *Science*, **275**, 322.
71 Fischer, E.W. (1993) *Physica A*, **210**, 183.
72 Tanaka, H. (2000) *Phys. Rev. E*, **62**, 6968.
73 Nelson, D.R. (1983) *Phys. Rev. B*, **28**, 5515.
74 Wales, D. (2003) *Energy Landscapes*, Cambridge University Press, Cambridge.
75 Gujrati, P.D., cond-mat/0412548, cond-mat/0412735.
76 Hammersley, J.M. (1957) *Proc. Camb. Phil. Soc.*, **53**, 642.
77 Frisch, H. and Hammersley, J.M. (1963) *J. Soc. Ind. Appl. Math.*, **11**, 894.
78 Stauffer, D. (1979) *Phys. Rep.*, **54**, 1.
79 Grimmett, G. (1980) *Percolation*, Springer-Verlag, New York.
80 deGennes, P.G. (1979) *Scaling Concepts in Polymer Physics*, Cornell University Press, Ithaca.
81 Fetter, A.L. and Walecka, J.D. (1980) *Theoretical Mechanics of Particles and Continua*, McGraw-Hill, New York.
82 Di Marzio, E.A. (1971) XXIII International Congress of Pure and Applied Chemistry, Butterworths London, vol. 8.
83 Kirkpatrick, T.R. and Wolynes, P.G. (1987) *Phys. Rev. A*, **35**, 3072; Kirkpatrick, T.R., Thirumalai, D., and Wolynes, P.G. (1989) *Phys. Rev. A*, **40**, 1045.
84 (a) Franz, S. and Parisi, G. (1997) *Phys. Rev. Lett.*, **79**, 2486; (b) Angelani, L., Parisi, G., Ruocco, G., and Viliani, G. (1998) *Phys. Rev. Lett.*, **81**, 4648; (c) Mézard, M. and Parisi, G. (1999) *Phys. Rev. Lett.*, **82**, 747; (d) Coluzzi, B., Parisi, G., and Verrocchio, P. (2000) *Phys. Rev. Lett.*, **84**, 306.
85 Kivelson, D. and Tarjus, G. (1998) *J. Chem. Phys.*, **109**, 234.
86 (a) Sastry, S., Debenedetti, P.G., and Stillinger, F.H. (1998) *Nature*, **393**, 554; (b) Sastry, S. (2000) *Phys. Rev. Lett.*, **85**, 590.
87 Adam, G. and Gibbs, J.H. (1965) *J. Chem. Phys.*, **43**, 139.
88 Gibbs, J.H. (1960) *Modern Aspects of the Vitreous State* (ed. J.D., Mackenzie), Butterworth, London.
89 Johari, G.P. (2000) *J. Chem. Phys.*, **112**, 8958.
90 Bouchaud, J.-P. and Biroli, G. (2004) *J. Chem. Phys.*, **121**, 7347.
91 (a) Cohen, M.H. and Turnbull, D. (1959) *J. Chem. Phys.*, **31**, 1164; (b) Cohen, M.H. and Grest, G.S. (1979) *Phys. Rev. B*, **20**, 1077.
92 (a) Chhajer, M. and Gujrati, P.D. (1998) *J. Chem. Phys.*, **109**, 9022; (b) Rane, S. and Gujrati, P.D. (2001) *Phys. Rev. E*, **64**, 011801.
93 Doolittle, A.K. (1951) *J. Appl. Phys.*, **22**, 1471.
94 Williams, E. and Angell, C.A. (1977) *J. Phys. Chem.*, **81**, 232.
95 Angell, C.A. (1997) *J. Res. Natl. Inst. Stand. Technol.*, **102**, 171.
96 Das, S.P. and Mazenko, G.F. (1968) *Phys. Rev. A*, **34**, 2265.
97 Götze, W. and Sjögren, L. (1987) *Z. Phys. B*, **68**, 415.
98 Cates, M.E. and Ramaswamy, S. (2006) *Phys. Rev. Lett.*, **96**, 135701.
99 Prigogine, I. and Defay, R. (1954) *Chemical Thermodynamics*, Longmans Green, New York.
100 Davies, R.O. and Jones, G.O. (1953) *Proc. R. Soc. Lond. A*, **217**, 26.

101 Goldstein, M. (1963) *J. Chem. Phys.*, **39**, 3369.

102 Gupta, P.K. and Monyihan, C.T. (1975) *J. Chem. Phys.*, **65**, 4136.

103 Bailey, N.P., Christensen, T., Jakobsen, B., Niss, K., Olsen, N.B., Pedersen, U.R., Schrø der, T.B., and Dyre, J.C. (2008) *J. Phys.: Condensed Matter*, **20**, 244113.

104 Ferrer, M.L., ALawrence, C., Demirjian, B.G., Kievelson, D., Alba-Simionesco, C., and Tarjus, G. (1998) *J. Chem. Phys.*, **109**, 8010.

105 Gujrati, P.D., cond-mat/0309143.

106 Gujrati, P.D., cond-mat/0404748.

107 Gujrati, P.D. (1981) *Phys. Rev. A*, **24**, 2096.

108 Gujrati, P.D. (1988) *Phys. Rev. A*, **38**, 5840.

109 deGennes, P.G. (1972) *Phys. Lett.*, **38 A**, 339.

110 Gujrati, P.D. (1985) *Phys. Rev. B*, **32**, 3319.

111 Gujrati, P.D. (1998) *J. Chem. Phys.*, **108**, 6952.

112 Lindemann, F.A. (1910) *Phys. Z.*, **11**, 609.

113 Jin, Z.H., Gumbsch, P., Lu, K., and Ma, E. (2001) *Phys. Rev. Lett.*, **87**, 055703.

114 Angell, C.A. (1995) *Science*, **267**, 1924.

115 Novikov, V.N., Rössler, E., Malinovsky, V.K., and Surovtsev, N.V. (1996) *Europhys. Lett.*, **35**, 289.

116 deGennes, P.G. (2002) *C.R. Physique*, **3**, 1263.

117 Toda, M., Kubo, R., and Saito, N. (1978) *Statistical Physics I*, Springer-Verlag, Berlin.

118 Thompson, C.J. (1988) *Classical Equilibrium Statistical Mechanics*, Clarendon Press, Oxford.

11
Liquid Crystalline Polymers: Theories, Experiments, and Nematodynamic Simulations of Shearing Flows

Hongyan Chen and Arkady I. Leonov

11.1
Introduction and Review

11.1.1
Low Molecular Weight and Polymeric Liquid Crystals

Low molecular weight liquid crystals (LC) have found large applications in many industries, especially in electronics. LCs display very specific properties characterized by such solid-like effects as orientation of a large group molecules under physical or stress fields coupled with flow [1–5]. Internal rotations, a specific Frank elasticity caused by mutual rotations of molecules, and anisotropic flows are common effects established for this type of liquids. Molecules in LC are arranged in an orderly manner, and there are three different types of their ordered structure: nematic, cholesteric, and smectic. The nematic structure where the molecules have a long-range orientation order but only a short-range positional order can be met more often because it has low ordered structure compared to other LC types. Remarkable, continuum type theories are the most common for description of both elastic and viscous LC properties. They are Frank theory for LC elasticity and Leslie–Ericksen–Parody (LEP) for viscous properties of LCs.

The quest for lightweight materials with great strength and stiffness has led to the synthesis of liquid crystalline polymers (LCPs) and occurrence of novel processes and theories to predict and control the LCP structures in final products. In LCPs, liquid crystalline properties are achieved either by insertion of rigid chemical fragments into the main polymer chain (main-chain LCPs) or by creating side branches (side-chain LCPs). To date, liquid crystalline polymers have found a variety of applications such as high-strength plastic fibers, bullet-proof garments, front panels of computers, cellular phones, electronic diaries, portable televisions, printed circuit boards, and so on [6]. Typically, LCPs possess outstanding mechanical properties at high temperatures, excellent chemical resistance, inherent flame retardancy, heat aging resistance, low viscosity, and good weather resistance. Such properties make LCPs ideal candidate for high-performance applications [7].

Modeling and Simulation in Polymers. Edited by P.D. Gujrati and A.I. Leonov

ISBN: 978-3-527-32415-6

Similar to the low molecular weight LCs, specific polymers display LC properties in a certain intermediate state between the solids and the liquids, which is called mesophase, with the combined properties of both crystalline and liquid states [5–7]. The molecular orientation order of chemical rigid fragments makes a polymeric material anisotropic and "crystalline," while the lack of strong positional order allows the material to flow like ordinary fluids. LCPs whose phase transition to the liquid crystalline phase occurs under change in temperature are called *thermotropic*, while a variety of LCPs exhibiting phase transitions by changing the polymer concentration in a solvent (as well as temperature) are called *lyotropic* liquid crystals. The unique feature of mesophase is that it is described by a positional order parameter because it is geometrically anisotropic in space. At present, the polymer nematics include liquid crystalline polymers and liquid crystalline elastomers (LCEs) [3]. Similar to LCs, the action of external field or flow causes the orientation of mesogens of LCPs or LCEs, which induces uniaxial anisotropy, with an additional degree of freedom – internal rotations. In certain cases, LCPs or LCEs possess partial flexibility, which displays macroscopically a common anisotropic molecular elasticity or viscoelasticity, and thus is important for the dynamics of these systems.

11.1.2
Molecular and Continuum Theories of LCP

Theoretical and experimental studies of nematic materials have over so many years produced an enormous amount of publications. Several excellent texts (e.g., see Refs [1–5]) are now available, presenting in historical perspective a balanced view of the most important experimental effects and their theoretical explanations, both of continuum and molecular types.

Two types of theories, continuum and molecular, attacked the problem of modeling nematodynamic properties of LCPs and LCEs. The continuum theories try to establish a general framework with minimum assumptions of molecular structures, involving, however, many material parameters related to both the basic properties of symmetry and interactions described by state variables. On the contrary, molecular approaches employ many particular assumptions and describe LCP properties with few molecular parameters. de Gennes [8] was first to attempt to extend the LEP theory to nonlinear case. Doi characterized this attempt as [9] "A way of generalizing the Ericksen–Leslie theory to the nonlinear regimes was suggested by de Gennes [8]. However, this phenomenological approach has not been pursued very far since the number of unknown parameters increases as more complexity is introduced. In this chapter, we describe a complementary approach, the molecular theory. Since this approach is based on a specific modeling of a system, it is less general, but it can give explicit results for rheological functions." Hopefully, the phenomenological and molecular theories will not be contradictory but complementary.

de Gennes and Prost [2] first developed the concept of *nematodynamics*. It was defined as a set of problems for deformation and flow of nematic systems under stress and external (magnetic and/or electrical) fields that can be solved or analyzed using specific macroscopic field equations. Research on polymer nematics has attracted long-standing academic and industrial interests for about three decades. Yet only few LCPs have been in use, which is mostly due to the lack of knowledge of the complicated behavior of these systems. Therefore, nematodynamic studies are an imperative for stimulating the progress in processing polymer nematics and predicting the properties of postprocessed products.

Larson and Mead [10] extended the Ericksen LC phenomenology to viscoelastic case, using instead of Ericksen's viscosities some linear viscoelastic memory functionals. Because these functionals were unknown, Larson and Mead exemplified their approach employing the linearized Doi theory [9, 11]. Using symmetry arguments, Volkov and Kulichikhin [12] proposed a continuum but nonthermodynamic approach to weak anisotropic viscoelasticity of Maxwell type with internal rotations. Pleiner and Brand [13, 14] developed a thermodynamic theory for linear anisotropic viscoelasticity of LCPs, using along with state variables their space gradients. However, the experience accumulated from many applications of continuum nonequilibrium thermodynamics clearly indicates that, except for very specific cases, extending the set of state variables with the use of their time–space derivatives leads to an awkward description, which typically involves a huge amount of material parameters without recommending how to fit them to experimental data. Rey [15, 16] applied a very particular thermodynamic approach to weak viscoelasticity of LCPs but obtained doubtful results of asymmetric stress. Terentjev and Warner [3, 17] developed a thermodynamic theory of solid viscoelasticity for LC elastomers, based on the Kelvin–Voigt type of nematic modeling (see also Ref. [18]). Leonov and Volkov [19–21] initiated thermodynamic studies of nonlinear nematic viscoelasticity for various polymer systems of different rigidity, such as LCPs, LCEs, and precursors of polymer nanocomposites. These approaches have met mathematical difficulties mentioned by Doi [9] and were overcome by creating a new mathematical tool, algebra of nematic operators [22]. Based on this tool a new nematodynamic theory [23] of LCP has been formulated.

A lot of effort was also undertaken to develop molecular theories that could model the lyotropic LCPs, starting from Doi theory [9, 11]. Marrucci and Greco [24], Larson and coauthors (see Ref. [5]), and Feng *et al.* [25] typically use and elaborate Doi's long rigid rod approach. Edwards *et al.* [26] applied a general Poisson bracket approach to LCP (see also the book by Beris and Edwards [27]). This approach is reduced to the Doi theory in homogeneous (monodomain) limit. The theories [24–27] employ the same state variables as in case of low molecular mass LC, that is, the temperature T, director $\underline{n}$ (or the respective second rank *order* tensor), and the director's space gradient $\underline{\nabla n}$. One should also mention the Rouse-like approaches to LCP developed by Volkov and Kulichikhin [28] and Long and Morse [29], which take into account the partial flexibility of LCP polymeric chains. Note that the approaches developed in Refs [28, 29] did not derive the closed set of nematodynamic equations. They also cannot be considered as purely molecular since they employed the phenomenology of Ref. [10] for describing the linear viscoelasticity in LCPs.

Along with the above nonequilibrium molecular theories, a lot of effort was made to develop equilibrium statistical mechanics of LCPs and LCEs. In particular, these theories present the scalar order parameter via molecular parameters of nematics. Some recent results in this field could be found in Refs [3, 30, 31].

11.1.3
Soft Deformation Modes in LCP

Golubovich and Lubensky [32] first predicted the general possibility of occurrence of deformation *soft modes* in anisotropic elastic solids. A remarkable feature of soft modes is that they bear almost no resistance in particular direction(s) of deformed anisotropic solids. Warner and coauthors found the soft/semisoft deformation modes for their particular molecular theory for LCEs (e.g., see Ref. [3]). Yet they were unaware whether the soft modes occurred in their theory by chance or because of a more fundamental reason. An attempt to justify the occurrence of soft modes by rotational invariance was made in Ref. [33]; however, it was not successful. The general physical idea, underlying the occurrence of soft modes in nematic solids, was proposed in Ref. [34]. According to this paper, large fluctuations typical of nematics in equilibrium, drive these systems almost to the boundary of their thermodynamic stability where the free energy is effectively minimized not only with respect to the state variables but also with respect to nematic material parameters. Thus, establishing the "marginal stability" conditions could be used as a theoretical tool for finding the soft deformation modes. This approach has been first introduced in Ref. [35] for nematic solids and in Ref. [36] for viscous nematic LC. In real situations, there always exist small energetic barriers caused by different physical reasons that create small deformation resistance in soft modes. These small barriers located near the boundary of thermodynamic stability, stabilize the behavior of nematic systems and constitute semisoft (close enough to the soft) behavior of nematics. Yet, really new statistical mechanics, similar to the Goldstone theory for magnetics, should be developed to justify the occurrence of the nematic soft modes for LCEs in equilibrium.

Using the marginal stability approach, the possible shearing and elongational soft/semisoft nematic modes were recently discovered for both weakly elastic and viscous nematics [35, 36]. In these systems, the rotational invariance of the shearing modes was found as a trivial consequence of marginal stability. Noticeable, the weak Warner elastic potential does not predict the soft elongation mode in the linear limit. Identical results of marginal stability analyses obtained for weakly elastic and viscous nematic theories [35, 36] are explained by the fact that the well-known LEP continuum theory of nematic LCs has a complete continual analogue for weakly elastic LCEs proposed first by de Gennes [37]. In this analogy, the Rayleigh dissipative function in LEP theory is similar to the de Gennes monodomain elastic potential. Since the minimum of free energy functional always exists for all elastic solids, this analogy was additionally justified by demonstrating the principle of minimum of dissipation functional for viscous nematics [38]. Note that in theories describing possible soft nematic deformation modes, the number of material parameters is highly reduced.

Notably, large fluctuations are typical features for the nematic systems [34]. Therefore, using attractively few parametric mean field molecular approaches yields poor predictions of experimental data for liquid crystals [2]. Consequently, nematic studies usually resort to continuum approaches based on fundamental principles of thermodynamics and symmetry.

As mentioned, a lot of theoretical and experimental studies have been performed to understand physics and rheological properties of lyotropic LCPs. The molecular Doi approach with many improvements and experimental tests is well presented in the literature (e.g., see Ref. [5]). But the thermotropic LCPs were poorly understood till recently, in spite of many attempts to develop either nematodynamic or molecular description of their flow properties. The beauty of continuum approach is that it can be applied to molecular nematics of both different types, as well as to the nonyielding suspensions with shaped particles. Yet, general nematodynamic theories are multiparametric. For example, the general LEP continuum LC theory contains five constitutive parameters [2]. Similarly, de Gennes potential proposed for the monodomain description of general weakly elastic behavior of LCE has also five parameters [37]. Because viscoelasticity is a combination of elastic and viscous effects, it is expected that even in easy theoretical schemes, the continuum approach to viscoelastic polymer nematodynamics should involve at least 10 constitutive parameters. This gives rise to the above pessimistic view [9] that the continuum theories of these systems are intractable.

As mentioned, there exists neither molecular nor continuum theory for describing complicated properties of thermotropic LCPs, although many experimental data for this type of LCPs have been accumulated. One of the objectives of the new continuum theory of weakly nonlinear viscoelastic nematodynamics [22, 23] is to interpret and simulate experimental data, and create models of processing for LPCs. New mathematical techniques [22] revealed the structure of the theory and were helpful in several derivations to present the theory in a simple form. The assumption of small transient (elastic) strains and transient relative rotations, employed in the theory, seems to be appropriate for most LCPs, which usually display a small macromolecular flexibility. This assumption has been used in Ref. [23] to simplify the theory to symmetric type of anisotropic fluid mechanical constitutive equations for describing the molecular elasticity effects in flows of LCPs. Along with viscoelastic and nematic kinematics, the theory nontrivially combines the de Gennes general form of weakly elastic thermodynamic potential and LEP dissipative type of constitutive equations for viscous nematic liquids, while ignoring inertia effects and the Frank elasticity in liquid crystalline polymers. It should be mentioned that this theory is suitable only for monodomain molecular nematics. Nevertheless, effects of Frank (orientation) elasticity could also be included in the viscoelastic nematodynamic theory to describe the multidomain effects in flows of LCPs near equilibrium. In the absence of external fields and neglecting the Frank elasticity, the simplified theory can employ less parametric description of LCP flows. An additional decrease in the number of parameters happens when viscoelastic soft/semisoft nematic modes are present [22, 23].

11.1.4
Specific Problems in LCP Theories

There are three additional specific problems that have to be addressed in any molecular theory for polymer nematics. The first one is a possible effect of the Frank elasticity in these systems. It was shown [3] that the Frank and molecular polymer elasticity for LCEs have well-separated space scales, with their crossover, the "characteristic scale" l_* evaluated as $l_* = \sqrt{K/G}$ [3]. Here, $G >\sim 10^6$ dyn/cm^2 is a typical rubber-like modulus and $K \sim 10^{-7}$ dyn is a typical value of the Frank modulus. This evaluation shows that $l_* <\sim 10^{-6}$ cm $= 10$ nm, that is, in the common macroscopic scales always larger than l_*, the effects of Frank elasticity on the nematodynamics of elastomers could be ignored.

The second problem is that the multidomain "textures" existing at rest in many nematic LCPs affect the slow (low Deborah number) flows of LCPs [5], or weak elastic deformations of LCEs [3]. Although the expression for l_* could also be applied to the molecular (or "instant") elasticity of LCPs, the nonequilibrium effects arising in textures do not allow to ignore the Frank elasticity effects for LCPs. It looks like the monodomain equations valid for the flow of LCPs in relatively strong stress/external fields acquire near the equilibrium some stochastic or periodic properties due to the action of Frank elasticity, ignored in monodomain theories [38]. Conversely, the polymer nematics usually forget their textures under action of higher stresses or external fields, which results in the monodomain description. A rough method of evaluating flow parameters based on mesoscale averaging of LEP equations over several monodomains, involving mesoscale averaged Frank elasticity terms has been discussed in the Ref. [5]. Complementary to the expression for characteristic space scale l_*, the scaling evaluation of characteristic time of spontaneous disorientation by forming a texture of characteristic domain size δ is $t_\delta = \delta^2\eta/K$. Here, η is a characteristic viscosity for LCP near the equilibrium. Using the value $\eta \approx 10^4$ dyn/cm^2 and $\delta \approx 10^{-4}$ cm for a characteristic size of domain in a typical multidomain texture, leads to the evaluation, $t_\delta \approx 103$ s, which seems to be realistic and validates the importance of the Frank elasticity in describing the slow disorientation process in some LCPs. Two timescales, one for fast stress monodomain relaxation and the other for the consequent slow texture formation reported in Ref. [39], seem to be very common for LCPs. The recent attempt to reveal the formation of texture based on a modified Doi rigid rod approach and related references can be found in Ref. [40].

The third problem is the possible effect of stress or external field on isotropic–nematic phase transition. In equilibrium, this phase transition is usually described by the well-known Landau phenomenology or more specifically (however, less reliably because of large fluctuations) by the Maier–Saupe mean field theory [2] (see also Refs [30, 31]). The assumption that the transition behavior of nematic elastomers is independent of stress was roughly confirmed while testing the LCE theory [3], where the parameters of anisotropy were assumed to be independent of stress. The possible dependences of scalar/tensor order parameter on stress/external field have been considered in molecular Doi theory [9, 11] or phenomenological approach by Ericksen [41].

Finally, the specific nematic kinematics caused both by the internal rotations and the macroscopic relations for the moment of momentum balance in external fields have been well understood and summarized in many publications (e.g., see Refs [1–4] and references therein). Inertial effects of internal rotations, commonly ignored in most cases for molecular nematics, might be important for nematic nonyielding suspensions and for such dynamic problems as propagation of sound in molecular nematics. In these cases, many generally small kinetic effects, such as internal spin [19] of nonnematic origin, could not be ignored too.

11.1.5
Experimental Effects in Flows of LCP

There are plenty of rheological experimental data, mostly obtained for liotropic LCPs, such as PBG and HPC solutions in simple shearing. They include measurements of shear viscosity and normal stress differences in steady shearing (including such features unusual for isotropic polymers as negative first normal stress difference), start-up transient simple shearing, and dynamic tests with small amplitude oscillations. These data, well presented in Ref. [5] demonstrated exciting success of the Doi theory in describing phase transition and the basic rheological properties for liotropic LCPs. It was, however, found that the above unusual rheological features predicted by the Doi theory have been observed in PBG and HPC solutions only within a range of medium concentrations: 10–25% for PBG and up to 50% for HPC. Surprisingly, the Doi rigid rod statistical theory is incapable of describing the rheological behavior of "rigid rod" polymers [42], such as PBZT [43, 44], PBO [45], and PPTA [46]. The typical thermotropic LCP, such as thermotropic HPC and thermotropic copolyesters do not demonstrate the above unusual rheological features found in the PBG and HPC solutions and are not described by the Doi theory.

The common commercial thermotropic LCPs, such as Titan (Eastman) with melting temperature $T_m = 330\,^\circ\mathrm{C}$ and Zenite 6000 (Dupont) with $T_m = 345\,^\circ\mathrm{C}$, are the random polyesters with rigid main-chain mesogenic groups. The degradation of commercial LCPs is the main problem faced in their rheological studies. Therefore, some LCP models were synthesized, introducing either (i) flexible spacers in the main chain or (ii) mesogenic side groups. Since the melting temperature of the model LCP are far below their degradation temperatures, the model nematic LCPs are convenient for rheological measurements. For example, the LCPs of (i) type have been widely used for rheological experiments [45–48].

Some specific experimental methods have also been recently elaborated and new reliable results have been obtained in rheological studies [49] of commercial thermotropic polymers, such as Titan and Zenith.

The whole sets of shearing rheological experiments for both model and commercial thermotropic LCPs demonstrated a good consistency when preshearing or using preliminary orienting magnetic field have been applied for eliminating the texture and other long memory effects. Yet, it is unclear what orientation the director acquires after preshearing. This uncertainty negatively affects the simulations of start-up LCP flows with data obtained in standard rheometric devices. Imposing

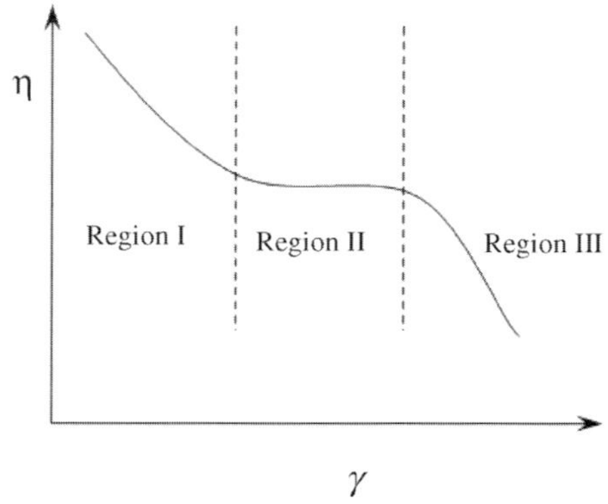

Figure 11.1 Schematics of three-region flow curve.

magnetic field in rheological experiments makes the initial orientation of director reliable, although it requires the use of nonstandard rheological equipment.

LCPs exhibit in simple shear steady-flow several rheological effects, uncommon to polymers with long flexible chains. The first is negative normal stresses observed in some LCPs. The second is the effect of "three-region" viscosity shown schematically in Figure 11.1.

This effect has mostly been observed for lyotropic LCPs, sometimes also for thermotropic ones. The existence of region I in Figure 11.1 is explained by the formation of "texture," a domain structure observed in many, mostly lyotropic LCPs. The texture occurs during relaxation when the stress levels are very low, that is, when approaching the rest state. Such a three-region flow curve was first observed in Ref. [50] and explained theoretically for lyotropic LCPs in Refs [51, 52] (see also Refs [4, 5, 53]). These theoretical descriptions are typically complementary to the more fundamental monodomain nematodynamic theories of both the molecular and the continuous types.

Next section employs the theory [22, 23] for simulations of steady and nonsteady shearing flows. The objective is to analytically and numerically simulate these basic flows of polymer nematics with possible aligning tumbling effects and compare the simulations with published experimental data. Hopefully, this approach will create a reliable and fundamental continuum framework for more detailed molecular theories for polymer nematics. It is also expected that the results of the present studies will significantly contribute in developing robust and reliable flow models that will be used to analyze processing and postprocessing properties of polymer nematic systems. The current lack of such models prevents the progress in processing of LCPs and LCEs.

11.2
General Equations and Simulation Procedures

The simulations in this and following sections are based on the continuum theory of weak viscoelastic nematodynamics [22, 23]. The closed set of constitutive equations includes the equation for the evolution of a unit vector $\underline{n}$ called director and anisotropic constitutive equation for the evolution of extra stress tensor $\underline{\underline{\sigma}}$. In general 3D case, the evolution equation for director is presented as follows:

$$\theta^*\left(\overset{\circ\circ}{\underline{n}} + \underline{n}\cdot\left|\overset{\circ}{\underline{n}}\right|^2\right) + \overset{\circ}{\underline{n}} = \mathbf{b}(\underline{n})\cdot\left(\lambda_e\theta^*\overset{\circ}{\underline{\underline{e}}} + \lambda_v\underline{\underline{e}}\right). \tag{11.1}$$

Here,

$$(\mathbf{b}(\underline{n})\cdot\underline{\underline{e}})_i = b_{ijk}(\underline{n})e_{kj} = (\delta_{ij}n_k - n_i n_j n_k)e_{kj},$$

and upper circles denote the Jaumann tensor time derivatives defined as

$$\overset{\circ}{\underline{n}} \equiv \dot{\underline{n}} - \underline{n}\cdot\underline{\underline{\omega}};\quad \overset{\circ}{\underline{\underline{e}}} \equiv \dot{\underline{\underline{e}}} - \underline{\underline{e}}\cdot\underline{\underline{\omega}} + \underline{\underline{\omega}}\cdot\underline{\underline{e}};\quad \overset{\circ}{\underline{\underline{\sigma}}} \equiv \dot{\underline{\underline{\sigma}}} - \underline{\underline{\sigma}}\cdot\underline{\underline{\omega}} + \underline{\underline{\omega}}\cdot\underline{\underline{\sigma}}.$$

Equation (11.1) is coupled with anisotropic equation for the evolution of extra stress tensor $\underline{\underline{\sigma}}$:

$$\begin{aligned}\theta_0\overset{\circ}{\underline{\underline{\sigma}}} + \underline{\underline{\sigma}} + (r_1-1)[\underline{nn}\cdot\underline{\underline{\sigma}} + \underline{\underline{\sigma}}\cdot\underline{nn} - 2\underline{nn}(\underline{\underline{\sigma}} : \underline{nn})] + 1.5(r_2-1)(\underline{nn} - \underline{\underline{\delta}}/3)(\underline{\underline{\sigma}} : \underline{nn}) \\ = \eta_0\left\{\underline{\underline{e}} + \alpha[\underline{nn}\cdot\underline{\underline{e}} + \underline{\underline{e}}\cdot\underline{nn} - 2\underline{nn}(\underline{\underline{e}} : \underline{nn})] + (\beta - 3/2)(\underline{nn} - \underline{\underline{\delta}}/3)(\underline{\underline{e}} : \underline{nn})\right\}.\end{aligned} \tag{11.2}$$

In Eqs. (11.1) and (11.2), η_0 is a characteristic nematic viscosity, $\theta_0 = \eta_0/G_0$ and $\theta^*(\sim\theta_0)$ are relaxation times, λ_e and λ_v are the elastic and viscous tumbling parameters, respectively, α, β, and r_1, r_2 are parameters characterizing anisotropy, $\underline{n}\,\underline{n}$ is dyadic with components $n_i n_j$, $\underline{\underline{e}}$ and $\underline{\underline{\omega}}$ are strain rate and vorticity tensors, respectively, commonly related to the gradient velocity tensor. The constitutive Eqs. (11.1) and (11.2) should be consistently used along with the momentum balance equation to determine both the stress and the flow fields. In few cases, however, the flow field is known. One of these flows, having many practical applications, is simple shearing. In this flow, solution of constitutive equations for weakly nonlinear viscoelastic nematodynamic can be simplified.

The shearing flows are commonly analyzed using a "standard" Cartesian coordinate system $\{\underline{x}\} = \{x_1, x_2, x_3\}$ where x_1 is directed along the flow and x_2 along the velocity gradient. In this coordinate system, the velocity vector $\underline{v}$ is $\underline{v} = \{\dot{\gamma}(t)x_2, 0, 0\}$, and the tensors of strain rate $\underline{\underline{e}}(t)$ and vorticity $\underline{\underline{\omega}}(t)$ for homogeneous shearing flows have the matrix forms:

$$\underline{\underline{e}}(t) = \frac{\dot{\gamma}(t)}{2}\begin{pmatrix}0 & 1 & 0\\ 1 & 0 & 0\\ 0 & 0 & 0\end{pmatrix},\quad \underline{\underline{\omega}}(t) = \frac{\dot{\gamma}(t)}{2}\begin{pmatrix}0 & -1 & 0\\ 1 & 0 & 0\\ 0 & 0 & 0\end{pmatrix}. \tag{11.3}$$

Here, the shear rate $\dot{\gamma}(t)$ is a given function of time. We will use below a common simplifying assumption that vector of director is located in shear plane, so the 2D expression for director is $\underline{n} = \{n_1, n_2, 0\}$. Substituting this expression along with (11.3) into evolution equation(11.1) yields the evolution equation for the longitudinal component n_1 of director located in the $\{x_1, x_2\}$ shear plane as

$$\theta^*\left(\ddot{n}_1+\frac{n_1\dot{n}_1^2}{n_2^2}\right)+\dot{n}_1-\frac{\dot{\gamma}n_2}{2}=\theta^*\ddot{\gamma}\left[\lambda_e(1-2n_1^2)+1\right]\frac{n_2}{2}/2$$
$$+\lambda_e\theta^*\dot{\gamma}^2n_1n_2^2+\frac{\lambda_v\dot{\gamma}n_2}{2}(1-2n_1^2). \tag{11.4}$$

Here, $n_2=\sqrt{1-n_1^2}$.

In simple shearing flow, Eq. (11.2) is rewritten in component form as

$$\theta_0\frac{d\sigma_{12}}{dt}+\theta_0\dot{\gamma}\frac{\sigma_{11}-\sigma_{22}}{2}+\sigma_{12}+(r_1-1)[\sigma_{12}+n_1n_2(\sigma_{11}+\sigma_{22})]$$
$$+(\sigma_{11}n_1^2+\sigma_{22}n_2^2+2\sigma_{12}n_1n_2) \tag{11.5a}$$
$$n_1n_2\left[\frac{3}{2}(r_2-1)-2(r_1-1)\right]=\frac{\eta_0\dot{\gamma}}{2}\left(1+\alpha+2n_1^2n_2^2\left(\beta-2\alpha-\frac{3}{2}\right)\right),$$

$$\theta_0\frac{d\sigma_{11}}{dt}-\sigma_{12}\theta_0\dot{\gamma}+\sigma_{11}+2(r_1-1)(\sigma_{11}n_1^2+\sigma_{12}n_1n_2)$$
$$+\left[\frac{3}{2}(r_2-1)\left(n_1^2-\frac{1}{3}\right)-2(r_1-1)n_1^2\right]$$
$$\times(\sigma_{11}n_1^2+\sigma_{22}n_2^2+2\sigma_{12}n_1n_2)=\eta_0\dot{\gamma}n_1n_2\left(\alpha(1-2n_1^2)+\left(\beta-\frac{3}{2}\right)\times\left(n_1^2-\frac{1}{3}\right)\right), \tag{11.5b}$$

$$\theta_0\frac{d\sigma_{22}}{dt}+\sigma_{12}\theta_0\dot{\gamma}+\sigma_{22}+2(r_1-1)(\sigma_{22}n_2^2+\sigma_{12}n_1n_2)$$
$$+\left[\frac{3}{2}(r_2-1)\left(n_2^2-\frac{1}{3}\right)-2(r_1-1)n_2^2\right]$$
$$\times(\sigma_{11}n_1^2+\sigma_{22}n_2^2+2\sigma_{12}n_1n_2)=\eta_0\dot{\gamma}n_1n_2\left(\alpha(1-2n_2^2)+\left(\beta-\frac{3}{2}\right)\left(n_2^2-\frac{1}{3}\right)\right). \tag{11.5c}$$

Equations (11.5a)–(11.5c) describe the time-dependent evolution and steady behavior of extra stress tensor in simple shearing.

In case of steady shearing analyzed in Ref. [23], the analytical expression for horizontal component of director n_1 was found as

$$n_1^2=\frac{1}{2}\left(1+\lambda_v\frac{1+|D\lambda_1/\lambda_v|\sqrt{D^2\lambda_e^2+\lambda_v^2-1}}{D^2\lambda_e^2+\lambda_v^2}\right). \tag{11.6}$$

Here, $D = \theta^* \dot{\gamma}$. In this case, a very awkward analytical solution of Eqs. (11.4) and (11.5a)–(11.5c) has also been obtained using Mathematica software. Formula (11.6) shows that the director position weakly depends on elastic tumbling parameterλ_e and strongly depends on the value of viscous tumbling parameterλ_v. The same aligning condition $|\lambda_v| > 1$ as in viscous case holds in viscoelastic nematodynamics. As compared to the Ericksen theory, the new fact here is the dependence position of director $\underline{n}$ on the flow Deborah number *De*. This dependence delays the onset of the rest state tumbling and tends to orient the director along or perpendicular flow lines when $\lambda_v > 0$ or when $\lambda_v < 0$, respectively. The dependence $\underline{n}$ on *De* mostly contributes to the shear thinning effect where the sign of the first normal stress difference N_1 depends on the sign of λ_v, with $|N_1|$ growing with increasing $\dot{\gamma}$.

In case of relaxation, the expression for horizontal component of director $n_1(t)$ was explicitly obtained in Ref. [23] as the solution of (11.4) using presentation $n_1(t) = \cos \varphi(t)$:

$$\varphi(t) = \varphi_0 + D\lambda_e(1-e^{-t/\theta^*}) \cos 2\varphi_0. \tag{11.7}$$

Here, the value $\varphi_0(D)$ is known from (11.6). In case of relaxation, the values of stress components are found in this chapter by numerical solution of the set of ODE (11.5a)–(11.5c), using formula (11.7).

In case of start-up flow from the rest state, a numerical solution of ODE set (11.4) and (11.5a)–(11.5c) should be obtained using the following initial conditions:

$$\underline{\underline{\sigma}}|_{t=0} = 0; \quad n_1|_{t=0} = n_{1r}; \quad \dot{n}_1|_{t=0} = n_{2r}\dot{\gamma}_0\left[\lambda_e\left(1-2n_{1r}^1\right) + 1\right]/2. \tag{11.8}$$

The first prerequisite in (11.8) is the natural (zero) initial condition for the stress tensor in start-up shearing flow from the rest state. To resolve the problem of choosing the initial condition n_{1r} for director in start-up flow, we preliminarily fitted the experimental data for stresses in steady shearing with following adjustment of parameters to also describe the stress relaxation. In this case, parameters of the evolution equation for director, along with its orientation in steady shearing, were also established. Calculating then the orientation of director during stress relaxation, we found its final orientation at the rest state, which was taken as initial director value $\underline{n}_r$ in the start-up flow. Thus the value of parameter n_{1r} is established as the fully relaxed value of director after relaxation. Using formula (11.7) in the limit $t \to \infty$, the value of n_{1r} is found as

$$n_{1r} = \cos \varphi_r; \quad \varphi_r = \varphi_0 + D\left[\lambda_e\left(1-2n_{1,0}^2\right) + 1\right]/2. \tag{11.9}$$

The second initial condition in (11.8) is derived from Eq. (11.4) using initial jump (or delta) conditions.

In the following discussion, we compare the results of numerical simulations with published shearing experimental data, generally presented as time-dependent plots for shear stress σ_{12}, apparent viscosity $\eta = \sigma_{12}/\dot{\gamma}$, and the "first normal stress difference" $N_1 = \sigma_{11} - \sigma_{22}$. The shearing flow simulations include the steady shearing, relaxation, and transient start-up flows. To analyze simple shearing flows for aligning thermotropic LCPs, we will use in the following the monodomain CEs (11.4) and (11.5a)–(11.5c) with above formulas (11.6)–(11.9) suitable for specific problems.

The ultimate goal of simulations was to show a possibility to describe experimental data along with determining constitutive parameters. These parameters established for steady shearing are then used for calculating the evolution of director, shear stress, and first normal stress difference during relaxation and start-up flow.

To resolve the problem of choosing the initial conditions for director in start-up flow, we preliminarily fitted the experimental data for stresses in steady shearing with the following adjustment of parameters to also describe the stress relaxation. In this case, parameters of the evolution equation for director, along with its orientation in steady shearing, were also established. Calculating then the orientation of director during stress relaxation, we found its final orientation at the rest state, which was taken as initial director value in the start-up flow.

11.3 LCP and their Parameters Established in Simulations

Common commercial thermotropic LCPs are the random polyesters with rigid main-chain mesogenic groups. These LCPs have very narrow time and temperature intervals between the beginning of crystal melting and the onset of polymer degradation, where the liquid crystalline phase exists. The existence of multi-domains in these commercial thermotropic LCPs has not been reported, and it seems improbable because of short time processing. The degradation of commercial LCPs is the main problem for their rheological studies. Therefore, some "model" LCPs have been synthesized in various laboratories. The main objective of the model LCP was to incorporate flexible spacers or pendent bulky group into the main chain and then reduce the melting point T_m and the clearing temperature T_{cl} of LCPs, which are far below their degradation temperature T_d. Using these techniques, it was possible to safely run the rheological measurements within the temperature intervals below T_d.

Our simulations of the above viscoelastic nematodynamics theory require the reliable and representative rheological data for LCPs, obtained for steady and transient shear flows and relaxation. We chose literature rheological data for two commercial LCPs, Titan and Zenith 6000 [49], as well as for two model polymers, a main-chain LCP, PSHQ9, and a side-chain LCP, PI-14-5CN [53].

According to Ref. [49] two commercial random copolyesters, Titan and Zenith 6000, were obtained from the Eastman and DuPont Chemical companies, respectively. These LCP have very narrow time and temperature intervals between the beginning of crystal melting and the onset of polymer degradation, where the liquid crystalline phase exists. The existence of multidomains in these commercial thermotropic LCPs has not been reported, and because of short time processing, it seems to be improbable.

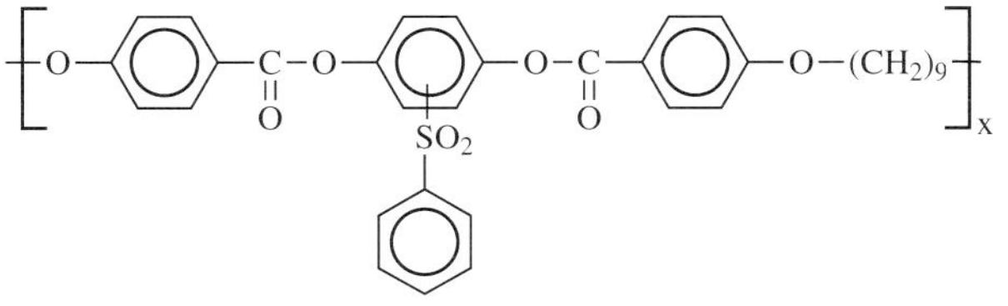

PSHQ9

Figure 11.2 Chemical structure of PSHQ9 [54].

Titan is a random copolyester probably composed of two random units, ethylene-terephthalate (PET) and hydroxybenzoic acid (HNA). Incorporating the PET unit in the main chain reduces the rigidity of the molecule due to the two methylene flexible spacers and in turn decreases the melting temperature of the material. On the other hand, Zenith 6000 is fully aromatic, where kinks were introduced by combining phenol and biphenol molecules by random copolymerization. As measured by DSC, the melting temperature T_m of Titan was between 325 and 335 °C, while for Zenith 6000, T_m was between 340 and 360 °C. TGA measurements indicated that the degradation temperature T_d of Titan is 450 °C, although it has been reported in Ref. [49] that the noticeable degradation begins at considerably lower temperatures.

One typical example of model LCP is PSHQ9, poly[(phenylsulfonyl)-*p*-phenylene nonanemethylene bis(4-oxybenzoate)] [54], whose chemical structure is depicted in Figure 11.2. This is a main-chain LCP, which has a glass transition temperature T_g of 84 °C, and a nematic-to-isotropic (N–I) transition temperature T_{NI} of 162 °C. This polymer has only nematic phase at temperatures between T_g and T_{NI}.

Another typical model LCP is PI-14-5CN. It is a nematic side-chain liquid-crystalline polymer (SCLCP), with the chemical structure [54] presented in Figure 11.3.

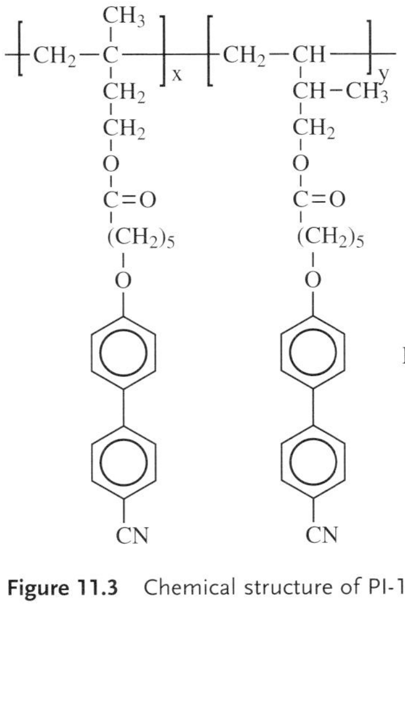

PI-14-5CN

Figure 11.3 Chemical structure of PI-14-5CN [54].

Table 11.1 Parameters of anisotropy and tumbling for industrial and model LCPs.

LCP	α	β	r_1	r_2	λ_e	λ_v
Titan, 340 °C	0.2	0.5	0.15	0.222	2.2	30
Zenith 6000, 360 °C	4	0.1	0.55	0.367	1.01	50
PSHQ9, 70, 80, 90 °C	0.7	1.3	0.5	0.9	2	12
PI-14-5CN, 130, 140, 150 °C	5	8	0.7	0.533	2	18

A nearly monodisperse PI-14-5CN was synthesized using anionic polymerization by grafting a liquid-crystalline monomer, 6-[(4-cyano-4′-biphenyl)oxy]hexanoic acid (5CN-COOH), onto a nearly monodisperse hydroxylated polyisoprene (PI). This polymer undergoes glass transition at 45 °C and N–I transition at 102 °C.

It is significant that almost all the model SC LCPs reported in literature have been synthesized by condensation polymerization, invariably giving rise to polydisperse SC LCPs. It was impossible to precisely determine the molecular weight of PI-14-5CN from the GPC because of a lack of information on its hydrodynamic volume. So the molecular weight was approximately calculated to be 7.19×10^4 g/mol using information on the degree of hydroxylation. The polydispersity of PI-14-5CN found from GPC was equal to 1.08.

Values of constitutive parameters for all tested LCPs, obtained with the mentioned intense fitting procedure are shown in Tables 11.1 and 11.2.

When comparing the values of constitutive parameters found for Titan and Zenith 600, one should recall that Titan is a random copolyester of ethylene-terephthalate and hydroxybenzoic acid with two methylene flexible spacers, whereas Zenith 6000 is a fully aromatic copolyester with kinks. Therefore, Zenith 6000 is expected to be much more rigid than Titan. Our simulations confirmed this. They demonstrate that the values of viscoelastic parameters θ_0 and η_0, parameter of viscous tumbling λ_v, and anisotropy parameters α, r_1, and r_2 found for Titan are smaller than those of Zenith 6000, while the values of elastic tumbling parameter λ_e and the anisotropic parameter β for Titan are greater than those for Zenith 6000.

It is also seen that all constitutive parameters for PSHQ9, except r_2 and λ_e, are larger than those for PI-14-5CN. We remind that PSHQ9 is a main-chain LCP, whereas PI-14-5CN is a side-chain one. The flexible main chain in PI-14-5CN makes

Table 11.2 Viscoelastic parameters for industrial and model LCPs.

LCP	Titan, 340 °C	Zenith 6000, 360 °C	PSHQ9 temperatures (°C)			PI-14-5CN temperatures (°C)		
			70	80	90	130	140	150
Relaxation time, θ_0 (s)	0.3	2.5	1.5	0.8	0.28	3	2.8	2.2
Viscosity, η_0 (kPa s)	0.34	1.55	200	9	2.9	220	70	10

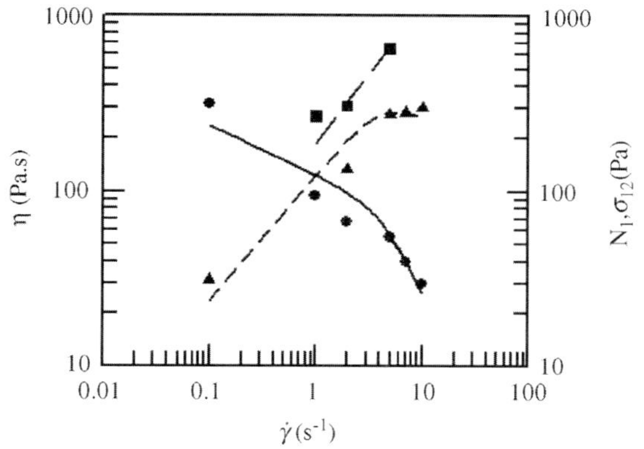

Figure 11.4 Experimental data for apparent viscosity η (●), shear stress σ_{12} (▲), and the first normal stress difference N_1 (■) versus shear rate $\dot{\gamma}$ for Titan at 340 °C. Fitting curves are shown by solid line for η, short dash line for σ_{12}, and by long dash line for N_1.

it much easier to relax and decouple mesogens from the main chain. In contrast, the main-chain LCP (PSHQ9) possesses mesogens in the main chain, which are tightly packed together and in turn have relatively low mobility and capability of relaxation.

11.4 Results of Simulations

11.4.1 Simulations of Steady Shearing Flows

Figure 11.4 demonstrates the experimental data and fitting curves for effective viscosity η, shear stress σ_{12}, and first normal stress difference N_1 versus shear rates $\dot{\gamma}$ for Titan at 340 °C.

Here, the constitutive parameters were found by the curve fitting procedure. As seen, there is a good agreement between simulations and experimental data. However, the fitting curve for apparent viscosity η does not exhibit the general three-region flow curve [8]. It is probably because only a narrow part of region II has been presented in the experimental data. The region I should appear at very low shear rates $\dot{\gamma}$ if there is a texture, which the current monodomain approach is incapable to describe. It should also be noted that the first normal stress difference is positive in the range of shear rates investigated.

Figure 11.5 demonstrates the flow curve for Zenith 6000 at 360 °C, presented as the logarithmic plot of apparent shear viscosity η versus shear rate $\dot{\gamma}$.

In Figure 11.5, the experimental results are shown by dots and the fitting curve by dashed line. Unfortunately, the normal stress data for this polymer were not reported. It is unclear whether the experimental plot is related to the transition from the region I to region II in the general three-region viscosity plot.

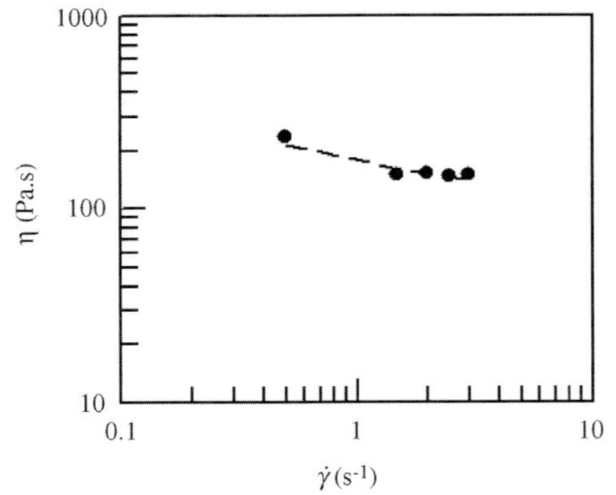

Figure 11.5 Experimental data (dots) and fitting curve (dashed line) for the logarithmic plot of viscosity η versus shear rate $\dot{\gamma}$ for Zenith 6000 at 360 °C.

Figure 11.6 shows steady shearing experimental data for PSHQ9 at different temperatures in the nematic region. Presented here are the logarithmic plots for apparent viscosity η and the first normal stress difference N_1 versus shear rate $\dot{\gamma}$. The viscosities of both experimental data and fitting curves exhibit regions II and III, and the Newtonian behavior becomes obvious as the temperature increases.

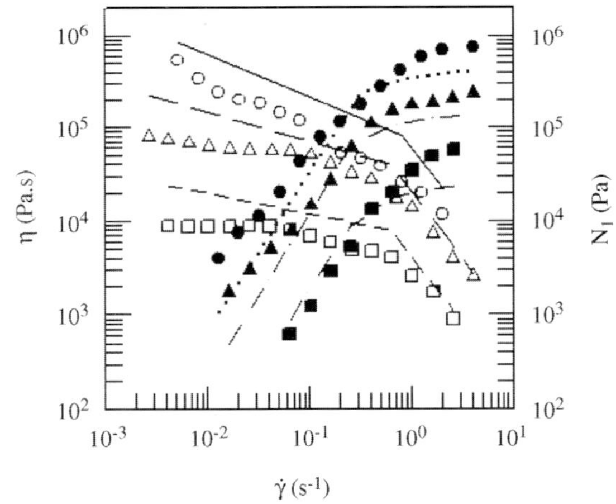

Figure 11.6 Experimental data for dependences of apparent viscosity η (open symbols) and first normal stress difference N_1 (filled symbols) on shear rate $\dot{\gamma}$ for PSHQ9 in steady flow at different temperatures in the nematic region: (σ, ●) 130 °C; (△, ▲) 140 °C; (□, ■) 150 °C. The fitting curves for plots of log η versus log $\dot{\gamma}$ are shown as 130 °C (solid line), 140 °C (long dashed line), and 150 °C (short dash line). The fitting curves for log N_1 versus log $\dot{\gamma}$ use the same curve symbols.

Both experimental and simulated data for N_1 are positive over the range of shear rates investigated. The viscosity of PSHQ9 exhibits a shear thinning at very low values of $\dot{\gamma}$, a Newtonian behavior at its intermediate values, and then, again, a strong shear thinning behavior at higher values of $\dot{\gamma}$. However, as the temperature increases from 130 to 140 °C, the region I in the η plot for PSHQ9 becomes very weak, and when the temperature increases further to 150 °C, the region I disappears. Since PSHQ9 is a polydisperse polymer, the low molecular fraction of PSHQ9 can transform into the isotropic state, forming a biphasic state before reaching T_{NI} (about 160 °C). Thus, the nematic property of PSHQ9 becomes progressively weaker as the temperature approaches T_{NI}. In such situations, region I in the η plot for PSHQ9 may not be observable.

Figure 11.7 shows the logarithmic plots of steady shearing experimental data (dots) for η and N_1 versus $\dot{\gamma}$ for PI-14-5CN at different temperatures. The corresponding fitting curves are shown by lines. It is seen that the fitting curves for both η and N_1 are in excellent agreement with experimental data, being better than those for PSHQ9. The viscosity of PI-14-5CN exhibits a Newtonian behavior at low $\dot{\gamma}$ values and then shear thinning behavior, while the first normal stress difference N_1 is positive over the entire range of $\dot{\gamma}$ tested at three temperatures in the nematic region. The absence of region I in the η plots might be due to the fact that PI-14-5CN is a side-chain LCP, and its nematic behavior is not sufficiently strong. The above observation supports the view [7] that region I in the η plots for various LCPs is associated with the existence of domain structure.

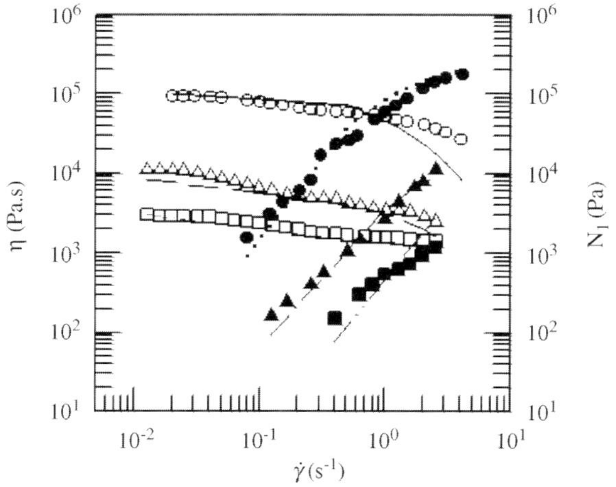

Figure 11.7 Experimental data for dependences of apparent viscosity η (open symbols) and first normal stress difference N_1 (filled symbols) on shear rate $\dot{\gamma}$ for PI-14-5CN in steady flow at different temperatures in the nematic: (σ, ●) 70 °C; (△, ▲) 80 °C; (□, ■) 90 °C. The fitting curves for plots of log η versus log $\dot{\gamma}$ are shown as 70 °C (solid line), 80 °C (long dash line), and 90 °C (short dash line). The same curve symbols are used for fitting plots of log N_1 versus log $\dot{\gamma}$ at different temperatures.

11.4.2
Simulations of Transient Start-Up Shear Flows

In the following discussion, we will use normalized stresses denoting by additional + symbol the time-dependent shear and normal stresses. The absence of + symbols indicates the stresses in steady flow.

The appearance of multiple overshoots in the first normal stress difference in transient shear flow can be qualitatively explained using molecular theory by Marrucci and Maffettone [9]. This monodomain theory is based on the Maier–Saupe potential, which is valid for low molecular weight thermotropic liquid crystals. Therefore, it may not be suitable for semiflexible main-chain LCP exhibiting flow-aligning behavior. On the other hand, the Larson–Doi mesoscopic model [10] describes the evolution of texture based on some experimental observations of decrease in domain size with increasing shear rate and growing domain size upon cessation of flow. Although the Larson–Doi mesoscopic model [10] for LCPs qualitatively predicts the experimental observations for the time evolution of shear $\sigma^{+}(\dot{\gamma},t)$ and normal $N_1^{+}(\dot{\gamma},t)$ stresses in start-up flows, the predicted magnitude of $\sigma^{+}(\dot{\gamma},t)$ is lower than that of $N_1^{+}(\dot{\gamma},t)/N_1$. This contradicts the experimental observations [20]. Furthermore, the models [9, 10] predict a much shorter transient time (strain) for $\sigma^{+}(\dot{\gamma},t)$ and $N_1^{+}(\dot{\gamma},t)$ variations compared to experimental data. The inadequacy of the Larson–Doi model to accurately predict the time evolution of $N_1^{+}(\dot{\gamma},t)$ in transient shear flow can be explained by the presence of long flexible spacers and bulky pendent side groups in LCPs. Both the long flexible spacers and the bulky pendent side groups in LCP macromolecules might directly suppress molecular rotations and perhaps, collective molecular rotations, or director tumbling. Although the Larson–Doi mesoscopic model is the only existing model that describes polydomain texture and distortional elastic effects, the model is based on Leslie–Ericksen equations, which is appropriate either for low molecular liquid crystals or for lyotropic systems that have short relaxation times.

It seems that the majority of thermotropic LCPs exhibit flow-aligning behavior. Thus, to describe the experimental observations for these polymers, the general viscoelastic nematodynamic theory [16, 17] is used in our simulations with aligning assumption.

Figures 11.8 and 11.9 demonstrate the start-up shear flow at $\dot{\gamma} = 6\ \mathrm{s}^{-1}$ for Titan at 340 °C. Figure 11.8 shows the evolution of normalized shear stress $\sigma^{+}(\dot{\gamma},t)/\sigma$ with strain ($\dot{\gamma}t$) and Figure 11.9 the normalized first normal stress difference $N_1^{+}(\dot{\gamma},t)/N_1$. Here, the experimental data are denoted by dots and simulated curves by dashed line. As seen, both the shear stress and the first normal stress difference display large and broad overshoots at $\dot{\gamma}t \approx 50$. Since Titan has very rigid macromolecules, the evolution of texture is slow, and in turn the overshoot occurs at high value $\dot{\gamma}t$.

Figure 11.10 demonstrates the evolution of shear stress σ with strain ($\dot{\gamma}t$) for Zenith 6000 at 360 °C in start-up shear flow at $\dot{\gamma} = 2\ \mathrm{s}^{-1}$. The experimental data are shown by dots and simulated curve by dashed line. The data for shear stress initially exhibit a small overshoot at $\dot{\gamma}t \approx 4$ and a large and broad overshoot at $\dot{\gamma}t \approx 76$. In the simulated curve, a small overshoot occurs at $\dot{\gamma}t \approx 16$ and a large overshoot at $\dot{\gamma}t \approx 50$.

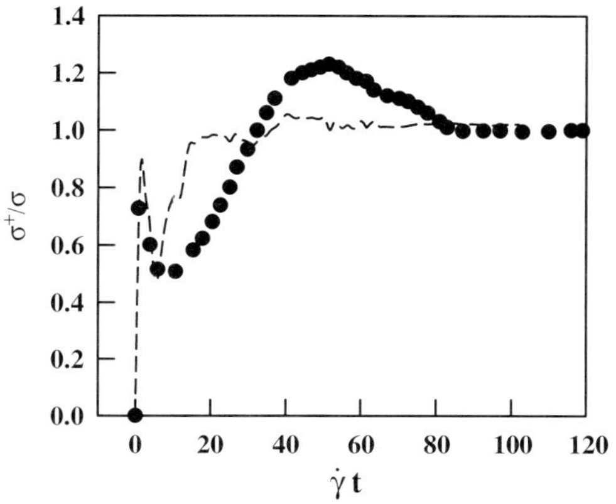

Figure 11.8 The evolution of normalized shear stress $\sigma^+(\dot{\gamma},t)/\sigma$ with strain ($\dot{\gamma}t$) for Titan at 340 °C in start-up shear flow at $\dot{\gamma} = 6\,s^{-1}$: experimental data are shown by dots and simulated curve by dashed line.

Figures 11.11 and 11.12 describe the evolution of normalized shear stress $\sigma^+(\dot{\gamma},t)/\sigma$ and first normal stress difference $N_1^+(\dot{\gamma},t)/N_1$ with strain $\dot{\gamma}t$ for PSHQ9 in start-up shear flow. The flow temperature was 130 °C and shear rate was $\dot{\gamma} = 1\,s^{-1}$. The experimental data are shown by dots and simulated curves by dashed line. The simulated overshoots for shear and normal stresses are the same as those for experimental data, but the overshoots in calculated curves occur at a relatively low $\dot{\gamma}t$ and are very narrow. One can attribute such large values of $\sigma^+(\dot{\gamma},t)/\sigma$ ratios, characteristic for liquid crystalline polymers, to the existence of a lot of polydomains in the nematic state when the start-up flow initiated. Recall that the theory used for simulation utilizes monodomain approach, whereas PSHQ9 exhibits polydomains in nematic state in start-up flow. Thus, deviation of simulated results from experimental data seems reasonable.

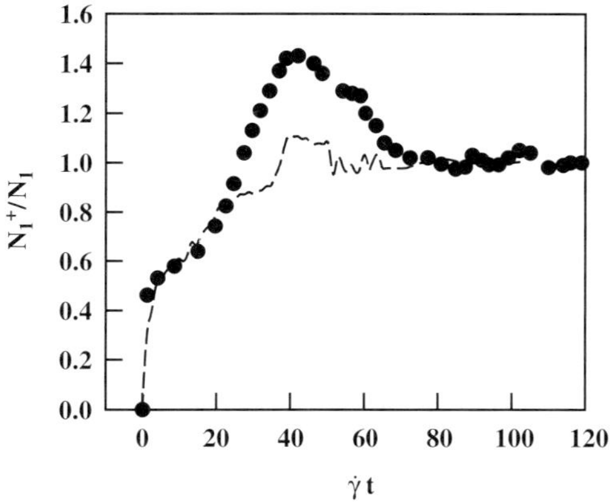

Figure 11.9 The evolution of normalized first normal stress difference $N_1^+(\dot{\gamma},t)/N_1$ with strain ($\dot{\gamma}t$) for Titan at 340 °C in start-up shear flow at $\dot{\gamma} = 6\,s^{-1}$; notations are the same as in Figure 11.8.

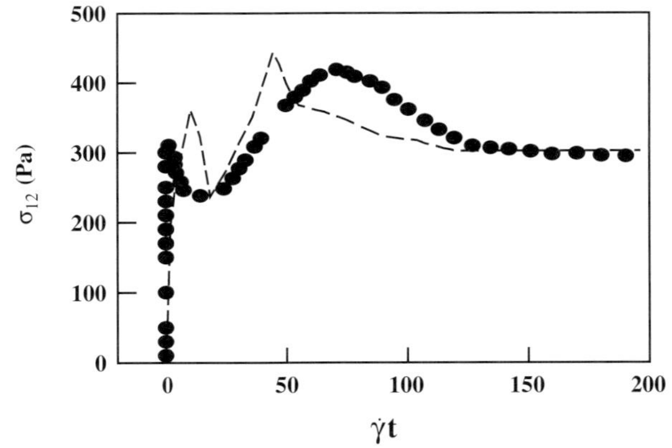

Figure 11.10 The evolution of shear stress (σ) with strain ($\dot{\gamma}t$) for Zenite 6000 at 360 °C in start-up shear flow at $\dot{\gamma} = 2\,s^{-1}$: notations are the same as in Figure 11.8.

Figures 11.13 and 11.14 describe, respectively, the evolution of normalized shear stress $\sigma^+(\dot{\gamma},t)/\sigma$ and first normal stress difference $N_1^+(\dot{\gamma},t)/N_1$ with strain $\dot{\gamma}t$ for PI-14-5CN at 70 °C in start-up shear flow at $\dot{\gamma} = 1\,s^{-1}$. In these figures, experimental data are shown by dots and fitting curve by dashed line. The experimental data for both the shear and normal stresses exhibit large overshoots and then decay to steady values. The simulated evolution of normalized shear stress with strain is fitted well enough with experimental data, whereas there is a relatively large difference between the simulated curve for normal stress and experimental data. This once again may be attributed to simulating the polydomain LCP using monodomain theory.

Compared to Figures 11.11 and 11.12, in start-up shear flow the overshoot peak for normal stress occurs much sooner in PI-14-5CN than in PSHQ9, while the overshoot

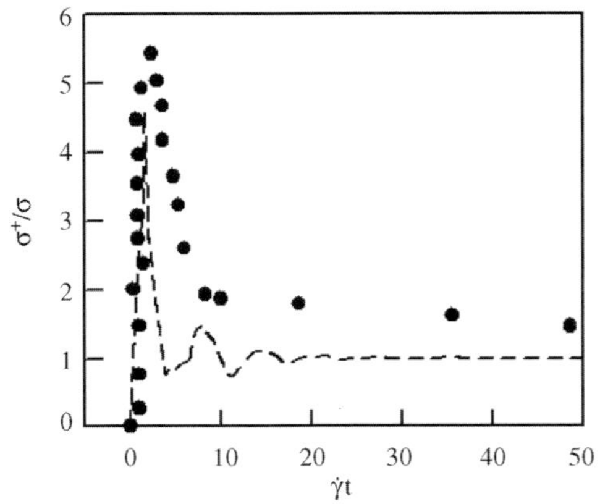

Figure 11.11 The evolution of normalized shear stress $\sigma^+(\dot{\gamma},t)/\sigma$ with strain ($\dot{\gamma}t$) for PSHQ9 at 130 °C in start-up shear flow at $\dot{\gamma} = 1\,s^{-1}$: notations are the same as in Figure 11.8.

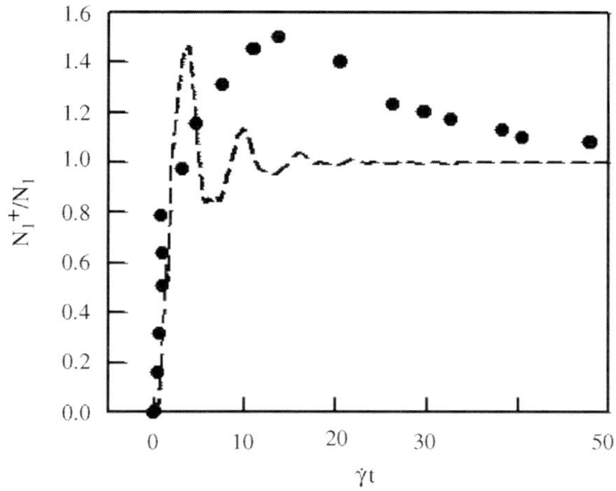

Figure 11.12 The evolution of normalized first normal stress difference $N_1^+(\dot{\gamma}, t)/N_1$ with strain ($\dot{\gamma}$) for PSHQ9 at 130 °C in start-up shear flow at $\dot{\gamma} = 1\,s^{-1}$: notations are the same as in Figure 11.8.

peak in shear stress occurs very quickly in both PI-14-5CN and PSHQ9. The overshoot peak value of fitted normal stress for PSHQ9 is slightly larger than that of PI-14-5CN, while the overshoot peak value of shear stress for PSHQ9 is about three–four times greater than that of PI-14-5CN. Thus, the transient responses in first normal stress difference and shear stress for side-chain LCP, PI-14-5CN are quite different from those for main-chain LCP, PSHQ9. In the latter case, 5CN-COOH mesogens are grafted onto the coil-like PI forming a polymer backbone through five

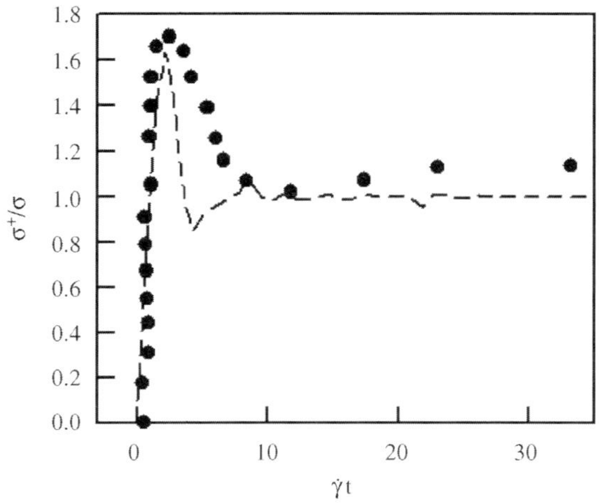

Figure 11.13 The evolution of normalized shear stress $\sigma^+(\dot{\gamma},t)/\sigma$ with strain ($\dot{\gamma}t$) for PI-14-5CN at 70 °C in start-up shear flow at $\dot{\gamma} = 1\,s^{-1}$: notations are the same as in Figure 11.8.

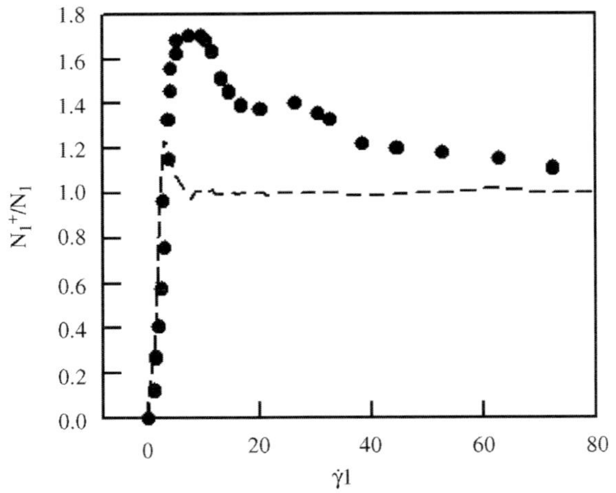

Figure 11.14 The evolution of normalized first normal stress difference $N_1^+(\dot{\gamma}, t)/N_1$ with strain ($\dot{\gamma}t$) for PI-14-5CN at 70 °C in start-up shear flow at $\dot{\gamma} = 1\,s^{-1}$: notations are the same as in Figure 11.8.

methylene groups as flexible spacer. So, the motions of the polymer backbone and the 5CN-COOH mesogens in PI-14-5CN may be regarded as being partially decoupled, making the 5CN-COOH mesogens mobile during shear flow [20]. Furthermore, each mesogen grafted onto the backbone of PI-14-5CN may move or orient, upon start-up of shear flow, slightly depending on other mesogens. On the other hand, the mesogens in PSHQ9 are directly linked to the polymer backbone, making motions of the mesogens and backbone of PSHQ9 strongly coupled during shear flow. Thus, each mesogen in PSHQ9 cannot act as independent, that is, all mesogens in PSHQ9 act in start-up of shear flow collectively or cooperatively.

11.4.3
Simulations of Relaxation after Cessation of Steady Flow

Figures 11.15 and 11.16 show the relaxation of shear stress σ_{12} and first normal stress difference N_1, respectfully, for Titan at 340 °C after cessation of steady flow with shear rate $\dot{\gamma} = 6\,s^{-1}$. Here, experimental data are shown by dots and fitting curve by dashed lines. The values of both σ_{12} and N_1 during relaxation drop abruptly and reach zero at the time of around 4 s. The simulated relaxation curve for σ_{12} has an excellent agreement with the experimental data, but this is not the case for N_1 when $t > 4$ s. This disagreement is seemingly attributed to the experimental error because the final value of N_1 during relaxation should reach zero.

Figure 11.17 presents the relaxation of normalized (a) shear stress and (b) the first normal stress difference upon cessation of steady shear flow at $\dot{\gamma} = 0.5\,s^{-1}$ for PSHQ9 at 130 °C. Here, experimental data are shown by dots and fitting curve by dashed line. The normalizing values σ_0 and $N_{1,0}$ are the steady values of shear stress

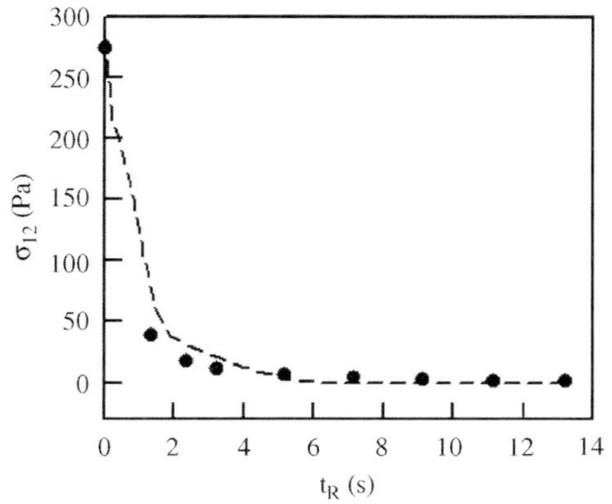

Figure 11.15 Relaxation of shear stress σ_{12} for Titan at 340 °C upon cessation of steady flow at shear rate $\dot{\gamma} = 6\,s^{-1}$: experimental data are shown by dots and simulated curve by dashed line.

and the first normal stress difference, respectively, just prior to flow cessation. As seen, the normalized shear stress approaches zero at $t_R > 7$ s, whereas the normal stress reaches zero when $t_R > 15$ s.

As the shear rate increases from $\dot{\gamma} = 0.5$ to $1.0\,s^{-1}$, as indicated in Figure 11.18, the relaxation rate of shear stress and first normal stress difference is getting a little slower.

Figure 11.19 demonstrates the relaxation of (a) normalized shear stress and (b) the first normal stress difference upon cessation of steady shear flow at $\dot{\gamma} = 0.5\,s^{-1}$ for PI-14-5CN at 70 °C. Experimental data are shown by dots and the fitting curves by

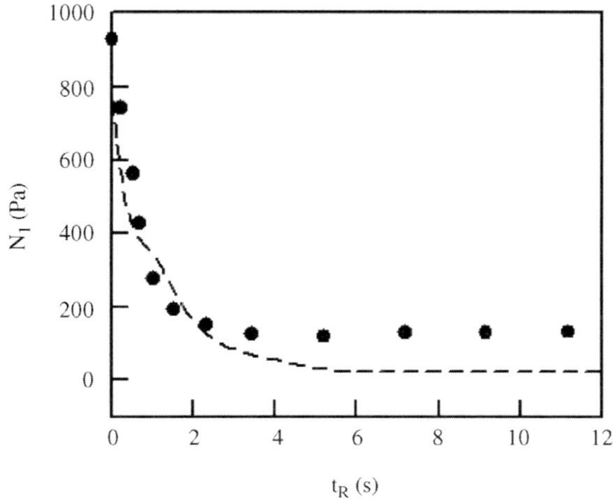

Figure 11.16 Relaxation of first normal stress difference N_1 for Titan at 340 °C upon cessation of steady flow at shear rate $\dot{\gamma} = 6\,s^{-1}$: notations are the same as in Figure 11.15.

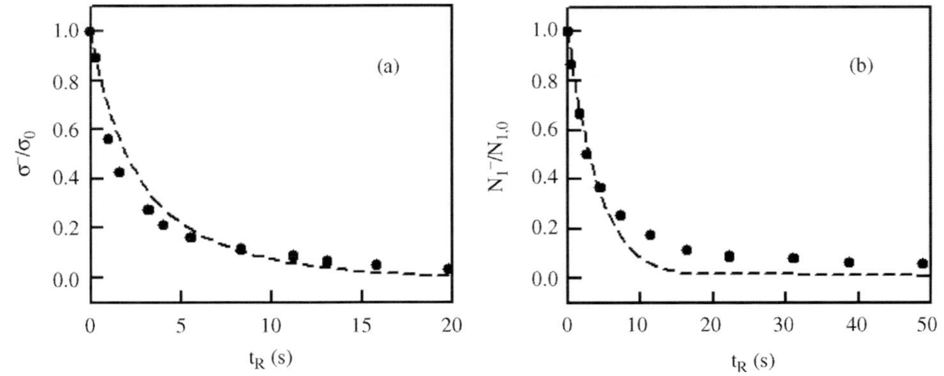

Figure 11.17 Relaxation of normalized (a) shear stress, $\sigma^-(\dot{\gamma},t)/\sigma_0$, and (b) first normal stress difference, $N_1^-(\dot{\gamma},t)/N_{1,0}$, upon cessation of steady shear flow at $\dot{\gamma} = 0.5\ \mathrm{s}^{-1}$ for PSHQ9 at 130 °C: experimental data are shown by dots and simulated curves by dashed lines. σ_0 and $N_{1,0}$ are the steady shear stress and first normal stress difference just prior to flow cessation.

dashed lines. Compared to Figure 11.15, the relaxation rates of shear stress and first normal stress difference are much faster for PI-14-5CN than for PSHQ9. This difference indicates how fast the recovery of the domain texture in side-chain LCP, PI-14-5CN is, after cessation of shear flow, compared to that in main-chain LCP, PSHQ9. The 5CN-COOH grafted on the coil-like backbone of PI, forming PI-14-5CN, might be very mobile and thus would relax rather quickly upon cessation of shear flow, compared to mesogens that are linked directly to the polymer backbone of PSHQ9.

As expected and seen from Figure 11.20, the relaxation of shear and normal stresses for PI-14-5CN are getting slower as $\dot{\gamma}$ increases from 0.5 to $1.0\ \mathrm{s}^{-1}$.

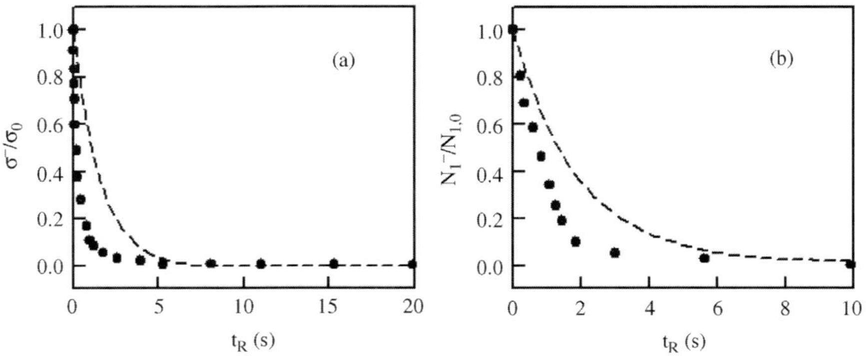

Figure 11.18 Relaxation of normalized (a) shear stress, $\sigma^-(\dot{\gamma},t)/\sigma_0$, and (b) first normal stress difference, $N_1^-(\dot{\gamma},t)/N_{1,0}$, upon cessation of steady shear flow at $\dot{\gamma} = 1\ \mathrm{s}^{-1}$ for PSHQ9 at 130 °C: notations are the same as in Figure 11.17.

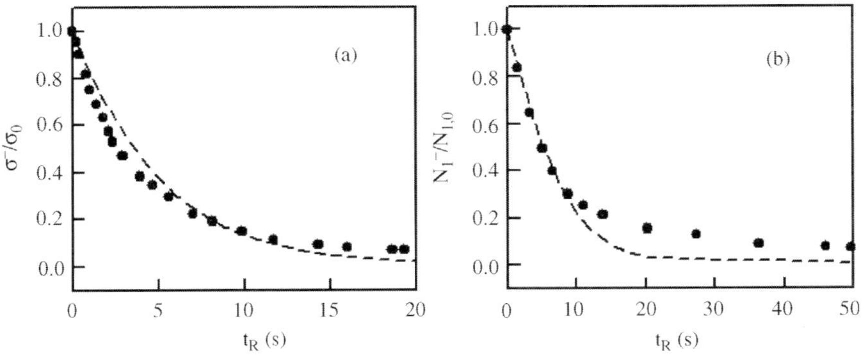

Figure 11.19 Relaxation of normalized (a) shear stress, $\sigma^-(\dot{\gamma},t)/\sigma_0$, and (b) first normal stress difference, $N_1^-(\dot{\gamma},t)/N_{1,0}$, upon cessation of steady shear flow at $\dot{\gamma} = 0.5\ s^{-1}$ for PI-14-5CN at 70 °C: notations are the same as in Figure 11.17.

11.4.4 On the Time-Temperature Superposition in Weakly Viscoelastic Nematodynamics

During the simulation of PSHQ9 and PI-14-5CN, the constitutive parameters α, β, r_1, r_2 characterizing the anisotropy of nematics, as well as the tumbling parameters λ_e, λ_v were assumed temperature independent. Thus, the only two parameters, characteristic relaxation time θ_0 and viscosity η_0, were assumed changing with temperature. In this case, the general equations of weak viscoelastic nematodynamics and their simple shearing specification (11.2) and (11.4) allow the scaling transformation:

$$\sigma_{ij} \to \hat{\sigma}_{ij} = \sigma_{ij}\theta_0/\eta_0, \quad t \to \hat{t} = t/\theta_0, \quad \dot{\gamma} \to \hat{\dot{\gamma}} = \dot{\gamma}\theta_0. \tag{11.10}$$

The transformation (11.10) demonstrates the time–temperature superposition scaling. Indeed, when the nondimensional variables denoted in (11.10) by overcaps

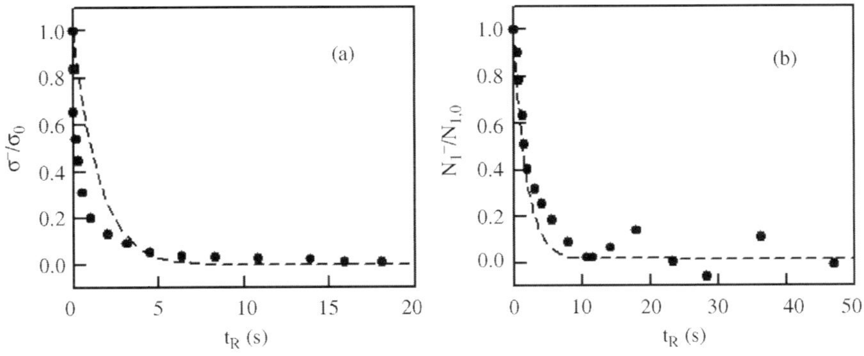

Figure 11.20 Relaxation of normalized (a) shear stress, $\sigma^-(\dot{\gamma},t)/\sigma$, and (b) first normal stress difference, $N_1^-(\dot{\gamma},t)/N_{1,0}$, upon cessation of steady shear flow at $\dot{\gamma} = 1\ s^{-1}$ for PI-14-5CN at 70 °C: notations are the same as in Figure 11.17.

are introduced in Eqs. (11.1) and (11.2) or in (11.4) and (11.5), these equations will describe isothermal, generally nonsteady shearing for various constant temperatures as temperature-independent curves. This is the time–temperature superposition principle for weak viscoelastic nematodynamics. Because in our modeling the fitted parameters of anisotropy were temperature independent, the time–temperature superposition principle does not need to be specially checked. There are no fundamental arguments, however, why it should be valid.

11.5 Conclusions and Discussions

The objective of this work was to analytically and numerically describe the rheological properties of thermotropic LCPs using recently developed thermodynamic monodomain theory of weakly viscoelastic nematodynamics [22, 23] and simulate the steady and unsteady shearing data. The Leslie–Ericksen theory, developed for low molecular weight liquid crystals, and the Doi theory for lyotropic LCPs are not suitable for description of flow properties of thermotropic LCPs. Therefore, with the use of these theories is difficult, if possible, to predict the rheological behaviors of LCPs in general. In contrast, viscoelastic nematodynamics proposes a new general approach to circumvent this problem. It should also be noted that in spite of a large number of constitutive equations proposed for many years for LCPs, no attempt for simulations of nonsteady shearing has been proposed for any type of LCPs. Moreover, there has been no theory to date that could be able to describe the complicated behavior of thermotropic LCPs.

In the absence of magnetic field, the theory exhibits viscoelastic transversally anisotropic behavior with symmetric stress tensor and orientation of director caused only by flow. Thus, this simplified approach has led to a closed set of two coupled anisotropic viscoelastic equations of quasilinear type for evolution of director and extra stress; the anisotropic properties in the set being described by viscoelastic evolution equation for director. Although this theory has been developed for low enough value of Deborah number, it is still possible to compare the simulations with experimental data.

Eight parameters are involved in the theory. They are three viscosities, three elastic module, and two tumbling (elastic and viscous) parameters. These constitutive parameters established for steady shearing were used for calculating the evolution of shear stress and first normal stress difference with corresponding evolution of director, during relaxation and start-up flow. The problem with initial conditions for director in start-up flow is resolved in the following way. We preliminarily fitted the experimental data for stresses in steady shearing with following adjustment of parameters for also describing the relaxation of stresses. In this case, parameters of evolution equation for director, along with its orientation in steady shearing, were also established. The orientation of director during stress relaxation was then easily calculated and its final orientation at the rest state was taken as initial value of director for the start-up flow.

The rheological shearing data for steady and transient shear flows, and relaxation after cessation of shear flow, were chosen for two industrial LCPs, Titan and Zenite 6000, as well as for two model LCPs, a main-chain LCP (PSHQ9) and a side-chain LCP (PI-14-5CN). This choice is justified by the most carefully made rheological measurements and their completeness. Specifically, Titan is random copolyester of ethylene-terephthalate and hydroxybenzoic acid with two methylene flexible spacers, whereas Zenite 6000 is a fully aromatic copolyester with kinks, Zenite 6000 is much more rigid than Titan. The absence of region I in the apparent shear viscosity plots might be due to the fact that PI-14-5CN is a side-chain LCP, and its nematic properties might not be sufficiently strong. Mesogens in the main-chain LCP PSHQ9 are directly linked to the polymer backbone, making motions of the mesogens and backbone of PSHQ9 strongly coupled during shear flow.

The most egregious deviations of our simulations from experimental data are observed for transitional start-up shearing flows. There might be several reasons for that.

The first is the weak viscoelasticity approach employed in the theory. For common polymers with long flexible chains, the weak viscoelasticity simply means the smallness of the Deborah number, $De = \theta\dot{\gamma} < 1$ where θ is the relaxation time averaged over the relaxation spectrum. In case of anisotropic LCPs where at least two relaxation times exist, the definition of Deborah number is not clear.

The second is how the fitting procedure was utilized in this chapter. We remind that the eight parameters were fitted to describe well enough the steady shearing data with adjustment for relaxations, for both shear and normal stresses and for each of four LCPs of different types. We use for fitting an exhausted computerized procedure of trials and attempts. Yet, we could not find the objective criterion for the quality of this fitting procedure.

The third and perhaps the more physically feasible reason for deviations is the inadequacy between the theory and the tested materials (or experimental procedures) selected for simulations. Yet, we should state that the materials and the experimental data chosen present the best choice made from a big pool of data. Simply, the better data do not exist.

The problem with start-up simulations of industrial LCPs Titan and Zenith 6000 is rather procedural. In order to obtain reliable data under relatively short operational time constraint (totally about 15min) preventing the samples from chemical degradation, a high-level preshearing procedure has been used [49]. Right after short relaxation period, the rheological measurements started, which include transient start-up flow, steady shearing and relaxation from the steady shearing level. To repeat the experiments, a fresh sample should be used. Using this procedure, the data showed to be reproducible. The problem with this procedure is that preshearing steady flow with following relaxation was not properly recorded. So, our way of establishing initial value n_{1r} of director in start-up flow is questionable for those materials.

Even worse, the complicated problem of polydomain behavior of model LCPs, PSHQ9 and PI-14-5CN, used in Ref. [54], cannot make reliable our simulations. Simply speaking, the monodomain theory we used is generally not suitable for the description of polydomain model LCPs. The polydomain effects are especially pronounced at the initial stage of start-up flow when the texture existed at rest is destroyed by the growing stresses. Because this initial stage may take a lot of time, developing the steady flow might be much longer than for the monodomain case. In addition, the procedure of finding initial value for director, established in our simulations, has no much sense in the polydomain case. On the other hand, in strong enough steady shearing flows and relaxation after their cessation, the polydomain effects are mostly insignificant. That is perhaps why the steady shearing and main relaxation processes (up to the beginning of the texture formation) have been well simulated using our monodomain approach. It is difficult to evaluate these effects without a good theory, which does not exist. So, the polydomain extension of the theory has to be developed to properly describe these data. Another way to rectify this problem is to prepare the monodomain samples. Although this is feasible, it needs a lot of experimental effort.

Nevertheless, with all these shortcomings, the simulations have demonstrated that these are at least in a semiquantitative agreement with the chosen experimental data. Moreover, it has been found that simulations of various LCPs were in accord with their different structural features.

References

1 Chandrasekhar, S. (1992) *Liquid Crystals*, 2nd edn, Cambridge University Press.

2 de Gennes, P.G. and Prost, J. (1993) *The Physics of Liquid Crystals*, 2nd edn, Clarendon Press, Oxford.

3 Warner, M. and Terentjev, E.M. (2003) *Liquid Crystal Elastomers*, Clarendon Press, Oxford.

4 Dierking, I. (2003) *Textures of Liquid Crystals*, Wiley-VCH Verlag GmbH, Weinheim.

5 Larson, R.G. (1999) *The Structure and Rheology of Complex Fluids*, Oxford Press, New York.

6 Collings, P.G. and Hird, M. (1997) *Introduction to Liquid Crystals Chemistry and Physics*, Taylor and Francis, London.

7 Donald, A.M. and Windle, A.H. (1992) *Liquid Crystalline Polymers* (eds A.M. Donald and A.H. Windle), University Press, Cambridge.

8 de Gennes, P.G. (1971) *Mol. Cryst. Liq. Cryst.*, **12**, 193.

9 Doi, M. (1987) Molecular theory for the nonlinear viscoelasticity of polymeric liquid crystals, IMA v. 5 in *Theory and Applications of Liquid Crystals* (eds J.L. Ericksen and D. Kinderlehrer), Springer-Verlag, New York.

10 Larson, R.G. and Mead, D.W. (1989) *J. Rheol.*, **33**, 185.

11 Doi, M. and Edwards, S.F. (1986) Chapters 8–10, in *The Theory of Polymer Dynamics*, Clarendon Press, Oxford.

12 Volkov, V.S. and Kulichikhin, V.G. (1990) *J. Rheol.*, **34**, 281.

13 Pleiner, H. and Brand, H.R. (1991) *Mol. Cryst. Liq. Cryst.*, **1991**, 407.

14 Pleiner, H. and Brand, H.R. (1992) *Macromolecules*, **25**, 895.

15 Rey, A.D. (1995) *J. Non-Newtonian Fluid. Mech.*, **58**, 131.

16 Rey, A.D. (1995) *Rheol. Acta*, **34**, 119.

17 Terentjev, E.M. and Warner, M. (2001) *Eur. Phys. J.*, **E. 4**, 343.

18 Fradkin, L.J., Kamotski, I.V., Terentjev, E.M., and Zakharov, D.D. (2003) *Proc. R. Soc., Lond.*, **A 459**, 2627.

19 Leonov, A.I. and Volkov, V.S. (2002) http://arxiv.org/pdf/ftp/cond-mat/papers/0202275.

20 Leonov, A.I. and Volkov, V.S. (2003) http://arxiv.org/ftp/cond-mat/papers/0203/0203265.pdf.

21 Leonov, A.I. and Volkov, V.S. (2003) *J. Eng. Phys. Thermophys.*, **76**, 498.
22 Leonov, A.I. (2008) *J. Math. Phys. Anal. Geom.*, **11**, 87–116.
23 Leonov, A.I. (2008) *Zeitschr. Angew. Math. Phys.*, **59**, 333–359.
24 Marrucci, G. and Greco, F. (1993) *Adv. Chem. Phys.*, **86**, 331.
25 Feng, J.J., Sgalari, G., and Leal, L.G. (2000) *J. Rheol.*, **44**, 1085.
26 Edwards, B.J., Beris, A.N., and Grmela, M. (1990) *J. Non-Newtonian Fluid Mech.*, **35**, 51.
27 Beris, A.N. and Edwards, B.J. (1999) *Thermodynamics of Flowing Systems*, Oxford University Press, Oxford.
28 Volkov, V.S. and Kulichikhin, V.G. (1994) *Macromol. Symp. Stat. Mech. Polym.*, **81**, 45.
29 Long, D. and Morse, D.C. (2002) *J. Rheol.*, **46**, 49.
30 Pickett, G.T. and Schweizer, K.S. (2000) *J. Chem. Phys.*, **112**, 4869.
31 Pickett, G.T. and Schweizer, K.S. (2000) *J. Chem. Phys.*, **112**, 4881.
32 Golubovich, L. and Lubensky, T.C. (1989) *Phys. Rev. Lett.*, **63**, 1082.
33 Olmsted, P.D. (1994) *J. Phys. II France*, **4**, 2215.
34 Lubensky, T.C. and Mukhopadya, R. (2002) *Phys. Rev. E*, **66**, 011702.
35 Leonov, A.I. and Volkov, V.S. (2004) *J. Eng. Phys. Thermophys.*, **77**, 717.
36 Leonov, A.I. and Volkov, V.S. (2005) *Rheol. Acta*, **44**, 331.
37 de Gennes, P.G. (1980) *Liquid Crystals in One- and Two Dimensional Order* (eds W. Helfrich and G. Kleppke), Springer, Berlin, pp. 231–237.
38 Leonov, A.I. (2005) *Rheol. Acta*, **44**, 573.
39 Odell, P.A., Unger, G., and Feijo, J.L. (1993) *J. Polym. Sci.: Polym. Phys.*, **31**, 141.
40 Yang, X., Forest, M.G., Mullins, W., and Wang, Q. (2009) *J. Rheol.*, **53**, 589.
41 Ericksen, J.L. (1991) *Arch. Rat. Mech. Anal.*, **113**, 97.
42 Guenther, G.K., Baird, D.G., and Davis, R.M. (1994) Liquid crystalline polymers, in *Proceedings of the International Workshop on Liquid Crystalline Polymers, WLCP'93, Capri, Italy, June 1–4, 1993* (ed. C. Carfagne), Elsevier, New York, pp. 133–142.
43 Einaga, Y., Berry, G.C., and Chu, S.-G. (1985) *Polymer*, **17**, 239.
44 Ernst, B. and Denn, M.M. (1992) *J. Rheol.*, **36**, 289.
45 Han, C.D. and Kim, S.S. (1993) *Macromolecules*, **26**, 3176.
46 Kim, S.S. and Han, C.D. (1993) *Macromolecules*, **26**, 6633.
47 Ugaz, V.M. and Burghardt, W.R. (1998) *Macromolecules*, **31**, 8474.
48 Han, C.D., Ugaz, V.M., and Burghardt, W.R. (2001) *Macromolecules*, **34**, 3642.
49 Belatreche, M.J. (2002) An Experimental and Theoretical Study of Liquid Crystal Polymers, M.S. Thesis, The University of Akron, Akron, OH.
50 Onogi, T.A. (1980) *Proceedings of the Seventh International Congress on Rheology*, vol. **I** (eds G. Astarita, G. Marrucci, and L. Nicolais), Plenum Press, New York, p. 127.
51 Marrucci, G. and Maffettone, P.L. (1990) *J. Rheol.*, **34**, 1217–1230.
52 Larson, R.G. and Doi, M. (1991) *J. Rheol.*, **35**, 539–563.
53 Han, C.D. (2007) *Rheology and Processing of Polymeric Materials*, vol. **I**, Oxford University Press, New York.
54 Lee, K.M. and Han, C.D. (2002) *Macromolecules*, **35**, 6263–6273.

Index

Modeling and Simulation in Polymers. Edited by P.D. Gujrati and A.I. Leonov

ISBN: 978-3-527-32415-6

k

q

r